Astrophysics and Space Science Library

Volume 387

For further volumes:
http://www.springer.com/series/5664

Giora Shaviv

The Synthesis of the Elements

The Astrophysical Quest for Nucleosynthesis and What It Can Tell Us About the Universe

Springer

Giora Shaviv
Department of Physics
Technion-Israel Institute of Technology
Technion City
32 000 Haifa
Israel

ISSN 0067-0057
ISBN 978-3-642-28384-0 e-ISBN 978-3-642-28385-7
DOI 10.1007/978-3-642-28385-7
Springer Heidelberg New York Dordrecht London

Library of Congress Control Number: 2012934662

Printed on acid-free paper

Springer is part of Springer Science+Business Media (www.springer.com)

Preface

Dol: *Are you sound?*
Have you your senses, masters?
Face: *I will have*
A book, but barely reckoning thy impostures,
Shall prove a true philosopher's stone to printers

Sub: *I have another work, you never saw, son,*
That three days since past the philosopher's wheel,
In the lent heat of Athanor; and's become
Sulphur of Nature

Ben Johnson (1573–1637), The Alchemist, 1612

A Few Words about What, Why, and How

A scientific theory is not a holy writ and sacred cows are routinely slaughtered in science. But do not hold this against science. On the contrary, it represents the perpetual struggle to reach the ultimate truth and understanding. As more details are discovered and experiments refined, scientific theories are improved or even revolutionized. And the theory of the synthesis of the chemical elements is no exception.

Recent discoveries in astrophysics set the amount of matter in the Universe, at least, matter of the kind we are familiar with, at around 4%. About 24–25% of the mass in the Universe is in the form of what is called 'dark matter', while the rest comes under the heading of 'dark energy'. Dark matter does not emit detectable photons and this is the reason for the annoying name. But in this book we shall discuss only the chemical composition of the visible matter. Visible means that it emits photons and can be seen or detected directly, by us. However, dark matter and energy affect our subject indirectly by influencing the formation of galaxies, and hence the formation of the first stars. A short digression into galaxy formation will thus be presented, to set the scene for those early stars.

In recent years we have witnessed a reversal in the line of research: from a quest for the formation of the chemical elements, how they came about, and why, the research has turned into attempts to understand what the various compositions tell us about the structure of the Universe. The change came about when attempts to explain the formation of the elements required hypotheses concerning global structure, in contradistinction to the investigation of pure nuclear physics processes and the way they lead to the synthesis of chemical elements.

Apart from a source of energy, everything needed for life can be obtained from this 4% of matter which is composed of ninety chemical elements found on Earth.

Two elements, technetium and promethium, are not found on Earth, while technetium is found in stars. On Earth, there are about 272 stable nuclei and 55 naturally occurring radioactive isotopes. This book describes how these elements were formed and where, how our understanding of their formation has evolved, and what the various compositions imply.

The question as to what constitutes a scientific discovery and who should be credited for it, or even when a discovery can be defined as such, does not have a definite answer: is it the first to make the discovery but is not believed, or is it the person who comes in at the end and clinches things? Is it the first who wrote the expression with an undefined constant or the one who took the new expression and fixed the constant? We are not going to answer this philosophical question, but instead attempt to describe how the idea has evolved up to the present day, identifying those who contributed to it, and conjecturing about what may happen tomorrow.

Element synthesis was born as a by-product of the problem of the energy source in stars. With the birth of nuclear astrophysics and the explanation of stellar energy sources came the saga of element synthesis in the stars. From its beginnings as a goal in itself, it soon turned out that the synthesis of the elements is intertwined and entangled with many fundamental processes in the cosmos. Today, element synthesis and nuclear astrophysics no longer stand alone, but are part of global processes that take place in galaxies, from galaxy formation and galaxy–galaxy collisions to mixing processes in galaxies and stars. Hence, as the title indicates, there is no question now of nuclear astrophysics remaining in splendid isolation. A global view is beginning to emerge, and it is the emergence of this panoramic picture which we shall try to depict and detail in this book.

It is not obvious that mankind will draw any practical benefits from knowing how the elements were formed. I find it appropriate at this point to cite what Roscoe[1] wrote in 1862, in his introduction to the papers by Kirchoff and Bunsen on spectroscopy, a discipline so essential for our subject here:

> It is unnecessary to insist, at the present day, upon the incalculable value of discoveries in the natural sciences, however abstruse they may be, or however far-distant may appear their practical application. If we put aside for the moment that highest of all intellectual gratifications afforded by the prosecution of truth in every form, the perception of which is one of the chief distinctions of humans from brute life, and if we look to the results of scientific discovery in benefiting mankind, we find so many striking examples of the existence of truths apparently altogether foreign to our everyday wants, which suddenly become points of great interest to the material prosperity and the moral advancement of

[1] Roscoe, H.E. Edinburgh Review, No **CXVI**, 295, (1862). *Researches on the Solar Spectrum, and the Spectra of the Chemical Elements* by Kirchoff and *Chemical Analysis by Spectrum Observations*by Bunsen and Kirchoff, Pogg. Annal. translated by Roscoe. Sir Henry Roscoe was a PhD student of Bunsen in Heidelberg, working on photochemistry, and in particular determining the effect of light on the reaction $H_2 + Cl_2$. This research led Bunsen to carry out the first attempt to estimate the solar constant, or as he called it, the radiant energy of the sun. However, this Bunsen–Roscoe collaboration was terminated abruptly by Bunsen when he discovered the power of spectroscopy and devoted all his time to it.

> the race, that we are less apt to utter the vulgar cry of 'cui bono' respecting any scientific discovery; and if we are not advanced enough to love science for the sake of her truth alone, we at least respect her for the sake of the power she bestows. [...] A great discovery in natural knowledge, for which no equivalent in direct benefit to mankind has as yet been found, but which nevertheless excites our liveliest interest and admiration, has lately been made in the rapidly advancing science of chemistry. [...] This is so delicate in nature, that, when applied to the examination of the substances composing our globe, it yields most new interesting, and unlooked-for information. At the same time it is so vast an application as to enable us to ascertain with certainty the presence in the solar atmosphere—at a distance of 95 000 000 miles—of metals, such as iron and magnesium, well known on this Earth, and likewise to give us good hopes of obtaining similar knowledge concerning the composition of the fixed stars. Here, indeed, is a triumph of science! The weak mortal, confined within a narrow zone on the surface of our insignificant planet, stretches out his intellectual powers through unlimited space, and estimates the chemical composition of matter contained in the Sun and the fixed stars with as much ease and certainty as he would do if he could handle it, and prove its reaction in the test tube. [...]
>
> How did two German philosophers, quietly working in their laboratory in Heidelberg, obtain this inconceivable insight into the process of creation? Are the conclusions which they have arrived at logical consequences of bona fide observations and experiment—the only true basis of reasoning in physical science—or do they not savour somewhat of that mysticism, for which our German friends are famous? Such questions as these will occur to all who hear of this discovery.

Roscoe went on to point out that:

> The further we penetrate into the secrets of nature, the more we find there remains to be learnt.

This is also the motto of this book about the chemical elements, a century and a half after spectroscopy was discovered as a tool. Indeed, not only was Roscoe's vision found to be correct, but it went well beyond expectations, and today, in retrospect, we can better appreciate his profound understanding. The present book is about just this.

The title of the book and the idea of 'element genesis' appeared for the first time in 1934, when the chemist Lewis (1875–1946)[2] suggested that a major part of the matter in the Universe might be composed chiefly of iron and nickel, like metallic meteorites. The idea was based on the fact that such a material is thermodynamically stable with respect to all spontaneous transmutations, except for very high temperatures. Lewis hypothesized that the bombardment of such a material by cosmic rays could generate all the elements seen in the crust of the Earth. Lewis claimed that the relative abundance of the main atomic species indicates 'a genetic relationship' between the cosmic rays and the crust of the Earth. The explicit idea that the elements are created by some cosmic physical process is probably due to Lewis, but present day views about the mechanisms have changed completely. It is the evolution of these ideas into present day thinking that will form the subject of the present book.

---

[2] Lewis, G.N.: *The genesis of the elements*. Phys. Rev. **46**, 897 (1934)

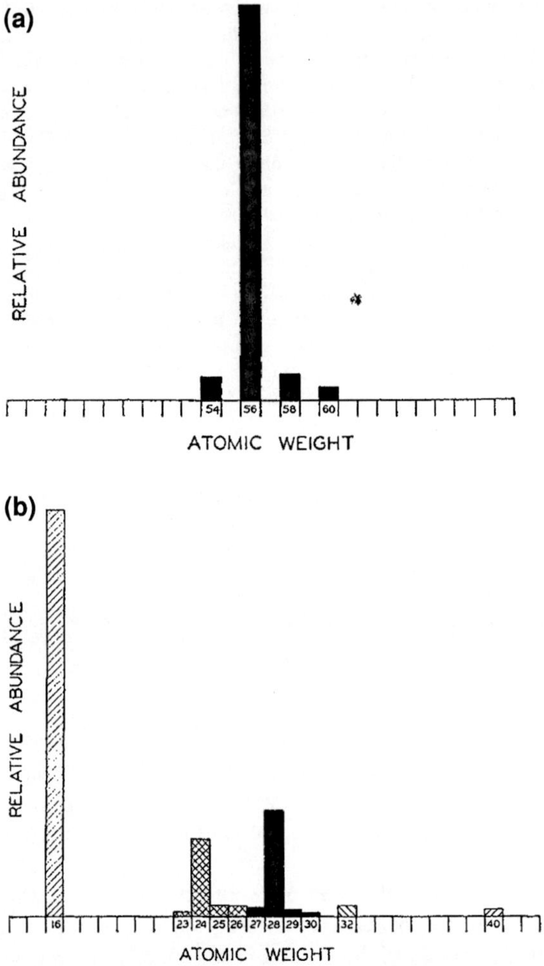

Fig. 1 The first attempt to hypothesize about the genesis of the elements as a result of bombardment of iron and nickel by cosmic rays. **a** Relative abundances of the major constituents of metallic meteors. **b** Relative abundances of the major constituents of stony meteors. Both panels give abundances as a function of atomic weight. After Lewis' paper 1934

For centuries scientists looked for ways to transform the chemical elements from one to the other, and in particular to produce gold out of sand. Stars materialize this long-standing dream of the alchemists, since they synthesize heavy elements out of lighter ones and thus create the chemical elements we see today. All chemical elements heavier than carbon (including carbon) were synthesized at some time in the past inside the hot cores of massive stars. The elements were removed from the furnace either by an explosion or a slow mass loss from the star.

So the process has two phases: synthesis deep within a star and removal into interstellar space without destruction. Chemical elements are a by-product of stellar energy sources which drive the cosmos.

The idea of element transmutation, which can be traced back to the Greek philosophers, prevailed amongst alchemists of the Middle Ages. People like Magnus, Bacon, and Vincent of Beauvais firmly believed in its feasibility. Such ideas about the transformability of matter were advocated even well into the seventeenth century, at least from the academic point of view. Among those who upheld them we find Newton and Leibniz, although they did not have a shred of evidence that this was really possible. This book describes the evolution of our ideas about the synthesis of the elements from the beginning right up to the present day, along with the controversies, discussions, and failures. Dramatic progress has been made in the past 80 years, but many problems remain. So this book will not be the 'last word'. The field is still evolving and represents an active field of research, promising many more surprises in the future.

Leading scientists are frequently immortalized by naming a lunar crater or a mountain after them. When this happens, the year of their death is followed by a small m.

Haifa, February 2011 Giora Shaviv

Acknowledgments

For years I had lengthy discussions with the late professor Rainer Wehrse, a leading expert in radiative transfer and spectra analysis, about science as well as about evolution, politics, and many other issues. It is a pleasure to acknowledge these great experiences and very sad that he cannot comment on this book.

It is my great pleasure to thank Dr. Smadar Bressler for proof-reading and raising a boundless range of interesting questions and comments.

Unique thanks are due to Stephen Lyle, who virtuously succeeded in editing the book and capturing the spirit of the original manuscript.

Contents

Chapter 1
Order in the Chemical Elements

1.1 An Early Start

The idea of atoms as forming the ultimate constituents of matter already arose in the great speculative or philosophical minds of the ancient world, although no exact chemical or physical atomic theory actually evolved from it at the time, and although, most importantly, there was no direct evidence for it. The fundamental questions were: What are the chemical elements? What are the basic units which carry the identity of a chemical element? Can matter be divided indefinitely while still preserving its identity?

Such questions already bothered Greek philosophers, including for example Aristotle (384–322 BCm), Leucippus (flourished circa 440 BCm), and Democritus (circa 460–370 BCm). According to the doctrines of Leucippus and Democritus, matter could not be divided indefinitely. Democritus believed in the existence of two types of reality: atoms and void. According to this hypothesis, the world consists of a collection of simple particles, of one kind of matter, and of indivisible smallness, as the name indicates. Indeed, the term 'atom' emerged from the Greek adjective *atomos* or *atomon*, which means 'indivisible'. According to this view, the atoms are infinite in number and have various sizes and shapes. They move about in an infinite void, interacting with one another to form clusters by mechanical means resembling hooks.

A variation on this idea was suggested by Anaxagoras (500–428 BCm), who proposed the *Homoiomeria*. According to this, material things consist of particles that are homogeneous in each kind of body, but dissimilar in different bodies. On the other hand, some 60 years later, Aristotle[1] suggested that matter might be indefinitely divisible. None of these ideas were put to a physical test, so they were just pure hypothesis. However, it is not clear what else they could have done, since the physical

[1] In some places it is claimed that Aristotle preached that we have to learn from experience. Experience is, however, not the same as an experiment which is carried out under controlled conditions. The ratio between the frequencies of string vibrations was probably found by Pythagoras and his students, and this is considered by many to constitute the first experiment.

G. Shaviv, *The Synthesis of the Elements*, Astrophysics and Space Science Library 387,
DOI: 10.1007/978-3-642-28385-7_1, © Springer-Verlag Berlin Heidelberg 2012

and technical basis for experimentation did not yet exist, even if they had wanted to do it.

With the lack of good tests, and centuries before Ockham's razor[2] was invented, an hypothesis was accepted or rejected according to the reputation of the philosopher who proposed it. In the present case, the later Aristotle came out best, and his hypothesis became a common conviction, irrespective of the fact that it was completely wrong and gave rise to more questions and problems than it solved.[3] However, Aristotle's reputation carried it through for centuries to come. The Greeks mastered logic, but logic without experiment or observation may lead us astray.[4]

1.2 Ideas in the Far East

Ideas about the ultimate indivisibility of matter evolved in China and India, but they were not expressed explicitly and must be inferred from contemporary writings. An atomic theory was proposed by Kanada, the founder of the Nyaya system of philosophy.[5] The technical term 'atom' or its equivalent was not invented, but can be inferred. According to Kanada, atoms are eternal but may compose more complex substances, although they themselves are simple and non-composite. Aggregates of atoms can be separated into the fundamental atoms.

1.3 Alchemy

It is an amazing fact that, through all the centuries of believing in and working on the possibility of transforming ordinary metals into gold, for example, none of the alchemists ever came up with a theory to explain how such a change could be brought about. Alchemists made innumerable attempts, and it is well known that all were in vain. And yet there was never a word, not a single idea about the fundamental constituents of matter they wanted to transform, or a theory that could be used as a guide. Every attempt was just a shot in the dark.

Despite its esoteric foundations, alchemy is in fact a precursor of practical chemistry, even though absolutely no new concept was invented over the many

---

[2] The so-called razor due to William of Ockham (1285?–1349) is the philosophical principle that, of all possible explanations of a phenomenon, the simplest is most probably the best one. Of course, this principle cannot be proven rigorously. We must 'take it or leave it'.

[3] Even today, an hypothesis is often accepted or rejected by popular vote, i.e., according to the number of scientists who believe it. See, for example, the idea that anthropogenic CO_2 causes global warming, while no evidence exists to support this quite common conjecture, and one group of scientists even tried to massage the data so as to 'prove' it. On the other hand, scientific revolutions are caused just when a single scientist attempts to disclaim a scientific dogma.

[4] For more detail on the ancient atomic theory, see Partington, J.R., Annals of Science **4**, 245 (1939).

[5] Ray, P.C., *A History Of Hindu Chemistry V1: From The Earliest Times to the Middle of the Sixteenth Century A.D.*, Calcutta, The Bengal Chemical and Pharmaceutical Works, Ltd., 1903. The book contains translations of relevant texts from oriental languages to English.

years during which 'alchemistry' was practised. Anyway we should not forget that Newton was a disciple of the alchemist movement, and that physics has already turned the old fantasy of element transmutation into a reality by transforming one nucleus into another.

It is instructive at this point to mention Ben Johnson's hilarious play *The Alchemist*, first performed in 1610, in which the author exploited the profession of alchemist as a platform on which to satirize about human greed and the partly scientific, partly pretentious language of the alchemist.[6]

1.4 Lavoisier and the Beginning of the Modern Era

The modern era of chemistry started in 1789 when the 'father of chemistry', Antoine-Laurent de Lavoisier (1743–1794m),[7] attempted to classify the elements.[8]

Lavoisier[9] defined what we call today a chemical element as *une substance simple n'étant susceptible d'être divisée par aucune méthode connue d'analyse chimique*. In other words, a chemical element was a substance that could not be further divided by any known method of chemical analysis, and this is the definition that we shall use here.[10] Lavoisier was unbelievably precise in his definition, almost predicting discoveries to be made 150 years on, when he restricted the definition to objects that were 'indivisible by analytical chemistry', as though suggesting that other methods could possibly further split the element. On the other hand, Lavoisier, who once and for all eliminated the imaginary phlogiston from chemistry, assumed the equally imaginary caloric to be a 'simple substance'.[11]

---

[6] Craig, D.H., *Ben Johnson: The Critical Heritage*, London, Routledge, 1999.

[7] Lavoisier, A., *Traité Elémentaire de Chimie*, Cuchet, Paris, 1789.

[8] As scientists could not make a living from doing research alone, Lavoisier, like Fourier after him, became a member of the *Fermiers Généraux*. This General Farmers association was a private company with 60 members which had at its disposal a fortune of about 100 million pounds. Every 6 years, the company negotiated with the royal treasury the right to collect tax and customs duties. Obviously, this company and its members were not very popular, to put it mildly. Lavoisier's immense scientific reputation did not save him on Judgement Day, and his job as tax collector cost him his life: he was guillotined in 1794, during the French revolution.

[9] There is a controversial claim regarding the extent to which Robert Boyle (1627–1691m) recognized the fundamental nature of a chemical element. See Knight, D., *Ideas in Chemistry: A History of the Science*, Cambridge University Press, 1992.

[10] One sometimes sees erroneous definitions of chemical elements as 'pure chemical substance' (see, for example, in Wikipedia). We adopted Lavoisier's definition here, although we did not define what exactly was meant by 'a method of chemical analysis'. The atoms of an element are complex systems that exist over a certain range of temperatures and densities, and in certain environments. So a chemical element can exist only if chemistry itself actually exists. In the cores of stars, for example, all atoms are crushed by the pressure, and there is no chemistry to speak of.

[11] On the other hand, it is not clear today whether the dark energy and dark matter postulated to explain the structure of galaxies and the expansion of the Universe are matter in the normal sense or merely an expression for our ignorance of some fundamental physical law.

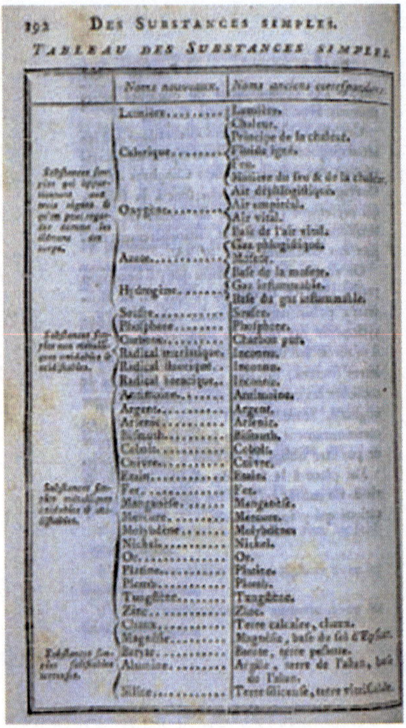

Fig. 1.1 Lavoisier's original classification of 33 'simple substances', including light (the first on the list) and heat (the second on the list) under the name 'calorique', which at that time was assumed to be a liquid that carries heat. Lavoisier, 1787

Lavoisier included all entities he could not split in his list of elements (see Fig. 1.1), including light, heat, and metal oxides. In addition, Lavoisier did not use the name 'element', but preferred 'simple substances'. Five of the substances listed by Lavoisier were discovered by Carl Wilhelm Scheele (1742–1786m), who was a German–Swedish pharmaceutical chemist. It was Scheele who discovered the chemical elements oxygen and nitrogen (in 1772–1773),[12] although it was Lavoisier who first named these gases.

1.5 Wenzel and Richter: Quantitative Chemistry

Quantitative chemistry was actually initiated by two German chemists, Karl Friedrich Wenzel (1740–1793)[13] and Jeremias Benjamin Richter (1762–1807),[14] who coined the word 'stöichiometrie' or 'stoichiometry' (from the Greek word *stoicheion* for

[12] Scheele, C.W., *Chemische Abhandlung von der Luft und dem Feuer*, Upsala und Leipzig, Verlegt von Magn. Swederus, Buchhändler; zu finden bey S.L. Crusius 1777.

[13] Wenzel, K.F., *Lehre von der Verwandtschaft der Körper*, Dresden, 1777.

[14] Richter, J.B., *Anfangsgründe der Stöchyometrie oder Messkunst chymischer Elemente*, Breslau und Hirschberg 1792–1793. Reprint Hildesheim 1968.

element). The main finding was that elements combine to form compounds in fixed mass ratios.[15] But the ideas of Wenzel and Richter apparently came too early, and involved some errors. As a result, the scientific community refused to recognize the discovery and it was mainly forgotten.[16]

1.6 Proust and Gay-Lussac Versus Berthollet: 1800–1808

Claude-Louis Berthollet (1748–1822) was a member of Lavoisier's entourage[17] and a leading French chemist in his own right. From the historical perspective, Berthollet is best known for his ideas on 'reverse chemical reactions', which led to Guldberg and Waage's 'law of mass action'.[18]

Berthollet's most remarkable contribution to chemistry was his *Essai de Statique Chimique* (Essay on Chemical Equilibria, 1803), which was the first systematic attempt to cope with the problems of chemical physics and chemical reactions. According to Berthollet's theory, the physical conditions surrounding reagents, including temperature and solubility, often offset the effect of affinities, thereby determining the direction of the reaction. Borrowing a physical term, he asserted that chemical reactions seek a 'static equilibrium', hence the title of his essay. Because it was thought that the molecules in a gaseous reaction were separated by caloric (a hypothetical weightless substance thought to account for the flow of heat),

---

[15] Note that we stress *mass ratios* and not *number ratios*. This is a fundamental point. Fixed mass ratios do not necessarily lead to the atomic hypothesis, but fixed ratios of integers do.

[16] Wenzel and later Richter published some inaccurate results and interpretations. This had to do with the conservation of electric charge in the dissociation of salts. See, for example, Wollaston, W.H., Phil. Trans. Roy. Soc. London **104**, 1 (1814). Consequently, they lost their credentials as chemists and it was only Berzelius who recognized their achievements about twenty years later. It is not absolutely clear to whom the credit for the discovery of the law of proportionality should go. Does the announcement of a correct law on the basis of incorrect experimental interpretation deserve full recognition? See *A History of Chemical Theory*, Wurtz, MacMillam and Co. London, 1869. But the idea was good so why was it ignored? Szabadváy (Szabadváy, F., J. Chem. Educ. **39**, 267, 1962) suggests that Richter began his book with an attempt to teach chemists simple arithmetic, e.g., explaining things like the simple plus sign. If you treat your fellow practitioners with such disdain, do not be surprised to find they ignore you.

[17] In contrast to Lavoisier, Berthollet served the revolutionary government as an adviser and emerged from the French Revolution unharmed. Later he joined Napoleon on his trip to Egypt, where he did the research reported later in his book *Essay on Chemical Equilibria.*

[18] Cato Maximilian Guldberg (1836–1902) and Peter Waage (1833–1900) suggested the law of mass action in 1864 (Forhandlinger: Videnskabs-Selskabet i Christiana **35**, 1864). In 1877, van 't Hoff (1852–1911m) (Berichte der Berliner Chem. Ges. **10**, 1877) rediscovered the law without knowing about the earlier discovery by Guldberg and Waage, which was published in Norwegian. In 1879, Guldberg and Waage decided to republish the paper in German (Erdmann's J. für Prac. Chem. **127**, 69, 1879) in order to claim the discovery. The law of mass action says that, when a chemical reaction is in dynamic equilibrium, e.g., $A + B \rightleftharpoons C$, then $K(T) = [A][B]/[C]$ is a constant which depends only on the temperature and pressure. Here $[X]$ denotes the concentration of the species X.

Berthollet insisted that they had to be carried out at certain temperatures in order to be successful.[19] Berthollet's thesis did not meet with general approval among his contemporaries, partly perhaps because he pushed it too far. For instance, he claimed that two elements might combine in varying proportions. Berthollet's experiments showed that the concentration of the species affects how the reaction ends, which is of course true for reactions in equilibrium. This result is at variance with the cases in which the affinity of the elements determines the outcome in an elementary reaction like $A + B \rightarrow C$.

1.7 Berthollet Was Both Right and Wrong

The confusion between reactions in equilibrium and regular chemical reactions, where the reacting species consume each other completely, worked against Berthollet and caused him to think that all chemical compounds exist in indefinite proportions in solution. When the acceptance of John Dalton's atomic theory discredited this thesis, some of Berthollet's contemporaries unjustifiably inferred that his views on mass action were erroneous. If partly wrong, then completely wrong! Others, however, notably Berzelius, distinguished between these two different situations. For several decades the prevailing belief was that Berthollet's mass action presented an alternative hypothesis to the affinities of elements as an interpretation for chemical reactions. The issue appeared chemically indeterminate because the relative quantities of each substance present in a solution could not be determined without applying reagents that themselves disturbed the equilibrium.

Joseph Louis Gay-Lussac (1778–1850m)[20] was quite prolific in his research interests, from the structure of the atmosphere to magnetism. However, Gay-Lussac is best known for the 'law of combining volumes': *The compounds of gaseous substances with each other are always formed in very simple ratios by volume.*[21] In a brilliant move, Gay-Lussac took Davy's (1778–1829m) results, in which the elements making up a compound were given as a ratio of weights (see Table 1.1),

[19] In conducting this research, Berthollet was assisted by his student Joseph Louis Gay-Lussac, with whom he was engaged in experiments in the laboratory as well as in a balloon flight to a then record height of 23,000 feet. Gay-Lussac and others, including Louis-Jacques Thenard, Jean-Baptiste Biot, Pierre-Louis Dulong, and ultimately Jacques-Etienne Bérard, constituted the active chemical group of the Société d'Arcueil, all inspired by the older academicians Berthollet and Laplace. The proceedings of the society were published in three volumes between 1807 and 1817 as *Mémoires de la Société d'Arcueil.*

[20] Gay-Lussac grew up during the French Revolution, and his education was disrupted when his private tutor ran away and his father was imprisoned. However, Gay-Lussac took advantage of the new order when he was chosen to attend the Ecole Polytechnique, the institution that the French Revolution designed to create the leading scientific and technical cadre, especially for the military. Gay-Lussac's mentors at the Polytechnique, Laplace and Berthollet, urged him to pursue a research-academic career, which he began shortly after graduation in 1800.

[21] Gay-Lussac, J.L., Mém. de la Soc. d'Arcueil **2**, 207 (1809)

Table 1.1 Davy's results for the relative masses of nitrogen and oxygen in various compounds of these two elements, and Gay-Lussac's conversion to volumes, 1809

Compound	Nitrogen mass à la Davy	Oxygen mass à la Davy	Nitrogen volume à la G-L	Oxygen volume à la G-L	Gay-Lussac assumed ratio
Nitrous oxide	63.30	36.70	100	49.5	2:1
Nitrous gas	44.05	55.95	100	108.9	1:1
Nitric acid	29.50	70.50	100	204.7	1:2

and converted them into volumes, thereby neutralizing the role of the weight of each atom.

No estimate of the error in the measurements was given by Davy. But Gay-Lussac understood the existence of an error in the measurements and ingeniously overlooked it. Gay-Lussac gave more examples before he summarized his conclusions:

> I have shown in this Memoir that the compounds of gaseous substances with each other are always formed in very simple ratios, so that representing one of the terms by unity, the other is 1, or 2, or at most 3.

But Gay-Lussac went on to conclude that:

> These ratios by volume are not observed with solid or liquid substances, nor when we consider weights, and they form a new proof that it is only in the gaseous state that substances are in the same circumstances and obey regular laws.

In 1805 Gay-Lussac and Alexander von Humboldt (1767–1835m) showed, for example, that hydrogen and oxygen combine in a ratio of 2:1 to form water.[22]

Gay-Lussac the scientist was uncompromising with regard to his mentor, pointing out that:

> The numerous results I have brought forward in this Memoir are also very favorable to the theory. But M. Berthollet, who thinks that combinations are made continuously, cites in proof of his opinion the acid sulphates, glass alloys, mixtures of various liquids, all of which are compounds with very variable proportions, and he insists principally on the identity of the force which produces chemical compounds and solutions.

Note that none of the counterexamples are gases. Did he mean that in the other cases Berthollet was right? Moreover, Berthollet spoke about mixtures while Gay-Lussac did not specifically eliminate this possibility.

Gay-Lussac turned against the view of his old mentor Berthollet, as it was expressed in the introduction Berthollet wrote to the French translation of the third edition of Thomson's (1773–1852) textbook, published in 1807.[23] In this book, Thomson presented Dalton's atomic hypothesis before Dalton published it,[24] and it is easy to imagine that Berthollet did not approve of it. Gay-Lussac referred in a very

[22] von Humboldt, A. and Gay-Lussac, J.F., Ann. der Physik **20**, 129, 1805.

[23] Thomson, T., *A System of Chemistry*, Edinburgh, 1807.

[24] Thomson had private communication with Dalton, in which Dalton provided him with a copy of *A New System of Chemical Philosophy*, which he published only a year later. Thomson gave all the credit to Dalton.

polite way to Berthollet's claims in the introduction, arguing that they did not agree with experiment:

> M. Berthollet has already strongly opposed it in the Introduction he has written to Thomson's Chemistry, and we shall see that in reality it is not entirely exact.

Berthollet did not allow the facts to confuse him again, claiming that the gaseous compounds were not the rule but the exception.

In 1785, after serving in different positions in Paris, Joseph Louis Proust (1754–1826) accepted a prestigious teaching position offered by the Spanish government. He spent the next twenty years in Madrid and Segovia, thereby completely 'missing' the French Revolution and the rise to power of Napoleon Bonaparte, with their various consequences for French science. While in Spain, Proust began studying different types of sugar, and was the first to identify glucose and study it.

Proust also studied copper carbonate,[25] the two tin oxides, and the two iron sulfides. In particular, he prepared artificial copper carbonate and compared it to the natural one. In this way he demonstrated that each compound has the same proportion of weights between the three elements involved (Cu, C, O). Moreover, Proust showed that no indeterminate intermediate compound exists between them. Consequently, Proust was able to formulate his law that compounds are formed by certain but fixed mass proportions. Although Proust published this paper in 1799, the law, let alone its implications, was not accepted by the scientific community until 1811, when Berzelius credited Proust with the discovery of the law and endowed it with his approval. No doubt, this was Proust's greatest achievement in chemistry. The law and its conclusions contradicted what Berthollet thought and wrote in his book, and unavoidably became the center of a long controversy between Berthollet and Proust which lasted from 1800 to 1808.

Proust owed nothing to Berthollet and expressed himself more critically than Gay-Lussac, although he remained friendly. For about eight years, Proust and Berthollet were immersed in a controversy over this issue, but in the end the truth of Proust's arguments prevailed. A frequently expressed view is that Berthollet treated mixtures, solutions, and compounds equally, and that this was the source of his error. However, things were a bit more complicated than that. Apart from discovering reactions in chemical equilibrium, Berthollet actually discovered the berthollites, that is to say, nonstoichiometric compounds in which the numbers of atoms of the elements present cannot be expressed as a ratio of integers.[26] It was ingenious on the part of

[25] Proust, J.L., Ann. de Chim. **23**, 101 (1797), ibid. **32** (1799).

[26] Nonstoichiometric compounds appear mainly among the transition elements which are used in solid-state electronic devices, such as rectifiers, thermoelectric generators, photodetectors, thermistors, and magnets useful in high-frequency circuits.

The existence of nonstoichiometric compounds which led Berthollet astray is related to the presence of defects in the lattice structures of crystalline substances. For example, a sodium chloride crystal in which one of the locations is occupied by a neutral sodium atom rather than a sodium ion. Another example is the FeO crystal with some vacancies (missing oxygen atoms at some location) leading to $Fe_{0.95}O$. Most nonstoichiometric compounds have compositions that are close to those of stoichiometric compounds.

Proust to disregard Berthollet's confusing non-true chemical compounds, although, like Gay-Lussac before him, he did not explain why the crystals, for example, do not obey his law. The problem for Proust was to demonstrate in a convincing way the difference between a compound and a mixture or an alloy. Still, Proust succeeded in showing[27] that the sulfurous compounds of copper were not true compounds, but mixtures of copper sulphide in sulphur.

Proust then went on to show how Berthollet was incorrect in his own chemical analyses, by showing that Berthollet treated some of the chemicals he used as oxides, when they were actually hydrates, i.e., they contained chemically-bonded water. This 'unpleasant discovery' completely pulled the rug from under Berthollet's claim.

On the other hand, Berthollet rejected the experiments carried out by Proust by stating that they did not have a general validity. Worse than that, Gay-Lussac, who worked on gaseous compounds under the directorship of Berthollet, managed to demonstrate that the gases combine in fixed ratios, but Berthollet simply refused to accept Gay-Lussac's results. Berthollet did not go as far as rejecting what he called 'Dalton's ingenious atomic idea', which was published during the ongoing controversy.

Berthollet also engaged in a long controversy with other chemists about the nature of the force acting between two atoms and making a compound, but this issue, known today as the chemical bond, goes beyond our scope here.

Berthollet insisted that Proust define a compound and a solution, definitions which Proust had difficulty in providing. He eventually defined a compound by saying that it was *a privileged product to which nature assigns fixed proportions*. But his definition was logically circular, hence useless.

Berthollet was famous for the controversy in which he was wrong, and Proust became famous for the controversy in which he was right. For more details about the Proust–Berthollet controversy, see Kapoor.[28] Berthollet's fundamental contributions which led to the law of mass action were unjustifiably overshadowed by the controversy. Here is that stupid law again: wrong once, wrong all the time. Similarly, Berthollet's contributions to the chemistry of dyes and bleaching agents, in particular, the suggestion that chlorine could be used as a bleaching agent, did not receive adequate recognition by his contemporaries.

1.8 Dalton and the Revival of the Atomic Theory

At the same time as the Berthollet–Proust controversy was taking place in France, Dalton proposed his seminal hypothesis about the structure of matter. It was in 1802 that John Dalton (1766–1844m),[29] professor of chemistry in Manchester,

---

[27] Proust, J.L., J. de Phys. **55**, 335 (1802).

[28] Kapoor, S.K., *Berthollet, Proust, and proportions*, Chymia **10**, 53 (1965).

[29] Dalton was also known for his extensive meteorological observations and the rediscovery of the Hadley circulation in the atmosphere, as well as research on color blindness.

resuscitated the old Greek atomic theory, but this time with good, or at least circumstantial, evidence. It must be admitted that similar ideas had been announced before by Wenzel and Richter.[30] Dalton had in hand the quantitative measurements of mass fractions of elements making up different compounds, as carried out by Lavoisier some two decades earlier, as well as the books by Richter and Wenzel. Atoms were defined by Dalton as:

> [...] the smallest parts of which bodies are composed. An atom therefore, must be mechanically indivisible and of course a fraction of an atom cannot exist.

Note the term 'mechanical', which is a physical term (as opposed to a chemical one). As we know today, Dalton's statement was too general (unless we add the clause 'for the energies available at the time').

In 1803, Dalton realized that carbon and oxygen can form two chemical compounds, CO and CO_2, and the ratio of oxygen between the two compounds is exactly 2.[31] It was this observation that led him to propose the 'law of simple proportions', which was later confirmed by Jöns Jacob Berzelius (1779–1848m). The key fact was that compounds are made of fixed proportions of the constituents. But at this point Dalton made a major step forward and interpreted the result as follows: each element has its unique atoms and the atoms of each element have a unique atomic weight. When an atom of element 1 combines with an atom of element 2 the mass ratio is always the mass ratio between atom 1 and atom 2 or a simple product of it. Questions such as how the chemical bond forms or why the properties of the thus formed molecule are not the average between the properties of the two atoms which composed it, were not asked. In the case of water it was clear that hydrogen and oxygen unite in no other proportions than two hydrogen atoms with one oxygen atom.[32] One may claim that the idea of atoms of matter can be traced back to the ancient Greek philosophers, but it was Dalton who based it on chemical facts. Moreover, Dalton essentially made a prediction, viz., in all chemical compounds, the elements that combine to form a compound combine in an integer(1):integer(2) ratio. There are no two elements which combine with a ratio of $\sqrt{2}:1$, for example. Assuming Dalton's hypothesis, the mass ratio of the atoms of oxygen to hydrogen, is 5.5:1 (see Table 1.2). If this is the mass ratio of the two atoms, then they combine either in a 1:1 or a 1:2 ratio, but not say in a 1:1.67 ratio.

In 1808, Dalton[33] published a new table in which all the weights were rounded off. The comparison between the two tables shows that the atomic weight of several

---

[30] Richter, J.B., *Ueber die Neuren Gegenstande der Chimie*, Breslau, 1791, *Anfangsgrunden der Stoichiometrie*, 1792–1794 (*Basics of Stoichiometry*).

[31] Dalton, J., Memoirs Literary Phil. Soc. Manchester, Second Ser. **1**, 271 (1805); reprinted in *Foundations of the Atomic Theory: Comprising Papers and Extracts by Dalton, William Wollaston, and Thomas Thomson*, Alembic Club Reprints No. 2, The Alembic Club, Edinburgh (1899) p. 15.

[32] Hydrogen peroxide, H_2O_2, was first discovered by Louis-Jacques Thenard (1777–1857) in 1818. However, it did not change the fact that the ratio of volumes is 2:1 or 1:1, i.e., always a ratio of two integers.

[33] Dalton, J., *A New System of Chemical Philosophy* **1**, Manchester, London, printed by S. Russel for R. Bickerstaff, 1808.

Table 1.2 Dalton's relative-to-hydrogen weights of the ultimate particles of gaseous and other bodies

Element	Dalton	Accurate
Hydrogen	1	1
Azote (nitrogen)	4.2	7
Carbon	4.3	6
Ammonia	5.2	5
Oxygen	5.5	8
Water	6.5	9
Phosphorus	7.2	15.5
Sulphur	14.4	14

From Dalton, read before the Philosophical Society of Manchester in 1803, but published in 1808. Our table here contains only the known elements. However, Dalton included many compounds, of which we include only water.

elements had 'changed' over the five years. The reason was the poor determination, impurities in the samples, and above all different methods used by various chemists. Furthermore, the comparison with present day values highlights the inaccuracies in the results. The admirable thing is that, despite the very poor available data for the atomic weights, the abstraction needed to get the law was nevertheless successfully made. Dalton disregarded the fact that two elements had the same atomic weight, and did not ask what made their atoms different.

A different classification of atoms was proposed by Berzelius,[34] namely, elementary atoms and compound atoms. The compound atoms divide into three types: atoms formed from two elementary substances united, which Berzelius called *compound atoms of the first order*, atoms composed of more than two elementary substances, which he found mostly to be organic compounds, so he called them *organic atoms*, and atoms formed by the union of two or more compound atoms, such as salts. These Berzelius called *compound atoms of the second kind*.

Most chemists accepted Dalton's hypothesis favorably. Thomson was even so enthusiastic as to argue that his own experiments were not accurate. Most of his papers were published in the journal he himself edited, which is quite convenient if you do not have accurate results. Thomson's estimates of atomic weights were worse than those of Dalton. Moreover, his estimates were influenced by Proust's hypothesis, something we might call theoretically biased experimental results.

Fundamental imperfections in Dalton's ideas were the lack of a definition of an 'ideal indivisible atom' (which caused Dalton to reject the work of Gay-Lussac) and some way to find the number of atoms in a given compound. The question of what the atoms were made of was not raised.

In the 1808 theory, Dalton rephrased Proust's law, calling it the 'law of multiple proportions'. Although it is unclear whether Dalton was directly influenced by Proust, the law of constant proportions provided evidence for Dalton's atomic theory, which in turn provided an explanation for Proust's observations.

---

[34] Berzelius, J.J., Annals of Phil. **2**, 443 (1813).

As Proust stated, after learning about Dalton's atomic hypothesis:

> According to Dalton's ingenious idea, that combinations are formed from atom to atom, the various compounds which two substances can form would be produced by the union of one molecule of the one with one molecule of the other, or with two, or with a greater number, but always without intermediate compounds. Thomson and Wollaston have indeed described experiments which appear to confirm this theory. Thomson has found that super-oxalate of potash contains twice as much acid as is necessary to saturate the alkali; and Wollaston, that the sub-carbonate of potash contains, on the other hand, twice as much alkali as is necessary to saturate the acid.

Dalton did not at first accept the results of Gay-Lussac concerning the fixed volume ratios of combining gases, and even claimed that he had had this idea earlier and found it untenable, despite the obvious support that Gay-Lussac's results provided to his hypothesis. He claimed that Gay-Lussac,[35] who was considered a very careful chemist and developed accurate methods, got inaccurate results. As an example, Dalton took HCl (or hydrogen chloride). If the supposition that equal volumes of gases contain equal numbers of particles is correct, equal volumes of hydrogen and chlorine should yield one volume of hydrogen chloride. But on the contrary, Dalton wrongly claimed, the yield is two volumes of hydrogen chloride.[36] It was clear that there was a missing link between the number of particles, the mass of particles, and the volume they occupy. It was Avogadro who showed how wrong Dalton was and how right Gay-Lussac was.

Berthollet eventually also attacked Dalton, who did not remain silent. Indeed, Dalton replied to Berthollet in the first part of his *New System of Chemical Philosophy*, which was published in 1808.[37] In Dalton's words, Berthollet's claim, *both in reason and observation, cannot fail to strike everyone who takes a proper view of the phenomenon*. In other words, Berthollet's conclusions, according to Dalton, had nothing to do with experiment. In the controversy, each protagonist referred to those experiments that agreed with his hypothesis. While Dalton knew of Berthollet's publication in France, he was apparently unaware of Proust's very correct publications in French. But summaries of Proust's papers were published in English,[38] so it is difficult to believe that Dalton was unaware of Proust's results. Furthermore, Proust's ideas appeared in Thomson's English textbook.[39] If the reader doubts whether Dalton read Thomson's description of Proust's results, then note that Dalton stressed in his book that Thomson failed to understand his hypothesis from 1802.[40] As for proper reference, there were so few chemists and publications in those days that it would have required a special effort to miss a publication!

---

[35] There is a long discussion and some surprises in Dumas' book *Lecons sur la Philosophie Chimique*, Bechet Jeune, Paris, 1837, based on his lectures at the Collège de France.

[36] Another source of conflict was the equation for water. Berzelius said H_2O, while Dalton assumed HO.

[37] Dalton, J., *A New System of Chemical Philosophy*, Manchester, 1808.

[38] Proust J.L., Ann. de Chim. **28**, 213 (1798), and the English abstract Nicholson J. **2**, 515 (1799).

[39] Thomson, T., *A System of Chemistry*, 4 vols. Edinburgh, 1802.

[40] Dalton, J., Phil. Mag. **14**, 169 (1802).

A burning question was: what is the weight of a hydrogen atom, or should the numbers be divided by the weight of oxygen? Various attempts by Thomson failed and led him to conclude that: *The above experiments* (to find the common divisor which yields integers) *will neither succeed with Dalton's, Wollaston's, nor Berzelius' numbers—a sufficient proof that none of them is absolutely correct.* Thomson therefore concluded that there was no proof for the atomic hypothesis.

1.9 Higgins and an Unjustified Claim for Priority

In 1814, William Higgins (1763–1825), who was a professor of chemistry in Dublin, published a book[41] in which he claimed priority for the atomic idea, giving as a reference his book from 1789.[42] The book dealt basically with the theory of phlogiston, which Higgins correctly claimed to be a wrong idea. Interestingly, Higgins wrote that:

> When I commenced my investigation [...] all the chemical philosophers were phlogistians. Under these circumstances, a very young man, such as I was at that time, might well be supposed to be intimidated, and even deterred from offering an opinion on the subject of chemical philosophy; however, the new view which I fortunately adopted furnished me with some degree of confidence.[43]

Regarding the atomic theory, Higgins wrote that *chemical attraction only prevailed between the ultimate particles of simple elementary matter, and also between compound atoms,* and he explained that *the term 'ultimate particles' means the last division of elementary matter.* This is more like an attempt at the chemical bond theory. Higgins continued: *Dalton repeats experiments formerly made by me, he does not even glance at the source from whence he derived his information.* But Higgins introduced the atomic hypothesis only as an additional argument against phlogiston and not as a basic argument on its own.

Dalton was probably unaware of Higgins' publication. A close examination shows that, even in 1814, Higgins assumed that all atoms have the same weight, while in Dalton's theory the atomic weight played the dominant role. According to Higgins,

---

[41] Higgins, W., *Experiments and Observations on the Atomic Theory and Electrical Phenomena*, Graisberry and Campbell, Dublin, 1814.

[42] Higgins, W., *A Comparative View of the Phlogistic and Antiphlogistic Theories: With Inductions.* To which is annexed, *An Analysis of the Human Calculus*, Publisher Murray, 1789. In this book, Higgins suggested marking the bonding interaction between atoms and within molecules by a straight line. Like Lavoisier, Higgins did not give up the caloric as he hypothesizes that *even when two atoms unite, the compound becomes surrounded with one common atmosphere of caloric* [...] and this description is continually mixed with the bonding of atoms.

[43] At this point Higgins cited the letter he got from the editor: *I shall be glad to see your book though I hope you have not taken Lavoisier's side of the question or else have defended it by arguments totally unlike any thing that has yet appeared. Dr. Priesley's late admirable experiments have in my opinion totally overset that doctrine and re-established the existence of phlogiston. Yours faithfully, Thos. Beddoes.* The editor was ready to publish only what he believed in!

atoms differ only in the *atmosphere of caloric which surrounds them.* But what does caloric have to do with the atomic theory? Thomson, who was a professor of chemistry in Glasgow, just on the other side of the Irish Sea, published right away in the journal he edited, Annals of Philosophy, a memoir[44] claiming that: *I have had that book* (he refers to Higgins 1789) *since the year 1798, and have pursued it carefully; yet I did not find anything in it which suggested to me the Atomic Theory. [...] and I still consider myself as the first person who gave the world an outline of the Daltonian theory.* Higgins complained that Dalton did not even bother to answer him. Only Thomson responded.[45] Alas, Higgins' attempts to shake Dalton's priority failed completely.[46] Dalton's ideas were not immediately accepted by all. Sir Humphrey Davy, for example, was among the notable objectors, yet he eventually yielded to the evidence.

1.10 If You Dislike a Theory, Reinvent it Under Another Name

On 4 November 1813 William Wollaston (1766–1828m) lectured before the Royal Society[47] and reviewed the work of his predecessors. Wollaston argued that:

> When the nature of any saline compound is proposed as the subject of inquiry to an analytical chemist, the questions that occur for his consideration are so varied and so numerous, that he will seldom be disposed of satisfying his inquiries, so long as he can rely upon the accuracy of those results that have been obtained by the labor of others, who have preceded him in this field of patient investigation.

In short, Wollaston was bothered by the fact that many published results were not identical, and even contradictory, which meant that it was a problem to sort out the results or repeat all the measurements. Indeed, Wollaston did not repeat all the measurements and a quick look at his table reveals that his results were not free of errors either (see Table 1.3).

So Wollaston devised a new scale which he called the synoptic scale. This scale was intended to *answer at one view all these questions*, and he meant questions concerning the properties of elements and compounds. You can see some hints of element classification. The rationale behind the need for a new scale was:

---

44 Thomson, T., Annal. Phil. **3**, 329 (1814).

45 Higgins, W., The Phil. Mag. **XLIX**, 241, 1817. The letter to the editor is amazing for its strange style of English: *It is stated that this paper*, referring to Dalton's paper, *was read before the Literary and Philosophical Society of Manchester in the year 1816. That should lie by since is not to be wondered at, as containing nothing new; it relate to a hackneyed subject, which chemists have lately gone over repeatedly. In that part of my paper which appeared in your excellent Magazine for December 1816, I observed that Dr. Thomson stepped forward repeatedly in a very unjust cause, which could never do him credit, as the advocate of Mr. Dalton, while the latter stood silent trembling as the bar of justice*. It could easily be part of a play.

46 A list of several tens of articles dealing with Higgins' claim can be found in Wheeler, T.S., Studies, An Irish Quarterly Review **43**, 327 (1954).

47 Wollaston, W.H., Phil. Trans. Roy. Soc. London **104**, 1 (1814).

Table 1.3 Wollaston's synoptic scale compared with accurate values of the atomic weights relative to oxygen = 10, as assumed by Wollaston in 1814

Element	Synoptic scale	Accurate
Hydrogen	1.32	0.625
Oxygen	10.00	10.00
Water	11.32	11.25
Carbon	7.54	7.51
Sulphur	20.00	20.0406
Phosphorous	17.40	20.609
Nitrogen	17.54	8.75
Chlorine	44.1	22.158
Magnesium	24.6	15.191
Strontium	69	54.763
Barium	97	85.829
Iron	34.5	34.903
Copper	40	39.719
Zinc	41	40.869
Mercury	125.5	125.369
Lead	129.5	129.5
Silver	135	67.418

> It it is requisite [...] to determine the proportions in which the different known chemical bodies unite with each other, and to express these proportions in such terms that the same substance shall always be represented by the same number.

Wollaston stated that the original idea was due to Richter.[48] Wollaston sorted out the results by his predecessors and came up with Table 1.3. The table gives the element and the synoptic scale of the element and the compound. As can be seen, Wollaston suggested taking oxygen as reference, and attributed a scale of 10 to it. A column has been added to show the correct atomic weight, assuming the atomic weight of oxygen to be 10. The comparison shows once again how poorly the atomic weights were known.

A simple glance at the table shows that:

- The synoptic scale[49] is nothing but the atomic weight.
- Many numbers in the table are in error.

The newly invented name was very promising, but as soon became clear it contained nothing but the atomic weights that Dalton had been talking about. Inventing a new name is not necessarily a discovery.

It is interesting to ask why the atomic weight of oxygen was used as a reference for the atomic weight. The reason Berzelius chose to relate all atomic weights to

[48] But I could find no reference for it in Wollaston's paper.

[49] Wollaston did not explain in the paper why he chose this name for his scale. It is implied in the paper that this number should provide a good explanation for the chemical properties of the elements and compounds.

oxygen was the fact that oxygen was capable of combining chemically with every other element. In fact, oxygen compounds were almost the only ones made use of at that time for the derivation of atomic weights. In Berzelius' own words: *Oxygen is the central point around which the whole of chemistry revolves.*

1.11 Avogadro: A Superb Insight, Perfect Timing, but Rejected

Shortly after Dalton suggested the atomic hypothesis, in 1811, the hitherto unknown Italian chemist Amedeo Avogadro (1776–1856m)[50] came up with a sensational hypothesis:

> The first hypothesis to present itself in this connection, and apparently even the only admissible one, is the supposition that the number of integral molecules in any gas is always the same for equal volumes, or always proportional to the volumes. Indeed, if we were to suppose that the number of molecules contained in a given volume were different for different gases, it would scarcely be possible to conceive that the law regulating the distance of molecules could give in all cases relations as simple as those which the facts just detailed compel us to acknowledge between the volume and the number of molecules.

Avogadro based his hypothesis on Gay-Lussac's demonstration *that gases always unite in a very simple proportion by volume*. Avogadro also reasoned that the distance between molecules is so large that their mutual attraction cannot be effective. He essentially found what we call today an ideal gas, i.e., a gas in which the interaction between the molecules of the gas is negligible. Avogadro needed this assumption about the mean distance between the molecules in order to safely assume that the imaginary caloric plays no role in the problem. Avogadro assumed that caloric is important only when the gases condense. So, to be on the safe side, Avogadro claimed that:

> [...] in our present ignorance of the manner in which this attraction of the molecules for caloric is exerted, there is nothing to decide us a priori in favor of the one of these hypotheses rather than the other, and we should rather be inclined to adopt a neutral hypothesis, which would make the distance between the molecules and the quantities of caloric vary according to unknown laws, were it not that the hypothesis we have just proposed is based on that simplicity of relation between the volumes of gases on combination, which would appear to be otherwise inexplicable.

As for Dalton, Avogadro claimed that:

> Dalton, it is true, has proposed a hypothesis directly opposed to this, namely that the quantity of caloric is always the same for the molecules of all bodies whatsoever in the gaseous state, and that the greater or lesser attraction for caloric only results in producing a greater or less condensation of this quantity around the molecules, and thus varying the distance between the molecules themselves.

The caloric was of course a red herring, but Avogadro did not know it, so to be on the safe side he considered those cases in which he thought that it could not be important.

---

[50] Avogadro, A., J. de Physique **73**, 58 (1811). Alembik Club Reprint no 4.

Avogadro's logic was very simple:

M. Gay-Lussac has shown in an interesting Mémoir [...] that gases always unite in a very simple proportion by volume, and that when the result of the union is a gas, its volume also is very simply related to those of its components. But the quantitative proportions of substances in compounds seem only to depend on the relative number of molecules which combine, and on the number of composite molecules which result. It must then be admitted that very simple relations also exist between the volumes of gaseous substances and the numbers of simple or compound molecules which form them. The first hypothesis to present itself in this connection, and apparently even the only admissible one, is the supposition that the number of integral molecules in any gas is always the same for equal volumes, or always proportional to the volumes. [...] For example, since the numbers 1.10359 and 0.07321 express the densities of the two gases oxygen and hydrogen compared to that of atmospheric air as unity, and the ratio of the two numbers consequently represents the ratio between the masses of equal volumes of these two gases, it will also represent on our hypothesis the ratio of the masses of their molecules. Thus the mass of the molecule of oxygen will be about 15 times that of the molecule of hydrogen, or, more exactly as 15.074 to 1. In the same way the mass of the molecule of nitrogen will be to that of hydrogen as 0.96913 to 0.07321, that is, as 13, or more exactly 13.238, to 1. On the other hand, since we know that the ratio of the volumes of hydrogen and oxygen in the formation of water is 2 to 1, it follows that water results from the union of each molecule of oxygen with two molecules of hydrogen.

Avogadro extended his theorem to gaseous molecules containing atoms from more than one element. But Avogadro made a grave mistake when, in his enthusiasm, he erroneously extended his theorem to non-gaseous bodies and consequently caused many to distrust the hypothesis and flatly reject it. It should also be pointed out that Avogadro had no reputation as a leading scientist, and nor did he have a prestigious affiliation—and in those days such things mattered. How about today?[51]

Three years later, in 1814,[52] André Marie Ampère (1775–1836m) rediscovered the hypothesis. Ampère was careful to distinguish between particles and the atoms which compose them, but it did not help to gain acceptance for the Avogadro hypothesis. It was just ignored. Progress was delayed for about half a century until Cannizzaro came along and rescued Avogadro's hypothesis from oblivion.

1.12 Prout and the Name of the Game: Numerology

As early as 1815, William Prout (1785–1850)[53] proposed that the atomic weight of any substance could be represented by an integer multiple of the atomic weight of hydrogen, as if hydrogen were the basic building block of all atoms. Prout was

---

[51] Avogadro was born into a well-known family of lawyers and as such, studied law and even started practising. But his soul was in the natural sciences, and in 1800, he began to study physics. In 1809 he started teaching in a high school and it was while serving as a teacher that he announced his discoveries.

[52] Ampère, A-M., Ann. de Chim., 1ère serie **xc**, 43 (1814). There is a mountain ridge called Mons Ampère on the moon.

[53] Anonymous, *On the relation between the specific gravities of bodies in their gaseous state and the weight of their atoms*, Annals of Phil. **6**, 321 (1815). The paper was published anonymously.

not sure whether his hypothesis was right, and wrote in a paper published with anonymous authorship:

> The author of the following essay submits it to the public with the greatest diffidence; for though he has taken the utmost pains to arrive at the truth, yet he has not that confidence in his abilities as an experimentalist as to induce him to dictate to others far superior that its importance will be seen, and that some one will undertake to examine it, and thus verify or refute its conclusions. If these should be proved erroneous, still new facts may be brought to light, or old ones better established, by the investigation; but if they should be verified, a new and interesting light will be thrown upon the whole science of chemistry.

Strange as it may be in science, the editor of the journal, Thomas Thomson,[54] a chemist and mineralogist, who through his chemistry textbooks did much to spread the atomic theory of Dalton, agreed to publish such a paper. This may have been because the idea appealed to him, as he was an ardent supporter[55] of the atomic idea.[56] Mind you, Thomson charmingly changed his mind about the claim that no such divisor could be found!

A year later, a second unsigned paper appeared in the Annals of Philosophy,[57] in which the anonymous author corrected some errors made in the first paper and improved the hypothesis. A year later, Thomson wrote a paper on the atomic hypothesis and exposed Prout as the anonymous author of the papers. Prout's original table is given in Table 1.4. As attested by the table, the atomic weights are extremely close to being integer products of the atomic weight of hydrogen. In the second paper Prout admitted that *if the view we have ventured to advance be correct, we may almost consider the πρωτηνλη*[58] *of the ancients to be realized in hydrogen*; *an opinion, by the by, not altogether new* (Fig. 1.2).

Although Dalton had already shown in 1805 that the gases in the atmosphere are a mixture and not a compound,[59] and even discussed the amount of water vapor in the atmosphere, Prout considered the atmosphere to be a compound made of one atom of oxygen and two atoms of nitrogen, whence it constituted a proof of the atomic theory.

(Footnote 53 continued)
This is a rare phenomenon. Most frequently you find cases were no citation to a previous discovery is made and the author wants to credit himself for it. Such behavior is a testimony to Prout's standards, i.e., if you are not sure, you publish anonymously.

[54] In 1813 Thomson founded and subsequently edited the Annals of Philosophy.

[55] Thomson, T., Annal. Phil. **2**, 32 (1813). Thomson announced Dalton's theory in his 1807 book, a year before Dalton himself published it.

[56] How times have changed! Can we imagine an anonymous publication today, motivated by the author's uncertainty regarding an idea?

[57] Anonymous, Annals of Philosophy **6**, 321 (1815).

[58] πρωτηνλη means 'raw material', from which the term protyle was coined for a hypothetical primitive substance.

[59] Dalton, J., Memoirs of the Literary and Phil. Soc. Manchester **1**, 244 (1805). Read before the society 12 November 1802.

Table 1.4 Prout's table of relative atomic weights of elements. Prout was courageous in ignoring the round-off errors in the inaccurate experimental results

Name	Sp. weight atmos. (air = 1)	Relative to hydrogen (accurate)	Prout's number rel. to H	Comment
Hydrogen	0.06944	1		Based on ammonia
Carbon	0.4166	6.003	6	Assumed weight
Nitrogen	0.9722	14.009	14	Based on air being a compound of O and N in 1:5 ratio
Phosphorus	0.9722	14.009	14	See comment in text
Oxygen	1.1111	16.010	16	Based on air being a compound of O and N in 1:5 ratio
Sulphur	1.1111	16.010	16	Weight assumed to be 20
Calcium	1.3888	20.012	20	Value 20.08 from Prout's numbers
Sodium	1.6666	24.014	24	
Iron	1.9444	28.017	28	Value 28.08 from Prout's numbers
Zinc	2.222	32.020	32	Value 31.85 from Prout's numbers
Chlorine	2.5	36.023	36	Assume that sp. gr. does not differ much from 2.5
Potassium	2.7777	40.025	40	
Barytium	4.8611	70.045	70	
Iodine	8.6111	124.079	124	Based on an iodine solution

The idea of error analysis did not exist at that time, and the difficulty in measuring the specific weight of a light gas like hydrogen was not expressed in the final result.[60] Checking Prout's calculations carefully, we find that he rounded his numbers without proper justification, sometimes giving numbers to 4 significant figures and then rounding at the 10% level! However, if it were not for this rounding, the atomic theory might have had to wait another century before the idea resurfaced again, and once again, the accuracy, or rather insufficient accuracy, would only lead to a vindication of Prout and the conclusions would still be wrong (see Sect. 1.36).

Why did Prout not have a good value for the atomic weight of phosphorus? As he explained in his paper:

> I have made many experiments in order to ascertain the weight of an atom of this substance; but, after all, have not been able to satisfy myself, and want of leisure will not permit me to pursue the subject further at present.

[60] As the molecular weight of hydrogen is 2, this means that 22.4 l of gas weighs 2 g. The weight of the container is then much greater, and manipulations were needed to determine the weight of the gas accurately. Comparison between the results of different researchers gives a good indication of the degree of accuracy in the weights obtained in those days. The fact that hydrogen gas is made up of molecules containing two hydrogen atoms confused the issue, since only one atom of hydrogen participates in some of the molecules, leading to 'half molecules'.

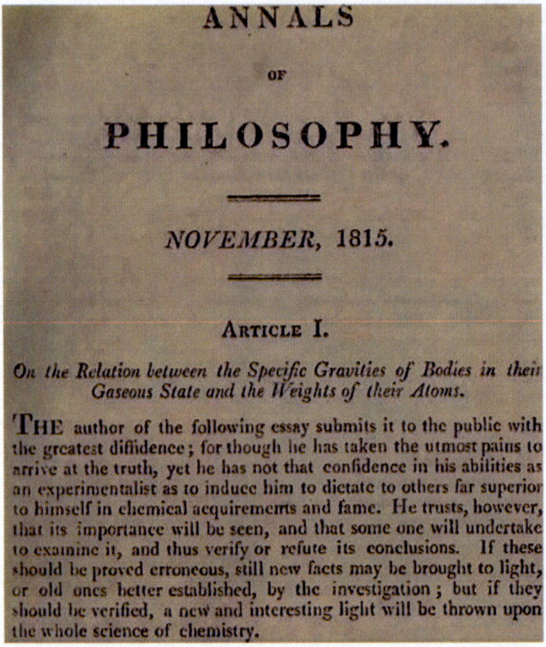

ANNALS

OF

PHILOSOPHY.

NOVEMBER, 1815.

ARTICLE I.

On the Relation between the Specific Gravities of Bodies in their Gaseous State and the Weights of their Atoms.

THE author of the following essay submits it to the public with the greatest diffidence; for though he has taken the utmost pains to arrive at the truth, yet he has not that confidence in his abilities as an experimentalist as to induce him to dictate to others far superior to himself in chemical acquirements and fame. He trusts, however, that its importance will be seen, and that some one will undertake to examine it, and thus verify or refute its conclusions. If these should be proved erroneous, still new facts may be brought to light, or old ones better established, by the investigation; but if they should be verified, a new and interesting light will be thrown upon the whole science of chemistry.

Fig. 1.2 Prout's first of two anonymous papers published in 1815 and 1816

It appears that in those days scientists appreciated rest more than the perpetual pursuit of new results.

Prout found that nitrogen and phosphorus should contain 14 hydrogen atoms, and likewise that oxygen and sulphur should contain 16 hydrogen atoms. So why was nitrogen different from phosphorus and oxygen from sulphur? No answer was given to this question, and nor did Prout comment on it. For Prout the issue was to get an integer number. Moreover, the fact that all numbers came out even, meaning that only an even number of hydrogen atoms was possible, did not cause any concern. It was not an oversight because, when he found that ammonia had been claimed to be composed of one atom of nitrogen and only 1.5 of hydrogen, he declared that this must be an error. A glance at the specific weights reveals that many of them were obtained by divisions of rounded, guessed numbers, in which the result was motivated by the hypothesis.

Worse than that, it is clear from Prout's own comments in the paper that he arbitrarily changed the atomic weights to make them fit his hypothesis. Prout himself was a physician, and his own investigations in chemistry were few in number and anything but conclusive, accurate, or methodical.

Table 1.5 The stable isotopes of chlorine. Atomic number 17

Atomic weight	Relative abundance
34.968 852 68	0.757 6
36.965 902 59	0.242 4

1.13 Berzelius and the End of Prout's Atomic Hypothesis

A few years after Prout put forward his hypothesis, Berzelius,[61] one of the most influential chemists of the 19th century,[62] changed his mind about Prout's atomic theory, although he continued to support Dalton's theory, i.e., that the atoms of different elements are different. Berzelius carefully examined[63] the measured atomic weights. The most damaging evidence against Prout's simple assumption was the atomic weight of chlorine, which was measured to be 35.47 times that of hydrogen.[64] The fact that an atomic weight was close to half an integer led some, in an attempt to salvage Prout's theory, to come up with the ad hoc suggestion that the basic unit was one half of a hydrogen atom. But to no avail. Further discrepancies surfaced which resulted in the hypothesis that 1/4 of a hydrogen atom was the basic unit. Berzelius' crusade against Prout's theory succeeded in driving Prout's hypothesis into oblivion.

Berzelius' liquidation of Prout's atomic theory brought a consensus among chemists. Spectroscopy, which was used as the exclusive method for identifying elements as of 1860, was not sufficiently accurate to distinguish the isotopic effect.

The solution of the chlorine conundrum was found in 1919–1920, when Aston discovered the existence of isotopes and it was clear that chlorine has just two stable isotopes (see Table 1.5) with atomic weights close to 35 and 37, respectively, and an abundance ratio of 3:1, yielding a chemical atomic weight close to 35.5, and thereby misleading all chemists into suspecting Prout's hypothesis. But as nickel and cobalt had the same atomic weight and were two distinct elements, it is strange that the opposite idea did not emerge, i.e., that chlorine has two types of atoms and that the chemist just observes a mixture of the two. Somehow it was simpler to believe in the existence of a fictitious fluid like caloric. The atomic weight of chlorine was the stumbling block upon which Prout's theory fell.

[61] Berzelius, J.J., Pogg. Ann. **vii**, 397 (1821).

[62] Berzelius discovered the elements cerium in 1803, selenium in 1817, silicon in 1824, and thorium in 1828. Lithium salt was discovered by Berzelius's student Johan August Arfwedson (1792–1841) in his laboratory in 1817 (pure lithium was isolated by Davy). Another student of Berzelius, Nils Gabriel Sefström (1787–1845), discovered vanadium. Berzelius also suggested the way we write chemical reactions today. Dalton for example, used geometrical symbols, while Berzelius suggested using the first letter of the element, for example, $C + O^2 \rightarrow CO^2$. Later the superscript became a subscript.

[63] Berzelius, J.J., *Essay upon the Theory of Chemical Proportions*, Ann. of Phil. **2**, 443 (1813).

[64] Berzelius actually got 17.735 times the weight of hydrogen for the atomic weight of chlorine.

1.14 Petit and Dulong's Law and Fading Hopes

The year 1819 brought a very important discovery in the atomic weight saga when Dulong (1785–1838) and Petit (1771–1820m)[65] demonstrated that the specific heat at constant volume of an element multiplied by its atomic weight is approximately constant. The number of atoms per unit volume is inversely proportional to the atomic weight. If the heat is expressed as vibrations of the atoms and does not depend on the mass of the vibrating atom, then the product of atomic weight times the specific heat at constant volume must be constant. However, the motivation for Dulong and Petit was formulated according to Dalton in the following way:

> The quantities of heat united to the elementary particles of the elastic fluid are the same for each. Hence we may, setting out from our knowledge of the number of particles contained in the same weight or the same volume of the different gases, calculate the specific heats of these gases. This M. Dalton has done, but the numbers which he obtained, and those likewise deduced from several other better founded hypothesis on the constitution of gases, are so inconsistent with experiment that it is impossible for us not to reject the principle upon which such determinations are founded.

Owing to these problems, Dulong and Petit decided to measure the cooling rates of solids and thereby deduce their heat capacity.

The most important conclusion reached by Dulong and Petit was this: *The atoms of all simple bodies have exactly the same capacity for heat.* They meant heat capacity at constant volume.[66]

Indeed, the first expectations were fulfilled and gave rise to the hope that a new and accurate method for measuring atomic weights, in particular for solids, might have been discovered. Dulong and Petit even applied their new discovery to correct some reported results of atomic weights. For example, while Berzelius obtained the atomic weight of lead as 2589 on the basis of O = 10, Dulong and Petit claimed that the value should be 1295, or half the value given by Berzelius. Hence, factors of two and four floating around in Berzelius's table were spotted very quickly. In short, they modified eleven out of the thirteen atomic weights given by Berzelius!

While Dulong and Petit hoped their discovery would be applied to the determination of atomic weights, the unexplained inaccuracy involved caused their hopes to fade away (see Fig. 1.3). However, the law helped in screening gross errors. In 1831, Franz Neumann (1798–1895)[67] discovered a similar relationship between

---

[65] Petit, A.T. and Dulong, P.L., Ann. Chim. Phys. **2**, 395 (1819). Dulong succeeded Petit as professor of physics, and his name is among the 72 mostly French names engraved on the Eiffel tower. Petit, on the other hand, has a mountain on the moon named in his honor.

[66] Almost a century later, in 1912, Debye discovered that every solid has a temperature, the Debye temperature, above which, and only above which, the Dulong and Petit law is valid. In their paper, Dulong and Petit did not provide any information about the temperature range of their measurements. But the Debye temperature, which they were unaware of, is very close in many cases to their supposed range of measurement. For example, $T_{\mathrm{Debye}} = 470\,\mathrm{K}$ for iron, whence the value they got for iron was doubtful. In other cases like gold, $T_{\mathrm{Debye}} = 165\,\mathrm{K}$ and Petit and Dulong could easily get an accurate measurement.

[67] Neumann, F., Poggendorff's Ann. Phys. **23**, 1 (1831).

Fig. 1.3 The results of Petit and Dulong 1819. The atomic weight multiplied by the specific heat is plotted for each element

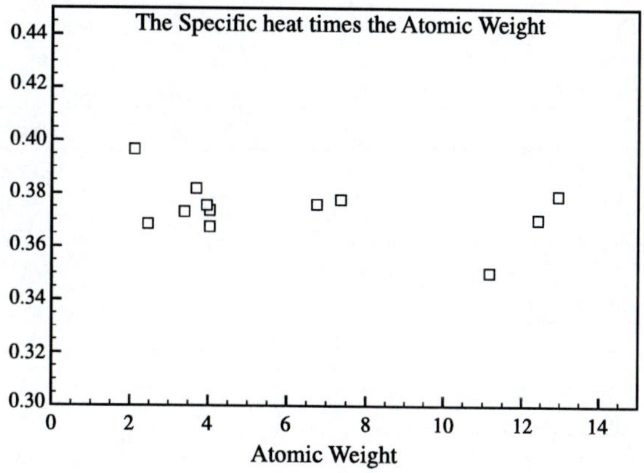

the specific heat and molecular weight of compounds, but again of limited use for the same reason: insufficient accuracy.

In 1828, Dumas made an attempt to advocate the use of the Petit and Dulong law to measure atomic weights, in particular after he cited experiments which he pointed out *were made with the greatest care* by La Rive (1801–1872) and Marcet (1803–1883)[68] and which confirmed the law, but he was not successful.

1.15 Thomson 1821: Prout Is Nevertheless Right

In 1821, Thomas Thomson (1773–1852)[69] published an attempt to confirm Prout's hypothesis. Thomson claimed to have shown that Berzelius' method to determine the atomic weights was *not precise enough for the determination of the weights of the atoms of bodies*. Since this was quite a blunt accusation, Thomson felt the need to add:

> I ought to observe, however, that Berzelius is undoubtedly a very great master of the analytical art; that his analysis approach upon the whole exceedingly near the truth; and exhibit a consistency which is highly satisfactory, and does a great deal of credit to the industry and sagacity of their authors. But unfortunately his mode of experimenting admits of no criterion by which the experimenter can determine whether the results be accurate or not; so he has not means of checking himself.

A compliment it was not. A glance at Tables 1.6 and 1.7 will explain the source of Thomson's comment. The comparison given in Table 1.6 shows how different

---

[68] La Rive, A.A. and Marcet, F., Ann. de Chim. et de Phys. **75**, 113, 119, 242 (1840). But their work was criticized by Weber, H.F., Phil. Mag. **49**, 161 (1875) as not being sufficiently accurate.

[69] Thomson, T., Annals of Phil. **I**, 1 (1821). This was the last volume of the journal edited by Thomson. In the editorial he explained that he was *beginning a laborious course of Chemical Lectures with scarcely any previous preparation*, and hence had to relinquish the position of editor.

Table 1.6 Typical results obtained by different chemists. After Thomson, 1821. While some of the results of each chemist are given to 4 or even 6 significant figures, the discrepancy between the various results is sometimes in the second significant figure

	Dalton	Wollaston	Berzelius	Prout	Thomson
Strontian	6.571	6.9	12.9460		6.5
Lime	3.428	3.536	7.1206	3.5	3.625
Magnesia	2.428	2.46	5.1272		2.5
Phosphoric acid	3.285	3.74	8.9230	3.75	3.5
Arsenic acid			14.4077	7.25	

Table 1.7 Thomson's 1813 table of atomic weights

Element	Atomic weight relative to oxygen	Atomic weight relative to hydrogen	Accurate
Oxygen	1.000	7.576	8.00
Hydrogen	0.132	1.000	1.00
Carbon	0.751	5.689	6.01
Nitrogen	0.878	6.652	7.03
Phosphorus	1.320	10.000	15.47
Sulphur	2.000	15.152	16.03

the results obtained by different chemists were. Thomson pointed out that dividing Berzelius' result by 2 provided a better agreement with the results of other investigators.

Berzelius did not remain in debt.[70] The report on the exchange between the two chemists was not signed. The title of the paper is: *Attack of Berzelius on Dr. Thomson's "Attempt to establish the First Principles of Chemistry by Experiment"*. It is clear from the text that the purpose of the anonymous author was to defend Thomson, saying that his *correspondence between his hypothesis and the results of his experiments* was *startlingly precise.* The author therefore asked *impatience* for Berzelius' opinion, which was *expressed in language extremely strong and extremely unusual*, to the extent that the author thought it *necessary to employ his own words.* He *accordingly translated a few passages from the Yares-Bericht for 1827.* And so wrote Berzelius:

> Someone had told Thomson that his whole work was of little value, because, in the fundamental experiment [. . .] Thomson openly replied that he had supposed chemists would have given him credit for a knowledge of the mode of separating oxide of zinc from acids, and had therefore omitted the details; but as he found this opinion erroneous, it became necessary to publish a full account of his process. [. . .] It appears to me that love for the real progress of science makes it imperative to detect quackery, and expose it to the judgement of every one as it merit.

[70] Anonymous, Phil. Mag. **4**, 450 (1828).

The author of this article went on to say that:

> We regret to see the dignity of science sacrificed by the intemperance of those who profess to be her advocates. It well becomes Berzelius to expose fallacy in argument, or detect error in analysis; but let him not pass beyond the limits of fair criticism; let him not arraign the character of an individual. [...] Intemperate attacks, such as this, reflect back upon their author, and indicate a mind inflamed by pique, jealousy, or some unworthy passion.

The name of the anonymous reporter remained anonymous.

Berzelius did not hesitate to express his view in textbooks.[71] In complete contrast to the way standard scientific texts are formulated, Berzelius criticized experiments from which, according to him, incorrect conclusions were drawn. Meyer[72] exemplifies Berzelius' approach with the following extract from a letter dated 3 April 1838, which Berzelius sent to Justus Leibig[73]:

> In the treatment of scientific subjects a man must have neither friends nor foes. He combats what appears to him to be an error without any reference whatever to the personality of the writer who advances it. Opinions are not individuals, and one can upset an opinion without finding in that a reason to treat its author as an enemy. Only in the case of palpable scientific plagiarism is one entitled to write sharply; and even then it is best to censure without the least sign of passion, for this awakens in the reader's mind the thought: *audiatur et altera pars*.

The meaning is that no person should be condemned unheard.[74]

In another paper, in the same year, Thomson[75] presented Table 1.8, which provides a nice demonstration of Prout's hypothesis. However, Thomson had problems with several metals, as can be seen from Table 1.9. These elements refused to obey Prout's law. Thomson got numbers like 3.39215 (relative to oxygen) for the atomic weight of iron, and similarly non-integer numbers for the other metals. So Thomson discussed each element separately to find the reason why it would not obey the law. Despite the disagreements with the metals, which were not resolved, Thomson did not abandon his faith in Prout's hypothesis. It was too beautiful and too symmetric to relinquish because of a couple of rebellious metal elements.

Thomson was happy with the resulting table. However, there was no question raised as to how 28 hydrogen atoms could yield both iron and manganese, and

---

[71] Berzelius, J.J., *Lehbuch der Chemie*, Dresden, 1820. The book first came out in Swedish in 1808–1818, and was later translated to German by Wöhler.

[72] von Meyer, E., *A History of Chemistry from Earliest Times to the Present Day*, translated from German by McGowan, MacMillan and Co., London, 1906. The original book was published in Leipzig in 1888.

[73] Leibig (1803–1873) was a well-known chemist, famous for his analysis of many organic compounds and his theory of organic radicals. He is generally considered to be one of the founders of organic chemistry. However, Leibig's name went before him in his appetite for scientific controversies. As of 1837, he held the powerful position of editor at *Anleitung zur Analyse organischer Körper*. So Berzelius' letter was sent hardly a year after Leibeig took the position as editor. The most famous controversies he was involved in were with Mulder, Wöhler, and Pasteur. We may say that Berzelius certainly sent the letter to the right person.

[74] The correspondence was published by Carriere, Pub. Lehmann, Munich and Leipzig, 1893.

[75] Thomson, T., Annals of Phil. **I**, 81, 241 (1821).

Table 1.8 Thomson's 1821 table of atomic weights

	O = 1	H = 1
Nitric acid	6.75	54
Sulphuric acid	5.0	40
Muriatic acid	4.625	37
Chromatic acid	6.5	52
Arsenic acid	7.75	62
Phosphoric acid	3.5	28
Potash	6.0	48
Soda	4.0	32
Barytes	9.75	78
Strontian	6.5	52
Lime	3.5	28
Magnesia	2.5	20
Silver	13.75	110
Lead	13.00	104

26 hydrogen atoms both nickel and cobalt. How could this be? It was not a question of oversight, because Thomson noted that all the numbers of hydrogen atoms were even. In retrospect, it is amazing how the atomic theory was 'proven' correct in the presence of such huge measurement errors.

The conflicting results from different chemists triggered comments by Thomson like[76]:

> My object in this paper is to show that the present state of our knowledge leaves no doubt whatever that Gay-Lussac's proportions are accurate and that Mr. Dalton has misled himself.

Here, Thomas refers to Dalton's rejection of Gay-Lussac's results, although these results provided more evidence for his theory.

Berzelius then set about determining atomic weights with more sophistication than Dalton had been able to do. In this project, Berzelius made use of the findings of Dulong and Petit and of Mitscherlich,[77] as well as Gay-Lussac's law of combining volumes. (He did not, however, use Avogadro's hypothesis.) Berzelius' first table of atomic weights, published in 1828, compared favorably for all but two or three elements, with today's accepted values.

Berzelius seemed to destroy *Prout's attractive suggestion* (attractive, because it reduced the growing number of elements to one fundamental substance, after the fashion of the ancient Greeks, and thereby seemed to increase the order and symmetry of the Universe) with more than one example. Thus, on a hydrogen-equals-1 basis, the atomic weight of oxygen is roughly 15.9, and oxygen can scarcely be viewed as being made up of fifteen hydrogen atoms plus nine-tenths of a hydrogenatom.

[76] Thomas, T., Annals of Phil. **I**, 322 (1821).

[77] Eilhard Mitscherlich (1794–1863) was a German chemist, best remembered for his law of isomorphism, which states that compounds crystallizing together probably have similar structures and compositions. This relationship was used by Berzelius in early attempts to assign relative masses to the elements.

Table 1.9 The number of oxygen and hydrogen atoms calculated by Thomson to compose metals. The last column is the accurate atomic weight as known today

	O = 1	H = 1	Accurate value found in nature
Copper	8.0	64	63.55
Zinc	4.25	34	65.39
Iron	3.5	28	55.845
Manganese	3.5	28	54.938
Nickel	3.25	26	58.6934
Cobalt	3.25	26	58.9332

The improved published tables of atomic weights and Berzelius' assertion that the atomic weights of the various elements were not integral multiples of the atomic weight of hydrogen was considered fully established and became an accepted dogma.

1.16 Berzelius and the Realisation that Imaginary Substances Are Not Chemical Elements

Lavoisier's table contained such entities as caloric and light which did not of course have any atomic weight. Chemists who came after Lavoisier correctly removed heat and light from the list of chemical elements without providing any explanation. It may be that caloric disappeared from the tables of elements precisely because it could not be weighed. However, it should be noted that, years later, Mendeleev suggested that the imaginary substance ether should be classified as an element with vanishing atomic weight.

It was Berzelius[78] who argued that light, heat, and magnetism were *properties of bodies* and not entities like elements, and hence had no place in the list of elements. But as late as 1827, Gmelin[79] still considered heat to be an imponderable substance. And as a matter of fact, heat was considered to be a chemical entity right up until 1842 when Julius Robert Mayer (1814–1878)[80] formulated the energy conservation principle and essentially moved heat away from chemistry and into physics.

1.17 Dumas 1828

In 1826, Jean Baptiste André Dumas (1800–1884)[81] developed a method for measuring the density of substances that are liquid or even solid at ordinary temperatures. The gaseous elements known at the time were hydrogen, nitrogen, oxygen, and

---

78 Berzelius, J.J., *Lehrbuch der Chemie*, Dresden und Leipzig, 1833.

79 Gmelin, L., *Handbuch der Theoretischem Chemie*, 1827.

80 von Mayer, J.R., Ann. der Chem. und Pharmacie **43**, 233 (1842).

81 Unlike Fourier before him, who served the emperor but continued to be creative in science, Dumas largely abandoned his scientific research and in 1848 went to work for the ministry of

chlorine, and only these gases could be used, for example, in the Gay-Lussac type of experiments,[82] unless one could use compounds that evaporated easily. Dumas' new method thus allowed a significant extension of the list of elements with measured atomic weights.

In 1826, Dumas[83] carried out experiments on Avogadro's hypothesis, which he called the Ampère hypothesis, without even mentioning Avogadro, and concluded at that time that the Ampère hypothesis was correct. Moreover, Dumas explained that the molecules that satisfy Ampère's law can be further divided. In this way, he explained how one volume of hydrogen combined with one volume of chlorine to produce two volumes of hydrogen chloride.[84]

A couple of years later, in 1837, Dumas published his lectures on the philosophy of chemistry in the form of a book.[85] One of the questions raised concerned the meaning of Dalton's hypothesis, and in particular the question: What are the atoms? Dumas thus wrote[86]:

> You may want proofs for the existence of atoms. But Dalton does not suggest any evidence. Dalton assumed the existence of atoms, but he does not prove their existence. The assumption about their existence helps to formulate the relations between chemical phenomena.

And so he claimed that the basis for the existence of atoms was never demonstrated: *C'était faire un cercle vicieux, et leur argumentation est demeurée sans authorité*, i.e., it was a vicious circle, and the argument remained without force. Dumas went on to argue that the atomic hypothesis was not required to explain observed chemical phenomena. For example, if a minimum mass was needed for a reaction to take place, one could do away with the atomic assumption, while the assumption about the indivisibility of the atom was not needed either. At this point, Dumas entered into a long, complicated, and farfetched astronomical argument ending with the claim that, since the atmosphere of the Earth is finite, this proved, à la Dumas, the infinite divisibility of matter. Strange logic indeed.

In the first volume of his earlier book[87] Dumas dealt with these questions in the introduction. It was in the introduction that he stated Avogadro's law, but without mentioning either him or Ampère. Dumas discussed the Dulong and Petit discovery. Although surprisingly Prout was not mentioned, Dumas wrote that *several English*

(Footnote 81 continued)
education under Napoleon III. Dumas became a member of the National Legislative Assembly, a senator, president of the municipal council of Paris, and master of the French mint. His administrative career came to its end with the collapse of the Second Empire.

[82] Dumas, J.B.A., Ann. Chim. Phys. **33**, 337 (1826), ibid. **50**, 337 (1832).

[83] Dumas, J.B., Ann. Chim. Phys. **33**, 337 (1826).

[84] Dumas realized that the reaction is $H_2 + Cl_2 \rightarrow 2HCl$ and not $H + Cl \rightarrow HCl$. In the first reaction, one volume of hydrogen and one volume of chlorine yield two volumes of HCl. The second reaction implies that one volume of hydrogen plus one volume of chlorine yields only one volume of hydrogen chloride.

[85] Dumas, J.B., *Leçons sur la Philosophie Chimique*, Bechet Jeune, Paris, 1837.

[86] This is the author's free translation of Dumas, p. 231.

[87] Dumas, J.B.A., *Traité de Chimie, appliquée aux arts*, Bechet Jeune, Paris, 1828.

chemists, among whom it should be noted Thomson, admitted that there exists a simple relation between the atomic weight of hydrogen and that of the other elements. Indeed, wrote Dumas, *One remarks that the weights agree in a striking form.* On the other hand, Dumas admitted that it was difficult to prove that an oxygen atom was composed of 8 hydrogen atoms. Clearly, there was no way chemists could demonstrate this in the laboratory. Dumas pointed out that the small deviations from integer values were of the order of the error in the measurements, and he thus justified rounding the numbers off to integers. In this way, Dumas vacillated from one extreme view to the other.

At the end of 1859, Dumas[88] published a Memoir on the 'equivalents of the elements'. As might be expected, this memoir produced a great sensation due to Dumas' leading position in chemistry. The illustrious French chemist, taking as basis M. de Marignac's researches on silver, found that all well known substances which he had had the occasion to examine conformed to Prout's principle, when certain modifications were applied to this principle. However, two elements were recalcitrant: chlorine and copper. Repeated experiments yielded 35.5 for the atomic weight of chlorine, while the atomic weight of copper was found to lie somewhere between 31 and 32. Thus, these numbers definitely did not agree with Prout's hypothesis. Normally, one example is sufficient to destroy a theory. However, Dumas concluded, in contrast to Berzelius, that the law must be accepted in the less general form, namely for chlorine the unit is 1/2 a hydrogen atom and for other elements the unit may be 1/4 of a hydrogen atom or even smaller.

1.18 Gaudin 1833 and the Correct But Ignored Explanation

It was Marc Antoine Gaudin (1804–1880) who realized in 1833[89] that the mysterious factors of two and four in the atomic weights had to do with the assumption that only a single atom can form a compound with another atom of another element. Consequently, he suggested that the number of atoms within a molecule might vary from element to element. In other words, the molecules of gaseous elements might contain two atoms (diatomic), while compounds of two or more elements might contain only one atom of each gas. It was not at all clear why the smallest particle in a pure hydrogen gas contains two atoms of hydrogen. Berzelius, for example, hypothesized that the atoms in a molecule were held together by electrical forces, whence two atoms of the same element should repel each other rather then attract. However, by then the entire atomic theory was in a mess, and all was confusion. Despite Gaudin being so right, the mood of the chemists was not appropriate for corrections to a poor theory.

---

[88] Dumas, J.B.A., Ann. Chim. Phys. **55**, 129 (1859).

[89] Gaudin, M.A.A., Ann. Chim. Phys. **52**, 113 (1833). See the strange history of Gaudin's papers in Cole, T.M., *Early Atomic Speculations of Marc Antoine Gaudin: Avogadro's Hypothesis and the Periodic System*, Isis **66**, 334 (1975).

Table 1.10 Marignac's table of atomic weights 1840–1850 in units of the atomic weight of hydrogen, and comparison with Stas and Prout

	Stas	Marignac	Prout
Silver	107.943	107.921	108
Chlorine	35.46	35.456	35.5
Potassium	39.13	39.115	39
Sodium	23.05		23
Nitrogen	14.04	14.02	14
Sulphur	16.037		16
Lead (from sulphate)	103.453		103.5
Lead (from nitrate)	103.460		103.5

It is interesting to note that Gaudin recapitulated Avogadro's hypothesis in this paper, and since the hypothesis was already out of favor with 'established opinion', this only added to the neglect of Gaudin's paper by the 'establishment'. With this reception of his ideas, and in particular Dumas' book in 1837, Gaudin lost interest in the problem and went to serve as a mathematician at the Paris Bureau of Longitudes.

1.19 Marignac and the Quest for Accurate Numerical Values

In 1842–1843, Jean Charles Galissard de Marignac (1817–1894) of Geneva, who was a student of Dumas, undertook a series of investigations, focusing in particular on chlorine, for the purpose of submitting Prout's law to a new and careful examination.[90] Marignac was fully aware of the experimental difficulties involved in further improving measurements of atomic weights and consequently was led to the conclusion that chlorine did agree with Prout's theory, but that the unit had to be redefined, and in particular that it should be halved. Moreover, Marignac concluded that future higher accuracy would not improve the agreement with Prout's law (Table 1.10).

1.20 Stas and Regrets for Earlier Thoughts

Jean Servais Stas (1813–1891) followed his teacher Dumas by developing accurate methods for measuring atomic weights and confirming the law of definite proportions to extreme accuracy. In 1860, Stas[91] commented cynically that:

> Prout had so little faith in the exactness of his hypothesis that he published his paper under the veil of anonymity. Whatever may be the opinion one holds about this hypothesis, whatever fate may be reserved for it in the future, it is impossible not to pay homage to the rare penetration of its author. Right or wrong, it permitted him to divine the specific weight of hydrogen forty years before it was determined experimentally with a degree of accuracy.

---

[90] Marignac, J.C., Pr. Ch. **26**, 461 (1842), Ibid., Compt. Rend. **14**, 570 (1842).

[91] Stas, J.S., Bull. de l'Acad. Roy. de Belge. **10**, 208 (1860).

And Stas continued to disprove Prout by stating that *there does not exist a common divisor for the weights of the elements which unite to form all definite compounds.* Indeed, this last statement by Stas is correct even today, while the wrong conclusion, i.e., that Prout was consequently in error, could be drawn only because too little was known, and the numbers were not accurate.

Stas' explanation as to why he returned to this dead-as-a-dodo hypothesis went as follows:

> In England the hypothesis of Dr Prout was almost universally accepted as absolute truth. The work executed by Professor Thomas Thomson of Glasgow in order to base it on analytical experiments greatly contributed to this result. Nevertheless, the same thing did not hold for Germany or France.[92] The immense prestige which surrounded the name of Berzelius, and legitimate confidence inspired by his work on the weights of the atoms were incontestably the cause of this. Even in England doubts arose: as early as 1833 Professor Turner, on the invitation of the British Association for the Advancement of Science, undertook a series of analyses, from which he drew the conclusion that the hypothesis of Dr Prout was not exact. In 1839, Professor Penny of Glasgow arrived at the same conclusion, although his results differed in some points from those obtained by Turner.[93] The determination of the atomic weight of carbon made by M. Dumas and myself in 1839–1840 called the attention of chemists anew to this subject. The atom of carbon deduced from our syntheses of carbonic acid agrees in fact entirely with Prout's hypothesis. The new syntheses of water which M. Dumas published in 1843, and which were confirmed by the work of MM. Erdmann and Marchand, led to the same result.

Consequently, Stas declared that:

> I dare affirm without fear of being contradicted that any chemist who has pursued research on the weights of the atoms, if he has varied his methods, if he has determined the weight of the atom of one substance as a function of the atom of two or three different substances, has encountered the same difficulties, the same contradictions. I accordingly experienced for long many a painful perplexity.

Similarly, Stas himself encountered a problem (with nitrogen), where the deviation was 0.02 out of 4, or 1/200, whence he inferred that the error was in the atomic weight of hydrogen, and consequently that all previous fits were fortuitous. He thus concluded that *the basis on which Dr Prout had erected in his law is itself without foundation.* Stas admitted that, when he started his research, he believed in Prout's hypothesis, but said that he changed his view when confronted with the facts.

Soon Stas was attacked by his compatriot Marignac.[94] While Stas believed that the numbers were sufficiently accurate to disprove Prout, Marignac did not think so. As an example, he claimed that Stas measured the atomic weight of silver by means of two methods, and got two different results, viz., 107.943 and 103.924, although the numbers should have been identical. Future accuracy, claimed Marignac, should be better. So which of Dumas' students, Marignac or Stas, would turn out to be right?

---

92 Note how the English Channel divided the world of chemistry, according to Stas.

93 Turner, E., Phil. Trans. Roy. Soc. London **123**, 523 (1833).

94 de Marignac, J.C., Bull. de l'Acad. Roy. de Belgique, 2nd Séries **x**, no. 8. Reprinted and translated in its entirety in Alembic Club Reprint no. 20, Prout's Hypothesis from Bibliothèque Universelle (Archives) **9**, 97 (1860).

1.21 Whewell: Philosophy a Bit Too Early

William Whewell (1794–1866) was one of the most important and influential figures in nineteenth century British science. An all round scholar, Whewell wrote extensively on a variety of subjects and was one of the founding members and an early president of the British Association for the Advancement of Science. Leading scientists turned to Whewell for advice on philosophical and scientific matters.[95]

In 1857, Whewell[96] argued essentially in the same way as Dumas before him, claiming that one cannot infer the existence of certain smallest possible particles from the simple fact that a certain chemical combines to form compounds in simple ratios. Whewell claimed that, strictly speaking, these experiments tell us only that, if there were such particles, they should be smaller than the smallest observable. *In chemical experiments, at least, there is not the slightest positive evidence for the existence of such atoms*. Both assumptions, *that the particles are not divisible or alternatively that they are divisible, explain the chemical fact equally well*, claimed Whewell. Note that this argument due to Whewell is exactly the opposite of the one given by Avogadro. This was written just three years before Kirchhoff and Bunsen demonstrated that the smallest particle of each element, the atom, has a unique spectrum which can be used for its identification.

1.22 How to Classify the Elements?

A crucial development for chemists took place in 1857, although it went largely unnoticed, when Dumas,[97] and soon after Strecker (1822–1871),[98] established the connection between atomic weight and chemical properties, and showed that chemically similar elements could be grouped into sets satisfying the following relation:

$$\text{atomic weight} = a + n \times 8\ ,$$

where n is an integer and a is the atomic weight of the first element in the sequence. However, most crucial was the finding that the atomic weight is the sole parameter that controls the properties of each set. Consequently, chemists searching for any kind of arrangement based their search on the atomic weight. Neither Dumas nor Strecker called it a period, but that is exactly what itwas.

---

[95] On the request of the poet Coleridge in 1833, Whewell invented the term 'scientist'. Before this time, the terms in use were 'natural philosopher' and 'man of science'.

[96] Whewell, W., *History of the Inductive Sciences*, in two volumes, New York, Appleton Comp., 1857.

[97] Dumas, J.B., Compt. Rend. **45**, 709 (1857), ibid. **46**, 951 (1858), ibid. **47**, 1026 (1858).

[98] Strecker, A., *Theorien und Experimente zur Bestimmung der Atomgewichte der Elemente*, Braunschweif, F. Vieweg, 1859.

1.23 The 1860 Karlsruhe Conference: The Fog Lifts

In the late 1850s, the situation with atomic weights of elements was in disarray. Two different scales were used. Dalton introduced hydrogen as the unit, and this was adopted by Gmelin and many others. Wollaston and Berzelius adopted oxygen, to which the former attributed the weight 10 and the latter the weight 100. Thomson, on the other hand, gave oxygen the weight 1. With such differences, life was still possible. But in Germany, carbon had atomic weight 6 and oxygen 8, while in France the corresponding values were 3 and 8, as though crossing the national border could affect the atomic weight. All this came on top of differences in results due to different methods and inaccuracies.

In 1858, a then relatively unknown chemist, Stanislao Cannizzaro (1826–1910) from the university of Genoa, published a pamphlet[99] about chemistry teaching, in which he demonstrated how to determine a consistent set of atomic weights. The paper was ignored.

In view of the disarray affecting the understanding of atomic weights, many voices were raised, calling upon the chemical community to dispel this confusion. The original idea was due to August Kekulé (1829–1896m).[100] He convinced Karl Weltzien (1813–1870) and Charles-Adolphe Wurz (1817–1884), who inherited Dumas' chair in Paris, of the need to organise a meeting. Support was secured from all the great names in chemistry. The invitation letter called *all chemists authorized by their work or position to express an opinion in a scientific discussion*. But did one have to have official authorization to express one's view on the problem? The declared purpose of the meeting was to reach *a common agreement* [...] *on some of the following points: the definition of important chemical notions, such as* [...] *atom, molecule,* and *in general terms, the differences that had arisen between the theoretical views of chemists and the urgency of putting an end to these differences by a common agreement*, as explained by Wurz.[101] Forty-five leading chemists signed the invitation and about 140 chemists responded and participated in the meeting.[102]

One of the central discussions in the meeting took place between Kekulé and Cannizzaro. The former claimed that one had to distinguish between a molecule and an atom and that there was a difference between a physical and a chemical molecule.

---

99 Cannizzaro, S., Il Nuovo Cimento **7**, 321 (1858). The origin was a letter Cannizzaro wrote to a colleague in Pisa, Sebastiano de Luca, a professor of chemistry. The paper in the journal is essentially Cannizzaro's letter. The title was: *Sunto di un corso di filosofia chimica fatto nella R. Università de Genova*, which means: Sketch of a Course in Chemical Philosophy at the Royal University of Genoa.

100 Friedrich August Kekulé von Stradonitz is known for discovering the structure of the molecule of benzene, C_6H_6, in 1865, long before the details of the chemical bond were known. Bull. Soc. Chim. Paris **3**, 98 (1865).

101 Wurz, C-A. *Account of the sessions of the International Congress of Chemists in Karlsruhe*, 3–5 September 1860.

102 The disarray went far beyond the question of atoms and molecules. There were problems in the structure of organic molecules as well as the naming of these molecules. The mid-nineteenth century saw a dramatic growth in the discoveries of organic molecules.

Cannizzaro, on the other hand, claimed that the Ampère–Avogadro law should be the basis of consideration for chemical molecules. In another session, Cannizzaro claimed that there was no difference between a physical and a chemical molecule. The situation was so bad that Wurtz had to stress the difference between an atom and a molecule in the summary of the conference.[103]

An important subject for us here is the session presided by Dumas. The original proposal was to debate the symbols of the elements and reactions, but the discussion went differently. The dominant speakers, usually expressing opposing ideas, were Cannizzaro and Kekulé. But the highlight of the conference was Cannizzaro's enlightening speech raising Avogadro's hypothesis from abeyance.

The Karlsruhe congress was both a failure and a great success. It was a failure because no practical decision was taken. As a matter of fact, Kopp (1817–1892)[104] and Erdmann (1804–1869)[105] argued that scientific problems could not be decided upon by a popular vote. It was a success because most participants took home Cannizzaro's pamphlet, which was essentially his two-year-old letter to Luca, and eventually overcame their bias against physical methods, agreeing to its content, and acting accordingly. This was probably Cannizzaro's greatest contribution to chemistry.[106]

1.24 Early Attempts to Classify the Elements

It is difficult to trace the origin of the idea of 'analogy between elements', which was never defined. In 1812, Davy[107] discussed the 'analogy' in properties between, for example, silver, palladium, antimony, and tellurium, or potassium, sodium, and barium, but did not define what was meant by an 'analogy'.

---

[103] Some idea of the personal quarrels that went on in the congress and why no agreement could be achieved can be obtained from Hartley, H., Notes and Records of the Royal Society of London **21**, 56 (1966).

[104] Kopp found that the molecular heat capacity of a solid compound is the sum of the atomic heat capacities of the elements composing it. The elements with atomic heat capacities lower than those inferred from the law of Petit and Dulong keep these lower values when they are in compounds. Kopp, H., Proc. Roy. Soc. London **13**, 229 (1863).

[105] Erdmann and Marchand measured atomic weights, in particular those of copper and silver Erdmann, O.L. and Marchand, R.F., Jour. f. Prakt. Chem. **xxxi**, 389 (1844).

[106] Italy was not a country of chemists like France, Germany, or England at that time. Thus, Cannizzaro, an unknown Italian chemist (before the meeting) revived the theory of another unknown Italian, Avogadro. In August 2004, Vincenzo Salerno wrote in the journal *Best of Sicily Magazine*: *His niche is a small but important one. Palermo-born Stanislao Cannizzaro is a special footnote to nineteenth-century physical science. He was, according to some, one of the scientists responsible for bringing chemistry and physics out of the realm of alchemy and into the modern world.* The title of the article is: *Stanislao Cannizzaro. A Political Physicist.*

[107] Davy, H., *Elements of Chemical Philosophy* **1**, 273, London: Printed for J. Johnson and Co., 1812. Apparently the second part never appeared. This is the only book on chemistry I have ever seen which discusses gravitation as part of chemistry.

Table 1.11 The triads discovered by Döbereiner, 1829

Element	Lithium	Beryllium	Boron	Arithmetic mean
Atomic weight	7.0	9.0	11.0	9.0
Element	Lithium	Sodium	Potassium	
Atomic weight	7.0	23.0	39.0	23.0
Element	Carbon	Nitrogen	Oxygen	
Atomic weight	12.0	14.0	16.0	14.0
Element	Calcium	Strontium	Barium	
Atomic weight	40.0	87.5	137	88.1
Element	Chlorine	Bromine	Iodine	
Atomic weight	35.0	80.0	127.0	80.6

In the year 1829, Johann Wolfgang Döbereiner (1780–1849),[108] a German scientist, was the first to classify elements into groups based on John Dalton's assertions. He grouped the elements with similar chemical properties into clusters of three called 'triads'. The distinctive feature of a triad was the atomic weight of the middle element. When elements were arranged in order of increasing atomic weight, the atomic weight of the middle element was approximately the arithmetic mean of the other two elements of the triad (see Table 1.11).

The problem with the triads was that a large number of similar elements could not be grouped into triads, e.g., iron, manganese, nickel, cobalt, zinc, and copper are similar elements, but could not be organised into triads. On the other hand, some quite dissimilar elements could apparently be grouped into such triads. In any case, Döbereiner was only able to successfully identify 5 triads, and since he failed to arrange all the then known elements in this form, his attempt at classification was not very successful, and consequently not very influential. In retrospect, we realize that lithium, for example, belonged to two triads. Moreover, while Döbereiner did correctly place lithium, sodium, and potassium (but also incorrectly beryllium and boron), he placed carbon, nitrogen, and oxygen in the same triad. Silicon was discovered by Berzelius in 1824, 5 years before Döbereiner came up with the idea of triads. It is therefore strange that it does not appear in Döbereiner's table, given that it is so similar to carbon. Further progress with the triad idea is due to Gmelin in 1843.[109] While manipulating the triads, he managed to arrange the elements, without really noticing, into an increasing order of atomic weight. But for some reason, Gmelin's role is hardly recognised.

It appears as if the subject of element classification lost favor with chemists, because it is only 33 years later that we encounter the next attempt. In 1862, the geologist de Chancourtois (1820–1886)[110] published a suggested arrangement of the chemical elements. The elements were organised into a quite complicated

[108] Döbereiner, J.W., Ann. Phys. **56**, 331 (1817), ibid. **57**, 435 (1817), ibid. Ann. der Phys. u. Chem. **15**, 301 (1829).

[109] Gmelin, L., *Hanbuch der Chemie*, 4th edn. Heidelberg, Winter, 1843.

[110] de Chancourtois, B., Compt. Rend. **54**, 757, 840, 967 (1862) ibid. **55**, 600.

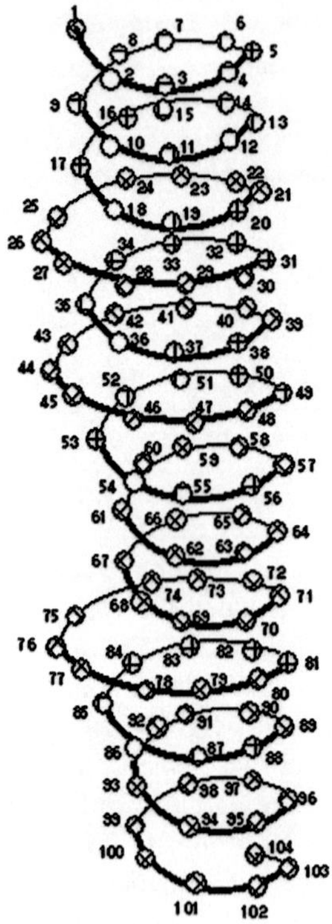

Fig. 1.4 The periodic spiral due to Chancourtois, 1862

three-dimensional spiral (see Fig. 1.4). De Chancourtois called his idea ‘the telluric spiral’, because the element tellurium came in the middle. The name for this particular arrangement was appropriate, being coined by a geologist, since the element tellurium is named after the Earth (from the word *Tellus* in Latin). De Chancourtois plotted the atomic weights on the surface of a cylinder in such a way that they would increase by 16 for each trip around the circumference of the cylinder. However, some of de Chancourtois correlations and arrangements were accidental, while others were wrong. Moreover, in certain cases he correlated atomic weights with prime numbers in a rather superstitious way. When he encountered a gap in the table, he filled it with chemical radicals or alloys. Thus about 23 elements were placed correctly in the table, while the rest, about 26, were all wrongly placed in one way or another. Today de Chancourtois retains to his credit the first usage of the term ‘period’. The figure showing the table as arranged by de Chancourtois was so complicated that chemists did not follow the basic idea and it was ignored, as well as being wrong.

1.25 Newlands Rediscovers the Octave Law

The next step in the development of the classification of the elements based on atomic properties was proposed by John Newlands (1837–1898) in a series of papers published in 1864–1866.[111] This was an extension of the results reported by Dumas in 1857, which presumably were not known to Newlands. Newlands had the advantage that several discoveries had since been made: cesium in 1860, rubidium[112] and thallium[113] in 1861, and indium[114] in 1863, thereby filling several 'gaps' in the table.

Since Newlands, who had not attended the Karlsruhe congress, arranged the elements according to atomic weight, he discovered a periodicity: the eighth element was similar to the first, a phenomenon he called the 'the law of octaves' (see Table 1.12). The problems began after calcium, where the periodicity became vague. No less damaging was the fact that there was little room left for unknown elements. In other words, Newlands blindly followed the atomic weight and ignored inconsistencies in the chemical properties. And in other places, two elements occupied the same location. The halogens, F and Cl, are in the same row as Co and Ni (which occupy the same location), and so on. The element Gl (the third element, sometimes denoted by G alone) is known today as beryllium.[115] The atomic weight of uranium is completely off. Newlands' greatest blunder was, however, to insist on a fixed length for the period and overlook chemical properties.

Newlands presented the 'new order' to the Chemical Society on 1 March 1866, and several very interesting comments were made, which reflect the thinking of his colleagues. For example, John Hall Gladsone (1827–1902) objected on the grounds that Newlands assumed that no elements remained to be discovered. This comment shows how Newlands adopted the atomic weight as the sole factor determining the location of the element, while ignoring its chemical properties. It was too much according to Dumas' principle, without any suspicion that something might be wrong. Carey Foster (1835–1919), who was on the publication committee of the Royal Chemical Society, *humorously inquired of Mr. Newlands whether he had ever considered examining the elements according to the order of their initial letters*, because he believed that any arrangement would present occasional coincidences in the chemical properties. On the other hand, he wondered why Mg was so far away from Cr, or Fe from Ni. Newlands response was that all other schemes he tried had failed, in particular one based on the specific weight.

---

[111] Newlands, J.A.R., Chemical News **7**, 70, (1863); ibid. **10**, 59 (1864); ibid. **10**, 94 (1864); ibid. *On the Law of Octaves* **12**, 83 (1865); ibid. **13**, 113 (1866).

[112] Bunsen, R.W., Ann. der Chemie und Pharmacie **119**, 107 (1861).

[113] Crookes, W., Chem. News **3**, 193 (1861).

[114] Reich, F. and Richter, H.T., J. fur Praktische Chemie **89**, 441 (1863), ibid. **90**, 172 (1863).

[115] Beryllium was discovered in 1789 by Nicolas Louis Vauquelin in the mineral beryl, whence its name beryllium. For some 160 years, the element was called glucinium (from the Greek *glykys*, meaning sweet), due to the sweet taste of its salts. But note that the metal and its salts are toxic, so it is not advisable to taste the salts, even though they are sweet.

Table 1.12 Elements arranged in octaves by Newlands, 1864

H	1	F	8	Cl	15	Co and Ni	22	Br	29	Pd	36	I	42	Pt and Ir	50
Li	2	Na	9	K	16	Cu	23	Rb	30	Ag	37	Cs	44	Os	51
Gl	3	Mg	10	Ca	17	Zn	24	Sr	31	Cd	38	Ba and V	45	Hg	52
Bo	4	Al	11	Cr	19	Y	25	Ce and La	33	U	40	Ta	46	Tl	53
C	5	Si	12	Ti	18	In	26	Zr	32	Sn	39	W	47	Pb	54
N	6	P	13	Mn	20	As	27	Di and Mo	34	Sb	41	Nb	48	Bi	55
O	7	S	14	Fe	21	Se	28	Ro and Ru	35	Te	43	Au	49	Th	56

Newlands' paper was rejected and the distinguished society failed to publish this seminal paper. The reason given by Odling, then the president of the Chemical Society, was that they made a rule not to publish theoretical papers, and this on the quite astonishing grounds that *such papers lead to a correspondence of controversial character*. So does this mean that chemists do not like to argue? Perhaps we might suggest instead that the refusal had more to do with Newlands' affiliation, with a private laboratory rather than a distinguished university.

In 1873, Newlands tried to establish his priority as discoverer of the periodic law, but his request to the Chemical Society to publish a note referring to his 1866 presentation was flatly rejected by Odling. Newlands, unhappy with the response, published a booklet in 1884,[116] in which his results and claims were detailed. Only upon his death, in 1898, did the Royal Society of Chemistry concede Newlands' contribution, and a plaque was placed on the wall at his birthplace which states: *Newlands, chemist and discoverer of the periodic law for the elements*. Better late then never, but how many people pass by West Square in south London and note the small plaque?

1.26 Odling 1864

William Odling (1829–1921) started to investigate the grouping of similar elements in 1857,[117] and was the first to take into consideration the valence concept in addition to the atomic weight.[118] In 1864,[119] he noticed that the difference in atomic weight between the lightest two or three elements of every group is approximately 16. The

---

[116] Newlands, J.A.R., *On the Discovery of the Periodic Law and on Relations among the Atomic Weights*, London, Spon, 1884.

[117] Odling, W., Phil. Mag. **13**, 423 (1857), ibid. 480.

[118] Each element has a chemical valence, which is an integer. This number expresses how many univalent atoms of each element are needed to form a chemical compound. For example, the valence of hydrogen is 1 and the valence of oxygen is 2, so two hydrogen atoms are needed for a single atom of oxygen to form a molecule of water.

[119] Odling, W., Phil. Mag. **27**, 115 (1864); Quarterly J. Sci. **1**, 642 (1864).

Table 1.13 Odling's suggested table of the elements in 1864. Note that all atomic weights are either integer or half integer

I		II		III		IV		V	
						Ro	104	Pt	197
						Pu	104	Ir	197
						Pd	106.5	Os	199
H	1					Ag	108	Au	196.5
				Zn	65	Cd	112	Hg	200
L	7							Ti	203
G	9							Pb	207
B	11	Al	27.5			U	120		
C	12	Si	28			Sn	118		
N	14	P	31	As	75	Sb	122	Bi	210
O	16	S	32	Se	79.5	Te	129		
F	19	Cl	35.5	Br	89	I	127		
Na	23	K	39	Rb	85	Cs	133		
Mg	24	Ca	40	Sr	87.5	Ba	137		
		Ti	50	Zr	89.5	Ta	138	Th	235
				Ce	92				
		Cr	52.5	Mo	96	V	137		
		Mn	55			W	184		
		Fe	56						
		Co	59						
		Ni	59						
		Ni	59						
		Cu	63.5						

next publication[120] already included a table which contained (a) missing elements and (b) unequal lengths of periods. The last point was crucial (Table 1.13).

1.27 Mendeleev 1869 and Meyer 1864–1870

Julius Lothar Meyer (1830–1895),[121] who was a student of Bunsen in Heidelberg, also attempted to arrange the chemical elements. The first publication in 1864 included only 28 out of about 80 elements known at that time. Dumas' and Stecker's influence was obvious, and the latter was actually cited by Meyer. Meyer's system

[120] Odling, A., *A Course of Practical Chemistry*, 2nd edn., Longmans, Green, London, 1865, ibid. 3rd edn., 1868.

[121] Meyer, L., *Die Moderne Theorien der Chemie*, Breslau, 1864. English version: Longman Pub., London, 1888.

evolved in time and the final version was only published in 1870,[122] not before Mendeleev's publication, and included some improvements.

Soon after the Karlsruhe meeting, which both Mendeleev and Meyer attended, many new atomic arrangements were published, culminating in the works of Meyer and Dmitri Mendeleev (1834–1907m). The influence of Dumas and Stecker is clear in both works. In the first edition of his book,[123] Meyer used atomic weights to arrange 28 elements into 6 families that bore similar chemical and physical characteristics, leaving blanks for the as-yet undiscovered elements. His one conceptual advance over his immediate predecessors was in seeing the valence of the element, namely the number that represents the combining power of an element (e.g., with atoms of hydrogen), as the link among members of each family of elements, and as the pattern for the order in which the families were organized. In his original scheme the valences of the succeeding families, beginning with the carbon group, were 4, 3, 2, 1, 1, and 2. All elements in a given row had the same valence. Meyer identified 9 groups bearing similarities with regard to these properties. Note that hydrogen did not appear in Meyer's table (see Table 1.14).

The only parameter chemists had at the time was the atomic weight. The atomic weight is a simple, well defined parameter and the idea was to arrange the elements according to increasing atomic weight and then look for periodicities. The problems were that:

- several elements had not yet been discovered,
- the periods appeared to be of variable length.

The valency was also known, but it was definitely not periodic and did not help in the mission of classification.

Meyer published his classic paper in 1870,[124] and described the evolution of his work since its beginning in 1864. This paper is particularly famous for its graphic display of the periodicity of atomic volume plotted against atomic weight (see Fig. 1.5). This was a very quantitative expression of the periodicity and the only one presented at that time! All other descriptions were entirely qualitative. While many chemists, including Bunsen,[125] had their doubts about the periodic law at first, these doubters were gradually converted by the discovery of elements predicted by the tabular arrangement and the correction of old atomic weights that the table cast in

[122] Meyer, L., *The Nature of the Chemical Elements as a Function of Their Atomic Weights*, Justus Liebig Annalen der Chemie, Suppl. **7**, 354 (1870).

[123] Meyer, L., *Die Modernen Theorien der Chemie und ihre Bedeutung für die chemische Statik*, Breslau, Verlag von Maruschke and Berebdt (1864).

[124] Meyer, J.L., Liebig's Annalen der Chemie, Suppl. **7**, 354 (1870).

[125] Meyer and Mendeleev were Bunsen's students, with a five year separation. Interestingly, Bunsen did not trust the work of his students. Bunsen, like most members of the German chemical community, was mainly interested in experimental work and analytical chemistry, as opposed to theoretical work like classification of elements, let alone prediction of the existence of unknown elements. Years later, Meyer commented on the German chemists' resentment of theoretical papers 'without any new data'. Brauner, who was also a student of Bunsen and an enthusiastic supporter of Mendeleev's periodic law, argued that the young members of the editorial board of the journal press for the publication of Mendeleev's 'no new data' paper.

Table 1.14 The periodic system according to Lothar Meyer (1870)

I	II	III	IV	V	VI	VII	VIII	IX
	B = 11.0	Al = 27.3				?In = 113.4	Tl = 202.7	
	C = 11.97	Si = 28				Sn = 117.8		Pb = 206.4
			Ti = 48		Zr = 89.7			
	N = 14.01	P = 30.9		As = 74.9		Sb = 122.1		Bi = 207.5
			V = 51.2		Nb = 93.7		Ta = 182.2	
	O = 15.96	S = 31.98		Se = 78		Te = 128?		
			Cr = 52.4		Mo = 95.6		W = 183.5	
	F = 19.1	Cl = 35.8		Br = 79.75		J = 126.5		
			Mn = 54.8		Ru = 103.5		Os = 198.6?	
			Fe = 55.9		Rh = 104.1		Ir = 196.7	
			Co = Ni = 58.6		Pd = 106.2		Pt = 196.7	
Li = 7.01	Na = 22.99	K = 39.04		Rb = 85.2		Cs = 132.7		
			Cu = 63.3		Ag = 107.66		Au = 196.2	
?Be = 9.3	Mg = 23.9	Ca = 39.9		Sr = 87.0		Ba = 136.8		
			Zn = 64.9		Cd = 111.6		Hg = 199.8	

Fig. 1.5 Atomic volume as a function of atomic weight, calculated and plotted by Meyer 1864. This was the first time that the periodicity of the properties received such a conspicuous quantitative demonstration

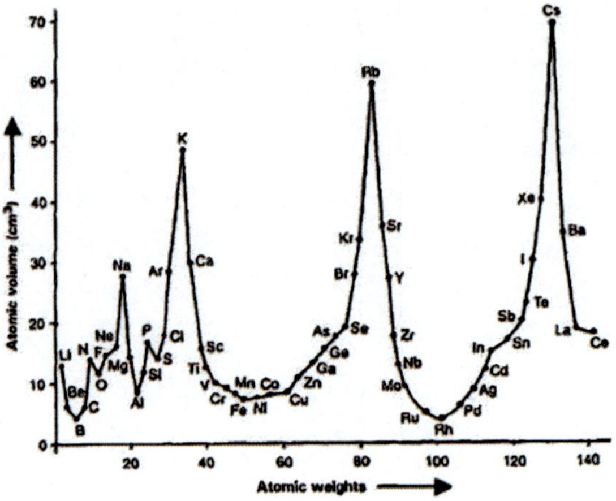

doubt. Meanwhile, Meyer[126] and Mendeleev[127] engaged in a long drawn-out priority dispute.

A glance at Meyer's table (see Table 1.14) reveals many empty locations. Were these as-yet undiscovered elements? Meyer did not specify, and refused to commit himself. The table looks more like a Swiss cheese than a complete table with a few missing elements. In any case, there were too many 'holes' in the table to justify Meyer's claim for priority in the discovery of the periodic table. The table essentially already appeared in his book in 1864.

In 1869, Mendeleev published[128] a two-page paper which had a profound impact on chemistry and physics. And so wrote Mendeleev:

> By ordering the elements according to increasing atomic weight in vertical rows so that the horizontal rows contain analogous elements, still ordered by increasing atomic weight, one obtains the following arrangement, from which a few general conclusions may be derived.

Note, however, that Mendeleev deduced the periodic table without any knowledge of the existence of electrons. [The electron was discovered by J.J. Thomson[129] (1856–1940) in 1897.] It was phenomenology at its best, without explanation or theory. The sole criteria were undefined chemical similarity and atomic weight. The original Mendeleev table is given in Table 1.15. Mendeleev was in the process of writing a

---

[126] Meyer, L., Zur Geschichte des priodischen Atomistik. Ber. Deuts. Chemi. Ges. **13**, 259 (1880), ibid. 2043 (1880).

[127] Mendeleev, D., Zur Geschichte des priodischen Gesetzes. Ber. Deuts. Chemi. Ges. **13**, 1796 (1880).

[128] Mendeleev, D.I., J. Russ. Chem. Soc. **1**, 35 (1869), ibid. 60; ibid. **3**, 25 (1871); Zeit. für Chemie **12**, 405 (1869).

[129] Thomson, J.J., Phil. Mag. **44**, 293 (1897). The discovery of the electron won him the Nobel Prize in 1906.

textbook on chemistry[130] when he got the idea about how to arrange the elements. It is puzzling as to why Mendeleev did not rewrite his book in order to make his breakthrough discovery the central theme.

Cassebaum and Kauffman[131] describe the interesting development of events concerning Mendeleev's discovery. Odling's book, in which his results appeared, was translated into Russian in 1867. However, in a footnote to his 1869 paper,[132] Mendeleev reported that he was referred by Savchenkov to Odling's table. Cassebaum and Kauffman show a picture taken during the November 1868 meeting of the Russian Chemical Society, in which one can see Mendeleev standing near Savchenkov, and wrote: *Yet, Mendeleev emphasized in the footnote in question that he had no previous knowledge of Odling's table.* So why did Mendeleev find it necessary to add his note? On the other hand, Odling had 5 columns, while Mendeleev found that the chemical properties require 6 columns. Hence, as a quick look at the two tables reveals, it would be impossible for the two schemes to relate the elements in the same way.

Mendeleev's claim was that the elements are arranged according to increasing atomic weight. However, this is not absolutely correct. The two elements nickel and cobalt appear to have the same atomic weight and share the same position. But if two elements share a position, there must be another parameter which distinguishes between them. The question as to what this extra parameter might be was not asked. Furthermore, rhodium (Rh) was wrongly classified. It belongs between ruthenium (Ru) and palladium (Pd). The element in-between, which years later would be called technetium, does not exist in nature. And the order of the last column is a bit confused.

The element J (atomic weight 127) is iodine. As Mendeleev published his paper in German, he used the German letter J for this element. Iodine is placed correctly after tellurium (i.e., with the halogens), despite having a lower atomic weight than tellurium. The idea of isotopes did not exist, and had to wait about five decades for Aston.

We note that there is no helium, and indeed no other noble gas in Mendeleev's table. Helium was discovered by Joseph Lockyer (1836–1920m),[133] in 1868, when he carried out spectroscopic observations of the Sun during an eclipse and discovered a yellow line never seen before in the laboratory. An independent discovery was made the same year by Pierre Janssen (1824–1907m).[134] Lockyer suggested that this might be a new element, unknown on Earth, and this explains the name he gave to it. It was the first element to be discovered by spectroscopy.

---

[130] Mendeleev, D., *The Principles of Chemistry*. The book was written 1868–1870 and translated into English in 1891.

[131] Cassebaum, H. and Kauffman, G.B., Isis **62**, 314 (1971).

[132] Mendeleev, D.I., J. Russian Chemical Soc. **1C**, 60 (1869).

[133] Lockyer is among the very few astronomers who turned stellar spectroscopy into a science. It led in the end to our understanding the composition of the stars.

[134] Janssen, M., AReg **7**, 107 (1869). The names Janssen and Lockyer appear together because they independently devised a method to see the corona in broad daylight without the need for an eclipse. Janssen described the spectra, but the courageous hypothesis about a new element is due to Lockyer. Janssen was too enthusiastic about the new instrument to be bothered by an unidentified new element.

Table 1.15 The original periodic table due to Mendeleev (1869). Numbers are atomic weights as known to Mendeleev

I	II	III	IV	V	VI
			Ti = 50	Zr = 90	? = 180
			V = 51	Nb = 94	Ta = 182
			Cr = 52	Mo = 96	W = 186
			Mn = 55	Rh = 104.4	Pt = 197.4
			Fe = 56	Ru = 104.1	Ir = 198
			Ni = Co = 59	Pd = 106.6	Os = 199
H = 1			Cu = 63.4	Ag = 108	Hg = 200
	Be = 9.4	Mg = 24	Zn = 65.2	Cd = 112	
	B = 11	Al = 27.4	? = 68	Ur = 116	Au = 197?
	C = 12	Si = 28	? = 70	Sn = 118	
	N = 14	P = 31	As = 75	Sb = 122	Bi = 210?
	O = 16	S = 32	Se = 79.4	Te = 128?	
	F = 19	Cl = 35.5	Br = 80	J = 127	
Li = 7	Na = 23	K = 39	Rb = 85.4	Cs = 133	Tl = 204
		Ca = 40	Sr = 87.6	Ba = 137	Pb = 207
		? = 45	Ce = 92		
		?Er = 56	La = 94		
		Yt = 60	Di = 95		
		?In = 75.6	Th = 118?		

But the idea that the stars might contain elements that do not exist on Earth was too revolutionary, and definitely too outlandish to allow them to be included in tables made by chemists. Lockyer himself had to wait until 1895, at which point William Ramsay (1852–1916m), a Scottish chemist who discovered almost all the noble gases,[135] identified helium as the gas released from the mineral cleveite. Radon (the radioactive daughter of radium) was discovered in 1900 by Rutherford and named 'radioactive emanation' by the Curies. The gas is radioactive and decays into polonium. The element was isolated in 1908 by Ramsay and Gray, who called it niton. Since 1923, it has been called radon. The chain of events was such that none of the noble gases were known before Mendeleev's discovery, but they were all discovered over rather a short period of time soon after the introduction of the periodic table. Mendeleev succeeded on several counts where Newlands failed:

- He allowed for periods of different lengths for the lanthanides, elements between which the chemical differences are smaller than usual.
- He ingeniously ignored the small details in order to see the overall picture.
- Above all, he had the audacity to predict the existence of unknown elements.

[135] Argon in 1894, neon in 1898, krypton in 1898, and xenon in 1898. All were found in liquid air. Ramsay, W., Nature **56**, 378 (1897).

This audacity has to be appreciated against a background in which the general ethos among chemists was not to publish papers 'without new data'.[136] The successful attempt to arrange the elements, or classify them, or find a general pattern, came against a background of failed attempts to classify the elements according to their physical properties. Clearly, he was helped by the fact that none of the noble gases was known at the time he published the periodic table, so he escaped further confusion on their count.

Not all scientists hailed Mendeleev's discovery with the same enthusiasm. Friedrich Beilstein (1838–1906),[137] who succeeded Mendeleev at the Technological Institute in St. Petersburg and may have been driven by personal conflicts with Mendeleev, claimed that:

> Mendeleev [. . .] has prophesied the existence of all sorts of new elements and believes that he needs only to conceive of them in order to have them immediately in the bag [. . .]. He is in general an odd chap.

Mendeleev may have been an 'odd chap', but imaginative and courageous, unlike some of his critics.

Mendeleev drew several interesting conclusions:

- The elements, if arranged according to increasing atomic weights, exhibit a periodicity of properties.
- Chemically analogous elements have either similar atomic weights (for example, Pt, Ir, Os), or weights which increase by equal amounts, as found already by many before him.
- The arrangement according to atomic weight corresponds to the valence of the elements and to a certain extent to the difference in chemical properties, for example Li, Be, B, C, N, O, F. Here Mendeleev realized that the chemical valence was a consequence of the trends or location in the table, something like a prediction, and not part of the criterion for classification.
- The most abundant elements in nature have small atomic weights.
- One can predict the future discovery of many new elements, for example, analogous to Si and Al, with atomic weights of 65–75.

So Mendeleev noted one of the important characteristics of element abundance in nature, namely, that the abundance decreases with increasing atomic weight. This was the first observation that the atomic weight might somehow be associated with the abundance of the element or the way it is synthesized. No one suspected in those days that the elements might not be primordial, or that they might have been naturally synthesized. Clearly, when Mendeleev wrote 'abundances', he meant abundances in

---

136 Winkler, who discovered germanium in 1886, the element predicted by Mendeleev as eka-silicon, called Mendeleev's three predictions 'bold speculation' in *The Discovery of New Elements During the Last Twenty Five Years*, in 1897 [34, p. 136].

137 Beilstein is known for compiling the first *Handbuch der organischen Chemie* **2**, 1880–1883, or *Handbook of Organic Chemistry*, which contained the chemical properties of about 15,000 organic compounds.

the crust of the Earth, as opposed to cosmic abundance. In any case, this particular conclusion about the abundances went unnoticed.

For Mendeleev[138] as a chemist, the 'chemical element' was a well defined *chemical* entity, characterized by its chemical reactions, with no need for a definition. The discovery of argon in 1894,[139] the first noble gas to be identified, caused havoc as its discovery was totally unexpected, and there was no room for it in Mendeleev's table. The new gas was found to be monatomic, unlike all the known gases, and to have an atomic weight of 39.8. Consequently, it did not fit at all into the periodic table. The shock was so great that several chemists suggested abandoning the idea of a periodic table. Some suggested that the newly discovered 'element' was not really an element, but a more complex molecule comprising several atoms. Even Mendeleev considered the possibility that 'argon' might be a compound made up of three nitrogen atoms.[140] Adding to the mystery and confusion among chemists was the fact that it did not react chemically—it was found to be inert, something we call 'noble' in today's parlance, since it refused to react with other elements, as though it had a vanishing valence. The discovery of the rest of the noble gases rather soon afterwards partially resolved the problem, because then a special group, the inert gases, could be introduced into the periodic table. So the confusion turned into a victory of the idea of the periodic table, to the point where Mendeleevsuggested

---

[138] In 1893, Mendeleev was appointed Director of the Bureau of Weights and Measures. In this capacity, he was directed to formulate new state standards for the production of vodka. The Vodka Museum in Moscow claims that, in 1894, Mendeleev determined the new standards for vodka as 38% proof. Distilled spirits were taxed according to their alcoholic strength, and consequently the percentage was rounded upwards, to 40%, for simplified taxation calculations. See http://www.russianfoods.com/cuisine/article00016/default.asp. Pravda has a slightly different version (see Pravda, 31 October 2009):

> In 1860–1870, private production of vodka in Russia became very popular. It was very difficult for the government to control the quality of vodka and fresh raw material. Soon Russia introduced the state monopoly for vodka production. Chemists got involved in the development of the production process to enable product standardization. The brand 'Moskovskaya Osobaya' is directly connected to the name of Dmitry Mendeleev, the inventor of the periodic table of elements. It took the scientist 18 months to find a perfect ratio for volume of spirit to water. Mendeleev published his findings in his doctoral thesis 'Essay on integration of spirit with water'. This is why Russian vodka is always 40% Alc. by Vol. Mendeleev's conclusions were appreciated and utilized in alcohol testing and vodka production. His work became the basis for the state comparison standard for the original Russian vodka established in 1894–1896.

[139] The discovery of argon is unique as it collected two Nobel prizes in the same year, 1904. Rayleigh (1842–1919) was awarded the Nobel prize in physics *for his investigations of the densities of the most important gases and for his discovery of argon in connection with these studies*, while Ramsay (1852–1916) was awarded the Nobel prize in chemistry *in recognition of his services in the discovery of the inert gaseous elements in air, and his determination of their place in the periodic system.*

[140] Mendeleev's view was reported in Nature **51**, 543 (1895) under the title *Professor Mendeleev on Argon*, and it is a report of what Mendeleev said in the meeting of the Russian chemical Society on 14 March 1895. The idea of a new set of elements constituting a new column in the periodic table appeared to Mendeleev too absurd.

in 1903 that the ether,[141] the elusive imaginary fluid that physicists invented to help them comprehend and 'explain' the propagation of light, should be included in the periodic table with atomic weight zero. But two years later Einstein put paid to the ether idea.

1.28 Loschmidt and the Value of the Avogadro Number

So far, the actual value of Avogadro's number was not known. In 1865, Loschmidt (1821–1895)[142] was interested in estimating the size of air molecules. It was during this work that he succeeded in deriving the first estimate for the size of a molecule by comparing the volume occupied by N molecules in the air, where they move freely, with the volume they have when the gas condenses to form a solid and the molecules touch each other.

The formula Loschmidt got was $d = 8C_f l$, where l is the mean free path of the molecules in the gas, C_f is the condensation coefficient as defined by Loschmidt (ratio of gaseous to solid volumes), and d is the diameter of the molecule. At that time liquefaction of air had not yet been achieved, so his gases were water, CO_2 and NO, NO_2 and NH_3.

Loschmidt applied Maxwell's[143] theory to calculate the mean free path of air from the value of the viscosity, and found that the mean free path of a molecule under standard conditions is 6.2×10^{-6} cm. The condensation coefficient for air was taken from the fact that air is 770 times lighter than water. A correction factor of 1.5 was assumed by Loschmidt, due to changes in temperature, etc. Consequently, the condensation coefficient of water is $1/(770 \times 1.5)$. Hence, substituting the numbers in the formula yields $d = 9.69 \times 10^{-8}$ cm. Once Loschmidt had the size of a molecule and the mean free path (from Maxwell), he could calculate the number of molecules in a given volume of the air. Loschmidt's estimate for the Avogadro number, the number of molecules in one gram mol,[144] was 4×10^{23}.

[141] Mendeleev, D., *Popytka khimicheskogo ponimaniia mirovogo efira* (in Russian). An English translation appeared as Mendeleff, D. (1904). G. Kamensky (translator): *An Attempt Towards a Chemical Conception of the Ether*, Longman, Green and Co., New York.

[142] Loschmidt, J., Proc. Acad. Sci. Vienna **52**, 395 (1865).

[143] Maxwell, J.C., Phil. Mag. **19**, 28 (1860).

[144] A gram mol is the molecular weight expressed in grams. In the case of oxygen, for example, it is $16 \times 2 = 32$ g.

1.29 Meyer 1888: A Historical Perspective

Historians of chemistry toward the end of the nineteenth century, such as Meyer (1847–1916),[145] were confident that Prout was utterly wrong and rejected his hypothesis, mainly due to the influence of Berzelius. Here is a typical passage from a history book regarding the advancement of Prout's hypothesis:

> This was one of those factors which materially depreciated the atomic doctrine in the eyes of many eminent investigators. On account of its influence upon the further development of the atomic theory, this hypothesis must be discussed here, although it has happened but seldom that an idea from which important theoretical conceptions sprang originated in so faulty a manner as it did.

In short, the idea was wrong and its consequences should be ignored. An incorrect idea had had too much influence.

Meyer also pointed to wrong chemistry textbooks, for example the one by Gmelin, who:

> [. . .] gave the mixture weights as far as possible in whole numbers, which he was assuredly not justified in doing after Berzelius' classical researches. Later still in 1840, Dumas and Stas [. . .] and Erdmann and Marchand [. . .] and Marignac. The predilection shown by many chemists for this conception, which led to such far-reaching deductions, helped to depreciate the whole atomic doctrine in the minds of thoughtful investigators.

Gmelin, the author of the most complete handbook of chemistry up to his time,[146] and the most influential as it went through many editions, adopted Wollaston's ideas, which were not so bad. Worse was that fact that he confused compounds and mixtures.

1.30 Certain Elements Are Not Forever

During the last decade of the nineteenth century, the radioactive elements were discovered, causing further confusion in chemistry, particularly with regard to the periodic table. Could there be such a thing as an unstable element? What could that mean? What were the characteristics of an element? What were its chemical properties, and what about the spectra?[147] The clarification of these and many other questions came as a bonus to the discovery of radioactivity.

Since there were now known to be unstable nuclei, the International Union of Pure and Applied Chemistry in conjunction with the International Union of Pure

---

[145] von Meyer, E., *A History of Chemistry from Earliest Times to the Present Day*, translated from German by McGowan, MacMillan and Co., London, 1906. The original book was published in Leipzig in 1888.

[146] Gmelin, L., *Handbuch der Theoretischen Chemie*, Frankfurt am Main, bei Franz Varrwntrapp, 1827.

[147] It became clear very early on that the light emitted by each element was characteristic of that element, rather like a 'finger print'. Thus, by heating the sample in question and carrying out spectroscopic analysis of the light it emitted, it was easy to identify the content of the sample.

and Applied Physics[148] established criteria that must be satisfied for the discovery of a new chemical element to be recognized. The first condition was that the nuclide should exist for at least 10^{-14} s. Thus, unstable nuclei with shorter lifetimes were not defined as 'chemical elements'. The reason for choosing this particular lifetime was that this is the timescale for electrons to arrange themselves in the various electronic shells.

1.31 Philosophy of a Discoverer When No Theory Is Yet Available

Twenty years after the historical paper about the periodic table, on Tuesday, 4 June 1889, Mendeleev was invited to deliver the Faraday lecture before the fellows of the Chemical Society in the theatre of the Royal Institution,[149] and allowed himself to wander into philosophy and the deeper meaning of his discoveries.[150] Ironically, the lecture was delivered not too long before radioactivity, the nucleus, and quantum theory were discovered, whence the picture changed completely, along with our understanding and the philosophy.

---

[148] Wapstra, A.H., Pure and Appl. Chem. **63**, 879 (1991). In 1991, a committee decided on criteria for the recognition of new chemical elements.

[149] Mendeleev, D., J. Chem. Soc. **55**, 634 (1889).

[150] Here are some selected quotes:

> The most important point to notice is that periodic functions, used for the purpose of expressing changes which are dependent on variations of time and space, have been long known. [...] This constituted such a novelty in the study of the phenomena of nature that, although it did not lift the veil which conceals the true conception of mass, it nevertheless indicated that the explanation of that conception must be searched for in the masses of the atoms [...].
>
> If we mark on an axis of abscissae a series of lengths proportional to angles, and trace ordinates which are proportional to sines or other trigonometrical functions, we get periodic curves of a harmonic character. So it might seem, at first sight, that with the increase of atomic weights the function of the properties of the elements should also vary in the same harmonious way. But in this case there is no such continuous change as in the curves just referred to, because the periods do not contain the infinite number of points constituting a curve, but a finite number only of such points [...].
>
> While connecting by new bonds the theory of the chemical elements with Dalton's theory of multiple proportions, or atomic structure of bodies, the periodic law opened for natural philosophy a new and wide field for speculation. Kant said that there are in the world

It is interesting to note that Mendeleev lived for almost 7 years after the first Nobel prize was awarded, but he never received it himself, owing to a personal vendetta.[151]

1.32 Is the Atomic Weight Variable?

Moreover, the confusion was so great that Paul Schützenberger (1829–1897)[152] along with Alexander Butlerow (1828–1886)[153] and Vogel suggested that the atomic weight might be variable! Even as late as 1887, Theodore Richards (1868–1928) from

---

(Footnote 150 continued)

'two things which never cease to call for the admiration and reverence of man: the moral law within ourselves, and the stellar sky above us'. But when we turn our thoughts towards the nature of the elements and the periodic law, we must add a third subject, namely, 'the nature of the elementary individuals which we discover everywhere around us'. Without them the stellar sky itself is inconceivable; and in the atoms we see at once their peculiar individualities, the infinite multiplicity of the individuals, and the submission of their seeming freedom to the general harmony of Nature.

From the foregoing, as well as from the failures of so many attempts at finding in experiment and speculation a proof of the compound character of the elements and of the existence of primordial matter, it is evident, in my opinion, that this theory must be classed amongst mere utopias. But utopias can only be combatted by freedom of opinion, by experiment, and by new utopias. In the republic of scientific theories freedom of opinions is guaranteed [...].

Unfortunately, this is not always so.

[151] Mendeleev's discovery, and his courageous predictions which were soon vindicated, were seminal. But traditional dirty politics was also quick off the mark and mixed itself in liberally with the science. In 1906, Mendeleev was elected by the Nobel prize committee to win the award, but the Royal Swedish Academy of Sciences interfered and the prize went to the Frenchman Henri Moissan (1852–1907) *in recognition of the great services rendered by him in his investigation and isolation of the element fluorine, and for the adoption in the service of science of the electric furnace called after him*.

The intervention was prompted by the Swedish chemist Svante Arrhenius (1859–1927m), who won the prize in 1903 for his theory of electrolytic dissociation. Mendeleev had been an outspoken critic of Arrhenius (correct!) theory, and the unforgiving Arrhenius played it ugly, seizing the opportunity to take revenge. Arrhenius' disguised argument was that Mendeleev had made his discovery too many years earlier. This illogical argument, which at best implies the failure of the committee to recognize Mendeleev's feat sooner, was nevertheless accepted by the prize committee. The attempt to nominate Mendeleev a year later was thwarted once more by Arrhenius. Animosity between chemists apparently has an infinite decay time. For more, see Friedman, R.M., *The Politics of Excellence: Behind the Nobel Prize in Science*, Times Books, New York (2001). Meyer and Mendeleev, who argued over priority, shared the Davy Medal in 1882.

[152] Schützenberger, P., Chem. News **46**, 50 (1882). See also Schützenberger, in a discussion of the variability of the law of definite proportions before the Chemical Society of Paris in 1883, as quoted in the Proc. Am. Acad. Arts Sci. May, 1888.

[153] Butlerow, A., Ber. **16**, 1559 (1882), ibid. *The Inconsistency of the Atomic Weights*, Am. Chem. J. **5**, 137 (1883).

Harvard[154] found it necessary to investigate whether the atomic weight of copper might depend on its origin. However, he reached a negative conclusion.

In 1906, about a hundred years after Lavoisier's original study, Hans Heinrich Landolt (1831–1910)[155] felt the need to reexamine the law of mass conservation. A special vessel was prepared so that two substances could be accurately weighed, and then a chemical reaction allowed to take place. A large number of experiments were carried out and the relative differences were less than 10^{-7}, which was considered to be the error in the experiment.[156]

The fact that the atomic weight of each element does not vary was essentially proven by Stas when he measured the atomic weight of silver in different compounds. Stas found 107.921–107.943 for the atomic weight of silver, and the difference was attributed to the error in the measurement. In 1911, Baxter and Thorvaldson[157] compared the atomic weight of meteoritic iron with that of iron of terrestrial origin and found them to be identical.

Crookes, Richards' boss in Harvard, suggested[158] that one might assume the presence of a few of what he called 'worn atoms' in the countless numbers of other atoms, which must come under consideration in any atomic weight determination. This meant the existence of atoms of the same element that had 'lost some mass'.

Eventually the issue was settled with the discovery of isotopes. The slightly different isotopic composition of various elements found in different natural ores with different physical and chemical history (like diffusion) gave rise to minute differences in the chemical atomic weights.

At the beginning of the twentieth century, Richards[159] taking fantastic precautions, produced atomic weight values that may well represent the ultimate possible accuracy using purely chemical methods.

If Berzelius's work had left any questions concerning the exact value of the atomic weights, that of Stas and Richards did not. The non-integer values of the atomic weights simply had to be accepted, and Prout's hypothesis seemed doomed. Yet, even as Richards was producing his remarkably precise results, the whole meaning of atomic weight suddenly had to be reevaluated, and Prout's hypothesis rose from its ashes.

---

154 Richards, T.W., Proc. Am. Acad. Arts and Sci. **XXIII**, 177 (1887).

155 Landolt, H.H., Sitzunsb. d. K. pr. Akad. Wiss. **16**, 354 (1908).

156 We know today that the chemical binding energy is smaller than $10^{-8}mc^2$, whence Landolt's careful experiment was not that accurate, although in 1906, a year after the theory of special relativity, this could have been figured out.

157 Baxter, G.P. and Thorvaldson, T., J. Am. Chem. Soc. **33**, 337 (1911), Baxter, G.P. and Parsons, L.W., J. Am. Chem. Soc. **43**, 507 (1921).

158 Crooke, J.P., Memoirs Am. Acad. Arts Sci. **5**, 23 (1854), ibid. Nature **34**, 423 (1886).

159 Theodore William Richards won the Nobel prize in chemistry in 1914, *in recognition of his exact determinations of the atomic weights of a large number of the chemical elements*. Generations of chemists had improved the measurements and, at long last, the one who was most accurate and made measurements for a sufficiently long time was finally Nobelized.

1.33 Moseley: At Long Last, an Accurate Table

The periodic table remained in the above form until 1913, when Henry Moseley (1887–1915m),[160] a student of Rutherford, investigated[161] the X-ray spectra of elements (see Fig. 1.6). Until that time, only visible light spectra had been used to identify chemical elements. The spectrum in the visible range was obtained by heating the sample and analyzing the visible light emitted by the element. Moseley extended the technique to the newly discovered X rays. As it was not possible to heat the elements to such high temperatures that the elements would emit X rays, other techniques had to be used, namely, accelerating electrons in cathode-ray tubes and letting the electrons bombard the element under examination. A description of these techniques would carry us outside the scope of this book.

Until Moseley's day, only the atomic weights were known, and the notion of atomic numbers did not exist. For this reason, Mendeleev and all his predecessors could only use the atomic weight to classify the elements, a fact that caused problems from time to time, as discussed above. What Moseley did was to arrange the elements according to their X-ray properties. Every element gave two main spectral lines, with a factor of five difference in intensity and about 10% difference in frequency.[162]

Moseley soon discovered that the frequency of the line was given by

$$\frac{\nu(X)}{R} = \frac{3}{4}(Z - 1)^2 ,$$

where Z is the location of the element in the table and R is the constant Rydberg number, obtained from attempts to find a phenomenological formula for spectral lines in the visible range (see Sect. 2.34). For this reason Moseley plotted the Z of the element as a function of the square root of the frequency. The meanings to be attributed to the formula and the coefficients were left to various guesses. No such number as Z was measured. The only relevant parameter to describe the element was its abstract location in a table of X-ray emissions.

However, Moseley quickly realized that he had generated a periodic arrangement of the elements in which the chemical properties of the element were related to the position in the table. Moseley[163] therefore defined the position in his table as a new quantity and called it the 'atomic number' of the element.[164] So the atomic number was an integer by definition, in contrast to the atomic weight, which was not. It soon became clear that the atomic number as defined by Moseley was really the

[160] One of the tragedies of science happened when the brilliant scientist Moseley was killed on the battle field of Gallipoli, Turkey, in 1915.

[161] Moseley, H., Phil. Mag. **26**, 1024 (1913), ibid. **27**, 703 (1914).

[162] Moseley looked for the highest X-ray frequency emitted by the element.

[163] Moseley, H.G.J., Phil. Mag. **XX**, 1024 (1913).

[164] In some textbooks you may find it claimed that, when Moseley arranged the elements according to the atomic number, he got a nice periodicity. The truth is, however, the other way round. Moseley invented the atomic number.

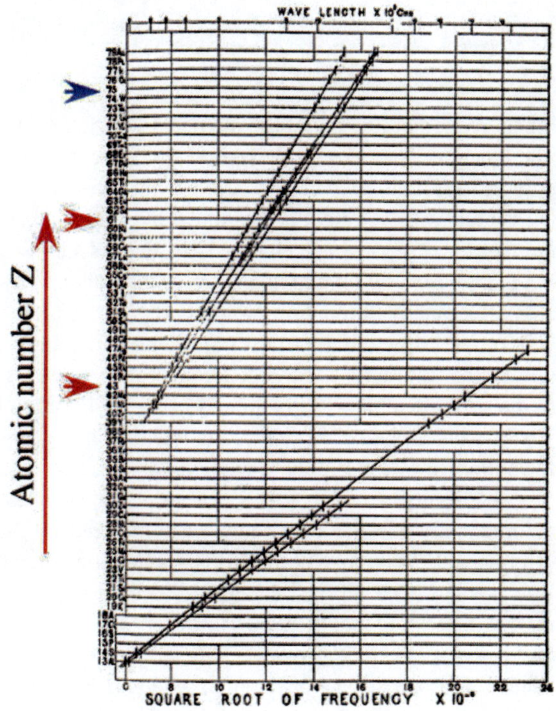

Fig. 1.6 Moseley's original 1913 figure. Note that the *horizontal axis* shows the square root of the frequency, while the *vertical axis* shows 'the location in the table'. Marked are $Z = 43$ and 61, which do not exist in nature, and $Z = 75$, which was discovered by Noddack in 1925. *Straight lines* indicate the perfect relation

number of electrons in the atom, as well as the position of the element in the periodic table. Moreover, and most importantly, it was the atomic number that determined the chemical properties of the element, and not the atomic weight.

In retrospect, we can understand the difficulties faced by chemists in the era prior to the discovery of electrons, neutrons, and so on. They had only the non-integer atomic weight at their disposal. In general, the atomic weight increases faster then the atomic number, so Mendeleev did not need to encounter problems arranging the elements according to their atomic weights. But alas, this is not always the case, and the most abundant isotope, which affects the atomic weight, is dictated by periodicities in the nucleus (see later). In any case, Moseley's new definition provided a unique integer number without reference to previous data.

Moseley left spaces for elements he did not know about, whenever the data did not fit the straight line, as he explained:

> The order chosen for the elements is the order of the atomic weights, except in the cases of A, Co, and Te, where this clashes with the order of the chemical properties. Vacant lines have been left for an element between Mo and Ru, an element between Nd and Sa, and an element between W and Os, none of which are yet known, while Tm, which Welscach has separated into two constituents, is given two lines. [...] This is equivalent to assigning to successive

> elements a series of successive characteristic integers. On this principle the integer N[165] for Al, the thirteenth element, has been taken to be 13. [...] This procedure is justified by the fact that it introduces perfect regularity into the X-ray spectra.

Moseley did not know that the elements with $Z = 43$ and 61 do not exist in nature. The location $Z = 75$ was also empty, because the element rhenium was discovered only in 1925.

At that point in time, the periodic table was complete (save for one element with number $Z = 75$, which changed nothing as far as the table was concerned),[166] but there were still no explanations for this table. Why should the properties of the elements vary with the atomic number? Why is there a periodicity? Why does the table come to an end? Why do the lanthanides have similar chemical properties, despite the fact that they have different atomic numbers? The first partial explanation of the periodic table was given in 1921 by Niels Bohr (1885–1962m, Nobel laureate 1922), who explained[167] the arrangement of the electrons in the quantum levels of the atom. He was also able to explain why the noble gases do not react chemically.

Even beautiful, clear, and clean results like the ones obtained by Moseley are not automatically accepted. Moseley was criticized by the mathematician Lindemann (1852–1939),[168] who denied Moseley's statement that his results strongly supported Bohr's view. As a matter of fact, Lindemann argued that Mosley's results *yield no information about the structure of the atom beyond confirming the views of Rutherford and van den Broek.*[169] However, Moseley and Bohr pointed out Lindemann's error. Lindeman assumed an arbitrary unit of length in the problem of the electric charge.[170] On the other hand, Moseley stated that he could not obtain any agreement with the prediction of Nicholson,[171] while the agreement with Bohr's theory was perfect (see Sect. 2.35).

---

[165] Moseley used N to denote the location in the table. Today we use Z to denote the number of elementary charges in the nucleus, and $Z \equiv N$.

[166] The element nipponium was discovered in 1908 by Ogawa. It later became clear that nipponium is actually rhenium. See Yoshihara, H.K., Proc. Jpn. Acad. Ser. B, Phys. Biol. Sci. **84**, 232 (2008).

[167] Bohr, N., Nature **107**, 104 (24 March 1921).

[168] Lindemann, F.A., Nature **92**, 631 (5 February 1914). Lindemann was the one who proved that you cannot 'square the circle' by a ruler and compass construction, since he proved that π is a transcendental number. This great proof did not prevent him from publishing several incorrect proofs of Fermat's last theorem See Bell, E.T., *Men of Mathematics*, Simon and Schuster, New York 1986.

[169] Moseley, H., Nature **92**, 554 (15 January 1914). Moseley and Bohr replied to Lindemann just a week later. Bohr's letter preceded Moseley's. Lindemann was wrong.

[170] It was strange to uncover such an error in the article by the scientist who was the first to prove that π is transcendental. Lindemann, who was essentially a mathematician, also worked on the theory of the electron, and had engaged in controversy over this with Arnold Sommerfeld (1868–1951), one of his PhD students. Sommerfeld was known as the Prince of German Physics. He was nominated a record 81 times for the Nobel Prize in Physics, and served as PhD supervisor for more Nobel prize winners in physics than any other supervisor before or since. However, he never actually got the award.

[171] See Shaviv, G., *The Life of Stars*, Magnes and Springer, 2009.

1.34 The Theory Behind Moseley's Discovery

When Moseley looked for a theory to explain his X-ray results, the only available theory with any degree of success was Bohr's. However, the theory was good for atoms with one electron only, i.e., the so-called hydrogen-like atoms. There was as yet no successful theory for atoms with many electrons.

Moseley found that the frequency of the X ray could be described by

$$\nu = A(N - b)^2 ,$$

where N is an integer and[172] A, which depends on the spectral lines, known as K_α and L_α lines, is given by

$$A(K_\alpha) = \left(\frac{1}{1^2} - \frac{1}{2^2}\right)\nu_0 , \quad \text{with } b = 1 ,$$

and

$$A(L_\alpha) = \left(\frac{1}{2^2} - \frac{1}{3^2}\right)\nu_0 , \quad \text{with } b = 7.4 ,$$

where ν_0 and b are constants. The problem was to know how the constant ν_0 varied from one element to the other. The particular form of the formula, with squares in the denominator, is a consequence of the Rydberg formula (see Sect. 2.34).

1.35 The Last Element to Be Discovered

The last chemical element to be discovered was rhenium, named after the Rhine river. Rhenium (Re) was discovered by two chemists, Ida (Tacke) Noddack (1896–1979) and Walter Tacke (1893–1960) in 1925. Rhenium, which is the eighty-first element to have stable isotopes, was detected in platinum ores. The atomic number is 75 and the atomic weights of the stable isotopes are 185 and 187. The specific density is very high, about 21 g/cm^3 at room temperature (only platinum, iridium, and osmium are denser), and at 3 453 K it has the third highest melting point after tungsten and carbon. With the discovery of Re, the periodic table was complete save for two unstable elements.

1.36 Aston and the Discovery of Isotopes

As pointed out above, the atomic number is monotonic in the periodic table (by definition), while the atomic weight is not. The solution to this mystery came

---

[172] Moseley, H.G.J., Phil. Mag., Ser. 6, **27**, 703 (1914).

in 1919 when Francis Aston (1877–1945)[173] managed to significantly improve his mass spectrometer and published his extensive research on the masses of the chemical elements. Aston examined a series of chemical elements and found that they were all composed of a mix of isotopes, and that each isotope had an integer atomic weight. On the other hand, Aston found that hydrogen has an atomic weight of 1.008 (relative to oxygen 16), in agreement with findings by chemical methods. For He, Aston found an atomic weight of 4.007.

Aston reached the conclusion that, apart from the two lightest elements, hydrogen and helium, all elements have an integer atomic weight, a result he called the whole-number rule. It became clear that the non-integer atomic weights found by the chemists were fortuitous and arose from mixtures of isotopes. In particular, the historical problem with the non-integer atomic weight of chlorine was solved, as chlorine was found to have two isotopes (with atomic weights of 35 and 37). So the 'evidence' raised by Berzelius to destroy Prout's hypothesis finally turned out to be wrong. The implications for atomic structure were enormous. In Aston's words:

> An elementary atom with mass m may be changed to one with mass $m + 1$ by adding a positive particle and an electron. If both enter the nucleus, an isotope results.

Aston's major discovery provided experimental evidence for accepting the assumption that all atoms were built out of the same standard building blocks.

But Aston faced a problem. The theory of relativity was already known as of 1905, although many of its consequences were not yet confirmed experimentally, and many physicists were slow, or even reluctant, to adopt it. So Aston knew about $E = mc^2$ (although he wrote that this formula derived from the electromagnetic theory,[174] rather than from relativity), which meant that the weights could not be additive. When two atoms fuse and there is a binding energy, the mass of the new nucleus should be less than the sum of the masses of the individual atoms, *but*, he argued, *it only becomes so when the charges are relatively distant from each other*. According to Aston, this appeared to be so in light nuclei. In the case of the heavier atoms, Aston argued that the charges were very close and that *the law* $E = mc^2$ *is not effective*.

But if the weights were indeed an exact product of the same building unit, there was no room for the formula $E = mc^2$! By pure chance, Aston's instrument at that time was not sufficiently advanced to see the effect of $E = mc^2$ on the atomic weight of nuclei, and hence he claimed incorrectly that the formula would not apply to nuclei. It is not clear from his paper why it should not apply. Apparently, Aston could not imagine that the nuclear binding energy would affect the mass to a degree below the accuracy of his measurements! Only 8 years later, and not before the accuracy was improved, were the effects of the binding energy discovered with the mass spectrometer.

---

[173] Aston, F.W., Nature **104**, 383 (1919) and Phi. Mag. Ser. 6, **39**, 611 (1920), ibid. Nature **104**, 393 (1919).

[174] There are several calculations in electrodynamics which yield $E = \alpha mc^2$, where α is of the order of unity.

Aston's discovery of isotopes won him the Nobel prize just three years later. In giving the committee's arguments for awarding Aston the 1922 Nobel prize in chemistry, professor Henrik Gustaf Söderbaum (1862–1933),[175] who was a chemist, stated that:

> All the masses so far measured [...] can be expressed by means of whole numbers in relation to oxygen 16 [...] and must be regarded as the expression of a natural law [...] which has been named the 'whole-number rule'.

In other words, Aston got the Nobel prize for showing the unity of matter by demonstrating that all elements were composed of the same units. In a sense, he proved Prout's hypothesis that the atoms of the elements are all made up of aggregations of a certain number of atoms of the lightest known element hydrogen. If Prout had been right, so claimed Söderbaum, then all elements should be exact multiples of hydrogen, but they are not. So Prout's hypothesis was proven with a twist, the fundamental unit à la Aston is the sum of a negative and a positive charge. Söderbaum did not mention that $E = mc^2$!

The impression one gets from reading Söderbaum's address is that he and the committee did not fully comprehend the repercussions of Aston's results—they left no room for binding energy and special relativity. After getting the Nobel prize, Aston improved the mass spectrometer and discovered the behavior of the binding energy with mass. He definitely deserved the prize for this discovery, but doubtfully for the earlier discoveries, which prima facie contradicted relativity and were misunderstood by the prize committee.

1.37 A Solution to a Long-Standing Problem

It was in 1828 that the botanist Robert Brown (1773–1858)[176] observed minute particles within vacuoles in pollen grains executing a continual jittery motion. He then observed the same motion in particles of dust, enabling him to rule out the hypothesis that the effect was due to pollen being alive. Brown did not provide a theory to explain the perpetual fluctuation. Jan Ingenhousz (1730–1799) had earlier reported a similar effect using charcoal particles in German and French publications of 1784 and 1785.[177] A year before Brown, another botanist, Adolphe-Théodore Brongiart (1801–1876),[178] also observed the phenomenon, but his contribution was

---

175 Söderbaum was a chemist, the secretary of the Swedish Academy of Sciences and a member of the Nobel committee for chemistry as of 1900. (He must therefore have been involved in the Mendeleev case.) In his later years, Söderbaum became interested in the history of chemistry and in particular the biography of Berzelius, about whom he wrote several books. It is an irony of history that the Swedish Söderbaum awarded the Nobel prize to the physicists who spoiled Berzelius' arguments against Prout's hypothesis.

176 Brown, R., *Miscellaneous Botanical Works* **1**, London, Royal Society. Ibid. Edinburgh New Phil. J. **5**, 358 (1828). Ibid. Phil. Mag. **4**, 161 (1828).

177 van der Pas, P.W., *The Discovery of Brownian Motion*. Scientiarum Historia **13**, 17 (1971).

178 Brongiart, A-T., Ann. Sci. Naturelles **12**, 41 (1827).

apparently completely missed by physicists, and the phenomenon is known today as Brownian motion.

The mystery of Brownian motion prevailed for close to forty years.[179] In 1889, Gouy (1854–1926)[180] found that Brownian movement was more rapid for smaller particles, while in 1900, Exner (1849–1926)[181] undertook the first quantitative studies, measuring how the motion depends on temperature and particle size. The mathematical treatment of Brownian motion involved new concepts from stochastic approaches to problems in physics.[182] However, we are interested here in the atomic nature of the disturbing agent and what can be inferred from it.

At long last, in 1905, it was Albert Einstein[183] who solved the problem, when he showed that the jittery motion is due to collisions of the pollen with molecules. Moreover, Einstein demonstrated that the mean distance squared the particle moves is given by $\bar{x}^2 = 2Dt$, where D is the diffusion coefficient and t the time. This was the same diffusion coefficient that Fick was talking about, although Einstein did not mention him.

Only then was Avogadro fully vindicated. In 1916, Perrin[184] used the phenomenon to determine the numerical value of the Avogadro number,[185] and his result was $(6.0\text{–}8.9) \times 10^{23}$.

---

179 In many writings on Brownian motion, it is claimed that, in 1877, J. Desaulx hypothesized that the phenomenon was due to thermal motion of molecules. The texts even supply a quote from Desaulx. However, I failed to find any article by an author with this name, let alone a hypothesis. See also the footnote on Einstein's paper.

180 Gouy, L.G, Compt. Rend. **109**, 102 (1889).

181 Exner, F.M., Ann. Phys. **2**, 843 (1900). Exner became famous because he was assisted for a while by Erwin Schrödinger.

182 Fick, A., Ann. Phys. **94**, 50 (1855). This is a purely phenomenological approach. There is no physics in it. Smoluchowski, M., Ann. Phys. **21**, 756 (1906). This paper was published shortly after Einstein's paper.

183 Einstein, A., Ann. Phys. **17**, 549 (1905), ibid. **19**, 289, 371 (1906). This was part of Einstein's dissertation at Zürich University. Einstein gave no reference in his paper, save one to Kirchhoff's lectures. In Ann. de Physik **19**, 371 (1906), there is a paper by Einstein in which he tells the reader that Seidentopf informed him that he (Seidentopf) and Gouy (J. de Phys. **7**, 561, 1888) carried out experiments on Brownian motion and had data. Einstein remarked that he found agreement between his theory and the experimental data of Gouy. There is no reference to Desaulx, and nor does such a reference appear in Exner's paper. Apparently, the heros of Brownian motion did not know about Desaulx.

184 Perrin, J., *The Brownian Motion in Atoms*, D. van Nostrand, 1916. Ibid. Compt. Rendu, May 1908. Perrin, J. and Donau, J., Fortschritt. über Kolloide u. Polymere **1**, no. 6–7, April 1910.

185 The accurate value is 6.022×10^{23}. Reasonable values were obtained in the late nineteenth century using the sedimentation equilibrium of colloidal particles. Millikan's oil drop experiment in the 1900s gave even better accuracy and was cited in most chemistry text books till about 50 years ago.

1.38 Is the Periodicity of the Chemical Elements Really So Obvious?

Is it so obvious that the properties of the chemical elements are periodic? In fact, it is not at all obvious without quantum theory. If we have a potential well with energy levels, then it can be filled with bosons or fermions. In the first case, all bosons will stay in the lowest energy level and all systems will look alike, irrespective of the number of bosons. In the second case, the fermions will fill all levels, and every time a level is full we expect to see similar behavior. Thus, when the particles filling the potential well are fermions, we should expect periodicity, although not necessarily of equal length, as a function of the number of particles in the system.

Chapter 2
Preparing the Ground for Delving into the Stars

2.1 The Long Road to Deciphering the Composition of the Stars

All our knowledge about the composition of cosmic objects is obtained via spectroscopy. Two key disciplines are required to extract this information from observations: the theory of radiative transfer through stellar material and the theory of atomic structure. Spectroscopy is as old as modern science. It began with Johannes Kepler (1571–1630m)[1] and later Isaac Newton (1643–1727m), who knew about the effect of the prism on sunlight.[2] When they cast the outgoing light of the prism on a screen, they discovered all the colors of the rainbow. Naturally, Newton used a circular aperture, and consequently his spectrum was not pure. Despite this early start, progress was slow at the beginning, and even after major breakthroughs, about 400 years were needed before reliable information about stellar composition could be obtained.

In 1752, Thomas Melvill (1726–1753)[3] began to study the spectra of salts placed in a flame. Melvill reported that heating table salt converted the flame to yellow. No explanation of the phenomenon was given.

After Kepler, about 150 years went by before William Wollaston, a London physician, invented the very narrow slit in 1802. This apparently trivial invention allowed him to obtain a pure spectrum by preventing the differently colored lights from over-

---

[1] Kepler, J., *Ad Vitellionem Paralipomena, Quibus Astronomiae Pars Optica Traditur*, Claudius Marnius, Frankfurt, 1604.

[2] Newton, I., *Treatise of the Reflections, Refractions, Inflections and Colours of Light*, London, 1704. The discovery was in 1666, according to Newton's letter to the secretary of the Royal Society dated 6 February 1672.

[3] Melvill, T., J. R. Astr. Soc. Can. **8**, 231 (1914). This is a reprint of the original paper from 1752. And so says the special introduction: *Had the ingenious author of this paper (who died December, 1853, at the age of 27) lived to put the finishing hand to it, he would probably, have added many things*. I could not discover who wrote the introduction, nor the occasion on which the paper was reprinted.

G. Shaviv, *The Synthesis of the Elements*, Astrophysics and Space Science Library 387,
DOI: 10.1007/978-3-642-28385-7_2,

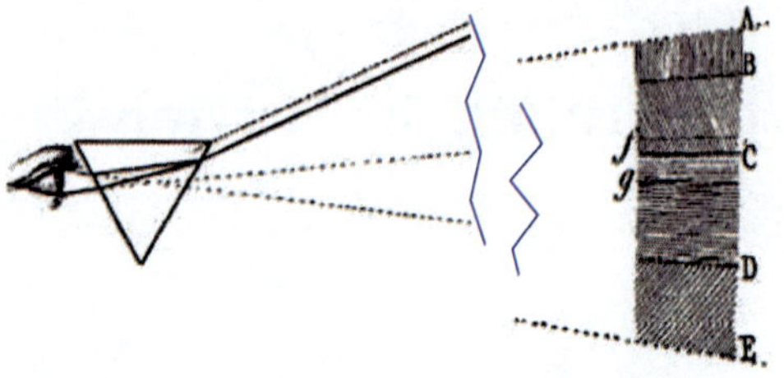

Fig. 2.1 The first description of what is known today as the Fraunhofer line, by Wollaston (1802). Note that the lines appear to separate different regions (what Wollaston called 'images' and 'separation lines') in the continuous spectrum. Wollaston marked the lines by A, B, ..., and this is the source of the notation D, for the famous *yellow* sodium line

lapping, as Henri Roscoe (1833–1915) described it in 1862.[4] The narrow slit is still in use today.[5] The fact that the spectrum was now pure enabled Wollaston[6] to make a seminal discovery. He noticed that the colors seen in the spectrum of the Sun (about 5) are separated by dark lines (see Fig. 2.1). At the same time, Wollaston discovered that the light emerging from a candle is not continuous, but exhibits distinct colored lines. Thus, hot gases do not emit continuous light, but lines. Wollaston offered no explanation. In his own words: *I cannot undertake to explain* the dark lines. He made no attempt to explore their origin. Florian Cajori (1859–1930)[7] claimed about a century later that Wollaston's 'explanation' *shows how a most plausible theory may be destitute of all truth*, though Wollaston admitted that he had no explanation.

The next major progress came in 1814, when Joseph Fraunhofer (1787–1826m)[8] added a small telescope of a theodolite after the prism and put the slit at a distance of 30 meters before the prism. As a result, he created a powerful spectrometer. When he examined solar light, he was able to observe a total of 576 dark lines crossing the colorful spectrum, the positions of which he recorded. It was from this moment on that these lines became known as the Fraunhofer lines. Fraunhofer noted that the intensity of the solar continuum radiation is not uniform and that the maximum intensity occurs in the yellow color. Next, Fraunhofer identified the location of the bright lines emitted by hot gases with some dark lines he saw in the spectrum of the Sun. In particular, he noticed that the D line (in Wollaston's notation), which appeared as two very close yellow lines in the spectrum of a hot gas, was present as one of the darkest lines in the spectrum of the Sun. Fraunhofer turned the system

---

[4] Roscoe, E.H., The Edinburgh Review **CXVI**, 295 (1862).

[5] Still, the question remains as to how Newton missed the discovery of the Fraunhofer lines in the solar spectrum. There is a claim (Johnson, A., Nature **26**, 572, 12 October 1882) that Newton had to rely on a young assistant with better eyesight, and it was the assistant who missed the lines.

[6] Wollaston, W.H., Phil. Tran. R. Soc. Lond. **92**, 365 (1802), read 24 June 1802. In a paper just after Wollaston's in the journal, Young (Young, M.D.; ibid., 387, read 1 July 1802) describes how he repeated Wollaston's experiment and got 'perfectly' identical results. In particular, he mentions 'the line of yellow'.

[7] Cajori, F., *A History of Physics*, The Macmillan Comp., London, 1899.

[8] Fraunhofer, J. von, Denkschriften de K. Acad. der Wissenschaften zu Munchen, Band **V**, 193 (1817).

of dark lines into standards for the calibration of achromatic lenses. He also tried to discover the source of the lines, but failed.

At first, Fraunhofer thought the dark lines were an artefact of interference caused by the slit. But further experiments convinced him that they were a genuine feature of the Sun's spectrum.[9] Fraunhofer measured the relative distance between the lines and discovered that these distances did not change with the location of the Sun in the sky, or even when the solar light was reflected from planets, like Venus. The positions of the lines were fixed! Moreover, he was able to observe three dark lines in the spectrum of Sirius, the brightest star seen from the Earth. Fraunhofer was thus convinced that the source of the dark lines was in the stars.

The scientific story of the Fraunhofer lines as described above camouflages a rivalry driven by national pride, which took place between the German and English academic communities over a long period of time.[10] The establishment of the Fraunhofer Optical Institute in Bavaria with the financial support of Napoleon's army, destined to produce superb optical instruments needed for the artillery, allowed Fraunhofer to develop the ultimate skill in glasswork and optical instrumentation. He used this to facilitate major discoveries in astronomy in general, and in the spectrum of the Sun in particular. England, on the other hand, decided to impose a tax on flint glass, and in this way strangled any technical progress. As an excuse, some of the leading English scientists began to argue that the art of manufacturing glass was not science. This was an exaggeration because the Fraunhofer calibration system, facilitated by the unrivaled quality of their instrumentation, was described by Babbage in his essay on why English science declined[11] as an example of what high precision instruments, craftsmanship, and above all observation can lead to.[12]

In 1824, Fraunhofer observed the coincidence of the yellow sodium lines with the double *D* line in the solar spectrum. Unfortunately, the inevitable conclusion that the

---

[9] Fraunhofer, Edinb. Phil. J. **9**, 299 (1823); ibid. **10**, 22 (1823).

[10] Jackson, M.W., Studies in History and Philosophy of Science Part A **25**, 549 (1994). An interesting account of the way short-sighted politics on the one hand and military needs on the other affected progress in science on both sides of the English channel.

[11] Babbage, C. *Reflections on the Decline of Science in England and on Some of Its Causes*, London, Printed for B. Fellowes, Ludgate street, 1830, p. 210. Charles Babbage (1792–1871m) was a mathematician.

[12] The irony is that the German scientists thought the same way and expressed contempt for experimental work. When the brilliant self-taught optical inventor Fraunhofer applied for membership to the Bavarian Academy of Sciences, his application was rejected (1819) because Bavarian academics were convinced that the discoveries had only technological significance (what a shame!). Indeed, Fraunhofer regarded himself as an optical engineer. But the telescope lenses produced by Fraunhofer were considered the best in the world. In 1838, Friedrich Bessel (1784–1846m) used a Fraunhofer telescope to determine the first parallax of a nearby star (Bessel, F.W., MNRAS **4**, 152, 1838). The star was 61 Cygni, with a mean annual parallax of 0.3135 arcsec. Johann Galle (1812–1910m) was using a Fraunhofer telescope when he discovered the theoretically predicted planet Neptune in the year 1846 (Galle, J.G., MNRAS **7**, 153, 1846). It was estimated that Fraunhofer's refractors of a given aperture were as effective as reflectors with an aperture three times as big. Better late than never, Fraunhofer was accepted as a member of the Bavarian Academy in 1823 and died three years later, before he turned forty.

Fig. 2.2 Testimony to discoveries made in Heidelberg around 1855–1860. *Left* the plaque in the main street of Heidelberg, on the building where Kirchhoff carried out solar spectroscopy for the first time. *Right* on the other side of the street, in a small garden, stands Bunsen's statue looking at the plaque commemorating Kirchhoff and Bunsen's discoveries. Photos by Shaviv

yellow lines and the dark *D* lines were produced by the same chemical element, but under different physical conditions, never crossed Fraunhofer's mind (Fig. 2.2).

2.2 Why the Sodium *D* Line Appears Everywhere

An annoying problem in spectroscopy was that the usual flames used for heating the substances contained sodium as a trace element from salt, and consequently every spectrum was contaminated with sodium lines. The problem was solved when Robert Bunsen (1811–1899m) invented the Bunsen burner, or in short the Bunsen, which produced a hot flame but without sodium.[13] Using the Bunsen, Joseph Swan (1828–1914) was able in 1857[14] to identify the *D* lines with sodium, because the clean flame simply did not show them. A long list of discoverers had suggested the coincidence between the Fraunhofer *D* line and sodium, but had not been able to verify it. With this discovery, Swan[15] confirmed the suspected discoveries by William Herschel (1738–1822m) in 1823, David Brewster (1781–1868m) in 1835, and Leon Foucault (1819–1868m) in 1849.

[13] Bunsen invented the Bunsen burner sometime in 1855. The goal was mainly to develop a better heat source for laboratory work. The standard flames used were smoky and produced a low heat intensity. Bunsen's breakthrough was simple: mix the gas with air before combustion instead of during combustion. Two years later in 1857, Bunsen and Roscoe described the new burner in Pogg. Ann. Phys. **100**, 84 (1857). Mixing the air and gas before burning left the salt outside the burning volume.

[14] Swan, J.W., Edinb. Trans. **21**, III, 411 (1857).

[15] Swan did not like the way Kirchhoff and Bunsen attributed the discovery to him, and found it necessary to write a letter to the editor stressing that he found it to be the case in *all flames*. Swan Phil. Mag. **20**, 169 (1860).

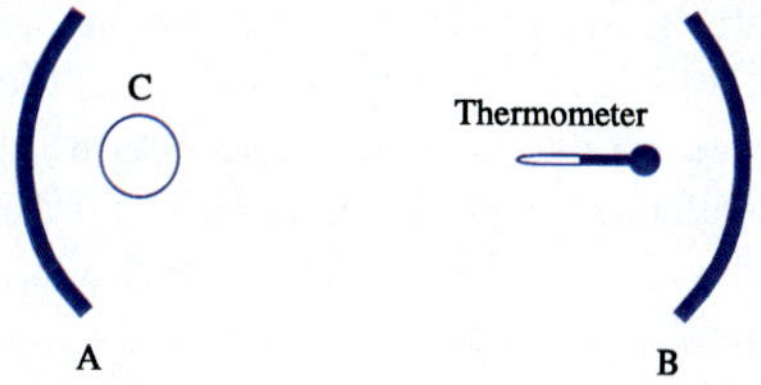

Fig. 2.3 The setup of Pictet's experiment. *A* and *B* are two mirrors. *C* is the hot/cold object. The thermometer and the hot/cold object are placed at the focal points of the two mirrors

It is fascinating to note how the strong line of sodium, present as a trace element almost everywhere, played such a crucial role in the development of spectroscopy.

2.3 The Concept of Equilibrium: Pictet and Prévost

In the middle of the eighteenth century, the concept of 'radiant heat' was still separated from the concept of light. Thomas Young (1773–1829m) wrote in 1803[16] that the first to concentrate radiant heat was Hoffmann, who used mirrors prepared by Wolfe. No further details were given by Young.

Marc-Auguste Pictet (1752–1825m)[17] used the possibility of concentrating heat rays by placing a hot blob of iron at the focus of a concave mirror and observing a thermometer placed at the focus of a second mirror placed about 3 meters away (see Fig. 2.3). After taking all the precautions to isolate the experiment from external perturbations, Pictet discovered that, as the iron blob cooled, the temperature of the thermometer rose. The experiment was conducted ten years before Herschel placed a thermometer behind a spectrometer at a location beyond the red color and discovered infrared radiation, demonstrating how the invisible 'radiant heat' and visible radiation behave and propagate in the same way. Pictet's experiment on the concentration and propagation of radiant heat was not the first in this story, but it was the first to stimulate interest among physicists and consequently incited them to provide a theoretical explanation.[18]

Pictet repeated the experiment with 'cold', i.e., he put snow at the focus of one of the mirrors, and found that the temperature at the other focus went down. In Pictet's language, the experiment demonstrated the 'reflection of cold'.

In the late eighteenth century, heat was assumed to be a kind of weightless fluid called caloric, which pervades everything. Those were the days when the similarly

---

16 Young, T., *A Course of Lectures on Natural Philosophy and the Mechanical Arts*, London, Taylor & Walton, new edition, 1846, p. 489. The original edition was published in 1803. The relevant report is by Wolfe (Phil. Tran. 4, 1769), who stated that the first to concentrate heat like light was Hoffmann, although Buffon (Histoire Naturelle, Supplement, 1774, I, p. 146) gave a more rigorous proof than Hoffmann did.

17 Pictet, M.A., *Essais sur le feu*, Geneva, 1790. Translated from French by W.B., printed for E. Jeffery, London, 1791.

18 See Cornell, E.S., Ann. Sci. **1**, 217 (1936) for a survey of previous experiments.

imaginary fluid phlogiston was invented to explain the chemical process during fire. Pictet believed in the caloric, and the two experiments he conducted differed from one another only in the direction of flow of the caloric. Pictet remarked that his explanation would not change if the heat were some vibrations in the elastic fluid of fire, because he argued that such vibrations would be reflected like sound. However, he did not specify who had proven such an effect. No matter how illogical it was, caloric served in those days as the best theory in town and many physicists subscribed to it.

When Pierre Prévost (1751–1839)[19] approached the problem in the 1780s, there were physicists who even believed that 'cold' was a different weightless fluid from caloric fluid, calling it 'frigoric'. Prévost argued, on the other hand, that cold was simply a lack of caloric.

In 1791, Prévost, who was Pictet's colleague at Geneva university, published[20] a seminal interpretation of Pictet's experiment. First, Prévost made the following assumptions:

> Heat is a discrete fluid. [...] And this is the effect of the movement of its particles. This movement is caused by the impulse of a much more subtle fluid whose effect upon its particles is determined to a certain extent by their form. It is so swift that when heat is freed, its translation from one to another appears instantaneous. [...] A discrete fluid whose particles radiate like those of light may be confined by barriers, but may not be confined by another radiant fluid nor, in consequence by itself.

We may wonder whether anyone really understood these sentences in full. It reflects how vague the notions of the physical meaning of heat and light were at that time. Yet despite such foggy notions, the important hypothesis put forward by Prévost was based on common sense. It could not have been based on thermodynamics, which did not exist at that time, nor on the as yet unknown microphysics.

Prévost conceived the first thought experiment to consider *two portions to be enclosed in an empty space, terminated on all sides by impenetrable walls*. It was this concept that developed into a cavity emitting black body radiation, although the terms 'cavity' and 'black body' did not exist in Prévost's lexicon.

Next, Prévost defined two types of equilibrium:

> Absolute equilibrium of free heat is a state of this fluid in a portion of space which receives as much as it allows to escape it. Relative equilibrium of free heat is the state of this fluid in two portions of space which receive from each other equal quantities of heat, and which are, moreover, in absolute equilibrium, or experience changes [that are] precisely equal.

---

[19] Prévost was the first known case of a lawyer who turned into a physicist. The other famous cases are Edwin Hubble, who studied law before changing his mind and pursuing a unique career in astronomy, and Lewis Rutherfurd, an attorney and amateur astronomer, who built an observatory at the center of New York city in 1856. Avogadro studied law but reached the conclusion that physics is more interesting.

[20] Prévost, P., J. Phys. **38**, 314 (1791).

The term 'free heat' was not defined by Prévost in this paper, but the definition can be found in his book published in 1809,[21] where it is defined as: *free heat as radiant fluid.* Not a terribly helpful definition.

Prévost conceived the following solution: the two bodies (the hot/cold blob and the thermometer) exchange caloric all the time and tend to a state in which each one absorbs and emits the same amount of caloric. This is the idea of dynamic equilibrium, in which heat flows permanently in both directions (with vanishing net flow), in contrast to static equilibrium, in which case there is no flow at all. According to Prévost's model, the cooling takes place because the hot body receives less caloric than it loses. Prévost hypothesized that, in equilibrium, the net flow vanishes, irrespective of the composition of the two radiating bodies. On the other hand, the emission and absorption do not vanish in equilibrium. It did not seem logical to Prévost that, when the temperature of a body equals that of its surroundings, the body would suddenly stop radiating. No mathematical formulation or proof was provided. Prévost's book was published in 1809, and as the author pointed out, after Herschel's *beautiful demonstration that caloric rays reflect and refract*, which implied that these rays behaved like light. This is the reason for the name Prévost chose for his book, viz., *Du Calorique Rayonnant*, which means, the radiating caloric. While Prévost rejected the explanation of his colleague, he provided an explanation as to why Pictet was led to his own erroneous explanation. There was no mention of Hoffmann and his priority.

Towards 1804, Rumford and Leslie discovered that the nature of the surface of the substances is not less important for absorption and emission. Benjamin Thompson, known as Count Rumford (1753–1814m), discovered that the heat losses are maximal when the substance is painted black and minimal when it is metallic. The particular property of black or blackened surfaces was already discovered in Rumford's first experiments on the nature of heat.[22] Rumford found that black or blackened surfaces cool fast, and hence are the best heat emitters, while polished metallic surfaces are the slowest to cool, and hence are the worst heat emitters.[23] We can trace the birth of the black body as best emitter to this discovery by Rumford.[24]

---

[21] Prévost, P., *Du calorique rayonnant*, Paris, Chez Paschoud, J.J., Libraire, Quai des Grands Augustins, no. 11, près du pont Saint-Michel, à Genève, chez le même libraire, 1809. The proof of the radiation law in the book is only verbal. Mathematics appears only in some examples, and the cooling law is expressed algebraically, and not as a differential equation, although the latter techniques had been known since Newton's times. Another interesting part of the book is Chap. VII, p. 298, where the author discusses the importance of the radiative heat exchange of the Earth in determining its global temperature, a key factor ignored by many in those days. This was at the time Fourier began his attempts to calculate the heat balance of the Earth.

[22] Rumford, *An inquiry Concerning the Nature of Heat, and the Mode of Its Communication*, 1804, in *Collected Works of Count Rumford*, Harvard Press, Cambridge, 1970.

[23] In 1796, on the basis of his research results for radiant heat, Rumford invented what is called today the Rumford fireplace. It reflects heat well and eliminates turbulence. The Rumford fireplace was popular until 1850.

[24] In *The Gentleman's Magazine*, p. 394, October 1814, we find the following anecdote about Rumford, who applied his own research results:

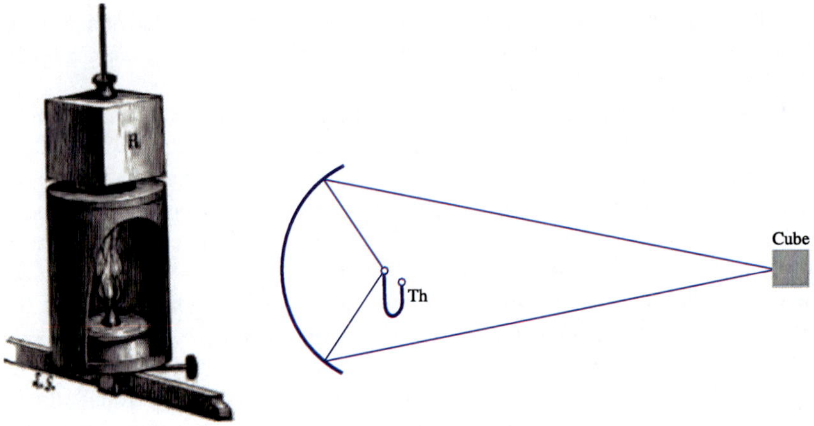

Fig. 2.4 *Left* the Leslie cube used by Leslie in his thermal radiation experiments. *Right* the arrangement applied by Leslie to investigate the emission of various substances. The parabolic mirror concentrates the radiation coming from the heated cube onto a thermometer. The cube is filled with boiling water and painted or coated with the substance to be tested

In 1809, Prévost called his hypothesis *l'équilibre mobile*, or dynamic equilibrium, and it is frequently referred to as *the theory of exchange*. Prévost was interested in heat transfer between bodies. He assumed that space is full of *radiant heat*, because every body radiates and absorbs heat continuously. But the dependence of the rate of cooling on the temperature of the body was unknown at that time.[25] So the implicit, unverified assumption was that the rate of cooling of a hot body is proportional to some monotonic function of the temperature. In equilibrium, the emission equals the absorption, and consequently the temperatures of the two bodies are equal. Prévost's idea was challenged by Rumford[26] in the realm of 'radiant heat', but this discussion would carry us too far afield and is less crucial for the physics of the Fraunhofer lines.

(Footnote 24 continued)

> Nor did any one follow (which is not to be wondered at) his whimsical winter dress, which was entirely white, even his hat. This he adapted agreeably to the law of nature, that more heated rays are thrown from a dark body than a light one; an experiment easily made, by taking two vessels of equal capacity, one blackened, the other white, and filling them with water heated to the same temperature: the water contained in the dark vessel will be found to arrive at the temperature of the surrounding bodies considerably sooner than the white, and vice versa.

The obituary is not signed. However, this is a good example of how physics may be important in fashion, even if it is wrong. Colors are meaningful only in the visible range. In the infrared, which is the relevant radiation in this case, there are no colors, and practically all materials behave the same way.

[25] This was even before 1817, when Dulong and Petit got the first result.

[26] Rumford, C., *The Complete Works*, Pub. Am. Acad. Arts Sci., Vol II, Boston, 1873.

In parallel with Prévost, but completely independently, John Leslie (1766–1832) carried out a long series of experiments and published them as a book[27] in 1804. The book describes 58 different experiments on the propagation of 'radiant heat', mostly carried out with the setup shown in Fig. 2.4. Qualitative conclusions are inferred after each experiment. The most important conclusion for our discussion here was already drawn after the fourth experiment:

> The power of absorbing heat, and the power of emitting it, seem always conjoined in the same degree.

However, he then added:

> [...] and this uniform conjunction clearly betrays a common origin, and discovers the evolution of a single fact, which assumes contrary but correlative aspects.

It is not clear what 'common origin' Leslie had in mind, but it is more plausible that the absorption and emission have the same origin and hence are proportional to each other, as was later demonstrated. Leslie's results are plotted in Fig. 2.5. We see that, because the results were related to lampblack[28] (the measurements were relative), Leslie effectively had two data points (two of his three data points were very close to each other). It was therefore daring to state, as he did, that the emission of a body is proportional to its absorption power. But Leslie was proven right, even though he never got the credit for this observation. We may say that Leslie discovered the existence of black bodies as perfect emitters and absorbers when he realized that surfaces coated with lampblack were the best emitters and absorbers he could lay his hands on.

2.4 Beyond the Visible Light

In 1800, Herschel[29] took the solar spectrum and put a detector beyond the place where the red light fell (see Fig. 2.6). He thus discovered what is known today as infrared radiation ('infra' means 'coming after'). Herschel also found that about half of the

---

27 Leslie, J., *An Experimental Inquiry into the Nature and Propagation of Heat*, printed for J. Mawman, London, 1804. In 1805, Leslie was elected to the Chair of Mathematics at Edinburgh. His unsuccessful competitor for the chair was backed by some Edinburgh clerics, members of the moderate wing of the Scottish church. This group sought to have Leslie's election overturned, invoking a clause in the university's statutes requiring the electors to take the advice of the Edinburgh clergy! As evidence of Leslie's unsuitability for the job, they cited a footnote from the book in which he agreed with David Hume's view of cause and effect (p. 521), saying that Hume's writings were *a clear model of accurate reasoning*. But Hume was hated by the church. Leslie's opponents objected because his views challenged *traditional arguments for the existence of God*. Leslie denied any connection to Hume, however. What saved him was the fact that the different clerical groups hated each other more than they hated Leslie, and did not want one group (in this case the moderate one) to win the battle to change the decision. (Price, H., *John Leslie*, Oxford, Clarendon Press, 2001).

28 Fine soot of incompletely burnt coal.

29 Herschel, W., Phil. Trans. R. Soc. Lond. **90**, 284 (1800).

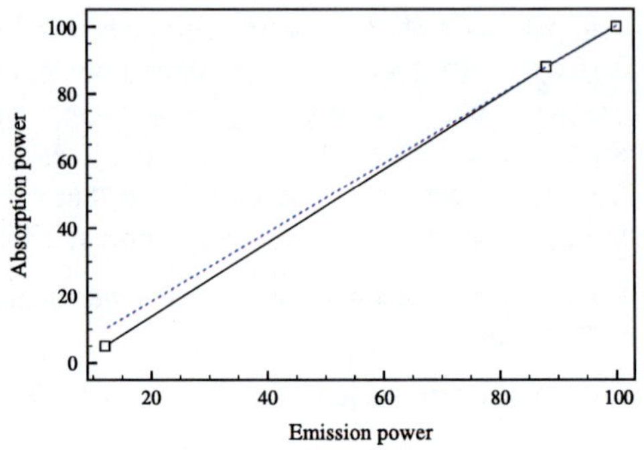

Fig. 2.5 Absorption as a function of emission from Leslie's data. *Squares* are the measured data for a list of materials, and the *continuous line* is the author's eye fit. Effectively, Leslie had three data points. The *straight line* is the *broken line*. Most of the substances at Leslie's disposal were either good emitters or bad ones

solar heating comes in the visible and half past the red color (in the IR). The violet, which is most refracted by the prism, had the least *efficacy*. Herschel concluded:

> If we call light, those rays which illuminate objects and radiant heat those which heat objects, it may be inquired, whether light be essentially different from radiant heat? [...] we are not allowed, by rules of philosophizing, to admit two different causes to explain certain effects, if they may be accounted for by one.

While Young[30] thought that Herschel had also had *the good fortune to discover the separation of the rays of heat from those of light*, as if it were a question of luck and not of a physical idea, he also considered that Herschel's discovery *must be allowed to be one of the greatest that has been made since the days of Newton, although the theories of some speculative philosophers might have led to it a few years earlier*. On the other hand, Leslie asked:

> What, then, is this calorific and frigorific fluid after which we are inquiring? It is no light, it has no relation to ether, it bears no analogy to the fluids, real or imaginary, of magnetism and electricity. But why have recourse to invisible agent?

Leslie thought that it was all an effect of heating the air, which then gave rise to hot currents. Leslie lost all his supporters of this idea (of hot air)[31] after Davy showed that the intensity of the radiation increased in vacuum and Ritter made his discovery of ultraviolet radiation (see below).

John Ritter (1776–1810) discovered that silver salts darken when exposed to sunlight. So a year after Herschel's discovery, in 1801, Ritter[32] repeated Herschel's experiment, but putting the silver salts beyond the blue solar light, and effectively discovered the ultraviolet range of the spectrum. At first these rays were called 'chemical rays', because it was soon found that they were chemically active and induced chemical reactions, in contrast to heat rays. The exact nature of these rays

---

[30] Young, T., *A Course of Lectures on Natural Philosophy and the Mechanical Arts*, printed for Taylor and Walton, London, 1845.

[31] The expression 'full of hot air' meaning nonsense or exaggerated, originated at some time during 1835–1845, but it is not clear to what extent Leslie's idea contributed to it.

[32] Ritter, J.W., Ann. Phys. **7**, 527 (1801).

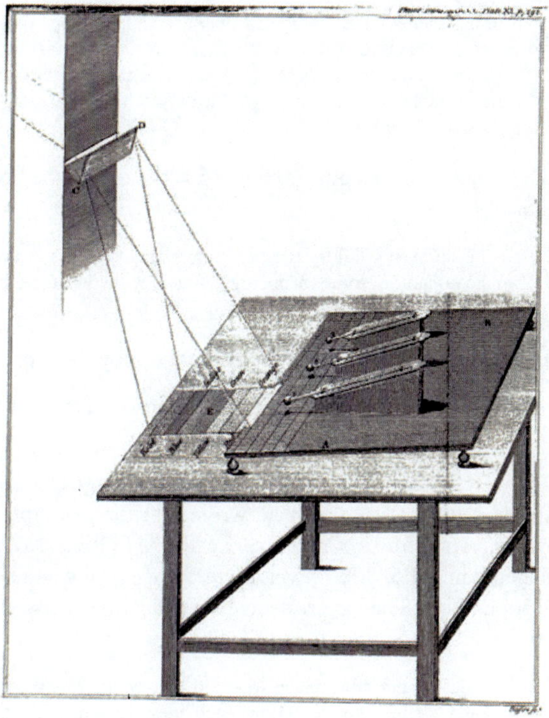

Fig. 2.6 The way Herschel discovered the infrared radiation by placing a thermometer beyond the place where the *red light* fell on the table

was at the center of a controversy, mainly between John Draper (1811–1882)[33] and his contemporaries. Draper thought that they were special type of rays that differed from visible light.

Returning to Leslie, while most of his conclusions were correct, some appeared strange. For example, experiment XX with the conclusion: *The impressions of heat or cold are, therefore, propagating through the air, with unequal degrees of diffusion.* Why should cold spread at a different speed to heat?

Unfortunately, the book contains no reference to any previous work, and in particular, Pictet and Prévost are not mentioned, although one of the experiments described in the book resembles Pictet's. Leslie made the admission: *I am free to confess that the propagation of heat is still a subject of immense difficulty*, then digressed to philosophy. Thus, 13 years after Prévost had stated his principle about emission and absorption on the continent, word had still not reached Edinburgh.

The book contains philosophical digressions which are relevant to the synthesis of the chemical elements, and for this reason we bring them into the discussion at this point. An example is:

[33] John was the father of Henry Draper (1837–1882m) from the Henry Draper (HD) catalogue, see Shaviv, *The Life of the Stars*, 2009. John William Draper took the first photograph of the Moon in 1840.

> Nothing seems more chimerical than to indulge a hope that the humans will ever be able to achieve the transmutation of earths into precious metals; but those hapless visionaries who consumed their days in the obscure search after the philosopher's stone, did not, like their fellow labourers who sought the perpetual motion, advance pretensions which involve a physical absurdity.

The connection is a puzzle. Finally, at the end of the book, there appears the conclusion:

> Heat is an elastic fluid, extremely subtle and active. Is it a new and peculiar kind of fluid, we are already in some manner acquainted with? Heat and light are commonly associated. Heat is only light in the state of combination.

Indeed, as Leslie wrote about his own discoveries (p. 115 in Leslie's book):

> We have thus deduced a train of phenomena which must be deemed equally novel and striking.

Many years later, in 1865, the Frenchman Paul-Quentin Desains (1817–1885) published a book[34] in which he wrote that the *principle of the equality between the emission and absorption powers* was due to René-Just Haüy (1743–1822), who first stated it in 1806.[35] However, no reference is given to a paper by Haüy except for his lecture notes. The experiments by Leslie are discussed, but the principle is credited to Haüy. Indeed, on p. 108 of Haüy's lecture notes, he wrote:

> The equilibrium takes place when all the affinities of the bodies for caloric are satisfied, [...] and when, at the same time, each body sends forth to others as much radiant caloric as it received; and this equal repartition continues so long as the system remains at the same temperature.

It is interesting that Haüy, who was a crystallographer, did not carry out experiments with heat, and consequently we may infer that he reached this conclusion logically, without bothering to publish the result. This is also why no reference was given. The only place I could find the name Haüy in this connection is in Desains' book.

In 1826, Henri Talbot (1800–1877)[36] made the important claim that, if his theory that certain bodies gave characteristic lines should prove to be correct, then a glance at the prismatic spectrum of a flame would suffice to identify substances that would otherwise require a tedious chemical analysis for their detection. In 1834, Talbot studied lithium and strontium, both of which paint the flame in red. He then wrote: *The prism betrays between them the most marked distinction which can be imagined.* However, the mission was not simple. In 1845, William Miller (1817–1870)[37] carried out a detailed analysis of the spectra of the alkali metals, but was unable to infer any specific characteristic lines, the reason being the use of impure flames. Moreover,

---

[34] Desains, P., *Leçon de Physique*, Tome second, Dezobry, E. Magdeleine et Co., Paris, 1865.

[35] Haüy, R-J., *Traité de Physique*, Berthollet, Daubenton, Paris, 1806. The translation into English by Gregory is from 1807. These are Haüy's lectures at the Ecole Normale, Paris.

[36] Talbot, H.F., Brewster, D., Sci. **5**, Phil. Mag. **3**, 33 (1833); ibid. **9**, 3 (1936).

[37] Miller, W.A., Phil. Mag. **27**, 81 (1845).

it was impossible to state with confidence that the spectral lines were unique for each element, and hence that the spectral identification was unique.

In 1830, the optician William Simms (1793–1860)[38] made a very important improvement in the spectroscope. Instead of merely using a prism and observing the slit with the naked eye, he placed a lens in front of the prism, so arranged that the slit was in the focus of the lens. The light which is allowed to pass through the slit is thus turned into a parallel beam hitting the prism. The beam emerging at each wavelength could then be magnified.

In 1847, Draper[39] concluded that all solids, and probably all liquids too, become incandescent, i.e., red hot, at the same temperature of 525°C. While below this temperature invisible rays are emitted, as the temperature rises above 525°C, rays of greater refrangibility (which in today's parlance means radiation of shorter wavelengths) are continually added. Moreover, all spectra of incandescent solids are continuous. Luminous gases, on the other hand, emit only bright lines. Draper's statement that only solids and liquids emit a continuum spectrum provided support for the then widely accepted conclusion that, as the Sun emits a continuous spectrum, it must be either a solid or a liquid, as 'can be inferred' from its mean density (1.45 g/cm^3). Scientists could not imagine gases at such high densities. Here is an example of the way imagination can sometimes fail us.

2.5 Stewart, the Forgotten Discoverer

Early in 1858, Balfour Stewart (1828–1887)[40] carried out a series of three types of experiment in which he compared the radiation emitted from different types of metal plate. Stewart was able to apply a sensitive radiation detector (see Figs. 2.7, 2.8), developed by Leopoldo Nobili (1784–1835)[41] and Macedonio Melloni (1798–1854),[42] who applied the thermoelectric effect discovered in 1822 by Thomas Seebeck (1770–1831)[43] to the problem of detecting radiation. According to Stewart:

> Heat rays have the same nature as light rays; these constitute a special class of the former. The invisible heat rays are distinguished from light rays only by the period of vibrations or the wavelength.

---

[38] Simms was an acclaimed family of opticians who contributed to improvements in spectroscopy and telescopes through the company Troughton & Simms.

[39] Draper, J.W., Phil. Mag., May, 345 (1847).

[40] Stewart, B., *An Account of Some Experiments on Radiant Heat, Involving an Extension of Prévost's Theory of Exchanges*, Trans. R. Soc. Edinb. **XXII**, PART I, 1 (March 1858) and Proc. R. Soc. Edinb. **6**, 93, session of 1857–1858.

[41] Nobili, L., Univ. Sci. et Art Genève **29**, 119 (1825).

[42] Melloni, M., Ann. Chim. Phys. **53**, 5 (1833); also Pogg. Ann. **35**, 112 (1835). His great work on thermal radiation was published in 1850 under the title *La thermochrose, ou la coloration calorifique*, Naples.

[43] Seebeck, T.J., Pogg. Ann. **6**, 1 (1823).

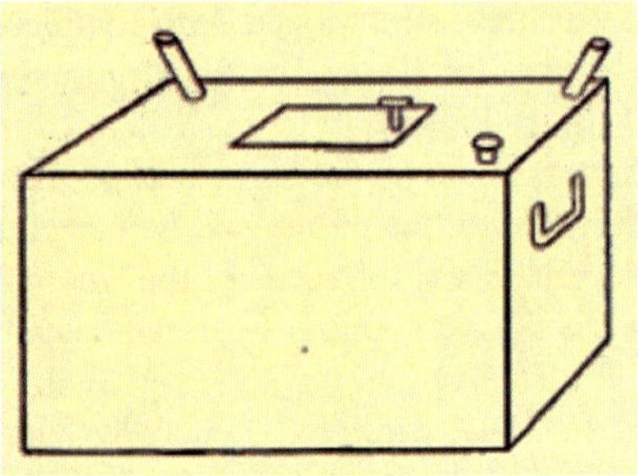

Fig. 2.7 The first cavity to serve as an effective *black body* (Stewart 1858)

Stewart repeated the basic experiments carried out by Leslie, Frédéric Provostaye (1812–1863), Desains,[44] and Melloni, but with greater accuracy. Inevitably, Stewart reached the conclusion already pointed out by Leslie, as Stewart asserted himself,[45] namely that:

> The absorption of a plate equals its radiation, and that for every description of heat.

It seems that 'for every description of heat' was intended to mean 'for every wavelength'. This important addition, however, did not appear in Leslie's conclusion. Neither Leslie nor Stewart applied filters to their experiments, and hence the experiment did not test what went on at particular wavelengths. The wavelength independence of the law came in the theoretical proof provided by Stewart. The novelty here lay in the fact that Stewart provided the first theoretical proof of the law. In this theoretical proof he demonstrated that the law held for every 'quality' of the heat wave, this being the term Stewart used for what we call wavelength today, as can be understood from the context. Finally, the above formulation of the law is incomplete. The correct formulation is this: at a given temperature, the emission of any substance is proportional to the absorption. The constant of proportionality changes with temperature.[46] This particular point was missing in Stewart's formulation.

In his second paper,[47] Stewart discussed the connection between the radiation emitted by a body and its temperature. Stewart found that the radiation of a naked

[44] For an extended description of these experiments, see the book by Desains, P., *Leçon de Physique*, Dezobry, E., Magdeleine et Cie., Lib-Éditeurs, Paris, 1860.

[45] In the textbook *Handbook of Natural Philosophy and Astronomy*, by Lardner, D., Second Course, Blanchard & Lea Pub., Philadelphia, published in 1854, we already find a table *showing the absorbing and reflecting power* of various substances, as found by Melloni in the 1830s. The interesting point is that the title of the column is *Radiating and Absorbing Powers*, and a single number is given. This confirms that Melloni showed the emission to be equal to its absorption, as is also explicitly stated in the text. However, although Melloni gave more data than Leslie, Stewart did not mention him. As a matter of fact, in a textbook on physics from 1837, *The Elements of Physics* by Webster, T., Scott, Webster, and Geary Pub., London, 1837, it is stated that, *in a state of equilibrium, the absorbing power is always equal to the radiating power* (p. 257).

[46] The coefficient of proportionality is given by the radiation density at equilibrium. See later.

[47] Stewart, B., *Researches on radiant heat*, Trans. R. Soc. Edinb. **XXII**, Part I, 59 (April 1859).

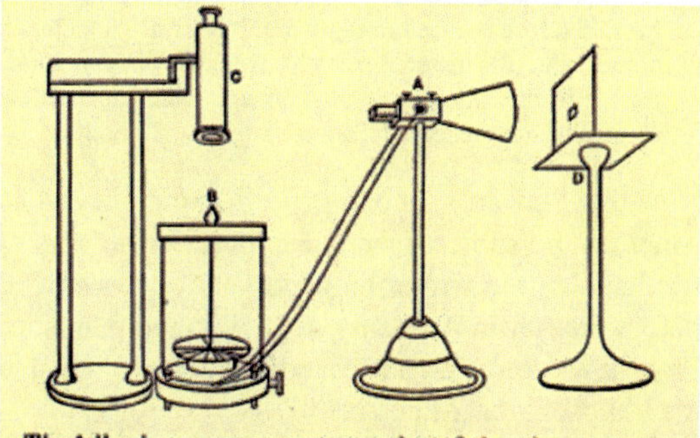

The following arrangement was adopted for the great mass of the experiments:

A. Is the sentient pile, with a polished brass cone attached to it, for collecting the rays of heat.

B. Is the galvanometer, the position of its needle being read to $\frac{1}{18}$th of a degree by the telescope C.

D. Is a screen placed before the mouth of the cone in which there is a small hole or diaphragm .65 inch square. The screen is covered with gilt paper, in order that, should it get slightly heated, it might radiate as little as possible.

Fig. 2.8 The tools used by Stewart in his radiant heat experiments. Note in particular item *A*, the sensitive detector of radiation (Stewart 1858)

thermometer increased somewhat less rapidly with temperature than that of a silvered thermometer. Dulong and Petit found no difference between the two.

The experiment on how plates of different thickness cool inspired Stewart to comment in the following way:

> What then, does Dulong and Petit's law express? The answer is, it expresses the law of radiation of indefinitely thick plates, and we have shown that it increases faster than the law of radiation of a material particle. [...] We have thus ascertained first that Dulong and Petit's law is not the law of radiation of a material particle, and second that this law increases less rapidly with the temperature than Dulong and Petit's law.

Stewart could not do better because his apparatus was not in vacuum, and heat conduction and convection by air constituted non-negligible cooling processes.

Furthermore, Stewart found that:

> The absorption of a plate equals its radiation and since roughening its surface does not influence the radiation it ought not to influence the absorption.

He also defined 'perfectly black bodies', or 'black bodies' for short, as bodies which *completely absorb all rays which fall upon them*, and the substance which was the best approximation for a black body was found experimentally to be lampblack.[48] At the end of the second paper, Stewart reached a pessimistic conclusion:

48 Note that lampblack is a perfect absorber in the infrared and was not really tested in the visible. However, in the visible, it has a black color. Colors have no meaning in the infrared as we cannot see in the infrared. So the term 'black body' really emerged from an object which behaved in the infrared as a perfect absorber and emitter, but has a black color in the visible range. Put another way, the name arose from an irrelevant property of the body, because it could have had any color in the visible.

> I am therefore induced to think that it is nearly hopeless to attempt to ascertain the true law of radiation of a material particle, as least by any method of experimenting depending upon the use of thin plates, or on the change which absorption may be presumed to cause in the amount of heat reflected from the surface of a body.

Note that Stewart's second paper[49] came out just after Kirchhoff published his first paper on the subject (see later), but Stewart was certainly unaware of Kirchhoff's results when he wrote the paper, and vice versa. Kirchhoff's first paper was read before the Berlin Academy in 1859[50] and published a few months later in the Ann. Phys. It was only in 1860 that the paper was translated into English and published in the Philosophical Magazine.

2.6 Kirchhoff's Law

During the years 1859–1860, Gustav Kirchhoff (1824–1887m) by himself and Kirchhoff together with Robert Bunsen made three independent major discoveries. Kirchhoff explained how the Fraunhofer lines form[51] (see Sect. 2.7) and proved his radiation law,[52] while Kirchhoff and Bunsen[53] promoted the idea of spectral analysis as a tool for identifying chemical elements and even certain compounds.

Kirchhoff proved that:

> The ratio between the emissive and the absorptive power is the same for all bodies at the same temperature in thermal equilibrium.

Note the additional clause 'at the same temperature'. Kirchhoff proved the law first for black bodies and then for general bodies. We describe these seminal discoveries in a logical rather than historical order, so as to emphasize the importance and explain the logic.

Kirchhoff's proof, like Stewart's, hinges on the assumption that there exist bodies which absorb and radiate energy only in a certain restricted wavelength range. In addition, the proof requires the assumption that perfect diathermanous substances exist.[54] The point would arise some thirty years later.

---

49 Stewart, B., Proc. R. Soc. Lond. **10**, 385 (1859–1860).

50 Kirchhoff, G., Monat. der König. Preussischen Akad. der Wissen. Berlin, October, 1858.

51 Kirchhoff, G., Ann. Phys. **184**, no. 12, 567 (1859); ibid. Ann. Phys. **185**, no. 1, 148 (1859).

52 Kirchhoff, G., Ann. Phys. **185**, no. 2, 275 (1860); ibid. Ann. Physik und Chemie **CIX**, 6, 275 (1860); ibid. Ann. Chim. Phys. **LXII**, 3, 160 (1861).

53 Kirchhoff, G. and Bunsen, R., Ann. Phys. und Chemie **CX**, 6, 161 (1860); ibid. **CXIII**, 7, 337 (1861); ibid. Ann. Chim. Phys. **LXII**, 3, 452 (1861); ibid. **LXIV**, 3, 257 (1862).

54 Diathermanous means permeable by heat waves, so it describes a body which transmits heat as electromagnetic radiation. The term was apparently invented by Melloni somewhere between 1830 and 1840. Melloni proposed to use 'diathermanous' for bodies that let heat pass easily, and athermanous for those that do not let heat pass. However, see W.D.L.L. Nature **7**, 242 (1873). In general, transparency in the visible and the property of being diathermanous are not related.

When heated, every chemical element or compound emits a set of emission lines and/or continuum radiation. The wavelengths at which these emission lines appear are characteristic of the chemical element. They are like a fingerprint of the element. Hence, we can say that the emission E is given by

$$E = \begin{cases} E(\lambda, T, \text{chemical composition}), & \text{for } \lambda = \lambda_i, \\ 0, & \text{for } \lambda \neq \lambda_i, \end{cases}$$

where the λ_i form a set of wavelengths corresponding to emission lines. Consequently, the spectral lines λ_i can be used for chemical identification. There was no theoretical proof that two different chemical elements cannot have the same set of emission lines. By tedious careful analysis of all known chemical elements and by cataloging the observed spectral lines, it was found empirically that each element has its own series of emission lines. Clearly, it was not known why there are spectral lines and not say, a continuum, let alone what determines the wavelengths of the lines. The question as to why there are lines was not asked in those days, and there was no hypothesis as to their origin. Only when quantum theory was invented and developed was it proven that the spectral lines are unique to each element.

Kirchhoff carried out the following experiment. He first observed the continuum radiation emitted from a very bright source, and then he observed the emission lines λ_i emerging from a faint source. Lastly, he placed the faint source in front of the bright source and discovered that the emission lines in the faint source became dark exactly at the positions λ_i, like Fraunhofer lines at the locations λ_i. There was no case in which one of the Fraunhofer lines did not correspond to an emission line and vice versa.

Kirchhoff formulated the result of the experiment by stating that, when you place a cooler gas in front of a hot one, the spectrum of the cooler gas is 'reversed', i.e., instead of an emission spectrum it becomes an absorption spectrum. The reversal phenomenon of spectral lines (reversal from emission into absorption) was, according to Kirchhoff, the origin of the Fraunhofer lines. This explains why the term 'reversing layer' is used with reference to the solar atmosphere. The hot solar interior is covered by a cooler atmosphere, and the Fraunhofer lines are formed in the cooler atmosphere, or 'reversing layer'. Note that we write 'cooler' rather then 'cold', because the gas may be at a high temperature and emit radiation itself, but it has to be at a lower temperature than the gas behind it.

The possibility of reversing the spectrum of a gas implied that the absorption a could be written as

$$a = \begin{cases} a(\lambda, T, \text{chemical composition}), & \text{for } \lambda = \lambda_i, \\ 0, & \text{for } \lambda \neq \lambda_i. \end{cases}$$

In other words, the absorption takes place at the wavelengths λ_j, and depends on the temperature and the identity of the matter. As the absorption spectrum is the reversed emission spectrum created under unique conditions, it is clear that it can also be used to identify chemical elements. The idea of 'reversing the spectrum' means that

$$E(\lambda=\lambda_j, T, \text{chemical composition}) \\ = \alpha(\lambda, T)a(\lambda=\lambda_j, T, \text{chemical composition}),$$

where $\alpha(\lambda, T)$ is some function of wavelength and temperature. Most importantly, the set of emission lines is identical to the set of absorption lines. What the above formula says is that the positions of the emission lines are identical to the positions of the resulting Fraunhofer lines, but the power of the emission is not necessarily equal to the power of the absorption. This is why α appears in the formula. It is impossible for an emission line to disappear in the reversed spectrum and vice versa. All emission lines are reversed and appear as Fraunhofer lines. This relation does not require equilibrium conditions.

In the words of Kirchhoff and Bunsen:

> Within the spectrum, an element absorbs the light at the exact location of the lines which it can emit.

They stated the basic law of elementary spectrometry which is:

> Each element has specific properties as regards the light it emits.

While the emission power (how much energy is emitted per unit time) of the gas depends on the wavelength, temperature, and composition, the absorption power (how much energy is absorbed per unit time) depends on the same parameters, and in addition the intensity of the radiation $J(\lambda)$. The total power absorbed by the gas is therefore given by

$$A = J(\lambda)a(\lambda = \lambda_j, \ldots, T, \text{chemical composition}).$$

In 1858, Stewart, following Leslie and Melloni, carried out experiments which showed that, in thermal equilibrium,

$$\text{total emission power} \equiv \int E(\lambda = \lambda_j, T, c)\mathrm{d}\lambda \\ = \int J(\lambda)a(\lambda = \lambda_j, T, c)\mathrm{d}\lambda = \text{total absorption power},$$

where c means 'composition'. The integration is carried out over all wavelengths. The function $J(\lambda)$ did not appear explicitly in Stewart's paper, but its existence was implied. The result makes a lot of sense. If more power were absorbed than emitted, the object would heat up, and vice versa. All the experiments implied was that the total power of emission and absorption, and not the power absorbed/emitted in a given spectral line, were equal. It was only in their theoretical proofs that Stewart and Kirchhoff established that the integral signs can be removed, whence, in thermal equilibrium,

$$E(\lambda = \lambda_j, T, c) = J(\lambda)a(\lambda = \lambda_j, T, c).$$

Kirchhoff's dramatic and momentous discovery was that, in thermal equilibrium,

$$\frac{E(\lambda = \lambda_j, T, \text{composition})}{a(\lambda = \lambda_i, T, \text{composition})} = J(\lambda_j, T) \equiv e(\lambda_j, T),$$

where $e(\lambda, T)$ is a universal function that does not depend on the material involved. The formula states that the emission at wavelength λ_j divided by the absorption at the same wavelength is given by a universal function $e(\lambda, T)$ evaluated at $\lambda = \lambda_j$. The function $e(\lambda, T)$ is the spectral distribution of the radiation in equilibrium. It is the same function for all materials, provided there is thermal equilibrium.

Let us formulate it slightly differently. The emission of the body depends only on its temperature. On the other hand, the absorption of the body depends on the temperature. However, the absorption power is a product of the properties of the matter and the incident radiation. It is the incident radiation as well as the emission which have, in the case of equilibrium, a universal shape and distribution. Kirchhoff realized how important the function $e(\lambda, T)$ is, but was unable to derive it theoretically because Maxwell's theory of light had not yet been born. Note that the law does not specify the temperature dependence of $E(\lambda = \lambda_j, T, c)$ or $a(\lambda = \lambda_j, T, c)$, but only the ratio between them.

Let us now cross the bridge before coming to it. Forty years later, Max Planck (1858–1947m) discovered the exact form of the function $e(\lambda, T)$, and it is usually denoted in present day textbooks by $B_\nu(T)$, and called the Planck function. In addition to the classical electromagnetic theory of light, the theoretical derivation of this function requires the fundamental assumption of quantum theory, namely the quantization of energy, and it thus opened the gate to quantum mechanics. In the view of the present author, it would be better called the Kirchhoff–Planck function, as Unsöld and Baschek do.[55]

About a year after publishing the first proof of his law, Kirchhoff published a second and more rigorous one.[56] The basic idea of the law emerged from general thermodynamic considerations. Kirchhoff's result was truly astonishing: why should all materials behave in the same way? The reason is that, in thermodynamic equilibrium, there is only one solution for the radiation field, irrespective of the materials out of which the enclosure is made. In stars, we define local thermodynamic equilibrium or LTE to be a situation in which the gas is in equilibrium with the radiation at a given point. Thus, whenever there is LTE in stars, and one can show that this assumption holds perfectly in the stellar interior, the distribution of the radiation field is $B_\nu(T)$. Naturally, such a sweeping discovery resulted in commotion, not so much about the validity of the result as about the priority of the discovery.

---

[55] Unsöld, A. and Baschek, B., *The New Cosmos*, 4th edn., Springer, 1991.

[56] Kirchhoff, G., Ann. Chim. Phys. **67**, 160 (1861).

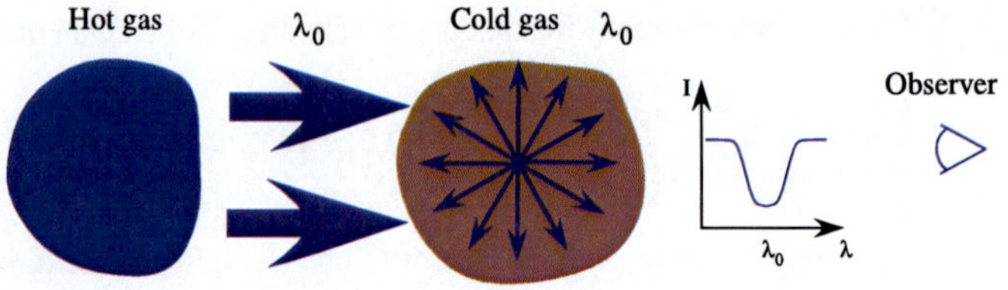

Fig. 2.9 The setup which leads to a reversed spectrum

2.7 A Modern Physical Explanation for the Reversal Phenomenon

Consider a hot gas which emits a continuum. Next consider the photons at wavelength λ_0, the wavelength at which the cold gas absorbs. Assume for simplicity that this is the only wavelength at which the cold gas absorbs.[57] Thus, the strong flux of radiation is absorbed by the cold gas. But the cold gas cannot absorb forever and must eventually re-emit the radiation. The emission takes place in all directions. Ignoring for a moment stimulated emission (to be discussed later), the atom forgets the original direction of the photon it absorbed and emits the newly born photon in any direction without preference. Consequently, the original intensity of the radiation is distributed in all directions and the intensity in the original direction, the direction of the observer, is reduced. When compared with the radiation intensity in wavelengths adjacent to λ_0, where no absorption took place, the intensity in $\lambda = \lambda_0$ is lower and hence appears darker, or more accurately, less bright. This is the Fraunhofer line. It is not that there is no radiation at all at this wavelength, but rather that it is significantly weakened in relation to the absorption/emission power of the hot gas.

The detailed process is more complicated than the above simple explanation and requires solution of the radiative transfer equation in which Kirchhoff's law plays a central role. The radiative transfer equation was not known in Kirchhoff's day, and nor was the atomic physics and detailed structure of the atomic levels, whose difference determines the wavelengths of the spectral lines. Moreover, the shape of the line, which appears very wide in the figure, is in fact quite narrow, and the details of its shape provide valuable information about the physical conditions under which the particular line formed. In the Sun, for example, not all lines form at the same location. A very strong line, which implies high absorption, forms high up in the atmosphere because a small amount of matter (counting from outside) is needed for its formation. On the other hand, a weak line requires a large amount of matter, and hence forms deeper in the star. By comparing the shape and intensity of various lines, the detailed temperature run can be traced through the outer layers of the star (Fig. 2.9).

[57] This approximation is called the two-level atom. This simplification, which is frequently used even today, was introduced by Milne, E.A., J. Lond. Math. Soc. **1**, 40, 1926.

2.8 Could the Coincidence Be Fortuitous?

Could the coincidence between the Fraunhofer dark lines and the bright emission lines emitted by the heated elements be fortuitous? Kirchhoff calculated, from the number of coincidences he found and from the degree of exactitude with which each coincidence could be determined, that the fraction of chance agreements was less than 10^{-18}. In other words, it was practically certain that these lines had a common origin. And if this were not sufficient, Kirchhoff pointed to the fact that the characteristic lines of elements occurred in groups, which of course appeared as such in the spectra of the Sun. Once quantum theory became available, it was found that the locations of the lines were not accidental but unique to every atom.

2.9 The Priority Debate

So much for the physics behind one of the most important laws in radiation theory. Soon after the German publication of Kirchhoff's radiation laws, several English scientists started to question, not Kirchhoff's scientific results, but the priority in his discoveries. The pages of the 1860 London, Edinburgh, and Dublin Philosophical Magazine and Journal of Science (with Brewster as one of the editors) **20**, July–December 1860, are testimony to the priority dispute, and contain practically no controversy about the physical content.

All sorts of priority claims were raised against Kirchhoff, which were tantamount to plagiarism: that 'Kirchhoff's law' was discovered first by Stewart; that several English researchers had long since proposed spectroscopic analysis as a tool for identification of chemical elements; and that the explanation of the way the Fraunhofer lines form was also put forward before Kirchhoff announced his own explanation.

2.10 The Radiation Law

The controversy over the priority for the proof of the radiation law took place directly between Stewart and Kirchhoff. Although Stewart discussed 'radiant heat' and Kirchhoff discussed 'visible light', they arrived at the same law (except for the clause about the temperature). So who had the priority? Once the dispute had come out into the open, Kirchhoff tried to establish his priority in the Annalen der Physik[58] and Stewart responded in the Philosophical Magazine.[59] In many textbooks today, Kirchhoff is credited with the priority, probably because Kirchhoff provided what was considered to be a more rigorous mathematical/physical proof of the law, while Stewart reasoned in a largely logical way and merely described the law.

---

58 Kirchhoff, G., Ann. Phys. **194**, 94 (1862). The English translation is: Phil. Mag. **25**, 250 (1863).

59 Stewart, B., Phil. Mag. **25**, 354 (1863).

Kirchhoff began the 1860 de facto reprint in English[60] of the paper first published in German with the following statement:

> A body placed within a covering whose temperature is the same as its own, is unaffected by radiation and must therefore absorb as many rays as its emits.

For the experimental verification, Kirchhoff cited de la Provostaye and Desains,[61] claiming that:

> This was proved for bodies emitting rays of one kind.[62]

But Kirchhoff stressed that:

> Whether the same law holds good when bodies emit rays of different kinds, has never hitherto been determined theoretically or by experiment.

Kirchhoff then argued that:

> The ratio of the radiating and absorbing powers of all bodies at the same temperature is the same.[63]

It is here that Kirchhoff defines the 'perfectly black' body as the perfect absorber. Years later, Wien[64] would redefine it more precisely and call it a black body, which is the term used today.

At the end of the paper, in a postscript, Kirchhoff turned to the argument over priority, writing:

> After the appearance of my paper in the Pogg. Annalen, I received information of a prior communication closely related to my own. The communication in question is by Mr Balfour Stewart [...]. The principle enunciated by Mr Stewart is, however, less distinctly expressed, less general, and not altogether so strictly proved as mine.

It is not difficult to imagine that this statement by Kirchhoff did not appease his rivals. Note that Kirchhoff himself was not happy with his first proof (the subject of the priority dispute) and provided another proof. As for Stewart, the proof he gave

---

60 Kirchhoff, G., Phil. Mag. Sect. 4, **20**, 1 (1860).

61 Kirchhoff wrote that de la Provostaye and Desains did the experiment, but did not provide any published reference. I suspect that he meant: de la Provostaye, F., and Desains, P., Ann. Phys. **150**, 147 (1848), which was a translation from Compt. Rend. **XXVI**, 212 (1848).

62 Kirchhoff mentions that, in the case of de la Provostaye and Desains, the rays were invisible, while he discussed visible light. The conclusions were nevertheless the same. In 1868, Desains wrote a review about French scientific achievements in *La Théorie de la Chaleur* and did not mention any 'rays of heat'.

63 Kirchhoff also argued incorrectly as follows: *The wavelengths which correspond to maxima of the radiating and absorbing powers are, as will be fully explained in another place, altogether independent of the temperature*. This was before the discovery of Wien's displacement law, which is a relation between the temperature of a black body and the wavelength at which its emission power reaches maximum. The existence of such a law became clear once the shape of the spectral distribution had been found. Such a law had been suggested earlier by Wilhelm Weber (1804–1891) in his 1888 paper.

64 Wien, W., Ann. Phys. **288**, 132 (1894).

in his book, published in 1866,[65] was also a new proof which relied on more general equilibrium considerations. We can summarize this dispute by stating that it began as a controversy about the rigour of the proofs, which was very quickly relinquished. The experimental evidence and the hypothesis for the law belong to Leslie almost fifty years earlier (or even Haüy slightly earlier). It was, however, Kirchhoff's analysis as a theoretical physicist that led to the formulation in terms of a universal function, and that was the wonderful physical essence of the law.

But this conclusion was not universally accepted. As late as 1901, Rayleigh (1842–1919)[66] raised the issue again. Rayleigh, in his fight for the priority of Stewart, admitted that the experiments did not strictly demonstrate what Stewart's conclusion implied. Yet the theoretical proof according to Rayleigh was perfect and no less rigorous than Kirchhoff's. Rayleigh's argument did not convince the entire community, and in 1902, Heinrich Kayser (1853–1940)[67] continued to argue that Stewart's proof was *not a strict demonstration*. However, Kayser did not supply an alternative proof which would be to his satisfaction.

The zeal to protect Stewart's priority continued even as late as 1925, by which time both Stewart and Kirchhoff had met their maker (both passed away in 1887). At this point, Arthur Schuster (1851–1934m) published an article in Nature *which may possibly leave with the ordinary reader an impression that Balfour Stewart's contributions to the establishment of the laws of natural radiation were slighter than was actually the case*. Larmor[68] immediately volunteered to straighten things out. What Larmor did was to direct the reader to Rayleigh's earlier discussion. The interesting fact, which tells us something about how high feelings were running, is that Schuster's article was published on 17 January and the letter Larmor wrote to the editor of Nature was dated 16 January. Note in this respect that Schuster was a German-born English physicist. He became an English citizen in 1875.

2.11 Stewart in Retrospect

In 1866, Stewart published the above-mentioned Rumford Medal winning book about 'Heat' and it is interesting to see his final remark on the subject. Stewart clearly had a hunch about Wien's law when he wrote:

> We have reasons to believe that as the temperature rises the spectrum of a black substance is extended in the direction of greatest refrangibility, so as to embrace more and more of the violet and photographic rays.

This should be contrasted with Kirchhoff's incorrect statement about the effect of the temperature of a body on its color. The formulation of the famous law was now:

---

[65] Stewart, B., *An Elementary Treatise on Heat*, Clarendon press, Oxford, 1866.

[66] Rayleigh, Lord, Phil. Mag. **6**, 1, 98 (1901); *Scientific Papers* **4**, 494.

[67] Kayser, H., *Handbuch der Spectroscopie*, Hirzel, Leipzig, 1900.

[68] Larmor, J., Nature **115**, 159 (16 January 1925).

> In an enclosure of constant temperature the heat reflected plus the heat radiated by any substance will be equal to the total lampblack radiation of that temperature.

So Kirchhoff's clause was added.

In Chap. II, in his discussion of the 'theory of exchange' on p. 199, Stewart wrote simply:

> This theory, since its proposal by Prévost, has been developed by Provostaye and Desains, and more recently by the author of this work and by Kirchhoff. [...] Bunsen and Kirchhoff, who have done much more than any one else to introduce and perfect this method of analysis.

Stewart, a man of honor, paid tribute to Kirchhoff:

> Before concluding this chapter we ought to allude to the beautiful discovery of Kirchhoff, by which it has been proved that substances with which we are here familiar exist also in the atmospheres of the Sun and stars.

Regarding the Fraunhofer lines, Stewart stated that:

> The inference naturally drawn from this experiment was that the lines of the solar spectrum do not denote rays originally wanting in the light of the Sun, but are due to the absorption of his light by some substance interposed between the source of light and the spectator. It was doubtful, however, whether this stoppage of light occurred in the atmosphere of the Sun or in that of our Earth, until the matter was finally settled by Kirchhoff, not however before the true explanation had been divined by Professor Stokes.

Stewart did not and could not give a reference to Stokes' discovery, because such a reference does not exist (see later).

2.12 Who Discovered the Source of the D_2 Lines?

Several of Kirchhoff's critics and quite a few textbooks claim that Foucault preceded Kirchhoff in identifying the D line of sodium in the spectrum of the Sun. Regarding this point, Kirchhoff wrote (see Fig. 2.10):

> The fact that the bright lines of the spectra of sodium and lithium flames may be reversed, was first published by me in a communication to the Berlin Academy, October 27, 1859. This communication is noticed by M. Verdet in the February number of the Ann. de Chim. et de Phys. of the following year, and is translated by Prof. Stokes in the March number of the Philosophical Magazine. The latter gentleman calls attention to a similar observation made by M. Leon Foucault eleven years ago, and which was unknown to me, as it seems to have been to most physicists. This observation was to the effect that an electric arc between charcoal points behaves, with respect to the emission and absorption of rays of refrangibility answering to Fraunhofer's line D, precisely as the sodium flame does according to my experiment. The communication made on this subject by M. Foucault to the Soc. Philom. in 1849 is reproduced by M. Verdet, from the Journal de l'Institut, in the April number of the Ann. de Chim. et de Phys.

He went on to say:

> Foucault's observation appears to be regarded as essentially the same as mine; and for this reason I take the liberty of drawing attention to the difference between the two.

Fig. 2.10 The typical voltaic arc used by Foucault, in which electrical sparks gave rise to spectra with *D* lines

Foucault discussed cold electric discharges while Kirchhoff discussed flames. We know today, but not in Kirchhoff's times, that the excitations of the gas which lead to the formation of the observed lines are different. And so alleged Kirchhoff:

> M. Foucault's observation does not afford any explanation of mine, and could not have led to its anticipation. My observation leads necessarily to the law which I have announced with reference to the relation between the powers of absorption and emission; it explains the existence of Fraunhofer's lines, and lead the way to the chemical analysis of the atmosphere of the Sun and the fixed stars. All this M. Foucault's observation did not and could not accomplish, since it related to a too complicated phenomenon, and since there was no way of determining how much of the result was due to electricity, and how much to the presence of sodium. If I had been earlier acquainted with this observation, I should not have neglected to introduce some notice of it into my communication, but I should nevertheless have considered myself justified in representing my observations as essentially new.

Foucault did apparently challenge Kirchhoff. It seems that Foucault's attention was focused on his experiments about the speed of light.

In view of the controversy about the priority of Kirchhoff's laws, Foucault's work was republished in 1860[69] and later translated in the Philosophical Magazine, publications which did not change the general perception that Foucault's experiment was different from Kirchhoff's, and no general conclusion was drawn.

---

[69] Foucault, L., Ann. Chim. Phys. **58**, 476 (1860).

According to Kirchhoff, he knew about the previous attempts to correlate the sodium emission line with the Fraunhofer *D* line, and wanted to either confirm or disprove the connection. However, he gave a reference to Swan 1857. Swan in turn, gave a reference to Auguste-Arthur de La Rive (1770–1834),[70] who wrote about Foucault's identification of the *D* line of sodium as early as 1849. Kirchhoff knew about the work of de la Provostaye and Desains[71] in which they discussed the issue.

2.13 Kelvin and the Reversal Phenomenon

Several times for too many years, William Thomson (1824–1907m), who later became Lord Kelvin, raised the priority issue concerning the discovery of the reversal of the spectrum. Thomson was a celebrity thanks to his contributions to thermodynamics (the second law), his theory about the energy of the Sun, the controversy with Darwin, and the dispute with the entire community of geologists about the age of the Earth.[72]

In his presidential address to the British Association in 1871,[73] Lord Kelvin stated his belief that the application of *prismatic analysis*[74] of light to stellar chemistry had never been suggested directly or indirectly by anyone when George Stokes (1819–1903m) taught it to him at Cambridge University, some time prior to the summer of 1852, and he set forth the conclusions, theoretical and practical, which he learnt from Stokes at that time, and which he afterwards gave regularly in his public lectures in Glasgow. And so wrote Kelvin:

> Professor Stokes mentioned to me at Cambridge some time ago (it was probably 1851 when Stokes apparently discussed a possible solution for the Fraunhofer lines), probably about ten years, that professor Miller had made an experiment testing to a very high degree of accuracy the agreement of the double dark line *D* of the solar spectrum with the double bright line constituting the spectrum of the spirit-lamp burning with salt. I remarked that there must be some physical connexion between two agencies presenting so marked a characteristic in common. He asserted, and said he believed a mechanical explanation of the cause was to be on some such principles as the following: vapours of sodium must possess by its molecular structure a tendency to vibrate in the period corresponding to the degree of refrangibility of the double line *D*. Hence the presence of sodium in a source of light must tend to originate

[70] De La Rive, A.-A., *A Treatise on Electricity*, London, Longman, Green, and Longman, 1853.

[71] de la Provostaye, F.H., and Desains, P., Compt. Rend. **38**, 977 (1854).

[72] Kelvin was famous for many provocative statements which turned out later to be wrong, in particular in the discussion with Darwin. See Shaviv, G., *The Life of Stars*, Springer, 2009.

[73] Thomson, W., MacMillan's Magazine, March, 1862, and Rep. Br. Assoc. **3**, 27 (1871).

[74] Kelvin used the term 'prismatic analysis' rather than 'spectroscopic analysis' to refer to spectroscopic analysis by means of a prism. Today, spectroscopic analysis can be better carried out with a grating. The first to construct and apply a grating was David Rittenhouse (1732–1796) in 1785, followed later by Fraunhofer in 1821. However, the prisms were still better in those days. The priority of Rittenhouse in the use of a grating was established by Babb as late as 1932. See Cope, T.D., J. Frank. Inst. **214**, 99 (1932). It was not until 1873 that Friedrich Nobert (1806–1881) perfected the grating to reach 9,000 lines per millimeter and eventually produce a superior resolution to prisms.

> light of that quality. On the other hand, vapours of sodium in an atmosphere round a source, must have a great tendency [...] to absorb and to have its temperature raised by light from the source, of the precise quality in question. In the atmosphere around the Sun, therefore, there must be present vapours of sodium. [...] I have the impression that some Frenchman did make this out by experiment, but I can find no reference on the point.

And he continued:

> I am not sure whether professor Stokes's suggestion of a mechanical theory has ever appeared in print. I have given it in my lectures regularly for many years always pointing out along with it that solar and stellar chemistry were to be studied by investigating terrestrial substances giving bright lines in the spectra of artificial flames corresponding to the dark lines of the solar and stellar spectra.

Kelvin claimed that:

> Although Professor Stokes unfortunately did not publish his theory (I say unfortunately because valuable time has been lost), the world was not long in ignorance of a matter of such general interest, for in 1853 the idea was published by the celebrated Ångström[75] who found that in many cases the Fraunhofer lines were an inversion of bright lines which he observed in the spectra of various metals.[76]

Kirchhoff's reply to Kelvin's letter does not appear in the paper. In any case, Stokes' ideas were apparently considered by him not to merit publication. On the other hand, this phenomenon later became the cornerstone of stellar spectroscopy. So the least one can say is that Kirchhoff recognized the impact of his discovery.

Not long afterwards, Whitmell,[77] who had been watching the ongoing controversy, obtained Stokes' consent to publish in Nature a letter that Stokes had written to him. And so wrote Stokes:

> Hence the sodium compounds [...] are transparent [...] and the absorbing vapour was that of sodium itself. Knowing the powerful affinities of sodium, I did not dream of its being present in a free state in the flames of a spirit lamp.

Stokes was confused by the signals from the molecule of salt (NaCl) and from the sodium atom. Finally, Stokes concluded his letter to Whitmell with:

> Reviewing my then thoughts by the light of our present knowledge, I see that my error lay in the erroneous chemical assumption that sodium could not be free in the flame of a spirit-lamp; I failed to perceive the extension of Prévost's theory, afterwards discovered by Stewart, nor perceived that the emission of light of definite refrangibility necessitated (and not merely permitted) absorption of light of the same refrangibility, which would have come in conflict with that error.

Stokes added to his letter to Whitmell a description of the conversation he had with Kelvin:

---

[75] Ångström, A.J., Optiska Undersokningar, Trans. R. Acad. Stockholm, 1853. Translated in Phil. Mag., fourth series, v, **IX**, 327.

[76] See Phil. Mag., fourth series, **XXIV**, 2, 3: Monatsberichte Akad. Wissen. Berlin, 1859, p. 662.

[77] Whitmell, C.T.L., Nature, p. 188 (January 1876).

> I mentioned to him [Kelvin] the perfect coincidence of bright and dark *D* [...] and described it using the dynamical illustration of the piano string. [...] I mentioned also, on authority of Sir David Brewster, another case of coincidence. [...] On hearing this Kelvin said something like: "Oh then, the way to find what substances are present in the Sun and stars is to find what substances give bright lines coincident with the dark lines of those bodies." I thought he was generalising too fast. [...] If, as I take it for granted, Kelvin is right as to the dates [1852] when he began to introduce the subject into his lectures at Glasgow [...] he must be mistaken as to the time when I talked with him about Foucault's discovery, for I feel sure I did not know it till 1855.

A most interesting remark by Stokes is:

> I have never attempted to claim for myself any part of Kirchhoff's admirable discovery, and cannot help thinking that some of my friends have been over zealous in my cause. As, however, my name has frequently appeared in print in connection with it, I have been induced to put on paper a statement for the views I entertained and talked about, though without publishing. In ascribing to Stewart the discovery of the extension of Prévost's law of exchange, I do not forget that it was rediscovered by Kirchhoff, who, indeed, was the first to publish it in relation to light, though the transition from radiant heat to light is so obvious that it could hardly fail to have been made, as in fact it was made, by Stewart himself. Nor do I forget that it is to Kirchhoff that we owe the admirable application of this extended law to the lines of the solar spectrum.

We are left to admire Stokes' academic honesty. As a matter of fact, Stokes had so many diamonds in his crown of discoveries that he did not need an additional one.[78]

As a matter of fact, Kelvin argued in favor of his compatriot Stokes, and in his zeal overlooked the priority of his fellow countryman and other English scientists like Brewster, who had identified even earlier[79] many lines that appeared in flames with lines that appeared in the solar spectrum. Obviously, Brewster could not have identified all the lines in the solar spectrum because he did not have the data on the spectra of many elements. Even so, Brewster concluded that the phenomenon was universal long before Swan clinched the identification of the *D* line with sodium.

As late as 1857, Swan[80] published a work on the spectra of carbon compounds. These compounds have a very large number of spectral lines and some coincided with solar lines (at least within the accuracy of the measurements). But Swan nevertheless concluded that the explanation for the reversal was not correct.

---

[78] Irony of fate, there exists a Kirchhoff–Stokes equation for sound attenuation. The name was given after both heroes had passed away. Kirchhoff, who is mostly known for his work on electricity and radiation, derived a formula for the absorption of sound due to conduction (Kirchhoff, G., Ann. Phys. **134**, 177, 1868) which is similar to the formula for viscosity deduced by Stokes (Stokes, G.G., Phil. Mag. **I**, 305, 1851) several years earlier.

[79] Brewster, D., Rep. Br. Assoc. **11**, 15 (1842).

[80] Swan, J.W., *On the Prismatic Spectra of the Flames of Compounds of Carbon and Hydrogen*, Royal Society of Edinburgh. Transactions, 1857.

2.14 Kirchhoff's Rebuttal

No doubt, Kirchhoff's proof was more complete than the one given by Stewart. This led Stewart to write to the editor Brewster and pose the dilemma: to what extent should priority go to the first who discovers a law or to the one who provides the more accurate proof? A problem that exists even today when it comes to attributing credit.

According to Kirchhoff,[81] Talbot had already guessed in 1826[82] that there was a connection between a substance and its spectral lines, but careful reading reveals that Kirchhoff did Talbot a great favor in this statement, because Talbot's papers contained many other hypotheses that were later found to be wrong.[83] Herschel on the other hand discussed the Fraunhofer lines in 1836,[84] but did not mention 'such a fantastic idea' of using the spectral lines for chemical analysis of stars. Could it be that Herschel actually missed this great idea?

A letter Kirchhoff wrote to Erdmann was transferred by the addressee to Roscoe.[85] And so reads the letter:

> The Sun possesses an incandescent gaseous atmosphere, which surrounds a solid nucleus having a still higher temperature. If we could see the spectrum of the solar atmosphere, we should see in it the bright bands characteristic of the metals contained in the atmosphere, and from the presence of these lines should infer that of these various metals. The more intense luminosity of the Sun's solid body, however, does not permit the spectrum of its atmosphere to appear, but reverses it, according to the proposition I have announced. So instead of the bright lines which the spectrum of the atmosphere by itself would show, dark lines are produced. Thus we do not see the spectrum of the atmosphere, but we see a negative image of it. This, however, serves equally well to determine with certainty the presence of those metals which occur in the Sun's atmosphere.

A statement valid even today.

---

[81] Kirchhoff, G., Pogg. Ann. **118**, 94 (1863).

[82] Talbot, W.H.F., Edinb. J. Sci. **5**, 77 (1826). Talbot discusses experiments carried out by Herschel.

[83] Talbot claimed that the *D* line was due to sulphur and sodium salt. He was correct about the contribution of the sodium to the line, but wrong about the sulphur. Moreover, if two elements can produce the same line, then the lines cannot be used to identify elements.

[84] Herschel, J.F.W., *A Treatise on Astronomy*, 3rd edn., Carey, Lea & Blanchard, Philadelphia, 1835.

[85] Roscoe forwarded part of the letter to the Editor of the Philosophical Magazine and Journal, on 1 February 1860, and wrote: *As it gives a later account of Kirchhoff and Bunsen's most important researches than has yet appeared in the English journal, I think it may be of interest to you and to your readers.* [...], *signed: Henry E. Roscoe.*

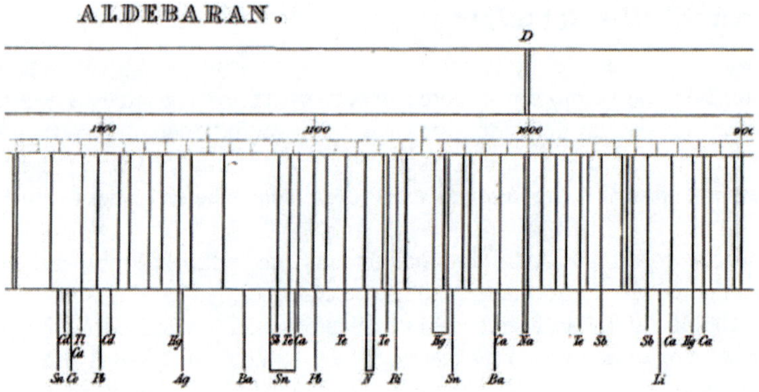

Fig. 2.11 The first ever spectra of the bright star Aldebaran taken by Huggins and Miller, 1862. At the *top* is the spectrum of sodium for comparison, while *below* are the identifications of the spectral lines of metals

2.15 Huggins' Particular View: How Stellar Spectroscopy Came into Being

In 1897, Huggins described how he started stellar spectroscopy in 1864[86]:

> It was just at this time [in 1862] that I happened to meet at a soirée of the Pharmaceutical Society, where spectroscopes were shown, my friend and neighbour, Dr. W. Allen Miller, Professor of Chemistry at King's College, who had already worked much on chemical spectroscopy. A sudden impulse seized me to suggest to him that we should return home together. On our way home I told him of what was in my mind, and asked him to join me in the attempt I was about to make, to apply Kirchhoff's methods to the stars. At first [...] he hesitated as to the probability of our success. Finally he agreed to come to my observatory on the first fine evening, for some preliminary experiments as to what we might expect to do upon the stars.

This is the story of the beginning of a long-standing friendship and collaboration, and it probably explains why Huggins felt compelled to defend Miller's priority in using spectral analysis for chemical identification (Fig. 2.11).

The relevant part for the priority Huggins claimed for Miller (1817–1870m) is based on what Miller[87] himself wrote in 1845, namely:

> It may be interesting to remark, in connexion with the speculation on the absorptive action of the Sun's atmosphere, that if solar light be transmitted through a flame exhibiting well-marked black lines, these lines reappear in the compound spectrum, provided the light of day be not too intense compared with that of the coloured flame: this may be seen in the red light of the nitrate of strontia, and less perfectly in the green of chloride of copper. It would therefore be that luminous atmospheres exist in which not only certain rays are wanting, but which exercise a positive absorptive influence upon other lights.

---

[86] Huggins, W., *The New Astronomy, A Personal Retrospect*, The Nineteenth Century **41**, 911 (1897).

[87] Miller, W.A., Phil. Mag. III, 27, p. 81, 1845.

The grounds for Huggins' claim are not very convincing. If the discovery was already so deeply appreciated, then why there were no follow-up papers?

In his influential textbook, the second edition of which appeared in 1860,[88] just before he met Huggins and before he embarked on his prolonged research on stellar composition, Miller addressed the problem of the Fraunhofer lines in the spectra of stars, writing:

> These lines are independent of the nature of the refracting medium, and occur always in the same colour, and at corresponding points of the spectrum.

While admitting their usefulness for opticians in accurately determining a substance's index of refraction, Miller added:

> No satisfactory explanation has yet been found for the cause of this phenomenon.

His long-standing interest in the Fraunhofer lines should have made Miller more receptive than most to the news of Kirchhoff's theory regarding their cause, but there was no mention of the new discoveries in his book.

2.16 Lockyer 1887: The Chemical Composition of the Sun

In 1887, Norman Lockyer[89] published a book on the chemical composition of the Sun,[90] in which he claimed that the solution for the Fraunhofer lines was found as early as 1823 by two English scientists, Brewster[91] and Herschel,[92] who realized that when a cool absorber is placed in front of a bright light source, dark lines will appear. It was Brewster, according to Lockyer, whose experiments convinced him that the dark lines offered a unique method of chemical analysis. It was also Brewster who claimed (erroneously) to have discovered nitrous acid in the atmosphere of the Sun. This was the first chemical compound to have been claimed to be identified in the solar spectrum.

But the reality or its rigorous interpretation, was a bit different. In 1835, Herschel himself[93] wrote as follows:

---

[88] Miller, W.A., *Elements of Chemistry: Theoretical and Practical*, Part I. Chemical Physics, 2nd edn. John W. Parker and Son, London, 1860, p. 146.

[89] Publisher Alexander Macmillan chose Lockyer as Nature's founding editor in 1869. As Ruth Barton suggests (Barton, R., *Lockyer's columns of controversy in Nature*, in *History of the Journal Nature*, Nature **16**, October 2007), Lockyer endorsed discussions of controversies. Should we understand his remarks on the priority in terms of his passion for controversy?

[90] Lockyer, N., *The Chemistry of the Sun*, MacMillan, 1887. See Chap. V.

[91] Brewster, D., Edinb. Phil. Trans. **9**, 433 (1823).

[92] Herschel, J.F.W., Edinb. Phil. Trans. **9**, 445 (1823).

[93] Herschel, J.F.W., *A Treatise on Astronomy*, 3rd edn., Carey, Lea, & Blanchard, Philadelphia, 1835, p. 203.

> The prismatic analysis of the solar beam exhibits in the spectrum a series of fixed lines totally unlike those of any known terrestrial flame. This may hereafter lead us to a clearer insight into its origin. But before we can draw any conclusion from such an indication, we must recollect that previous to reaching us it has undergone the whole absorptive action of our atmosphere, as well as of the Sun's. Of the latter we know nothing, and may conjecture everything. [...] It deserves inquiry whether some or all of the fixed lines observed by Wollaston and Fraunhofer may not have their origin in our own atmosphere. [...] The absorptive effect of the Sun's atmosphere, and possibly also of the medium surrounding it (whatever it be), which resists the motion of comets, cannot be thus eliminated.

Two points are worth noting. We observe the Sun through a telescope, so obviously everything we see can be either on the Sun, or in the space between the Sun and the Earth, or in the Earth's atmosphere. But suggesting these three possibilities is not a solution or an explanation of the Fraunhofer lines. Furthermore, everything we see may be used to understand structure and composition, and Lockyer's interpretation of a naive phrase like 'clear insight into its origin' as a prediction of the powerful future methods of spectroscopic analysis, is to my mind, stretching the meaning a bit too far.

In 1840, Herschel published his *Outline of Astronomy*.[94] In 1871, in the 11th edition, he wrote:

> The reference of the dark lines in the solar spectrum to absorptive action in the Sun's atmosphere has of late received a most unexpected confirmation, and it may now be considered as almost certain that they owe their origin to the presence in that atmosphere of the vapours of metals and metalloids identical with those which exist here on Earth. These vapours, or many of them, have been shown by Kirchhoff, Bunsen, and Fizeau to possess the singular property when present in an unburnt (or metallic) state in a flame, of destroying in the spectrum of that flame rays of precisely the refrangibilities of those which they themselves when burning emit in peculiar abundance. Though there is something so enigmatical as almost to appear self-contradictory in the facts adduced, the conclusion, especially as applied to the most conspicuous of all the lines (one double one in the yellow, marked *D* by Fraunhofer, and which owes its origin to sodium) seems inevitable. The spectra of some of the stars seem to indicate the presence of chemical elements not identifiable with any terrestrial ones.

We learn that, even ten years after Kirchhoff's major discoveries, Herschel was quite reserved about the possible scope of spectral analysis and the explanation of the Fraunhofer lines, and in particular he did not mention the explanation attributed to him.

The details in the case of Brewster[95] are not that different. In 1837, Brewster wrote about *the coloured bands of the reflected spectrum*, and did not mention any explanation.

In 1859, a year before Kirchhoff published the English version of his German paper in the journal Brewster edited, Brewster and John Gladstone (1827–1902)[96]

---

[94] Herschel, J., *Outline of Astronomy*. The first edition came out in 1840 and the 11th edition was issued in 1871. The 'new edition' came out in 1893, London, Longmans, Green, 1893.

[95] Brewster, D., Phil. Trans. R. Soc. Lond. **127**, 245, 1837.

[96] Brewster, D. and Gladstone, J.H., *On the Lines of the Solar Spectrum*, Proc. R. Soc. Lond. **150**, 339, 1859.

wrote: *The origin of these fixed lines and bands in the solar spectrum is a question still unresolved.* The possibilities mentioned were as follows:

- That the light when emitted from the photosphere (the region in the Sun from which the radiation we see emerges) is itself deficient in these rays. This was evidently Fraunhofer's idea.
- That they are due to absorption by the Sun's atmosphere.
- That they are due to absorption by the Earth's atmosphere.

As noted before, these were the three obvious possible solutions. *The first supposition scarcely admits a positive proof,* the authors wrote. The second supposition implied that there should be a difference between spectra emitted from the center of the Sun and the limb. However, Brewster and Gladstone cited Gladstone, who tried to observe this effect during the eclipse in March 1858:

> Unfortunately clouds prevented the experiment. [...] However, by other contrivances, each of the authors came independently to the conclusion that there is no perceptible difference in this respect between the light from the edge and that from the centre of the solar disk.

With some twist of the exact meaning, one can ignore the first and the third explanations and state that this paper provided the explanation for the Fraunhofer lines. Even so, this solution at best identified a configuration that might produce the lines, but it by no means offered an explanation.

Brewster, however, made an interesting experiment. He passed the solar light through vapors of nitrous acid (HNO_2) and discovered *a multitude of lines crossing the spectrum in the same direction as those observed by Fraunhofer*. Miller tried to repeat Brewster's experiment and did not find any gases that would produce the same lines.[97] He also noticed that the number of lines increased as the gas was compressed. In 1833,[98] Brewster discovered that, when the Sun was low above the horizon, additional Fraunhofer lines appeared in the spectrum, thus favoring a local source for the lines.

At about the same time, John Forbes (1787–1861)[99] was determined to take advantage of the opportunity presented by the forthcoming eclipse to examine the spectrum of the edge of the Sun without any disturbance from the central rays. He argued that, if the absorption was due to the solar atmosphere, it ought to be more marked at the edge of the Sun, because the light has to pass through a thicker stratum of atmosphere than at the centre, from which the rays proceed vertically and through a minimum of atmosphere. The general darkening of the Sun's limb was a well-known fact that could be observed at any time by examining the Sun through a dark glass (see Fig. 2.12[100]). Forbes was probably aware of this, and expected to find the limb-darkening accompanied by an increased selective absorption, i.e., stronger Fraunhofer lines.

[97] Miller, W.A., Phil. Mag., Ser. III, **27**, 81, published in German Ann. Phys. **145**, 404 (1846).

[98] Brewster, D., Phil. Trans. Edinb., 1833.

[99] Forbes, J.D., Phil. Trans. R. Soc. Lond. **126**, 453 (1836).

[100] Pierce, A.K. and Slaughter, C.D., Solar Phys. **51**, 25 (1977).

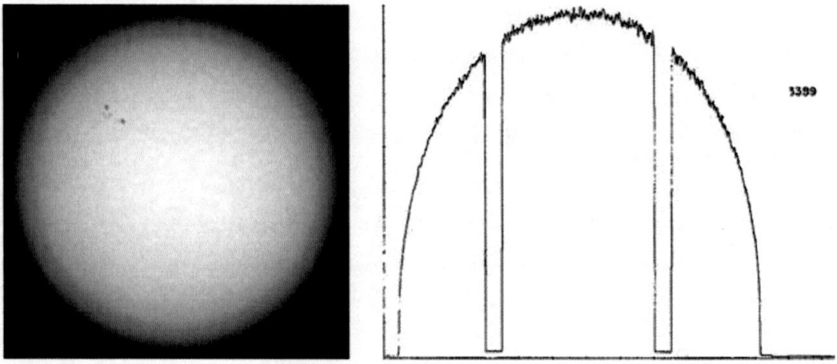

Fig. 2.12 *Left* Limb-darkening. Note how the brightness decreases towards the limb of the Sun. *Right* a modern result for limb-darkening of the Sun at a wavelength of 3,389 Å, after Pierce and Slaughter, (1977)

The result of the observation was that, as the eclipse progressed and the proportion of lateral to central light consequently increased, no change was observed in the number, position, or thickness of the lines, and from this observation Forbes concluded that[101]:

> This result proves decisively that the Sun's atmosphere has nothing to do with the production of this singular phenomenon.

Lockyer dismissed Forbes' observations and wrote that his conclusion *was at variance with that held by his predecessors*, which is not a scientifically valid argument, *and we now know that Prof. Forbes' conclusion was wrong*. He also presented it as an example illustrating *how near one may be to a most important discovery and yet miss it*, because of poor weather.

---

[101] In a footnote to his paper Forbes wrote:

> I do not know with whom the idea of the absorptive action of the Sun's atmosphere originated. The editors of the London and Edinburgh Phil. Mag. have, however, referred me to the mention of Sir John Herschel's writings, particularly his *Elementary Treatise on Astronomy*, from which I extracted the following remarkable passage: "The prismatic analysis of the solar beam exhibits in the spectrum a series of fixed lines totally unlike those of any known terrestrial flame. This may hereafter lead us to clearer insight into its origin. But before we can draw any conclusions from such an indication, we must recollect that previous to reaching us it has undergone the whole absorptive action of our atmosphere, as well as of the Sun's. [...] It deserves inquiry whether some or all of the fixed lines observed by Wollaston and Fraunhofer may not have their origin in our own atmosphere. [...] The effect of the Sun's atmosphere, and possibly also of the medium surrounding it (whatever it be), which resists the motion of comets, cannot be eliminated."

If we continued in this way, we would conclude that Herschel even predicted the effect of the solar wind on comets.

But the situation was not as trivial as Lockyer had depicted it. In 1887, the same year as Lockyer published his book, Clerke[102] published a popular book on astronomy in which she claimed that the problem of the formation of Fraunhofer lines across the solar disk *still remains an anomaly of which no satisfactory explanation has been offered*. Indeed, she was quite right.

In 1902, Very (1852–1927)[103] attempted to measure the absorption coefficient of the solar atmosphere, about which practically nothing was known. He measured the intensity across the disk (see Fig. 2.13). The results allowed him to check various models and to find that none of them was particularly successful. So he suggested that irregularities in the photosphere might explain his results, namely that:

> The photosphere is made up of brilliant 'rice-grains' and their component 'granules', separated by a relatively dark reticulation in which the light having come from greater depths suffers larger absorption than where it proceeds from the summits of the photosphere clouds, or granules.

A comparison between the predictions of such a model and the observations turned out to be good, according to Very.

In 1902, Schuster attacked the problem of limb-darkening. He found that:

> The radiation received from different portions of the solar disk is known to diminish from the center towards the limb in a manner which is generally considered not to be consistent with the assumption of a uniformly absorbing solar atmosphere. [...] The difficulty is easily removed. It is only necessary to place the absorbing layer sufficiently near the photosphere and to take account of the radiation which this layer, owing to its high temperature, must itself emit.

The end equation Schuster got was

$$A = (I - F)z^{\sigma} + F,$$

where A (not to be confused with the absorption discussed earlier) is the radiation leaving the absorbing layer in the direction of the Earth, I is the intensity of the radiation which is incident on the absorbing shell, and F is the radiation of a perfect black body at the temperature of the shell. z and σ are two geometrical factors associated with the Sun.[104] The equation can be solved if the radiation is measured at three points across the solar disk, because there are three unknowns: I, F, and z.

Two years later, Very[105] was unhappy with Clerke's statement that the problem had not yet been solved and claimed that he had offered a solution. He repeated Forbes' argument that *a critical examination of the appearances of the Fraunhofer*

---

[102] Agnes Mary Clerke (1842–1907) was an astronomer and a well-known writer, mainly in the field of astronomy. She wrote a popular book on astronomy (Clerke, A.M., *A Popular History of Astronomy During the Nineteenth Century*, Edinburgh, A. & C. Black, 2nd edn. 1887).

[103] Very, F.W., Astrophys. J. **16**, 73 (1902).

[104] $z = e^{kt_0}$, $\gamma = r/R_{\odot}$, and $\sigma = 1/\sqrt{1-\gamma}$, where t_0 is the thickness of the absorbing layer, k the coefficient of absorption, r the perpendicular distance between any point on the Sun and the line drawn from the Sun's surface towards the observer on the Earth, and $R_{\odot}$ the radius of the Sun.

[105] Very, F.W., Astrophys. J. **19**, 139 (1904).

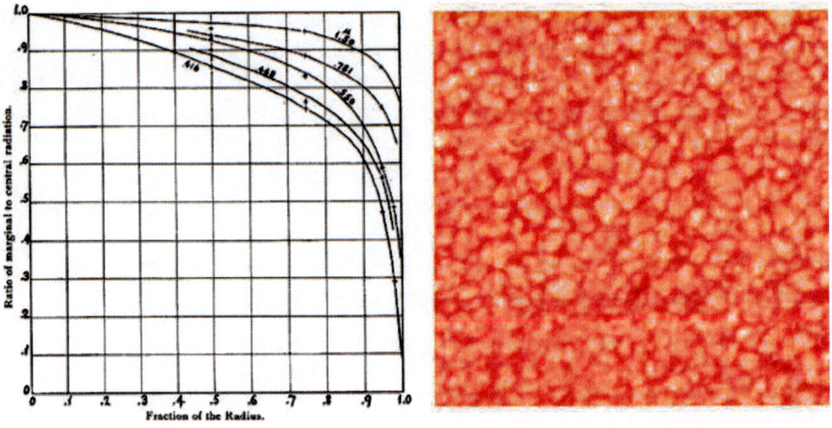

Fig. 2.13 *Left* Very's observation of the change in the solar intensity as a function of the distance from the center and the wavelength. After Very (1902). *Right* a close-up view of the solar surface shows the granules and the dark boundaries between them

lines from different parts of the solar disk discloses the remarkable fact that they do not vary appreciably at any point of the unspotted surface, and can have nothing to do with the progressively increasing selective absorption, with the obvious (but incorrect) conclusion about how the dark lines form.

Very's claim, as presented in the previous paper, was:

> The failure of the Fraunhofer lines to become intensified at the Sun's limb is to be attributed chiefly to the corrugation of the photospheric surface, and to the fact that the efficient absorbing layer is of a depth not great in relation to the vertical dimension of these irregularities.

Then Very attacked Schuster's solution[106] which asserted that:

> Solar radiation comes from an absorbent and radiating layer, distinct from and immediately above the photosphere, and the apparent change of absorption at different distances from the Sun's limb is explained as due to the varying relative preponderance of the two sources—photosphere and atmosphere—in producing the radiation.

Very referred to certain observations, claiming that *Schuster does not avail himself of this evidence*. As for Schuster's assumptions, he assumed a coefficient of absorption which was independent of wavelength. But, so claimed Very:

> Anyone who has compared the strongly contrasted colors—blue at the center and reddish-brown at the limb of the Sun—will recognize that, if this explanation is to be accepted, the radiation at the limb must come almost entirely from the red-hot particles of the envelope, while the blue-hot photosphere is mainly in evidence at the center.

The discussion by Very is quite long and completely rhetorical, without any modeling. Moreover, he claimed to find an error in Schuster's analysis.

---

[106] Schuster, A., Astrophys. J. **16**, 320 (1902).

The story about solar limb-darkening is compounded with the question: what is the solar radius? The final answer to this question appeared to require a satisfactory solution and attracted astronomers even as late as the beginning of the twentyfirst century.[107]

2.17 More Creditors?

Among the English scientists on whose behalf credit was requested for discoveries in spectroscopy was Charles Wheatstone (1802–1875).[108] As early as 1835, Wheatstone made some important observations about the origin of the lines as well as their position, and also established that all the elements could be identified by their spectral lines. It would have been virtually impossible for Kirchhoff and Bunsen to know of this, for the lecture in which Wheatstone publicly announced the results of his investigation was held in 1835 and was not published until 1861, when Crookes, in the wake of Kirchhoff's publication, found it appropriate to straighten out the story. Only a summarized extract was published in 1835. According to Huggins, Wheatstone's words in that lecture were:

> We have here a mode of discriminating metallic bodies more readily than by chemical examination, and which may hereafter be employed for useful purposes.

In another publication during that year, Wheatstone[109] described the spectra of Cd, Sn, Pb, Hg, and Zn. No statement about the future or the potential of the method was made in this single reference by Wheatstone.

Similarly, we should mention Anders Ångström (1814–1874m)[110] who reported a relation between absorption and emission. Volkert van der Willigen (1822–1878)[111] discovered in 1859 that the same metal produces the same spectrum whether it is in the form of nitride or chloride. It is thus possible to identify the metal even if it is in a chemical compound.[112]

As in all other branches of science, the discovery of Kirchhoff and Bunsen did not come out of the blue, but rested on significant layers of experiments, ideas, failures, thinking, etc. Few today would argue with the statement that Kirchhoff and Bunsen's

---

107 See for example, Livingston, W.C., Milkey, R., and Sheeley, N., Jr., AAS meeting, no. 211, **159.06**, 2008; Neckel, H., Solar Phys. **229**, 13 (2005).

108 Wheatstone, C., *Prismatic decomposition of electric light*, Reports to the British association for the Advancement of Science **5**, 11 (1835); Crookes, Chem. News **3**, 198 (1861).

109 Wheatstone, C., Phil. Mag. **7**, 299 (1835).

110 Ångström, A.J., Pogg. Ann. **117**, 290 (1862); Phil. Mag. **9**, 327 (1855).

111 Willigen, V.S.M., Ann. Phys. **182**, 610 (1859).

112 Roscoe (Roscoe, H.E., *The Edinburgh Review or Critical Journal* for July 1862 to October 1862, p. 295) gives the following example about the unbelievable way in which scientific discoveries are interpreted by laymen, and suggests that it could be *an interesting branch of study to the psychologist*: Kirchhoff and Bunsen got a letter form a Silesian farmer who thanked them for proving his theory that no inorganic materials should be added to plants as all the required minerals exist in solar light.

research was imperative in converting spectroscopic analysis into a paramount tool for analytical chemistry. However, it was only after Kirchhoff and Bunsen recognized the enormous potential of spectroscopy that the community started to appreciate the power of the method.

2.18 An Unimaginable but True Story: No Life on the Sun

Soon after the publication of the great papers, Kirchhoff published a memoir on the *Solar spectrum and the spectra of the chemical elements*, which was quickly translated into English by Roscoe.[113] Kirchhoff credited Swan, Brewster, Gladstone, and Miller, and stressed that his discoveries were due to the excellent apparatus he obtained from the company Steineil-Söhne in Munich.[114]

It is interesting to note that in this first analysis of the solar spectra, Kirchhoff inverted the logic. Hence, having problems with measuring the exact location of certain lines seen in the laboratory, he claimed that:

> The dark lines of the solar spectra afford invaluable assistance in determining the position of the bright lines of the various elementary bodies.

Thus, instead of finding the composition of the Sun, the solar spectrum was used to provide accurate spectroscopic data for chemical elements on Earth. This situation was repeated several times in the following years (like the discovery of helium on the Sun).

To explain the formation of the dark lines Kirchhoff assumed that:

> [...] the solar atmosphere encloses a luminous nucleus, producing a continuous spectrum, the brightness of which exceeds a certain limit. The most probable supposition [...] is that it consists of a solid or liquid nucleus, heated to a temperature of the brightest whiteness, surrounded by an atmosphere of somewhat lower temperature. This supposition is in accordance with Laplace's celebrated nebular theory respecting the formation of our planetary system.

So, on top of explaining the origin of the dark lines, Kirchhoff concluded that the composition of the Solar System and the Sun must therefore be alike, and argued that geology tells us that the Earth was once hot and liquid. Hence, he concluded that all bodies in the Solar System were hot in the past and were now cooling. Small objects like the Moon and the Earth would cool relatively fast compared to the age of the Earth, while big objects like the Sun would cool more slowly. So far, Kirchhoff's assumption was in agreement with Kelvin's incorrect theory of the Sun.

---

[113] Kirchhoff, G., *Solar Spectra and the Spectra of the Chemical Elements*, MacMillan, Cambridge, 1862.

[114] Recall the power of the excellent optical equipment produced by Fraunhofer. Similarly, the Steineil-Söhne company was founded in 1855 by Carl (1801–1870) and his son Adolf (1832–1893) Steineil, and excelled in spectrographs.

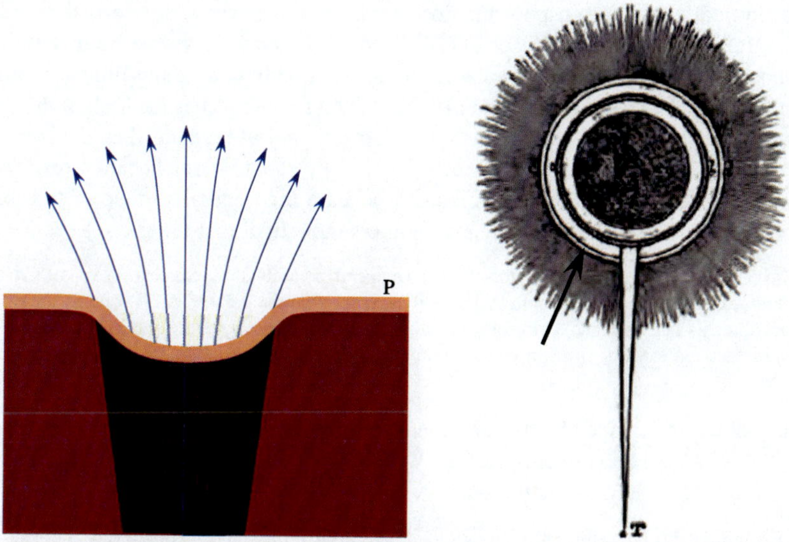

Fig. 2.14 *Left* the Wilson depression was discovered by Wilson in 1769. The photosphere is the region from which the radiation we see emerges from the Sun. In other words, the region where the probability of a photon escaping from the star is high. The *dark area* is a sunspot. Wilson did not know that sunspots contain strong magnetic fields. *Right* Arago invented the term 'photosphere' in his book *Popular Astronomy*. The caption in Arago's book is *Formation of a spot without a nucleus*. The *arrow* marks the photosphere. The 'solid' nucleus of the Sun is seen through the spot without nucleus

Like Galileo before him, Kirchhoff observed the spots on the Sun and accepted Galileo's explanations that they were clouds in the solar atmosphere. But the cloud theory for sunspots was relinquished by many astronomers on account of some peculiarities in the spots which were brought to light by further observations. According to François Arago (1786–1853m),[115] the Sun consisted of a dark nucleus surrounded by an opaque and reflecting atmosphere, this being enclosed by a luminous atmosphere or photosphere,[116] which was in its turn surrounded by a transparent atmosphere. Arago declared:

> If I were asked whether the Sun can be inhabited by beings organized in a manner analogous to those which people on our globe are, I should not hesitate to reply in the affirmative. The existence of a central obscure nucleus, enveloped in an opaque atmosphere, far beyond which the luminous atmosphere exists, is by no means opposed, in effect, to such a conception.

[115] Arago, F., *Popular Astronomy*. Translated by Smyth and Grant, Longman, Brown, Green and Longman, London, 1955. From 1813 and until 1845, Arago gave very popular non-technical lectures on astronomy. Chapter XXIX of the book is entitled: *Is the Sun inhabited*?

[116] The term photosphere was invented by Arago in *Popular Astronomy*, p. 411:

> All the phenomena of which we have just been speaking, may be explained in a satisfactory manner, if we assume that the Sun is an obscure body surrounded to a certain

The physical nature of sunspots remained a topic of controversy for nearly three centuries. In 1774, Alexander Wilson (1714–1786) observed the asymmetric appearance of sunspots when seen near the solar limbs. The phenomenon, called Wilson depression (see Fig. 2.14) was discovered in 1769. When a spot reaches the limb of the Sun due to solar rotation, the penumbra[117] of a spot appears wider than when observed at the center of the Sun. The entire effect is very small,[118] but Wilson became convinced that the depressions were regions located beneath the general surface of the Sun, in contrast to mountains, as many astronomers thought at that time:

> Is it not reasonable to think that the great and stupendous body of the Sun is made up of two kinds of matter, very different in their qualities; that by far the greater part is solid and dark and that this immense and dark globe is encompassed with a thin covering of that resplendent substance from which the Sun would seem to derive the whole of his revivifying heat and energy?

Herschel was convinced by this argument, and his first paper on the Sun, published in 1794, was based on this supposition. By 1801, Herschel concluded that there were even two types of clouds on the Sun:

> The solid body of the Sun beneath these clouds appears to be nothing else than a very eminent, large and lucid planet, evidently the first, or in strictness of speaking, the only primary one of our system.

And so he was led to claim that: *We need not hesitate to admit that the sun is richly stored with inhabitants*. As hallucinatory as this idea may seem in our view today,

---

> (Footnote 116 continued)
> distance by an atmosphere, which may be compared to the terrestrial atmosphere when the latter is occupied by a continuum stratum of opaque and light reflecting clouds. If moreover, we place above this first stratum a second luminous atmosphere which will assume the name of a photosphere, this photosphere, more or less remote from the interior cloudy atmosphere, determines by its contour the visible limits of the body.

Since then, this term has been adopted to describe the 'last luminous visible layer of the Sun'.

[117] Penumbra literally means dim light, in this case, the outer filamentary region of a sunspot.

[118] The existence of magnetic fields in sunspots was demonstrated in 1908, shortly after the Zeeman effect was discovered. The Zeeman effect, the splitting of spectral lines in the presence of a magnetic field, was discovered by Zeeman (1865–1943m) in 1896 (Zeeman, P., Phil. Mag. [5], **43**, 226, 1897; Astrophys. J. **5**, 332, 1897). He observed in the laboratory how the two sodium *D* lines (the *D* lines played an important role once again) broaden when the flame is placed between the magnetic poles of a strong electromagnet. Hale used the Zeeman effect to identify the magnetic field in sunspots in 1908 (Hale, G.E., PASP **20**, 287, 1908; Astrophys. J. **28**, 315, 1908). In the wake of this discovery, Hale asked Zeeman to send his views about the discovery to Nature. While Hale was rather cautious about the discovery and its implication, Zeeman stated that:

> Hale has given what appears to be a decisive evidence that sunspots have strong magnetic fields, the direction of these fields being mainly perpendicular to the Sun's surface.

The effect was expected by Faraday (Maxwell, J.C., Collected Works, II, 790, Cambridge Press, 1890), who tried in vain to detect it as early as 1862. The temperature in the spot is about 4,000 K. The magnetic fields in the spots are about 1,000 times greater than the mean solar magnetic field, and may reach 1,000–4,000 gauss, while the magnetic field of the Earth is 0.3–0.6 gauss.

some regard it as the first attempt to provide a coherent explanation for isolated pieces of data seen on the surface of the Sun. Several historians of astronomy claim that this is almost the only point on which Herschel was first mistaken,[119] but later corrected.

Kirchhoff claimed that this theory:

> [...] appears to me to stand in such direct opposition to certain well established physical laws, that in my opinion it is not tenable even supposing that we were unable to give any other explanation of the formation of sunspots. This supposed photosphere must, if it exists, radiate heat towards the Sun's body as well as from it. Every particle of the upper layer of the lower or opaque atmosphere will, therefore, be heated to a temperature at least as high as that to which it would be raised if placed on the Earth in the focus of a circular mirror exposed to the Sun's rays, whose surface seen from the focus is larger than a hemisphere.

Kirchhoff argued that the atmosphere keeps the nucleus of the Sun at least as hot as the temperature of the atmosphere (heated to the point of incandescence). So Kirchhoff rejected this theory and returned to the cloud hypothesis:

> A local diminution of temperature must give rise to the formation of clouds, only that the solar clouds will be of a different chemical composition from terrestrial ones. When a solar cloud is formed, all the portions of the atmosphere lying above it will be cooled down, because a portion of the rays of heat which are emitted from the incandescent surface of the Sun are cut off by the cloud. [...] The temperature of the cloud sinks below the point of incandescence, it becomes opaque, and forms the nucleus of the solar spot.

Kirchhoff claimed that he could fully explain Wilson's apparent depression of spots. But Kirchhoff noticed that sunspots are observed only close to the equator and admitted that *this fact cannot be explained by my theory*.

Despite Kirchhoff's rather convincing arguments, several years later, we find people like George Stoney (1826–1911)[120] who had other ideas about the Sun, describing it in the following way:

> The true surface of the Sun is the outer boundary of this enormous atmosphere. [...] Within this luminous film there is a dark body, glimpses of which are occasionally seen as the umbrae of spots.

Stoney dedicates a special section to the *clouds in the outer atmosphere*, which were made, according to his hypothesis, of carbon. This section contains an extensive description of the clouds, their size, and what exists between them.

2.19 Final Comments on the Kirchhoff Saga

It is difficult to detach the controversy around Kirchhoff's discoveries from parallel discussions which took place across the English channel at about the same time. Consider, for example, the controversy between the Scotsman Peter Tait (1831–1901)

---

[119] Macpherson, H., *Century Progress in Astronomy*, William Blackwood and Sons, Edinb. and Lond. 1906.

[120] Stoney, G.J., Proc. R. Soc. **XVII**, 1, 1867.

and the German Herman Helmholtz (1821–1894m)[121] about the contributions to the basic laws of thermodynamics, and the claim for priority by the Englishman James Joule (1818–1889) against the claim by the German Julius Mayer (1814–1878) that he had established the energy conservation law,[122] or the dispute about the contributions to thermodynamics of the Irishman Lord Kelvin (1824–1907) and the German Rudolf Clausius (1822–1888m) to thermodynamics.

As for Kirchhoff himself, it is important to note that he was not so satisfied with his own proof of the radiation law.[123] So, only a few weeks later, Kirchhoff apparently changed his view that a general proof could be attained by the simple theoretical considerations he had invoked. In January 1860, he submitted a second, much more involved proof without initially commenting on the fate or rigour of the first. Two years later, he published a structurally improved version of this second derivation, which the editors chose for his collected works.[124] In this revision, he commented on the supposition that bodies emit or absorb only at one specific wavelength[125]:

> The necessary completion of the proof may easily be given when a plate is supposed to exist, having the property of transmitting undiminished rays whose wavelength lies between l and $l + dl$ and whose plane of polarization is parallel to the plane a; but which completely reflects rays of other wavelengths or of an opposite polarization.

He now regarded this supposition as inadmissible. Instead, he relied on an even more intricate object:

> [A] plate is possible which, of the rays striking it at the same angle, transmits and reflects them in different degrees according to their wavelengths and plane of polarization. A plate, which is so thin that the colors of thin films are visible and which is placed obliquely in the path, shows this.

In 1877, John Tyndall (1820–1893m),[126] who succeeded the great Faraday, gave a series of 6 lectures in the USA, and the fifth lecture was devoted to spectral analysis.

---

[121] Knott, C.G., *Life and Scientific Work of Peter Guthrie Tait*, Cambridge University Press, 1911.

[122] Shaviv, G., *The Life of the Stars*, Springer, Heidelberg, 2009.

[123] Kirchhoff, G., *On the relation between emission and absorption of light and heat*, which was presented to the Berlin Academy of Sciences on 15 December 1859: Gustav Kirchhoff, *Ueber den Zusammenhang von Emission und Absorption von Licht und Wärme*, Akad. der Wissen. Berlin, pp. 783, 784, 786, reprinted in Gustav Kirchhoff, *Untersuchungen über das Sonnenspektrum und das Spektrum der chemi schen Elemente und weitere ergänzende Arbeiten aus den Jahr en 1859–1862, Osnabrück, 1972, ed. Kangro. The more detailed paper was published in Ann. Phys., January 1860.*

[124] Gustav Kirchhoff, Untersuchungen über das Sonnenspektrum und die Spektren der chemischen Elemente (2nd edn., Berlin, 1962), appendix, *Über das Verhältnis zwischen dem Emissionsvermögen und dem Absorptionsvermšgen der Körper für Wärme und Licht*, 22–39; also in Gesammelte Abhandlungen. Vol. 1 (Leipzig, 1882) 571, English trans. in D.B. Brace, ed., *The Laws of Radiation and Absorption: Memoirs by Prévost, Stewart, Kirchhoff, and Kirchhoff, and Bunsen*, New York, 1901, p. 75.

[125] Note that Kirchhoff's one-wavelength plate is a perfect mirror for all radiation with a wavelength different from the specified one.

[126] Tyndall, J., *Six Lectures on Light Delivered in America in 1872–1873*, Appleton and Comp., New York, 1877.

So who did the Irishman Tyndall, who worked in England, consider to have explained the Fraunhofer lines? Tyndall's version was as follows:

> The explanation of these lines was, as I have said, a problem which long challenged the attention of philosophers, and to Kirchhoff, Professor of Physics in the University of Heidelberg, belongs the honour of having first conquered this problem.

By 'lines' Tyndall meant the Fraunhofer lines. It should be mentioned that Tyndall did not adopt the English version in other priority squabbles that took place at the same time.

2.20 Epilogue

The radiation law practically ceased to be a point of contention between the English and German scientific communities when, in 1862, Kirchhoff won the Rumford medal. The citation was: *For his researches on the fixed lines of the solar spectrum, and on the inversion of the bright lines in the spectra of artificial light*. The citation failed to mention his law of radiation. When Balfour Stewart got the Rumford medal in 1868, the citation said: *For his researches on the qualitative as well as quantitative relation between the emissive and absorptive powers of bodies for heat and light, published originally in the Transactions of the Royal Society of Edinburgh, and the Proceedings of the Royal Society of London, and now made more generally accessible by the publication in 1866 of his treatise on heat*. It is not sure that Kirchhoff would concur.[127]

In 1868 Kirchhoff was elected Fellow of the Royal Society of Edinburgh, and in 1875 Fellow of the Royal Society. As a further recognition by the English scientific

---

[127] The Rumford Medal is awarded by the Royal Society every other year for *an outstandingly important recent discovery in the field of thermal or optical properties of matter made by a scientist working in Europe*. The medal is based on a donation by Rumford, who was the first to be awarded the prize! Of the names mentioned so far, the following were winners:

establishment, a year before his death, Kirchhoff was awarded a gold medal by Queen Victoria in 1887.[128]

The doltish non-scientific controversy driven by nationalistic feelings between the English and German scientific communities did not end with the recognition of the German Kirchhoff by the English scientific establishment, and even perdured beyond World War I. The details, though interesting historically, are beyond the scope of this book (for details see, for example, Heilbron[129]).

2.21 The Late Kirchhoff Grilled by His Compatriots

Naturally, the debate in the German scientific community was not about priority. Kirchhoff's priority was obvious in Germany. However, claims about the rigour of the proof were raised. Forty years after the declaration of Kirchhoff's law, Voigt[130] still found it necessary to prove it theoretically and experimentally. It just so happened that the criticism of the rigour of Kirchhoff's proof by German scientists was no

(Footnote 127 continued)

1804 Leslie	For his Experiments on Heat, published in his work, entitled, An Experimental Enquiry into the Nature and Propagation of Heat.
1816 Davy	For his Papers on Combustion and Flame, published in the last volume of the Philosophical Transactions.
1834 Melloni	For his discoveries relevant to radiant heat.
1838 Forbes	For his experiments on the polarization of heat, of which an account was published in the Transactions of the Royal Society of Edinburgh. Not for spectroscopy.
1842 Talbot	For his discoveries and improvements in photography. Not for heat research.
1852 Stokes	For his discovery of the change in the refrangibility of light.
1872 Ångström	For his researches on spectral analysis.
1874 Lockyer	For his spectroscopic researches on the Sun and on the chemical elements. Not for the helium discovery (it was too 'risky').
1876 Janssen	For his numerous and important researches in the radiation and absorption of light, carried on chiefly by means of the spectroscope. Not for discovering helium!
1880 Huggins	For his important researches in astronomical spectroscopy, and especially for his determination of the radical component of the proper motions of stars.
1886 Langley	For his researches on the spectrum by means of the bolometer.

Note those who did not get the prize, in particular, Herschel, Prévost, Pictet, Wheatston, Miller, de la Provostaye, Desains, and Bunsen. (Source: Rumford archives, 1800–1898.) Prize committees are driven by internal politics.

[128] The following story is amusing. Kirchhoff (Smithsonian Report, p. 537, 1889) once told his banker about the discovery of terrestrial metals on the Sun. The banker responded somewhat indifferently, with: *Of what use is gold on the Sun if I cannot get it down to Earth*? Later, after Queen Victoria of England had presented Kirchhoff with a medal and a prize in gold sovereigns for work on the solar spectrum, he took the gold sovereigns to the banker and retorted: *Here is some gold from the Sun*!

[129] Heilbron, J.L., *The Dilemmas of an Upright Man*: *Max Planck and the Fortunes of German Science*, Harvard University Press, 1996.

[130] Voigt, W., Ann. Phys. **67**, 366 (1899).

different from Kirchhoff's criticism of the accuracy of Stewart's proof. But by then, the early 1890s, Kirchhoff was long dead.

In 1894, Wilhelm Wien (1864–1928)[131] thought he had found a flaw in Kirchhoff's assumptions.[132] In his proof, Kirchhoff had assumed the existence of a material with optical properties that the experimentalist Wien argued could not exist. Even as late as 1909, Wien[133] claimed that Kirchhoff's proof was 'artificial' or contrived.

In 1894, Friedrich Paschen (1865–1940m)[134] used a spectrobolometer[135] to measure the total intensity of the two *D* lines emitted by salt placed in the flame of a Bunsen burner. All other things being equal, he then measured the corresponding intensity of a selected region in the spectrum of a black body, a region completely contained within the two *D* lines. Assuming incorrectly an excessive value for the width of the sodium lines, Paschen then calculated a maximum value for the intensity of the part of the spectrum of the black body corresponding to the lines. The value he found was not even half the value found with the lines given by the burner. Paschen reached the inevitable conclusion that Kirchhoff's law was inapplicable. Paschen conjectured that the brightness of these lines was due, at least for a large part, to a phenomenon of luminescence. Using a different method, Gustav Wiedemann (1826–1899)[136] arrived at the same conclusion. Thus, even the experiments were problematic.

The status of Kirchhoff's law was reviewed by Aimé Cotton (1869–1951) in 1899.[137] The need for such a review forty years after the discovery reflects the difficulties in accepting this far-reaching physical law. Cotton attempted to prove Kirchhoff's law by starting with Prévost's assumption (see Sect. 2.3), and complained that *we have no method of studying a radiation without causing it to disappear*. Next he assumed that *the emission depends solely on the temperature, and conversely, the absorbed radiation is converted wholly into heat*. Cotton pointed to the fact that Kirchhoff's first proof was not sufficiently general while the second proof was much better. However, as Cotton remarked, the proofs by assuming certain plates with unique optical properties did not make a theory because it was not connected with a theory of light. Hence, Cotton tried to combine the law with the properties of the ether. Cotton ended his paper with the statement:

> The law which connects together so many experimental facts brings an important contribution to the theoretical study of the relationship between ether and matter, which is still so mysterious.

---

[131] Wien, W., *Temperatur und Entropie der Strahlung*, Ann. Phys. **288**, 132 (1894).

[132] Wien was very nationalistic, and a defender of German science, so for him to raise such a claim was no a trivial matter.

[133] Wien, W., Encyklopadie del mathematische Wissenschaften **5**, 282 (1909), Leipzig.

[134] Paschen, F., Wied. Ann. **51**, 41 (1894).

[135] A spectrometer which splits the light, combined with a bolometer which measures the power emitted in a certain wavelength range.

[136] Wiedemann, G.H., Wied. Ann. **37**, 180 (1893). In 1877, Wiedemann became the editor of the Annalen der Physik und Chemie, succeeding Johann Christian Poggendorff. Consequently, the journal is frequently cited as Wiedemann Annalen.

[137] Cotton, A., Astrophys. J. **9**, 237 (1899).

Similarly, Ernst Pringsheim (1859–1917)[138] requested further proof in 1903, but neither offered nor performed any new experiments.

2.22 A Mathematical Proof (if Needed): Hilbert

At the beginning of the twentieth century, and in particular after the death of Henri Poincaré (1854–1912m), David Hilbert (1862–1943m) became the leading mathematician in mathematical physics.[139] Hilbert was interested in establishing the whole of physics on axiomatic grounds, and the problem of Kirchhoff's law was no exception to this.

The delicate question of the axiomatic approach versus a genuine physical approach was the subject of a debate between Hilbert who, as a mathematician, favoured the axiomatic approach, and a group of physicists, headed by Pringsheim (and including Kayser and Wien) who, as physicists favored the physical approach. But this debate, which is very important for physics, had no effect on the application of Kirchhoff's law either to the transfer equation or to abundance determinations of chemical elements in stars. For this reason, we shall leave the discussion at this point.[140]

In 1912, Hilbert[141] proved Kirchhoff's law using an axiomatic approach to radiative transfer. But Kirchhoff's law had become a standard assumption in radiative transfer well before Hilbert's proof, with almost no questions raised about its applicability or validity. Physicists, even theoreticians like Karl Schwarzschild, took it for granted. The foundations of radiative transfer were secured at long last.

2.23 The French View

In 1860, Desains published his book *Leçon de Physique*,[142] and in Chap. VIII discussed *Chaleur Rayonnante*. It is interesting to see the French version of the above events. According to Desains, the experiments on the transmission of radiant heat were carried out by François Delaroche (1781–1813) in 1811.[143] Pictet is not

---

[138] Pringsheim, E., *Herleitung des Kirchhoffschen Geset zes*, Zeit. f. Wiss. Photographie **1**, 360 (1903).

[139] The two volume book by Courant, R. and Hilbert, D., *Methods of Mathematical Physics*, first published in 1924 and updated in 1953, Interscience Pub., served for many years as the 'bible' for mathematical physicists.

[140] For more details on this interesting and important debate, see Schirrmacher, A., *Experimenting theory*: *The proofs of Kirchhoff's radiation law before and after Planck*, Historical Studies in the Physical and Biological Sciences **33**, 299 (2003).

[141] Hilbert, D., Phys. Zeit. **13**, 1056, 1912; ibid. 1885 **14**, 592 (1913).

[142] Desains, P.Q., *Leçon de Physique*, Dezobry, E., Magdelenine et Cie., Lib-Éditeurs, Paris, 1860.

[143] Delaroche, F., J. Phys. **25**, 201 (1812); ibid. Ann. Phil. Lond. **2**, 100 (1813).

mentioned. Desains emphasised all the results of de la Provostaye and himself which confirmed the equality of the absorption and emission. The heroes were Melloni and de la Provostaye and Desains. Melloni studied in France, but returned to his home country, Italy, after completing his studies.

Desains attributed the discoveries in radiant heat to de la Provostaye and his collaborator.[144] However, the reader will easily recognize that the formulation is Kirchhoff's, although Desains wrote that he repeated the proof by de la Provostaye. The omission of Kirchhoff may be just about justified by the fact that Kirchhoff discussed light, while this book discussed heat. But what about Stewart? After all, the controversy between Stewart and Kirchhoff was run out on the pages of the best scientific journals of the time. And these are not the only inaccuracies in the book. For example, according to Desains, Kirchhoff repeated Foucault's experiment. But Kirchhoff never used a voltaic arc.

2.24 One Can Never Foresee the Future

The discoveries by Fraunhofer did not impress the French philosopher August Comte (1798–1857), who claimed in 1835[145] that it would never be possible to discover the composition of the stars. This 'prophecy' was repeated in the second edition of the course of lectures, which appeared in 1864, after the publication of Bunsen and Kirchhoff's discoveries and Kirchhoff's paper about the solar composition.

2.25 Getting the Black Body Function: Maxwell

The equations of electromagnetism were discovered by various authors and over many years. These appeared as separate equations for the magnetic and electric fields, and for the interaction between them. It was the Scotsman James Clerk Maxwell (1831–1879m) who unified the electric and magnetic fields into the electromagnetic field and formulated modern electrodynamics. Maxwell showed that light is a propagating electromagnetic field composed of inseparable and varying electric and magnetic fields. It seems that Faraday had entertained this idea before Maxwell expressed it in writing, but it was Maxwell who found the theoretical expression and calculated the resulting speed of propagation. The theory of light as an electromagnetic field provided the basis for the forthcoming discovery by Planck. Assuming that light propagated in the 'luminiferous aether', Maxwell derived its properties about 40 years before Einstein buried the idea of the hypothetical aether, which had been assumed to permeate all matter.

---

144 de la Provostaye, *Considérations théoriques sur la chaleur rayonnante*, Ann. Chim. Phys., série 3, **67**. This is the reference given by Desains. The full reference is: p. 5, 1863.

145 Comte, A., *Cours de Philosophie Positive* **2**, 6, Pub. Baillière, Paris, 1864.

It is interesting to note that Maxwell wrote his equations in a form that was consistent with the special theory of relativity, long before Einstein discovered his theory in 1905. Einstein's special theory of relativity did not change anything in Maxwell's theory.

The basic Maxwell equations are:

- An equation for the connection between the electric charge and the electric field it produces.
- An equation for the connection between the magnetic charge and the magnetic field.
- An equation for Faraday's law of induction, which describes how a changing magnetic field can create an electric field.[146]
- An equation for Ampère's law[147] (with Maxwell's correction) which states that magnetic fields can be generated in two ways: by electrical current (this was the original Ampère law) and by changing electric fields. This is the inverse equation of the previous one. However, Maxwell realized that the equation had to be corrected.

Maxwell's theoretical correction to Ampère's law is described in *A Dynamical Theory of the Electromagnetic Field*, where he commented that[148]:

> The agreement of the results seems to show that light and magnetism are affections of the same substance, and that light is an electromagnetic disturbance propagated through the field according to electromagnetic laws.

Maxwell's equations can be found in almost present day notation in Vol. 2 of Maxwell's book.[149]

2.26 Attempts to Obtain Kirchhoff's Function

Towards the end of the eighteenth century several physicists attempted to find the complete behavior of the universal function $e(\nu, T)$ of black bodies. During the years 1897–1899, Otto Lummer (1860–1925) and Pringsheim[150] accurately measured the energy distribution as a function of wavelength emitted by a black body. The first realization of a black body, and the once and for all disposal of the lampblack, was carried out by Lummer and Wien,[151] using Kirchhoff's idea that a cavity is a good

146 The equations for the electric field $\mathbf{E}$ and magnetic field $\mathbf{B}$ are $\nabla \cdot \mathbf{E} = \rho/\varepsilon_0$, $\nabla \cdot \mathbf{B} = 0$, $\nabla \times \mathbf{E} = -\partial \mathbf{B}/\partial t$, $\nabla \times \mathbf{B} = \mu_0 \mathbf{J} + \mu_0 \varepsilon_0 \partial \mathbf{E}/\partial t$, where ε_0 and μ_0 are the permittivity of free space and the permeability of free space, respectively, and ρ is the total charge density.

147 For historical justice, we mention, Ampère, Weber, and Robert Thomson (1822–1873).

148 Maxwell, J.C., *A Dynamical Theory of the Electromagnetic Field*, Phil. Tran. Lond. **155**, 459 (1865).

149 Maxwell, J.C., *A Treatise on Electricity and Magnetism*, MacMillan, 1873.

150 Lummer, O.R., and Pringsheim, E., Wied. Ann. **63**, 395 (1897).

151 Lummer, O.R. and Wien, W., Wied. Ann. **56**, 451 (1895).

approximation for a black body. Kirchhoff slightly confused the problem when he wrote:

> In an enclosure or a cavity which is enclosed on all sides by reflecting walls, externally protected from exchanging heat with its surroundings, and evacuated, the condition of 'black radiation' is automatically set up if all emitting and absorbing bodies at the walls or in the enclosure are at the same temperature.[152]

2.27 Getting the Integral of an Unknown Function First

The first attempts to find the radiative cooling law were by Dulong and Petit in 1817.[153] They heated a thermometer in a vacuum and watched how it cooled (see Sect. 1.14). The problem was that the air pressure in the vacuum was 2–3 mm of mercury, and this is not sufficient to suppress heat losses by air conduction relative to radiation losses. Dulong and Petit found that they could approximate their results with a cooling law of the form $E = m \times a^T$, where T is the temperature and m and $a = 1.0077$ are constants.

It was Draper who, in 1847,[154] discovered that, as the temperature increases, the emission takes place at shorter and shorter wavelengths. This observation was confirmed by Desains in his book.

The experiment was repeated with some improvements by Provostaye and Desains,[155] who managed to reduce the air pressure until no changes in the cooling were measured, and in this way obtained the 'cooling in vacuum'. Provostaye and Desains claimed that the cooling, as given by Dulong and Petit, should be corrected by adding a constant which depended on the radiation. Moreover, they remarked that the numerical coefficient in the cooling law changed with the dominant wavelength of the cooling body. However, the variations were not given.

Alexander-Edmond Becquerel (1820–1891)[156] first confirmed Dulong and Petit's result, and then discovered that the emission at a certain given wavelength increased with temperature. Becquerel found that the constant a in Dulong and Petit's law varied with color as shown in Table 2.1.

---

[152] If matter is not completely evacuated from the cavity, the squared index of refraction of the matter enters the law. Here we leave this point aside and assume unity for the index of refraction (vacuum). If the walls are perfect reflectors, the radiation does not interact with the walls. If the radiation is to 'feel' the temperature of the walls, they must absorb at least part of the radiation (and then re-emit it). As for stars, the 'enclosure' is not empty, but full of matter and there are no walls to speak of.

[153] Dulong, P.L. and Petit, A.T., Ann. Chim. Phys. **vii**, 225, 237 (1817).

[154] Draper, J.W., Phil. Mag. **XXX**, 345 (1847). This is basically the displacement law formulated by Wien in 1894.

[155] Desains, P., *Leçons de Physique*, Tome second, Dezobry, E. Magdeleine et Co., Paris, 1865, pp. 704–705.

[156] Becquerel, A.E., J. Phys. **VII**, November, 1878. *La Lumière*, Vol. I, 61. His father was Antoine César Becquerel (1788–1878), a pioneer in the study of electric and luminescent phenomena, and his son was Henri Becquerel (1852–1908), the discoverer of radioactivity.

Table 2.1 Becquerel's radiation law

a	Color	Wavelength
1.01180	Red	6,700 Å
1.01371	Green	5,260 Å
1.01660	Blue	4,600 Å
1.00770	Total	Dulong and Petit

In 1878, André Crova (1833–1907)[157] tried to use the published formulae to obtain an expression which included the wavelength and the temperature. However, it was just an interpolation between known results. The matter and the radiation were treated as ensembles of oscillators, the simplest physical system. Each oscillator has a different frequency or wavelength. The cooling takes place through each oscillator radiating away its energy. The question was, in Crova's words, how much would each oscillator of wavelength λ radiate at a given temperature? Oscillators with different wavelengths radiate different amounts of energy.

The next known attempt was by Francesco Rossetti (1833–1885),[158] who repeated Leslie's cube experiment. He essentially found that the rate of cooling goes as $aT^2(T-\theta)+b(T-\theta)$, where T is the absolute temperature of the body, θ that of the surroundings (lampblack enclosure), and a and b are two constants for the body.

Only in 1879 did Joseph Stefan (1835–1893m)[159] correct the Dulong and Petit result for the disturbing convection and found that $E=\sigma T^4$ provides a better approximation. Dulong and Petit's results extended between the temperatures of 273 and 573 K, which is a relatively small range. To check how the approximation works at higher temperatures, Stefan took the results of Tyndall[160] for platinum at two temperatures, viz., 273 + 525 K and 273 + 1,200 K. The observed ratio of emissions was 10.4, while the ratio of the fourth powers of the temperature was 11.6. Though 10.4 is not equal to 11.6, Stefan considered the agreement satisfactory. But now contemplate the irony of fate: the fact that platinum is a poor black body at this temperature combined with the errors in Tyndall's experimental results (the emission ratio is 18.6 and not 11.6) played in Stefan's favour: all the errors compensated each other to produce the right result! Stefan did not evaluate the constant σ and his result was not accepted because of the crude way it was obtained, until it was proven theoretically about 5 years later by Boltzmann.[161]

[157] Crova, A., J. Phys. **7**, 357 (1878).

[158] Rossetti, F., Phil. Mag. **viii** (1879).

[159] Stefan, J., Math. Naturw. Akad. Wiss., Wien, Classe Abteilung 2 **79**, 391 (1879).

[160] Tyndall, J., *Heat Considered as a Mode of Motion*, Longman, Green, Longman, Roberts and Green, London, 1865. The book contains the experimental results.

[161] Boltzmann was a PhD student of Stefan and got his PhD in 1866, long before Stefan found his law empirically.

The first theoretical attempt to prove the law was by Adolfo Bartoli (1851–1896), who, in 1875, published a monogram and a short article[162] in which he used pure thermodynamics reasoning and analysis of Crookes' experiments to reach the fundamental conclusion that the radiation in an enclosure operates like a Carnot thermodynamic machine and that light must exert pressure in the direction of propagation. Bartoli's result was overlooked by the community. In contrast, the astronomer Henry Eddy (1844–1921) devised a thought experiment from which he concluded that radiant heat might not obey the second law of thermodynamics.[163] Moreover, Eddy claimed that the prediction of the thermal death of the Universe was wrong.

By pure chance,[164] Bartoli's results became known to Ludwig Boltzmann (1844–1906m).[165] Boltzmann showed, using Maxwell's kinetic theory of gases, that if Stefan's law is valid and the radiative energy losses $\phi(T)$ go as $\phi(T) \approx T^{\alpha}$, then the radiation pressure in thermodynamics goes as $p(T) = (\pi/3c)\phi(T)$. The radiation obeys Maxwell's relation between the energy density and pressure in thermodynamic equilibrium, namely $p = u/3$, where p is the radiation pressure and u the energy density of the radiation. Thus, while the rate of cooling at every wavelength was not yet known, the functional dependence of the total losses (sum over all wavelengths) was already known, and given by what is known today as the Stefan–Boltzmann law. In other words, the integral of Kirchhoff's function $e(\nu, T)$ over all frequencies was known without knowing the shape of the function.

Once the Stefan–Boltzmann law had been established, Stefan could calculate the surface temperature of the Sun. The amount of energy received at the Earth (1,366 W/m^2) was measured by Samuel Langley (1834–1906) in 1884. Langley's result was 2,903 W/m^2, or about a factor of two higher than the present day value. Langley must have erred in the data reduction, since his assistant, Charles Abbot (1872–1973m), who used Langley's original data, found a value of 1,465 W/m^2. The distance to the Sun was known, so it was a simple matter to calculate its emissivity. Use of the new law gave a temperature range of 5,600–11,000K, depending on the exact value of σ, while previous estimates had been significantly lower. The details of the solar spectra were first obtained by Langley (1834–1906m) in 1886.[166] Langley also invented the bolometer in 1878, and this allowed measurement of the total energy arriving at the Earth from stars.[167]

---

[162] Bartoli, A., *Il calorico raggiante e il secondo principio di termodynamica*, Nuovo Cimento **15** 196 (1876); ibid. *Sopm i movimenti prodotti dalla luce e dal calore e sopra il radiometro di Crookes*, Firenze, 1876.

[163] Eddy, H.T., J. Franklin Inst. **115**, 182 (1883).

[164] Boltzmann wrote that the editor Wiedemann drew his attention to Bartoli's paper, which was then translated from Italian.

[165] Boltzmann, L., Wied. Ann. **22**, 31 (1884); ibid. 291 (1884).

[166] Langley, S.P., Ann. Chim. Phys. **9**, 433 (1886). Previous attempts had been made by Müller (Müller, J.H., Ann. Phys. **11**, 337, 1858) and Lamanky (Lamanky, S.I., Ann. der Phys. **146**, 200, 1872).

[167] After Langley's death, Abbot continued to measure the solar constant and searched for places with clear sky. Abbot discovered that the southern mountainous region of the Sinai Peninsula enjoyed excellent weather conditions all year round and established an observatory on Mount St.

In 1894, Wien[168] carried out the following thought experiment. Consider a cavity with a piston. Let the cavity be closed so that no radiation can escape from it. As the piston moves slowly inward, the radiation reflected by the slowly moving piston undergoes a small Doppler shift.[169] Calculating the effect on the energy of the enclosed electromagnetic waves with frequency ν, Wien succeeded in showing that the universal function $e(\nu, T)$ must have the form

$$e(\nu, T) = \frac{\nu^3}{c^2} G\left(\frac{\nu}{T}\right),$$

where c is the speed of light in vacuum. Regarding the function G, Wien could only state that it must be a function of ν/T. Obviously, the function G could not increase forever, so at higher and higher frequencies (at a given T), the body must radiate more and more, and consequently must reach a maximum, or else the total emitted radiation would be infinite. Hence, there had to be a value of ν/T for which the maximum emission (per unit frequency) was reached. If the maximum is reached for $\nu_{\max}/T$, then

$$\nu_{\max} = c_1 T,$$

where c_1 is a constant. So although Wien did not know the exact form of the function G, he could predict that it reached a maximum value, the location of which could be measured. This simple law, which quantifies Draper's observations in 1847, became known as Wien's displacement law.

Next, assuming the validity of Maxwell's law for the distribution of velocities among the molecules of gas,[170] Wien succeeded in showing that

$$G\left(\frac{\nu}{T}\right) = \alpha\, \mathrm{e}^{-\beta\nu/T},$$

where α and β are constants. So when Wien combined the two results, he found

$$e(\nu, T) = \alpha \left(\frac{\nu^3}{c^2}\right) \mathrm{e}^{-\beta\nu/T},$$

Katherine, only a few kilometers from the famous Santa Katherina monastery. One can still see the stairs with the sign 'to the observatory'. The mountain on which, according to tradition, the ten commandments were announced provides an excellent place for measuring the solar constant! The observatory, established in 1931, operated for about 6 years and was then shut down. The reason for the consistent effort to measure the solar constant accurately was the belief by Langley and Abbot that the energy flux from the Sun dictates the weather on the Earth, so that an exact knowledge of this quantity would be a prerequisite for Earthly weather predictions. Langley suspected the solar radiation of varying periodically.

[168] Wien, W., Wied. Ann. **52**, 132 (1894).

[169] The Doppler effect, namely the shift in wavelength due to the relative velocity between the source and the observer, was discovered by Doppler in 1842.

[170] According to Maxwell, the energy distribution of the molecules in a gas is given by an exponential law, namely, $\mathrm{e}^{-E/kT}$, where k is a constant.

which is Wien's famous radiation law.

Very soon afterwards, these results of Wien and Stefan–Boltzmann were confirmed by the experiments of Lummer and Pringsheim,[171] but only for high frequencies or short wavelengths. A systematic deviation from Wien's law was discovered at small frequencies (long wavelengths).[172] The mere existence of deviations between theory and observation was controversial. Paschen[173] insisted that he had succeeded in proving the universality of Wien's radiation law. Planck[174] on the other hand, rederived Wien's law from the theory of irreversible processes and provided *a more rigorous proof*. Planck's basic idea was to consider the radiation field as a collection of harmonic oscillators[175] exchanging energy with the walls of a cavity. The idea of presenting the radiation field as a collection of oscillators was crucial.

Planck's proof went in two steps. In the first, he showed that the unknown function was given by

$$e(\nu, T) = \frac{\nu^2}{c^2}\overline{U},$$

where $\overline{U}$ is the mean energy of an oscillator. In the next step, he proved that $\overline{U}$ was given by

$$\overline{U} = \alpha\nu\, \mathrm{e}^{-\beta\nu/T},$$

and in this way he proved Wien's law. Planck was pleased with the result.

While Wien's radiation formula was the only one derived on the basis of classical physics, several phenomenological formulae were proposed as well. One of the better known amongst these was invented in 1888 by Weber,[176] who found a simple interpolation formula which described a wide range of experiments. The formula was

$$\frac{c}{\lambda^2}\exp\left(aT - \frac{1}{b^2\lambda^2T^2}\right).$$

Mysteriously at that time, this empirical formula described some of the available data better than the theoretically derived Wien formula.

---

[171] Lummer, O.R. and Pringsheim, E., Wied. Ann. **63**, 395 (1897).

[172] Lummer, O.R., and Pringsheim, E., Verh. Dtsch. Phys. Ges. **1**, 215 (1899).

[173] Paschen, F., Berliner Ber. 405 and 959 (1899).

[174] Planck, M., Wied. Ann. **57**, 1 (1896). There exists a report on the same subject a year earlier: Sitzunsber. Berliner Akad. Wiss., 21 March, 289 (1895).

[175] A harmonic oscillator is any physical system that behaves like a spring attached to a mass or a pendulum, i.e., a system which, when displaced slightly from its equilibrium position where it would remain without motion, feels a restoring force proportional to its displacement from equilibrium. This is the simplest mechanical system one can conceive of, and many physical systems behave this way upon sufficiently small perturbation from their steady state. Hence, it was natural to assume the simplest possible system for the emitters and absorbers of radiation.

[176] Weber, H.S., Phys. Rev. **2**, 112 (1894), Sitzungsberichte der Akad. Wiss. Berlin 933 (1888).

2.28 A Black Body Does Not Necessarily Look Black

The term 'black body' may wrongly imply that it looks black, since it absorbs all incident light. However, a black body is also a perfect emitter, and its color changes with temperature according to the Wien displacement law. As the temperature reaches about 500 K the body starts to glow with a red color. This was what Draper had already found in 1847. As the temperature of the black body continues to rise, the color changes gradually, passing through all the colors in the spectrum, until at very high temperatures all such bodies shine with a blue–white color. Stars behave to a good approximation like black bodies, and the color they exhibit is an indication of their surface temperature. Paradoxically, one may say that there are black bodies which appear red (when they are at the relevant temperature) and there are yellow black bodies (our Sun for example!), and even blue black bodies (stars with surface temperatures above 15,000 K).

2.29 Buried in the Finer Details

But the experimentalists Lummer and Pringsheim were not happy with the result of the theoretician Planck. In 1900, they returned to experiment[177] and showed convincingly that Wien's law deviates more and more from observation at short wavelengths. It was only then that Planck went back to the problem[178] and came up with his seminal paper, in which the energy of the oscillators was assumed to be quantized. It is extremely instructive to see how this physics-shaking idea was born.

Planck proceeded as follows. He imagined that the walls were composed of small oscillators which radiate. The energy of each oscillator was divided into a discrete number of 'energy quanta' of magnitude ε and it was supposed that these energy quanta were distributed randomly among the individual oscillators. We are not concerned here with which particular energy lies in which particular oscillator, but with the number of energy quanta that each oscillator has.[179] Then Planck calculated the number of ways that each particular distribution of energy quanta could be realized (following the standard procedure in statistical mechanics to find the distribution). From this point on it only required the standard mathematics of permutations to obtain $\overline{U}$. Using Boltzmann's relation between the entropy and the probability for the realization of a state, Planck obtained

$$\overline{U} = \frac{\varepsilon}{e^{\varepsilon/kT} - 1},$$

177 Lummer, O.R. and Pringsheim, E., Verh. Dtsch. Phys. Ges. **x**, 163 (1900).

178 Planck, M., Verh. Dtsch. Phys. Ges. **x**, 202, 237 (1900). The derivation of Wien's law already appeared in the first report, but was not accompanied by explanations.

179 Ehrenfest, P. and Kamerlingh-Onnes, H., Ann. Phys. **46**, 1021 (1915).

where ε is the energy of the oscillator. But from Wien's displacement law and the expression for $e(\lambda, T)$, it follows that

$$\overline{U} = \nu G\left(\frac{\nu}{T}\right).$$

How could Planck provide a bridge between the two results? By inspection, they can be made to agree with one another if one assumes that ε is proportional to ν. In other words, to get agreement with Wien's law, Planck had to assume that

the energy of an oscillator with frequency ν is $\varepsilon = nh\nu$,

where h is known today as the Planck constant[180] and n is a positive integer. In other words, an oscillator with natural frequency ν can only have an energy that is a whole number multiple of $h\nu$. The natural frequency of an oscillator does not depend on temperature, but the number n does. The complete formula then becomes

$$e(\nu, T) = \frac{h\nu^3}{c^2}\frac{1}{e^{h\nu/kT} - 1},$$

and this is the function Planck derived in 1900, which started the quantum revolution.

The result was soon confirmed experimentally by Heinrich Rubens (1865–1922) and Ferdinand Kurlbaum (1857–1927)[181] for long wavelengths and by Paschen[182] for short wavelengths. Further confirmations were carried out by Warburg and his associates.[183]

But even success left skeptics. Walther Nernst (1864–1941m) and Theodor Wulf (1868–1946)[184] insisted that the accuracy of the experiments (estimated at 7%) was not sufficient to convince them of the correctness of the new revolutionary theory. As a matter of fact, many, including Planck himself and Einstein, had difficulty swallowing the new theory with its unimaginably far-reaching implications.

Attempts by Max Thiesen (1849–1936)[185] and Eugene Jahnke (1863–1921)[186] to improve agreement with experimental results by playing around with Wien's formula failed.

---

180 Planck, M., Ann. Phys. **4**, 553 (1901) found that $h = 6.548 \times 10^{-27}$ erg s.

181 Rubens, H. and Kurlbaum, F., Sitzunsber. Berliner Akad. Wiss. 929 (1900), Ann. Phys. **4**, 649 (1901).

182 Paschen, F., Ann. Phys. **4**, 277 (1901). Paschen had by now changed his mind and accepted the new result.

183 Warburg, E., with 3 coauthors, Ann. Phys. **40**, 609 (1913); Warburg, E. and Müller, C., Ann. Phys. **48**, 410 (1915).

184 Nernst, W. and Wulf, T., Berliner Dtsch. Phys. Ges. **21**, 294 (1919).

185 Thiesen, M., Verh. Dtsch. Phys. Ges. **2**, 65 (1900).

186 Jahnke, E., Ann. Phys. **3**, 283 (1900).

2.30 Rayleigh: Give Classical Physics a Chance

It was only after Planck got his incredible result that Rayleigh[187] set out to discover what traditional physics had to say about the problem. Rayleigh therefore calculated the energy distribution in a cavity, assuming Maxwell's classical theory. The crucial assumption Rayleigh made was that the energy of the small oscillators in the walls, those oscillators which radiate the energy, varied as kT, where k is a constant.[188] In other words, the energy varies continuously and is proportional to the temperature, as classical physics dictates. Rayleigh's result for $e(\nu, T)$ was

$$e_{\text{class}}(\nu, T) = \frac{\nu^2}{c^2} kT.$$

A comparison with Planck's formula shows that this result corresponds to the limit of long wavelengths, i.e.,

$$e_{\text{class}}(\nu, T) \quad (\text{for } kT \gg h\nu) = \lim_{kT \gg h\nu} \left(\frac{h\nu^3}{c^2} \frac{1}{e^{h\nu/kT} - 1} \right) = \frac{\nu^2}{c^2} kT.$$

Note that the new constant h, which was introduced by Planck, disappears from the Planck formula in this limit. So there was no sign of a new constant in what is known as the classical limit. Thus, Planck's formula combines nicely with Wien's law on the one hand and Rayleigh's result on the other. However, it would be a grave mistake to consider Planck's formula as an interpolation between two formulae. The actual situation is quite the other way round: Wien's and Rayleigh's results should be considered as approximations to the full, true result given by Planck's law.

2.31 Jeans: No Way of Saving the Classical Picture

Planck's idea was so radical that it was too much even for physicists to swallow. So many attempts were carried out to escape from Rayleigh's result, which was based on the pure classical theory of electricity and statistical mechanics. The most famous of these attempts was the one by Jeans,[189] carried out five years after Planck's publication. Jeans dropped the idea of radiating oscillators in the wall and assumed that the radiation energy inside the cavity was distributed among the 'degrees of

[187] Rayleigh, Lord, Phil. Mag. **49**, 539 (1900).

[188] According to Maxwell's classical theory of a gas of molecules, or any collection of particles or systems, the distribution of energy of the molecules is exponential ($e^{-E/kT}$) and the mean energy is $3/2kT$. This may sound complicated, but it can be shown that, if we bring two systems together and the energy of the new bigger system is the sum of the energies of the individual systems, then this law must follow.

[189] Jeans, J.H., Phil. Mag. **10**, 91 (1905).

freedom of the radiation'. The radiation was considered as an infinite collection of waves.[190] The degrees of freedom were simply the amplitudes of the waves.[191] In other words, while Rayleigh considered the radiators in the wall, Jeans considered the radiation in the cavity. However, in both cases one had oscillators: in the first case material oscillators and in the second radiative oscillators. But the classical treatment of the oscillator's energy was the same in Rayleigh's and in Jeans' calculations. The fundamental difference was that, in classical physics, the energy of a wave depends on the amplitude squared, while in quantum physics the energy depends on the frequency of the oscillator, whatever it is (material or radiation), and as described in Sect. 2.29, only insertion of the frequency dependence by Planck resulted in an expression that agreed with observation.

2.32 Classical Physics is Ruled Out for Microscopic Phenomena

But to no avail, Jeans recovered the Rayleigh result once again. Less well known attempts were carried out by Hendrik Lorentz (1853–1928),[192] Albert Einstein (1879–1955) and Ludwig Hopf (1884–1939),[193] Adriaan Fokker (1887–1972),[194] and even Planck.[195] Physicists were reluctant to accept such a revolutionary step. All attempts to evade the new hypothesis led to the same Rayleigh result. Note that Einstein and Hopf's paper was published five years after Einstein published his explanation for the photoelectric effect in 1905,[196] which required as a prerequisite Planck's idea of discrete energies. Although Einstein implemented the idea, he was unhappy with it.

---

190 Strictly speaking, the wavelengths were discrete and of the form $L/2n$, where L was the size of the cavity and n an integer. But these fine details make no difference for the final result and were added here for physical accuracy only.

191 A degree of freedom of a physical system is a parameter needed to determine the state of the system uniquely. In the case of a monochromatic wave, there is one degree of freedom, the amplitude, and for a general wave there can be an infinite number of degrees of freedom.

192 Lorentz, H.A., Proc. Kon. Akad. v. Wet., Amsterdam 666 (1903). See also *The Theory of Electrons*, Teubner, Leipzig, 1909, Chap. 2.

193 Einstein, A. and Hopf, L., Ann. Phys. **33**, 1105 (1910).

194 Fokker, A.D., Ann. Phys. **43**, 810 (1914).

195 Planck, M., Ber. d. Berliner Akad. Wiss. 8 July (1915).

196 The irony is that much of the experimental work on the photoelectric effect on which Einstein based his theory was discovered by Lenard (1862–1947). Lenard, who got the Nobel prize in 1905, was an adamant promoter of the 'Deutsche Physik' idea and a declared anti-semite. As such, he did not believe in the 'Jewish physics' as reflected in the special theory of relativity. For him it was a bitter pill to watch Einstein being awarded the Nobel prize in 1921 for explaining the data obtained by an Aryan. Add to this his annoyance over the fact that the effect was named after Einstein and himself, since he did the experiment, and you will appreciate the rumor that he was even ready to declare his results wrong.

The final death blow to the idea that there might be an alternative assumption that could salvage classical physics came in the 1911 Solvay congress, when Lorentz[197] provided a general proof that the classical concepts of energy distribution among particles, or degrees of freedom as Jeans would have it, lead unavoidably to the Rayleigh result. There was no way to escape from Planck's dramatic assumption, if one wanted to obtain the Planck distribution.

2.33 The Photoelectric Effect: A Further Success for the Quantum Theory

A dramatic confirmation of Planck's result came when Einstein[198] attacked the problem of the photoelectric effect.[199] He realized that, in order to explain the effect, he had to assume that *radiation, when propagated through a vacuum or any medium, possesses a quantum-like structure*. Einstein retreated from the wave description of light back into the massless particle picture. But like Planck, Einstein did not reach the idea of quanta in a straightforward way. Consider a volume v_0 filled with molecules. The probability that *n* molecules will be found at a certain moment in time in volume $v < v_0$ is given by

$$w = \left(\frac{v}{v_0}\right)^n.$$

Einstein found that, if the radiation satisfies Wien's radiation law, then the probability that the entire radiation converge into a volume $v < v_0$, leaving no radiation in the volume $v - v_0$, is given by

$$w = \left(\frac{v}{v_0}\right)^{E/h\nu}.$$

A comparison with the previous classical result shows that the results agree if

$$n = \frac{E}{h\nu},$$

where *n* is an integer, since it is the number of particles. But this is exactly what Planck discovered.

---

[197] Lorentz, H.A., *Die Theorie der Strahlung und der Quanten*, Abhandlungen d. Deut. Bunsen-Ges. no. 7, 10 (Eucken, 1914). This is a summary of the first Solvay meeting which took place from 30 October to 3 November 1911.

[198] Einstein, A., Ann. Phys. **17**, 132 (1905).

[199] In the photoelectric effect, a metal is illuminated by monochromatic light. Only when the frequency of the light is above a certain value are electrons emitted from the metal. Each metal has a different threshold frequency. A high intensity of light, but at a frequency below the threshold, does not release electrons.

2.34 Are the Positions of the Fraunhofer Lines Random?

Spectral lines seemed to appear in quite chaotic positions. Lockyer, for example, thought that for some mysterious reason every element had its own lines, and that they would not change whatever happened, like human fingerprints. Moreover, it was not clear at all why two different elements could not have the same set of spectral lines. However, there was no attempt to explain the positions of the lines.

So the obvious thing people tried to do was to find some systematics. The first attempt was made by Paul Lecoq de Boisbaudran (1838–1912)[200] for the case of nitrogen lines. But his conclusions, which were based on insufficiently accurate line positions, were not confirmed by others.[201] Stoney[202] was the first to observe some regularity. He found that the wavelengths of the hydrogen lines, denoted by H_α, H_β, and H_γ, were related to each other in the ratios $1/20{:}1/27{:}1/32$, to very high accuracy. It was not clear what these 'nice' fractions could mean.

Well-known investigators like Osborne Reynolds (1842–1912),[203] Charles Soret (1854–1904),[204] and others tried various combinations of formulae, until Schuster[205] showed in 1881 that, even if there were absolutely no connection between the positions of the lines, the chances were in favor of finding a harmonic relationship if the spectrum was rich in lines. Consequently, the interest in finding a connection between the positions of the lines faded away. How statistics can be misleading!

In 1885, Johann Balmer (1825–1898m)[206] made a travesty of statistics when he discovered a formula that described the then known spectral lines of hydrogen to very high accuracy. The formula was

$$\lambda = A\frac{m^2}{m^2 - 2^2}, \quad \text{or} \quad \nu = \frac{4}{A}\left(\frac{1}{2^2} - \frac{1}{m^2}\right) \quad \text{for } m = 3, 4, \ldots,$$

where A is a constant and m an integer greater than 2. Balmer had the problem that the wavelengths of the hydrogen lines were not very accurately known, so he had to compare the results obtained by different experiments and use the average values of the latter for comparison between the formula and observation. Yet the errors in the

[200] Lecoq de Boisbaudran, P.E., Compt. Rend. **69**, 610, 694 (1869).

[201] Thalen, T.R., Svenska. Vetensk. Akad. Handl. **8**, 1 (1869). See also, Landauer, J., *Spectrum Analysis*, Wiley, 1898, Chap. I.

[202] Stoney, G.J., Phil. Mag. **41**, 291 (1871).

[203] Reynolds, O., Phil. Mag. **42**, 41 (1871).

[204] Soret, C., Phil. Mag. **42**, 464 (1871).

[205] Schuster, F.A., Proc. R. Soc. Lond. **31**, 337 (1881).

[206] Balmer, J.J., Verh. Naturforsch. Ges. Basel **7**, 548 (1885); ibid, 750. The name 'Balmer series' was given to this series because Balmer discovered the phenomenological law that yields the frequencies of the lines, and not because he discovered the lines themselves as was the case for the names of the other hydrogen series.

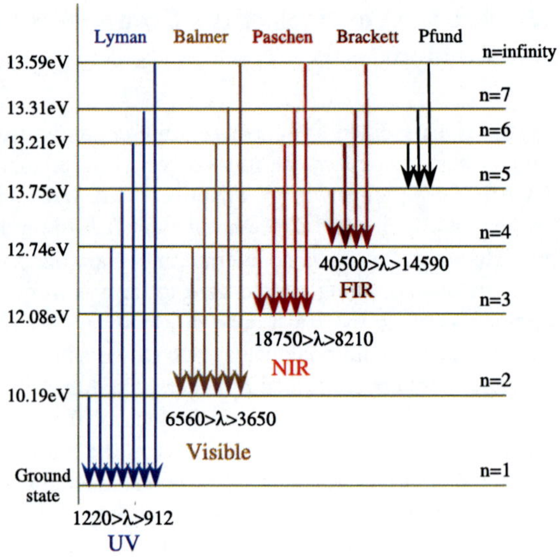

Fig. 2.15 The hydrogen series. Note the ranges in which the series are observed. The energies of the levels are not to scale. There are no energy levels between 0 and 10 eV and all levels lie between 10 and 13.56 eV. There is an infinite number of levels crowding together towards 13.56 eV

fit were about 3 Å, which were considered to be very small in those days. The known lines at the time corresponded to $m = 3$–11. Balmer wrote the equation in the first form (the one on the left), but to display the similarity with Rydberg's formula (see below), it has been rewritten here in the second form. The spectral lines represented by this formula are known today as the Balmer series. Once Balmer had published his formula, Marie Cornu (1841–1902)[207] discovered a similar relation for aluminum and thallium, and later Henri Deslandres (1853–1948m)[208] found additional elements that satisfied this rule. After the success with hydrogen, Balmer attempted to find a formula for the helium lines,[209] but luck did not strike twice (Fig. 2.15).

In 1887, Heinrich Kayser (1853–1940)[210] and Carl Runge (1856–1927m)[211] started their investigation and managed to find various formulae which 'reproduced' the positions of the lines. First, Runge found that

$$\nu = A + \frac{B}{m} + \frac{C}{m^2}, \qquad m = 3, 4, \ldots,$$

provides good agreement with the spectrum of lithium. Then Runge and Kayser found that the following formula provides excellent agreement with the observed positions of the lines of the alkali metals (lithium, sodium, potassium, rubidium, and cesium):

---

207 Cornu, M. A., Compt. Rend. **100**, 1181 (1885).

208 Deslandres, H.A., Compt. Rend. **103**, 375 (1886); ibid. **104**, 972 (1887).

209 Balmer, J.J., Astrophys. J. **5**, 199 (1897).

210 The physical unit of wave number was formerly called the kayser.

211 Kayser, H. and Runge, C., Ann. Phys. **41**, 302 (1890).

$$\nu = A + \frac{B}{m^2} + \frac{C}{m^4}, \qquad m = 3, 4, \ldots.$$

Janne Rydberg (1845–1919m) attempted to find an empirical expression for the lines of the alkali metals. These were the days after Mendeleev's discovery of the periodic table in which the alkali metals occupied a well defined group. The spectra of the alkali elements were relatively simple when compared for example with that of iron. So it was natural to start with their spectra. At the time it was not known that these metals have a single electron in the outermost shell, and for this reason resemble hydrogen. What Rydberg saw were many lines in the visible. Today we know that this arises because the last electron in the alkali metals is bound to the nucleus with an energy of less than 5 eV. Consequently, most of the spectral lines are in the visible range and had already been observed at that time. The binding energy of the electron in hydrogen is 13.5 eV, so many of the lines lie in the UV and hence were not known at the time Rydberg began his investigation.

Balmer's discovery led Rydberg to attempt in 1889[212] the complete formula (all values of n and not just $n = 2$), which includes all the lines of the alkali metals. Rydberg realized that Balmer's formula was a special case of the following expression:

$$\nu = R_y \left(\frac{1}{n^2} - \frac{1}{m^2} \right), \qquad m \geq n = 1, 2, \ldots, \quad m \text{ integer},$$

which applied nicely to the alkali metals. R_y is a constant known today as the Rydberg constant.

It was a mystery why the hydrogen lines required just $n = 2$, and it was not known what the other values of n implied. Interestingly, Rydberg himself was not sure of the general validity of his formula. In 1897,[213] Rydberg participated in a debate about the nature of the hydrogen lines observed in stars, and claimed that:

> The two series of hydrogen are to be represented by two distinct formulae, even if it may be possible to unite them with great approximation in a single equation.

Moreover, the spectra were classified into two types, those with a clear structure following Rydberg's formula (class I, like the alkali metals) and those which did not follow such a structure (class II).[214]

The enlightenment came when spectroscopy extended its domain to the UV and the far IR. In 1906, Lyman[215] discovered a new set of hydrogen lines in the UV. These lines are 'predicted' by the Rydberg formula if one inserts $n = 1$. In 1908, Paschen[216] discovered a new series of hydrogen lines in the far IR. This series corresponds to $n = 3$. These three series were known to Bohr when he modelled the structure of the

---

212 Rydberg, J.R., K. Svenska Vetensk. Akad. Handl. **23**, 1 (1889).

213 Rydberg, J.R., Astrophys. J. **6**, 233 (1897).

214 Rydberg, J.R., *Rapports Présentés au Congrès International de Physique, Paris* **2**, 200 (1900).

215 Lyman, T., Astrophys. J. **23**, 181 (1906).

216 Paschen, F., Ann. Phys. **332**, 537 (1908).

hydrogen atom in 1913. When Brackett (1897–1974m) discovered the $n = 4$ series in 1922 and Pfund (1879–1949) discovered the $n = 5$ series in 1924, the existence of these series could be predicted by Bohr and constituted an outstanding victory for the new theory.

2.35 The Structure of the Atom: Only Complete Solutions Are Acceptable

Once Einstein's and Planck's ideas were in the air, it became clear that quantization should prevail. The problem was the source of the spectral lines. It was evident that atoms should be treated as oscillators, but the idea of simple oscillators did not help. According to Planck, simple oscillators radiate at a single frequency ν and emit an energy quantum of $h\nu$. If atoms behaved like such oscillators, they would yield a single spectral line with frequency ν, which is obviously not the case. Hence, various attempts were carried out to formulate an atomic oscillator that would oscillate at the observed frequencies of the spectral lines. Among these attempts, probably the best known is the one by Arthur Haas (1884–1941).[217] Haas adopted the Thomson pudding model for the atom in which the negatively charged electrons are immersed in a positively charged jello, and tried to quantize it. This was a year before Rutherford showed, in 1911, that there is a small condensed nucleus and the electrons move around it, like planets around the Sun.

Haas assumed that the maximum frequency of the electron in an atom occurs when it revolves around the atom along its surface (note the contradiction, since the electrons do not move in Thomson's model). Next, he assumed that this negatively charged electron had one energy quantum and revolved around a positively charged sphere. Haas' result for the maximum frequency was

$$\nu_{\max} = \frac{4\pi^2 e^4 m_e}{h^3},$$

where m_e is the mass of the electron. He identified this maximum frequency with the series limit, i.e., he substituted $m \to \infty$ in Balmer's formula

$$\nu = R_y \left(\frac{1}{2^2} - \frac{1}{m^2}\right),$$

from which it follows that

$$\upsilon_{\max} = R_{\text{Haas}} = R_y/4.$$

[217] Haas, A.E., Jahrb. d. Rad. u. El. **vii**, 261 (1910). Other attempts were made by Schidlof, A., Ann. Phys. **340**, 90 (1911); Wertheimer, E., Phys. Zeitschr. **xii**, 409 (1911), Verh. Dtsch. Phys. Ges. 1912, p. 431; Lindemann, F.A., Verh. Dtsch. Phys. Ges. **482**, 1107 (1911); Haber, F., Verh. Dtsch. Phys. Ges. **482**, 1117 (1911); and Nicholson, J.W., MNRAS **130**, 49 (1912).

Had Haas taken the Lyman series, he would have obtained $R_{\text{Haas}} = R_y$. But there was no way he could reproduce the following lines in the series. Deriving the Rydberg constant approximately may appear to be a success, but the only parameters in the problem are the charge e and mass m_e of the electron and the Planck constant. The simplest unit of frequency (inverse time) that one can build from these three constants is $R_{\text{dim}} = e^4 m_e / h^3$, so it was no wonder that playing around with the problem would yield the value R_{dim} times a constant of the order of unity for the Rydberg constant. The problem did not lie here, but rather with the relation between the frequencies of the lines!

In contrast to Haas' model, Nicholson's model took the force between the particles to vary inversely as the square of the distance and related the energy of the particles to Planck's theory. The atoms were supposed to consist of a ring of a few electrons surrounding a positive nucleus of negligibly small dimensions. The ratios between the frequencies corresponding to the lines in question were compared with the ratios between the frequencies corresponding to different modes of vibration of the ring of electrons. As Bohr wrote[218]:

> Excellent agreement between the calculated and observed values of the ratios between the wavelengths in question seems a strong argument in favour of the validity of the foundation of Nicholson's calculations.

So what was the problem? Bohr answered:

> In Nicholson's calculations the frequency of lines in a line spectrum is identified with the frequency of vibration of a mechanical system, in a distinctly indicated state of equilibrium, [...] but systems like those considered, in which the frequency is a function of the energy, cannot emit a finite amount of homogeneous radiation [as is observed].

In other words, a certain principle has to be violated or a new one invented.

Why can atoms only radiate at certain frequencies? As well as failing to predict the sequence of spectral lines, all models which assumed atoms in which the positive charge was distributed over the volume completely failed to explain the Geiger (1882–1945m) and Marsden (1889–1970)[219] experiment with scattering of α particles. In this experiment, Geiger and Marsden bombarded a piece of gold foil with α particles. They discovered that:

- The vast majority of the α particles passed through the gold as though through a vacuum.
- Those very few particles that were deviated from the straight line were scattered through a very large angle.

If as Thomson assumed the electric charge is smeared over the whole atom, then the probability that the α particle will hit a constituent of the atom is high, in contradiction with the experiment. The experiment indicated primarily that most of the atom is empty.

---

[218] Bohr, N., Phil. Mag. **26**, 1 (1913).

[219] Geiger, H. and Marsden, E., Proc. R. Soc. Ser. A **82**, 495 (1909).

Rutherford[220] explained the experiment with a new model. In Rutherford's model, the positive charge was concentrated at the center and the electrons moved freely through the volume of the atom. The atom as a whole is neutral, but inside the atom the electric field of the nucleus prevails and causes those α particles that pass close to the nucleus to scatter in a special way. Rutherford succeeded in nicely predicting the distribution of the scattered particles. Today the experiment is called the Rutherford scattering experiment, although the experiment was conducted by Geiger and Marsden, while Rutherford provided the theoretical breakthrough regarding the atomic structure.[221]

According to van den Broek,[222] the charge of the nucleus is Ze, where Z is half the atomic weight of the atom. The electron moves in a Keplerian orbit around the nucleus. The atoms are neutral, and hence the number of electrons is equal to the charge of the nucleus.[223]

Rutherford's model explained the scattering but not the emission lines. Moreover, the atom according to Rutherford could not emit a sharp spectral line, thereby introducing the problem that all previous models tried to avoid, namely that an accelerating electron must radiate continuously and hence lose energy and eventually collapse into the nucleus. In short, the atom was unstable.[224]

Bohr adopted the Rutherford model and overcame the problems by revolting against classical physics. He took the liberty of making three bold assumptions which contradicted the notions and spirit of classical physics. His postulates were[225]:

- The electrons cannot revolve around the nucleus in arbitrary orbits, but only in certain discrete orbits determined by quantumtheory.

---

220 Rutherford, E., Phil. Mag. **6**, 21 (1909).

221 Note the coincidence of facts that helped Rutherford. The atomic weight of gold is 200 while that of the α particle is 4. Thus, when the α particle is scattered by the nucleus of the gold atom, the recoil of the nucleus is very small and the assumption that the scatterer has an infinite mass is a good one. Under this assumption, Rutherford's calculation was simple, while taking the recoil into account would have greatly complicated the theoretical modelling. Gold was chosen because it can be stamped into very thin foils in order to have the smallest number of atomic layers in the target.

222 van den Broek, A., Phys. Zeit. **14**, 32 (1913).

223 There are claims that van den Broek suggested in 1911 [Nature **87**, 78 (1911)] that the number of an element in the periodic table corresponds to the charge on its nucleus. But in fact, what van den Broek said was that *if the charge is equal to half the atomic weight, than one can infer from the existence of uranium that there are* $238/2 \sim 120$ *elements*. This claim was repeated in Nature **92**, 372 (1913), the year when Mosley published his correct results with 92 elements.

224 As the Rutherford atom resembles the Solar System and the central force behaves in much the same way, the reader may ask whether the Solar System is stable. The answer is that, according to the general theory of relativity, the Earth emits gravitational wave radiation exactly like the electron which moves around the nucleus. The planets thus lose energy and gradually approach the Sun. In due time, they will therefore collapse into the Sun. However, the rate of energy loss is so small that we can forget this process on the scale of the age of the Universe. But in close binary stars, where the mass of the companion is large and the separation between the two stars is quite small, the collapse happens on a time scale shorter than the age of the Universe. In all likelihood, the collapse triggers a supernova and/or the formation of a black hole.

225 Bohr, N., Phil. Mag. **26**, 1 (1913).

- If we restrict the motion to circular orbits, then only orbits with angular momentum (mass times radius times circular velocity) equal to a whole multiple of $h/2\pi$ are permitted. This assumption restricts the possible orbits to those which relate to each other like the squares of integer numbers (namely, 1:4:9:16, and so on). These allowable orbits are stationary.

Bohr did not explain why there was a ground state, let alone why the ground state was stable and allowed the electron to stay in it forever, but simply assumed this to be the case. If Bohr was correct, then clearly Maxwell's electrodynamic theory would have to be revised in a substantial way to permit the existence of such orbits, since they violate classical electrodynamics.

- The electrons can jump from one orbit to the other, and when the electron jumps from a level with energy E_2 down to a level with energy E_1, it emits a photon[226] (a quantum of electromagnetic energy) with frequency

$$\nu = \frac{E_2 - E_1}{h}.$$

  This is called the Bohr frequency condition. What happened during the jump from one energy level to the other remained unknown.

These three assumptions allowed Bohr to derive the Rydberg formula for the hydrogen atom, along with the accurate value of the Rydberg constant. The theory also explained the spectral lines of ionized helium, namely an atom with double the charge but only one electron. The problem of atoms with more than one electron remained unsolved.

Although Bohr's theory introduced new unsolved questions, the victory in the explanation of the existing hydrogen line series, and the prediction of those series that were discovered shortly after, was sweeping.

2.36 Einstein: An Explicit Expression for the Kirchhoff Law

A dramatic change in the attitude to Kirchhoff's law was brought about by Einstein in 1916–1917.[227] By now, Bohr had the explanation for the discrete frequencies at which the hydrogen atom emits radiation, including a full explanation for the position of the emitted lines. So for simplicity, Einstein considered an atom with two energy levels (see Fig. 2.16).

---

[226] The term 'photon' appeared for the first time in Lewis, G.N., Nature **118**, 874 (1926) and was quickly adopted. It is surprising that, despite the fact that the concept of a particle of light was known to Newton and was the subject of a centuries long controversy between the supporters of wave theory and the advocates of particle theory, it was such a long time before a proper name was invented.

[227] Einstein, A., Verh. Dtsch. Phys. Ges. **18**, 318 (1916); Mitt. Phys. Ges. **16**, 47 (1916); Phys. Zeit. **18**, 121 (1917).

Let A_{21} be defined by

$$A_{21} = \frac{\text{transition probability per unit time}}{\text{for spontaneous emission per second}}$$

That is to say, an electron staying in the upper level eventually jumps back to the lower level and this process is spontaneous. It is not known when the electron will jump, but it does so with a certain probability per unit time. Einstein simply assumed that the 'decision' to jump down is statistical. Then define the absorption by

$$B_{12}J = \frac{\text{transition probability per unit time}}{\text{for absorption}}$$

where J is the mean intensity. The spontaneous jump of the atom from the high to the low energy level does not depend on the radiation that the atom is immersed in. On the other hand, the rate of absorption depends on the mean intensity of radiation and is also probabilistic. The assumptions about these two processes were not sufficient to yield the Planck distribution. Worse than that, they yield Wien's law.

So Einstein found that he had to assume the existence of an additional term, namely, that on top of the spontaneous downward jump of the atom from the high to the low energy level, the transitions are stimulated by the external radiation. Hence, the additional term is

$$B_{21}J(\nu) = \frac{\text{transition probability per unit time}}{\text{for stimulated emission}}$$

In other words, there is an additional probability to jump down, and this probability is proportional to the intensity of the radiation. The existence of stimulated emission was another major discovery by Einstein. Radiation of a certain frequency induces the oscillator with the same frequency to radiate more. In the simple case, with probability A_{21}, once the electron jumps down, the photon can be emitted in any direction. However, the photon produced by the stimulated electron jump is always emitted in the direction of the stimulating photon. Stimulated emission is not predicted by classical physics, and is a pure quantum phenomenon.

In the steady state, we have

$$n_1 B_{12} J = n_2 A_{21} + n_2 B_{21} J,$$

where n_1 and n_2 are the numbers of atoms/oscillators in the corresponding levels.

From here Einstein rederived Planck's function. To obtain the Planck function, Einstein had to assume that $B_{12} = B_{21}$, a result that does not depend on the properties of the matter. Moreover, Einstein required that $A_{21} = (2h\nu^3/c^2)B_{21}$. This result implies that the coefficient of spontaneous emission is always related to the absorption in the opposite direction, and for all possible materials. In particular, there is no

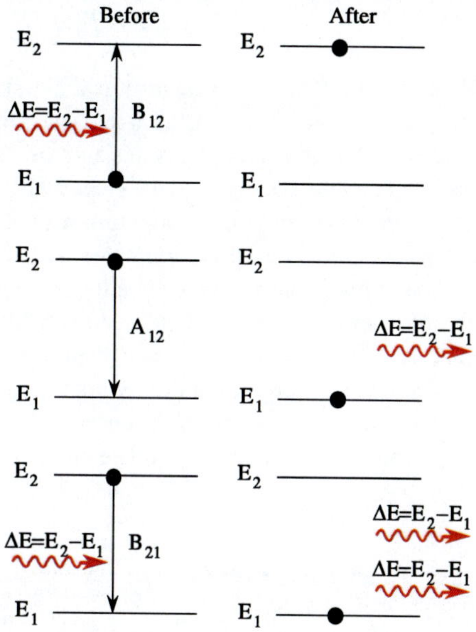

Fig. 2.16 Radiation absorption and emission according to Einstein's phenomenological theory. The *top* process is simple absorption of a photon, with transition of the electron to the upper level. The *second* process is spontaneous emission of a photon, with transition of the electron to the lower level. The *third* process, discovered by Einstein, is stimulated emission. A photon induces the transition of the system to the lower level. Thus for any incoming photon, two photons come out

temperature dependence in this relation. The stimulated photon is coherent with the photon that stimulated the emission.

From the above relation, it follows that the mean radiation intensity is given by

$$e(\nu, T) \equiv J = \frac{n_2 A_{21}}{n_1 B_{12} - n_2 B_{21}}.$$

The strict Kirchhoff relation is

$$e(\nu, T) \equiv J = \frac{n_2 A_{21}}{n_1 B_{12}},$$

namely, without the stimulated emission. If we want to recover Kirchhoff's relation, we must correct the absorption for the stimulated emission. In this way Einstein proved and corrected Kirchhoff's law. But why did Kirchhoff and the others not discover stimulated emission during 40 years of research, and why was it left to Einstein to discover it? There are in fact two reasons. Firstly, Kirchhoff spoke only of the effective absorption and not the theoretical probability of light being absorbed. And secondly, the stimulated effect becomes important only at very short wavelengths, such as UV, and UV spectroscopy was in its infancy or non-existent in Kirchoff's day.

To proceed, Einstein needed the ratio n_1/n_2 between the numbers of atoms in the two levels. He assumed that the ratio satisfied the Boltzmann relation $n_2/n_1 = \exp\left[-(E_2 - E_1)/kT\right]$ and this was sufficient to prove that

$$e(\nu, T) \equiv J \equiv B_\nu(T).$$

Note that the Boltzmann relation is a classical expression, and it is not clear a priori why it should hold in the quantum domain. Thus, the proof was a kind of a mixture between classical physics and the new quantum theory. Moreover, Einstein assumed that the transition of the electron between the energy levels was a probabilistic process, long before the quantum wave function was discovered by Schrödinger and its statistical interpretation given by Born.

The connection between the atomic energy levels, the frequency of the emitted photons, and Kirchhoff–Planck's universal function was finally established. Note that, while the coefficients A and B introduce the probabilistic element of quantum mechanics, Einstein did not draw upon any other results from quantum mechanics. In particular, the structure of the energy levels of the atom, e.g., the Rydberg formula, played no role in the derivation. The matter out of which a cavity is made is irrelevant!

2.37 A Physical Note

The description so far is ideal. Consider a cavity with temperature T and a substance composed of certain kind of two-level atom. Moreover, assume that the cavity is also composed of two-level atoms, but a different kind, so that the two frequencies are different. Clearly, such a system cannot reach thermal equilibrium, because there is no process which converts the radiation energy into thermal energy. Does this mean that there are theoretical systems that cannot reach equilibrium? In principle it does, although in practice it probably does not. If we allow some dissipation to take place, like recoil of the absorbing atom, the cavity will eventually approach thermal equilibrium with the body inside. How long it would take to reach such an equilibrium is another question that will not be discussed here.

2.38 Not All Are Quick to Accept

Not everyone was happy with the new discoveries about the radiation mechanisms. Gilbert Lewis (1875–1946m) was one of the most influential American chemists around the turn of the nineteenth century. As late as 1925, Lewis stated[228] that the number of photons was conserved. He was led to this conclusion from considerations of detailed balance between absorption and emission. Lewis himself asserted that the principle contradicted existing notions of light absorption. The idea was that, in the emission process, only one photon should appear. Likewise, Lewis[229] advocated *the law of entire equilibrium*, which he formulated as:

---

[228] Lewis, G.N., Nature **118**, 874 (1925).

[229] Lewis, G.N., PNAS **11**, 179 (1925).

> Corresponding to every individual process there is a reverse process, and in a state of equilibrium the average rate of every process is equal to the average rate of its reverse process.

This appears to be what had been demonstrated long before, but just in case the reader was not clear about it, Lewis wrote:

> I believe that some of the ideas contained in this paper have been suggested by the work of Einstein, but he has not proposed this law of equilibrium.

Einstein simply wrote the equation without the accompanying fanfare.

A couple of pages after Lewis' paper in Nature, Ralf Fowler (1889–1944) and Edward Milne (1896–1950)[230] referred to this paper with the remark:

> It may [...] be helpful to the reader of Lewis' note to call attention to the considerable amount of recent work in physics, under the name of the 'principle of detailed balancing'. It seems unnecessary that the relevant parts of these investigations should be worked through anew.

In short, the chemist should look at what the physicists had been doing before. On this occasion Fowler and Milne cited a long list of researchers who had investigated this principle: Owen Richardson (1879–1959),[231] Einstein,[232] James Franck (1882–1964),[233] Günther Cario (1897–1984),[234] Cario and Franck,[235] and many others. This is one of the rare cases in which the journal Nature bothered to correct a paper published in its own pages.

2.39 New Elements in the Sun

The discovery of helium in the Sun was an impressive manifestation of Kirchhoff's and Bunsen's discovery that each atom and molecule has its own characteristic spectrum. However, the contribution of luck cannot be ignored.

Indeed, helium was first discovered on the Sun, for which reason it was named after the Greek word *helios*. In 1868, Pierre-Jules Janssen went to India to study the total solar eclipse of 18 August 1868. During the eclipse, he was able to take a spectrum of the prominences, which are usually too faint to be observed at full

---

230 Fowler, R.H. and Milne, E.A., PNAS **11**, 400 (1925).

231 Richardson, O.W., Phil. Mag. **27** (1914).

232 Einstein, A., Phys. Zeit. **18**, 121 (1917).

233 Franck, J., Phys. Zeit. **9**, 289 (1922).

234 Cario, G., Phys. Zeit. **10**, 185 (1922).

235 Cario, G., and Franck, J., Phys. Zeit. **11**, 161 (1922).

D_1 D_2 D_3

FIG. 42.—Line D_3 (yellow), with radial slit.

Fig. 2.17 The *upper panel* is the famous D_3 line as depicted by Lockyer in 1887 in his book. The *lower part* is the laboratory spectrum attached to the telescope and used to calibrate the observed spectra. The D_1 and D_2 sodium lines are marked as well. Note that the D_3 appears in emission

Sun. And in the spectra of these prominences[236] and the corona, he discovered an unfamiliar single(!) yellow spectral line in emission (see Figs. 2.17, 2.18).[237]

The setting was perfect for Kirchhoff's explanation of the Fraunhofer lines, i.e., a very bright source of radiation, the Sun, surrounded by a faint corona. Yet the observed yellow line in the prominences and the corona, was not seen as a dark Fraunhofer line, but as an emission line. So what was it? Lockyer measured the location of the lines so accurately that there was no room for confusion. The reason, as we know today, is a streak of luck. Helium needs high temperatures (of the order of 2.5×10^5 K) to excite the electron so as to produce spectral lines by jumping back to a lower energy level. The temperature of the corona is about 2×10^6 K, and at this high temperature, a few helium lines remain possible. One of them is the yellow line. At a lower temperature such as 5,800 K, which prevails in the reversing layer of the Sun, the electron of the helium is not excited at all and no spectral line appears. On the other hand, the corona is so much hotter than the solar surface that no reversal of the line takes place. The line was designated D_3 because of its proximity to the sodium D_2 lines.

Janssen informed Lockyer[238] in England about the mysterious line. On 20 October 1869, Lockyer pointed his 6-inch telescope at the Sun and verified Janssen's discovery

---

[236] A prominence is a large, bright torus extending outward from the Sun's surface. Prominences are anchored to the Sun's surface in the photosphere, and extend outwards into the Sun's corona. A prominence forms over timescales of about a day, and particularly stable prominences may persist for several months. It is believed that, when the prominences break the energy released heats the corona.

[237] Janssen, P.J.C., Compt. Rend. **67**, 838 (1868).

[238] Lockyer also won a crater on Mars.

Fig. 2.18 The famous sodium D_1 and D_2 lines, and the D_3 line of helium which was identified by Lockyer and Janssen in the Sun. The conditions in the solar corona allow only for the D_3 line to appear

of a yellow line. Edward Frankland (1825–1899) and Lockyer[239] tried to discover the element in the laboratory, but to no avail.

Janssen and Lockyer announced the discovery of a heavenly element never seen before. This was too much for the 'sophisticated' scientific community. Lockyer and Janssen were ridiculed in scientific circles[240] for many years. The idea that there might be elements in stars that do not exist on Earth seemed absurd.

In 1874, Lockyer[241] gave the Bakerian talk, in which he discussed the spectrum of the Sun and explained how there could be emission lines, observed during the eclipse, which had no corresponding Fraunhofer lines formed in the 'reversing layer'. His working hypothesis was that:

> The so-called elements not present in the reversing layer of a star will be in course of formation in the coronal atmosphere and in course of destruction as their vapour densities carry them down.

They would thus be effectively invisible. But 'helium' was not mentioned!

In later years Lockyer referred to helium indirectly. For example, in 1878,[242] he concluded that:

> The substances which give us the non-reversal line in the chromosphere [...] termed the coronal line, are other forms of hydrogen.

He decided to devote a special paper to the subject.

Still in 1877, Draper[243] wrote:

> The case of the D_3 line strengthens the argument in favor of the apparent exemption of certain substances from the common law of the relation of emission and absorption, for while there can be no doubt of the existence of an ignited gas in the chromosphere giving this line, there is no corresponding dark line in the spectrum of the solar disc.

---

239 Frankland, E. and Lockyer, N.J., Proc. R. Soc. **17**, 288 (1869); ibid. **18**, 79 (1869). Lockyer was hesitant about announcing the discovery. Frankland strongly disapproved. But Kelvin did so at the 1871 British Association meeting. In 1886, Copeland, the Astronomer Royal for Scotland, discovered a single helium line in the Orion nebula.

240 Lockyer, T.M. and Lockyer, W.L., *Life and Work of Sir Norman Lockyer*, Macmillan, London, 1928.

241 Lockyer, J.N., Phil. Trans. R. Soc. Lond. **164**, 479 (1874).

242 Lockyer, J.N., Proc. R. Soc. Lond. **28**, 157 (1878–1879).

243 Draper, H., Proc. Am. Phil. Soc. **17**, no. 100, 74 (1877).

Kirchhoff's partial explanation of the formation of the Fraunhofer lines confused observers. In 1893, there was a solar eclipse which Lockyer observed. Alas, Lockyer, the scientist, reported to the Royal Society only in 1896 that:

> D_3 was absent [...] and the reason given suggested that its recorded appearance in 1882 was simply a photographic effect.

Was Lockyer ready to withdraw his claim? Definitely not. But a later attempt by Herbert Turner (1861–1930)[244] to check Lockyer's claims during the total solar eclipse of 29 August failed as well!

Luigi Palmieri (1807–1896), a well-known vulcanologist, reported that, while investigating an eruption of mount Vesuvius in 1881,[245] he found the D_3 line which Lockyer had identified as helium in the spectroscopic analysis of samples of ejected matter from the volcano. Perplexingly, Palmieri did not save his samples or collect the emitted gas, thus forestalling verification of his claim. Today we know that helium is released in vulcanic eruptions. This is a testimony to radioactive decay taking place in the Earth's solid outer layers. But when Palmieri announced his discovery, it went more or less unnoticed.

In March 1895, William Ramsay (1852–1916) published in Nature[246] the results of his analysis of the spectrum of gases emanating from a uranium mineral called clevite, and discovered an unfamiliar yellow line, as well as:

> [...] four specially characteristic lines in the helium which are absent from that of argon: they are a brilliant red, the D_3 line of a brilliant yellow, a peacock-green line, and a brilliant violet line.

Ramsay did not provide the wavelengths of the spectral lines, but sent gas samples to both Lockyer and Crookes. Within a week, Crookes confirmed that the gas was the same as the one Lockyer had observed on the Sun. The Sun, however, showed just a single line! It took a quarter of a century for Lockyer to celebrate his hard-won victory over those who had mocked him for such a long time. Ramsay's announcement in the journal was followed by a note from Lockyer.[247]

In 1892, before the identification of helium was confirmed, the French Academy minted a medal with the faces of Lockyer and Janssen, who were still alive, to commemorate this outstanding discovery. The French Astronomical Society was more conservative and it was only in 1960 that it established the *Jules Janssen prize*.

---

[244] Turner, H.H., Phil. Trans. R. Soc. Lond. **180**, 385 (1886).

[245] Palmieri, L., Rendiconto dell'Academia delle Scienze Fisiche e Matematiche **20**, 150 (1881).

[246] Ramsay, W., Nature **52**, 55 (1895).

[247] Lockyer, J.N., Nature **52**, 55 (1895). Although this was Lockyer's great victory, which came after years of laughter at his expense, his comments were very reserved. Dignity prevailed. Lockyer was then the chief editor of Nature.

Chapter 3
Probing the Stars from Afar

3.1 Spectral Classification and Abundance Uniformity

As soon as stellar spectra were analyzed spectroscopically in the second half of the nineteenth century, it became clear that these spectra exhibit a bewildering variety. The need for some order, or rather some 'classification', became urgent. Of the many schemes suggested for stellar classification, the Harvard scheme, proposed by Pickering, was accepted in 1910 as the standard.[1]

In Fig. 3.1, typical spectra of main sequence stars are shown one above the other, aligned in such a way that the wavelengths in all spectra are identical. One can easily see how the spectral lines change their shape from one spectrum to another. Therefore, in principle, stars may be classified according to how specific lines vary. Such a classification would require a rather involved calibration of each spectrum. To simplify, the classification is based on the ratio of intensities of selected prominent spectral lines in the spectrum.

The classification of stars into distinct groups or classes is clearly not continuous. The original classes were arranged according to the Latin alphabet, i.e., A, B, ..., but by shear ingenuity, Miss Cannon of Harvard realized that the resulting order of the various classes was not 'smooth'. She thus suggested, not to change the definition of the classes, but instead to change from the originally proposed alphabetical order to O, B, A, F, G, K, M.[2] In this way, the intuition of Miss Cannon led to the bizarre notation and order, but correct order of physical classification.

---

[1] See details in Shaviv, G., *The Life of the Stars*, Springer-Verlag, Heidelberg and Magnes, Jerusalem (2009).

[2] The order, which appears illogical, gave rise to stories, and some interesting mnemonics, such as: Oh, Be A Fine Girl Kiss Me Right Now, or Only Boys Accepting Feminism Get Kissed (which probably describes better the situation in Harvard when Miss Cannon worked there). For the correct historical description, see: Cannon, A.J., The Henry Draper Memorial, J. Roy. Ast. Soc. Canada **9**, 203 (1915), where she tells how she discovered the need to change the order of the classes and thereby destroy the original alphabetical order.

G. Shaviv, *The Synthesis of the Elements*, Astrophysics and Space Science Library 387,
DOI: 10.1007/978-3-642-28385-7_3,

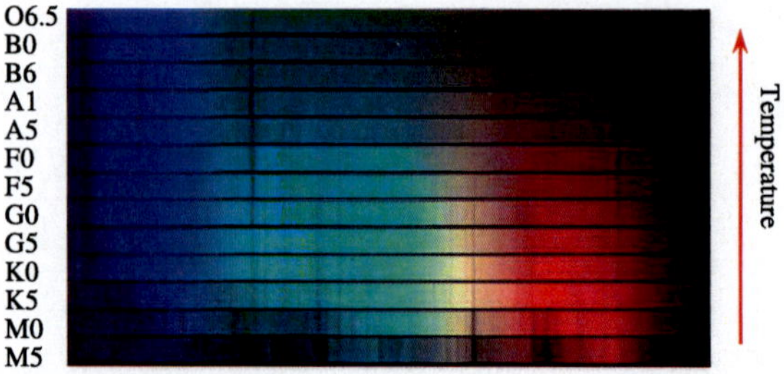

Fig. 3.1 The spectra of stars with different surface temperatures. The *letters on the left* are the astrophysical symbols for the different classes. The temperature increases from the *bottom*, where the star is *red*, upward to where the star is *blue*. O type stars have surface temperatures of 30,000–60,000 K, B stars 10,000–30,000 K, A stars 7,500–10,000 K, F stars 6,000–7,500 K, G stars 5,000–6,000 K, and K stars 3,500–5,000 K, while M stars have temperatures less then 3,500 K. The Sun is a G type star and hence appears *yellow*. Note that the strength of each line varies with the spectral class or temperature

Table 3.1 An example of absorption line ratios serving as criteria for spectral classification in the O spectral class

Spectral type	Standard star	Characteristic line ratio		
		HeI 4471 / HeII 4541	HeII 4541 / Hγ 4340	SiIV 4089 / NIII 4097
O5	ζ Pup	0.0	0.6	0.3
O6	λ^1 Cep	0.8	0.5	0.5
O7	*S* Mon	1.4	0.4	0.8
O8	λ Ori	2.0	0.3	1.0
O9	ι Ori	2.7	0.2	1.4

Note that two ratios are between lines of different elements. Such a ratio can serve as a spectral class indicator so long as the composition is identical

A tacit assumption regarding the classification of stars, seldom spelt out explicitly, is that all stars have identical composition. We distinguish between a line ratio between lines belonging to the same element and a line ratio between two lines each belonging to different elements. An example of how O-type stars are classified is shown in Table 3.1. In the first case, the line ratio depends primarily on the physical conditions prevailing in the star, while in the latter, besides the physical conditions, the ratio depends also on the relative abundances. And it is no trivial matter to distinguish between variations in composition and variations in physical conditions. The original and simplest assumption was that all stars had the same composition and that there was a universal abundance shared by all cosmic bodies. If so, variations in the spectra should imply variations in physical conditions. We will see how deviations from this assumption led to the recognition that the Universe and its stars evolve.

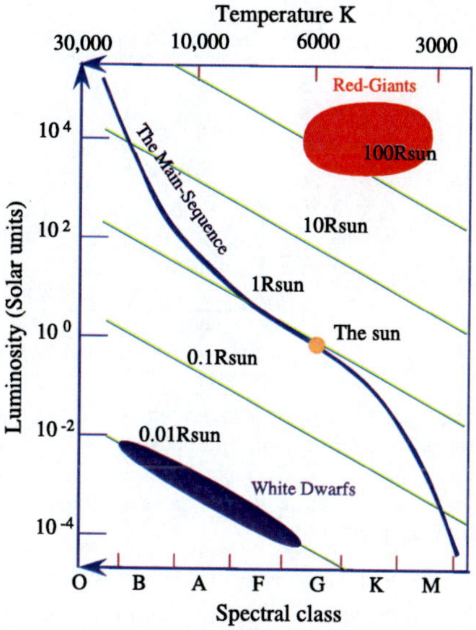

Fig. 3.2 The basic correlation between luminosity and spectral classes of stars, known as the main sequence. Note the large variation in luminosity versus the smaller variation in surface temperature (and radius)

3.2 Stellar Correlations: The HR Diagram

The Hertzsprung–Russell diagram[3] displays a correlation between the type of spectrum exhibited by the star (spectral class, denoted by Sp) and the luminosity (see Fig. 3.2).

Megh Nad Saha (1893–1956m)[4] managed to demonstrate that, if we assume that all stars have the same composition, the changes in the absorption lines seen across the main sequence result from the gradual change in the surface temperature taking place along the main sequence, as the mass of the star increases. As a matter of fact, Henry Norris Russell (1877–1957m) already hypothesized in 1914[5] that *stars of similar spectrum are at least approximately similar in surface brightness*. Russell was unable to prove this supposition and interpreted the main sequence of stars as a cooling sequence, so that stars were born hot and bright and then cooled. Hence, stars did not stay at the same point in the HR diagram but moved down the main sequence. This interpretation assumes implicitly that the composition of all stars is the same and does not change as the stars cool. The energy released was assumed to occur without change in the composition.

[3] For details see, Shaviv, G. *The Life of the Stars*, Springer-Verlag, Heidelberg, and Magnes, Jerusalem, 2009.

[4] Saha, M., Phil. Mag. **40**, 479 (1920); Proc. Roy. Soc. London **99A**, 135 (1921).

[5] Russell, H.N., PA **22**, 331 (1914). This is the detailed paper. Precursors with fragmentary ideas are: Obs. **36**, 324 (1913), in which the idea of giant and dwarf stars is discussed, and Obs. **37**, 165 (1914), in which Russell hypothesized that stars form hot, white–blue, and bright at the top of the main sequence and then cool, decrease in luminosity, and become redder and redder until they fade away.

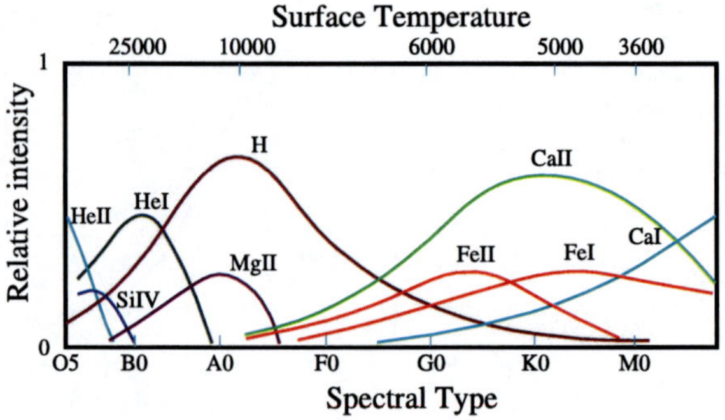

Fig. 3.3 The spectral classification is determined according to the relative intensity of selected strong spectral lines of different species. The figure shows how the relative intensity changes with the spectral class. The identification of the spectral class as a proxy for the temperature, as shown on the *upper axis*, came only in the early 1920s with the discovery by Saha that atoms ionize at high temperature

Saha showed how the intensity of the lines varies with temperature and density as can be seen in Fig. 3.3. For all lines, the intensity is small at low temperatures and increases with temperature. As the temperature rises, the bulk of the electrons move to higher and higher levels or are released completely from the atom. Hence, the line intensity reaches a maximum. This maximum line intensity depends on the particular energy levels between which the electrons jump, and as the energies are different, each line reaches its maximum at a different temperature. Thus, the spectral line ratios provide excellent approximations to a temperature scale for the stars, provided the composition is identical.

However, Russell concluded that:

> The differences in brightness between the stars of different spectral classes, and between the giants and the dwarf stars of the same class, do not arise (directly at least) from differences in mass. Indeed, the mean masses of the various groups of stars are extraordinarily similar.

This inference by Russell was found to be in error when Hertzsprung and Eddington discovered the mass–luminosity law. More accurately, the mass of the stars varies from $\sim 0.1\text{–}100 M_\odot$, while the luminosity varies by a factor of 10^6. Hence one can say that, relative to the luminosities, the masses of the stars hardly change. Russell's error was due to the fact that the luminosity is very sensitive to the mass.

3.3 Stellar Abundance Uniformity: Before the Revolution

Kirchhoff did not detect hydrogen on the Sun. Hydrogen was discovered on the Sun by Ångström in 1862,[6] when he mapped the Fraunhofer lines it produced. However, there was no way to know how much hydrogen there was on the Sun. The observation

[6] Ångström, A.J., Ann. Phys. **193**, 290 (1862). This is a bit strange. Hydrogen did not appear in Ångström's list of detected elements. It was Ångström's associate, Thalén, in a separate memoir, Nova Acta, Upsala, 1868, who listed 4 lines of hydrogen as observed on the Sun. In addition, Thalén

was not trivial. Huggins failed to observe hydrogen on the Sun,[7] although Johnstone Stoney (1826–1911)[8] argued in 1867, in a paper entitled *Constitution of the Sun and Stars*, that hydrogen was to be expected, so Huggins admitted: *I acknowledge that I ought to have found the lines*. Lockyer[9] tried hard to detect hydrogen on the Sun and failed, even when he observed the solar prominences. However, Lockyer stated in his 1874 Bakerian lecture[10] that he had observed hydrogen on Sirius A (the brightest star for an Earth-based observer).

In 1867, Stoney wrote[11]: *Hydrogen seems to be a very large constituent of the Sun's atmosphere*. Moreover, just from the observation of the spectral lines and without any ability to quantify the statement, as there was no theory of spectral lines available, Stoney went on to say:

> It appears from the analysis which has been made that none other of the gases in the solar atmosphere that extends as far as the stratum from which iron lines come, can compare in quantity with hydrogen and iron. [...] Hydrogen and iron are accordingly the principal ingredients of the part of the Sun's atmosphere which extends beyond the photosphere.

Stoney did not provide any reference to an observation which supported these statements. The declaration about hydrogen, just out of the blue, was correct, but the one about iron was wrong. It is interesting that astronomers and theoreticians (like Eddington) ignored or did not believe in Stoney's statements, probably because no theoretical support was available, but also because Stoney identified sunspots as clouds and cloud motions on the solar photosphere. The various observed elements were placed in separate layers, not mixed as we think today, just as Lockyer hypothesized.

In 1897, Lockyer published his book *On the Chemical Constitution of the Stars*,[12] in which he wrote:

> While I think that we shall all admit that different stars are in different stages of development, and that hydrogen stars will ultimately approach more nearly to the state of our Sun, it would be unwise to push the argument of uniformity too far, and to say that every star will pass exactly through the same stages.

So there was no hydrogen in the Sun and we had an explanation for it. Arthur Schuster (1851–1934m)[13] did not like Lockyer's conclusions and argued that they were not based on solid evidence. Moreover, argued Schuster:

(Footnote 6 continued)
removed zinc and barium from the list of elements that Kirchhoff claimed to have discovered. Kirchhoff's evidence depended on the identification of two lines and *these were doubtless thought insufficient*, as Lockyer put it. But recall that Lockyer identified a new element with only one line!

[7] Huggins, W., Nature **7**, 320 (1873).

[8] Stoney introduced the idea of an 'electron' (and the name) as the 'fundamental unit quantity of electricity' in 1891.

[9] Lockyer, N., Phil. Trans. Roy. Soc. London **163**, 253 (1873); ibid. 639 (1873).

[10] Lockyer, N., Phil. Trans. Roy. Soc. London **164**, 479 (1874).

[11] Stoney, J., *On the Physical Constitution of the Sun and Stars*, Proc. Roy. Soc. **XVII**, 1 (1867).

[12] Lockyer, N. *On the Chemical Constitution of the Stars. And Additional Remarks*, Proc. Roy. Soc. London **61**, 209 (1897).

[13] Schuster, A., Proc. Roy. Soc. London **61**, 209 (1897).

> I am not convinced that the difference in the spectra of stars, which Mr. Lockyer ascribes to ascending and descending temperatures, are not due to a real difference of constitution.

In other words, Schuster allowed for different stars to have different compositions. The theoretical explanation of the observer was right, and that of the theoretician wrong.

Last but not least, Lockyer envisaged the solar atmosphere as composed of different layers of elements, as shown in Fig. 3.4.

3.4 The Revolution

Cecilia Payne (1900–1979) took Eddington's course in theoretical astrophysics in Cambridge, England, became fascinated with astronomy and decided, under Eddington's strong recommendation, to go to Harvard, USA,[14] to prepare a Ph.D. thesis under Shapley's supervision.[15] Shapley's original suggestion for her thesis topic was the photometry of variable stars, along the lines of Miss Leavitt's research (see below). But it so happened that Payne was attracted by Miss Cannon's research on stellar spectroscopy and classification. Payne prepared her dissertation with the title *Stellar Atmospheres: A Contribution to the Observational Study of High Temperature in the Reversing Layers of Stars* during the academic year 1924–1925.[16] The long title did not reveal its revolutionary content.

---

[14] Cecilia Payne realized that England offered no chance for a woman to get involved in scientific research at that time and hoped that the situation in the USA would be different.

[15] Payne attended Eddington's lecture about the 1919 expedition to observe the solar eclipse and to provide observational proof for the theory of general relativity, became captivated, and later used to say that this was the moment she decided to devote her life to astronomy. The idea of going to the USA came when Payne heard a talk by Harlow Shapley (1885–1972m) from Harvard College Observatory. The flamboyant Shapley impressed Payne to the point that she decided to go and prepare a thesis in Harvard. After completion of her studies, she was not awarded a degree, as Cambridge, England, did not grant degrees to women at that time, but she was deluded if she really believed that the USA was ripe for women scientists.

[16] Payne, C., The Dyer's Hand: An Autobiography (1979), in Katharine Haramunda (ed.), p. 157. The finished Ph.D. thesis was submitted to the physics department, since no such degrees were offered in astronomy at that time. The Chairman of Physics, Theodore Lyman, would not approve her thesis for a physics Ph.D. because she was a woman, and in those days the physics department was 'men only'. However, he agreed to sign it for 'another department'. In a strange 'passing of the buck', they called it an astronomy Ph.D., thus creating a de facto Harvard astronomy department. Payne then worked in astronomy at Harvard without a position for eleven years, before finally obtaining the title of astronomer. Cecilia Payne, who was called Mrs G, after her husband's name, became Harvard's first female tenured professor in 1956, and later the first woman to serve as department chair.

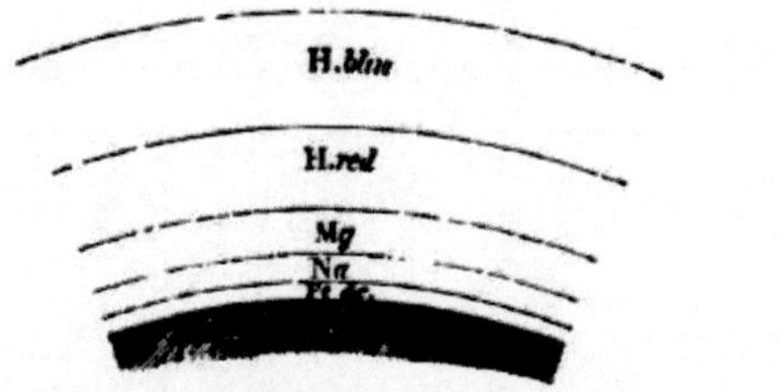

FIG. 69.—Imagined stratification of the solar atmosphere (1873). (H = Hydrogen; Mg = Magnesium; Na = Sodium; Fe = Iron.)

Fig. 3.4 Lockyer's solar model, with layers of different isotopes, one placed on *top* of the other. This stratified solar atmosphere was soon abandoned in favor of a well mixed atmosphere

Payne joined the extraordinary group of women researchers in Harvard,[17] and like the others, worked unofficially and unacknowledged at Harvard College Observatory until 1938, when she was granted the title of astronomer. She later asked to have this title changed to Phillips Astronomer, and in 1938 was appointed Phillips Astronomer in the Harvard College Observatory, without limit of time, and lecturer in astronomy, becoming the first woman to receive tenure in the Faculty of Arts and Sciences. None of the courses she taught at Harvard appeared in the catalogue until 1945. It was 'underground teaching'. In 1958, Payne-Gaposchkin was named the Phillips Professor of Astronomy.

The lack of official status did not prevent Payne from producing a unique Ph.D. thesis. It was an attempt to translate the observed Fraunhofer lines in the Sun into real relative abundances. The most important conclusions of Payne's thesis were as follows:

- The stars have uniform composition. As we saw, this conclusion was assumed by theoreticians and resulted in predictions that were not in contradiction with the observations. *Different stellar spectra result from physical conditions in stars, not from chemical abundance variations*, claimed Payne, and confirmed Saha's theory.
- The relative element abundances found by Payne are given in Table 3.2. The conspicuous feature is that most of the matter is in the form of helium and hydrogen. The abundance of elements with atomic numbers higher than 2 is significantly lower. This is a very important conclusion since many astrophysicists believed at that time that main sequence stars were liquid, because of the high mean solar density. As no gas with such a high density was known at the time, the default was that the stars were liquid. The 'accepted' source of energy was radioactive decay. If so, the abundances of the heavy elements which appear at the beginning of the radioactive chain should be high. The stars were assumed to be homogeneous and well mixed,[18] so that the products of radioactive decay should be observed in the

[17] Four outstanding women excelled in the Harvard Observatory in those days: Williamina Fleming (1857–1911m), who discovered the Horsehead nebula, Henriette Leavitt (1868–1921m), who discovered the period–luminosity law for Cepheids, Annie Cannon (1863–1941m), who contributed to stellar classification as we know it today, and Antonia Maury (1866–1952m), who discovered the giants.

[18] Stars were considered fully mixed by the meridional currents which result from rotation, as 'proven' by Eddington in 1929. Only in 1950, when Sweet (and Öpik) had shown that the

Table 3.2 The first table of relative abundances in stellar atmospheres

Z	Atom	[A]	Z	Atom	[A]
1	H	11	19	K	3.5
2	He	8.3	20	Ca	4.8
2	He^+	12	20	Ca^+	5.0
3	Li	0.0	22	Ti	4.1
6	C^+	4.5	23	V	3.0
11	Na	5.2	24	Cr	3.9
12	Mg	5.6	25	Mn	4.6
12	Mg^+	5.5	26	Fe	4.8
13	Al	5.0	30	Zn	4.2
14	Si	4.8	38	Sr	1.8
14	Si^+	4.9	38	Sr^+	1.5
14	Si^{+++}	6.0	54	Ba^+	1.1

Payne's Ph.D. thesis, 1925. H and He were omitted from the PNAS publication. The notation is $[A] \equiv \text{Log}A$. All abundances are relative to hydrogen, which is 10^{11}

atmospheres of the stars and dominate the abundances. Thus, Payne's observation appeared to shake 'well established truths'. Without mentioning Lockyer explicitly, Payne banished his complex and artificial picture of the stratified elements in the solar atmosphere, and assumed that all the elements were well mixed.

Note that helium was found to be more abundant than hydrogen by a factor of about 10 in number or by a factor of about 41 in mass. Thus about 2% of the mass is hydrogen and the rest is practically all helium. The result was amazing. Recall that the low-mass stars show no signs of helium,[19] because their surface temperature is too low to ionize helium. On the other hand, the high-mass stars have such high surface temperatures that the hydrogen is completely ionized and no neutral hydrogen atoms are left. Hence, hydrogen is not observed. This means that one has to infer from the observed spectral lines the abundance of what is not observed at all and composes most of the matter.

So unimaginable were the results that Payne wrote in her thesis:

> The outstanding discrepancies between the astrophysical and terrestrial abundances are displayed for hydrogen and helium. The enormous abundance derived for these elements in the stellar atmosphere is almost certainly not real. Probably the result may be considered, for hydrogen, as another aspect of its abnormal behavior, already alluded to; and helium, which has some features of astrophysical behavior in common with hydrogen, possibly deviates for similar reasons. [...] The observations on abundances refer merely to the stellar

(Footnote 18 continued)
meridional currents induced by rotation were extremely slow, did it become clear that stars were not mixed, and that their evolution leads to composition stratification. For the six orders of magnitude error in Eddington's calculation, see Shaviv, G., *The Life of Stars* Springer, Heidelberg, and Magnes, Jerusalem (2009).

[19] The ionization energy of helium is very high, about 24.59 eV, while the ionization energy of hydrogen is 13.6 eV. All other elements have lower ionization energies and hence are already ionized in stars with low surface temperatures.

> atmosphere, and it is not possible to arrive in this way at conclusions as to internal composition. But marked differences of internal composition from star to star might be expected to affect the atmosphere to a noticeable extent, and it is therefore somewhat unlikely that such differences occur.

In the first part of the penultimate sentence, Payne found a way out: H is present only on the surface. In the second sentence she retreats and admits that it cannot be a big effect. So even Payne had problems in digesting her revolution.

When Payne's manuscript was presented to Russell, he wrote that her ideas concerning the prevalence of hydrogen were 'impossible'. Shapley, Payne's supervisor, trusted the famous Russell and convinced Payne to dilute her conclusion substantially. This is the reason for the odd formulation: *almost certainly not real*. As a result, when Payne published,[20] she omitted the results for hydrogen and helium and explained that:

> Hydrogen and helium are omitted from the table. The stellar abundance deduced for these elements is improbably high, and is almost certainly not real.

So incredible was the result that Compton and Russell[21] suggested a peculiar behavior of the hydrogen lines as a way to explain why the observation was wrong. Plaskett[22] found similar excuses not to trust the results. It turned out that there was no need to concoct an explanation.

The theory used to infer the composition from the spectral lines was indeed extremely poor. Payne just used the Saha equation to calculate the relative strength of the lines, without any reference to the radiation. First, the 'theory' did not take into account the transfer of radiation through the atmosphere, nor the structure of the atmosphere, largely because there was no real theory in the first place. Second, the properties of the spectral lines of the different elements were not known, so the simplest assumption to use was that all atomic species behave the same way. Yet an error as high as a factor of 10^6 appeared unlikely. It took about 25 years for the theory to reach a state in which the error in the abundance determination could be reduced to about 10–20%.

Payne's Ph.D. was published as a book and won several reviews hardly a year later. For example, Paul Merrill[23] wrote a review of the new Ph.D. dissertation, but did not mention the dramatic discovery that hydrogen and helium were so abundant. A critical review[24] by Otto Struve (1897–1963), then a young lecturer at the Yerkes Observatory,[25] pointed to a series of problems with the thesis. However, what was probably the most important conclusion derived by Payne, namely, the *uniformity of composition of the stellar atmosphere* and the preponderance of hydrogen and helium

[20] Payne, C., PNAS **11**, 192 (1925).

[21] Compton, K.T. and Russell, H.N., Nature **114**, 86 (1924).

[22] Plaskett, H.H., Pub. Dom. Ast. Obs. **1**, 108 (1922).

[23] Merril, P.W., PASP **38**, 33 (1926).

[24] Struve, O., Ap. J. **64**, 204 (1926). Struve was the last member of a dynasty of astronomers and a great-grandson of the noted astronomer Friedrich von Struve.

[25] Struve himself received his Ph.D. in 1923 from the University of Chicago.

in stars, were not mentioned by Struve. But this did not prevent Struve and Zebergs from stating over three decades later, in 1962, that Payne's Ph.D. dissertation was *the most brilliant Ph.D. thesis ever written in astronomy*.[26] How the wheel of appreciation turns. It is amazing that, while Struve and Zebergs defined the dissertation as 'brilliant' and Lang and Gingerich included a condensed version of the thesis in their Astronomy and Astrophysics source book,[27] there are only about 23 citations to the thesis in the professional literature (as of January 2008), and there is only one citation to the PNAS paper based on the thesis.[28]

3.5 First Attempts to Derive the Shape of the Fraunhofer Lines

Payne's results gave the required push to the theoreticians to try to derive the shape of the Fraunhofer lines from theory, and the first attempts were carried out by Harald von Klüber (1901–1978)[29] and Marcel Minnaert (1893–1970)[30] as early as 1927.

In 1927, Unsöld[31] used a new method he had developed to calculate the shape of the Fraunhofer lines on the basis of Milne's[32] radiation theory. Unsöld was aware of the first attempts by Minnaert to derive the shape of the spectral lines. While the agreement with the observed shape of the lines was considered as acceptable for sodium, as can be seen from Fig. 3.5 (left), the discrepancy for hydrogen, shown in Fig. 3.5 (right), was disturbing. In particular, Unsöld's result deviated from Payne's by a factor of 5. Unsöld was disturbed by the failure to obtain the correct shape of the line and referred to Rosseland[33] and Freundlich[34] who demonstrated (wrongly) that, if one included the effect of the electrostatic force between the particles of the stellar plasma, then one could improve agreement between observation and theory for the hydrogen lines. In summary, the very high theoretically derived concentration bothered the theoreticians who were looking for reasons why the theory failed.

---

26 Struve, O. and Zebergs, V., *Astronomy of the 20th Century*, New York, 1962, p. 220.

27 Lang, K.R. and Gingerich, O., ed., *A Source Book in Astronomy and Astrophysics, 1900–1975*, Harvard University Press, 1979.

28 When Shapley prepared his *Source Book in Astronomy, 1900–1950*, Harvard University Press, 1960, he decided not to include Payne's thesis on the grounds that it was too technical. Similarly, Shapley did not include Bethe's 1939 Nobel prizewinning paper about the CN nuclear cycle (see later), but did include Russell's popular talk about Bethe's discovery.

29 Klüber, H., Zeit. f. Phys. **44**, 481 (1927).

30 Minnaert, M., Zeit. f. Phys. **45**, 610 (1927).

31 Unsöld, A., Zeit. f. Phys. **44**, 793 (1927).

32 Milne, E.A., Phil. Trans. Roy. Soc. London A **223**, 201 (1923).

33 Rosseland, S., MNRAS **85**, 541 (1925).

34 Freundlich, E.F., Erg. d. Exact. Natur. **6**, 35, Berlin, 1927.

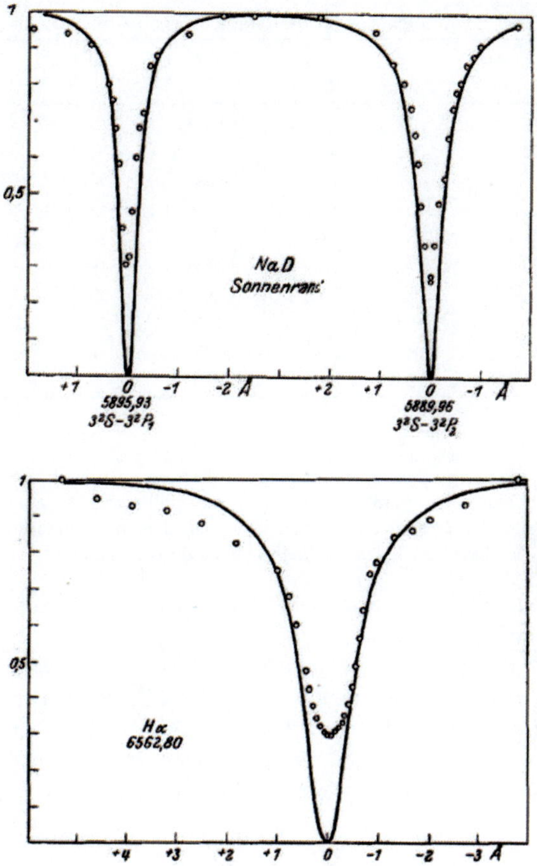

Fig. 3.5 *Top* Unsöld's calculation of the shape of the sodium *D* line. *Bottom* Unsöld's calculation of the shape of the hydrogen H_{α} line, where the agreement was unacceptable

3.6 Confirmation, and by Whom!

In 1929, Russell published a thorough analysis of the composition of the solar atmosphere,[35] and reached the conclusion that:

> The calculated abundance of hydrogen in the Sun's atmosphere is almost incredibly great.

Russell produced a table (see Table 3.3) in which he compared his results for the Sun with those of Payne for main sequence stars. Russell had to convert Payne's results to make them comparable with his own,[36] and for this reason the numbers in this table differ from those in Payne's. Russell found that:

---

35 Russell, H.N., Ap. J. **70**, 11 (1929).

36 The reference point of the abundances is arbitrary, and Payne's differed from Russell's. So Russell noted that, in order to relate the two results to the same reference point, a constant of 1.9 should be added to Payne's results as they appeared in her Ph.D. thesis.

Table 3.3 Comparison between Russell's and Payne's results

Element	Log*A* Payne	Log*A* Russell	Difference	Element	Log*A* Payne	Log*A* Russell	Difference
H	12.9	[11.5]	(+1.4)	Ca	6.7	6.7	0.0
He	10.2	–	–	Ti	6.0	5.2	+0.8
Li	1.9	2.3	−0.1	V	4.9	5.0	−0.1
C	6.4	7.4	−1.0	Cr	5.8	5.7	+0.1
C	8.0	9.0	−1.0	Mn	6.5	5.9	+0.6
Na	7.1	7.2	−0.1	Fe	6.7	7.2	−0.5
Mg	7.5	7.8	−0.3	Zn	6.1	4.9	+1.2
Al	6.9	6.4	+0.5	Sr	3.5	3.3	+0.2
Si	6.9	6.4	+0.5	Ba	3.0	3.3	−0.3
K	5.3	6.8	−1.5				

> This is very gratifying agreement, especially when it is considered that Miss Payne's results were determined by a different theoretical method, with instruments of a quite different type, and even on different bodies—a long list of stars, almost all of which are giants.

While Russell confirmed Payne's results, he did not believe that the numbers were correct.

Russell realized that the theory (and the observations) were far from mature, and pointed to the first steps taken by Unsöld[37] towards a reliable theory of line formation in stellar atmospheres, a theory which enables the derivation of relative abundances.

Russell calculated the abundances separately and took as a free parameter the pressure of the free electrons. What he found was:

> The electron pressures calculated from the degree of ionization and from the number of metallic atoms and ions are discordant.

Put simply, there had to be an unobserved element which supplied the electrons that made the difference in pressure. A similar discrepancy had been found by Unsöld a year earlier, and by Payne and Hogg.[38] Russell therefore reached the conclusion that the extra electrons must be supplied by helium, the abundance of which was:

> [...] very hard to estimate even in the stars, for its lines appear to be abnormally strong, like those of hydrogen.

In this way he arrived at the universal abundance of hydrogen and helium as given in Table 3.3.

As is clear from the numbers, the conclusion had to be that hydrogen was by far the most abundant element in the atmospheres of stars. However, these dramatic results did not resolve all remaining questions. As Russell noted (see Table 3.4):

[37] Unsöld, A., Zeit. fur Physik **46**, 765 (1928).

[38] Payne, C.H. and Hogg, F.S., HarCi **334**, 1 (1928).

Table 3.4 Probable composition of the Sun's atmosphere according to Russell (1929)

Element	By volume	By mass
Hydrogen	60 parts	60
Helium	2?	8?
Oxygen	2	32
Metals	1	32
Free electrons	0.8	0
Total	65.8	132

> Hydrogen must be extremely abundant in the atmosphere of the red giants, for its lines are stronger in their spectra than in that of the Sun. With any reasonable allowance for the effect of the lower temperature in diminishing the proportion of excited atoms, the relative abundance of hydrogen, compared with the metals, comes out hundreds of times greater than in the Sun. If this is true, the outer portions of these stars must be almost pure hydrogen, with hardly more than a smell of metallic vapors in it.

A new feature of scientific deduction appeared in Russell's analysis, namely, the *significance of the absences* (Russell 1929). Russell referred to elements that had excitation energies in a range that should be excited in the Sun, i.e., at a temperature of about 5,800 K, and hence should exhibit the relevant spectral lines if they were present, but which were not in fact observed.[39] Some of these elements had quite large relative abundances and yet their spectral lines were not observed. Another issue concerned elements that would not be excited at the relatively low temperature of the Sun, notably helium, whence their abundance could not be inferred directly from the spectra, despite their very high relative abundance.

In 1932, Eddington tried to resolve the problem of the giant star Capella. All attempts to model it assuming a homogeneous model had failed. So, when Eddington discussed *the hydrogen content of the stars*,[40] he discovered that the discrepancy between the theory of stellar structure and what was observed in Capella could be resolved if one assumed that the star contained 20% hydrogen by mass. But such a high abundance of hydrogen was at that time implausible (Eddington apparently did not trust Payne's results). Consequently, he *did not at that time think that the abundance of hydrogen was the actual explanation of the discrepancy*. But in 1934 Eddington changed his mind and claimed:

> This is no longer a matter of speculative curiosity; such determinations are needed to compare with and check the determination of abundance of hydrogen in stellar atmospheres made by Russell and others.

Disturbing as it was, Payne's work was not mentioned at all, although Payne did inform Eddington in person about her result during a visit to England.

Actually, the situation was even stranger. Even in 1919, Eddington had noticed the difference in mass between four hydrogen atoms and a helium atom and claimed that this difference could be the source of energy in the stars. At the time of his

---

[39] Examples are B, F, Ne, P, Cl, A, Se, Br, Kr, Te, I, Xe, Cs, Ta, Os, Au, Hg, Bi, Ra, Th, and U.

[40] Eddington, A.S., MNRAS **92**, 471 (1932).

famous address before the British Association for the Advancement of Science, the common wisdom was that stars contain very small amounts of hydrogen, and hence the process was considered possible but unrealistic. But then Payne came along and found that hydrogen is actually rather abundant in stars (at least in the envelope of the stars), and yet Eddington remained indifferent. Even after he himself discovered that hydrogen was so abundant in stars, he remained unmoved.

In retrospect, Eddington's argument using Capella is interesting, because we know today that Capella is not homogeneous. So in the spirit of Eddington himself, we can say that Eddington used an incorrect model to get the correct result.

The lines of hydrogen are exceptionally strong in the Sun, and yet it took Unsöld[41] three years to confirm, by independent means, that hydrogen is indeed the most abundant element in the solar atmosphere. Unsöld's conclusion was later confirmed by William McCrea (1904–1999).[42] At that time Russell did not believe the abnormally high result for hydrogen, and suspected that it was an artifact of a strong radiation pressure,[43] particularly on hydrogen, which gave rise to this anomaly.[44] However, while he was working on the problem, Russell visited Donald Menzel (1901–1976) at the Lick Observatory. Menzel was Russell's student in Princeton, obtaining his Ph.D. in 1924. This fact did not prevent the master from coming back to learn from his excellent past student. Menzel showed Russell his work on the solar spectrum and convinced Russell that hydrogen is indeed the most abundant element in it. Russell used Menzel's results before Menzel published them.[45] Menzel himself published his important results in 1931.[46]

It was only after the establishment of hydrogen as the most abundant element in the solar atmosphere that Elis Strömgren (1870–1947m)[47] extended Eddington's calculation to a full solar model (not just the atmosphere), and demonstrated that the entire Sun, not merely the atmosphere, is predominantly composed of hydrogen.

3.7 Finding the Abundances: The Curve of Growth

The spectra of stars contain Fraunhofer lines of different strengths. For example, in Fig. 3.6, we see the spectrum of solar sodium and its lines with different strengths. If an atom absorbs strongly at a certain wavelength, a small number of atoms would be needed to absorb the entire radiation at this wavelength. Hence, a strong line is formed up in the atmosphere at the lowest temperature. On the other hand, a weak

[41] Unsöld, A., Zeit. fur Physik **46**, 756 (1928).

[42] McCrea, W.H., MNRAS **89**, 483 (1929).

[43] The absorption of photons by an atom creates pressure on the atom, which is later transferred to all other non-absorbing atoms.

[44] Adams, W.S. and Russell, H.N., Phys. Rev. **68**, 9 (1928).

[45] Russell trusted his ex-student more than he trusted Shapley's student!

[46] Menzel, D.H., Pub. Lick Obs. **17**, 1 (1931).

[47] Strömgren, B., Zeit. für Astrophysik **4**, 118 (1932); ibid. **7**, 222 (1933).

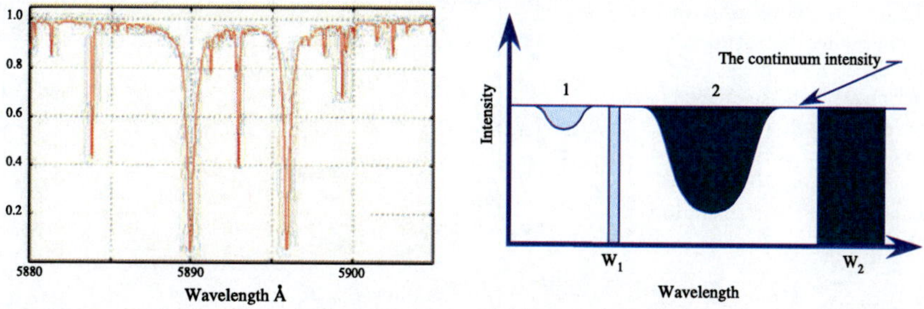

Fig. 3.6 *Left* typical sodium lines in the solar spectrum. *Strong* and *weak* lines are observed. *Right* the equivalent width W is essentially the area covered by the absorption line relative to the continuum around the line. The continuum is considered as unity. The *weak* line marked by 1 generates an equivalent width W_1, while the *strong* line 2 generates the equivalent width W_2

line requires many atoms for its formation, and so forms deep in the atmosphere, where the temperature is high. The strength of the line is expressed by the equivalent width, which is illustrated in Fig. 3.6. The equivalent width is the total area of the line.

The goal of the theory is to find the connection between the number of atoms with certain absorption properties and the observed equivalent width. The curve of growth describes how the number of atoms per unit volume of the given species increases (see Fig. 3.7). Each line of every element yields a point in the $(\mathrm{Log}(W/\lambda), \mathrm{Log}N)$ plane. The position of the point depends first of all on the number density of the species, but also on the structure of the stellar atmosphere and the physical properties of the line. Thus, by repeating the procedure for as many lines as possible, one can reconstruct the curve of growth. Clearly, there is only one free parameter, the number density of the species.

3.8 The Theory Gradually Improves

The theory of the curve of growth for absorption spectra was first developed by Minnaert and Slob in 1931,[48] when they showed how the intensity of a stellar absorption line, measured by its equivalent width, depended on the 'number of active atoms' involved in the electron transition. The theory was extended by Menzel[49] in 1936 and later by Unsöld in 1938.[50]

[48] Minnaert, M. and Slob, C., Proc. Kon. Akad. van Wetens. Amsterdam **34**, 542 (1931).

[49] Menzel, D.H., Ap. J. **84**, 462 (1936).

[50] Unsöld, A., Physik der Sternatmosphären, p. 264, Springer, Berlin, 1938.

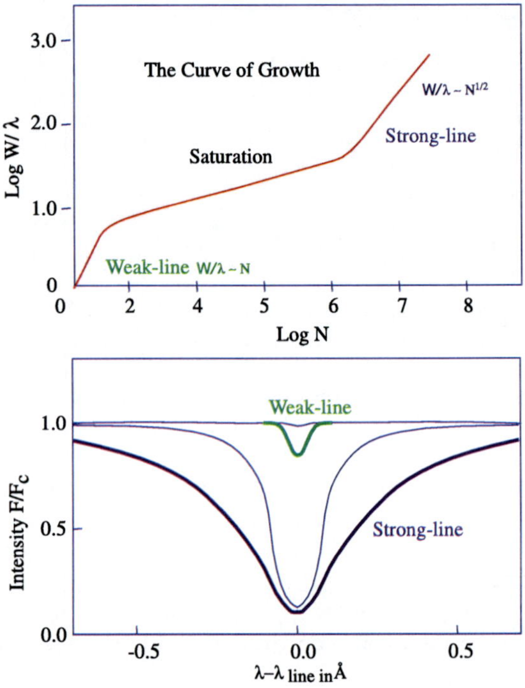

Fig. 3.7 The curve of growth which yields the relative abundance of elements. For each element and star there is a curve of growth which relates the equivalent width of the spectral lines and the relative abundance of the element in the star

The first attempts at detailed practical applications to the Sun and stellar spectra were carried out by Menzel, Baker, and Goldberg,[51] Greenstein,[52] and Wright.[53] The initial attempts to obtain the abundances yielded quite inaccurate results, mainly because the structure of the radiation field in the atmosphere, where the spectral lines formed, was poorly understood. Once Subrahmanyan Chandrasekhar (1910–1995)[54] had found the exact solution of the radiative transfer problem through the spectral line, in the late 1940s, Marshal Wrubel (1924–1968)[55] was able to provide accurate expressions for the connection between the equivalent width and the relative abundances, thereby clearing the way for reliable stellar abundance determinations. Consequently, reliable data could be obtained only after 1950. To be more accurate, the data and the theory of element abundances continues to improve even today.

[51] Menzel, D.H., Baker, J.G., and Goldberg, L., Ap. J. **87**, 81 (1938).

[52] Greenstein, J.L., Ap. J. **95**, 161 (1942).

[53] Wright, K.O., Pub. A.A.S. **10**, 34 (1940).

[54] Chandrasekhar, S., Ap. J. **106**, 145 (1947).

[55] Wrubel, M.H., Ap. J. **109**, 66 (1949); ibid. **112**, 157 (1950); Pub. Goethe Link Obs. Indiana University **13**, 117 (1954); Ap. J. **119**, 51 (1954). Wruber was a gifted pianist, who graduated from the Julliard School of Music and decided to give up a promising musical career in favor of astrophysics. Wrubel's illustrious career ended suddenly when he suffered a fatal heart attack and died at the age of 44.

The first attempts to determine the relative abundance suffered from several problems in both the stellar (observational and theoretical) and the atomic disciplines. On the one hand, the observational results were not smooth, but showed a bothering scatter around the mean value, because the stellar spectra were not measured sufficiently accurately. On the other hand, the theory was oversimplified, because the mathematical equations describing how the radiation passes through the layer where the lines form were not yet solved. Furthermore, the line intensities calculated on the basis of the atomic theory of spectral lines gave poor results for complex atoms. As a consequence, various fitting formulae were employed rather than accurate results.

Figure 3.6 shows the spectral line and the definition of the equivalent width W of the spectral line. The equivalent width is the width of the rectangle which has the same area as the spectral line. As is obvious, the value of W depends mostly on the shape of the wings of the line, and these in turn depend on how the electron in the high level decays, the damping of the spectral line, and the temperature. The damping depends in turn on the density. Moreover, the width depends on the temperature via the Doppler effect. The radiation from atoms moving away is shifted to the red while that from atoms moving toward the viewer is shifted to the blue. The result is a broader line. Consequently, it is via the shape of the wings that the stellar structure affects the shape of the line. Add to this the fact that the absolute value of the area of the line is sensitive to the temperature and we get an idea of the complexity of the problem, explaining why it took so many years to solve.

3.9 Conceptual Problems with Uniform Composition

If all stars had the same composition, this would imply two consequences, since the stars are well mixed and homogeneous, as was demonstrated by Eddington:

- The energy source of the stars does not change the composition, nor does it depend on it.
- The elements were formed before the stars, and all stars were formed from matter which already contained all the elements.

Clearly, at least one of the assumptions must be wrong.

Chapter 4
Is Physics the Same Everywhere?

4.1 Are the Laws of Physics Universally Valid?

As early as 1916, Eddington[1] proclaimed that the stars obey the same laws of physics as we discover in the laboratory. The known universe was small in those days, but today we can observe much further away and look back many more millions of years in time. The obvious question to ask is: were the laws of physics in general, and spectroscopy in particular, valid, say, 5 billion years ago? How confident can we be in the derived abundances of the elements, in objects that are ten billion light years away, using the physical laws we find today?

Two basic changes can take place: the laws of physics themselves can change, or the fundamental constants which enter into these laws can change. It is quite inconceivable, although we have no proof for it, that the nature of a physical law could change from one location to another. For example, it would be unreasonable for the gravitational force to have the known form, namely, inversely proportional to the distance squared, at one point, and another form, say, inversely proportional to another power, somewhere else. Consequently, the discussion is usually restricted to possible changes in the constants which enter the laws, like the gravitational constant. From our point of view here, the question is whether the spectral lines which we observe to have been formed a few billion years ago were formed according to the laws we discover in the laboratory today, so that we can trust our abundance determinations and what follows from their application to extremely remote objects.[2]

In 1967, George Gamow (1904–1968) wrote[3]:

> Since the works of Sir Arthur Eddington, it has become customary to discuss from time to time the numerical relations between various fundamental constants of nature. Although until today such discussions have not led to any practical result—that is, to any valuable road signs towards further development of the theory of the still unclear fundamental facts in

[1] Eddington, A.S., MNRAS **77**, 16 (1916).

[2] Note the paradox in the idiom: 'variability of the physical constants'.

[3] Gamow, G., PNAS **59**, 313 (1968).

G. Shaviv, *The Synthesis of the Elements*, Astrophysics and Space Science Library 387,
DOI: 10.1007/978-3-642-28385-7_4, © Springer-Verlag Berlin Heidelberg 2012

physics—it may be of some interest to survey the present status of this 'clairvoyant' branch of science.

One can consider the constants of physics as parameters, which may vary in value, but which have the known measured values in our particular universe at this point in time and location. If so, the question is how many fundamental parameters there are, i.e., parameters from which all other constants can be derived. The size of the set depends on the particular model we adopt.[4] The details are beyond our scope here, and the problem is only mentioned to alert the reader to the fact that this is still an open question. Our operative definition of a parameter is a physical constant which cannot be calculated from some known and established (!) physical theory.

4.2 The Connection Between the Micro and the Mega Cosmos

The idea that the constants of physics may depend on time arose from Paul Dirac's (1902–1984) hypothesis about large numbers,[5] which in turn emerged from the hypothesis by Hermann Weyl (1885–1955m)[6] twenty years earlier.

In 1917, when Einstein considered[7] a model for the entire Universe, the expansion of the Universe was not yet known. Moreover, there was controversy as to whether the nebulae, or some of them, were extragalactic or not.[8] In those days many thought that the Universe comprised only our Milky Way and its stars. Consequently, Einstein considered the stars and found that their speeds were much smaller than the speed of light. He thus concluded that the Universe must be static. But the equations of general relativity yielded only dynamic solutions. Being unable to reconcile his solution for the Universe with observation, Einstein invented the cosmological constant Λ, which was added ad hoc to the equation of general relativity, so as to force the solution to be static. The result was that

$$\frac{1}{R^2} = \frac{1}{2}\rho = \Lambda,$$

where R is the radius of curvature of the Universe, ρ is the mean mass density of the Universe, and Λ is the new constant Einstein found it necessary to introduce.

4 Weinberg, S., Roy. Soc. (London), Phil. Trans. Ser. A **310**, 249 (1983).

5 Dirac, P.A.M., Nature **139**, 323 (1937); Proc. Roy. Soc. London **A 165**, 199 (1938).

6 Weyl, H., Annal. Phys. **54**, 117 (1917); ibid. **59**, 101 (1919); Naturwiss. **22**, 145 (1934). See also, Gorelik, G., *Hermann Weyl and large numbers in relativistic cosmology*, in *Einstein Studies in Russia*, Einstein Studies **10**, ed. by Balashov and Vizgin, Birkhäuser, Boston, Basel (2002) p. 91.

7 Einstein, A., Sitzungsber. Berlin, 8 February, 142 (1917).

8 Telescopes were not sufficiently powerful to resolve individual stars, even in the nearest galaxy, and consequently extragalactic galaxies appeared no different from gaseous clouds, although they had peculiar shapes that the gaseous clouds did not have.

An estimate of the mean density of the Universe provides an estimate for the radius of curvature R.

In the same year, Willem de Sitter (1872–1934m)[9] solved Einstein's equations for the Universe, rederived Einstein's solution, and found that there were other solutions in which the relation between the cosmological constant and the radius of curvature was different, e.g.,[10]

$$\frac{3}{R^2} = \Lambda,$$

He then used observational data to find the actual value of the radius of curvature R, or equivalently, the radius of the Universe. Since extragalactic objects were not known, de Sitter could use only the density of observed stars in the Galaxy. With this limitation, he found that $\rho > 10^{-27}$ g/cm^3, and consequently $R < 10^{27}$ cm (about 10^9 light years). These are the numbers Weyl used in his estimates.[11] Once this number was available, people could play with it and find 'suggestive relations'.

Weyl's idea was as follows. Consider the radius of the electron and the radius of the Universe. Approximately, then,

$$\frac{\text{radius of Universe as found by de Sitter}}{\text{classical electromagnetic radius of electron}} = \frac{r_U}{r_e} \sim 10^{42},$$

where r_U and r_e are the radii of the Universe, taken as R, and the classical electromagnetic radius of the electron, respectively.

Let m_e be the electromagnetic mass of the electron, that is, the mass one measures for the electron in electrical experiments, and let r_e be the radius detected for the electron in electromagnetic experiments. Then one can define the gravitational radius of the electron as follows. First define a gravitational mass m_G by

$$\text{rest mass energy} = m_G c^2 = \frac{G m_e^2}{r_e} = \text{gravitational energy},$$

where m_e is the mass of the electron. This equates the gravitational energy of the electron (assuming that the radius of the electron is the classical electromagnetic

---

9 de Sitter, W., MNRAS **78**, 3 (1917).

10 The different solutions arise from assuming a different line element for the Universe. de Sitter also got a solution with $\Lambda = 0$ for $\rho = 0$, i.e., empty space.

11 Weyl pointed out that *the cosmological term that Einstein first added to his theory is a natural consequence of our original principle*. In other words, Weyl derived Einstein's cosmological constant from his principle. The irony of fate is that, when Hubble discovered the expansion of the Universe, the need for the cosmological constant disappeared, only to reappear about 60 years later when observations indicated that the expansion of the Universe is actually accelerating.

As a matter of fact, Weyl believed so strongly in his hypothesis that he criticized Einstein's theory of gravitation as leading to discordant gravitational and electrical radii of electrons. At the same time, Einstein was also critical of Weyl's theory, but his criticism did not convince Weyl, who did not provide a critical test for the theory.

radius) with the rest mass energy of a mass m_{G}. The classical radius of the electron is defined by

$$r_{\mathrm{e}} = \frac{e^2}{4\pi\varepsilon_0 m_{\mathrm{e}} c^2},$$

where ε_0 is the dielectric constant (needed to have the right units, but cancels out and has no effect). Next, we define the gravitational radius of the electron (using the same dielectric constant) by

$$r_{\mathrm{G}} = \frac{e^2}{4\pi\varepsilon_0 m_{\mathrm{G}} c^2} .$$

With this definition of r_{G}, it follows that

$$\frac{r_{\mathrm{G}}}{r_{\mathrm{e}}} \sim 10^{42}, \quad \text{while we also have} \quad \frac{r_{\mathrm{U}}}{r_{\mathrm{e}}} \sim 10^{42} .$$

The fundamental question that Weyl asked was whether the similarity in the numerical values of the two ratios could have any physical meaning.

This ratio turns up in many different ways. For example, it is the ratio between the Coulomb and the gravitational forces of the electron:

$$\frac{f_{\mathrm{Coul}}}{f_{\mathrm{grav}}} = \frac{e^2}{4\pi\varepsilon_0 G m_{\mathrm{e}}^2} \sim \sqrt{N} \sim 10^{42},$$

where N is Eddington's number and f_{Coul} and f_{grav} are the Coulomb force and the gravitational force between two electrons.

If the relation between the cosmic and the microscopic radii is valid, and if it is constant and not a provocation of nature, and this is a big 'if', then it follows that, as the Universe expands and changes its radius, the microscopic quantities should change as well. In other words, according to this far-reaching hypothesis, the expansion of the Universe induces a time variability in the structure of atoms. The structure of the macro-scale and the micro-scale are then closely tied together. And since the structure of the atom is inferred from the spectral lines, the location of the spectral lines should vary in time.

Weyl found in Eddington a great supporter of his theory. Eddington[12] believed that the *coincidence of large numbers* did indeed follow from Weyl's theory. When it became evident that Weyl's idea was a non-starter, Eddington did not abandon the hope that micro and mega physics might be connected, but substituted tempestuous hypotheses for well-grounded physics.

Obviously, changing physical constants is at least in part in contradiction with Eddington's statement about the laws of physics being applicable to stars in the

[12] Eddington, A.S., *Space, Time and Gravitation. An Outline of the General Relativity Theory*, Cambridge Press, 1920.

strong sense. Hence, Eddington did not believe in changing physical constants. In particular he rejected suggestions that the speed of light changes in time, as suggested by Peter Wold (1881–1945).[13]

Arnold Sommerfeld (1868–1951m) introduced the fine structure constant in 1916, when he expanded Bohr's theory of the hydrogen atom to comply with Einstein's theory of special relativity. The fine structure constant α was defined as $\alpha = 2\pi e^2/hc = 7.2977 \times 10^{-3} = 1/137.029$, or the ratio of the velocity of the electron in the first circular orbit of the relativistic Bohr atom to the speed of light in vacuum, and it expresses the magnitude of the effects induced by the relativity theory in Bohr's Newtonian quantum mechanics. The fine structure constant controls the size of the splitting in the energy level, and hence the spectral lines, in hydrogen-like atoms, due to first order relativistic effects. The energy levels in Sommerfeld's theory are given by

$$E_n = -\frac{2\pi^2\mu e^4}{h^2}\frac{Z^2}{n^2}\left[1+\frac{\alpha^2 Z^2}{n}\left(\frac{1}{k}-\frac{3}{4n}\right)\right],$$

where Z is the charge of the ion, μ is the reduced mass of the electron, $n = 1, 2, \ldots,$ is the principal quantum number, k is the azimuthal quantum number ($k = 1, 2, \ldots, n$), and $\alpha = 2\pi e^2/hc = 1/137$ is the fine structure constant. The complete formula contains terms to higher orders in α which are omitted here. The term in α^2 expresses the effect of relativity on the energy levels, as found by Sommerfeld. Since for every n, k takes integer values up to n, every Bohr level with a given n splits accordingly into n substates. For us here, the fine structure constant is the coupling between relativity (expressed in c) and quantum mechanics (expressed in h).

In the late 1930s, the best experimental value for the fine structure constant α was $1/136$. The fact that this number came out to an integer (to a very good approximation) excited the imagination of many physicists. It is extremely unusual for a dimensionless number to come out as an integer. Eddington[14] began by arguing from aesthetic and numerological considerations that α should be exactly $1/136$. He then gave a *proof* (!) that the number of protons in the Universe is exactly

$$N_{\text{Edd}} = \frac{1}{\alpha} \times 2^{256} = 136 \times 2^{256} \sim 1.57 \times 10^{79}.$$

---

[13] Wold, P., Phys. Rev. **47**, 217 (1935).

[14] Eddington, A.S., *New Pathway in Science*, Messenger Lectures, University Press, 1935. Eddington was a Quaker with deep commitments to the Society of Friends. He participated in the Quaker Guild of Teachers, whose mission was to help its members understand and integrate their faith and their intellectual life into their occupation. The Quakers believe that the 'inner light' enlightens every human being, and emphasize the personal experience of God. Eddington's philosophical writings can be traced to his frequent (and mostly correct) reliance upon his scientific intuition, combined with his inner conviction as a Quaker, based on spiritual practice and commitment. See also: Eddington, A.S., *Religion, causation, science and mysticism*, in *The Nature of the Physical World*, MacMillan, 1928.

When the accuracy improved and a better value for α became $1/137$, Eddington provided a new proof, this time that α must be exactly $1/137$.[15] However, Eddington's numerology failed in the most striking way possible when accurate measurements of α yielded $\alpha = 1/137.035999070$ and it stopped being a nice integer about which such pleasant stories could be woven.[16]

Eddington's explanation of the number 137 was as follows:

$$137 = \frac{1}{2}(16^2 - 16) + 16 + 1.$$

As a matter of fact, if you consider a symmetric matrix in $16 = 4 \times 4$ dimensions, then the number of independent terms in this matrix is $(16^2 - 16)/2 + 16 = 136$. The unit was added artificially[17] to the above formula after Eddington heard that a better experiment had given $\alpha = 1/137$. It is difficult to find physicists who would subscribe today to Eddington's arguments and his 'proof'.

It did not take a long time for physicists to get upset by such mystical arguments, and Beck, Bethe, and Riezler[18] published a travesty paper 'proving' that the absolute zero temperature must be -273°C.[19] To their dismay, the mockery was misunderstood, and on 6 March 1931, they had to publish a 'correction' and explain their motivation.

In 1936, Milne[20] discussed *the inverse square law of gravitation in the presence of a substratum or smoothed out universe*, and added the positive statement that the constant of gravity *depends on the epoch t*. The time dependence Milne got was $G \propto t$, i.e., it increases with time. Milne estimated that:

---

[15] This kind of reaction is reminiscent of a Jewish tradition of Gimatria, wherein each letter has a numerical value associated with it, so that every word corresponds to a numerical value given by the sum of the values of all its letters. In this way numerology enters Jewish mysticism like Kabbalah. If two words have the same numerical value, this must imply 'something'. For example, the words 'wine' and 'secret' have the same numerical value, viz., 70. Hence the proverb: *When wine enters, the secrets come out*, or *In vino veritas*. By the way, the numerical value of the Hebrew word Kabbalah is 137!

[16] The fine structure constant baffled Wolfgang Pauli (1900–1958), to the extent that he even involved the psychologist Carl Jung (1875–1961), with whom he entertained extensive correspondence, in the questions raised by this constant and its particular value. Pauli succumbed to cancer in 1958. When his assistant visited him in hospital, Pauli asked him: *Did you see the room number?* It was 137. Does this mean that even great physicists like Pauli can be superstitious? See Miller, A.I., *Deciphering the Cosmic Number: The Strange Friendship of Wolfgang Pauli and Carl Jung*, Norton, 2009.

[17] The addition of unity may equally be thought of as a kind of 'bakshish', claiming that the theory requires it.

[18] Beck, G., Bethe, H., and Riezler, W., Die Naturwissenschaften **19**, 39 (1931), published 9 January 1931.

[19] The spoof went as follows. They assumed 'according to Eddington' that every electron has $1/\alpha$ degrees of freedom and concocted the formula $T_0 = -(2/\alpha - 1)$ K. Then if $\alpha = 1/137$, this yields $T_0 = -273$ K.

[20] Milne, E.A., Proc. Roy. Soc. A **156**, 62 (1936).

> If G increases by one part in 2×10^9 per year, the sidereal year should be shortening, and the Earth in its orbit should be getting ahead of the position calculated for a constant G.

Milne argued that:

> Such an effect is observed and is usually described as the secular acceleration of the Sun.

Next, the length of the year should shorten by 10^{-9} parts of a year per year. Milne cited several observers who had estimated the change in the solar acceleration and found it to be consistent with his prediction.

In 1937, Dirac[21] discussed the cosmological constant Λ introduced by Einstein and also Eddington's thesis. Regarding Eddington, Dirac wrote:

> The significance of these numbers [the fundamental constants] has excited much interest in recent times, and Eddington has set up a theory for calculating each of them purely deductively. Eddington's arguments are not always rigorous, and, while they give one the feeling that they are probably substantially correct in the case of the smaller numbers [like the fine structure constant $1/137$], the larger numbers, namely the ratio of the electric to the gravitational force between electron and proton, which is about 10^{39}, and the ratio of the mass of the Universe to the mass of the proton, which is about 10^{78}, are so enormous as to make one think that some entirely different type of explanation is needed for them.

Thus, we do not speak about one theory, but two.[22]

Dirac then noted that the first number $t = 10^{39}$ is actually the square root of the second one 10^{78}, and hypothesized that we may have discovered a series of the form $t, t^2, \ldots$, i.e., the fact that fundamental dimensional constants appear in a set of order 10^{40n}, where $n = 0, 1, 2, \ldots$, is no accident. There must therefore be a physical reason behind it. The argument used by Dirac is reminiscent of Dirac's own criticism of Eddington, when Dirac claimed that:

> Present-day physics, both theoretically and experimentally, provides no evidence in favor of such an increase, but is much too imperfect to be able to assert that such an increase cannot occur, as it is so small; so that there is no reason to condemn our theory on this account.

The consequence, so argued Dirac, is that the constant of gravity must decrease with time as $G \propto 1/t$.

How did Dirac obtain the time dependence of the gravitational constant? The estimated age of the Universe was at that time 2×10^9 years. If one expresses this time in units of atomic constants, i.e., units of e^2/mc^3, one obtains a value of about 10^{39}, which is accidentally (?) close to the ratio between the Coulomb and gravitational forces given above:

> This suggests that the above-mentioned large numbers are to be regarded, not as constants, but as simple functions of our present epoch, expressed in atomic units.

---

21 Dirac, P.A.M., Nature **139**, 323 (1937).

22 Have a look at Farmelo, G., *The Strangest Man: The Hidden Life of Paul Dirac, Mystic of the Atom*, Basic Books, 2009, to appreciate Dirac's thoughts.

The reasoning continued by assuming that the 'gravitational power', defined by Dirac as the mass times the gravitational constant, increases with time and *this is to some extent equivalent to Milne's cosmology*. Would it be fair to claim that Dirac's argument was no more rigorous than Eddington's, the one that Dirac had been so critical about?

In 1939, Fritz Zwicky (1809–1974m),[23] while discussing the hypothesis of neutron stars, resorted to the Weyl and Dirac hypotheses, but not to Eddington's or Milne's which had been put forward just a few years earlier.[24]

In 1943, a year before his death, Eddington returned to the large numbers hypothesis and was ready to claim that N_{Edd} was the number of protons and electrons in the Universe. However, Eddington attributed a much greater significance to it, claiming that it implied the ratio of the electrical and gravitational forces between elementary particles. His explanations were as follows:

> If we consider a distribution of hydrogen in equilibrium at zero temperature, the presence of the matter produces a curvature of space, and the curvature causes the space to close when the number of particles contained in it reaches a certain total; this total is N.

In 1948, Edward Teller (1908–2003)[25] discussed the idea of variable physical constants and showed that it could easily get into conflict with geological evidence. Teller remarked that Dirac's hypothesis could be discussed either as a variable electronic charge or as a variable gravitational constant. However, Teller ends by stating that the geological evidence is not sufficienty accurate to disprove the hypothesis.

4.3 The Possible Variability of the Gravitational Constant

In 1961, Brans and Dicke[26] attempted to modify the general theory of gravitation in such a way as to explain what they claimed Einstein's theory failed to explain. A new dimensionless parameter ω was introduced into a generalized theory which boiled down to Einstein's theory for $\omega \to \infty$. One of the consequences of the new theory was the time variability of the gravitational constant.

---

[23] Zwicky, F., Phys. Rev. **55**, 726 (1939).

[24] Zwicky discussed *the difficulties with the classical theory of gravitation* when he considered the mass and radius of the neutron star and found that the critical mass for a neutron star should be $80M_{\odot}$. Zwicky completely ignored the theory of white dwarfs advanced by Chandrasekhar and the theory of neutron stars based on the same idea of degenerate matter as advanced by Landau, which were well established by then, despite Eddington's vocal protests.

[25] Teller, E., Phys. Rev. **73**, 801 (1948).

[26] Brans, C. and Dicke, R.H., Phys. Rev. **124**, 925 (1961). The first idea was due to Jordan [Jordan, P., Nature **164**, 637 (1949), *Schwerkraft und Welttall*, Vieweg, Braunschweig, 1955] and the proper credit should be given by calling this the Jordan, Brans, and Dicke theory. However Jordan's contribution was usually ignored by most researchers (but not by Brans in his PhD thesis), probably because of Jordan's affiliation with the NSDAP (the Nazi party). The irony was that Jordan was considered unreliable by the Nazi party, because of his former Jewish friends.

The problem of the variability of the physical constants has become very fashionable since several cosmological theories allow for such variability (see DeWitt,[27] Gamow,[28] and Dyson[29]). An exhaustive report would carry us too far astray from our main aim, which is to provide support for the claim that the abundance measurements in the most distant objects, those formed shortly after the beginning of the Universe, are in fact completely trustworthy.[30]

In 1973, Dirac[31] demonstrated that the large numbers hypothesis leads not only to the time variability of the gravitational constant, but also to the continuous creation of matter. Moreover, Dirac inferred that, if continuous creation takes place uniformly through the entire space, the Universe is flat and originated from a Big Bang.

4.4 Modern Attitude and Hypotheses

Until about 1980, the question as to whether the fundamental constants change or not appeared rather esoteric, like attempts to extend or replace the general theory of relativity. While observations did not require such an extension, some theories did. The suggested time variations had been based, at best, on numerology. However, various developments in cosmology and theoretical physics revived theories like that of Kaluza and Klein[32] and supergravity, theories which attempted to unify gravitation and electromagnetism. The Universe in these theories is not composed merely of our classical four-dimensional world. The 'actual Universe' in these theories comprises the classical four-dimensional world and a part with a higher number of dimensions in which the dimensions above four have today collapsed and become extremely small. These theories assume that all dimensions were alike when the Universe had just started. However, the extra dimensions evolved in time differently from the classical four-dimensions we are familiar with.

---

[27] DeWitt, B.S., PRL **13**, 114 (1964).

[28] Gamow, G., PRL **19**, 759 (1967).

[29] Dyson, F.J., in *Aspects of Quantum Theory*, ed. by Salam and Wigner, Cambridge University Press, 1972.

[30] Several authors attribute the suggestion about the possible variability of the fine structure constant to Landau (1955). It is interesting to note that different authors attribute to different scientists the invention of the idea, which so far has not been found to be true, that the fine structure constant may not actually be a constant. However, note that the review by the Russians Varshalovich and Potekhin (Varshalovich, D.A. and Potekhin, A.Y., Space Sci. Rev. **74**, 259, 1995) does not mention their fellow Russian Landau at all (Landau, L.D., *On the Quantum Theory of Fields. Niels Bohr and the Development of Physics*, ed. by Pauli, Pergamon Press, London, 1955). The paper discusses how to develop a perturbation theory for the strong interaction.

[31] Dirac, P.A.M., Proc. Roy. Soc. London A **338**, 439 (1974).

[32] The Kaluza–Klein theory is a model that seeks to unify the two fundamental forces of gravitation and electromagnetism. The theory was first published in 1921.

Such theories allow for a variability in the constants as seen in our 4D world (see, e.g., Okun[33] for a review). For example, in 1980, Chodos and Detweiler[34] derived $\alpha/G \propto t$ from such a theory, but this was already known to be wrong.

The discussion of such theories goes beyond our scope. Here we are interested in whether any of the constants which affect the determination of the abundances can vary in time and thereby confuse the meaning of the observed results, an experimental issue.

4.5 What Does Nature Say? The Oklo Phenomenon

A natural nuclear fission reactor is a sufficiently large natural uranium deposit for a chain fission reaction to be able to take place naturally in it. Uranium contains three isotopes, ^{238}U, ^{235}U, and ^{233}U. Only the last two can support a fission chain reaction because they emit more than two neutrons per fission and these neutrons can fission the next generation of uranium nuclei. ^{238}U converts into ^{239}Pu upon absorption of a neutron. The abundance of ^{235}U is usually 0.0072, and consequently natural uranium cannot support a chain reaction. In most cases the uranium ore is not pure and contains neutron absorbers (called poisons). Hence, this scenario always looked somewhat academic.[35]

A mix of ^{238}U and ^{235}U at the relative abundances which appear on Earth today cannot become critical in the sense of supporting a chain reaction in which the number of neutrons remains constant. However, if by some geological process the relative abundance of ^{235}U increases to something close to 3%, then criticality is possible. Alternatively, the presence of water which moderates[36] the neutrons is required. The issue of what exactly allows criticality is still not settled.[37] The idea that a natural reactor might occur in nature was predicted in 1956 by Paul Kuroda (1917–2001).[38] But in 1972, the phenomenon was discovered by the French physicist Francis Perrin in Oklo, Gabon. Several ore deposits at Oklo were found in which a chain fission reaction had taken place approximately 1.84 ± 0.07 billion years ago.[39] The reactor was critical for about $(2.29 \pm 0.7) \times 10^5$ years and produced a neutron flux of the

[33] Okun, L.B., Sov. Phys. Usp. **34**, 818 (1991).

[34] Chodos, A. and Detweiler, S., Phys. Rev. D **21**, 2167 (1980).

[35] The lifetimes of the uranium isotopes are 5×10^9 for ^{238}U and 7×10^8 for ^{235}U. The relative abundance of the two isotopes can be explained by their different lifetimes, provided they are synthesized at the same rate.

[36] If the energy of the neutrons is reduced, the probability of absorption increases, and a mix which cannot be critical without neutron moderation can then become critical.

[37] Ball, P., Nature, 15 May (2008).

[38] Kuroda, P.K., J. Chem. Phys. **25**, 781, 1295 (1956). Kuroda assumed that geological processes could cause uranium enrichment.

[39] It was a serendipitous discovery. The Oklo mines supply uranium mainly to the French nuclear industry. During the 1970s, a shipment of uranium from Oklo was found to be depleted in the fissionable isotope ^{235}U. The mystery was resolved when the operation of a natural reactor a billion years ago was discovered. The reactor would have consumed the fissionable uranium and produced

Table 4.1 Comparison of the samarium abundance in Oklo and universally

	Oklo (%)	Universal (%)
$^{149}Sm_{62}$	3.9	13.9
$^{150}Sm_{62}$	21.4	17.5

order of $(1-2) \times 10^{21}$ neutrons cm^{-2}. The amount of ore was sufficiently large to support the chain reaction for a few hundred thousand years, producing about 100 kW of power during the time of operation.[40]

In 1976, Shlyakhter[41] proposed to look at $^{149}Sm_{62}$, which is present in nature[42] at an abundance ratio of 13.8%, but should be depleted in the reactor zones because it is a strong neutron absorber through the reaction $n + {}^{149}Sm_{62} \rightarrow {}^{150}Sm_{62} + \gamma$.

What makes this reaction so unique is the fact that it is dominated by a nuclear level that lies as low as $E_r = 0.0973$ eV above the ground state and has a width of 0.063 eV. Typical nuclear energy levels are usually of the order of MeV above the ground state, so this particular level is, for some particular reason, extremely close to the ground state. In other words, the energy level is about seven orders of magnitude 'too low'. This must be due to a nearly perfect cancellation between two forces which determine the location of the energy level: the repulsive Coulomb force, which is proportional to the fine structure constant α, and the attractive nuclear force which depends on the strong interaction coupling constant.

Suppose now that one of the force constants, say for our purpose here α, varies even slightly. Then the strength of the Coulomb energy will also change slightly, and so will the energy of the relevant nuclear level. The probability of the nuclear reactions varies exponentially with the energy, and hence even a very small absolute change implies a large relative change in the energy level, and a concomitant change in the rate of the reaction and the relative abundance of the reaction products. This is the essence of Shlyakhter's idea about how to use the Oklo phenomenon[43] to establish bounds for the time variability of α over the past 2 billion years since the time the reaction took place.

The relative abundances of samarium are compared in Table 4.1, where the relative depletion of Sm^{149} is obvious. The neutrons emitted during the fission are very

(Footnote 39 continued)
radioactive nuclear waste like technetium. The longest half-life of any technetium isotope is 4.2 million years, and only extremely small amounts of technetium were found. To the surprise of the investigators, it was not a sophisticated theft, as might have been thought, but a natural nuclear reactor that did it.

40 The neutrons released by the fission of ^{235}U were also absorbed by the ^{238}U and converted it to plutonium and all its isotopes. However, the lifetime of Pu^{239} is about 24,000 years, so only extremely small amounts would be left in the ore after 2 billion years.

41 Shlyakhter, A.I., Nature **264**, 340 (1976).

42 The other stable isotopes of samarium and their relative abundances are ^{144}Sm 3.07, ^{147}Sm 14.99, ^{148}Sm 11.24, ^{149}Sm 13.82, ^{150}Sm 7.38, ^{150}Sm 7.38, ^{152}Sm 26.75, and ^{154}Sm 22.75%. The isotope under discussion, namely ^{149}Sm, plays an important role in the building of heavy elements in the slow neutron capture process.

43 The Oklo Phenomenon, I.A.E.A. Symposium proceedings, Vienna, 1975.

energetic (1−2 MeV), and as such have a low probability of being absorbed. The neutrons in man-made nuclear reactors are slowed down to thermal energies of the order of $kT = 0.024\,\mathrm{eV}$. In Oklo, the energetic neutrons were slowed down inefficiently by various isotopes so as to reach the resonance energy of ^{150}Sm, which lies at thermal energies. Once there, they were preferentially absorbed by ^{149}Sm forming ^{150}Sm.

Shlyakhter found that the abundances in Oklo imply that

$$\left|\frac{\Delta\alpha}{\alpha}\right| \leq 2\times10^{-8},\quad \text{or equivalently |relative change in } \alpha| \leq 10^{-17} \text{ per year.}$$

The most recent result from the Oklo reactor is by Gould, Sharapov, and Lamoreaux[44]:

$$-0.11\times10^{-7} \leq \text{relative change in } \alpha \leq 0.24\times10^{-7} \text{ since about 2 billion years ago,}$$

while Petrov et al.[45] found

$$-4\times10^{-17} \leq \text{relative change in } \alpha \text{ per year} \leq 3\times10^{-17},$$

which means that, if α changes, it does so on a timescale longer than 10^{17} years, or 10^4 times longer than the age of the Universe. The results are perfectly consistent with no change at all.

4.6 α Variability in the Distant Past?

If α changes in time, then the ratios of wavelengths emitted by atoms should change as well. By observing distant quasars, it is possible to set limits on the amount of variability in time of α. Quasars are the ideal objects because they are very luminous. This means that they are observed at very large distances, or equivalently, at much earlier times of the Universe.[46]

In 1956 Savedoff[47] used observations by Minkowski and Wilson[48] and Lilley and McClain[49] on a set of spectral lines observed in two colliding galaxies known as

---

[44] Gould, C.R., Sharapov, E.I., and Lamoreaux, S.K., APS, 73rd Annual Meeting of the Southeastern Section of the APS, 9–11 November 2006, abstract #BC.001, 11/2006.

[45] Petrov, Yu.V., Nazarov, A.I., Onegin, M.S., and Sakhnovsky, E.G., Phys. Rev. C **74**, 064610 (2006).

[46] In a simple Einstein–de Sitter universe, the relation between distance and redshift z is given by $r = (2c/H_0)\left[1 - 1/\sqrt{(1+z)}\right]$, where H_0 is the Hubble constant and $z = \Delta\lambda/\lambda$, where λ is the wavelength in the laboratory and $\Delta\lambda$ the shift in wavelength due to the expansion of the Universe.

[47] Savedoff, M.P., Nature **178**, 688 (1956).

[48] Minkowski, R. and Wilson, O.C., Astrophys. J. **123**, 373 (1956).

[49] Lilley, A.E., and McClain, E.F., Astrophys. J. **123**, 172 (1956).

Cygnus A, seen at a distance of 3×10^8 years. They derived

$$\frac{\alpha^2(\text{Cygnus A})}{\alpha^2(\text{local})} = 1.0036 \pm 0.0032.$$

Note that the estimated error is just slightly smaller than the deviation from unity. A more stringent upper limit was set by Bahcall, Sargent, and Schmidt[50] and Bahcall and Schmidt,[51] who used the observed fine-structure splitting of the emission lines of OIII and NeIII[52] to find that

$$\frac{\alpha(\text{at redshift 1.95})}{\alpha(\text{today})} \equiv \frac{\alpha(2.7 \times 10^9\,\text{years ago})}{\alpha(\text{today})} = 0.97 \pm 0.05.$$

Assuming that the quasars are indeed situated at cosmological distances, Bahcall and Schmidt found that

$$\frac{\alpha(\text{at redshift 0.2})}{\alpha(\text{today})} = 1.001 \pm 0.002.$$

In 1967, ignoring the previous negations of such a change listed above, Gamow[53] suggested that G is constant but α should vary with time according to $\alpha \propto t$. Gamow was aware of the above-mentioned limits, but argued that the spectral lines used for the determination of the upper limits were not quasar lines formed so many years ago, but rather spectral lines somehow produced more recently in intervening galaxies. At that time controversy was raging over the location and nature of quasars. The bone of contention was whether the quasars really are located at enormous cosmological distances, as implied by their unusually large redshifts, and consequently produce unprecedented amounts of energy, or whether they are 'local', in fact close to our Galaxy, or even inside it, and do not generate a problem regarding their energy source. Gamow was aware of the previous results which limited the variability of α, and moreover, he apparently realized that the limits were obtained by assuming that the quasars were where their redshift implied, namely at cosmological distances. So adopting the side which claimed that the quasars were local, Gamow could safely suggest that e^2 had changed much more recently. But then the time span over which the possible change is measured would be much shorter.

---

[50] Bahcall, J.N., Sargent, W.L.W., and Schmidt, M., Astrophys. J. **149**, L11 (1967).

[51] Bahcall, J.N. and Schmidt, M., PRL **19**, 1294 (1967). The authors refer in the paper to Bahcall and Salpeter [Bahcall, J.N. and Salpeter, E.E., Astrophys. J. **142**, 1677 (1965)] as setting a limit on the variation of α. However, Bahcall and Salpeter have a footnote in which they point out that they assume that *all physical constants are independent of time*.

[52] OIII and NeIII refer to twice ionized oxygen and neon, respectively, while the neutral species are denoted by OI and NeI.

[53] Gamow, G., PRL **19**, 759 (1967).

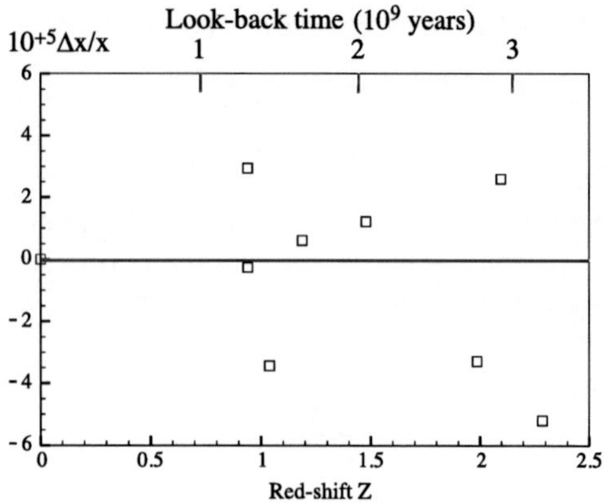

Fig. 4.1 The results for $\Delta x/x$ (in units of 10^{-5}) as found by Tzanavaris et al. (2005). The results from eight quasars are shown. The error in the measurements has been suppressed so as to better appreciate the spread in the results. The distance is given in redshift and in units of 10^9 light-years

Gamow's paper got a fast response by many critics, to the point that the editor of the PRL had to limit the number of rebuttals. How many times does one have to kill a dead idea?

Chitre and Pai[54] adopted an idea due to Wilkinson in 1958[55] according to which a change in e causes a change in the radioactive decay time. They applied a geological method based on the α decay[56] of U^{238}, which has a decay constant of 6.49×10^9 years, and the decay of K^{40}, which has a time constant of 1.72×10^{10} years. The idea was that, if the basic charge e of the protons changes, then the Coulomb barrier of the nucleus changes, and with it the energy of the emitted particles and the decay time. The two effects operate in opposite directions but the net effect is to decrease the decay probability and increase the decay time. By comparing data from meteorites, they could place limits on the decay time of uranium and potassium and set the limit $\Delta e^2/e^2 \leq 5 \times 10^{-13}$ years^{-1}. In words, this means that, 10^9 years ago, e^2 differed from its present value by less than 5×10^{-4} of the present day value (Fig. 4.1).

While so far all experimental data produced upper limits for a possible change in α, the first positive indication of such a change came in 1999, when a team led by John Webb of the University of New South Wales claimed the first detection of a variation in α.[57] Using the Keck telescopes[58] and a data set of 30 quasars at redshifts $0.5 \leq z \leq 1.6$, Webb et al. found that their spectra were consistent with a slight increase in α over the last 10^{10} years. Specifically, they found for the whole sample

[54] Chitre, S.M. and Pal, Y., PRL **20**, 278 (1968).

[55] Wilkinson, D.H.W., Phil. Mag. **3**, 582 (1958).

[56] This is the α particle model of decay and not the decay of the fine structure constant, also denoted by α.

[57] Webb, J.K. et al., PRL **82,** 884 (1999); Murphy, M.T. et al., MNRAS **327**, 1208 (2001); Webb, J.K. et al., PRL **87** (2001); Murphy, M.T. et al., MNRAS **345**, 609 (2003).

[58] The Keck telescope in Hawaii has the largest mirror in the world, with a diameter of 10 m and composed of 36 smaller mirrors.

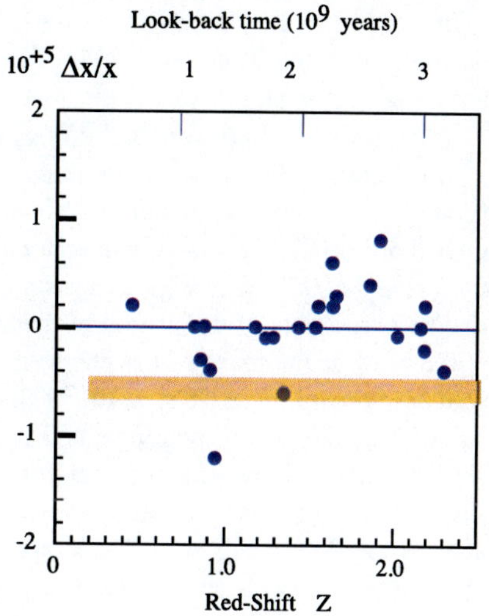

Fig. 4.2 Measured relative change $\Delta\alpha/\alpha$ in the fine structure constant α as a function of the redshifts of quasars, equivalent to the distance of the quasar or look-back time. The *yellow band* presents the Oklo phenomenon result plus the quoted error. The new results are inconsistent with the claimed change. European Southern Observatory, Press release 24 January 2010, based on Srianand et al. (2004)

that

$$\text{relative change in } \alpha = (-1.1 \pm 0.4) \times 10^{-5}.$$

More specifically, for the redshifts $z > 1$, they found that

$$\text{relative change in } \alpha = (-1.9 \pm 0.5) \times 10^{-5},$$

while for closer objects they found that

$$\text{relative change in } \alpha = (-0.2 \pm 0.4) \times 10^{-5}, \quad \text{for redshift } z < 1.$$

In 2001, Murphy et al.[59] measured the location of spectral lines of Mg, Al, S, Cr, Fe, N, Zn, and H and announced that the fine structure had shown a relative change of $(-0.72 \pm 0.18) \times 10^{-5}$, and with a high statistical degree of confidence, as the small error indicates. This degree of change was quoted without correcting for possible systematic effects. However, the authors argued that the only significant systematic effect so far identified by them, if removed, would only increase the relative change in α. Naturally, since the paper implied a revolutionary result, it received a warm response in the form of 175 citations by 2010.

Could the positive results found by Murphy and collaborators be consistent with the negative Oklo conclusion? Attempts to reconcile the measurements from quasars

[59] Murphy, M.T., and 7 coauthors, MNRAS **327**, 1208 (2001).

with those from the Oklo reaction were made by Fujii,[60] who suggested that α might oscillate in time. Of course, having just two numbers to hand, you can fit almost anything you want.

More recently, improved technology has made it possible to probe the value of α at much greater distances and to much greater accuracy. In 2005, Tzanavaris et al.[61] used spectra from quasars at redshifts between 0.24 and 2.04 (implying distances of up to 3 billion light-years), obtained by different astronomers, to measure the product $x = \alpha^2 g_p m_e/m_p$, where g_p is a proportionality constant that relates the observed magnetic moment of the proton to the appropriate angular momentum, and m_e/m_p is the ratio of the mass of the electron to the mass of the proton. The result found was $\Delta x/x = (1.17 \pm 1.01) \times 10^{-5}$. The estimated error was smaller then the result, so no change is out of the range, and the implication is thus that α has changed.

A recent result by Srianand et al.[62] based on spectra obtained from distant quasars set the limits of variability at $\Delta\alpha/\alpha = (-0.06 \pm 0.06) \times 10^{-5}$, consistent with no change at all (see Fig. 4.2). Thus, spectroscopy is safe and the determinations of abundances, as well as the physics, are valid. Obviously, there is no evidence to support the idea of any variability in the fine structure constant over the look-back time of 9.7×10^9 years.

Yet, in 2007, Murphy, Webb, and Flambaum[63] found errors in the analysis of Chand and his associates, which caused the null result to be ignored. Moreover, Webb and Murphy joined forces with King and Mortlock,[64] and used sophisticated statistical methods to reassure themselves that the change in α they had discovered previously was real and not a mere statistical fluke.

4.7 Present Day Laboratory Tests

Prestage, Tjoelker, and Maleki[65] applied atomic clocks to obtain an upper limit on the relative change in α that was smaller than 3.7×10^{-14}/year.

In 2008, Rosenband et al.[66] used the frequency ratio of singly ionized aluminum and singly ionized mercury in atomic clocks to obtain an upper limit on the possible present day change in α, namely,

$$\text{relative change in } \alpha = (-1.6 \pm 2.3) \times 10^{-17} \text{ per year.}$$

---

60 Fujii, Y., in Lecture Notes in Physics **648**, 167, 2004.

61 Tzanavaris, P., and 4 coauthors, PRL **95**, issue 4, id. 041301, 07/2005; arXiv e-print (arXiv:astro-ph/0412649).

62 Srianand, R., Chand, H., Petitjean, P., Aracil, B., PRL **9**, issue 12, id. 121302 (2004).

63 Murphy, M.T., Webb, J.K., and Flambaum, V.V., PRL **99**, 239001 (2007); ibid. MNRAS **384**, 1053 (2008).

64 King, J., Mortlock, D., Webb, J., and Murphy, M., Mem. S.A. It. **1**, 1 (2008); arXiv:0910.2699 [astro-ph.CO] 14 October 2009.

65 Prestage, J.D., Tjoelker, R.L., and Maleki, L., PRL **74**, 3511 (1995).

66 Rosenband, T., and 15 coauthors, Science **319**, 1808 (2008).

This quite stringent upper limit for present day possible change does not exclude a larger change in the past and hence does not eliminate the importance of cosmological checks.

A review on testing the stability of the fine structure constant in the laboratory can be found in Kolachevsky et al.[67]

Thorsett[68] used the change in the rate of rotation of neutron stars to derive the upper limit

$$\text{relative change in } G < (-0.6 \pm 4.2) \times 10^{-12} \text{ per year.}$$

However, this limit is theory-dependent. On the other hand, the result agrees perfectly with no change at all.

4.8 Upper Limits for the Variability of the Gravitational Constant

The present day limit on the variability of the gravitational constant was derived by Isern, García-Berro, and Lorén-Aguilar[69] from the rate of cooling of white dwarfs.[70] The result is

$$-(1 \pm 1) \times 10^{-11} < \text{relative time variability of } G < 0.$$

It is amazing that G is known experimentally only to four significant digits while the upper limit for a change in G is much smaller than the error in the measurement.[71] Note that no change at all lies within the range of possibilities. When white dwarfs cool, the only available energies are thermal and gravitational. When G changes, the gravitational energy of the star changes. When G increases, this means that the star becomes more tightly bound and releases its energy to the Universe, and vice versa. An upper limit to the change in G can be derived by comparing the cooling curve of such stars with observation.

---

[67] Kolachevsky, N., and 8 coauthors, Sp. Sci. Rev. 08/2009.

[68] Thorsett, S.E., PRL **77**, 1432 (1996).

[69] Isern, J., García-Berro, E., and Lorén-Aguilar, P., Mem. S.A. It. **80**, 820 (2009).

[70] Two properties make white dwarfs ideal for this problem. First, the only energy available for these stars is the internal energy of the ions and gravitational energy. So when the white dwarfs (Footnote 70 continued)
cool, only these two sources of energy play a role. Second, the radius of the white dwarfs is very small, about 1/100 the radius of the Sun. Consequently, the gravitational energy given by GM^2/r is very large. So if G changes in time, the change in the gravitational energy is very large, and so hopefully might be detected by changes in the cooling rate.

[71] $G = (6.693 \pm 0.027) \times 10^{-8}\,\text{cm}^3\text{g}^{-1}\text{s}^{-2}$ [Fixler, J.B. et al. Science **315**, 74 (2007)].

4.9 No Detectable Variability in the Physical Constants (So Far?)

In summary, despite incessant attempts to find changes in α and G, these two constants scorn the physicists and astronomers who have desperately sought a change of some kind and only obtained upper limits (see, for example, Flambaum,[72] who presents a summary in which all results are consistent with no change). The extensive literature about possible changes in α and/or G can be summarized as follows:

- So far the most consistent evidence is that they are constant.
- If there is any variability, the upper limit is so stringent that it has no effect on the abundance measurements, even in the oldest objects.
- Large variability cannot be ruled out for times earlier than the formation of the first atoms, in which the spectral lines are measured.

Modern theories like string theory may still lead to changes in these constants, but these should be restricted to extremely early times for which we have so far no data. The certainty of the abundance information that we derive from the earliest known stars is therefore secure.

What Gamow said in 1967 is still valid today.

[72] Flambaum, V.V., Eur. Phys. J. Special Topics **163**, 159 (2008).

Chapter 5
Towards the Bottom of the Nuclear Binding Energy

Once it became known that hydrogen was the most abundant element in the Universe (in 1931), and once the binding energy per nucleon could be obtained as a function of the atomic weight (in 1935) (see Fig. 5.1), it was quite natural to assume that the fusion of light elements into heavier ones was the source of stellar energy. But the mechanism remained mysterious. There is a basic nuclear difference between hydrogen and heavier elements in their neutron to proton ratio. While in hydrogen there are no neutrons, helium has two neutrons and two protons, and if four protons fuse to form a helium nucleus, then two protons must β-decay. We say that $Y_e = n(p)/[n(p)+n(n)]$ changes from $Y_e = 1$ for hydrogen to $Y_e = 1/2$ for He and heavier nuclei. How does this change take place? Inside nuclei or during a collision? This is the basic difference between the CN cycle and the proton–proton chain.[1]

Figure 5.1 shows how the binding energy per nucleon varies with the number of nucleons (protons and neutrons). The binding energy is negative, but is usually plotted as a positive number. The graph implies that the ^{56}Fe nucleus is the most tightly bound. Consequently, fusion of nuclei lighter than iron in stars provides them with energy sources, while the synthesis of elements heavier than iron requires energy and cannot be done in large amounts. In fact, elements heavier than iron can only be formed in minute quantities. Nuclear stellar evolution follows a descending trend along the binding energy curve, and this is the reason for the title of this chapter. Nuclear stellar evolution ends when the core of the star has been synthesized to iron, or in other words, when the matter has reached the minimum of the nuclear binding energy. As we shall see, stars with an iron core continue to contract until the radiation field disintegrates the iron and the result is collapse, followed by a gigantic explosion. Details are given below.

[1] In the CN cycle, the conversion of a proton into a neutron takes place inside a nucleus which absorbed an extra proton. In the pp chain, the conversion takes place 'in flight', while the two protons move one towards the other. A detailed description of the idea of nuclear reactions as energy source in stars can be found in Shaviv, G., *The Life of Stars. The Controversial Inception and Emergence of the Theory of Stellar Structure*, Springer, Heidelberg, and Magnes, Jerusalem (2009), so only a concise summary is presented in this chapter.

G. Shaviv, *The Synthesis of the Elements*, Astrophysics and Space Science Library 387,
DOI: 10.1007/978-3-642-28385-7_5,

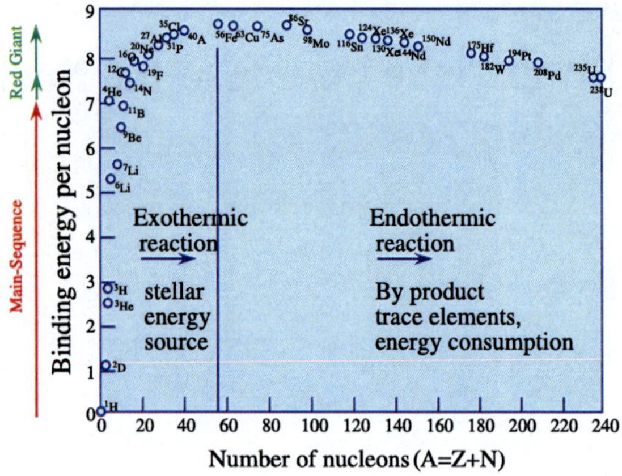

Fig. 5.1 Binding energy per nucleon (in MeV) as a function of $A = N + Z$, the number of protons and neutrons in the nucleus. *Arrows* on the *left-hand side* indicate the stellar phase when the relevant fusion takes place. Most of the nuclear energy is released during the stellar main sequence phase

Historically, astrophysicists discovered how stars die before it was understood how they live or how they are born. As early as 1919, observing the mass difference between 4 protons and a helium nucleus, Eddington suggested that the synthesis of hydrogen to helium might be the stellar energy source, but he had no idea how that could work. Recall that the neutron was discovered in 1932, when the helium nucleus was assumed to contain 4 protons and 2 electrons, so as to have a net charge of two. Furthermore, no nuclei with atomic masses of 2 or 3 were known to exist in nature. Hence, the only way Eddington's hypothetical synthesis could go would be for 4 protons and 2 electrons to come together simultaneously, release the binding energy, and stay together as a He nucleus, a hopelessly complicated and extremely rare process.

5.1 The Light at the End of the Tunnel

A major breakthrough in our understanding of radioactive decay came in 1929 when George Gamow (1904–1968m),[2] and Ronald Gurney (1898–1953) and Edward Condon (1902–1974)[3] discovered the phenomenon of quantum tunneling, i.e., the fact that particles in a potential well can escape even if classical physics predicts that they are bound. Gamow's paper was sent to the editor of Zeitschrift für Physik on

[2] Gamow, G., Zeit. f. Phys. **51**, 204 (1928). In his first attempt, Gamow assumed the α particle to be moving freely in the Coulomb field of the nucleus and got a continuous spectrum of emitted particles, in contrast with experiment. In his next attempt, Gamow combined the Coulomb and the nuclear forces to obtain an effective barrier.

[3] Gurney, R.W. and Condon, E.U., Phys. Rev. **33**, 127 (1929).

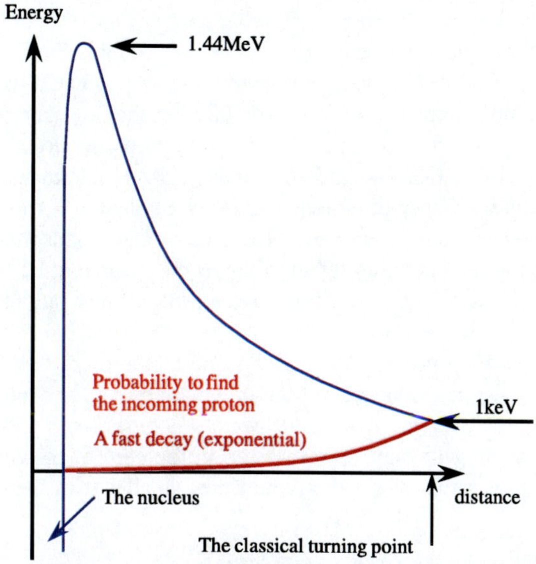

Fig. 5.2 Coulomb barrier between two approaching protons and the energies involved. The probability of a proton with energy smaller than the peak energy of the Coulomb barrier penetrating the classically forbidden region decreases as the wave–particle approaches the nucleus. The *red curve* gives the probability of entering the potential. The energy axis is not to scale. The peak of the Coulomb potential is 1,440 times higher than the energy of thermal protons in the Sun. The radius of the nucleus and the classical turning point are not to scale. The actual classical turning point is over 10^3 times greater then the radius of the nucleus

29 July 1928 and was accepted on 2 August, which is considered an unusually fast appreciation by the editor.[4]

The motivation of all physicists involved was to explain natural radioactivity: why certain nuclei are not stable and decay; why in some cases it takes billion of years for unstable nuclei to disintegrate, while in others the decay occurs in a fraction of a second; why a nucleus from a collection of atoms suddenly disintegrates, while the others follow, but each at a different time. At a certain time, radioactive decay had been considered as a possible candidate for the stellar energy source, but by the late 1920s, it was already clear that this could not be the case.

It would be hard to overestimate the importance of the tunneling phenomenon in so many disciplines, and above all in astrophysics. This discovery essentially instigated nuclear astrophysics. One can safely state that the tunneling phenomenon is one of the crucial factors for the existence of life in the cosmos, providing energy *and* a long time-scale for its release. It is amusing to note that Condon, who eventually explained how nuclear reactions can take place in stars, had believed several years previously

[4] Gurney and Condon's paper was received on 20 November and published in the February issue of Phys. Rev. However, the authors presented the results in several conference meetings before journal publication.

that mass annihilation was the source of stellar energy,[5] and that this explained the long lifetime of the stars. Since he assumed mass annihilation, the stellar ages he got were of the order of 10^{13}–10^{14} years, provided that they radiated like the Sun. But no reason was found to explain why mass annihilation should generate energy at this rate. Tunneling explains both the energy and its rate of production.

In October of 1928, Gamow and the young Fritz Houtermans[6] published an extension of Gamow's theory of radioactive decay. Soon after it was found out how α particles can escape from the nucleus, Houtermans met Atkinson and the two got the idea of applying the tunneling effect to the inverse problem, i.e., the problem of how stellar particles with energy well below the peak of the Coulomb barrier could pass through that barrier.[7]

Robert d'Escount Atkinson (1893–1981)[8] and Houtermans (1903–1966)[9] were quick to work out the consequences of the tunneling effect for nuclear reactions in stars and the paper was submitted on 19 March 1929. Figure 5.2 shows the parameters of the tunneling in the p + p reaction in the core of the Sun. The energy of the colliding protons is over a thousand times smaller than the peak energy of

---

[5] Condon, E., Proc. Natl Acad. Sci. U S A **11**, 125 (1924).

[6] Gamow, G. and Houtermans, F.G., Zeit. f. Phys. **XX**, 496 (1929).

[7] The Coulomb barrier is due to the mutual repulsion by the positive charges of the nucleus. In stars the atoms are stripped of their electrons. The electrons move like free particles and form a sea of negative charges. The effects of the free streaming electrons were later considered by Salpeter, E.E., Aust. J. Phys. **7**, 373 (1954), who found that they reduce the effective peak of the barrier and increase the tunneling.

[8] Atkinson, who was an ex-student of Rutherford, went to study physics in Germany, where he married a German woman who knew Fritz Houtermans from high school. So when the Atkinsons and Houtermans moved from Göttingen to Berlin-Charlottenburg to accept new jobs, the two families met and the collaboration started.

Charlotte Riefenstahl, a PhD student in physics at the University of Göttingen in 1927, was courted by Robert Oppenheimer and Houtermans. During a physics conference at the Black Sea resort of Batumi, Riefenstahl and Houtermans got married, with Wolfgang Pauli and Rudolf Peierls as witnesses to the ceremony.

Houtermans was a Communist and a member of the German Communist Party. When Hitler came to power in 1933, Charlotte insisted that they leave Germany, and in 1935, after a short stay in England, they emigrated to the Soviet Union. In the Great Purge, Houtermans was arrested by the NKVD, tortured, and forced to confess to being a Trotskyist and German spy. After the Hitler–Stalin Pact of 1939, Houtermans was turned over to the Gestapo and imprisoned in Berlin (out of the frying pan into the fire). Through efforts of (the anti-Nazi) Max von Laue, Houtermans was released in 1940, and in 1944 accepted a position as a nuclear physicist at the Physikalisch-Technische Reichsanstalt. During this period he discovered that neptunium and plutonium are fissionable.

While working in Forschunsinstitut Manfred von Ardenne, Houtermans showed that transuranic isotopes, such as neptunium and plutonium, could be used as fissionable fuels in substitution for uranium. During the war, Houtermans sent a telegram from Switzerland to Eugene Wigner, warning the USA of extensive German research on fission: *Hurry up. We are on the track.* After Khriplovich, I.B., *The Eventful Life of Fritz Houtermans*, Phys. Today **45**, 29 (1992).

[9] Atkinson, R.d'E., and Houtermans, F.G., Zeits. f. Physik **54**, 656 (1929). The original paper had an error. Due to a lack of experimental data, the authors took the classical theoretical probability for the nuclear reaction, rather then the quantum value. This is a contradiction in terms, and it was Gamow who corrected them.

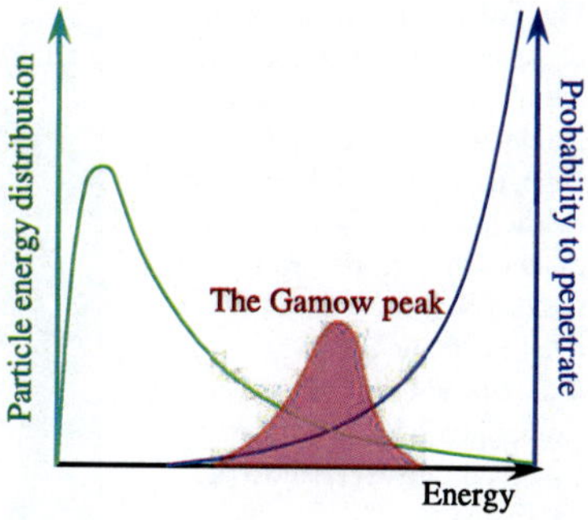

Fig. 5.3 The so-called Gamow peak. The result of the product of increasing and decreasing functions

the barrier. The results of Atkinson and Houtermans are shown in Fig. 5.3.[10] The mathematical expression Atkinson and Houtermans derived for the tunneling was quite complicated, and Gamow and Teller found it necessary, *because of subsequent development of nuclear physics*, to rederive the formula in a different way.[11] The expression they got, whose main components are described below, is simpler and is the one used today.

The probability of penetrating the Coulomb potential between the two charged particles behaves as $P \sim \exp(-b/\sqrt{E})$, where b is a constant that depends on the charges and the atomic weight of the colliding particles.[12] On the other hand, the energy distribution of the colliding particles is given by $\exp(-E/kT)$. The rate of the nuclear reaction in the star is therefore given by the product of the number of particles in each energy and the tunneling probability. This product of an increasing function with a decreasing function results in a rather narrow peak called the Gamow peak (see Fig. 5.3).[13] At low energies the probability of penetration vanishes, and at high energies the number of particles vanishes. In-between there is a small peak.

[10] Here is Houtermans's own account of the denouement: *That evening, after we had finished our essay, I went for a walk with a pretty girl. As soon as it grew dark the stars came out, one after another, in all their splendor. Don't they shine beautifully? cried my companion. But I simply stuck out my chest and said proudly: I've known since yesterday why it is that they shine.* von Buttlar, in *Leonium und andere Anekdoten um den Physik professor Dr. F.G. Houtermans*, Bochum (1982). Strictly speaking, this self-glorification, rather typical of physicists, came slightly too soon!

[11] Gamow, G. and Teller, E., Phys. Rev. **53**, 608 (1938).

[12] In fact, $b = 31.28 Z_1 Z_2 A^{1/3}$, where Z_1 and Z_2 are the charges of the colliding particles and A is the reduced atomic weight $A = A_1 A_2/(A_1 + A_2) = \mu/M_u$. The units are $\mathrm{keV}^{1/2}$ and M_u is the mass of 1 atomic mass unit.

[13] In 1936, Atkinson called the tunneling expression the Gamow formula (Astrophys. J. **84**, 73, 1936). As far as could be traced, Hoyle, F., Astrophys. J. Suppl. **1**, 121 (1954) used *the well-known Gamow peak*, although it does not appear in the literature before. Subsequently, one finds the Gamow peak mentioned in A.G.W. Cameron's series of lectures *Stellar Evolution, Nuclear Astrophysics, and Nucleosynthesis* CRL-41, Atomic energy of Canada Ltd, 1957. Cameron derived the peak without giving any reference whatsoever, and added: *This peak in the integrand will be called the Gamow peak.* No reason was given as to why the peak was not called the Houtermans, Atkinson, and Gamow peak. Independently, Satio Hayakawa, S., Hayashi, C., Imoto, M., and Kikuchi, K., Prog. Theor. Phys. **16**, 507 (1956) treated the Gamow peak as the accepted term and discussed

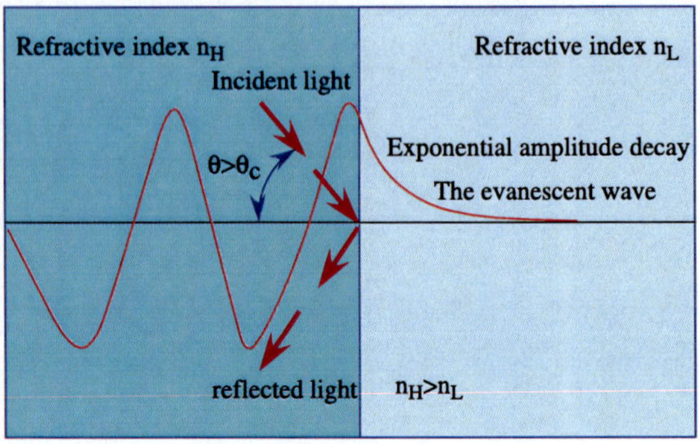

Fig. 5.4 Quantum tunneling of particles is analogous to the classical phenomenon of an evanescent wave which appears when light is incident on the interface between two media with different refractive indices. The evanescent wave, which propagates 'where it should not exist', decays exponentially and can be observed only very close to the interface

In the case of the *pp* reaction, the probability of two protons with energy of 1 keV penetrating the Coulomb barrier is 1.8×10^{-7}. The Gamow peak in the core of the Sun is at $\sim 5kT$ or ~ 5 keV, so the penetration probability increases by a factor of about 5,200. However, the number of particles decreases by a factor of 0.018.

The quantum tunneling phenomenon, in which a particle behaves like a wave, has a classical analog: the existence of an evanescent wave when light is incident, at an angle of incidence which is greater than the critical angle, on the interface between two media with different refractive indices. Full reflectance is expected in this case (see Fig. 5.4).

5.2 Stellar Energy Balance

If E_{grav} is the gravitational energy of a star and E_{therm} is its total thermal energy, then the binding energy of the star is given by $E_{\text{bind}} = E_{\text{grav}} + E_{\text{therm}} < 0$, where E_{grav} is negative and E_{therm} is positive. The temperature of the cores of stars is determined by the balance between the gravitational attraction and the gas pressure. Since the gravitational energy determines the gravitational force and the thermal energy determines the gas pressure, it is obvious that there exists a relation between the two. If the gas in the star behaves like an ideal gas, then $2E_{\text{therm}} + E_{\text{grav}} = 0$, a result which is known as the virial theorem.[14] The theorem is extremely simple and

(Footnote 13 continued)
modifications due to resonances, ignoring the fact that Gamow and Teller had already done just that in 1938.

[14] Poincaré, H., *Lecons sur les hypothèses cosmogoniques*, Librarie Scientifique, A. Hermann, 1811. This theorem is also the basis for the negative effective specific heat of stars. Only at this point do we need the Clausius connection between kinetic energy and temperature [Clausius, R.J.E., Phil. Mag. **2**, 1 (1851); ibid. 102; ibid. **12**, 81 (1856)]. The temperature is given by the kinetic energy divided by the Boltzmann constant. If the star loses energy L, it must contract, i.e., reduce its radius, and consequently lower its negative gravitational energy E_{grav}. The kinetic energy T is then more positive, so the temperature rises. In this way the star can lose energy and increases its temperature, in contrast to normal matter. This is one of Eddington's famous paradoxes about stars: they lose energy and heat up.

can be derived in just 6 lines.[15] The total thermal energy is proportional to the mean temperature of the star times the mass of the star, so the implication of the theorem is that the mean energy E of a particle in a star is related to the gravitational potential GM/R, where M is the mass of the star and R the radius. The mean gravitational potential of the Sun is 1 keV, and this is the mean energy of a proton in the Sun.[16] This is the connection between the macrophysics, i.e., the stars, and the micronuclear physics, i.e., the rate of the nuclear reactions.

But this is not the entire story. The star must be in hydrostatic and thermal balance. The energy produced by the nuclear reactions must be radiated away by the surface. The radiation emission at the surface (as well as the transfer of the radiation from the core to the surface) is determined by atomic physics. We can say that the gravitational field determines the core temperature at which the energy produced in the core can be radiated by the surface. This is where the microatomic physics comes in. The effect of the atomic physics on the radiating power of the surface depends on aspects of the chemical composition, such as the amount of hydrogen and helium there is relative to heavier elements. Consequently, the equilibrium state into which the star settles depends on the chemical abundance. The relative chemical abundance affects the location of the star in the HR diagram.

Let us reflect for a moment on these discoveries. The temperature of the stars is sufficiently low (!) for the nuclear reactions to proceed slowly enough to provide a long time scale for the energy supply needed for planetary evolution, and indeed for the evolution of life. If the temperature in the cores of stars had been higher, the reactions would have gone faster, the energy generation would have been greater, and stellar lifetimes shorter. Although they have central temperatures in the millions of degrees, stars on the main sequence are cool from the point of view of nuclear reactions, and for this reason have a long nuclear lifetime. But once the temperatures in stars increase by a factor of ~500, the physical scenario changes completely.

Checking the possible rates of all available nuclear reactions, Atkinson and Houtermans discovered that significant probabilities of penetration exist only in proton collisions with the first 7 elements. In fact, the lifetime of these elements against absorption of a proton turned out to be less than a few billion years. The penetration probability decreases quickly with increasing nuclear charge Z because the potential barrier is too high for the physical conditions in stars. They thus correctly concluded that, under the conditions prevailing in the Sun (a temperature of 40 million degrees was estimated at that time), proton reactions with heavy nuclei would be totally negligible. On the other hand, they did not discuss the proton–proton reaction, probably because deuterium was not yet known to exist in nature. The temperatures required for fusion were still too high for the Sun, and consequently many were skeptical about whether the Sun was sufficiently hot for fusion to take place in its core.[17]

---

[15] See Jeans, *Astronomy and Cosmogony*, Cambridge University Press, 1929, p. 67.

[16] For comparison, the potential energy of a proton on the surface of the Earth is 9.7×10^{-6} keV.

[17] To the skeptic, Eddington retorted: *If you don't think the center of the Sun is hot enough, go and look for a hotter place.* That being said, once tunneling had been discovered, it is a puzzle as to why Eddington did not attempt to see how his original hypothesis from 1919 could be realized.

And this is still not the full story. In the initial state we have a bound nucleus and a free particle. Once the colliding particle penetrates the Coulomb barrier and gets inside the nucleus, it reacts with the other particles and forms a nucleus with surplus energy, i.e., an excited nucleus. Unless the extra energy is removed from the excited nucleus, the nucleus cannot remain in this bound state. The simplest relaxation process, and best in some sense, is for the excited nucleus to emit the extra energy in the form of a γ ray. But there are other possibilities: the nucleus may eject a particle in such a way that the nucleus left behind is stable. But the question of what happens to an excited nucleus with extra energy lies in the domain of nuclear physics, a subject that was only in its infancy when Atkinson and Houtermans made their discovery. It was at this point, therefore, that nuclear physics was summoned.

5.3 The Birth of Nuclear Astrophysics

Many would identify the discovery of the neutron in 1932 as the birth of nuclear physics. However, the present author considers that nuclear astrophysics started a year earlier with two papers by Atkinson,[18] published in 1931. In these papers, the connection was established between nuclear reactions, stellar structure, and stellar evolution, and the first attempt was made to explain the relative abundances of elements as a consequence of element synthesis from hydrogen. The three papers published by Atkinson, two in 1931, and the last one in 1936, do not contain actual calculations and can be considered only as hypotheses and scenarios, probably because the requisite nuclear physics was not yet known or ready.[19]

From the outset, Atkinson admitted that he did not intend to derive a detailed theory that could explain, for example, the observed abundances of the elements. Atkinson adopted the observation about the predominance of hydrogen in stars and generalized it:

> It seems very reasonable to assume that in its initial state any star, or indeed the entire Universe, was composed solely of hydrogen.

The situation was, as Atkinson remarked, that so much observational data had accumulated that it was no longer possible to *construct an arbitrary hypothesis without producing a contradiction*. Notwithstanding, the recognition that hydrogen was the first and sole chemical element at the beginning, the difficulties in finding how fusion

(Footnote 17 continued)
Furthermore, there was apparently no reaction from Eddington to Bethe's discovery of the CN cycle, which vindicated his 20 year old hypothesis.

[18] Atkinson, R.d'E., Astrophys. J. **73**, 250, 308 (1931); Astrophys. J. **84**, 73 (1936).

[19] In 1960, Atkinson was awarded the Eddington Medal by the Royal Astronomical Society in recognition for the idea of *a regenerative process, having the essential property of the CN cycle later proposed by Bethe and von Weizsäcker*. Redman, R.O., President RAS, QJRAS **1**, 26, 1960. It is interesting to note that the seminal papers which won the Eddington medal were cited less than 10 times until 2010. Explanations of the regenerative process appear in the next section.

could start with just pure hydrogen, led all researchers to look for alternatives, and all alternatives assumed the pre-existence of heavy elements. There was no explanation as to how these heavy elements were formed and how the paradox of the need for heavy elements to synthesize them should be resolved.

5.4 How to Jump from H to He

The straightforward reaction $4^1\text{H}+2e^- \rightarrow ^4\text{He}+Q$, which 'jumps' over the barrier of the non-existence of a nucleus with two protons, *is almost certainly so improbable a process*, concluded Atkinson, that it can be ignored.

Since all combinations failed, Atkinson invented a new concept whereby helium could nevertheless be synthesized out of hydrogen. The essentials of the new idea were as follows: protons are captured successively by the light elements. In this way heavier nuclei are built one after the other. The proton absorption process continues until the product nucleus becomes unstable and disintegrates by emitting an α particle, i.e., a helium nucleus:

> The nuclei act as a sort of a trap and a cooking-pot combined, catching four protons and two electrons in such sequences and at such intervals as may prove practicable, fettering them by emitting as radiation most of the surplus mass brought in [...] combining its captives into an α particle, and emitting this after a delay.

Suppose that, after several absorptions of protons, an α particle is emitted. Then clearly a nucleus with $A-4$ (and a helium nucleus) appear, and this nucleus will now absorb four protons and reform the nucleus A that disintegrates. In this way a cycle is formed in which the nuclei A, $A+1$, $A+2$, $A+3$ act as catalysts, absorbing protons and forging them into a helium nucleus inside the nucleus $A+4$. The first three nuclei must emit a γ and the last nucleus must be unstable against the emission of a helium nucleus. As an example, Atkinson cited ^{56}Fe, which absorbs 4 protons to become ^{60}Zn. It is clear that the essence of the idea is the unstable nucleus, which disintegrates to the original nucleus and to a helium nucleus. The ^{60}Zn disintegrates into ^{56}Ni, which then absorbs two electrons and returns to ^{56}Fe. The total amount of these nuclei does not change during the process, although the relative amount may change because the speed of proton absorption may be different. As a result, the amount of the slow absorber will be high and vice versa. The breakdown of the abundances of the catalysts is a signature that such a process has in fact taken place. For this reason, Atkinson called it a 'regenerative process'. The rationale behind the process was that, in natural radioactive decays, an α particle is emitted and not a proton. This is the case with heavy elements. Atkinson hypothesized that it holds also for light elements.

To better appreciate the idea, let us quote Eddington[20] on this very subject:

> Indeed the formation of helium is necessarily so mysterious that we distrust all predictions as to the condition required. The attention paid to temperature, so far as it concerns the

[20] Eddington, A.S., *The Internal Constitution of the Stars*, Cambridge, p. 301.

cookery of the helium atom, seems to neglect the adage 'First catch your hare ...'. How the necessary materials of 4 mutually repelling protons and 2 electrons can be gathered together in one spot, baffles imagination. One cannot help thinking that this is one of the problems in which the macroscopic conception of space has ceased to be adequate, and that the material need not be at the same place (macroscopically regarded) though it is linked by a relation of proximity more fundamental than the spatial relation.

How desperate Eddington was a few years before pure quantum tunneling was discovered!

The experimental data on the masses and the binding energies of nuclei could not provide Atkinson with any clue as to which element stood at the end of the process, because the data were not sufficiently accurate and detailed at that time. So Atkinson resorted to the recently published nuclear stability theory by Gamow,[21] published just before Atkinson started his research. The theory provided good results, which nicely reproduced the measured numbers for the nuclei with atomic weights $4n$, where n is an integer smaller than 5. But for higher values of n, namely A greater than 25, the results were poor. Hence, Atkinson was unable to maintain accuracy when he assumed a very high Z (around that of iron).

Worse than that, since Atkinson was unable to exactly determine the 'last nucleus', he was compelled to assume a very high nuclear charge, making the reaction with the proton very difficult at the relatively low stellar temperatures, and therefore also making the process extremely slow. Inevitably, Atkinson proposed an untenable scenario with the remark:

It is only with reluctance that we introduce such a hypothesis, since there is no a priori justification for it at all, as far as is known.

It is a pity that, due to the lack of proper data, Atkinson had to retreat from his original ingenious and correct idea.

5.5 The Discovery of the Neutron

In the early 1930s, several major discoveries were made which changed nuclear physics and nuclear astrophysics. In particular, the neutron was discovered by Chadwick[22] in 1932. This discovery caused a revolution in nuclear physics and made it possible for the first time to apply the new quantum theory to nuclear physics.

Then deuterium was discovered following a theoretical prediction. It was well known that the atomic mass of hydrogen, when measured chemically, was found to be 1.00777 ± 0.00002, while Aston found 1.00756 ± 0.00015 using the mass spectrometer. So the difference between the two results was greater than the error. In 1931, Menzel[23] speculated that the difference might be due to an isotope of hydrogen of mass 2, with relative abundance $^1H/^2H = 4{,}500$, pointing out that:

---

[21] Gamow, G., Proc. R. Soc. **126**, 632 (1930).

[22] Chadwick, J., Nature, 27 February, 312 (1932). The full paper is: Proc. R. Soc. A **136**, 692 (1932).

[23] Birge, R.T., Menzel, D.H., Phys. Rev. **37**, 1669 (1931).

It should be possible to detect such an isotope by means of spectra.

Hardly a year after this speculation was announced, Harold Urey (1893–1981m),[24] Ferdinand Brickwedde (1903–1989), and George Murphy discovered deuterium in water,[25] and with about the predicted abundance.

Soon after the discovery of the $A = 2$ nucleus, tritium, the $Z = 1$, $A = 3$ nucleus, was produced independently in various experiments carried out by different groups of scientists: Rutherford and Cockroft, and Lawrence, Alvarez, and Libby.[26] The positron was discovered by Anderson in 1933.[27]

5.6 The Return of Atkinson

In 1936, Atkinson returned to reexamine his scenario from 1931 in view of the important recent discoveries. Atkinson considered the relevance of the neutron and deuterium for the physics of the energy sources in stars. Clearly, reactions with neutrons do not have the problem of a Coulomb repulsion and hence can operate at any temperature. The question was whether neutrons could be produced in sufficient amounts in stars. A check of all possible neutron production reactions revealed that they were very slow and that no significant quantities of neutrons could be formed. For example, the reaction $\mathrm{p} + \mathrm{e}^- \to \mathrm{n}$, namely, the absorption of an electron by a proton resulting in a neutron, was examined in the laboratory,[28] and the results were all negative. The only alternative left, and suggested by Atkinson, was to generate neutrons by first producing plenty of deuterium via the reaction $\mathrm{p} + \mathrm{p} \to {}^2\mathrm{D} + \mathrm{e}^+$, where e^+ is a positron, and then splitting the deuterium.

In this way, Atkinson discovered the first reaction which leads to what is known today as the pp chain, namely, the fusion of hydrogen to helium starting from pure hydrogen without any other element. However, Atkinson called for this reaction to produce deuterium and from it the neutrons. Atkinson expected it to be easy to

---

[24] Urey was awarded the Nobel Prize for Chemistry in 1934 for the discovery of deuterium. The high-minded Urey publicly acknowledged the crucial role played by Brickwedde and Murphy in the discovery, and gave each of them one-quarter of the Nobel prize money.

[25] Urey, H.C., Brickwedde, F.G., and Murphy, G.M., Phys. Rev. **39**, 164 (1932). Brickwedde produced the first sample of hydrogen in which the spectrum of deuterium was observed. The idea of a heavy hydrogen isotope was rejected by Aston who, in 1927 [Aston, F.W., Proc. R. Soc. Lond. A **115**, 487 (1927)], used mass spectrometric evidence to set an upper limit for the deuterium abundance ratio of ${}^2\mathrm{D}/{}^1\mathrm{H} < 1/5{,}000$. The spectroscopic evidence was not sufficient to convince the scientific community of the existence of a heavy hydrogen nucleus, although it was extremely clear. The history with the discovery of helium on the Sun was not sufficiently convincing for the existence of a new isotope in water, and Urey and Brickwedde were obliged to distill the water and provide a 'real proof'. This concise description does not do justice to the story, which is a chronicle of trial and error, false ideas, and misleading opinions by well-known scientists. See Brickwedde, F.G. *Harold Urey and the Discovery of Deuterium*, Phys. Today **35**, 34 (1982).

[26] The discovery was not credited because of misidentifications by this list of distinguished scientists.

[27] Carl Anderson won the 1936 Nobel Prize for Physics for the discovery of the positron.

[28] Livingood, J.J. and Snell, A.H., Phys. Rev. **48**, 851 (1935).

measure the reaction in the laboratory. But he could not have made a bigger mistake than to hold such an expectation. This is the most famous nuclear reaction in stars, one which cannot be measured in the laboratory because of its extremely low yield.[29] As a matter of fact, there is a mistake in the $p+p$ reaction as suggested by Atkinson, because the Gamow–Teller selection rule for β-decay was not yet known. According to the Fermi β-decay transition rule, the only transition known to Atkinson, this reaction cannot take place (detailed explanation to follow).

Atkinson realized that his previous hypothetical regenerative process, which reached elements with an atomic number as high as $Z = 28$, was probably not correct, but found comfort by stating that *there are not yet enough data*. He was right on that count.

When all attempts to generate neutrons by light elements had failed, Atkinson considered the possibility that neutrons might be formed by means of catalysts in a two-step regenerative process of the kind $M + {}^1H \rightarrow N$ and $N + e^- \rightarrow M + n$, where N and M are two nuclei which carry the reaction, and the transformation of a proton into a neutron takes place inside the nucleus N. This is actually what Hans Bethe (1906–2005) assumed to happen, following Atkinson's 'regenerative process' idea. The problem was that the two processes had to occur rather fast. If either of the reactions required a few billion years, the suggested process could still operate in stars, but the rate of energy generation would become too small to be relevant.

The possibility of deuterium disintegration, and recombination of a neutron and a proton, was treated theoretically by Bethe and Peierls[30] in 1935, a year before Atkinson came up with the idea of forming and splitting deuterium. They assumed that, during the merger of the particles, the interaction resembled an electromagnetic interaction. Fermi[31] was quick to show that Bethe and Peierls' theoretical results did not agree with experiment. According to Fermi, the reaction could not take place unless some additional interaction could bring the theory into agreement with experiment. But Fermi did not go as far as Gamow and Teller went two years later.

5.7 Weizsäcker: A Good Idea, but the Wrong Calculation

In 1937, Carl Weizsäcker (1912–2007) published two papers that led many to credit him with the invention of the CN cycle.[32] The first paper examined various reactions, none of which could serve as a solution to the problem, and Weizsäcker, becoming desperate about the situation emerging in stars, wrote:

[29] Willy Fowler used to give the following example in his lectures on nuclear astrophysics. With a 10 MW beam of protons in an accelerator which produces this power and accelerates 1 MeV protons before causing them to impinge on other protons, the result will be on average one reaction per year of operation. Now note that the Gamow peak for the pp reaction in the Sun is at 4–5 keV, so the rate of this reaction in the Sun will be significantly smaller.

[30] Bethe, H., and Peierls, R., Proc. R. Soc. A **148**, 146 (1935).

[31] Fermi, E., PRL **48**, 570 (1935).

[32] Weizsäcker, C.F., Physik. Zeits. **38**, 176 (1937) Paper I; ibid. 633 (1937) Paper II.

> We cannot exclude the possibility that the elements were created before the stars were formed, and presently the stars generate energy by creating minor changes in the composition of the stars.

The second paper was qualitative, like the first one. In it, he discussed and rejected various possibilities. Then Weizsäcker proposed the CN cycle, which is a concrete version of what Atkinson called a 'regenerative process'. However, he was unable to write down the relevant reaction rates. The set of reactions according to Weizsäcker is:

$$^{12}\mathrm{C} + {}^{1}\mathrm{H} \rightarrow {}^{13}\mathrm{N}\,, \quad {}^{13}\mathrm{N} \rightarrow {}^{13}\mathrm{C} + \mathrm{e}^{+}\,, \quad {}^{13}\mathrm{C} + {}^{1}\mathrm{H} \rightarrow {}^{14}\mathrm{N}\,,$$

$$^{14}\mathrm{N} + {}^{1}\mathrm{H} \rightarrow {}^{15}\mathrm{O}\,, \quad {}^{15}\mathrm{O} \rightarrow {}^{15}\mathrm{N} + \mathrm{e}^{+}\,, \quad {}^{15}\mathrm{N} + {}^{1}\mathrm{H} \rightarrow {}^{12}\mathrm{C} + {}^{4}\mathrm{He}\,.$$

The two conversions of protons into neutrons are assumed to take place inside the nucleus, and the extra positive charge is emitted as a positron. The nucleus $^{12}\mathrm{C}$ returns to its initial state. A crucial issue is why the pattern does not continue, and why $^{15}\mathrm{N} + {}^{1}\mathrm{H} \rightarrow {}^{16}\mathrm{O} + \gamma$ does not occur? Or rather, why it happens that, once in a thousand times, the excited 4α nucleus decays into a stable nucleus and not into a 3α nucleus and a free α particle.

5.8 The Answers Are in Bethe's Bible

During the years 1936–37, Bethe published a trio of synopses which covered everything known in nuclear physics, and very soon these bulky reviews became known as Bethe's Bible.[33] Of particular importance for us here is the table given by Livingston and Bethe, which summarizes reactions of the type $^{A}\mathrm{Z} + \mathrm{p} \rightarrow {}^{(A-3)}(\mathrm{Z}-1) + \alpha$, i.e., proton absorption with emission of a helium nucleus. The first part of the table is shown in Table 5.1. All the reactions can serve as the last reaction in the cycle if the product is one of the isotopes that appears in the leftmost column. One has to check that there is no leakage to nuclei that are not part of the cycle.

Bethe's Bible was published too late for Weizsäcker, although the data compiled in the Bible had all been published by others previously. But Weizsäcker chose not to look for the rates and made a crucial mistake when he assumed that the CN cycle was in dynamic equilibrium, an assumption for which he needed only the masses of the nuclei, which were already available. As the energy difference between the different nuclei is $\sim$10 MeV, the temperatures required for the process to be in equilibrium are $T \sim 10\,\mathrm{MeV}/k_{\mathrm{B}} = 1.1 \times 10^{11}$ K. More accurately, Weizsäcker found a temperature of 2.3×10^{11} K, which was way beyond what stellar models predicted to exist in stars. Consequently, Weizsäcker realized that his CN cycle could not operate in stars. Had

[33] Bethe, H.A. and Bacher, R.F., *Nuclear Physics A. Stationary States of Nuclei*, Rev. Mod. Phys. **8**, 82 (1936); Bethe, H.A., *Nuclear Physics B. Nuclear Dynamics, Theoretical*, Rev. Mod. Phys. **9**, 69 (1937); Livingston, M.S., and Bethe, H.A., *Nuclear Physics C. Nuclear Dynamics, Experimental*, Rev. Mod. Phys. **9**, 245 (1937).

Table 5.1 Summary of proton capture—α emission reactions, as given by Livingston and Bethe in 1937

Z	Isotope	Product	Q (theoretical)	Q (observed)
3	^{6}Li	^{3}He	3.76	3.72
	^{7}Li	^{4}He	17.25	17.13
4	^{9}Be	^{6}Li	2.25	2.28
5	^{11}B	^{8}Be	8.60	8.60
	^{11}B	^{4}He	8.72	8.7
6	^{13}C	^{10}B	−4.15	–
7	^{14}N	^{11}C	−3.3	–
	^{15}N	^{12}C	4.79	–
8	^{16}O	^{13}N	−5.4	–
	^{17}O	^{14}N	1.3	–
	^{18}O	^{15}N	2.82	–
9	^{19}F	^{16}O	8.14	–

Q is given in MeV

Weizsäcker calculated the rates of the reactions as given by Livingston and Bethe, he could have solved the problem. It suffices to identify the slowest reaction, and it is this that determines the energy generation rate.

5.9 Hans Bethe and the First Calculation of the pp Reaction

Hans Bethe was the leading light of nuclear physics, with a unique reputation as a world expert in all matters of nuclear reactions. Bethe published two papers almost at the same time. In these two seminal papers on stellar nuclear energy sources, Bethe reached the pinnacle of a twenty-year ascension which had started with Eddington's subatomic energy hypothesis and culminated in the solution. Eddington was alive to see the confirmation of one of his most important hypotheses, while Bethe's achievement won him the Eddington medal in 1961 (a year after Atkinson was awarded the same medal) and the Nobel prize in Physics in 1967.

Charles Critchfield (1911–1994) was a PhD student in George Washington University under the guidance of Teller and Gamow. The subject of his thesis, suggested by Gamow, was to calculate the rate of the p + p reaction in stars. When Critchfield finished the calculation, Gamow suggested that he should present the calculation to the 'high priest' of nuclear physics, Bethe himself, and try to get his approval. It was in 1938 that Critchfield presented his calculations to Bethe, who found them to be basically correct (apart from some factors of 2 here and there and a somewhat more powerful method for calculating the deuteron wave function), and in 1938 Bethe and Critchfield[34] published their calculation. The authors gave credit

[34] Bethe, H.A. and Critchfield, C.L., Phys. Rev. **54**, 248 (1938).

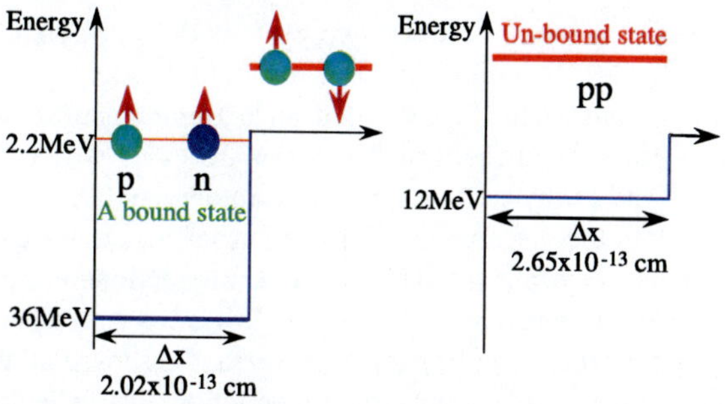

Fig. 5.5 The nuclear potential well between a neutron and a proton which gives rise to deuterium, and the nuclear potential acting between two protons which has no bound state. The Coulomb repulsion between the two protons is not shown. A potential of 36 MeV is needed to get just one energy state. A shallower potential would not allow for an energy level and would render the pn system unstable, hence non-existent in nature. As the protons repel each other, in order to have a bound state and allow the existence of ^{2}He, the nuclear potential well between the two protons should be much deeper than 12 MeV. After Preston, M.A., *Physics of the Nucleus*, Addison-Wesley, 1962

to Weizsäcker,[35] but not to Atkinson (to whom Weizsäcker himself gave credit), although it was Atkinson who had come up with the idea of the pp reaction, and not Weizsäcker. They wrote:

> There seems to be a general belief that the reaction $\mathrm{H}+\mathrm{H}=\mathrm{D}+\mathrm{e}^{+}+0.42\,\mathrm{MeV}$ is too rare to account for any appreciable fraction of the energy production in stars and that it can serve only to start the evolution of elements in a star which will then be carried on by other, more probable processes. It is the purpose of this paper to show that this belief is unfounded but that the reaction $\mathrm{p}+\mathrm{p}$ gives an energy evolution of the correct order of magnitude for the Sun.

The paper was published while Bethe was working on the CN cycle. So he added:

> We do not want to imply that the pp reaction is the only important source of energy. [...] The capture of protons by carbon and nitrogen will also play an important role.

The question mark in the calculation of the pp reaction was on the nuclear part: once the two protons are together, how fast will one of them decay into a neutron so that the neutron and the proton can capture each other to form deuterium? Due to the very low energy of the colliding protons in the Sun, only states with no angular momentum (s-waves) contribute significantly. One can consider it as a head-on collision, so that angular momentum plays no role. Consequently, the total angular momentum is the sum of the spins, and the spins alone control the reaction. Because of Pauli's exclusion principle, the incoming protons must have opposite spins. On the other hand, in the only bound state of deuterium, the spins of the neutron and proton are aligned. Hence a spin flip must take place (see Fig. 5.5). The strength of the nuclear force which holds the neutron and the proton together depends on the spin of the particles. The force between an aligned proton and neutron is sufficient to give a

[35] von Weizsäcker, C.F., Physik. Zeits. **38**, 176 (1937).

bound state, but the interaction between two protons does not yield a bound state under any circumstances. Deuterium has only one bound state.

In 1937, Gamow and Teller[36] postulated an extremely important addition to Fermi's β-decay theory. They realized that there are cases where the Fermi theory fails to explain the decays. Consequently, Gamow and Teller proposed an ad hoc solution to explain the discrepancy. For our purposes here we can simplify the difference between the Fermi and the Gamow–Teller interactions as they are expressed in the reaction relevant to stars, namely $p+p \to {}^2D+e^+ +\nu$. In a Fermi interaction which converts a proton into a neutron and vice versa, the sum of all the spins of the particles does not change. In a Gamow–Teller interaction, the total spin must change by one unit.

What Gamow and Teller actually discovered was that the weak force, which is responsible for the β-decay, has two different components which behave and act differently and have different strengths. A good example is the electromagnetic force which can appear as a Coulomb force between electric charges or as a magnetic force acting on moving charges. The electric and magnetic components behave differently. Fierz[37] generalized the theory by combining the Fermi and the Gamow–Teller conditions into a unified theory of this complicated force. The strength of the two components of the force, when compared with the force acting between the protons and the neutrons for example, known as the strong force, is very small, hence the name weak force.

Figure 5.5 shows the potential well of deuterium. The actual potential well looks only slightly different. In this particular case only a single bound state exists. The effective nuclear potential in the case of the deuterium is simple because there is no Coulomb potential acting between the proton and the neutron. From Fig. 5.6, which shows the wave function of deuterium, we learn that the probability of finding the distance between the neutron and the proton greater than the range of the force (given by the absolute value squared of the wave function) is quite high. The reason for this is the finite depth of the potential, which leads to an exponential decay outside the potential well.

It is clear that element synthesis in a pure hydrogen environment could not have started without deuterium having at least one bound level. The deuterium is barely bound! If the nuclear force acting between the neutron and the proton had been slightly weaker, the potential well of the deuterium would have been a bit shallower and there would have been no bound state at all, and no deuterium nuclei would have existed in nature. If on the other hand, the potential well had been much deeper, that is, if the nuclear force had been much stronger, then we would have seen many bound energy levels inside the deuterium potential well, with many more options for the reaction to proceed.

This situation should be contrasted with the nuclear potential acting between two protons (see Fig. 5.5). The Coulomb repulsion between the two protons is not drawn. Here the nuclear potential is significantly shallower and consequently there is no

---

[36] Gamow, G., and Teller, E., Phys. Rev. **49**, 895 (1936).

[37] Fierz, M., Zeit. für Phys. **194**, 553 (1937).

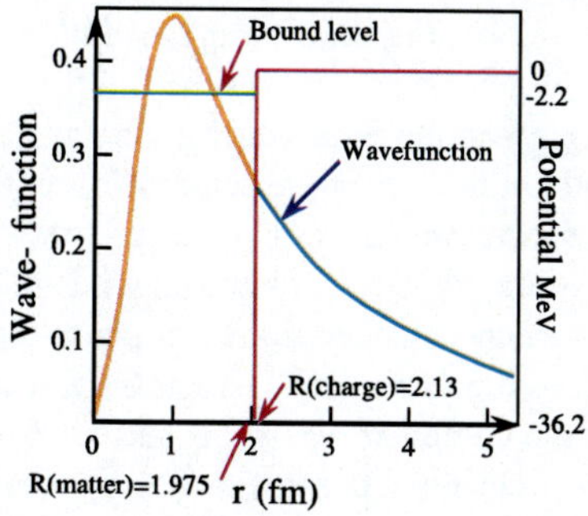

Fig. 5.6 The spherically symmetric part of the deuterium wave function. Also shown are the square potential well (*pink*), the bound level (*green*), and the charge radius and matter radius, found experimentally, which are not identical. The wave function is shown *yellow–brown* inside the potential well and *blue–green* outside. 1 fm $= 10^{-13}$ cm and the units of energy are MeV. After Garcon, M. and Van Orden, J.W., Adv. Nucl. Phys. **26**, 293 (2001)

bound state. Because of the Coulomb repulsion, the nuclear potential between two protons must be even deeper than the potential between a neutron and a proton for a bound energy level to exist. This is the reason why the nucleus ^{2}He does not exist in nature.[38] Had it existed, it would have been the first step in the synthesis of the elements in stars.

To provide additional support to their theory of how two colliding protons can be converted into deuterium, Bethe and Critchfield cited Goldhaber,[39] who investigated a similar process (in terms of the need for a flip of the spins), namely, the fast decay $^6\text{He} \rightarrow {}^6\text{Li}$, which is a β-decay with similar properties.

As Bethe and Crtichfield pointed out, *the transition is therefore allowed only if the Gamow–Teller form of the β theory is used.* At the time of publication, the Gamow–Teller theory was hardly two years old and consequently had not yet acquired sufficient credibility, so Bethe and Crtichfield pointed to various experimental data which confirmed the Gamow–Teller theory. Bethe and Critchfield commented that, if the original Fermi theory were used instead of the Gamow–Teller theory, the reaction would still have proceeded, but it would have been suppressed by a factor of

---

[38] Similarly, in the cases of ^{5}Li, ^{5}He, and ^{8}Be, the effective potential wells of the protons and the neutrons is not sufficiently deep to allow for a bound state, and these nuclei do not have a stable state and consequently decay quickly. On the other hand, ^{4}He has the largest binding energy per nucleon and only one bound state, the ground state.

[39] The reference given is Goldhaber, Physical Review, to be published. However, no such paper by Goldhaber could be found in the Phys. Rev. A year later, Margenau [Phys. Rev. **55**, 1173 (1939)] investigated the structure of ^{6}He, and Grönblom at Cornell University, where Bethe was based, investigated the β-decay of ^{6}He [Phys. Rev. **56**, 508 (1939)]. Bethe was aware of Grönblom's results since he suggested the problem to him. As a matter of fact, Grönblom found that the strength of the interaction between the protons, the factor which determines the speed of the reaction, was significantly stronger than what Bethe and Critchfield had assumed, and hence the rate of energy production was greater. Bethe and Critchfield estimated their data from the reaction $^{13}\text{N} \rightarrow {}^{12}\text{C} + e^+ + \nu$.

about 10^5, and hence *in this case, then, the energy evolution in the Sun due to proton combination would be negligibly small.* As for the full process, they calculated that it would yield about 2 erg/g/s in the Sun, which is the average value of the energy production in the Sun and not the rate at the center. Hence, they concluded that this was *the energy source for stars smaller than the Sun.*

After the publication of the Physical Review paper, someone[40] suggested to the authors to extend the symmetry considerations leading to the process, so Bethe and Critchfield added a short communication.[41] At the time, it was not clear what type of interaction it was, so the comment was short and far from inclusive. However, several years later it became clear that this is a unique interaction which distinguishes between an image and its mirror image. This is like the difference between the right and the left hand. When a physical system is invariant under a certain transformation, a conservation law must follow. If we cannot tell whether an experiment is seen in the laboratory or in the mirror, the system must conserve what is known as parity. The conservation of parity dictates certain properties of the system. The Gamow–Teller interaction was found not to conserve parity.

After the difficult formation of deuterium, the road is open and all reactions go fast. The reason for the difficulty is the weak force, and the reason for the speed of the subsequent reactions is that they are controlled by the strong force. Hence, soon after the formation of deuterium, it will be consumed by the fast reaction $D + H \rightarrow {}^3He + \gamma$, leaving only a tiny amount of deuterium. A short while afterwards, Alvarez and Cornog[42] discovered that ${}^3He_2$ is stable. Helium has two isotopes, one extremely stable, regular helium in the form of ${}^4He$, and a light isotope ${}^3He$. When Bethe devised the CN cycle, it was not known that helium had this light stable isotope, so he only considered the formation of deuterium. The discovery of ${}^3He$ had a significant impact on the pp reactions, as will be seen later.

When the formation reaction is very slow but the destruction reaction very fast, the constituent disappears quickly and only small amounts are left. Consequently, the concentration of deuterium in the Sun is about 10^{-17}! Generally observed relative-to-hydrogen abundances of deuterium are of the order of 10^{-5}–10^{-6}, and hence a sample with this deuterium concentration could not have been found inside stars.

The conclusion is that the pp reaction gives an energy evolution of the right order of magnitude for the Sun. But, argued Bethe and Critchfield:

> It seems that there must be another process contributing somewhat more to the energy evolution in the Sun. This is probably the capture of protons by carbon.

While the paper by Bethe and Critcfield was in print, Gamow[43] evaluated the effect of the newly calculated pp rate on stellar evolution. First Gamow derived the mass–luminosity law on the basis of Critchfield's new result as $L \sim M^{5.5}$, which agreed

---

40 On the basis of a footnote in the Physical Review paper, one could guess that the commentator was Robert Oppenheimer.

41 Bethe, H.A. and Critchfield, C.L., PRL **54**, 862 (1938).

42 Alvarez, L.W. and Cornog, R., Phys. Rev. **56**, 379 (1939).

43 Gamow, G., PRL **53**, 907 (1938).

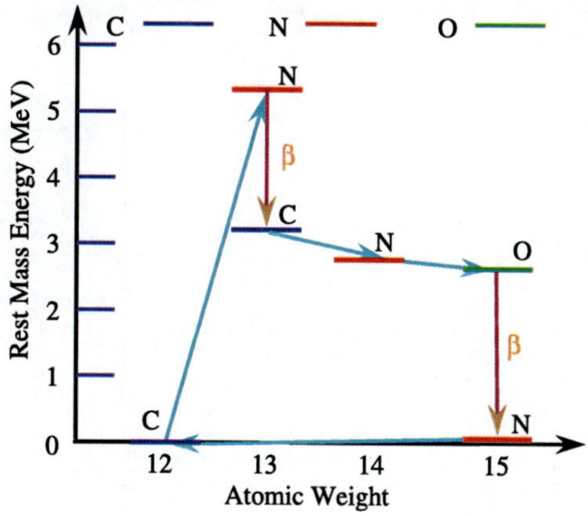

Fig. 5.7 The CN cycle in the atomic weight–energy plane. Carbon isotopes are *blue*, nitrogen isotopes are *red*, and oxygen isotopes are *green*. Proton capture is shown in *light blue* and beta-decays are shown in *pink*

quite well with observation. Then he calculated the evolutionary tracks and found that they hardly differed from what he, Gamow, had calculated a year earlier. In short, the new expression for the pp reaction rate did not have any major impact on the agreement between theory and observation.

5.10 Hans Bethe and the CN Cycle

Two reactions in the CN cycle are the most important, namely, the last and the first.[44] The last reaction is $^{15}\mathrm{N} + {}^{1}\mathrm{H} \rightarrow {}^{12}\mathrm{C} + {}^{4}\mathrm{He} + 4.79\,\mathrm{MeV}$. The reaction appeared in Livingston and Bethe's table (see Table 5.1), but was not measured at the time the CN cycle was proposed.[45]

As can be seen from Fig. 5.7, this is the only reaction where the atomic weight decreases (due to the release of the helium nucleus). With no experimental data concerning the reaction, Bethe claimed that he estimated the rate by analogy with other reactions, but did not specify exactly which reaction he compared it with. Fortunately, the exact value of the rate of this reaction is not so important, because the rate of the entire cycle depends on the slowest reaction in the cycle, and if one assumes that the slowest reaction is not this one, the results hardly change.[46] The

44 Legend has it that Bethe attended a conference in Washington organized by Teller and Gamow on stellar energy sources, and solved the CN cycle in the train on his way back to Cornell. Bethe made a point of diffusing this story, but admitted that figuring out the workings of the CN cycle required two weeks. *Hans Bethe and His Physics*, ed. by Brown, G.E. and Lee, C-H., World Scientific, 2006.

45 Burcham, W.E., and Smith, C.L., Nature **143**, 795 (1939) investigated short-range α particles from oxygen, nitrogen, and fluorine bombarded with protons, and did not bother to derive the probability for the reaction.

46 This is true only if the cycle operates at equilibrium, which is a good assumption when the star is on the main sequence. But, in an explosion, for example, this is not the case.

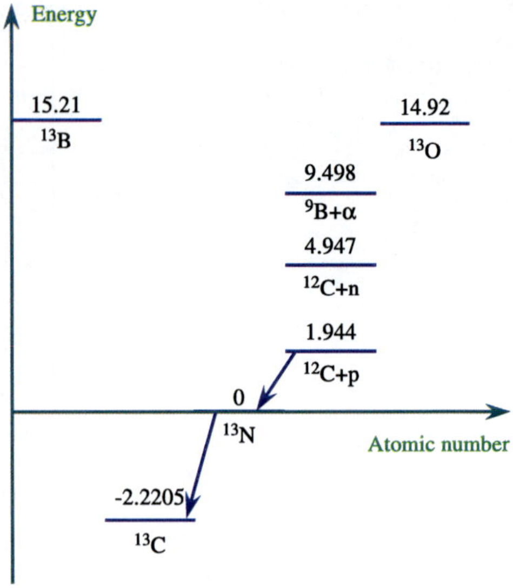

Fig. 5.8 The energy diagram for atomic weight 13

slowest reaction, which determines the rate of the cycle, is $^{14}N+^{1}H \to {}^{15}O$, and has a time scale of 5×10^7 years, while the reaction in question was estimated (at this stage) to occur on a time scale of only 2,000 years. Hence, the fact that this reaction was not measured exposed no problem. Once the paper was published, Holloway and Bethe measured the reaction in Cornell in 1940.[47]

Let us examine how the nuclear parameters of the nuclei allow for such a cycle. Consider the first step, namely a proton capture on ^{12}C, which yields ^{13}N. Of all nuclei with atomic weight 13, ^{13}C has the lowest energy, and hence ^{13}N decays into ^{13}C. All other options, like emission of an α particle or a neutron, require more energy and are excluded (see Fig. 5.8). The second step is $^{13}C + p$, which is 7.5506 MeV above the ground state of ^{14}N. The option $^{10}B + \alpha$ requires 11.6125 MeV above the ground state of ^{14}N, and is thus excluded. The third step is $^{14}N + p \to {}^{15}O$. The situation is similar to the first step, and ^{15}O decays quickly to ^{15}N.

The last step is the reaction $^{15}N+p$, which, energetically speaking, can go either to $^{12}C+\alpha$ or to $^{16}O+\gamma$. Here the α reaction, which ends the cycle, requires less energy than the p reaction, and is therefore energetically possible. This is the first time in the cycle that the p absorption reaction is more energetic than α emission/capture. It so happens that the ratio between the two rates is 10,000:4, so in most cases the first option wins. An examination of the structure of the ^{16}O nucleus (see Fig. 5.9) shows that $^{15}N + p$ has an energy of 12.1276 MeV above the ground state of ^{16}O, where two energy levels with the right quantum numbers lie very close. Furthermore, ^{15}N and p each have spin 1/2. As they collide at low energy, they come towards each

47 Holloway, M.G. and Bethe, H.A., Phys. Rev. **57**, 747 (1940). Even today the investigation of the reaction continues. For a recent paper, see Mukhamedzhanov, A.M., and 15 coauthors, Journal of Physics Conference **15**, 202, 012017 (2010).

Fig. 5.9 The structure of the ^{16}O nucleus. The important levels are marked with *thick lines*. Unimportant levels are marked with *thin lines*. The energy and the spin of the levels are marked on the *right*

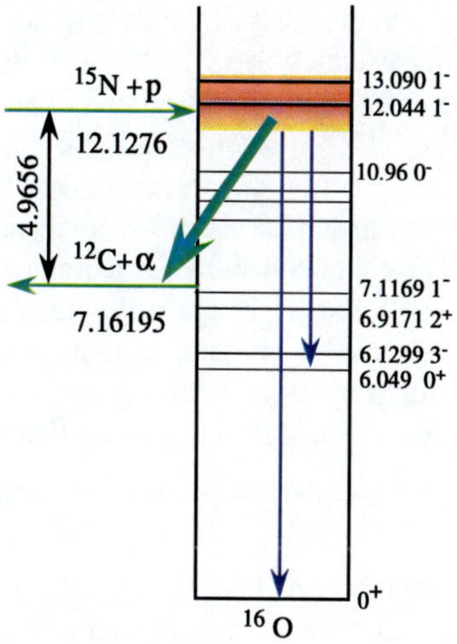

Table 5.2 Typical times of the reactions in the CN cycle under solar conditions, according to Bethe (1939)

H + H	→	$^{2}D + e^{+} + \nu$	1.2×10^{11} years
$^{12}C + H$	→	^{13}N	2.6×10^{6} years
^{13}N	→	$^{13}C + e^{+} + \nu$	870 s
$^{13}C + H$	→	^{14}N	5×10^{4} years
$^{14}N + H$	→	^{15}O	5×10^{7} years
^{15}O	→	$^{15}N + e^{+} + \nu$	870 s
$^{15}N + H$	→	$^{12}C + {}^{4}He$	2,000 years

other with no orbital angular momentum (s-wave), and hence can populate levels with spin 1. Just where $^{15}N + p$ forms the excited ^{16}O nucleus, there are two levels with the right quantum numbers. The fate of the reaction now depends on how these levels decay. Do they decay by disintegrating the excited nucleus into ^{12}C plus an α and an energy of 4.97 MeV, or do the excited levels emit one or two γ and decay into the ground state of ^{16}O, in which case no helium is produced? It so happens that the excited state prefers to disintegrate and in this way complete the cycle.

The above simple reasoning was already available to Bethe in 1935,[48] when he arranged the masses of the nuclei. This is the reason why the last arrow (Fig. 5.7) goes backwards, that is, from right to left. However, this fact went unnoticed. Similarly, the possibility of the ^{20}Ne–^{24}Mg cycle, which is very similar to the CN cycle but with heavier catalysts, was already evident to Livingston and Bethe in 1937.

Bethe's estimate for the time scales for the various reactions is given in Table 5.2. The times are calculated for the conditions in the Sun. The time scale of the pp reaction has been added. The CN cycle is faster than the pp chain. This is because the β-decays are much faster, since they take place inside the nuclei rather than in flight, when the two protons are within nuclear distance from one another.

[48] Bethe, H.A., Phys. Rev. **47**, 747 (1935).

On the basis of the assumption that $^{12}C + p$ is the slowest reaction in the cycle, Bethe concluded that the energy produced in the Sun by the CN cycle is much greater than the energy produced by the pp chain. But improved measurements of the $^{14}N + p$ reaction indicated that this was the slowest reaction in the cycle, and consequently that the cycle was slower than Bethe had previously thought, whence our Sun derives about 94% of its energy from the pp and the rest from the CN cycle.[49] The old conclusion of Friedman and Motz thus received further support.

Captive of the prevailing idea that giants precede the main sequence stars, Bethe examined the possibility that the giants might still be breaking down lithium, beryllium, and boron, while the main sequence has already started the carbon cycle. Bethe soon reached the conclusion that[50]:

> It seems, however, doubtful whether the energy production in giants is due to nuclear reactions at all.

The reaction $^{7}Li + H \rightarrow 2^{4}He$ was known to be 'improbable'.[51] These were among the first signs that something was wrong with the hypothesis that stellar evolution goes from the giants to the dwarfs, with continuous contraction of the star.

5.11 Schatzman 1951: From White Dwarfs to Main Sequence

In 1949, Schatzman investigated the problem of novae. Novae are white dwarfs which erupt periodically, and Schatzman's idea was[52] that the surfaces of these stars somehow contain plenty of ^{3}He, and that the reaction between two ^{3}He nuclei triggers the eruption. Later, in 1951, Schatzman realized that the same reaction can operate in main sequence stars.[53] Independently, Fowler[54] proposed in 1951 that *the proton–proton cycle is most probably completed with the* $^{3}He + {}^{3}He$. (At this point Fowler still called the pp reactions a cycle. Later the name became a chain.) The problem with the reaction

$$^{3}He + {}^{3}He \rightarrow {}^{4}He + 2p + 12.8\,MeV$$

was to estimate its rate. Fowler suggested an estimate which implied that only one out of $\sim 10^4$ helium nuclei in the Sun was ^{3}He. But the rate was so high as to compensate for the low abundance. Another basic question concerned the extent to which the competing process $^{3}He + {}^{3}He \rightarrow {}^{5}Li + {}^{1}H + 11\,MeV$ was faster or not.

[49] Caughlan, G.R. and Fowler, W.A., Astrophys. J. **136**, 453 (1962).

[50] Here Bethe cited a private communication with Gamow.

[51] Goldhaber, M., Proc. Camb. Phil. Soc. **30**, 560 (1934).

[52] Schatzman, E., Ann. Ap. **12**, 281 (1949).

[53] Schatzman, E., Compt. Rend. **232**, 1740 (1951); Ann. Ap. **16**, 162 (1953).

[54] Fowler, W.A., Phys. Rev. **81**, 655 (1951). This was a brief report at the annual meeting of the American Physical Society.

This question was not considered by Schatzman and Fowler, because they assumed that ^{5}Li is unstable and does not exist in nature.

As a matter of fact, there was no experimental data for this reaction. It was first suggested by Fermi and Turkevich as a solution to the $\alpha\beta\gamma$ idea of hot early universe synthesis (see Chap. 7), but rejected because the density in the Big Bang was too low for this reaction to go at an appreciable rate. Schatzman realized that stellar densities are significantly higher than the densities in the Big Bang, so the reaction that was rejected for the Big Bang could operate in stars.

The discovery of this reaction aroused interest in stellar nuclear reactions starting from pure hydrogen, an interest which was quenched by the success of Bethe's CN cycle and Hoyle's steady claim that all elements were formed in stars. The problem of how stars without CN can generate energy was ignored. Soon it became clear that this reaction is the fastest reaction in the pp chain and, most importantly, releases the largest amount of energy per interacting particle.

When Bethe and Critchfield[55] considered pure hydrogen fusion, they did not consider the ^{3}He reaction. At that time it was not clear whether ^{3}He is more stable than ^{3}H or not. If it were more stable, Bethe and Critchfield argued, it would be destroyed rather quickly by the more abundant ^{4}He in the reaction $^3\mathrm{He} + {}^4\mathrm{He} \rightarrow {}^7\mathrm{Be} + \gamma$. Hence, they dismissed the idea of $^3\mathrm{He} + {}^3\mathrm{He}$.

5.12 Salpeter 1952: So It Should Be

Soon after the suggestion by Schatzman and by Fowler, Salpeter attempted to calculate the rate of the reaction. There were no measured rates. Good et al.[56] observed the reaction and confirmed its existence, but did not measure the rate. On the other hand, the analogue reaction $^3\mathrm{H}+{}^3\mathrm{H} \rightarrow {}^4\mathrm{He}+2\mathrm{n}+11.4\,\mathrm{MeV}$ was observed and measured by Angew et al.[57] In the absence of better data, Salpeter[58] assumed that the reaction in question resembles the measured reaction between two tritium (^{3}H) nuclei, and calculated the rate in this way.[59] The rate Salpeter got was 1.5 times higher than the approximate rate given by Fowler. Thus the reaction turned out to be even faster than what Fowler had found. This was not the last word. When the reaction was properly measured, the rate found was even faster. The latest experimental data[60] are about 5.5

[55] Bethe, H.A. and Critchfield, C.L., Phys. Rev. **54**, 248 (1938).

[56] Good, W.M., Kunz, W.E., and Moak, C.D., PRL **83**, 845 (1951).

[57] Agnew, H.M. and 6 coauthors, Phys. Rev. **84**, 862 (1951).

[58] Salpeter, E.E., Phys. Rev. **88**, 547 (1952); Astrophys. J. **116**, 649 (1952).

[59] Salpeter's great idea was as follows: consider helium 3 as composed of 2p + n and tritium as 2n + p. In the mutual collision, the extra particle of the second nucleus is captured to form the ^{4}He nucleus in both reactions. The experiment was carried out by Allen, K.W., and 4 coauthors, PRL **82**, 262 (1951). Among others, this experiment showed that a di-neutron does not form in this experiment. Two neutrons, like two protons, do not have a bound state.

[60] Junker, M., Phys. Rev. C **57**, 2700 (1998); Arpesella, C., et al., Phys. Lett. B **389**, 452 (1996).

Table 5.3 The complete pp chain according to Salpeter (1952)

$p + p$	$\rightarrow$	$^2D + e^+ + \nu + 0.42\,MeV$	4×10^9 years
$p + {}^2D$	$\rightarrow$	$^3He + \gamma + 5.5\,MeV$	4 s
$^3He + {}^3He$	$\rightarrow$	$^4He + 2p + 13\,MeV$	3×10^5 years

times the original estimate of Fowler and a factor of 1.7 greater than what Salpeter estimated theoretically (see Table 5.3).

To assess the importance of the reaction, Salpeter calculated the rate of this reaction and various competing reactions as summarized in Table 5.4. The comparison shows that all competing reactions are slower and hence can be neglected in comparison with the $^3He + {}^3He$ reaction.

Since the deuterium and ^3He are destroyed so quickly, it can be assumed that both reactions are in equilibrium. Under this assumption, one can easily calculate the concentrations of deuterium and ^3He, and the results are $n_D = 3 \times 10^{-17} n_H$ and $n_{^3He} = 4 \times 10^{-5} n_H$. Salpeter commented that the calculated concentration of ^3He was *not in disagreement with spectroscopic observations on stellar atmospheres.*[61] In particular, Salpeter commented that the abundance of deuterium on the Earth is *enormously* greater and concluded:

> This seems to suggest that the material from which the Earth was formed could not have come from the interior of a main sequence star.

Salpeter was right, with the knowledge available at that time, and he would have been even more right had he replaced the word 'material' with 'deuterium'. To be more accurate, Fermi and Turkevich, as reported by Alpher and Herman,[62] found that the deuterium abundance in the Big Bang was about 10^{-2}. However, Salpeter did not refer to this result. Today we know that deuterium was created in the Big Bang with an abundance of a few times 10^{-4}, close to the concentration of deuterium on the Earth. On the other hand, the heavy elements found on the Earth, say iron and heavier, must have been synthesized in stars. We can conclude that some mixing between primordial gas and material processed in stars must have taken place before the formation of the Earth.

Two years later, after Salpeter had carried out his theoretical estimates, Good, Kunz, and Moak[63] got the first experimental numbers for this reaction. They also demonstrated that ^5Li does not form at all in the reaction. They mentioned Schatzman's original idea and Fowler and Lauritsen's estimate (as a private communication) for this reaction and stated that: *The essential correctness of their assumption seems established.* No reference was given to Salpeter, who actually calculated the rate of this reaction.

---

[61] Greenstein, J.L., Astrophys. J. **113**, 531 (1951).

[62] Alpher, R.A. and Herman, R.C., Rev. Mod. Phys. **22**, 153 (1950).

[63] Good, W.M., Kunz, W.E., and Moak, C.D., Phys. Rev. **94**, 87 (1954). The paper was published on 1 April 1954.

Table 5.4 Salpeter's estimate of the mean lifetime of ^{3}He against interactions with different species, establishing the dominance of the ^{3}He $+^3$ He reaction (1952)

Reaction	Mean lifetime (years)	
	Temperature 5×10^6 K	Temperature 30×10^6 K
$^3He + ^3He \rightarrow ^4He + 2p$	6×10^{10}	240
$^2D + ^3He \rightarrow ^4He + p$	7×10^{12}	3×10^5
$^4He + ^3He \rightarrow ^7Be + \gamma$	1×10^{18}	8×10^3
$^1H + ^3He \rightarrow ^4He + e^+ + \nu$	5×10^{14}	2×10^8

In 1953, unaware of Salpeter's work, Friedman and Motz[64] recalculated the pp chain. The work was published after Salpeter's, but was based on Friedman's Ph.D. thesis, and a preliminary report was presented at the May 1951 meeting of the American Physical Society. This is the reason why there is no mention of Salpeter's paper, which was published a year earlier. Friedman and Motz's conclusion concerning the role of ^{3}He was identical to Salpeter's. However, Friedman and Motz concluded, in contrast to Bethe's 1939 result, that the pp chain is the dominant contributor to the energy production in the Sun, rather than the CN cycle. They thus confirmed the conclusion reached by Epstein[65] three years earlier, namely, that in the Sun the pp chain energy production exceeds that of the CN cycle by a factor of 12. This is very close to the present day value of CNO contributing just 6%.

5.13 The Discovery of the ^{3}He + ^{4}He Reaction: A New Branch

The second historical event occurred in 1958 when Holmgren and Johnston from the Naval Research Laboratory reported at the 1958 meeting of the American Physical Society that they had succeeded in measuring the rate of the $^3He + ^4He \rightarrow ^7Be + \gamma$ reaction and found it to be about 2,500 times higher than what Salpeter and Bethe had predicted. The news was quick to spread, and Willy Fowler and Al Cameron were prompt to appreciate the tremendous potential by suggesting that, if ^{7}Be is stable and produced in significant amounts in the Sun, then it could capture a proton and yield ^{8}B. The ^{8}B nucleus is unstable and decays into two helium nuclei, a positron (e^+), and an antineutrino ($\bar{\nu}$) during the conversion (in the newly formed nucleus) of the proton into a neutron. The exciting thing about this decay was that the emitted antineutrino is very energetic (the maximum energy is 14.04 MeV), while all other neutrinos emitted by the Sun have significantly lower energies (0.42–2 MeV). Neutrino detectors are extremely sensitive to the energy of the neutrinos and consequently the ability to detect such a particle increases very steeply with its energy. As a matter of fact, Fowler and Cameron realized that the discovery by Holmgen

64 Friedman, E.A. and Motz, L., Phys. Rev. **89**, 648 (1953).
65 Epstein, I., Astrophys. J. **112**, 207 (1950).

and Johnston had the potential of opening up a new field of what is known today as neutrino astronomy, which greatly increases the opportunities for investigating the interior of the Sun directly.[66]

Holmgren and Johnston reported the discovery at the meeting[67] and started to prepare the paper for publication. Recognizing the importance of this reaction, Fowler reacted with a publication in the Astrophysical Journal,[68] and so did Cameron.[69] In his paper Fowler made the prediction:

> If energetic neutrinos are released in the ^{8}B decay in the Sun, they may be observable at the surface of the Earth [...] using the techniques developed by Davis at Brookhaven.

This conditional prediction paved the way to neutrino astronomy and to the direct confirmation that the Sun synthesizes helium, providing a way of looking directly into the core of the Sun.

The issue at stake was the lifetime of ^{7}Be against capture of an electron, because if it is very fast the beryllium will have no chance to absorb a proton and become ^{8}B, thereby yielding the much wanted energetic neutrinos. What Fowler did was to rely on Bethe's theory and estimate the lifetime of ^{7}Be against electron capture. This solved half of the problem. The second half was to estimate the rate of the competing reaction: $^7\mathrm{Be} + \mathrm{p} \rightarrow {}^8\mathrm{B} + \mathrm{e}^+ + \bar{\nu}$. In the lack of an experimental value, Fowler was unable to provide an accurate answer to the question of how many high energy neutrinos the Sun emits, and for this reason the prediction was conditional.[70]

5.14 Where is the pp Reaction Crucial?

At the time the pp chain was discovered, the problem of the first generation of stars, namely, how the first stars formed and how they produced their energy, had not yet been posed. Bethe for example, did not know how the CN species he needed for energy production were synthesized. The outcome of the Big Bang is an extremely low amount of CNO and heavier elements. If the heavy elements form in stars, then clearly the question about the composition and formation of the first stars emerges. Lacking any heavy elements in significant amounts leaves the pp chain as the only energy source for the first generation of stars.

---

[66] The neutrinos created in the core escape from the Sun without any interaction with the solar matter. The Sun is practically transparent to neutrinos. In contrast, photons released in the core of the Sun interact with the matter and diffuse out gradually, losing all memory of what happened in the core.

[67] Holmgren, H.D., and Johnston, R.L., Bull. Am. Phys. Soc. Ser. II **3**, 26 (1958).

[68] Fowler, W.A., Astrophys. J. **127**, 551 (1958), sent for publication 24 February 1958.

[69] Cameron, A.G.W., Bull. Am. Phys. Soc. II **3**, 227 (1958); Ann. Rev. Nucl. Sci. **8**, 299 (1958); Chalk River Report CRL-41, 2nd edn., 1958, unpublished.

[70] The story of the solar neutrinos is unfolded in Shaviv, G., *The Life of Stars. The Controversial Inception and Emergence of the Theory of Stellar Structure*, Springer, Heidelberg, and Magnes, Jerusalem (2009).

Chapter 6
The Composition–Age–Velocity Connection

We stressed that the observed composition of stars is uniform. All stars have, to a very good approximation, the same composition. The justification for this fundamental result was the success of the Harvard spectral type classification, based on spectral line intensity ratios. The observations demonstrated time and again that, to a very good approximation, the overwhelming majority of stars (at least those nearby) exhibit similar line ratios for the same effective surface temperatures, thus indicating similar compositions. This general tenet survived until the beginning of the 1950s, when it became clear that this was only approximately correct. The problem was then completely turned round, to ask what was the source of the small difference in the abundances of the heavy elements.

Similarly, the velocities of the stars were known to follow a general distribution found by Schwarzschild,[1] and called the Schwarzschild velocity ellipsoid. But observation soon began to indicate that there are fundamental differences in the velocity distribution of stars. Even before Schwarzschild's theory was invented, Eddington[2] analyzed the results of Jacobus Kapteyn (1851–1922m) and Willem de Sitter (1872–1934m),[3] and discovered that the stars do not have random motions, but that there are two systems of stars, or 'drifts', as Eddington called them. These two drifts are in motion relative to one another. In 1908, Eddington[4] tried to compromise between Schwarzschild's and Kapteyn's apparently contradictory results. He concluded:

> It does not at present seem practicable to decide by a crucial test between the two proposed laws of distribution of stellar velocities, as neither law pretends to be more than an approximation to the truth.

But Eddington ends with:

> It is relatively unimportant that we are unable to agree that his explanation is physically so simple and free from difficulties as that afforded by the two-drift hypothesis.

---

[1] Schwarzschild, K., Göttingen Nachr. 614 (1907).

[2] Eddington, A.S., MNRAS **67**, 34 (1906).

[3] Kapteyn, J.C. and de Sitter, W., PGro. **19** D1, 1908. This is a complete catalogue of velocities (proper motions). Eddington based his analysis on Kapteyn, Brit. Assoc. Report, 1905.

[4] Eddington, A.S., The Obs. **31**, 119 (1908).

G. Shaviv, *The Synthesis of the Elements*, Astrophysics and Space Science Library 387, DOI: 10.1007/978-3-642-28385-7_6,

In parallel, it became clear that there are different groups of stars. In fact, there are groups which contain very bright blue stars, while others contain only bright red stars. The properties of the various groups were correlated with their location in the Galaxy: in the halo, in the disk, or in the center. The different groups of stars exhibited different HR diagrams.

This was certainly mind-boggling. Why should the kinematical properties of stars be associated with their appearance in the HR diagram? How can internal properties be connected to the motion of the stars? A long trail of discoveries would be required before the grand picture would crystallize and the connections could be uncovered.

Four pivotal stories unfolded at the same time. The first story concerned the way the elements were formed (the question of nuclear reactions and stellar energy sources). The second story concerned the reasons for the differences in abundances, although small in absolute values, between certain groups of stars. The third story accounted for the post-main sequence evolution of stars and the transformation from the homogeneous Eddington model to a inhomogeneous Milne model, which we call the stellar onion model. Finally, the fourth story was the discovery, by means of 21 cm radio waves, of the gas in the galactic spiral arms, where most of the stars in the disc of the Galaxy actually form.

The unification of the above stories into the saga of cosmic chemical evolution is described in this chapter. The synthesis of the chemical elements and the distribution of the elements in the space of the Galaxy, which ultimately brought the heavy elements to the relatively young Sun, cannot be separated from this saga. At the same time the chemical elements served as tracers of galactic evolution.

6.1 The Stars are Never at Rest

The so called 'fixed' stars are not really fixed, and as a matter of fact they cannot be fixed in space. They are in perpetual motion. The stars move under the attractive gravitational force of the Galaxy, which is in turn composed of the sum total attraction by all stars in the Galaxy.

Observations of the exact location of stars in the sky and their change in time, observations which extended over many years, gave information about the transverse velocities (the component perpendicular to the line of sight), although one needs to know the distance to the star in order to get the velocity from the apparent motion in the sky. The professional term for this velocity is proper motion. As soon as the Doppler effect became known (see Sect. 6.2), astronomers used the effect to measure the stellar velocities along the line of sight. In this way two components of the velocity were measured, and with them the absolute value of the velocity.

The velocity of a star depends on its location in the Galaxy and the gravitational field of the Galaxy as well as on its total energy. Stars with high kinetic energy can reach greater distances from the galactic center, and vice versa. By measuring the velocity of the stars, one can depict the gravitational field of the Galaxy (and the distribution of mass in the Galaxy). Typical velocities of stars in the solar neighborhood of the Galaxy are 10–40 km/s relative to the Sun.

6.2 The Discovery of the Doppler Effect

On 25 May 1842, the mathematician Christian Doppler (1803–1853m) presented a paper[5] before the Royal Bohemian Society of Sciences, hypothesizing that the color of a source of light as seen by a distant observer depends on the relative velocity between the source and the observer. When the source and the observer move towards each other, the color of the source appears bluer, but when they move away from one another, the color becomes redder. The inspiration was Doppler's observation, as described in his paper, of how ships meet waves in the sea. When the ship moves towards the waves, it encounters more waves per hour and vice versa. The number of waves encountered per unit time is just the observed frequency of the wave. Hence, when moving towards the ship, the observer on the ship will count more waves per unit time and hence the observed frequency will increase. The opposite occurs when the ship moves in the same direction as the waves.

Doppler derived his principle in just a few lines, assuming that both light and sound are longitudinal waves (which means that the oscillation of the wave is in the direction in which the wave propagates[6]). Moreover, he assumed the light to be waves in the ether, in accordance with the then widespread belief, just as sound consists of waves in matter.

Doppler demonstrated the correctness of his principle by an attempt to explain the observed colors of the stars. The stars emit white light, so postulated Doppler, and the color is a consequence of their motion relative to us. Doppler had a valid principle but the wrong explanation for the colors of double and single stars. The effect on the observed color of stars moving in the Galaxy is negligible because the velocities of the stars in the Galaxy are much too small relative to the speed of light, being of the order of just tens of km/s. Doppler ended his presentation with a prediction that astronomers would use the principle to measure the velocities of stars. Until that time astronomers could only follow changes in the position of stars in the sky and infer the transverse velocities from that.

The dutch meteorologist Christophorus Buys-Ballot (1817–1890m) criticized Doppler in his 1844 Ph.D. thesis statement, and in 1845,[7] set out to refute Doppler's theory. Consequently, he organized a famous experiment with musicians, who played a fixed note in trains going from Amsterdam to Utrecht, only to discover that Doppler was right. Yet, Doppler was nervous and replied to Buys-Ballot,[8] claiming the correctness of his theory, even though Buys-Ballot had just proved him right!

---

[5] Doppler, C., *Über das farbige Licht der Doppelsterne und einiger anderer Gestirne des Himmels*, Abhandlungen der k. Böhm. Gesellschaft der Wissen. **11**, 465 (1842); *On the Colored Light of Double Stars and Certain Other Stars of the Heavens, Beitrage zur fixsternenkunde*, Prag. Druck von G. Haase sohne, 1846.

[6] The simple classical sound waves are longitudinal, while light waves are transverse: the magnetic and electric fields oscillate perpendicularly to the direction of propagation.

[7] Buys-Ballot, C.H.D., *Akustische Versuche auf der Neiderländische Eisenbahn nebst gelegentlichten Bemerkungen zur Theorie des Hrn. Prof. Doppler*, Pogg. Ann. B. **LXVI**, 321 (1845).

[8] Doppler, C., Pogg. Ann. B. **LXXXI**, 270 (1846); Pogg. Ann. B. **LXXXVI**, 371 (1846); Ann. der Phys. **144**, 1 (1846).

Six years after Doppler's announcement of his principle, on 23 December 1848, Hippolyte Fizeau (1819–1896m) gave a lecture before the Société Philomatique (The Scientific and Philosophical Society of Paris) under the title *Des effets du mouvement sur le ton des vibrations sonores et sur la longueur d'onde des rayons de lumière*.[9] For some reason which is not quite clear, the lecture was only published in volume 19 of the Annales de Chimie et de Physique, in 1870.[10] It was a special case for Fizeau, who worked on the propagation of light, to discuss acoustic waves. Unfortunately, his derivation of the change in frequency and the final formula is wrong.[11] Yet the effect is sometimes called the Doppler–Fizeau effect.

6.3 The Doppler Effect at the Service of Astrophysics

Initial observations of fixed stars by Father Benedict Sestini (1816–1890)[12] seemed to confirm Doppler's theory about what caused the colors of stars. On the other hand, attempts by Klinkerfues (1827–1884)[13] to measure the effect were criticized by Sohncke,[14] who was a crystallographer and not an astronomer, claiming that the results were inconceivable. As a matter of fact, Klinkerfues carried out the first experiments which paved the way to the famous Michelson–Morley experiment.

Some twenty years later, in 1868, Father Angelo Secchi[15] claimed that he could not detect any temporal change in the colors of the components of a double star which move one around the other with velocities of the order of tens of km/s. The same null result had been obtained by Huggins and Miller a few years earlier.[16] Their short paper did not report any actual observations of the effect. As a matter of fact, they had an upper limit of 196 miles/s for the velocities of some of the brightest stars in our sky, but this upper limit was published by Huggins only in his next paper.

The first, almost successful, attempt to measure stellar velocities using the new effect was carried out in 1868 by Huggins.[17] He criticized Doppler as follows:

---

[9] The effect of motion on the tone of acoustic vibrations and the wavelength of light waves.

[10] One can guess that the idea of measuring the speed of light, for which he got the value 313,000 km/s, attracted his attention.

[11] The correct formula for the Doppler effect relating the observed frequency ν_{obs} to the natural frequency ν_{nat} of the source is

$$\nu_{obs} = \nu_{nat}\frac{1 - v/c}{\sqrt{1 - v^2/c^2}},$$

while Fizeau got the formula without the square root. Here, v is the relative velocity and c the velocity of light or sound in the medium.

[12] Sestini, B., *Memoria sopra i colori delle stelle del catalogo de Baily*, Roma 1847.

[13] Klinkerfues, W., A.N. **1538**, 17 (1865); A.N. **1582–83**, 337 (1866).

[14] Sohcke, L., A.N. **1646**, May (1867).

[15] Secchi, A., Compt. Rendu. **66**, 396 (1868).

[16] Huggins, W., and Miller, W.A., Phil. Trans. Roy. Soc. Lond. **154**, 437 (1864).

[17] Huggins, W., Phil. Trans. Roy. Soc. Lond. **158**, 529 (1868).

> That Doppler was not correct in making this application of his theory is obvious from the consideration that, even if a star could be conceived to be moving with a velocity sufficient to alter its colour sensible to the eye, still no change of colour would be perceived, for the reason that beyond the visible spectrum, at both extremities, there exists a store of invisible waves which would be at the same time exalted or degraded into visibility, to take the place of the wave which had been raised or lowered in the refrangibility by the star's motion. No change of colour, therefore, could take place until the whole of those invisible waves of force had been expended, which would only be the case when the relative motion of the source of light and the observer was several times greater than that of light.

Huggins was right to claim that the change in the color due to relative motion arises only if the distribution of intensity in wavelengths is not uniform. We speak of Doppler shift to the red or to the blue, and this means that the wavelengths of all spectral lines becomes shorter (bluer) or longer (redder) relative to the wavelengths measured at rest. The color, on the other hand, is determined by the ratio of intensities in two wavelengths and not by the wavelength of a single spectral line. Huggins was wrong about velocities greater than the speed of light.

In the same paper, Huggins reported his observations of Sirius and how he succeeded in measuring a recession velocity of −41.4 miles/s. As Huggins knew that Sirius moves in an ellipse (because it had been suspected of being a binary system for over twenty years[18]), he calculated the velocity due to this motion and found it to be 43.2 miles/s. So he concluded that:

> It may be that in the case of Sirius we have two distinct motions, one peculiar to the star, and a second motion which it may share in common with a system of which it may form a part.

This time Huggins was right. Today we know that the Sirius system moves in space with a velocity of −8 km/s (which is −4.7 miles/h).

All the above observations were carried out by eye and consequently were of limited accuracy. There were no proper devices to register the observations. A significant improvement in the accuracy of the velocity measurements came when Cornu (1841–1902) suggested, as late as 1890,[19] using photographic plates to measure the effect in stars, and the first such measurement was published by Vogel[20] without waiting for Cornu's suggestions. The technique developed by Vogel and already implemented in 1887 paved the way to accurate measurements of stellar velocities, which quickly began to accumulate, and opened the way to a new field of investigation, namely, the study of stellar velocities and its far-reaching impact on our understanding of the structure of the Galaxy.

---

[18] Bessel discovered the particular motion of Sirius and suggested that Sirius was a binary star, but at that time the secondary could not be discovered. See Shaviv, G., *The Life of the Stars*, Springer, Heidelberg, and Magnes, Jerusalem (2009).

[19] Cornu A., *Sur la method Doppler–Fizeau permettant la détermination, par l'analyse spectrale, de la vitesse des astres dans la direction du rayon visuel*, Gauthier-Villars et fils, Paris, 1890.

[20] Vogel, H.C., *On the Spectroscopic Method of determining the Velocity of Stars in the Line of Sight*, MNRAS **52**, 87 (1891); A.N. **125**, 305 (1890), # 2995. Even earlier publications are cited in the MNRAS paper.

By 1907, Schwarzschild[21] was able to demonstrate that the velocity distribution of the stars in the solar neighborhood follows a simple Gaussian shape, exactly like the velocity distribution of molecules in a gas. However, Schwarzschild discovered that, although the distributions in the three perpendicular directions are Gaussian (like molecules), the mean velocities in the three directions are not equal, in contrast to the case of a gas in a container. The stars do not after all move like molecules in a container.[22] The stars are affected by the force acting on them, which is not equal in all directions, while molecules in a container are affected only by the walls of the container. But Eddington refused to abandon his rival two-drift theory.[23]

6.4 Reality is More Intricate

The first hint that the situation might be more complex already appeared in Eddington's publication. This is where Eddington's so-called rabbit diagrams, which demonstrate the deviation from spherical symmetry, appeared for the first time (see Fig. 6.1).

Eddington's 1911 analysis[24] of the data of Hough and Halm[25] and Campbell's data[26] indicated deviations from Schwarzschild's simple ellipsoid in the form of 'streams of stars'. Eddington explained that:

> Campbell's results do not show the phenomenon so pronouncedly as might at first be expected; but there is little doubt that the effect is partly obscured in his figures by another cause, namely the prevalence of early type stars (which have small individual velocities) in the neighborhood of the vertices.

In other words, the 'streams of stars' observed by Hough and Halm were not seen by Campbell because his sample contained many low-mass stars with low velocities, and low-mass stars must have low velocities, in contrast to massive stars.

Soon after, in 1913, Carl Vilhelm Ludwig Charlier (1862–1934m)[27] discovered an excess of stars with large velocities giving rise to a skewed distribution (too many high velocity stars move in a given direction). This was the first sign that the Schwarzschild's original ellipsoidal velocity distribution was not universally valid.

In 1914, Walter Adams (1876–1956m) and Arnold Kohlschütter (1883–1969m)[28] discovered a few stars (with the peculiar names Lal. 1966 and Lal. 15290 from the

---

[21] Schwarzschild, K., Göttingen Nachr. **614** (1907).

[22] Because the velocities in the three perpendicular directions are not identical, the distribution is called ellipsoidal, in contrast to spherical.

[23] Eddington, A.S., MNRAS **71**, 4 (1910).

[24] Eddington, A.S., Obs. **34**, 355 (1911). See also MNRAS **73**, 346 (1913), where two types of stars are examined, A and K types, and different kinematical properties were discovered.

[25] Hough, S.S. and Halm, J., MNRAS **70**, 578 (1910).

[26] Campbell, W.W., Lick Obs. Bull. 211, **7**, 19 (1912).

[27] Charlier, C.V.L., MNRAS **73**, 486 (1913).

[28] Adams, W.S., and Kohlschütter, A., Astrophys. J. **40**, 385 (1914).

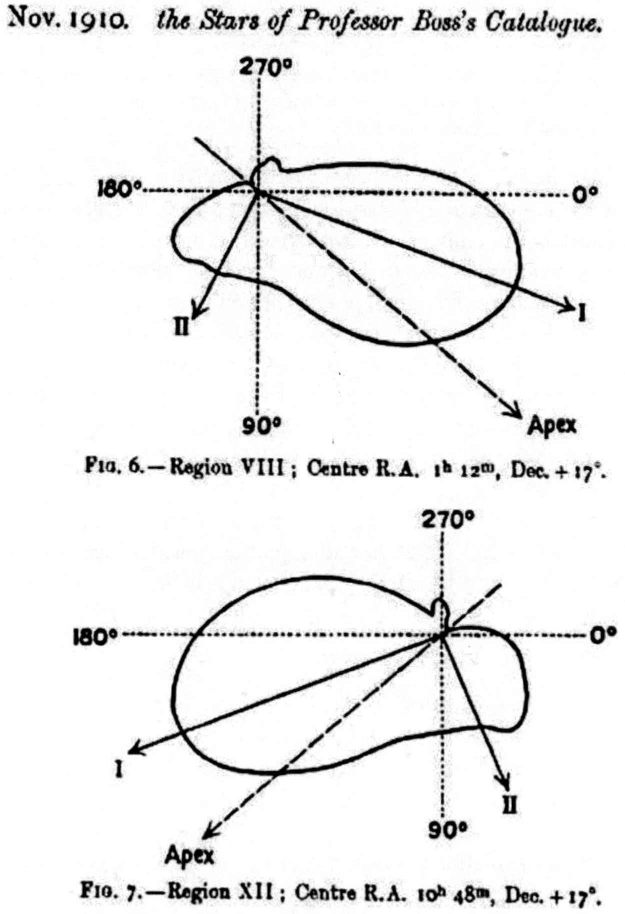

Fig. 6.1 Distribution of stellar velocities with direction. The non-spherical symmetry and the preponderance of stars moving in certain directions is clear. The fine details are strange and obscure. These unique diagrams won the name of 'Eddington rabbits' and were among the first to indicate that the kinematics of stars in the Milky Way is complicated. There was no explanation for the 'ears of the rabbits'

Lalande (1732–1807m) catalogue of fixed stars) with very large radial velocities (of 325 and 242 km/s, respectively), and they were all negative (moving towards us). By galactic standards, these are very high velocities. Moreover, they noticed that there are more stars with negative velocities than stars with positive velocities, as if most stars were falling onto the Sun. Interestingly, Adams and Kohschütter made a note that the two stars with the largest velocities had peculiar spectra, and in particular the spectral lines of magnesium were *either absent or extremely faint*. However, they remarked that:

> The remarkably high and nearly equal velocities for two stars of nearly identical but peculiar spectra, form a singular coincidence. The stars are widely apart in the sky, and the apices of their motion are quite different. The existence of high radial velocities among stars having what is generally considered an early type of spectrum is shown by these results, although there can be no doubt that such cases are rare.

They cited Campbell,[29] who found no stars with a constant velocity[30] exceeding 40 km/s in an investigation of 337 stars having spectra between type B and F4, while Adams and Kohschütter claimed to have found just one such star. In the following paper,[31] they paid more attention to the connection between spectral type and velocity, and inferred that stars with small proper motion have weak ultraviolet emission. They hypothesized that this was due to special conditions in the atmosphere of these stars. However, just what these conditions were they did not specify, which is a default assumption, or an excuse for not specifying them.[32] In summary, the situation was quite confusing, and astronomers were looking for specific reasons to explain deviations from the Schwarzschild ellipsoid distribution and from the general spectral classification of stars.

The hints observed by Adams and Kohlschütter were strengthened by Kapteyn and Adams,[33] who concluded that the data supported the idea that there were streams of stars. However, the most interesting conclusion is their last one, namely:

> Some indications have been found of a change of radial velocity with brightness, the brighter stars moving more slowly than the faint stars.

To comprehend and appreciate the strangeness of the situation, assume for simplicity that all stars, irrespective of their distance from us and spectral type, move with the same velocity and have the same intrinsic brightness. In this case, the far away stars will appear less bright and with smaller proper motion. However, the opposite was observed!

So far the various relations between spectra and velocity concerned the radial velocity, i.e., the velocity along the line of sight. In 1915, Adams showed[34] that the stars with large radial velocities also have large transverse velocities, so it is not true to say that the fast-moving stars move towards the Sun (Fig. 6.2).

Adams noted that the distant stars showed hardly any change in velocity with spectral type:

---

[29] Campbell, W.W., *Stellar Motions*, Yale University Press, 1913, p. 198.

[30] The 'constant' means that Campbell also observed stars with variable radial velocity, PASP **30**, 353 (1918).

[31] Adams, W.S., and Kohlschütter, A., Astrophys. J. **40**, 305 (1914).

[32] Just hypothesizing that it is due to the atmosphere of the star is not terribly helpful for understanding the phenomenon.

[33] Kapteyn, J.C., and Adams, W.S., PNAS **1**, 14 (1915).

[34] Adams, W.S., Astrophys. J. **42**, 172 (1915).

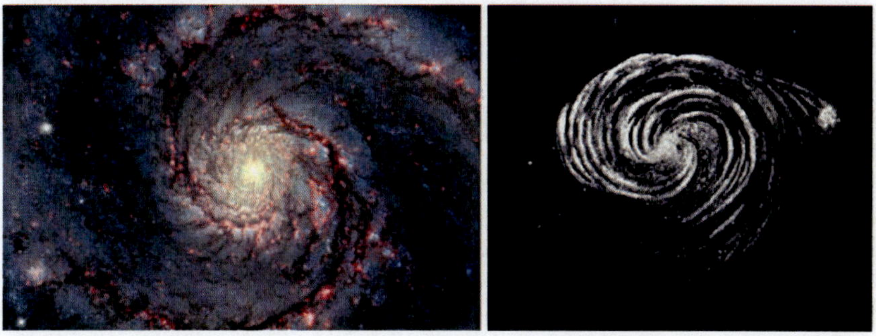

Fig. 6.2 *Left* the Whirlpool Galaxy M51. The differences in color between the core and the spiral arms. *Blue* implies a preponderance of massive stars and *red* a preponderance of low mass stars. Credit: NASA and the Hubble Heritage Team. *Right* the nebula M51 as observed in 1845. The image was hand drawn by Parsons before astronomical photography was invented by Draper. The comparison with the previous image is a testimony to what has been achieved in astronomy in a century and a half

> This would be in agreement with the hypothesis put forward by Eddington,[35] but later entirely disproved, as he considered, by the evidence of the A stars,[36] that the relation between velocity and spectral type might be a relation between velocity and distance.

In 1916, Frederic Seares (1873–1964m)[37] observed the spiral nebulae M51, M94, and M99.[38] At that time it was not yet known that these were galaxies like our own, and it was impossible to resolve the nebulae and see the individual stars out of which they are composed. Seares discovered differences in the color between the nuclear part of the nebula and the spiral arms. Today we know that what Seares discovered without realising it was that the stellar populations of the nuclear part and the spiral arms are different. It was Baade, over 28 years later, who finally clarified this.

Interestingly, in his address to the American Association for the Advancement of Science in New York City on 26 December 1916, William Campbell (1862–1938m) described the two greatest problems of the Universe. The first was the *evidence of matter in space obscuring the passage of light*. Campbell concluded that:

> There are many lines of evidence in support of the hypothesis that invisible matter exists in abundance within the stellar system.

Campbell named several unexplained phenomena, and one of them was associated with the enigma of the high velocity stars. The essence of the problem was that these stars appeared to move faster than the gravitational attraction of the observed mass of the stellar system could allow.[39] These were the roots of the discovery, announced

---

35 Eddington, A.S., Brit. Assoc. Report, 1911.

36 Eddington, A.S., *Stellar Movements*, p. 161.

37 Seares, F., PASP **28**, 123 (1916).

38 The letter M refers to the Messier catalogue of nebulae.

39 When a star moves faster than the gravitational field allows, the star overcomes the gravitational attraction and escapes. Hence, if we observe stars moving in the gravitational field of the Galaxy

only many years later, which raised one of the most important issues in astrophysics, namely, the nature of 'dark matter', i.e., unobserved matter which is supposed to supply the extra gravitational attraction needed to explain the motion of the stars in the Galaxy. Today we know that about 96% of the mass in the Universe is in the form of 'dark matter' and 'dark energy'. This discovery relegated hadronic matter (protons, neutrons, electrons, etc.) to a mere 4% of the matter in the Cosmos.

As early as 1915, Kapteyn[40] and Adams[41] showed that the radial velocities of a considerable number of stars depended upon the proper motion and mean parallax[42] in way that suggested that the intrinsically brighter stars were moving more slowly than the fainter stars!

Kapteyn and Adams ended their paper by suggesting that either the velocity depends on the brightness of the star, i.e., faint stars move faster than bright stars, or nearer stars move faster than distant stars. In 1917, Adams and Gustaf Strömberg (1882–1962)[43] succeeded in finding the relationship between stellar motion and brightness, thereby confirming the first possibility. They reached the conclusion that the radial velocity is a function of absolute brightness for about 1,300 stars they examined. The relation is simple: for every decrease by a factor of 2.512 in brightness, the velocities increase by 1.5 km/s. On the face of things, this looks like a crazy idea. Why for heaven's sake would the brightness of a star depend on its spatial velocity? Or vice versa, why should the velocity depend on the brightness?

A year later, Strömberg[44] extended the measurements and provided more evidence confirming these results. Soon Benjamin Boss (1880–1970)[45] followed this with the discovery that the velocities of high velocity stars were confined to particular directions in the Galaxy. A similar observation was made by Adams and Alfred Joy (1882–1973m).[46] A short time later Strömberg[47] confirmed Boss' discovery.[48] An interesting feature of the results, so remarked Adams and Joy, was that the high

and not escaping, this means that there must be enough mass to generate the gravitational field required to keep the system bound.

[40] Kapteyn was a Dutch astronomer who published a huge catalogue of stars (450,000 in total) and repeated Herschel's star count by sampling various parts of the heavens. On the basis of these star counts, he confirmed Herschel's view that the Galaxy has a lens shape (which is true) with the Sun at the center (which is false). His estimate of the size of the Galaxy was off by a factor of two.

[41] Kapteyn, J.C., and Adams, W.S., Contribution from the Mount Wilson Obs., # 210, 1915, PNAS **1**, 14 (1915).

[42] Parallax is the angle at which the orbit of the Earth around the Sun is seen by the star. The star with largest parallax, i.e., the nearest star, is α Centauri or Proxima Centauri, for which the value is 0.772 arcsec, corresponding to a distance of 4.22 light years. All other stars have smaller parallaxes.

[43] Adams, W.S. and Strömberg, G., Astrophys. J. **45**, 293 (1917).

[44] Strömberg, G., Astrophys. J. **47**, 7 (1918).

[45] Boss, B., Astrophys. J. **31**, 121 (1918).

[46] Adams, W.S., and Joy, A.H., Astrophys. J. **49**, 179 (1919).

[47] Strömberg, G., Astrophys. J. **56**, 265 (1922).

[48] It is interesting to note that Adams and Joy used Boss and Russell Porter's (1871–1949) catalogue of proper motions, but did not compare their results with those of Boss, although they confirmed his results.

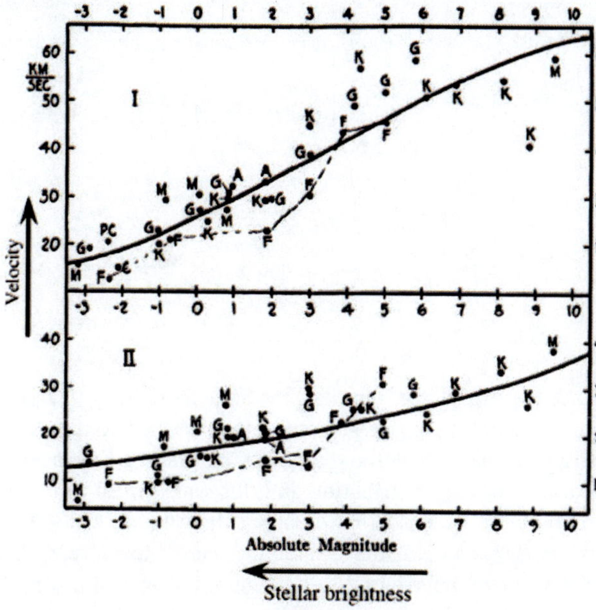

Fig. 6.3 The velocity–brightness relation which was the main result due to Adams et al. (1921)

and low luminosity stars seemed to move in different directions. However, the most important of their results was this:

> For the average velocities of the stars of high and low luminosity, the latter stars move more rapidly.

In 1921, Adams, Strömberg, and Joy[49] produced the results shown in Fig. 6.3, from which the relation between brightness and velocity emerges clearly. They found that the most probable velocity of giant stars was 18.8 km/s, while that of dwarfs was 40.8 km/s. The brightness–velocity relation was finally considered as established. However, Eddington and Douglas[50] apparently had a hard time swallowing such a relation and carried out an analysis which seemed to indicate an improper reduction of data by Adams and his collaborators. The quick response by Adams and Strömberg[51] was that, even when the required corrections claimed by Eddington and Douglas were taken into account, the relationship between brightness and velocity did not disappear.

In 1925, Strömberg[52] was the first to separate the stars in the Galaxy according to their velocities. It was not a matter of few stars here or there, but two distinct groups

[49] Adams, W.S., Strömberg, G., and Joy, A.H., Astrophys. J. **54**, 9 (1921).

[50] Eddington, A.S., and Douglas, A.V., MNRAS **83**, 112 (1923).

[51] Adams, W.S., and Strömberg, G., MNRAS **83**, 474 (1923).

[52] Strömberg, G., Astrophys. J. **61**, 363 (1915).

Fig. 6.4 Hubble's 1926 galaxy classification. The drawings do not appear in the original paper and were made later

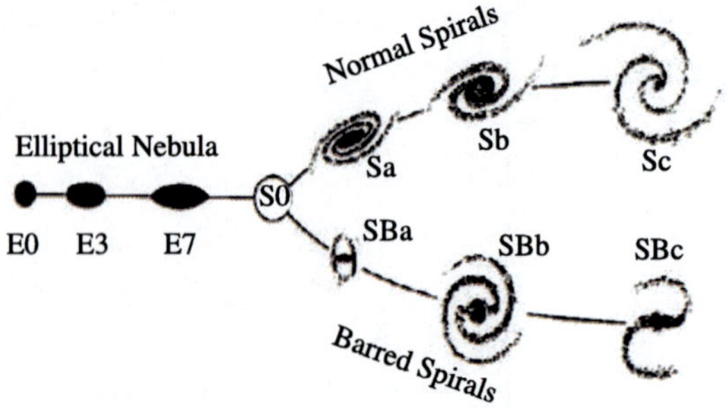

were identified. Strömberg's paper provided the basis for Oort's 1927 papers[53] about the discovery and general theory of galactic rotations, thus confirming Lindblad's (1895–1965m) hypothesis[54] about the rotation of the Galaxy. Strömberg discovered an asymmetry in the velocity distribution and this was critical to the discovery of galactic rotation by Oort. The assumption about the rotation of the Galaxy nicely explained the asymmetry. As a matter of fact, Poincaré[55] had already hypothesized that the lenticular shape of the Milky Way was caused by rotation, and estimated the rotation period to be about 3,700 years (based on the difference between the mean density of the Sun and the mean density of matter in the Galaxy). But the Milky Way does not rotate like a rigid body. In fact, the period of rotation varies with the radius. The period of rotation of the Sun around the center of the Milky Way is 220 million years, quite different from Poincaré's estimate.

6.5 Unintentional Evidence from Other Galaxies

Seares[56] and Seares and Shapley[57] were the first to point out a color difference between the arms (blue) and the unresolved nuclear regions (red) of spiral nebulae. At that time the nature of the spiral nebulae was not known and in particular the nebulae were not known to be composed of gas and stars. In the same paper, Seares discussed the observations of two gaseous nebulae which, as we know today, are pure gas and contain no stars, so he in fact treated gas clouds and galaxies on an equal footing.

The first nebula to be resolved into stars was the great nebula in Andromeda, which is a spiral nebula. However, the stellar nature of the elliptical nebulae

---

[53] Oort, J.H., BAN **3**, 275 (1927); ibid. **4**, 79 (1927).

[54] Lindblad, A., Arkiv. f. Mat. Astr. o. Fysik, Bd. **19A**, # 21, 27, 35 and Bd. **19B**, No 7 (Upsala Meddelanden # 3, 4, 6, 13).

[55] Poincaré, H., Bull. Astron. de la Société Astron. de France, April 1906.

[56] Seares, F.H., PANS **2**, 553 (1916).

[57] Seares, F.H., and Shapley, H., PASP **28**, 123 (1916).

remained unknown. In his famous 1926 paper about the classification of extragalactic nebulae,[58] Hubble classified the elliptical nebulae as extragalactic despite the fact that *they show no evidence of resolution* (see Fig. 6.4). The only hint Hubble had that elliptical nebulae might contain stars was an image of the M87 galaxy, which showed some faint images of what were *apparently stars*. One of the images was Belanowsky's nova[59] of 1919.[60] This particular observation is interesting because Hubble already remarked in this paper that M87 was *a non-galactic nebula of the amorphous type*. Obviously, by comparing the maximum brightness of this nova with other novae observed in our own Galaxy, Hubble could infer that its distance was way beyond the size of the Galaxy, whence it must be extragalactic. To make sure, Hubble included a specific discussion about whether it was a pure coincidence that the nova was observed on the background of M87, or whether it really belonged to this nebula.

By 1929, the inability to resolve the elliptical nebulae led Jeans[61] to consider elliptical nebulae as possible gaseous spheres, and he worked out their dynamic properties under this assumption. In 1930, Ten Bruggencate (1901–1961m)[62] suggested that the surface irradiation of elliptical galaxies was due to scattering of light from a bright central region by a thick, opaque cloud of dust, with no observed stars outside the nucleus.

6.6 Doubts

Charles Perrine (1867–1951)[63] was puzzled by the discovery of a correlation between radial velocity, brightness, and distance. Perrine claimed that he investigated the apparent relation, particularly the dependence upon apparent brightness, and:

> [...] accepted the evidence of these relations until quite recently, [when he found that] if the ellipsoidal motion was taken into account simultaneously, practically all of the other peculiarities could be satisfied upon the hypothesis that the more distant stars, as inferred from their proper motions, were in general moving more slowly than the near stars.

However, the result that remote stars move more slowly than nearby stars is quite surprising, to say the least. On top of this, as an alternative argument, Perrine argued

---

[58] Hubble, E., Astrophys. J. **64**, 321 (1926). Hubble's original paper includes only the description of the nebulae and not the famous figure showing the shapes of the galaxies.

[59] A nova is a star whose luminosity suddenly rises by a factor of up to 10^5 for several weeks and then fades away, and in most cases is too faint to be observed. The peak luminosity of a nova can be used as a standard candle, and hence, as a distance indicator.

[60] Hubble, E., PASP **35**, 261 (1923). For the discovery of the nova see, Belanowsky, I., AN No. 5144.

[61] Jeans, J.H., *The Universe Around Us*, Macmillan, New York, 1929.

[62] Braggencate, P. ten, *Der physikalische Zustand eliptischer Nebel*, Naturwiss **19**, 455 (1931).

[63] Perrine, C.D., Astrophys. J. **47**, 179 (1918); ibid. **46**, 278 (1917).

Fig. 6.5 The strange galaxy NGC 1247A in the Fornax cluster. The galaxy contains many young, massive, recently born blue stars, and is similar to the one observed by Baade, Hubble, and Shapley in 1939. Credit: Hubble Heritage Team, AURA, ESA, NASA

that Adams and Strömberg had not reduced their data properly and had not included all known effects.

Perrine argued as follows:

> The spectral condition of a star depends chiefly upon its size and mass and the external conditions of density of cosmical matter and the relative velocities of star and matter. This matter is assumed to be most dense in the relatively distant regions in the direction of the Galaxy.

The external matter had to slow down the stars of higher velocities:

> The operation is so obvious in principle that no more detailed explanation seems called for at this time.

In other words, the interstellar matter through which stars move slows them down. It was as simple as that.

If this were true then the next question would be: have stars grown up by accretion from very small beginnings, and their velocities gradually decreased in time? In a present day formulation one might say that the stars were born while moving fast, whence fast-moving stars are young, and with time they slow down. Perrine's explanation was not obviously wrong, and as a matter of fact would later be taken up as an argument (Fig. 6.5).

6.7 Oort

In 1926, Oort[64] reached two important conclusions for our discussion here:

- The group of high velocity field stars did not contain any highly luminous massive stars which belong to the upper branch of the ordinary HR diagram.
- The mean intrinsic brightness of low mass stars of a given spectral type seemed to be the same for the high and low velocity stars.

This means that globular clusters must have a main sequence whose stars begin only as early as middle F spectral type (stars of mass 1.4–1.6$M_{\odot}$), because high velocity field stars were confined to such types and later. In other words, the globular clusters do not possess high mass stars. The power of the telescopes available at that time was insufficient to observe the faint lower main sequence part (low mass stars) of globular clusters. Oort's claim was therefore a prediction.

Let us summarize the picture as of 1930. The stars in the Galaxy seemed to be divided into two categories. The distant stars in the Galaxy were fast-moving and faint and belonged to late type spectral classes. On the other hand, all nearby stars were low velocity, and this group contained bright massive stars. The mass of the stars and the dynamics of the Galaxy appeared to be connected, but the nature of the connection was obscured.

6.8 The Enigma of the Nebulae

The first two decades of the twentieth century were witness to one of the greatest controversies in the history of astronomy. This controversy centered on the nature of the nebulae. On the one hand, telescopes were sufficiently powerful to see distant stars in our Galaxy, while on the other, they were too feeble to resolve stars even in the nearest nebula. It was pure coincidence that the interesting objects were just below the resolving power of the telescopes. Consequently, star counts, namely the number of stars as a function of the apparent brightness, could serve to measure the dimensions of our own Galaxy, but there was no tool to measure distances to the nebulae, let alone their physical size. The observed nebulae did not have any object or feature that astronomers could adopt as a standard candle, so it was impossible to measure the distance to these objects.

Suppose all stars have identical intrinsic brightness. Count the number of stars as a function of their apparent brightness (see Fig. 6.6). Clearly, the further away the star, the fainter it is. By measuring the brightness at which the number of stars diminishes to zero, astronomers can determine the size of the Galaxy, provided they can observe the most distant stars. The assumption that all stars have the same intrinsic brightness is called the standard candle assumption. These measurements can be refined by counting only stars of a certain spectral type, comprising a more

64 Oort, J.H., Groningen Pub., No. 40, 1926.

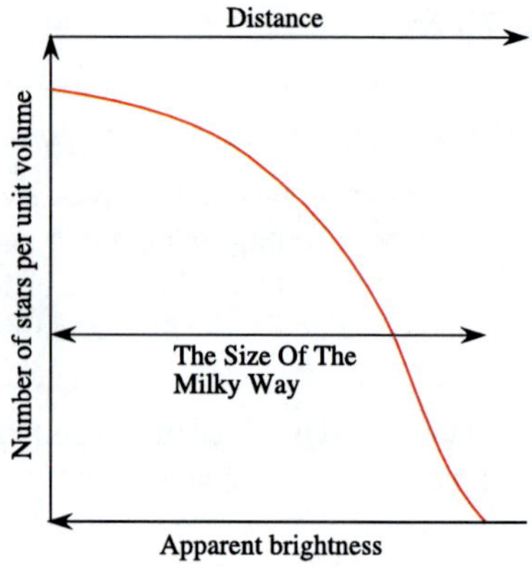

Fig. 6.6 A schematic luminosity function: the number of stars per unit volume as a function of the apparent brightness. Note the direction of the x axis. The fainter the star, the further away it is

homogeneous group of objects. However, the principle as described above remains the same. This simple and very reliable method allowed astronomers to determine first the size of the Galaxy and later the size of the Universe.

6.9 The Great Debate

The mysterious nebulae sometimes exhibited a regular shape with spiral arms and sometimes an irregular amorphous form like clouds in the Earth's atmosphere. However, the pivotal question concerned the location of these objects: were they inside or outside our Milky Way? This was a major question about the size of the Universe. Does it end where the Milky Way ends? Is our Galaxy the only one in the Universe? Is there anything else? Can we see objects that are beyond the outskirts of our Galaxy?

The great debate between Harlow Shapley (1885–1972m) and Heber Curtis (1872–1942m) about the nature of the nebulae took place in 1920.[65] At that time, Shapley's measurements implied a Milky Way diameter of about 60,000 light years, while Curtis found a smaller Milky Way, with a diameter of just 30,000 light years. Kapteyn and Pieter van Rhijn (1886–1960m)[66] got a diameter of 50,000 light years by counting stars. Shapley claimed that the nebulae were objects inside our Galaxy, while Curtis argued that they were independent objects outside the Milky Way, and so remote that our telescopes could not resolve them, which was why they looked

[65] Berendzen, R., Hart, R., and Seeley, D., *Man Discovers the Galaxies*, Science History Publications, New York, 1976; Curtis, H.D., *The Scale of the Universe*, Bull. Nat. Res. Coun. **2**, 171 (1921); Shapley, H., *The Scale of the Universe*, Bull. Nat. Res. Coun. **2**, 194 (1921); Hoskin, M., *The 'Great Debate': What Really Happened?*, J. Hist. Astron. **7**, 169 (1976); Smith, R.W., *The Expanding Universe. Astronomy's 'Great Debate' 1900–1931*, Cambridge University Press, 1982; Struve, O. and Zebergs, V., *Astronomy of the twentieth Century*, McMillan, New York, 1962.

[66] Kapteyn, J.C., and van Rhijn, P.J., BAN **1**, 37 (1922).

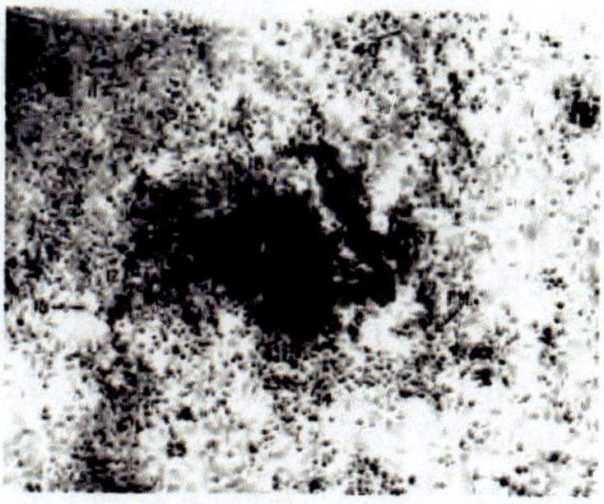

Fig. 6.7 Hubble's image of M33, using a 90 min exposure time with the 100 inch telescope and especially sensitive photographic plate. *Numbers* mark the identified Cepheids. Note that they all appear in the outskirts of the Galaxy, because otherwise it was impossible to resolve the stars. Hubble 1926

hazy or cloudy. It is puzzling to note that the Shapleys[67] assumed that the globular clusters were extragalactic, but refused to accept that the spiral nebulae might also be extragalactic. Indeed, they were quite right about the globular clusters, since they do indeed lie outside the plane of the Galaxy for the main part.

Fourteen different topics germane to the distance measurements were debated. In retrospect, each of the contenders was right on about half of the points and wrong about the other half. It was clear that a decisive new piece of data was needed to resolve the issue. At the end of the controversy, it was not clear who was right, and there was no declared 'winner'. It seems, however, that the show put up by the flamboyant Shapley won the sympathy of the audience, and he was the undeclared winner in this sense, although unjustifiably so.

With hindsight, two points should be made. First, Öpik[68] used Pease's observations of the rotation velocities in the galaxy M31. He assumed that the mass to luminosity ratio was the same as in our Galaxy to estimate the distance at 1.5 million light years, much bigger than the Milky Way. However, Öpik's result was practically overlooked.

Secondly, Curtis was right, although the final proof came only 4 years later with Hubble's[69] victorious measurement of the distance to the Great Nebula in Andromeda (M31), showing that its distance was indeed much greater than the size of our own Galaxy. The breakthrough came when Edwin Hubble (1889–1953m) succeeded in identifying Cepheids inside the Andromeda nebula. He used them as standard candles and in this way determined the distance to the Andromeda nebula (see Fig. 6.7).

It is interesting to note that, in the course of a search for novae inside the nebula M33 in 1920, John Duncan (1882–1967)[70] reported the discovery of a *suspected nova* in the Andromeda galaxy, as well as three variable stars[71] the exact nature of which had not been established. By securing photographic plates taken at several observatories between 1899 and 1922, he confirmed the variable nature of the stars

[67] Shapley, H. and Shapley, M.B., Astrophys. J. **50**, 107 (1919).

[68] Öpik, E., Astrophys. J. **55**, 406 (1922).

[69] Hubble, E.P., Obs. **48**, 139 (1925).

[70] Duncan, J.C., PASP **30**, 255 (1918).

[71] Duncan, J.C., PASP **34**, 290 (1922).

but was unable to determine the period of variation exactly, something which Hubble did succeed in doing.[72] In addition, the data about the maximum brightness of novae was not sufficiently accurate to use them as standard candles. Hubble knew about the existence of variable stars in M33.[73] He used a special technique to sort them out and later measure their periods. In this way he discovered the Cepheids in two nearby nebula. All in all, Hubble discovered 36 variable stars and 46 novae in M31, along with a similar number in M33. He then used Shapley's[74] period–luminosity law for Cepheids to obtain the intrinsic brightness. He could thus determine the distance by comparison with the observed brightness. The distance to Andromeda as found by Hubble was 930,000 light years, which was about a factor of 10 greater than the diameter of the Milky Way, irrespective of who measured it. In this way Hubble brought the controversy to an end: the spiral nebulae were extragalactic and the Universe was at least a factor of 10 times bigger in size than our Milky Way.

It soon became clear that the general term 'nebula' refers to two different objects, whence these objects had to be split into two quite different classes. The first kind are indeed gaseous objects in our own system of stars, the Galaxy, as Shapley had advocated. These objects kept the name 'nebulae'. The second kind of nebulae are huge islands of stars and gas in the Universe, just as Curtis had advocated. At first they were called extragalactic nebulae, but in the present-day terminology they are just galaxies.[75]

Obviously, it was the use of the Cepheids as distance indicators that allowed this dramatic discovery. It is worth noticing that, even in 1912, long before the heated debate took place, Maximilian Wolf (1863–1932m)[76] had discussed the properties and distances of spiral nebulae under the assumption that the nebulae themselves were standard candles, and found distances much in excess of the diameter of the Milky Way.

---

[72] Duncan could not set a lower limit on the variability and derived a minimal distance which was much greater than the size of the Milky Way. Had he done so, he could have scooped Hubble by about five years.

[73] Hubble cited Duncan and the existence of suspected variables.

[74] Shapley, H., Astrophys. J. **48**, 89 (1918).

[75] The two contenders of the debate won immortality by having lunar craters named after them. Curtis who was right, got a crater 3 km in diameter, and Shapley, who was wrong, got a crater 23 km in diameter. The craters are quite close to each other (14.6N, 56.5E and 9.4N, 56.9E). Having said that, one should stress that Shapley made many other contributions to astrophysics and well deserved a big crater. This is mentioned only so that the reader can appreciate that big names make big discoveries as well as big mistakes. When Edwin Hubble arrived at Mount Wilson in 1917, after a short military service, he acted like a major, as if his military rank of captain also applied in the observatory. This mode of conduct did not appeal to his new colleagues on the mountain, and in particular to Harold Shapley, who was the director and whose supposition about the nature of the nebulae Hubble was about to disprove using Shapley's own period–luminosity law.

[76] Wolf, M., A.N. **190**, 229 (1912).

6.10 Fornax and Sculptor Provide Hints

In 1925, Hubble drove the spiral galaxies out of the Milky Way. So the next natural question concerned the way the galaxies were distributed in space. Was the distribution uniform? How far did the galaxies extend? Do we see the end of the galaxy distribution, just as we see the edge of our Milky Way? The answer to the first question was given about 10 years later, when Bart Bok (1906–1983m)[77] discovered that galaxies group into giant clusters of galaxies. Half a year later, Jason Nassau (1892–1965m) and Louis Henyey (1910–1970m)[78] identified the cluster of galaxies in the constellation of Ursa Major. These two discoveries opened the way to a thorough investigation of these objects. The clusters of galaxies are the largest known gravitationally bound cosmic systems. Here we shall touch upon only what is relevant to the discovery of variations in composition throughout almost any galaxy.

Sculptor and Fornax are two clusters of galaxies. Each galaxy cluster may contain hundreds or even thousands of galaxies, moving under the mutual gravitational force of all members in the cluster. As the density of galaxies is high, they move quite close to each other and affect each other significantly. Inevitably, one can find very peculiar objects in such a cluster, such as the object NGC 1427A shown in Fig. 6.5. Inside this object, there are many blue stars. These are massive, newly born stars, and the preponderance of such stars indicates a considerable rate of star formation. This prodigious rate is in turn caused by the action of neighboring galaxies on the relatively small one. In a sense, the small galaxy is just the prey of the bigger one, and soon the big galaxies will swallow all the smaller ones. The merger of two galaxies would take some 10^8 years, and for a long time, during and after the collision, the mix of the two stellar populations of the merged galaxies remains evident. However, the idea of galaxy collision was only put forward in the early 1950s (see later).

In 1938, Shapley[79] discovered two stellar systems belonging to one of the many galaxies in each field of view. He pointed out that:

> The two systems are of the size of an average galaxy, but have low total intrinsic brightness and are devoid of supergiant stars.

For this reason Shapley described these objects as *stellar systems of a new kind*. This very point was noted by Baade and Hubble,[80] who were attracted by Shapley's use of the term 'new kind' and decided to observe these systems. They concluded that, by their paucity of massive stars, the objects discovered by Shapley resembled the unresolved members of the Local Group,[81] i.e., the companion galaxies of

---

[77] Bok, B.J., Harvard College Observatory Bulletin # 895, 1 May 1934.

[78] Nassau, J.J., Henyey, L.G., Astrophys. J. **80**, 282 (1934).

[79] Shapley, H., Nature **142**, 715 (1938); also PNAS **25**, 565 (1939).

[80] Baade, W., and Hubble, E., PASP **51**, 40 (1939).

[81] The Local Group is the group of galaxies that includes the Milky Way and some thirty other galaxies, with a gravitational center located between the Milky Way and the great Andromeda galaxy. The diameter of the this small group of galaxies is about 10^7 light years. The two most

Andromeda. But how come these systems were devoid of such stars? This was the next conundrum.

In 1939, Baade and Hubble investigated these objects and found that the total flux of blue light emitted by such systems was lower than the flux of blue light emitted by any previously known star system. The overall light emitted by these systems was therefore redder than usual. They soon realized that this change in the integrated color of the system was due to the absence of supergiant stars thus confirming Shapley's findings. Having identified this peculiarity of these stellar systems, the most interesting point in their article is a footnote:

> As a working hypothesis, it has been assumed that, in a statistical sense, supergiants are lacking in the nebula described as 'elliptical', in the central regions of 'early'-type spirals, and in irregular nebulae of types represented by M82 and NGC 3077.

Baade and Hubble realized already at this point that there were variations between different classes of galaxies. Among the most conspicuous differences was the fact that certain types of star (the massive ones) are missing in one and not in the other, just as Baade would demonstrate five years later.

6.11 Not all Variables are Alike

We know today about quite a large number of different types of variable star. Here we describe briefly only the pulsating types which exhibit regular periodicity. By pulsating, we mean stars whose light variations are due to internal causes, in contrast to systems where they vary due to external causes, like eclipsing binaries. There are many other classes of star which erupt like a supernova and do not display periodicity.

- δ Cepheids or classical Cepheids. The period is 1–70 days and the amplitude of variation varies from a factor of 1.1 to a factor of 6 in luminosity. The classical Cepheids obey a period–luminosity law. These are massive and very luminous stars.
- W Virginis or Cepheids of the second class. The period is 0.8–35 days and the amplitude about a factor of 1.3–3. These variables obey another period–luminosity law. At the same period, the W Virginis stars are fainter by about a factor of 2 relative to the classical Cepheids. The details of the light curve differ from those of δ Cepheids by having humps on the descending branch.

(Footnote 81 continued)
massive members of the group are the Milky Way and the Andromeda galaxy. These two spirals each have a system of satellite galaxies that move around them. The Milky Way's satellite system consists of Sag DEG, the Large Magellanic Cloud, the Small Magellanic Cloud, Canis Major Dwarf, Ursa Minor Dwarf, Draco Dwarf, Carina Dwarf, Sextans Dwarf, Sculptor Dwarf, Fornax Dwarf, Leo I, Leo II, Tucana Dwarf, and Ursa Major Dwarf. Andromeda's satellite system consists of M32, M110, NGC 147, NGC 185, And I, And II, And III, And IV, And V, Pegasus dSph, Cassiopeia Dwarf, And VIII, And IX, and And X. When the collection of galaxies is small, just a few dozen in all, it is called a group. The term 'cluster of galaxies' is reserved for much more numerous collections of galaxies.

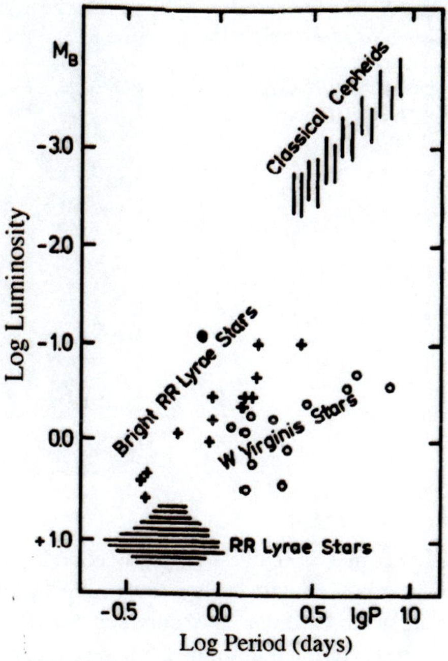

Fig. 6.8 Comparison of the period–luminosity laws of the various pulsating variables. The W Virginis stars are marked by *open circles*, while *crosses* mark variables with observed luminosity in globular clusters. From Kukarkin (1975)

- RR Lyrae. The period is 0.2–1 days and the amplitude of variation is from a factor of 1.3 to a factor of 6 in luminosity. The stars obey a period–luminosity law. The stars are much older than the Cepheids and have a lower mass (Fig. 6.8).
- RV Tauri. The period is 30–100 days and the amplitude can reach a factor of 16 in luminosity.
- Long period variables, known also as Mira variables. The period is 80–1,000 days and the amplitude (in the visible light) varies from a factor of 10 to a factor of 100. These stars are supergiants.
- Semi-regular variables. The period is 30–1,000 days and the amplitude varies between a factor of 2 to a factor of 4 in luminosity.

As a rule, the smaller and denser the star, the shorter is the period. The above definitions do not uniquely define the variables and subgroups that have been identified.[82]

6.12 Star Clusters as the Rosetta Stone of Stellar Evolution

Star clusters are groups of stars bound by their mutual gravitational attraction, and hence all the stars in the cluster are bound to the cluster, much like the nucleons in a nucleus, which are bound by their mutual nuclear attraction. There are two major types of star cluster: globular clusters (GC) and open or galactic clusters (OC). The globular clusters are very densely populated with stars at their center. The distribution

[82] For more details, see Kukarkin, B.V., IAUS **67**, 511 (1975).

Fig. 6.9 The old globular cluster M2 in Aquarius. The diameter is about 150 light years. Credit: NASA and the Hubble Heritage Team (STScI/AURA)

of stars in a typical GC is spherical with a radius of about 30 light years, and such a cluster will contain up to a few million stars. In comparison, the average density of stars in the Galaxy is about one star per 10 cubic light years. Hence, the mean density in GCs is about 8,000 times the mean density of stars in the Galaxy. As can be seen from Fig. 6.9, the density is maximal as the center. In atomic nuclei, by contrast, the density is constant as a function of radius and does not vary from one nucleus to the other, the main reason being that, at very short distances, the nuclear force changes from being attractive to strongly repulsive, thus preventing the nucleons from coming too close to one another. This difference between the nuclear and the gravitational forces allows for extremely high densities in stellar clusters (and the formation of black holes). Open clusters contain a few hundred stars and have an amorphous shape.[83]

There are over a hundred globular clusters known in our Galaxy. It was Shapley[84] who pointed out in 1918 that almost all globular clusters are to be found in the part of the sky which contains the galactic center. The globular clusters are evenly spread around the center of the Galaxy. Observations of other spiral galaxies shows that they also contain GCs, and that they are distributed symmetrically around the center. The brightest stars are red giants. One of the most famous globular clusters is M2 (see Fig. 6.9).

The OCs or galactic clusters are restricted to the plane of the Galaxy and as a rule are found inside the spiral arms. The open clusters do not have a definite shape (see below). The total number of stars is generally less than 10,000, and usually just a few hundred. The volume in space occupied by open clusters is about the same

[83] It will be shown that the two types of cluster have distinct dynamic properties as well as typical compositions, but this was not known until Baade's work.

[84] Shapley, H., A series of 12 papers on stellar clusters, Contributions from the Mount Wilson Observatory/Carnegie Institution of Washington **151–157** (1918).

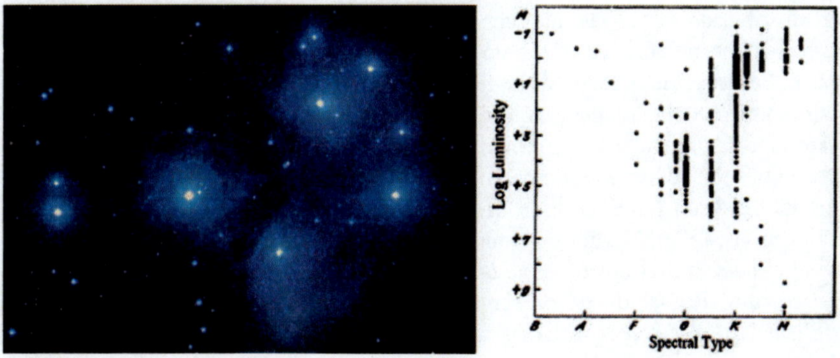

Fig. 6.10 *Left* the young open cluster M45, known as the Pleiades. The estimated age is just 10^8 years and the distance about 425 light years. The total mass of the cluster is about $800M_{\odot}$. Credit: NASA/ESA/AURA/Caltech. *Right* the HR diagram of the high velocity stars, as depicted by Miczaika (1940)

as the volume occupied by the globular clusters. However, since they possess far fewer stars, their mutual attraction is weaker. The stars are thus loosely bound and can more easily escape from the cluster. The brightest stars are blue supergiants. As these objects are young, in some particularly young OCs, one can still observe remnants of the initial gas cloud out of which the stars were formed. One of the most prominent open clusters is the Pleiades (Subaru in Japanese), also known as M45 (see Fig. 6.10 left).

The stars in a cluster were formed at about the same time from the same interstellar cloud of gas. There is a large age difference between globulars on the one hand and open clusters on the other. The globulars are as old as the Milky Way, while the open clusters are very young, about 1/10th to even 1/100th the age of the Galaxy. Simultaneous formation also means that all stars have the same composition but differ mainly in their total mass. There are other stellar parameters that vary from star to star (like rate of rotation, strength of magnetic field), but these are much less important to the evolution of the star and the synthesis of the elements. By observing how the stars in such a group occupy the HR diagram, we can learn a great deal about the evolution and structure of stars and the role of their composition. Indeed, the clusters can be considered as the Rosetta Stone of stellar evolution and element synthesis, as will become clear below.

6.13 Baade: Stellar Populations

In 1940, in the middle of World War II, Miczaika[85] investigated the *stars with large velocities*, by which he meant larger than 63 km/s. Miczaika compiled a catalogue and drew the HR diagram of the high velocity stars. He did not include the low velocity stars in the diagram (see Fig. 6.10 right). Miczaika observed that the spec-

[85] Miczaika, G., A.N. **270**, 249 (1940).

tral distribution of the fast-moving stars is different from that of normal stars, in particular by the absence of the upper part of the main sequence. Miczaika, who was in Berlin at the time, published his results during WW II in the German journal Astronomische Nachrichten. His paper remained basically unknown outside Germany, and in particular to the American scientific community working on the problem, until 1955. Most astronomers were drafted in to help with the war effort, and certainly did not have the time or the motivation to look at German astronomy publications. In 1953, Miczaika moved to the USA and started to publish in American journals. It is strange to note that he spoke of the Russell diagram and not the Hertzsprung–Russell diagram, even though Hertzsrpung, who was of Danish origin, published his discovery of the diagram in 1911 when he was in Potsdam, at the great observatory near Berlin.[86]

In 1944, Baade (1893–1960m) published two papers[87] which were printed in the Astrophysical Journal one after the other. In the first paper, he described how, using red sensitive plates, he succeeded through his ingenious capabilities as an observer in resolving the central portions of M31 and its companion galaxies M32 and NGC 205. As Baade wrote in the introduction:

> The two companions of the Andromeda nebula—M32 and NGC 205—and the central region of the Andromeda nebula itself have always presented an entirely nebulous appearance. Since there is no reason to doubt the stellar composition of these unresolved nebulae—the high frequency with which nova occur in the central region of the Andromeda nebula could hardly be explained otherwise—we must conclude that the luminosities of their brightest stars are abnormally low, of the order of 100 times less compared with the brightest stars in our own Galaxy and for the resolved members of the Local Group.

Baade converted a null observation into a positive statement:

> Although these data contain the first clear indication that in dealing with galaxies we have to distinguish two different types of stellar populations, the peculiar characteristics of the stars in unresolved nebulae remained, in view of the vague data available, a matter of speculation; and since all former attempts to force a resolution of these nebulae had ended in failure, the problem was considered one of those which had to be put aside until the new 200 inch telescope should come into operation.

So what happened? World affairs, ingenuity, luck, and the 'collaboration' of the stars joined forces to push science forward. The German citizen Baade came to the USA in 1931 as a staff member of Mount Wilson. When WW II broke out and most of the scientific staff were mobilized to the war effort, Baade was excluded, being classified as an enemy alien. For some unknown reason, his request for naturalization never went through. Somehow the papers were lost. So as an enemy alien, he was confined to Los Angeles County. Consequently, Baade found the 100 inch telescope dedicated to his research, and he made good use of it. The 100 inch telescope on Mount Wilson overlooks Los Angeles and the lights from the fast-growing city disturbed astronomers. But the wartime blackout order played into Baade's hands by providing good observing conditions.

---

[86] Publikationen des Astrophysikalischen Observatorium zu Potsdam, 1911, Nr. 63.

[87] Baade, W., Astrophys. J. **100**, 137 (1944); ibid. 147 (1944).

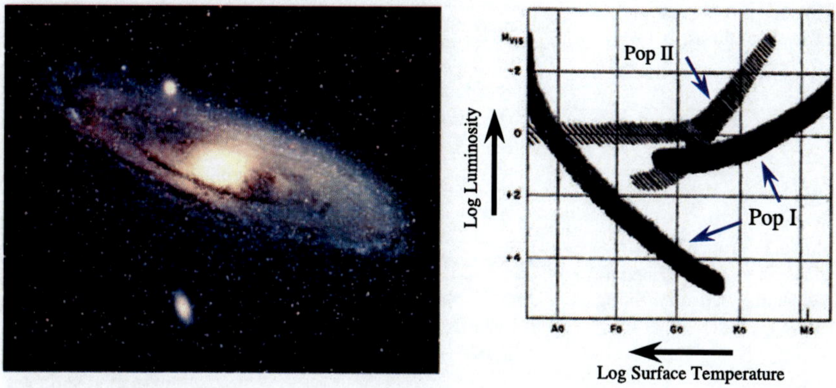

Fig. 6.11 *Left* the great spiral nebula in Andromeda. Baade and Hubble's research domain. Note the two small companion galaxies above and below the giant Andromeda galaxy. Credit: Hubble Heritage Team, AURA, ESA NASA. *Right* the original HR diagram of the two stellar types, Baade 1944. The Pop II stars do not exhibit the main sequence. The *horizontal axis* shows the spectral type, which is equivalent to the surface temperature, increasing *from right to left*

By sheer good fortune, Baade discovered that photographic plates taken with the 100 inch telescope two years earlier *revealed for the first time unmistakable signs of incipient resolution in the hitherto apparently amorphous region*. But how had this happened? Without going into the technical details of highly sensitive photographic plates, suffice it to say that the stars cooperated by having the unique property which made them detectable on the most sensitive red plate. Still, what made this discovery possible was the combination of long exposure times, at least 4 h, good seeing (stable atmosphere), the 'collaboration' of the stars themselves in changing their color, world politics, and above all a first-rate astronomer who was able to overcome all the 'small details'.

Baade discovered that the brightest stars in the newly resolved systems were not blue stars, like the brightest stars in the spiral arms of our Galaxy, but rather red stars, in fact red giants similar to those found in globular clusters in our own Milky Way. The discovery that there are different kinds of stars in different locations of the Galaxy led Baade directly to the hypothesis of two populations of stars. The main result, the HR diagram of the two stellar populations, is shown in Fig. 6.11 (right).

The cautious Baade concluded that:

> Although the evidence presented in the preceding discussion is still very fragmentary, there can be no doubt that, in dealing with galaxies, we have to distinguish two types of stellar populations, one which is represented by the ordinary HR diagram (type I), the other by the HR diagram of the globular clusters (type II).

This is Baade's important result. On the basis of the location in the HR diagram, Baade identified two classes of stars. It is not that the stars are spread over greater regions in the HR diagram. The stars belonging to type I actually differ in their properties from the stars of type II. The globular clusters are too far away and the faintest stars in a

Table 6.1 Properties of stellar populations, as described by Baade in 1944

Population I	Population II
Open (galactic) clusters	Globular clusters
Highly luminous O and B type stars	Short period Cepheids
Intermediate type nebulae	Elliptical and Sa nebulae
	Intermediate type nebulae

Table 6.2 Properties of variable stars as described by Baade at the Vatican meeting in 1958

Population I	Population II
Type I Cepheids	Type II Cepheids
β Can Maj. stars	RV Tauri stars
T Tauri	Cluster type variables
	Long-period variables
	Ordinary novae
Supernova Type II	Supernova Type I

globular cluster were too faint for the telescopes of the day to resolve them. Hence the story behind these faint stars was not yet known at that time. In particular, the globular clusters (the Pop II stars) showed no main sequence. The characteristics of the two types are summarized in Tables 6.1 and 6.2. Baade realized that none of the galaxies was a pure type. Every galaxy contained mixtures of populations. Table 6.2 shows the typical variable stars which belong to the corresponding stellar population. Note that Baade still referred to the galaxies as nebulae (and to the Milky Way as a galaxy).

The final note by Baade is most intriguing. Recalling Oort's 1926 paper, Baade identified the type II stars with the high velocity stars of our Galaxy (which he called *our type II*) and Oort's slow moving stars with type I, these stars being the majority in the solar neighborhood. Note the logic: Baade supported Oort's conclusion about the existence of two types of star by observing extragalactic objects. He was unable to measure the velocities of these stars relative to their galaxy (he could hardly even observe them, and to ask for their velocities was too much). So Baade identified the stellar populations observed in nearby galaxies with the dynamic properties of the stars in our own Galaxy.

In the second paper, Baade discussed the observations of two newly discovered members of our Local Group of galaxies, NGC 147 and NGC 185. The discovery that these galaxies were members of our small group of galaxies is less important for our discussion here. The crucial issue is that Baade discovered that the stellar populations of these galaxies had intermediate properties between type I and type II, indicating that the separation between the two types was not sharp, but in fact rather smooth. The idea of two populations was not the end of the story, a fact available to Baade from the very beginning.

The 200 inch telescope was completed in 1948 and the scientific dedication took place on 1 July of that year. Baade was invited to give the scientific dedication at the

Mount Palomar observatory,[88] and from this talk it is clear that he did not know the reasons for the differences between the two populations of stars. As an astronomer, Baade discovered the difference between the two populations, but refrained from hypothesizing about its meaning.

The first to provide Baade with part of the correct explanation was George Gamow, who sent him a note on a postcard to the effect that there was only a difference in the age of the stars.[89] In 1948, Russell[90] tried to explain the difference in brightness as due to age. Russell's logic was as follows. Stars form from gas. The massive stars are the bright ones and have the shortest lifetime, so they never succeed in moving far away during their lifetime, but remain immersed in the gas from which they formed. In contrast, the low mass stars have a long lifetime and succeed in moving away from the gas clouds they arose from. The old low mass stars are the faintest. Consequently, concluded Russell, it was a difference in age that caused populations I and II to differ.

This explanation described how the difference in age implied a difference in the main sequence, but it did not explain why the spectra were different. Gamow himself published his version of the explanation in 1951,[91] without reference to Russell. Gamow remarked that the upper part of the HR diagram of Pop II was absent. This fact suggested that the Pop II stars were old, and consequently that the massive stars, which have a short lifetime, had already disappeared. The Pop I group comprises young stars and so still possesses its massive stars. However, because his calculations for low mass stars were carried out assuming the CN cycle, the rest of the paper does not agree with observation.

In that year, Baade set out to discover the main sequence of globular clusters, since he believed that:

> There is every indication that, beginning with spectral type F, the dwarf branches of the Pop I and II coincide.

This statement accords with Russell's explanation, but we know today that it is not accurate. The difference in metal abundances[92] leads to a different atmosphere structure, so the lower part of the main sequence, like all other parts of the HR diagram, is not identical for the two populations.

As soon as the 200 inch telescope entered routine service in November 1949, Baade assigned the problem of the main sequence of globular clusters to his two

---

[88] Sandage, A., *The First 50 Years at Palomar*, Ann. Rev. Astron. **37**, 86 (1999).

[89] Osterbrock, D.E., IAUS **164**, 21 (1995).

[90] Russell, H.N., PASP **60**, 202 (1948).

[91] Gamow, G., Nature **168**, 52 (1951).

[92] In astrophysics, the word 'metal' refers to all elements with atomic number greater than 2. Hence, there is hydrogen (X), then helium (Y), and all the other elements are lumped under 'metal' (Z). The reason is that the ionization potential of hydrogen and helium is very high compared to kT_{surface}, where T_{surface} is the stellar surface temperature, while the opposite holds for the rest of the elements. Consequently, the electrons which affect the spectra are contributed by all elements save hydrogen and helium. Hence the significance of the total amount of heavy elements Z, while the details of the breakdown are less important.

graduate students, Halton Arp and Allan Sandage. Their story will be told subsequently.

6.14 The Details are Important

When Shapley updated the period–luminosity law in 1918,[93] he compiled a list of typical variable stars with clearly determined periods of less than forty days. The stars were obtained from the archives of the Harvard observatory as well as Hartwig's annual catalogue. Shapley was happy with the data, but added with caution that:

> The material is not extensive, and its sufficiency for this problem might well be questioned.

So Shapley took all the data available at the time and generated a plot of the intrinsic brightness against period. His result, based on 230 Cepheids, is shown in Fig. 6.12. To obtain this large number of Cepheids, Shapley combined data from 7 different stellar systems, and in this way, not realizing the possibility of Cepheids with dissimilar properties in different stellar systems, or more accurately not dreaming of the possibility that stars might contain different heavy element abundances or that this could affect the period–luminosity relation, lumped all the data together. It was this relation, based on all the available data, that Hubble used to derive the distance of the Andromeda galaxy. Shapley's problem was that he could not determine the intrinsic brightness of the variable stars in the globular cluster. The reason was that, at that time, he did not have any alternative way to measure the distance to the remote globular clusters and thereby find the period–luminosity law for these variables.

The warning signs that something might be wrong could already be seen from Shapley's plot of the period–luminosity diagram shown in Fig. 6.12, where the deviation from a straight line is seen at short periods. But this clue was overlooked. The flattening of the curve implies that the amplitude does not change for short periods, as if the law breaks down.

After some 30 years, astronomers began to realize that something was not quite right about the distance determination based on Cepheids. When Baade carried out his seminal work on stellar populations, he was already aware that:

> It is now quite certain that Hubble's value of the distance is somewhat too small.

Baade questioned Hubble's value for the distance to the Andromeda galaxy on the basis of a comparison between the brightness of the brightest stars he could see in Andromeda and the brightness of the brightest stars seen in our own Galaxy. The correction to the luminosity Baade found was a factor somewhere between 1.38 and 1.79, so the distance to Andromeda had to be corrected by a factor of 1.17–1.33. A similar situation existed with the stars in globular clusters in our Galaxy. Here the correction factor for the brightness was as high as 3.16.

---

[93] Shapley, H., Astrophys. J. **48**, 279 (1918).

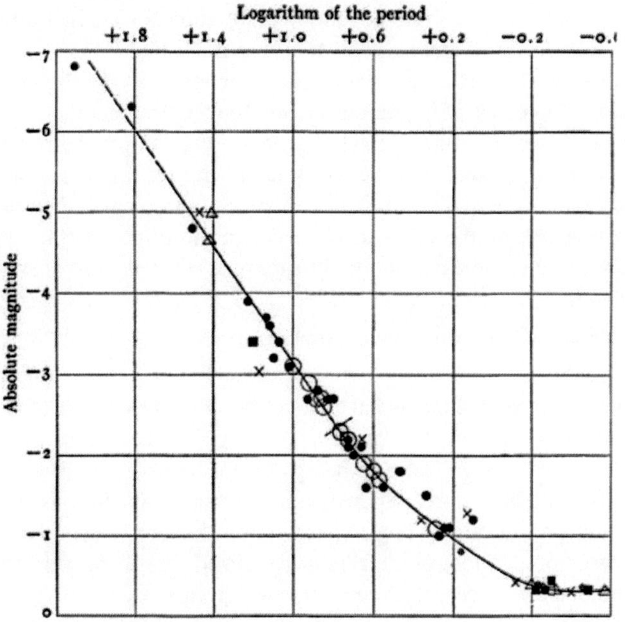

Fig. 6.12 The log luminosity–log period relation, as plotted by Shapley (1918) using all available data

6.15 Dispersing the Fog Around the Variable Stars

As late as 1949, Joy[94] examined the variable stars in globular clusters and found, in addition to the RR Lyrae class, four additional groups of variable stars that had properties differing from those of the classical Cepheids in the form of RR Lyrae. The last sentence in Joy's paper is:

> The occurrence of large numbers of RR Lyrae variables in systems where Cepheids are practically non-existent shows that the two classes of variables are not so closely related as has hitherto been assumed.

In other words, one should not mix Cepheids with RR Lyrae as Shapley had done. Still in 1949, Roscoe Sanford (1883–1958m)[95] examined the variable star W Virginis, which is the prototype of a special class of variable stars. Sanford noticed that, during certain parts of the period, all the absorption lines doubled, asserting that:

> None of the classical Cepheids has shown such doubling of its absorption lines.

The star was considered by Baade to belong to population II because it was located well above the galactic plane in the halo, and had high velocity. Sanford claimed

---

[94] Joy, A.H., Astrophys. J. **110**, 105 (1949).

[95] Sanford, R.S., PASP **61**, 134 (1949); ibid. Astrophys. J. **109**, 208 (1949).

therefore that the variables observed among population II stars had different properties from those observed among population I stars. Actually, Sanford claimed that even the prototype star RR Lyrae (which is a member of a GC) shows a somewhat analogous doubling of its hydrogen lines (for a limited time during the period).

A few years layer, in 1953, Shapley[96] returned to the problem of the period–luminosity relation. At that time he knew about the results of Joy and Sanford, and came up with three possibilities to explain the discrepancy between distance measurements based on the comparison between the brightest stars and distance measurements based on the period–luminosity relation. The third possibility raised by Shapley was that:

> The period–luminosity relation for the classical cepheids, in the Magellan Clouds and elsewhere, is wrong.

Realizing this possibility, Shapley calculated that, if the brightness of the Cepheids was off by a factor of 3.31, thus nearly doubling the distance of the Magellanic Clouds and the Andromeda Nebula, the agreement between the distance measurement methods could be restored. Shapley noticed that the W Virginis stars appeared in globular clusters in our Galaxy. These stars, Shapley claimed, are *somewhat abnormal variables*. At that point, Shapley realized that there were two types of Cepheid, and decided to separate them into two groups, each with its own period–luminosity relation, to discover that the difference between the luminosities of the two classes was just a factor of 4. Consequently, Shapley claimed that the distance to the Magellanic Clouds was 150,000 light years.[97] Next, claimed Shapley, the distance to the great nebula in Andromeda had to be 1,500,000 light years, almost twice the distance found by Hubble.

The correction of the period–luminosity relation solved an additional problem that Shapley mentioned, namely, the fact that radioactive dating of the Earth gave a greater age then the age of the expanding Universe. But now, since the Universe had become almost twice as large overnight, this contradiction disappeared. This particular problem, namely the conflict between the age determination from the expansion of the Universe and the age determination from the oldest stars, was recurrent in later years.

Thus, 35 years after overlooking the different groups of variable stars and fighting in favor of the idea that the nebulae were inside the Milky Way, Shapley corrected the basic tool for measuring cosmic distances, only to find out that the Great Nebula in Andromeda was even further away.

In 1956, Reddish[98] succeeded in explaining that the difference between the two groups of variable stars was a consequence of the difference in the abundances of the heavy elements,[99] thereby confirming the connection between heavy element abundance, type of variable, stellar population, and age.

---

[96] Shapley, H., PNAS **39**, 349 (1953).

[97] Hertzsprung, the first to think of using the period–luminosity law to measure distances, got just 33,000 light years.

[98] Reddish, V.C., MNRAS **116**, 533 (1956).

[99] The term 'heavy element' means here anything heavier than helium.

The RR Lyrae Cepheids belong to population II and have a low abundance of heavy elements. On the other hand, the Cepheids have a high abundance of heavy elements. Shapley was unaware of stellar populations and what a change in abundance would do to the period–luminosity relation.

Upon reflection, it was lucky that the error in the period–luminosity law was not in the opposite direction. It so happened that the measured distance to Andromeda was not as great as the true distance, but sufficiently large to fall outside the Milky Way by a safe margin, so as to convince the most skeptical astronomers that the Andromeda Nebula was indeed an extragalactic object.

6.16 It is the Chemical Elements: First Signs

A good spectroscopist has a sharp eye and can tell in seconds whether the spectrum of a star is normal or unique. The first signs that there were problems with the spectral classification appeared to spectroscopists as early as the mid-1930s.

In 1935, Adams et al.[100] investigated the relation between spectral lines and the brightness of a star. They found groups of stars which appeared *somewhat fainter than normal giants*, but they were not sure whether being fainter expressed a genuine distinct property of this group of stars or whether it was just an artifact of some sort. The authors also noted 3 stars which had especially large velocities, and from the point of view of brightness formed a kind of middle group, since they were exceptionally faint.

In view of the many variants of spectral types, it became necessary to document all the possibilities. This mission was carried out in 1943 by Morgan, Keenan, and Kellman, who produced an atlas of stellar spectra.[101] These spectroscopists remarked that the spectrum of the Cepheid variables in the RR Lyrae class was peculiar and could not be located uniquely in a spectral classification system, because the hydrogen lines were much too weak for the spectral type as indicated by the intensity of the calcium lines. Münch and Terrazas were intrigued by this remark and carried out a detailed analysis of these peculiar stars.[102] They confirmed the peculiarity of the Cepheids of the RR Lyrae type, but gave no explanation as to the possible source of the peculiarity. The reader may wonder what is so special about calcium, among all the heavy elements. The reason is that spectral lines of calcium fall just in a region of the spectra in which the energy flux emerging from these variable stars is high. Other stars with high energy fluxes in other regions of the spectrum would show lines of other elements, not necessarily calcium.

---

[100] Adams, W.S., Joy, A.H., Humason, M.L., and Brayton, A.M., Astrophys. J. **81**, 187 (1935).

[101] Morgan, W.W., Keenan, P.C., and Kellman, E., *An Atlas of Stellar Spectra with an Outline of Spectral Classification*, Chicago University Press, Chicago, 1943.

[102] Münch, G., Terrazas, L.R., Astrophys. J. **103**, 371 (1946).

In 1950, the Schwarzschilds[103] decided to examine in detail the idea that the differences between the high and low velocity stars could be put down to the abundances of heavy elements. With this goal in mind, they decided to compare only stars of the same spectral class, opting for a comparison of F type dwarfs (in the Hertzsprung terminology). The stars contained three high velocity stars and three low velocity stars. The Schwarzschilds measured the characteristics of thirteen lines and compared the results. The conclusion they reached was that there were *no differences between the high velocity and the low velocity F dwarfs*. However, the only difference between the high and low velocity stars was that the abundance of carbon relative to the metals was *probably* about 2.5 times greater in the high velocity stars. Note that the difference concerns the abundance of carbon relative to metals and not the absolute value.

Hardly a month and a half after the question had been raised by the Schwarzschilds, Roman,[104] who was unaware of their results, noted similar differences in 94 stars from other spectral classes, in particular that as a rule the high velocity stars had weaker spectral lines of heavy elements. No suggestion was offered as to the source of this difference.

Care must be taken to avoid a trap. The spectral type is determined by the ratio of line intensities assuming all stars have strictly the same composition. Thus, when two stars have the same spectra, surface temperature, and abundance, they must have the same line ratio and vice versa. However, if there are differences in abundance between the stars, there will be a difference in line ratio, even when the surface temperatures are presumed identical. Consequently, the different line ratio may be interpreted as the same abundance but different spectral class.

Indeed, in 1951, Schwarzschild (1912–1997), Spitzer (1914–1997), and Wildt (1905–1976m)[105] suggested a *tentative* explanation, namely that there is a general reduction in the abundance of the heavy elements in high velocity stars relative to low velocity stars and that it is this reduction which causes the spectral peculiarities. Thus, the idea that the abundance of the heavy elements was not the same in all stars was thrown up in the air. We will return to this paper when we discuss the idea put forward to explain how these differences came about.

A quite esoteric idea was suggested by Weizsacker in 1951,[106] namely, that population I stars are not *newly formed but rather rejuvenated, and that even this process of rejuvenation needs peculiar external conditions*. No details or mechanism were suggested. The origin of this speculation is Weizsacker's theory of galaxy formation, on the basis of which he concluded that there should not be any star formation in the Galaxy at the present time. It would seem that the argument should have been the other way round.

One may ask why the topic of peculiarities in the spectra did not sooner attract more astronomers and astrophysicists. Maybe the idea and dramatic impact of poten-

[103] Schwarzschild, M. and Schwarzschild, B., Astrophys. J. **112**, 248 (1950).

[104] Roman, N., Astrophys. J. **112**, 554 (1950).

[105] Schwarzschild, M., Spitzer, L., and Wildt, R., Astrophys. J. **114**, 398 (1951).

[106] von Weizsacker, C.F., Astrophys. J. **114**, 165 (1951).

tial implications of abundance differences just did not cross anyone's mind. Stellar classification was considered a dull subject. However, the situation changed dramatically a decade later, when the connection between composition and dynamics became evident.

6.17 Confirmation: It is the Chemical Elements

A major twist in the mystery and confusion came in 1951 when Chamberlain and Lawrence Aller (1913–2003)[107] analyzed in great detail the composition of three specific stars.[108] The reader may wonder why these particular stars were chosen of all the stars in the sky, or how they knew which were the 'right stars' to choose. If it looks as though Chamberlain and Aller just picked three stars out of the catalogue and analyzed them, this is a very misleading impression. The first two stars appeared in the Mount Wilson Spectroscopic Parallax Catalogue, prepared by Adams et al.[109] back in 1935, as being brighter than regular white dwarf stars but less bright than main sequence stars, so they were already put down as special in this catalogue. It was due to this peculiarity that they posed a problem for the astronomers preparing the Henry Draper catalogue, and this was exactly the point which attracted Chamberlain and Aller. The third star was apparently a normal A type star lying slightly above the main sequence.

In order to elucidate the reasons underlying the particular appearance of the spectra, Chamberlain and Aller considered various options. They cited Greenstein (1909–2002),[110] who had encountered a similar phenomenon. Greenstein was so perplexed by the results that he suggested that the Saha equation might be inaccurate, predicting a lower ionization than was actually the case. Another suggestion was made by Rudkjobing,[111] who surmised that the difference might be due to a different structure of the stellar atmosphere and argued (without any numerical calculations to support his claims) that differences in the structure of stellar atmospheres might give rise to a false impression of low or high abundances. He mentioned that there were different theories for stellar atmospheres (citing in particular the theories of Biermann and Unsöld) and asserted that:

> This difference in model is—contrary to other differences—characterized by the lack of intermediate models. Therefore this difference will be more marked in the HR diagram than other differences in model.

---

[107] Chamberlain, J.W., and Aller, L.H., Astrophys. J. **114**, 52 (1951).

[108] HD 19445 and HD 140283 in the Henry Draper catalogue, and 95 Leonis which is HD 103578. The star was recently observed by Royer et al. [Astron. Astrophys. **393**, 897 (2002)] who found it to rotate with a speed of 16 ± 4 km/s, a fact which complicates the analysis. On top of that, it is a binary system and the velocities of the stars are 11 and 8 km/s.

[109] Adams, W.S., Joy, A.H., Humason, M.L., and Brayton, A.M., Astrophys. J. **81**, 187 (1935).

[110] Greenstein, J.L., Astrophys. J. **109**, 212 (1949).

[111] Rudkjobing, M., Ann. Astrophys. **13**, 69 (1950); ibid. Zs. f. Astr. **21**, 254 (1942).

Once again, no numerical calculations were provided to substantiate this suggestion.

As Chamberlain and Aller were not convinced by the above explanations and suggestions, and for good reasons, they concluded:

> At any rate, the stars treated in the present paper are undoubtedly quite different from the metallic-line stars, and it may well be that super ionization (as suggested by Greenstein) cannot explain the apparent abundance.

So the authors reached the conclusion that a large hydrogen to metal ratio (i.e., a smaller amount of heavy elements relative to hydrogen) could possibly exist in type II population objects in general. This was the first indication that the population type had something to do with differences in the total amount of the heavy elements in the star. Chamberlain and Aller were bold enough to revoke the universally accepted assumption that there is a universal abundance and that all stars, irrespective of their location, mass, or age, have the same composition.

Sandage[112] tells the following fantastic story about Chamberlain and Aller's breakthrough paper. The first version of the paper claimed a reduction of the heavy elements by a factor of a 100 (which turned out to be the correct value). This result seemed so far out of line with the 'establishment' that the authors themselves, either independently before submission to the Astrophysical Journal or upon a suggestion of the referee, changed the physical parameters (at the cost of the quality of the fit) to obtain a reduction by a factor of only 10–30, the value ultimately published.

However, even this 'compromise' was questioned by Unsöld, the leading authority on stellar atmospheres. In 1938, Unsöld published the book *Physics of Stellar Atmospheres*,[113] which became the 'bible' for calculations of spectral line intensities and the abundances of the elements. It was during WW II that Unsöld published the first ever detailed spectral analysis of a star based on a consistent stellar atmosphere.[114] This was the star τ Scorpii, which is a hot star and consequently shows spectral lines of He, C, N, and O, not easily accessible in the relatively cooler Sun (τ Scorpii is a B0 star with an effective surface temperature of 32,800 K). For the first time the abundance ratio of helium to hydrogen could be measured directly in the same star, and the ratio found by Unsöld was $He/H = 0.21$.[115]

In 1957, Unsöld was honored with the George Darwin lecture,[116] which he devoted to quantitative analysis of stellar spectra, and finding the abundances of the elements.

---

112 Sandage, A., Ann. Rev. Astron. Astrophys. **24**, 421 (1986).

113 Unsöld, A., *Physik der Sternatomsphären*, Springer, Berlin.

114 Russell's famous determination of the solar abundances could not include the structure of the atmosphere, because a suitable theory did not yet exist. The theory was worked out during the 1930s and summarized in 1938 by Unsöld in his book.

115 Unsöld, A., Ziet. f. Ast. **21**, 1 (1942); ibid. **22**, 356 (1943); ibid. **23**, 75 (1944). Just before the outbreak of WW II, Unsöld spent a few months in the USA, and together with Otto Struve obtained high resolution spectra of τ Scorpii right after the inauguration, under the directorship of Struve, of the new McDonald Observatory in Texas with its 82 inch telescope. It was these spectra that kept Unsöld busy during his long hours of service as a meteorologist in a German air field during WW II. Unsöld's paper with Struve was published in the Astrophys. J. in 1940, after the outbreak of WW II, but before the US entered the war.

116 Unsöld, A., MNRAS **118**, 3 (1958).

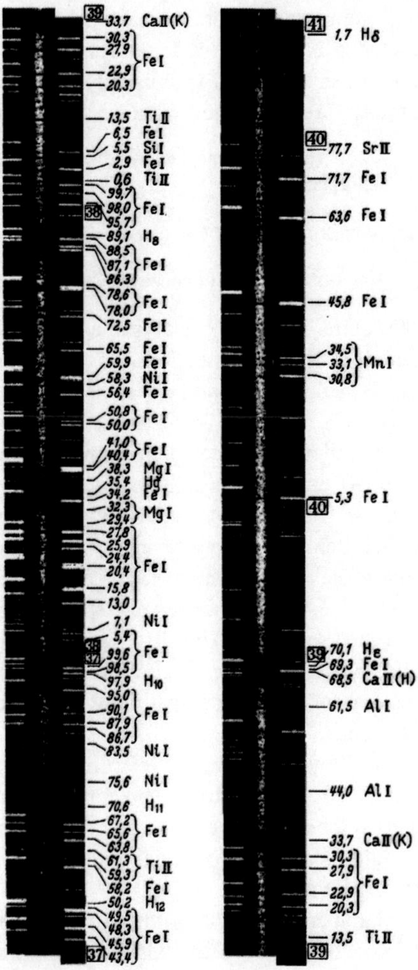

Fig. 6.13 The spectrum of HD 140283 with the identified spectral lines. This was the first spectrum indicating diffrences in the abundances of metals in stars

In his talk, Unsöld was skeptical of Chamberlain and Aller's dramatic results, because their analysis was 'quite coarse', as they themselves had admitted.

At that time, Bodo Baschek began his Ph.D. thesis at the University Christian-Albrechts-Universität zu Kiel under Unsöld's supervision, and the topic was the composition of the star HD 140283, one of the Chamberlain and Aller stars. Baschek applied the best available theory, as produced in Kiel to find the stellar abundances. During Baschek's work on his Ph.D., Aller visited Kiel and naturally the main topic of discussion was the analysis of the HD 140283 spectrum (see Fig. 6.13). Baschek realized that he might be scooped by Aller, and Aller's feelings were no different.

Upon his return to the US, Aller urged Greenstein to publish their analysis, while Baschek prepared his publication. Baschek's paper was the first to come out.[117] It

[117] Baschek, B., Ziet. f. Ast. **48**, 95 (1959). See also ibid. **56**, 207 (1962).

was submitted to the journal on 8 June 1959 and published on 30 September. It included a detailed model atmosphere. Aller and Greenstein submitted their paper on 25 January 1960 (published November 1960), but without the required model atmosphere. In the addendum to the paper, they apologized that:

> It would be difficult now to revise the present computation on the basis of [Baschek's] model atmosphere.

They acknowledged Baschek's priority, but claimed that:

> Abundances agree systematically, though the methods of analysis are completely different.

Not quite. Baschek's analysis, as would be typical for Unsöld's school, contained detailed calculations of the atmosphere and was much more reliable than the analysis performed by Aller and Greenstein.[118]

Baschek's result was that the metal abundances were lower by a factor of 200, while Aller and Greenstein found a factor of 30. Only after his student managed to show that HD 140283 was metal poor did Unsöld accept that there were some stars with a different composition, so deep was the conviction that the abundances of all elements were universal, and all stars alike. However, even this result due to Baschek did not change Unsöld's general view, and he considered the particular star HD 140283 to be a very special case. On the other hand, the acceptance of Baschek's result by Unsöld, one of the founding fathers of stellar spectral line theory and abundance determination, silenced all skeptics vis-a-vis this particular star. The general view was now that there were exceptions to the universal abundance of the elements.

A few years later, in 1969, Unsöld[119] finally accepted this value, but not before Aller and Greenstein[120] had repeated the observations and the calculations, and settled the question beyond any doubt. The abundances of all elements with atomic weight greater than or equal to that of carbon were down by a factor of 100 relative to Pop I stars.

Another nail was unwittingly hammered in the coffin of 'universal abundance' by Miss Nancy Roman in 1955.[121] when preparing a catalogue of high velocity stars. She pointed out that:

> The spectra of many high velocity stars are not normal and some compromise must be made in their classifications.

The intensity of the ultraviolet light from the high velocity stars is stronger than that from low velocity stars of the same spectral class. The strange spectrum is not a special property of one or two strange stars (as Chamberlain and Aller had asserted),

---

[118] Baschek's detailed and self-consistent paper accumulated about 59 citations up until 2009, while Aller and Greenstein's paper collected 76 citations. Is this because Baschek published in German and Aller and Greenstein in English? Or is there another excuse?

[119] Unsöld, A., Science **163**, 1015 (1969).

[120] Aller, L.H., and Greenstein, J.L., Astrophys. J. Suppl. **5**, 139 (1960).

[121] Roman, N.G., Astrophys. J. **59**, 307 (1954); Astrophys. J. Suppl. **2**, 195 (1955).

but actually appears in the majority of high velocity stars. The fact that most high velocity stars are *not normal*, and bluer than normal, was the major finding by Roman.

In 1955, Sandage and Walker[122] observed the globular cluster NGC 4147 and discovered an ultraviolet excess which amounted to a factor of 1.6 relative to what was observed in regular stars. Sandage and Walker, who knew of the suggestion about the differences in heavy element abundances between the stellar populations of Schwarzschild, Spitzer, and Wildt from 1951, proposed that the UV excess was due to the low abundance of heavy elements. As Sandage and Walker correctly stated:

> Proof of this qualitative suggestion must, of course, come from model atmosphere computations.

But these were not available at the time. Baschek's definitive result came only 4 years later.

This is not such a trivial result. If the heavy elements are synthesized in the stars, as was already quite clear at that time, the alternative is that all stars start their life with the same composition, and that, for some unknown reason, the difference in abundance develops in time. Stars with high abundances of heavy elements should then be older than stars with low abundances of heavy elements. At the same time Oort emphasized that it was by no means certain from the data then available whether the differences between the globular clusters merely represented differences in age or whether a more revolutionary assumption was needed, e.g., that some appreciable change in the chemical composition might have occurred in the early stages of the Galaxy.

In 1954, Reiz took the Chamberlain and Aller result one step further. Reiz set out to explain Roman's result concerning the UV excess and tried to build a stellar model assuming negligible amounts of heavy elements throughout the star, rather than just in the atmosphere of the star. The model was homogeneous. Reiz found that such a model has a lower luminosity than a normal star with the same surface temperature. A star with no metals therefore falls below the main sequence of normal stars, or exactly where the high velocity stars appear. In present day language, we would say that Reiz found that the location of stars in the lower part of the main sequence depends on the total amount of heavy elements. The lower the amount of heavy elements, the fainter the star for the same surface temperature.

In 1955, Schwarzschild, Howard, and Leonard Searle (1930–2010)[123] observed main sequence stars and sub-main sequence stars. (For unexplained reasons the authors used the Hertzsprung terminology, namely dwarfs and subdwarfs, rather than Eddington's term 'main sequence'.) The differences between the two classes is that a subdwarf with the same brightness as a dwarf but a lower level of metals is bluer, exactly the effect Roman found. The authors proposed to adopt Reiz's suggestion[124] that the two classes differ *not in the helium content, but only in having, in contrast to*

[122] Sandage, A.R., and Walker, M.F., Astrophys. J. **60**, 230 (1955).

[123] Schwarzschild, M., Howard, R.F., and Searle, L., Astrophys. J. **122**, 353 (1955).

[124] Reiz, A., Astrophys. J. **120**, 342 (1954).

the ordinary stars, a negligible abundance of the heavier elements. No consequences of this suggestion were worked out at that time.

The apparent conclusion that chemical composition and kinematic properties might be connected was quite puzzling. How come the motion of the star affects the composition? But this was the wrong question, as if Perrine's objections had been revived.

6.18 Theoretical Understanding of Stellar Populations

Milne[125] developed his core–envelope model, assuming identical composition but a different state of matter in the core and the envelope. However, very few researchers referred to Milne's seminal work, even though it had laid the theoretical grounds for the model of giants and for all phases of stellar evolution past the main sequence, by postulating small and very dense cores surrounded by very extended and highly tenuous envelopes with uniform composition. Milne's work was probably neglected because it came before its time, when the problems it eventually solved did not yet exist, or because it was embroiled in the controversy with Eddington, and Eddington's style was more flamboyant than Milne's. (Unlike Eddington, Milne did not write any popular books.)[126]

6.19 Early Successful and Failed Attempts to Explain the Red Giants

The first to calculate inhomogeneous models was Öpik.[127] He also discussed the problem of how to obtain the red giants and put forward the explanation that the giant phase follows the main sequence phase, in complete contrast to the prevailing view.

At that time, the year being 1938, the theory of strong mixing by meridional circulation as proven by Vogt and Eddington[128] was largely dominant. Hence, Öpik, in an unorthodox move, ignored the ongoing von Zeipel theory, according to which stars are mixed on a short time scale and hence must be homogeneous. Well aware of the literature, Öpik knew about von Zeipel's theorem but claimed that:

---

[125] Milne, E.A., MNRAS **91**, 4 (1930). See also Shaviv, G., *The Life of Stars*, Springer, Heidelberg, and Magnes, Jerusalem (2009).

[126] Milne had the humility and simplicity of character that often goes with scientific genius, and Eddington is one of the best examples to the contrary. Milne suffered from personal misfortunes in his family, while Eddington never married.

[127] Öpik, E., Pub. Obs. Tartu **30/3**, 1 (1938); **30/4**, 1 (1938); **31/1**, 1 (1943); Armagh Obs. Contr. # 2 (1949); # 3 (1951). It appears that Öpik did not like referees, as many of his unique papers were published in unrefereed observatory publications.

[128] See Shaviv, G., *The Life of the Stars*, Springer, Heidelberg, and Magnes, Jerusalem (2009).

> Nevertheless the mixing of the material may be incomplete.

Öpik cited Eddington's calculation that the Sun must be fully mixed by the fast mass flows. On that account, the governing idea at the time was that stars were fully mixed and homogeneous. On the other hand, Öpik did not have any good reason to assume that this was not the case, except for his own attempt to model giant stars. He therefore chose to ignore mixing, and the justification was left to the results. In 1951, Öpik returned to the problem of meridional circulation[129] and demonstrated (independently of Sweet[130]) that Eddington was wrong. The idea of mixing by fast meridional circulation was a red herring.

The next question was whether the non-homogeneity was intrinsic or developed in time. Two possibilities were put forward:

- Stars are formed inhomogeneous, and if mixing is too slow, they remain inhomogeneous. The idea went something along the following lines. The initial phase of stellar formation is the condensation of a core of heavy elements, or more accurately, a core enriched with heavy elements. Thus the core can contain any amount of hydrogen from zero to the initial amount. After condensation, the star undergoes a period of accretion from interstellar space. This accreted matter is enriched with hydrogen. Inspired by cloud formation in the atmosphere of the Earth, which requires seeds to start growing, Öpik suggested that stars also required a nucleation seed with a given composition to start forming, and then grew by accretion of matter that had a different composition. However, astronomical evidence and theoretical considerations show that newly born stars are homogeneous because they are completely convective and hence well mixed.
- Öpik assumed correctly that if energy production is concentrated in the core then the core must be convective. Consequently, stars must have a small convective core surrounded by a radiative envelope. If the star produced energy by conversion of hydrogen into helium, it would cause a gradual change in the molecular weight of the core with time. Thus, the secular evolution of stars gradually converts them from fully homogeneous to inhomogeneous. This was the first explicit assumption that changes in composition drive stellar evolution in the HR diagram.

What Opik actually discovered, in the wake of Milne's models,[131] was that, when the star is inhomogeneous, the relation between the central temperature and the

---

[129] Öpik, E.J., MNRAS **111**, 278 (1951). Submitted for the first time on 1 December 1949. Öpik probably had difficulties with the referee and the paper was resubmitted on 28 September 1951. See Shaviv, G., *The Life of Stars*, Springer, Heidelberg, and Magnes, Jerusalem (2009).

[130] Sweet, P.A., MNRAS **110**, 548 (1950), submitted on 13 October 1950, or two weeks after Öpik's submission of his revised paper. Yet Sweet's paper was published an entire volume ahead of the publication of Öpik's paper. Beats me!

[131] Öpik added the following note:

> We avoid the term 'centrally condensed', as it has been attached by Milne to a certain mathematical model which is a mathematical, and probably also a physical impossibility, leading to the central singularity which can be removed only by an appeal to an unknown physical property created ad hoc.

radius changes completely and the central temperature becomes proportional to the outer radius. So the bigger the radius, the higher the central temperature. What happens physically is that the core of a star initially devoid of energy sources contracts under the pressure of the outer layers and releases gravitational energy. According to the Ritter–Kelvin–Helmholtz theory, this energy is absorbed by the envelope which consequently expands.[132] It took astrophysicists a couple of years to realize that Öpik had the correct answer to the structure of the giants. But Öpik's results were almost completely ignored by the community, perhaps because the very long paper contained several errors,[133] along with many good ideas that eventually proved to be correct. But is it justifiable to throw away a good idea if a paper includes some ideas that are incorrect. Apparently it is often felt that all the ideas expressed in the paper must then be wrong. This was the case with Opik's paper. He had good ideas and bad ideas, and as a result, the whole paper was completely ignored!

In the same year as Öpik published his unrecognized seminal paper, Gamow published his version for the evolution of stars, assuming that they are fully mixed.[134] Figure 6.14 (left) summarizes Gamow's theory. As the low mass star exhausts its hydrogen, the luminosity increases due to contraction and the radius decreases. The star reaches maximum brightness before it starts cooling and moving towards the white dwarf region of the HR diagram. The massive stars (see Fig. 6.14 right) are first seen as giant stars which derive energy from the Li+H reaction. During their contraction towards the main sequence, the stars undergo a phase of pulsation like a Cepheid and, after exhausting the hydrogen on the main sequence, the star resumes its perpetual contraction until it explodes. Because the mass is above the limiting mass of a white dwarf, Gamow hypothesized that the remnant of the explosion would be a collection of white dwarf stars. The incorrect assumption of homogeneous stars led Gamow to his erroneous picture of their evolution.

(Footnote 131 continued)

What Öpik did, however, was not that different from what Milne did. You can create models with a very dense core and an extended envelope by increasing the space of possibilities, i.e., replacing a uniform composition with a non-uniform one, as Öpik did, or change the condition at the surface as Milne did. This is an example of how astrophysicists failed to appreciate Milne's results.

[132] How could this happen? Here is a simple physical explanation. Consider a simple model composed of a core with uniform composition and an envelope composed of another uniform composition. The star is in pressure equilibrium, i.e., at each point, the gravitational pull is balanced by the pressure of the gas. The latter is proportional to the number of particles. Consider now the conversion of hydrogen into helium. In this process, the star converts four protons into one helium nucleus, and hence the number of particles in the core decreases. The result is a decrease in the gas pressure, so that the unchanged gravitational pull wins out and the core contracts. As the core contracts it releases gravitational energy which is absorbed by the envelope. Hence, while the core contracts, the envelope expands. Öpik, however, did not provide a physical explanation as to why the composite model has a larger radius than the homogeneous one.

[133] Such as Öpik's hypothesis of the dissociation of nuclei or his rejection of stellar collapse.

[134] The problem of mixing is not mentioned by Gamow, and nor is there any reference to Eddington, as justification for this assumption. However, the assumption is tacitly implied throughout the entire paper.

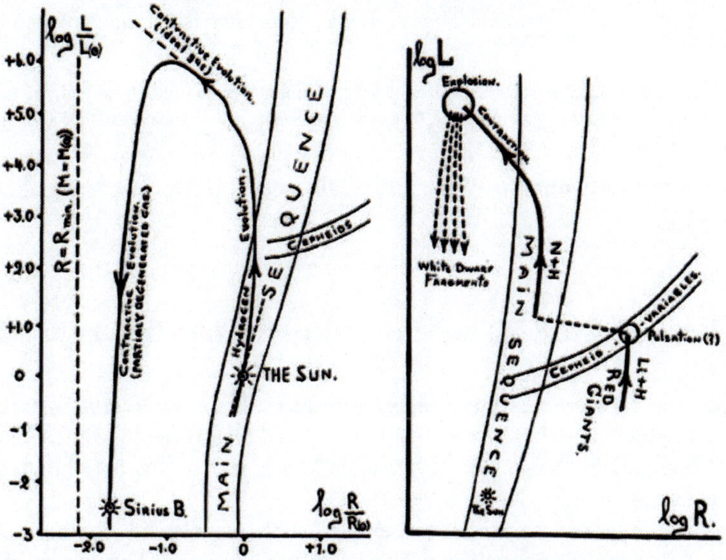

Fig. 6.14 *Left* the evolution of low mass stars under the assumption of full mixing, as hypothesized by Gamow in 1938. *Right* the evolution of massive stars as hypothesized by Gamow in 1938

In 1939, Critchfield and Gamow[135] constructed a model of a star with a 'shell source'. The concept was based on Gamow's idea that the nuclear reactions might have a resonance.[136] Since the temperature changes throughout the star, being maximal at the center and decreasing outwards, there should be only one point in the star where the temperature is exactly right for the resonance. In this case, argued Critchfield and Gamow, the burning will take place in a very narrow shell, and not in the core. Öpik suggested a 'shell source', but only after the exhaustion of hydrogen in the core, whence it was a 'shell source' within an inhomogeneous star.

6.20 Who is Right?

In 1940, Wasiutynski (1907–2005)[137] tried to compare Gamow's (homogeneous) theory, Critchfield and Gamow's (shell model) theory, and Öpik's (inhomogeneous) theory. He repeated the calculations of the inhomogeneous model and included the effect of the radiation pressure, which had been ignored by the above authors. The results were disappointing, as all models gave large changes in the luminosity of the model but very small changes in radius. The models simply did not predict

---

135 Critchfield, C.L. and Gamow, G., Astrophys. J. **89**, 244 (1939).

136 Gamow, G., Astrophys. J. **87**, 206 (1938); Phys. Rev. **53**, 595 (1938).

137 Wasiutynski, J., MNRAS **100**, 362 (1940).

expansion in the way expected. The result was quite surprising and, as Wasiutynski claimed:

> Though in homogeneous stars, as well as in several inhomogeneous cases studied, the trend of evolution is similar to that indicated by Gamow, this can scarcely be considered a decisive proof of its validity for real stars.

In other words, according to Wasiutynski, all models led to practically the same result.

6.21 Hoyle & Lyttleton and Schönberg & Chandrasekhar (1942)

A major step forward was made independently by Fred Hoyle (1915–2001) and Raymond Lyttleton (1911–1995),[138] who submitted their paper to MNRAS in June of 1942, and by Schönberg (1914–1990) and Chandrasekhar,[139] who submitted their paper to Astrophys. J. in September of 1942.

Hoyle and Lyttleton constructed a series of static models with composition differences between the convective core and radiative envelope. The authors did not discuss mixing but assumed that the composition was constant, both in the core and in the envelope. Thus they tacitly assumed strong mixing in a convective core and none at all in the envelope. By changing the composition of the *outer part of the star*, they managed to get an increase in the radius by a factor of 3.55. At this point they mentioned that Öpik had suggested earlier that the giants could be modelled with:

> [...] a convective core of higher molecular weight than the radiative envelope exterior. [However], according to Wasiutynski, the change in radius thereby produced is negligible. This is in general agreement with the findings of the present paper which indicate that large changes of radius occur through non-uniform composition of the atmosphere only.

The authors cited Wasiutynski's wrong criticism of Öpik's correct result. By 'atmosphere' they meant the outer radiative zone. If so, they put forward the hypothesis that the star was first formed from a high atomic weight material and later accreted hydrogen, the molecular weight of the envelope changing during the motion through interstellar space.

Here is a related but not directly relevant story. Eddington[140] had already considered the idea of accretion in the context of the energy source of stars. Assuming the interstellar gas to have a mean density of about one particle per cubic centimeter (which is a good approximation) and to behave as free particles attracted by stars, he estimated the rate of mass accretion by the Sun to be 4.8×10^8 g/s. This may sound like a large mass, but when integrated over 10^{10} years it amounts to just $10^{-8} M_{\odot}$. Hence, this mass is unimportant for the energy problem, but may be important in polluting the surface. However, low mass stars have an extensive convective zone

---

[138] Hoyle, F., and Lyttleton, R.A., MNRAS **102**, 21 (1942).

[139] Schönberg, M., Chandrasekhar, S., Astrophys. J. **96**, 161 (1942).

[140] Eddington, A.S., *The Internal Constitution of the Stars*, Dover, 1959, p. 391.

which is significantly more massive than the total mass accreted, so the accreted matter would have no effect on the appearance of such stars.

Unlike Eddington, who assumed accretion by individual particles, Hoyle and Lyttleton[141] assumed that the interstellar gas behaves as a fluid, and calculated how much mass could be accreted by a star moving through interstellar space. The results were criticized by Atkinson,[142] but Hoyle and Lyttleton responded to this criticism,[143] and derived an accretion rate proportional to $1/v^3$, where v is the velocity of the star through the interstellar medium. As can be seen, such a formula yields prohibitively high rates of mass accretion for very small velocities, and it is this high rate which motivated the conclusion about the importance of accretion. On top of this, Hoyle and Lyttleton assumed that the gas density in the solar neighborhood was as high as 500 atoms/cm^3, while the actual value is about 1 atom/cm^3. The problem of accretion was revisited by Bondi and Hoyle,[144] who found the same basic result. Only in 1952, when Bondi revisited the problem again,[145] was it discovered that the rate of accretion is actually proportional to $1/(v^2 + c^2)^{3/2}$ where c is the speed of sound in the gas. In this way, the extremely high accretion rate from interstellar space disappeared for good.

The problem of accretion by an old star moving randomly in the Galaxy is extremely important because of the following question: are the abundances observed at the surface of an old star a testimony to the composition of the matter in the galactic regions it has visited or does the star preserve the pristine composition it had at the moment of formation? The present theory is that the original composition of the stellar surface is preserved, apart from modifications by internal processes like diffusion.

At the same time, but on the other side of the Atlantic Ocean, Mario Schönberg (1914–1990) and Chandrasekhar investigated the general picture of stellar evolution on the hypothesis of Gamow,[146] and rejected it. Schönberg and Chandrasekhar assumed fast mixing to take place only in the convective core and none at all in the envelope,[147] whence Gamow's assumption was not valid. In other words, Schönberg and Chandrasekhar realized, exactly like Öpik before them (but without any reference to him), that the assumption of fully mixed stars had to be abandoned. They did not provide any argument as to why full mixing does not take place in stars, but they were right in abandoning it. What they found therefore was that, after the exhaustion of the hydrogen in the core, when no more energy is released, the core becomes isothermal (constant temperature) because there is no energy to carry outward.

---

[141] Hoyle, F., and Lyttleton, R.A., Proc. Camb. Phil. Soc. **35**, 405, 592 (1939).

[142] Atkinson, R.d'E., Proc. Camb. Phil. Soc. **36**, 313 (1940); MNRAS **100**, 500 (1940).

[143] Hoyle, F., and Lyttleton, R.A., MNRAS **101**, 227 (1941).

[144] Bondi, H., and Hoyle, F., MNRAS **104**, 273 (1944).

[145] Bondi, H., MNRAS **112**, 195 (1952).

[146] Gamow, G., Phys. Rev. **53**, 59 (1938); Nature **144**, 575 (1939).

[147] The authors did not explicitly discuss the mixing and the need to neglect it, but in practice they did indeed ignore it. They merely stated that: *The later stages of the evolution would be the same as in the case of slow mixing.*

Once they had freed themselves from the constraining assumption of full mixing, Schönberg and Chandrasekhar returned to the Cowling model,[148] as developed in more detail by Chandrasekhar.[149] These are frequently wrongly cited references for the first calculation of composite models involving a convective core and a radiative atmosphere. However, neither Chandrasekhar nor Cowling treated the case of different composition in the core and in the envelope. Like Öpik and Hoyle and Lyttleton, Schönberg and Chandrasekhar assumed non-homogeneity, but they went a step further and analyzed the stability of the model. In doing so they discovered that, if the convective core of the star is above a certain limit, which is known today as the Schönberg–Chandrasekhar limit,[150] then as soon as the nuclear burning in the core extinguishes, the core collapses and heats up and the star moves quickly to another point in the HR diagram. The composite model has a large radius, and the authors remarked that mixing eliminates this phenomenon. They proved that, for stars in which the core is beyond the limit, the star is not stable. Schönberg and Chandrasekhar formulated their result a bit differently:

> There is an upper limit ($\sim$10%) to the fraction of the total mass of hydrogen which can be thus exhausted.

The implication is that there is a limit to how much hydrogen can be converted into helium when the star is on the main sequence.

As for what happens beyond this point, the authors just speculated and were found, years later, to be correct. The meaning of a stellar instability is that the star cannot remain in the same place in the HR diagram. The internal stellar structure changes quickly and the star moves to another location in the HR diagram. The space in the HR diagram between the old and the new locations does not contain stars. Hence, a gap in the HR diagram is a region in which stars cannot remain. The observation of such a gap is therefore an excellent confirmation of the theory.

A year later, in 1943, as if Schönberg and Chandrasekhar's important result had never been published, Gamow[151] calculated the evolution of a massive star, still assuming a homogeneous model and still ignoring the old Öpik paper and the more recent Schönberg and Chandrasekhar paper. As can be seen in Fig. 6.15 from Gamow, the massive star continues to contract past the main sequence (the solid line). Gamow could not obtain a model of a giant star. On the contrary, as the hydrogen was exhausted, the model contracted. Adherence to the assumption of full mixing misled Gamow.

---

[148] Cowling, T.G., MNRAS **91**, 92 (1931).

[149] Chandrasekhar, S., *An Introduction to the Study of Stellar Structure*, Chicago, 1939, p. 352.

[150] The exact limit depends on the composition of the star and is 0.10–0.15 of the mass of the star. The source of the instability is the following. A star with a fixed temperature (isothermal) must have an infinite radius. If the isothermal envelope is replaced by a non-isothermal one, a finite mass and radius can be found. The process can continue only until the limiting core mass is reached, because there is no envelope mass to stabilize the star above this limit.

[151] Gamow, G., Astrophys. J. **98**, 498 (1943).

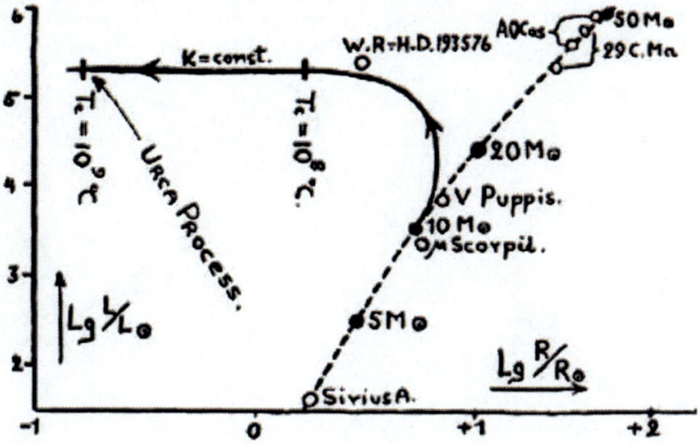

Fig. 6.15 The evolution of a contracting massive stars as hypothesized by Gamow in 1943. The evolution is expected to end with an explosion which spreads the synthesized elements around the Galaxy

6.22 Gamow Acquiesces: 1945

It was in 1945 that Gamow and Keller[152] were finally persuaded to assume good mixing inside the core and tacitly assume, without alerting the reader, no mixing in the envelope. Naturally, they then reproduced the result that, as hydrogen is exhausted in the core and an inhomogeneity develops, the radius increases. Following Schönberg and Chandrasekhar, Gamow and Keller calculated the evolution of three inhomogeneous stellar models. They found that a star with a mass of $0.1M_\odot$ evolves by decreasing its radius and increasing in luminosity, a star with a mass in the range of 0.4–$1M_\odot$ evolves by expanding and increasing in luminosity, and massive stars evolve by expanding and hardly changing in luminosity (see Fig. 6.16 left). The conclusion was:

> Sufficiently massive stars may lead to a very large increase of stellar radius, thus bringing the star into the region of the HR diagram occupied by the giant.

This was the first identification of the giants as composite models after the exhaustion of hydrogen in the core. In particular, Gamow and Keller revoked all Gamow's previous hypotheses by stating:

> In fact, it is not possible to consider stars of the red giant branch as still being in the stage of gravitational contraction since in this case their radii would be decreasing at a faster rate than is consistent with the observational evidence.

After several decades and not a moment too soon, the correct composition-induced evolution of the giant stars was finally discovered.

---

[152] Gamow, G., and Keller, G., Rev. Mod. Phys. **17**, 125 (1945).

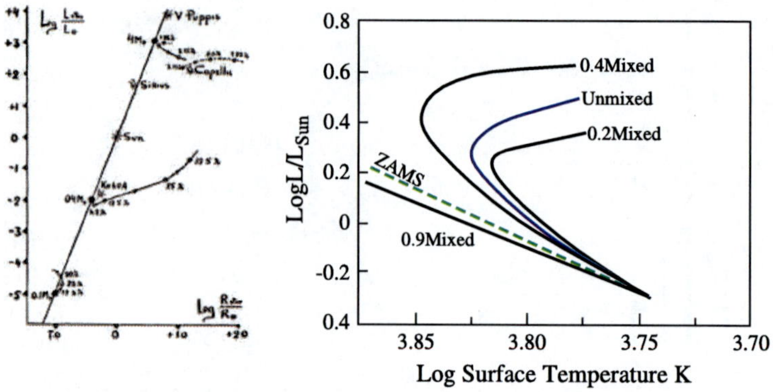

Fig. 6.16 *Left* evolution off the main sequence according to Gamow and Keller (1945), based on Schönberg and Chandrasekhar (1942). *Right* evolution of the Sun under the assumption of partial mixing (Shaviv and Salpeter 1971). The unmixed Sun evolves along the *blue line*. Almost fully mixed stars evolve to higher luminosities, but with decreasing radius, to below the main sequence. ZAMS stands for zero age main sequence

Gamow and Keller continued to explain that Gamow and Teller's[153] attempt to explain the red giants as burning light elements like Li, Be, or B failed because it could not explain the distribution of these stars in the HR diagram. As Gamow and Keller stated:

> A look at the general position of the red branch, especially in the case of Baade's stellar population of type II, suggests on the other hand that most red stars represent evolutionary stages subsequent to the main sequence; in fact, only in such a case would the brighter, faster evolving stars get farther away from their main sequence position.

This time they were absolutely correct.

A decade after the important paper by Schönberg and Chandrasekhar, Sandage and Schwarzschild[154] followed the evolution towards the red giant, invoking a shell source. After hydrogen exhaustion in the core and during the contraction of the core now devoid of fuel, the temperature of the core increases but remains too low for helium ignition. On the other hand, the envelope heats up and soon reaches a point where the temperature at its base is sufficient to ignite hydrogen. The consequence is that hydrogen burns just at the bottom of the envelope. The burning zone does not extend over the entire envelope because (a) the temperature is mostly low and insufficient to ignite hydrogen and (b) as soon as the energy release by the burning adjusts to the energy flow from the surface, an equilibrium is reached and there is no reason for the star to extend the burning any further. For this reason the restricted region just outside the burnt core is called a burning shell. Since no mixing was assumed, we can also refer to this mode as cigar burning, in the sense that the burning takes place at one end of the cigar and slowly propagates (out from the base in this case).

---

[153] Gamow, G. and Teller, E., PRL **55**, 791 (1939).

[154] Sandage, A. and Schwarzschild, Astrophys. J. **116**, 463 (1952).

6.23 When Do Stars Expand and When Do They Contract

As discussed previously, Sweet and independently Öpik proved in 1950–1951 that meridional circulation is very slow and can be ruled out as a mixing mechanism for stars. Still there remained a question as to whether other mechanisms, like diffusion, a fast-rotating core, and a slow envelope, etc., could operate to cause various degrees of mixing in the stars, thereby affecting their evolution.

To clarify this point, Shaviv and Salpeter[155] calculated a series of solar models under the assumption of various degrees of forced mixing by whatever agent. The question was: can mixing be ruled out? The results are shown in Fig. 6.16 (right) and appear to be more complicated than expected. The unmixed Sun evolves by expansion. Just modest mixing, which mixes the innermost part, leads to a smaller change in the luminosity and about the same change in radius. As can be seen, a model which is partly mixed evolves towards the giants, while mixed models with cores greater than 90% of the mass contract. The conclusion agrees with what was implemented in all previous calculations of clusters of stars, i.e., fast mixing can be ruled out.

The subject of mixing by element diffusion was revived recently when Noerdlinger,[156] Wambsganss,[157] Bahcall and Pinsonneault,[158] Bahcall and Loeb,[159] and Kovetz and Shaviv[160] revisited the problem. The effects of the diffusion are small, but they do lead to important changes in the composition, increasing the surface abundance of hydrogen (by 2.5%) and reducing the surface abundance of the heavy elements by the same amount. Very accurate helioseismology[161] can trace small composition changes and verify the effects of diffusion near the surface of the Sun. However, the most important result is that the surface composition of very old stars differs from the primordial value by a non-negligible amount.

6.24 Failures in Modelling Red Giants

In 1947, Reiz[162] checked the model suggested by Schönberg and Chandrasekhar by attempting to construct a model for the giant star Capella. This star has a yellow color which indicates that its surface temperature is close to that of the Sun, but it

155 Shaviv, G. and Salpeter, E.E., Astrophys. J. **165**, 171 (1971).

156 Noerdlinger, P.D., A and A **57**, 407 (1977); Astrophys. J. Suppl. **36**, 259 (1978).

157 Wambsganss, J., A and A **205**, 125 (1988).

158 Bahcall, J.N. and Pinsonneault, M.H., Astrophys. J. **395**, L119 (1992).

159 Bahcall, J.N. and Loeb, A., Astrophys. J. **360**, 267 (1990).

160 Kovetz, A. and Shaviv, G., Astrophys. J. **426**, 787 (1994).

161 Helioseismology is the science of deciphering the structure of the Sun from observations of solar oscillations. As in seismology on the Earth, one can find the structure of stars by tracing the propagation of sound waves. Today it is possible only in the case of the Sun, but soon the method will be extended to other stars.

162 Reiz, A., Ann. Astrophys. **10**, 301 (1947).

has a radius about 10 times greater. Capella is a modest giant. Supergiants have radii as large as 1,000 times the radius of the Sun. The 'moderate' success in constructing such a model misled later researchers into thinking that giant stars derive their energy from hydrogen burning. Three years later, Reiz[163] improved his model by assuming that the central temperature was 33 million degrees instead of 20 million, and indeed the agreement improved, but the deviation between theory and observation was still much too large.

As a matter of fact, the (wrong) idea that the energy source in red giants is hydrogen burning had already been suggested by Bethe in 1939,[164] when he gave a table with the energy generation by hydrogen fusion to helium in several stars including Capella. Bethe arranged the stars according to increasing energy production and compared his calculations of the central temperature with those of Eddington. Eddington's calculation was based on solving the governing equations of the star while Bethe's calculation was based on generating the total luminosity of the star. In general, as Bethe pointed out, the comparison between the results obtained by the two independent methods agreed quite well, except for the case of Capella, where there was a large discrepancy. Thus it was evident that hydrogen fusion was not the energy source for giants. However, both Eddington and Bethe erred by assuming Capella to be homogeneous.

6.25 A Byproduct: The First Inhomogeneous Solar Model

In 1948, Ledoux (1914–1988)[165] was the first to obtain a static, inhomogeneous model of the present day Sun, and used this model to estimate its age. Ledoux knew about von Zeipel's theorem and the fast mixing that rotation seemed to induce. His calculation was carried out before Sweet and Öpik finally killed the idea that fast meridional circulation mixes the stars. However, he provided an argument which claimed that the large scale circulation which engulfed the entire star had to break down to a circulation in the convective core and a circulation in the envelope. There were therefore two independent circulations in the Sun, and for this reason:

> Even in the case of a rotating star, there is a possibility of a difference in composition between the convective core and the radiative envelope.

Once no mixing between the core and the envelope was assumed, or once it had been ruled out, Ledoux could generate a sequence of models with different amounts of

---

[163] Reiz, A., Ar. A. **1**, 1 (1950).

[164] Bethe, H.A., Phys. Rev. **55**, 449 (1939).

[165] Ledoux, P., Ann. Astrophys. **11**, 174 (1948). See also C&T **64**, 221 (1948). In the last publication, which is a summary of a conference, Ledoux remarks that Milne made a grave mistake in his model of inhomogeneous stars, as he neglected the condition that the temperature and the radiative flux must be continuous across the boundary between the core and the envelope. This condition was implemented by Ledoux in his calculation of the Sun.

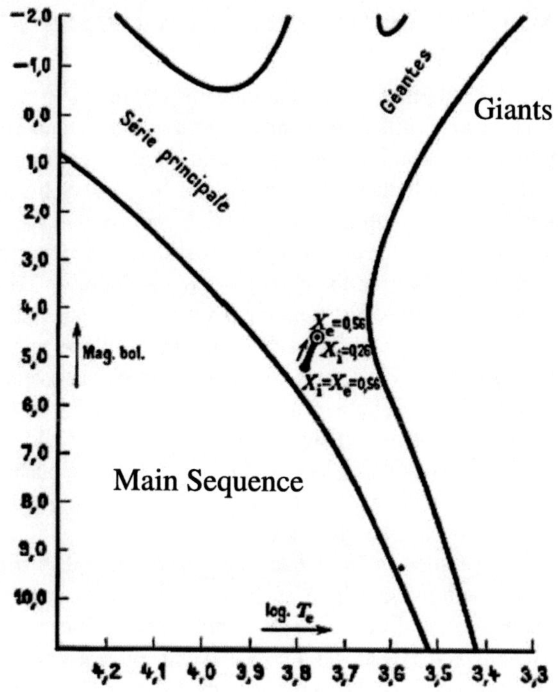

Fig. 6.17 The first calculation of the evolution of the Sun, by Ledoux in 1948. The arrow of the effective temperature points in the wrong direction. The evolution of the Sun is expressed by the motion of the HR position with changing hydrogen abundance X. The values given for X in the figure are the hydrogen mass fractions when the model is at that point

hydrogen in the core. He then compared the luminosity of the model with that of the Sun in order to get the present day values.

Now the idea was simple. The model of the Sun contained a core with one abundance of hydrogen and an envelope with another, which was the original hydrogen abundance. The difference between the two abundances yields how much hydrogen the Sun has consumed since its formation, assuming of course that the Sun started its life as a fully homogeneous star. If the luminosity of the Sun did not change appreciably throughout its steady burning of hydrogen, then the lifetime is given by the total energy content of the hydrogen consumed divided by the rate of consumption, which is the solar luminosity. The age Ledoux obtained was 5 billion years, very close to the present day estimate of 4.54 billion years for the age of the Earth. Ledoux actually performed the first ever calculation of the evolution of the Sun, as can be seen from Fig. 6.17.

Finally, Ledoux noticed that the motion of the Sun in the HR diagram was very slow, and that, during the past 5 billion years, it has moved less than the 'width' of the main sequence. He thus concluded that stars spend *a significant part* of their life on the main sequence. More accurately, stars spend most of their life (up to 90%) fusing hydrogen into helium in the core.

6.26 The Giants: Alternative Solutions?

Are there alternative explanations for why the giant stars are so big? In 1948, Robert Richardson (1902–1981) and Schwarzschild[166] chose to ignore the result from composite models and attempted to build homogeneous models of giant stars comprising three regions rather than two. Richardson and Schwarzschild sought a new mechanism, as they implied that *models based on the classic equilibrium conditions do not apply to the red giants*, and postulated therefore that there was a third region in the star, between the core and the envelope, which was isothermal. This ad hoc postulate required justification and they argued that such a region between the inner convective core and the outer radiative envelope might exist because of the mode of energy transfer in the convective zone. It was a very uncompelling argument and imaginative, too, but wrong.

Not surprisingly, Richardson and Schwarzschild succeeded in building such a model with a suitably large radius and fitting the observed parameters of red giants. They concluded that they could reproduce very big red giants, but admitted that their assumption was *speculative*.[167]

A year later, in 1949, Hen and Schwarzschild[168] tried a model with a convective core, a radiative envelope, and a chemical inhomogeneity which did not coincide with the boundary of the convective core. Usually, the strong currents inside the convective core give rise to full mixing, so that convective cores have uniform composition. On the other hand, radiative regions do not mix, and may have non-uniform composition. What Hen and Schwarzschild assumed was that the inner part of the radiative zone was fully mixed like the convective core. They thus had three regions in the star (Fig. 6.18):

- A fully mixed convective core.
- A radiative zone having the same composition as the convective core.
- A radiative envelope.

The rationale behind Hen and Schwarzschild's model was that inhomogeneity can arise due to accretion of the star from the interstellar medium (the Hoyle and Lyttleton model). If the stars are not mixed and the hydrogen abundance in the interstellar medium is different from what the star was born with, then accretion creates a third region in the star in which the hydrogen abundance differs from that in the region below it. Can such models possess radii as large as those observed?

Hen and Schwarzschild succeeded in getting models with *stellar radii sufficiently large for moderate red giants*, but not large enough for the largest observed giants.

[166] Richardson, R.S. and Schwarzschild, M., Astrophys. J. **108**, 373 (1948). Robert Shirley Richardson was a science fiction writer, who used the pen name Philip Latham when he wrote fiction, e.g., The Xi Effect, Five Against Venus, Disturbing Sun, and more. His papers with Schwarzschild were, however, real science.

[167] The way to build stellar models in those days was to start a solution from the core and another from the surface and fit them together somewhere inside the star. Adding a third region provided plenty of freedom for this solution fitting, and hence allowed for giants.

[168] Hen, L. and Schwarzschild, M., MNRAS **109**, 631 (1949).

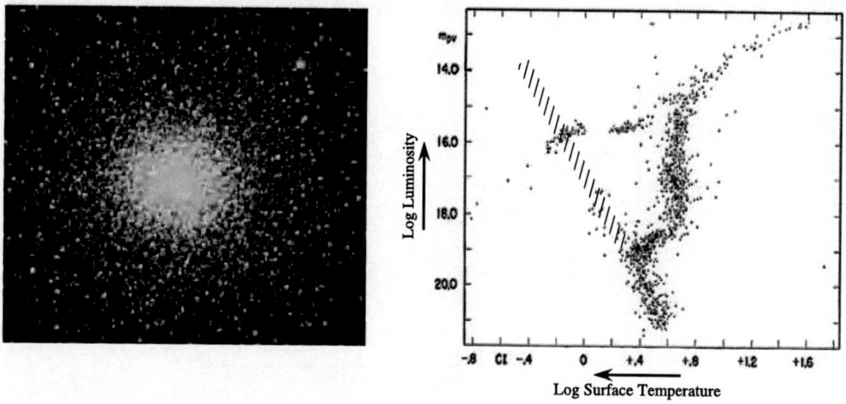

Fig. 6.18 *Left* the globular cluster Messier 3 (NGC 5272) in the constellation Canes Venatici. Discovered by Charles Messier (1730–1817) in 1764. The distance is 30,600 light years. The cluster contains about a half a million stars and is noted for its large number of variable stars. The central parts are so dense that it is impossible to resolve individual stars. *Right* the HR diagram of M3 as discovered by Sandage in 1953, using the new 200 inch telescope. For the first time, the faint main sequence is observed. To facilitate recognition of the main sequence, the *hatched region* has been added to Sandage's figure to indicate its missing continuation. The *original axes* are given in astronomical units and the physical meaning is included

The model they got for a $3M_{\odot}$ star had a total radius of only $20R_{\odot}$. In view of this moderate success, the authors speculated that a smaller mixed zone might lead to larger radii. However, they never published such a model.

In 1950, Herman Bondi (1919–2005) and Christine Bondi[169] constructed a series of inhomogeneous models and demonstrated that there was no need to assume three regions of different composition in order to explain the large radii of giants, as Hen and Schwarzschild had assumed. All observed data could be explained with just a core and an envelope. They criticized Hen and Schwarzschild as having solutions that were *unrepresentatively distributed and confined to the less interesting part of the HR Diagram*. In other words, of little relevance to observation. Forgetting the contribution of Öpik, they attributed the basic mechanism for generating giant stars, the non-homogeneity of the star, to Hoyle and Lyttleton in 1942,[170] and also ignored Schönberg and Chandrasekhar.

6.27 Inhomogeneous Models: Comparison with Observation

In 1952, John Oke and Schwarzschild[171] returned to the simple inhomogeneous model of a core with one abundance of hydrogen and an envelope with another (see Fig. 6.19). By calculating models with several masses they succeeded in estimating

[169] Bondi, C.M. and Bondi, H., MNRAS **110**, 287 (1950).

[170] Hoyle, F., and Lyttleton, R.A., MNRAS **102**, 218 (1942).

[171] Oke, J.B., Schwarzschild, M., Astrophys. J. **116**, 317 (1952).

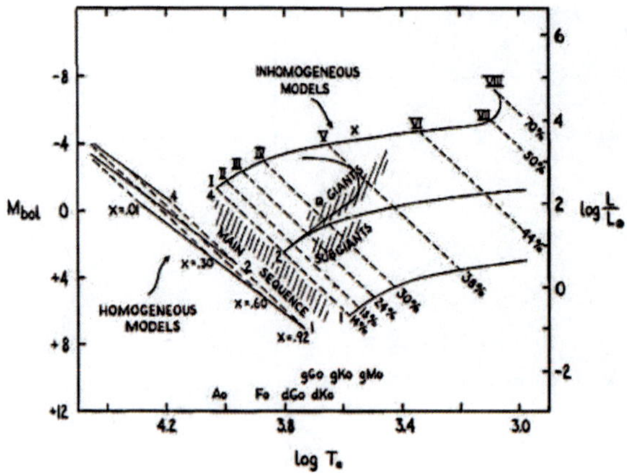

Fig. 6.19 Evolution of main sequence stars off the main sequence towards the red giants. Oke and Schwarzschild (1952). X is the hydrogen mass fraction. The main sequence is a contour of homogeneous stars. The giants are all inhomogeneous. A change in the initial mass fraction of hydrogen causes a small change in the location of the main sequence. A much greater change takes place by depletion of the hydrogen in the core

the mass of two well known red giants: Capella and Zeta Aurigae. The theoretical result of Oke and Schwarzschild for Capella's mass was 2.5$M_\odot$, and they compared it with the measured value of 2.7–3.2$M_\odot$ given by Struve,[172] which gave a tolerable although not fantastic agreement.[173]

Struve pointed out that Capella showed spectroscopic anomalies that were not observed in single stars, implying the need for extra caution in the comparison between observation and theory. Consequently, inferred Struve, the anomalies had to be caused by the mutual interactions of the two giant stars composing the binary system. These interactions disturb the atmosphere of Capella and hence deprive this system of much of its merit as an observational standard to which theoretical predictions should be compared. The same was true with Eddington's mass–luminosity relation, in which Capella played a significant role. Hummel et al.[174] described the binary as separated by only 0.73 astronomical units and having a period of 104.0 days. The stars are therefore extremely close to each other.

Zeta Aurigae is a binary system composed of a very hot massive and luminous blue star and an even more luminous orange supergiant star. The plane of the orbit

[172] Struve, O., PNAS **37**, 327 (1951).

[173] Capella is a binary system and Struve could not determine the mass ratio between the two components because their spectra are so close and similar. He therefore provided two estimates for the mass of Capella. For the first he assumed a mass ratio of 1.2 and then the resulting masses were 3.9$M_\odot$ and 3.2$M_\odot$. When Struve assumed that the mass ratio was 1 he got a mass of 2.7$M_\odot$. Thus, the quality of the agreement depended on what was assumed for the mass ratio, but in no case could the mass be less then 2.7$M_\odot$, so the discrepancy remained.

[174] Hummel, C.A., Armstrong, J.T., Quirrenbach, A., Buscher, D.F., Mozurkewich, D., Elias, N.M. and Wilson, R.E., Astrophys. J. **107**, 1859 (1994).

is exactly in the line of view and consequently the stars move in such a way that they eclipse each other periodically. Because of the huge difference in radii and the fact that the larger star is the more luminous one, only the eclipse in which the giant orange star comes in front of the blue star is observed. The suspicion that the star was not single was already raised by Miss Maury,[175] who claimed that the spectrum was composite, implying that there were two different spectra which appeared as one; in other words, two different, but very close stars. Maury simply could not disentangle the spectrum of the two stars.

Later, in 1908, Campbell[176] discovered that it is an eclipsing binary. Besides the masses of the stars, such a binary system provides plenty of information about the radius of the stars, because they eclipse each other periodically. Since the binary period is 2.7 years, which is long, the stars are sufficiently far apart not to distort each other's atmosphere, and present a 'cleaner' system for comparison with theory than Capella does.

In 1935, Christie and Wilson (1909–1994)[177] carried out an extensive series of observations, in particular, during the eclipse, and estimated the masses to be $15.5M_\odot$ and $8.3M_\odot$ on the basis of lengthy orbital measurements. The result of Oke and Schwarzschild was $5M_\odot$, which, to their disappointment, was too far off. Worse then that, it was not clear what had gone wrong.

Years later astronomers discovered that Zeta Aurigae produces a wind, like the solar wind, and it became a popular subject for research. Over 400 observations of the star were carried out by 1983. During this extensive research,[178] improved measurements yielded masses of $8M_\odot$ and $6M_\odot$. These masses are much closer to the theoretical value obtained by Oke and Schwarzschild, but still the theoretical value was too low compared with what was observed. Oke and Schwarzschild decided to follow Eddington's maxim:

> It is also a good rule not to put overmuch confidence in the observational results that are put forward until they are confirmed by theory.

So Oke and Schwarzschild decided to ignore the measurements of the mass of the star and deduced its mass from the mass–luminosity relation. In this case they found that the mass should be $5M_\odot$, in perfect agreement with theory. Time has shown that Oke and Schwarzschild were only partially right in this case, and the 'measured masses' decreased towards the theoretical value but did not meet it. However, the fact that these stars have a strong wind blowing off their atmosphere, a phenomenon not known to Oke and Schwarzschild, invalidates the mass–luminosity estimate, and hence the observed value.

In summary, the early attempts to understand the structure of the giants by comparing the theory of composition-driven evolution with observations of single stars (in binary systems) were not very successful.

---

[175] Maury, A., Harv. Ann. **28**, 99 (1897).

[176] Campbell, W.W., Cont. Mount Wilson Obs. # 519, 1908.

[177] Christie, W., and Wilson, O., Astrophys. J. **81**, 426 (1935).

[178] Darling, D., Astronomy **11**, 66 (1983).

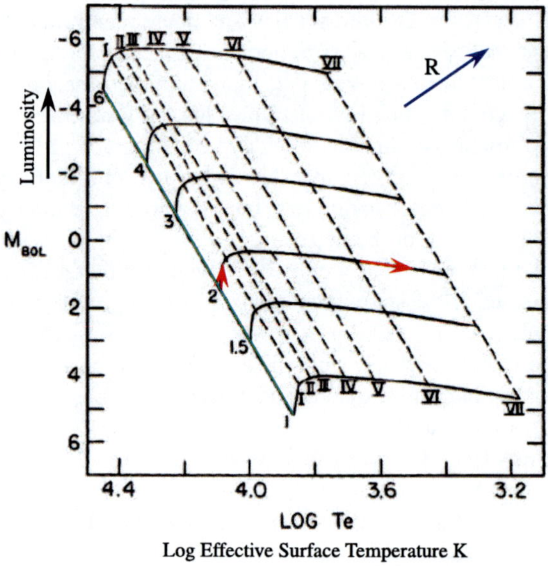

Fig. 6.20 Theoretical results of Sandage and Schwarzschild (1952). *X* is the hydrogen mass fraction. The main sequence is a contour of homogeneous stars. The giants are all inhomogeneous. *Numbers* along the main sequence indicate the total mass of the model

6.28 The Key Is Again Provided by the Globular Clusters

A major breakthrough in the understanding of the globular clusters came when the new Mount Palomar telescope supplied wonderful new data about the faint stars. One of the most important theoretical papers explaining the HR diagram of globular clusters was published in 1952 by Sandage and Schwarzschild,[179] who improved upon Oke and Schwarzschild by taking into account the gravitational contraction of the core when free from nuclear energy production. In principle, Sandage and Schwarzschild provided the physical explanation as to why the inhomogeneous stars expand so tremendously. In their calculations they found that the gravitational energy released by the core when it is devoid of hydrogen contributes only 4% of the total luminosity of the star. This implies that most of this released energy goes into expanding the envelope, puffing the star up to gigantic dimensions. The results are summarized in Fig. 6.20. The leftmost green line represents homogeneous stellar models for the zero age main sequence and uniform composition. Roman letters denote the location of non-uniform models with different degrees of hydrogen depletion in the core, which are non-uniform in composition.

[179] Sandage, A.R., and Schwarzschild, M., Astrophys. J. **116**, 463 (1952).

6.29 Observations of Globular Clusters

In 1953, Sandage[180] published the first HR diagram of the GC M3,[181] in which the faint part of the main sequence was detected for the first time. Baade was proven right in his speculation that GCs have lost the upper part of the main sequence. As can be seen from Fig. 6.21, the Pop II stars which lived in the top part of the main sequence and have a short lifetime are either on the giant branch or absent because they are dead. If the birth rate of massive Pop II stars is the same as the birth rate of massive Pop I stars, then it is clear just from the observed number of Pop II giants relative to massive Pop I stars that many massive Pop II stars have disappeared, probably exploded, spreading their nuclear ashes all over the Galaxy. The heavy elements out of which the younger Pop I stars were formed were synthesized in the missing part of the main sequence of Pop II stars. The claim that we are 'stellar dust', or that the heavy elements in our body came from the stars, expresses the fact that the atoms of the heavy elements in our body were synthesized by those massive stars which occupied the upper part of the main sequence of the now gone Pop II stars. The atoms out of which we are composed were synthesized by a now defunct generation of stars.

One can now very easily determine the age of globular clusters, provided of course that the stars do not lose mass during their life on the main sequence. The age of the cluster is the lifetime on the main sequence of the star which has just left the main sequence. Each stellar mass has a definite effective temperature along the main sequence. Once the effective temperature of the turning point is observed, the mass and the age can be determined. Of course, the time needed for the collapsed stars to reach the main sequence should be taken into account, but this time is so short relative to the time the star spends actually on the main sequence that it can be comfortably neglected.

By comparing theoretical calculations and observations of GCs, Sandage showed that:

> Öpik's initial suggestion of chemically inhomogeneous and Hoyle and Lyttleton's subsequent independent suggestion of chemically inhomogeneous stellar models to explain the large radii have given rise to many papers on the problem of the giants. It now seems probable that chemical inhomogeneity does lead to stars with large radii and high central temperature.

But be warned! The above researchers assumed inhomogeneous initial models, while Sandage was referring to homogeneous main sequence stars which develop an inhomogeneity. No doubt the stellar clusters provided observers and theoreticians with a fantastic victory.

---

[180] Sandage, A.R., Astrophys. J. **58**, 61 (1953).

[181] It is interesting to note how Messier described this globular cluster:

> Nebula discovered between Bootes and one of the Hunting Dogs of Hevelius [Canes Venatici], it does not contain any star, its center is brilliant, and its light is gradually fading away. It is round.

From Messier, C., in *Mémoirs de l'Académie Royale des Sciences*, 1779.

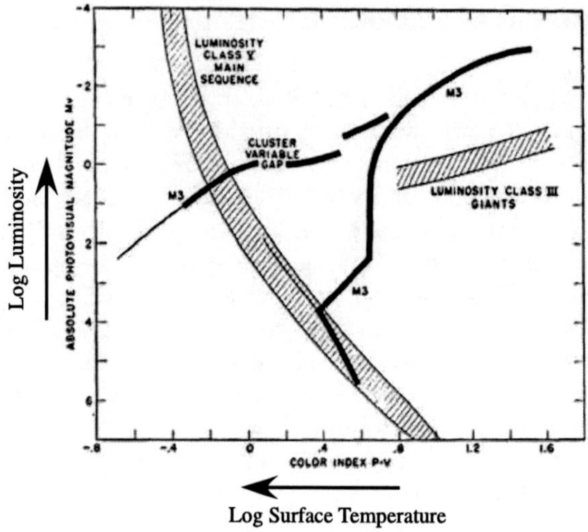

Fig. 6.21 The schematic HR diagram of Pop I and Pop II stars, after Sandage (1953). The point where the main sequence breaks is called the turning point

The age Sandage found for M3 was 5 billion years. Sandage did not notice that this age for Pop II stars was identical with the age Ledoux had found for the Sun, which is a Pop I star. The discrepancy was eliminated when more accurate stellar models were developed. The best present day measurements of the age of M3 is 10.3 ± 1.6 billion years.[182] There are older GCs like M92, with an age of 14.2 ± 1.7 billion years. On the other hand, applying the same method to OCs, ages of less than half a billion years were discovered.

To summarize, Sandage discovered the chronology of stars and the history of element synthesis in the Milky Way.

6.30 Galaxy Collisions and Composition Differences

A seminal idea regarding the origin of the abundance differences between groups of stars was put forward by Lyman Spitzer (1914–1997) and Baade in 1951.[183] The basic idea was that galaxies collide with one another and consequently may merge and mix their different stellar populations. In other words, the stellar populations are the remnant of a major catastrophe in which two galaxies collided with one another. The idea of collision between galaxies was actually suggested by Fritz Zwicky (1898–

182 Paust, N., Ph.D. Thesis, United States, Dartmouth College, 2006. Publication Number: AAT 3219697. Source: DAI-B 67/05, November 2006.

183 Spitzer, L. and Baade, W., Astrophys. J. **113**, 413 (1951); ibid., Astrophys. J. **55**, 183 (1950).

1974m) as early as 1937.[184] However, he just threw up the idea as a possibility and did not carry out any estimates to show that it might really occur. Zwicky was interested in a completely different problem which had to do with the structure of clusters of galaxies. He noted the difference between stellar collisions and galaxy collisions and invoked the latter to explain the structures of clusters of galaxies.

Spitzer[185] and Baade considered collisions between galaxies as a mechanism to explain the existence of different stellar populations of stars inside the same galaxy. There are several differences between stellar and galaxy collisions. First the frequency of collision. The mean distance between stars is about 50,000–100,000 times larger than the typical radius of stars. Hence, the probability of encounter between stars is quite small—most of the Milky Way is empty. On the other hand, the distance between galaxies today is just 10–20 times the size of the galaxies, and hence encounters between galaxies are much more likely. Indeed, Spitzer and Baade estimated that, if the age of the Universe is 3×10^9 years, then the number of collisions between galaxies in a cluster of galaxies ranges from 20 to 150.

The idea was that the stars in different galaxies may have dissimilar compositions and, when a collision takes place between galaxies, the two types of stars mix, while many other phenomena are instigated by the collision. Spitzer and Baade were also interested in explaining the shape of the galaxies as well as the lack of dust in some of them. They hypothesized that the elliptical galaxies contained only stars of population II. Collisions sweep away the gas, leaving the galaxies without gas and subsequently preventing the formation of the next generation of stars, the population I stars. It was already known from the Hubble galaxy classification that elliptical galaxies contain no gas and possess predominately population II stars. On the other hand, they realized that there is correlation between the amount of gas possessed by a galaxy and the type of stars. In particular, galaxies with large amounts of gas possess population I stars.

Spitzer and Baade did not discuss composition, or why stars in different colliding galaxies should have different compositions. They simply replaced one assumption by another. They did not discuss the possibility that a collision between galaxies might induce formation of stars that would carry the signature of the heavy elements in the gas at that moment. However, we can assume that if the colliding galaxies had different histories of collisions, the compositions might indeed be different.

6.31 Observations of Galaxy Collisions

When Spitzer and Baade came up with the idea of galaxy collisions, it seemed quite inconceivable. But reality was more imaginative. The resolution of radio antennas was not very great in the early years of radio astronomy, in contrast to the high

[184] Zwicky, F., Astrophys. J. **86**, 217 (1937).

[185] The Spitzer Space Infrared Telescope is named after Spitzer. It was launched on 25 August 2003 and expected to operate for 5 years.

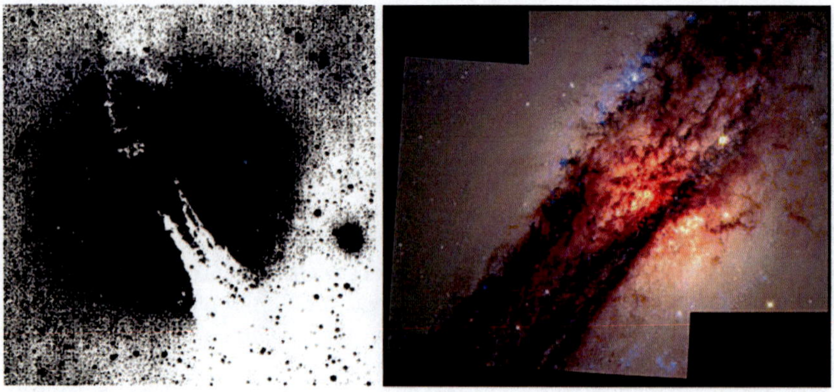

Fig. 6.22 *Left* the galaxy Centaurus A, photographed in 1954 by Baade and Minkowski, who identified it as a galaxy collision. *Right* the innermost part of Centaurus A, photographed by the Hubble Space Telescope

resolution of optical telescopes. When attempts were made to identify a radio source with some object in the sky, the field of view of the radio telescope contained a large number of objects and the identification of the object in the visible range with the radio source was very tricky. Strong observational evidence in favor of collisions between galaxies was supplied by John Bolton (1922–1993), Gordon Stanley (1921–2001), and Bruce Slee,[186] who identified one of the strongest radio sources with the very peculiar galaxy Centaurus A. This was the first detection of an extragalactic object as a radio source. In 1954, Baade and Minkowski confirmed the identification.[187] The nebula exhibits a heavy absorption band due to absorbing dust and has a quite breathtaking appearance (see Fig. 6.22).

As stressed above, the radio telescopes of the day had very low resolving power and the identification of radio sources with visible sources was a far from trivial problem. Thus, the hunt for the visible counterparts of radio sources was a continuing process. In 1954, Baade and Minkowski managed to identify discrete radio sources as remnants of supernovae, galactic nebulosities *of a new type*, *peculiar extragalactic nebulae*, and normal extragalactic nebulae. Of special interest here is the third kind of objects. They confirmed the findings of Bolton et al. and added a few more objects. No explanation was offered, and nor was there mention of galaxy collisions, although they describe various signs which hint at this possibility. Only in a later publication,[188] which was less formal and committing, did they dare to state clearly that:

> The available evidence suggests that it may represent two galaxies in close interaction or in collision.

---

[186] Bolton, J.G., Stanley, G.J., and Slee, O.B., Nature **164**, 101 (1949).

[187] Baade, W., and Minkowski, R., Astrophys. J. **119**, 215 (1954).

[188] Baade, W., and Minkowski, R., Obs. **74**, 130 (1954).

Another example which we have already encountered is the galaxy M51 shown in Fig. 6.2. The small galaxy is colliding with the big galaxy to its left and in a few million years will be completely swallowed by the big galaxy. Since one of the colliding galaxies is quite small, no dramatic features are observed. Collisions between galaxies, while violent, are far from the most violent events in the Universe.

6.32 Unifying Chemical Element Synthesis and Galaxy Formation

As can be imagined, attempts to form a coherent picture which explained the profusion of phenomena were made as early as possible. Two of the most prominent attempts were even made in the early 1950s by Hoyle[189] and by Schwarzschild and Spitzer.[190]

In 1953, Hoyle came up with a key idea that aimed to explain the plethora of observations we have discussed so far. It connected galaxy formation, stellar populations, and element synthesis, and in many senses laid the foundations for the theory of galactic chemical evolution. The basic idea is that the clusters of galaxies formed out of the *extensive* intergalactic cloud of gas as soon as it became unstable against gravitational collapse. (This problem was first treated by Jeans.[191]) The initial gas is assumed to contain a very low, but not vanishing, abundance of heavy elements. The collapsing cloud fragmented into a cluster of fragments each of which had the size of a galaxy. Each fragment continued to fragment and in this way a hierarchical structure formed. Hoyle followed the fragmentation of the Galaxy into Pop II stars and deduced that the last fragment would have a mass of about 0.3–1.5$M_\odot$. The entire process starting from the initial cloud to the last stars would then take about 3 billion years, which was an uncomfortably long time.

Next, Hoyle speculated about the formation of the Pop I stars, after the formation of the galactic disk had taken place. The basic point to make here is that no explanation was given as to why the stars which formed during the collapse of the Galaxy had a lower abundance of heavy elements than the stars formed in the disk, although it was one of the most fundamental observations. So Hoyle returned to the problem a year later[192] to suggest how the heavy elements from carbon to nickel were formed by the massive Pop II stars and subsequently got into the young Pop I stars. A summary is given in Fig. 6.23.

At this point, Hoyle, a staunch supporter of element synthesis in stars rather than in cosmic explosions, introduced the idea that the massive Pop II stars generated

[189] Hoyle, F., Astrophys. J. **118**, 513 (1953).

[190] Schwarzschild, M., and Spitzer, L., The Obs. **73**, 77 (1953).

[191] Jeans, J.H., Phil. Trans. Roy. Soc. Lond. **213**, 457 (1914).

[192] Hoyle, F., Astrophys. J. Suppl. **1**, 121 (1954). This is the famous paper where Hoyle predicted the nuclear structure of ^{12}C. The paper is discussed in connection with the synthesis of carbon in stars.

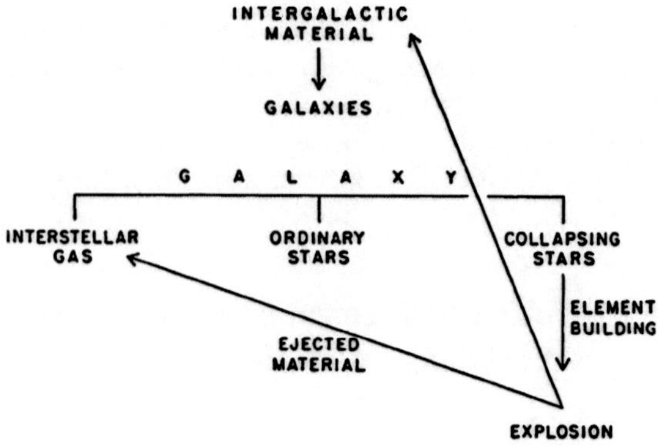

Fig. 6.23 Hoyle's original description of the circulation of the heavy elements in the Galaxy. Hoyle (1954)

the heavy elements either during slow evolution or during their final explosion, and spread them all over the collapsing galaxy. The stellar ejecta, rich in heavy elements, mixed with the prevailing gas to increase the abundance of the heavy elements in the interstellar gas. (Again we have a problem of mixing efficiency and time scales. Inefficient and slow mixing would result in stars having the same age but different abundances.) The collapse time must be sufficiently long for the massive Pop II stars to end their life on the main sequence and disappear before the formation of the disk of the Galaxy. The products are spread everywhere. A part escapes from the Galaxy but the other part remains in the Galaxy to enter into the next generation of stars, the Pop I stars. This idea of cycling matter from one generation of stars to the next leads to a definite prediction that was not known at the time Hoyle wrote his paper, namely that, since the collapse of the Galaxy and stellar evolution took place simultaneously, one should expect to see how the abundance of the heavy elements changes during the collapse, provided the mixing is fast. Indeed, this phenomenon was observed not long after.

Hoyle suggested that present day Type I supernovae occur in Pop II:[193]

> That same star in which exceptional nuclear reactions occur at very high temperatures may well be the stars that by explosion come to scatter their material into space. It is suggested that it is by this process that elements other than hydrogen and helium are built up and distributed in the Universe.

To see whether this hypothesis has any chance of being correct, assume that the number of massive Pop II stars formed is similar to the number of massive Pop I stars. According to Hoyle, this implies that the number of dead stars in our Galaxy should be 3×10^8, and if each star ejects just $1M_\odot$ into the galactic volume, then there will be enough heavy elements to enrich all Pop I stars. This number is roughly what is needed to explain the heavy elements in Pop I, and also implies a prohibitive

[193] Today SN Type I are observed in elliptical galaxies, where the stellar population is old Pop II. On the other hand, SN Type II are observed in galaxies where Pop I stars are abundant. Current SN paradigms state that SN Type I take place on small mass stars while SN Type II occur in massive stars. Thus, in past times, the massive Pop II stars exploded as SN Type II and not as SN Type I.

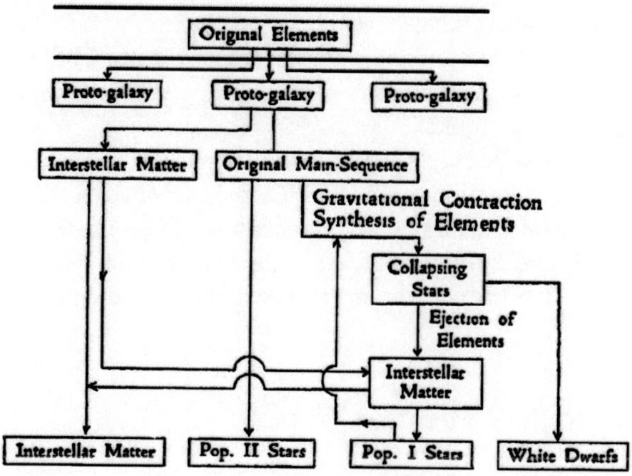

Fig. 6.24 The extension of Hoyle's theory by Taketani, Hatanaka, and Obi in 1956

rate of SN and a huge amount of energy poured into the Galaxy during the collapse phase. According to Hoyle, if the rate of SN Type I is about 1 per 300 years in a galaxy like ours, then only $1/3 \times 10^8$ SN can explode during the lifetime of the Galaxy, which he estimated as 10^{10} years. Hoyle realized the conflict with his steady state theory (to be elaborated in Chap. 7) and argued that:

> Strictly speaking, the concept of 'first' galaxies can be applied only in cosmologies that assign a finite age to the Universe. In the steady state theory of the Universe, on the other hand, there are no 'first' galaxies in the present sense, since, in this theory, the Universe has an infinite past. Consequently, at all times the extragalactic material must contain elements that were ejected from previously existing galaxies. Accordingly, all galaxies should possess heavy elements at the time of their birth.

Hoyle supported the idea that all elements were formed in stars rather than in the Big Bang. But he had a problem:

> SN Type I are not adequate to explain the total abundance of all elements heavier than helium. The SN Type I are indeed adequate to explain the total abundance of elements heavier than neon, but not that of carbon, oxygen, and the neon group, which exceeds the total abundance of heavier elements by a factor of about 10, this being just the order of the discrepancy revealed by the present discussion.

Hoyle stretched the parameters to their limit to obtain agreement with observation. Years later, when more accurate data became available and models of stellar evolution matured, it turned out that Hoyle's basic idea was partially right (Fig. 6.24).

Taketani, Hatanaka, and Obi[194] extended Hoyle's scheme and attempted to form a complete picture. The new idea was that the massive stars evolved to the collapsing stage and ejected most of their material into the surrounding space rather early in the lifetime of the Universe. The new material mixed with the original. The authors

[194] Taketani, M., Hatanaka, T., and Obi, S., Prog. Theor. Phys. **15**, 89 (1956).

did not provide any estimate whatsoever about whether the number of massive stars would be sufficient or whether their extension required an additional assumption about the rate of birth of massive stars. The authors were not aware of Salpeter's paper discussing just this problem.[195]

Taketani et al. did not cite Hoyle's 1953 paper, the paper which laid the grounds for the population–galaxy evolution connection. There is, however, a citation of Hoyle's 1954 paper, in which he did not cite his own previous paper. On the other hand, the paper by Taketani, Hatanaka, and Obi was not cited in the West, but only by Japanese researchers. Hoyle's 1953 paper was and still is abundantly cited in the West.

6.33 Can Alternative Theories Work?

If the heavy elements are not formed in stars, then how did the abundance of heavy elements increase during the collapse of the Galaxy? Stars are not mixed, and if the stars do synthesize the heavy elements, they do so in the cores and nothing should be seen on the surface of the stars. What is observed is the original abundance of heavy elements, which has little to do with the composition of the core. Could the heavier elements arrive from outside, say by accretion? This assumption means that we know about another place where the heavy elements are synthesized, say the beginning of the Universe, or other giant explosions.

Having no place to form the heavy elements, Schwarzschild, Spitzer, and Wildt[196] considered the possibility that the amount of heavy elements in the cosmos was fixed and that concentration mechanisms operated in the Galaxy, leading to a gradual increase in the abundance of the heavy elements, in particular in the objects from which stars form. Now stars form out of dense clouds of gas, and these gas clouds are spread throughout the Galaxy. But why should the heavy element abundance in the denser clouds be higher? The authors invoked radiation pressure:

> The force of radiation pressure from the general galactic light will push the surrounding grains into these regions. Essentially, a grain in the shadow cast by a dense, opaque cloud will be propelled towards the cloud by radiation coming from the opposite direction and will diffuse through the hydrogen cloud. Since the grains contain mostly heavy elements, with relatively little hydrogen, this concentration of grains in the dense clouds will increase the abundance of the heavy elements relative to hydrogen in such clouds.

The time needed for such a process was estimated at 10^7 years. The trigger for this supposition was the authors' conviction at the time that:

> According to current physical theory, there is no way in which these differences (in abundance of heavy elements) can result from nuclear processes inside a typical star. One must therefore seek some process by which more heavy elements are brought into a type I star than into a star of type II.

---

195 Salpeter, E.E., Astrophys. J. **121**, 161 (1955).

196 Schwarzschild, M., Spitzer, L., and Wildt, R., Astrophys. J. **114**, 398 (1951).

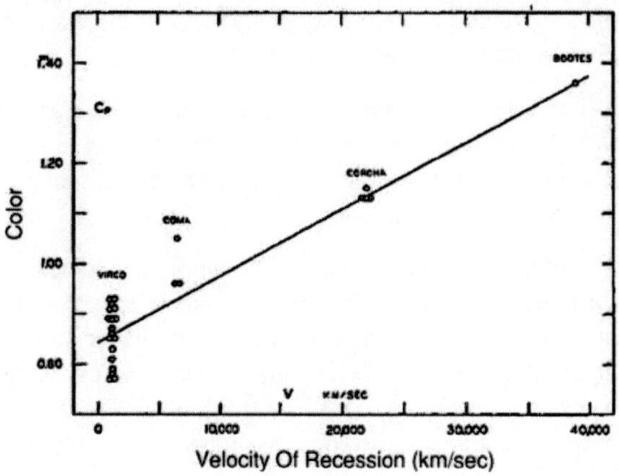

Fig. 6.25 Correlation between red excess and distance (Stebbins and Whitford 1948). Names refer to clusters of galaxies inside which the observed elliptical galaxies reside

The year was 1951. Bethe had solved the problem of the energy by hydrogen fusion into helium using the CN elements as catalysts 12 years earlier. No breakthrough regarding the way helium could be synthesized into heavier elements like carbon had been discovered. It appears that the abstruse assumptions that the heavy elements were formed before the stars were, and that the differences in abundances were created by other processes only after the formation of stars and the elements, were called upon because the authors lost hope in the existence of a conventional solution.

The authors estimated that the process could enhance the abundance of the heavy elements by at most a factor of 4, and hence argued that:

> There is good evidence that at least some of the type I stars have formed recently from interstellar clouds, while the type II stars are all old, having been formed by some different process about three billion years ago. If one assumes that all type I stars have condensed from interstellar clouds since the formation of the Galaxy—an assumption which is consistent with this evidence from stellar velocities—it appears that this difference in chemical composition between stars of the two populations can be explained.

At the time of publication of the paper, it was not yet known that the difference in the observed abundance of the heavy elements could be as large as a factor of 100, and such a factor lies way beyond what concentration by grains can achieve.

6.34 Spitzer and Schwarzschild (1953)

A coherent hypothesis for the evolution of a galaxy was put forward in 1953 by Schwarzschild and Spitzer.[197] The authors put forwards scenarios for the evolution

[197] Schwarzschild, M., and Spitzer, L., Obs. **73**, 77 (1953); Astrophys. J. **118**, 106 (1953). The paper was received by the editor on 26 February while Hoyle's paper was received in June of the same year.

of stars and the chemical elements in the early phases of a galaxy. Three fundamental observations were on the table:

- The very low abundance of the metals in the earliest Pop II stars.
- The high frequency of the white dwarfs.
- The red excess of distant elliptical galaxies discovered by Stebbins and Whitford[198] a few years earlier.

The term 'red excess' means that the objects looks redder than usual. Let us digress for a moment to describe Stebbins and Whitford's main result. As was already well known, the galaxies recede from us and hence look redder. Stebbins and Whitford discovered that the galaxies under consideration looked redder than expected if they were only receding according to Hubble. The excess reddening was found to be proportional to the distance: the further away, the redder the galaxy appeared (see Fig. 6.25).

It was quite implausible, argued Stebbins and Whitford, that the phenomenon was due to an age difference, namely, the fact that the Milky Way and nearby galaxies look older than the distant observed galaxies, because we see the latter when they were younger by the light travel time, i.e., about 2×10^8 years, which is a short time relative to the age of the galaxies or the Universe. (This possibility was mentioned by Hubble and Tolman.[199]) They then mentioned the discussion at the meeting of the American Astronomical Society in December 1947, when Schwarzschild suggested to the authors that:

> Since our measures were of elliptical nebula only, we are dealing with population II and these objects may once have contained large numbers of high luminosity red giants. These stars would have consumed their energy at a high rate and have faded in the time since the light left the distant ones. In population II there is no interstellar material out of which a continuing supply of giants may be condensed.

Indeed, Schatzman[200] discussed the effect of the short life of the massive stars on the apparent luminosity and color of galaxies. If the massive hot blue stars disappeared, the galaxy would look redder than normal. Russell's idea[201] was identical. In summary, if the massive stars evolve in just 2×10^8 years, then Stebbins and Whitford provided the proof for the stellar evolution which changed the color of the elliptical galaxies. The idea that massive stars end their life in an explosion which spreads the newly synthesized heavy elements in space had not yet been raised, because at that time it was not clear at all that the heavy elements could be synthesized in stars.

Schwarzschild and Spitzer, giving up the idea of grain concentration, looked for another possibility. By comparing the number of Pop I stars and the total mass of heavy elements in the Galaxy, they reached the conclusion that, contrary to prior

[198] Stebbins, J., and Whitford, A.E., Astrophys. J. **108**, 413 (1948). This was a collaborative work with Baade and Hubble.

[199] Hubble, E., and Tolman, R.C., Astrophys. J. **82**, 302 (1935).

[200] Schatzman, E., Ann. Astrophys. **10**, 14 (1947).

[201] Russell, H.N., PASP **60**, 202 (1948).

estimates, a large fraction of heavy elements now found in the Galaxy could actually have been produced by massive stars after the first stars were born.

A similar difficulty was posed by the high number of white dwarfs.[202] These stars cannot contain hydrogen. Consequently, they concluded that the high number of white dwarfs implied that at least part of the heavy elements found in them were the product of nuclear reactions occurring during the lifetime of the Galaxy:

> If so, one might assume, as has often been proposed[203] that the death of heavy, fast living stars is the cause of both.

On the one hand, a simple estimate showed that one needed about 10^8 stars of type A (mass 1.5–3$M_\odot$) that lived and died in the last 3×10^9 years to supply the heavy elements seen in Pop I:

> A similar order of magnitude is obtained on the basis of the present frequency of supernovas, if one assumes that the death of every such relatively massive star produces a supernova. On the other hand, the total number of WD is estimated at 3×10^9 and the total amount of heavy elements in Pop I stars may amount to about $10^9 M_\odot$.

Hence, concluded Schwarzschild and Spitzer:

> It appears that the present rate of star death is probably not sufficient to explain the number of WD[204] and the amount of heavy elements now found in the Galaxy.

In other words, it was improbable that the entire amount of the heavy elements was produced in stars, because more than can be produced by stars is buried in WDs. To aggravate the situation, the authors assumed that all stars with masses above the Chandrasekhar mass of 1.44$M_\odot$ become WDs, and thereby exaggerated in their estimate. Today we know that only stars with masses less than 8$M_\odot$ become WD, so that the yield is smaller and the problem bigger.

A possible way out argued by the authors was to assume that:

> During the formation of stars, stars of different masses are produced. However, the fraction of massive stars formed was significantly larger in the past and hence the estimates are not very accurate.

This ad hoc assumption solves the problem. Support for it was provided by the observations of Stebbins and Whitford. By the way, this phenomenon contradicts the simple idea of a stationary universe (all galaxies should look the same irrespective of their distance from us). When we observe distant galaxies, we look back in time, and if they are different it means that there is evolution. The lack of gas in elliptical galaxies implies that no new stars are born in these galaxies and hence the stellar population ages and does not remain the same. This is in contrast to spirals, which contain gas and in which new stars continue to form.

---

202 White dwarfs (WD) are stars which have exhausted their nuclear fuel and have masses below the critical mass of Chandrasekhar. The contraction of the star supplies the energy as the stars cool. WDs cannot contain any hydrogen, save tiny amounts on the surface. The rest must be heavy elements.

203 Baum, W.A., Astrophys. J. **57**, 222 (1952).

204 Schatzman, E., Symposium on Stellar Evolution, Rome, September 1952.

6.35 Successes

In 1955, Johnson and Sandage[205] published the HR diagram of the GC M67 (see Fig. 6.26). Besides observing the lower part of the main sequence, they also identified the Schönberg–Chandrasekhar gap. The cluster must have the right age, so the particular star with the right size of core must leave the main sequence to reveal this fact.

New numerical techniques to solve the complex equation of stellar structure and evolution were developed in the early 1960s.[206] With the advent of new powerful computers, this allowed static and dynamic evolution calculations. An excellent theoretical calculation of the gap, allowed by such methods, was provided by Iben's[207] extensive analysis of stellar evolution, as can be seen from Fig. 6.26. First note the line of constant radius. Next note the movement of the 1.25 and $1.5M_\odot$ models from points 3 to 4 in their evolution. This part of the evolution takes a very short time due to the Schönberg–Chandrasekhar instability. The model with one solar mass does not exhibit this instability, because its core is smaller than the Schönberg–Chandrasekhar limit.

6.36 Salpeter and the Invention of the Initial Mass Function: Stellar Birth Statistics

In 1954, when stellar evolution calculations started to show how stars evolve off the main sequence, Salpeter[208] came up with the following idea. Assume that (a) stars move off the main sequence after burning about 10% of their hydrogen mass, and (b) stars have been created at a uniform rate in the solar neighborhood for the past 5 billion years. Using these two assumptions, he was able to calculate the rate of star formation from the observed luminosity function, namely the number of stars as a function of their absolute luminosity. The result was

$$N(M) \approx \left(\frac{\mathrm{M}}{\mathrm{M}_\odot}\right)^{-1.35}.$$

The result is called the Salpeter initial mass function (IMF) and it has far-reaching consequences for galactic evolution, as it tells us how many stars are formed at each mass. The IMF shows that massive stars are less frequent. For example, the

205 Johnson, H.L. and Sandage, A.R., Astrophys. J. **121**, 616 (1955).

206 Henyey, L.G., Wilets, L., Böhm, K.H., LeLevier, R., and LeLevée, R.D., Astrophys. J. **129**, 628, 1959; Henyey, L.G., Forbes, J.E., and Gould, N.L., Astrophys. J. **139**, 306 (1964); Rakavy, G., Shaviv, G., and Zinamon, Z., Astrophys. J. **150**, 131 (1967).

207 Iben, I., Astrophys. J. **147**, 624 (1967).

208 Salpeter, E.E., Astrophys. J. **121**, 161 (1955).

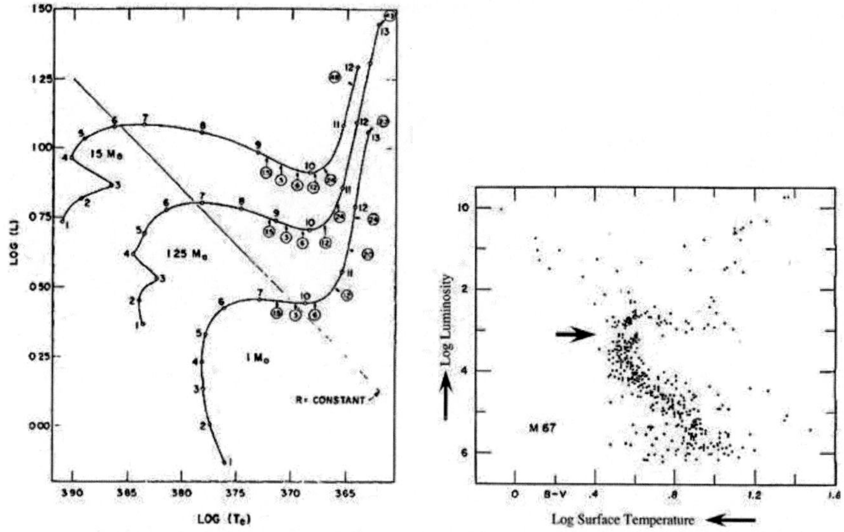

Fig. 6.26 *Left* the evolution of three stellar models (Iben 1967). The Schönberg–Chandrasekhar instability is seen in the 1.25 and 1.5$M_{\odot}$ models, in steps 3 to 4. *Right* the HR diagram of M67 as observed by Johnson and Sandage (1955). The *arrow* marks the location of the Schönberg–Chandrasekhar instability gap. Over the following years, exact measurement of the gap provided information about the size of the core and the interface between the core and the envelope

probability for the creation of a 10$M_{\odot}$ star is only 0.045 of the probability of forming a 1$M_{\odot}$ star. More massive stars have still lower probabilities of formation.

Salpeter could now calculate how many stars had died in the Galaxy and compare this with the number of observed white dwarfs. The numbers were found to be comparable if all stars lighter than 8$M_{\odot}$ end their life as white dwarfs after extensive mass loss. Salpeter's crucial assumption was that all massive stars lose most of their mass to the interstellar medium during the last stages of their evolution and that the small remnant becomes a white dwarf. However, since the number of stars with masses greater than 8$M_{\odot}$ is very small relative to the number of stars with masses between 1.44 and 8$M_{\odot}$, the error in Salpeter's estimate was negligible. Indeed, about 10% of all existing stars should be white dwarfs. This implies that the rest of the stellar mass was poured into the interstellar medium and only a small fraction will be buried forever in white dwarfs.

Moreover, Salpeter wrote:

> If all the assumptions of this paper should turn out to be at least roughly correct, it would then appear likely that an appreciable fraction of the interstellar gas present has at some time passed through the interior of stars. This conclusion, if true, would have some bearing on Hoyle's (1946) theory, which claims that all the chemical elements (except hydrogen) found in the Universe today were formed in the interior of stars.

Salpeter found that most stellar mass is reprocessed!

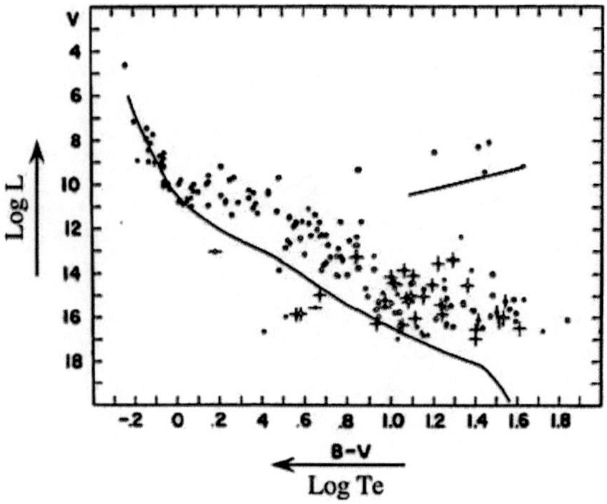

Fig. 6.27 HR diagram of an extremely young galactic cluster of stars. Walker (1956). The *continuous line* is the location of the main sequence of Pop I stars. The *horizontal line* is the location of the giants

6.37 Stars Are Born Homogeneous: 1956

In the 1950s, Walker carried out several very detailed examinations of the particularly young OC NGC 2264.[209] The results are shown in Fig. 6.27. The continuous line is the main sequence of homogeneous stars. We can see that the top of the main sequence is already populated by stars, while the lower part is not yet. There are two possible explanations:

- The massive stars reach the main sequence much faster than the low mass stars. The stars above the main sequence are homogeneous stars contracting towards the main sequence.
- The stars along the lower part of the main sequence are on their way to becoming red giants as they develop an inhomogeneity.

However, the second possibility contradicts the observation of GCs in which the upper part of the main sequence is not seen since it is the first to leave the main sequence. We are left with the first possibility. Thus, if the stars are above the main sequence, they have large radii. If the age is very small, these stars did not ignite any nuclear fuel. Hence, their large radii is a result of the fact that they have not yet contracted sufficiently to reach the main sequence and ignite nuclear reactions. Consequently, the newly formed stars must be fully homogeneous.

The estimated age of the cluster, as can be derived from the most massive star seen at the top of the main sequence, is merely 3×10^6 years. These observations supported the recent theoretical predictions of Salpeter and of Henyey, Lelevier, and Levée.[210]

---

[209] Walker, M.F., Astrophys. J. Suppl. **2**, 365 (1956); Astrophys. J. **125**, 636 (1957).

[210] Henyey, L.G., Lelevier, R., and Levée, R.D., PASP **67**, 154 (1955).

6.38 The Vatican Meeting 1957

The importance of the existence of stellar populations led to a unique conference on stellar evolution that took place in the Vatican. On 20–28 May 1957, the Pontifical Academy of Sciences and the Vatican Observatory sponsored a conference[211] in Rome to discuss stellar populations in the light of the theoretical and observational knowledge which had accumulated by then. Many provocative papers were presented. All in all, only 21 astronomers were invited.

While Baade's original idea of two populations agreed with almost all the available data, the skies were not completely bright. Two small clouds in the form of papers by Keenan and Keller[212] and by Nicholas Mayall (1907–1993),[213] now a few years old, hovered over the problem of stellar populations. What Keenan and Keller had done was to look for stars in the solar neighborhood with velocities greater than 85 km/s. Since this is considered to be a high velocity by galactic standards, the expectation was to see an HR diagram similar to that of Pop II stars. Instead, they found one that resembled the diagram for Pop I stars, reminiscent of the results of Adams et al.[214] for low velocity stars in the solar neighborhood.

Mayall measured the velocities along the line of sight of 50 GCs and discovered that 19 had positive velocities and 31 had negative velocities, ranging from +291 to −360 km/s. This in itself means that the GCs move around the center of the Galaxy. Such rotation is the natural explanation for the difference between the positive and negative velocities. The annoying problem raised by Mayall's observations was the following. Mayall measured the composite spectral type of the GCs and found them not to be of a single type. The implication is that the stellar content is different. In other words, not all Pop II stars have the same composition. William Morgan (1906–1994)[215] repeated this part of Mayall's work and confirmed the results. The beautifully consistent picture seemed to collapse.

Even Baade realized that the picture was more complex. At the conference, he presented a paper entitled[216] *The population of the galactic nucleus and the evidence for the presence of an old population pervading the whole disk of our Galaxy.* Here are just three pieces of Baade's observational evidence:

- The distribution of well known and established old objects like novae and planetary nebulae. Despite being old there are also such objects in the young disk of the Galaxy.
- Several GCs were found in the plane of the Galaxy, close to the center.

[211] *Stellar Populations*, Specola Atronomica Vaticana, 1958, ed. by D.J. O'Connell, North Holland, Amsterdam.

[212] Keenan, P.C., and Keller, G., Astrophys. J. **117**, 241 (1953).

[213] Mayall, N.U., Astrophys. J. **104**, 290 (1946).

[214] Adams, W.S., Joy, A.H., Humason, M.L., and Brayton, A.M., Astrophys. J. **81**, 187 (1935).

[215] Morgan, W.W., PASP **68**, 509 (1956).

[216] Baade, W., R. A. **5**, 303 (1958).

Table 6.3 Schwarzschild's interpretation of stellar population as age sequence

Population	Typical member	Age (billion years)	Z	Velocity (km/s)	Subsystem
Young Pop I	Galactic cluster	0–0.1	0.04	10	Disk
Intermediate Pop I	'Strong line' stars	1–3	0.03	20	
Old Pop I	'Weak line' stars	3–3.5	0.02	30	Intermediate
Mild Pop II	Oort's 'high velocity' stars	5.5–6	0.01	50	
Extreme Pop II	Globular clusters	6–6.5	0.003	130	Spherical

Table 6.4 A simplified version of Oort's classification at the Vatican Meeting of 1957

	Halo Pop II	Intermediate Pop II	Disk pop	Older Pop I	Extreme Pop I	Pop I
Mean height (light year)	6,400	2,200	1,400–1,000	500	380	
Distribution	Smooth	Smooth	Smooth	Spiral arms	Spiral arms	In the plane
Mass fraction of heavy elements Z	0.003	0.01	?	0.02	0.03	0.04
Age (billion years)	6	6.0–5.0	5	1.5–5	0.1–1.5	≤ 0.1

- Observations of other galaxies have shown that pulsating variables of the population II type (RR Lyrae variables) can be found near the central bulge of the Galaxy.

So what was going on? The ideal picture of a halo with one composition and a disk with another composition did not agree with the observations. The unavoidable solution, as was realized at the conference, is that the stars cannot be divided into just two distinct classes. In other words, there is a continuous variation in the properties of the stars. The heavy element abundance, for example, varies gradually between the different populations. Consequently, Father O'Connell, the organizer of the conference, proposed a compromised scheme of 5 populations: extreme populations I, Older Pop I, Disk Pop, Intermediate Pop II, and Halo Pop II. More recently the names of the intermediate stellar populations have been changed from halo and disk to the somewhat more complex Young (thin) Disk, Old (thin) Disk, Thick Disk, and Halo. But this refined classification did not solve the problems raised by Keller and Mayall.

Schwarzschild[217] suggested that the stellar populations might exhibit an age sequence, as can be seen from Table 6.3, which he presented at the conference. Schwarzschild did not discuss the spatial location of the different populations. On the other hand, Oort presented his Table 6.4, in which the emphasis was on location (although the velocity did not appear in Oort's table).

217 Schwarzschild, M., R. A. **5**, 204 (1958).

Oort[218] did not provide an explanation as to how or why the concentration of heavy elements in the collapsing gas increased with time. The explanation was provided by Strömgren,[219] who was experienced in building stellar models. Strömgren argued that the internal structure of a star of a given mass, radius, and luminosity depends on its chemical composition. The latter influences the equation of state, the opacity, and the energy production:

> In a general way it is therefore expected that composition differences between stellar populations will cause corresponding differences in stellar properties as expressed through the distribution of stars in the HR diagram.

Strömgren assessed the variation of the spectral properties with population (the HR diagram), but did not hypothesize about the enrichment mechanism.

Actually, Strömgren asked the provocative question as to whether or not there was any evidence for the assumption *of a unique relation between initial chemical composition and epoch of formation of a star*. Strömgren found some support for this assumption, but the data available to him at the time was not terribly good.

At the Vatican meeting, Baade[220] confirmed that:

> Galaxies containing only population II stars are free from dust and gas or contain at best insignificant amounts. In contrast, the population I is found only in the presence of dust and gas.

Hence, argued Baade, gas and dust provide the first hint that the two populations differ in age. There are old galaxies in which all the stars are old and no new stars form (like elliptical galaxies) and there are galaxies which today exhibit a mixture of young and old stars as well as gas out of which new stars continue to form even now (spiral galaxies.)

6.39 Further Evidence for Enrichment During Collapse

Gradually, clues accumulated about the variation in the amount of the heavy elements in different locations in the Galaxy. In 1962, Eggen, Lynden-Bell, and Sandage[221] set out to explore the formation and evolution of the Milky Way, and they discovered a connection between the amount of iron (relative to hydrogen) and the kinematics of the high velocity stars. The lower the amount of iron, the greater the height above the plane of the Galaxy that the star reaches. They proposed that the Galaxy started out as an extended body and collapsed. They found that the collapse of a rotating galaxy would take a few 10^8 years with gradual enrichment of the intergalactic space as the collapse proceeds. This was to a large extent a very refined version of Hoyle's

---

218 Oort, J.H., Vatican Meeting 1957, R. A. **5**, 415 (1958), ed. by O'Connell, D.J.

219 Strömgren, B., R. A. **5**, 245 (1958).

220 Baade, W., R. A. **5**, 3 (1958).

221 Eggen, O.J.M., Lynden-Bell, D., and Sandage, A.R., Astrophys. J. **136**, 748 (1962).

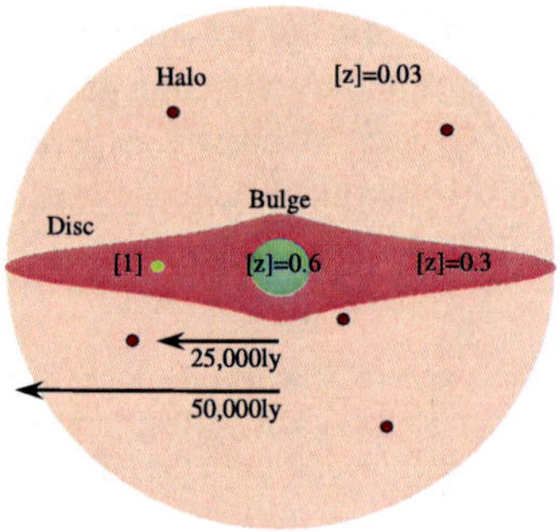

Fig. 6.28 Distribution of stellar populations throughout the Milky Way. *Numbers* indicate the total heavy element abundances relative to the Sun. *Small black circles* are the locations of typical globular clusters. The Sun is marked with a *yellow circle*

1953 or Spitzer and Schwarzschild's 1953 suggestions, but this time it included a genuine mathematical formulation capable of providing falsifiable predictions. The authors were criticized over selection effects in obtaining their data, but the general framework remained. Since then many refinements have been made, but the overall picture of an initial collapse of the Galaxy has been confirmed by many observations. The approximate situation in our Galaxy is depicted in Fig. 6.28.

6.40 Direct Observational Evidence for Mixing in Stars

A 'smoking gun' was discovered in 1952, when Paul Merrill (1887–1961m)[222] identified the spectrum of the element technetium (^{98}Tc) in a spectrum of the S-type red giant star R Andromeda.[223] Particularly good spectra of this star, and several other S-type stars, were obtained by Bowen and analyzed by Merrill. The spectroscopic plates were obtained during the testing and adjusting of the new 200 inch telescope and spectrograph.

The element technetium is special in that it is not found in the crust of the Earth because it has no stable isotopes! Even its most stable isotopes, technetium 97 and 98 have half-lives of only about 2.6×10^6 years and 4.2×10^6 years, so that, in an old Solar System, this element would have vanishing abundance.

---

[222] Merrill, P.W., *Spectroscopic observations of stars of class S*, Astrophys. J. **116**, 21 (1952). The boring title Merrill chose for his paper hides the great discovery. In today's way of doing things, the title would have been *At long last: The first ever discovery and proof of nuclear reactions in stars*. Merrill detected four spectral lines: 403.1, 423.8, 426.2, 429.7 nm, and this was sufficient to identify the element ^{98}Te.

[223] S-type stars are those whose spectrum displays bands from zirconium oxide in addition to titanium oxide which is characteristically exhibited by K and M class giant stars. Other s-process elements, for example yttrium oxide and technetium, are also enhanced, clearly indicating that the s-process has operated in the star. Most of these stars are long-period variables. A full discussion will be given in Chap. 12.

Technetium, which has atomic number 43, was discovered in 1937 by Carlo Perrier (1886–1948) and Emilio Segrè (1905–1989)[224] in molybdenum that had been irradiated with deuterium. The authors suggested the name technetium because it was the first element to be prepared artificially.

Since the S-type stars live significantly longer than 4.2 million years, this discovery implied that nuclear reactions within the stars must be producing it. Moreover, by some process, the newly made isotopes are carried from the hot interior, where they are formed, to the surface of the star where they can be observed. The discovery was the first direct evidence that nuclear reactions do indeed take place in stars.

More recently, van Assche[225] has put forward arguments that Noddack, Tacke, and Berg should be credited with the discovery of technetium. Tacke and Noddack[226] were searching for the missing elements eka-manganese when, in 1925, together with Berg, they isolated the element with $Z = 75$ and called it rhenium, from Rhenus, the Latin name for the Rhine. So far so good. It was during the process of distillation of the molybdenite ore that they also claimed to have discovered traces of an element with $Z = 43$, which they named masurium, after Masurien, the region south-east of Ost-Preussen where W. Noddack was born. The identification of the unknown element was carried out by Berg, who examined the X-ray emission along the lines of Moseley's previous discoveries. However, the discovery of masurium was not confirmed by others and the authors' credibility was tarnished.

In 1934, Fermi attempted to synthesize elements heavier than uranium by means of neutron bombardment. Ida Tacke[227] challenged Fermi's conclusions that he had produced a transuranium element artificially. Although several scientists sided with Fermi, Noddack suggested that Fermi had actually fissioned the uranium nucleus into lighter but known elements.[228] With her tarnished credentials and no empirical data to support her suggestion, it was duly ignored. It was then left to Meitner, Frisch, Hahn, and Strassman to discover fission by means of neutrons in 1939 and thereby vindicate Ida Noddack. In retrospect, the elusive technetium, which does not really exist in nature, probably appeared to be correlated with uranium as a product of the rare spontaneous fission of uranium, not known to exist at the time. A list of known Te isotopes is presented in Table 6.5.

---

[224] Perrier, C. and Segrè, E. Nature **140**, 193 (1937). Segrè visited Berkeley in 1936, where Ernest Lawrence gave him a sample of molybdenum that had been irradiated for several months by deuterium nuclei. It was in this sample that the new element was discovered. Early suggestions for the name of the element, including 'trinacrium', after Trinacria (an old name for Sicily), were resisted. It was actually only named after WW II. Its nomenclature is a reference to its being the first artificial element. According to the Best of Sicily Magazine, 2004, the unofficial version of the naming is that the University of Palermo officials pressed scientists to call the new element 'panormium' (Palermo in Latin), but the discoverers preferred to start from the Greek word *technètos*, meaning 'artificial', because it was the first artificially produced element.

[225] van Assche, H.M., Nuc. Phys. A **480**, 205 (1988).

[226] Noddack, W., Tacke, I., and Berg, O., Die Naturewissenschaften **13**, 567 (1925).

[227] Noddack-Tacke, I., Zeit. fur Angewande Chemie **37**, 653 (1934). Tacke married Noddack sometime after the beginning of their collaboration.

[228] Years later Segrè admitted that *they did not take her seriously, to their loss*, Segrè, E., Ann. Rev. Nucl. Part. Sci. **31**, 1 (1981).

Table 6.5 The isotopes of technetium and their half-lives

Isotope	Mass	Half-life
^{93}Tc	92.91	2.73 h
^{94}Tc	93.91	4.88 h
^{95}Tc	94.91	20.0 h
^{96}Tc	95.91	4.3 day
^{97}Tc	96.91	2.6×10^6 year
^{98}Tc	97.91	4.2×10^6 year
^{99}Tc	98.91	2.13×10^5 year

The spectrum of technetium isotopes was investigated by Meggers and Scribner[229] just a year before Merrill needed the information, so this enabled Merrill to identify the spectral lines. The observation of technetium in stars showed that nucleosynthesis was not confined to past times, but was in fact ongoing! But Merrill's seminal discovery[230] did not even win a place in the abstract of his own paper.

6.41 The Stellar Onion Model

We have already discussed how important it is to know whether or not the stars mix internally. Any new chemical element synthesized in the core does not appear on the surface, but remains in the core. We have also reviewed the observational evidence, which proves directly that mixing processes other than convection are extremely slow in gaseous stars, and hence can be safely neglected.

As the hydrogen burns into helium, the star develops a helium core surrounded by a hydrogen envelope. It is therefore to be expected that when helium later burns to give carbon, and carbon to give magnesium, and so on, each new species creates its own core, so that an onion-like structure gradually emerges. The star becomes composed of layers of different chemical elements, where the heaviest and latest to be synthesized are closest to the center (see Fig. 6.29).

The idea of the onion model and non-mixing is crucial for the synthesis of the elements in stars and in the Universe. Suppose the stars evolve completely mixed.

[229] Meggers, W.F, Spectrochim. Acta **4**, 317, (1951); Meggers, W.F. and Scribner, B.F., J. Res. Nat. Bur. Stand. **45**, 476 (1950).

[230] Merrill, P.W., Astrophys. J. **116**, 21 (1952).

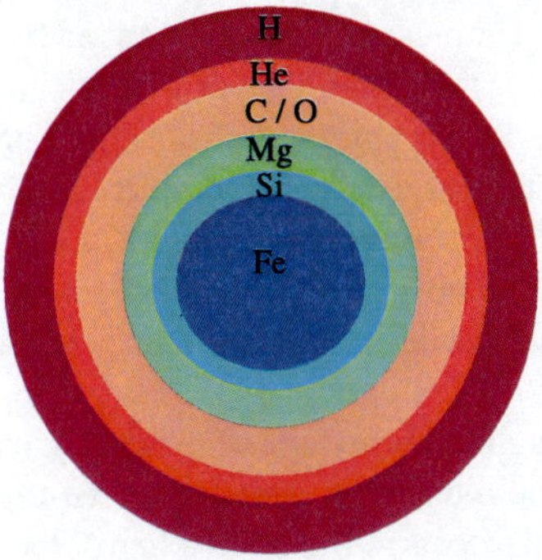

Fig. 6.29 The evolution of non-mixed stars leads to the onion model. Fully mixed homogeneous stars are ruled out by observation. The onion picture is crucial for post-MS evolution of stars. Only inhomogeneous models can develop the high temperatures needed for the nuclear synthesis of high atomic number elements. The very small degree of mixing in stars is crucial for the synthesis of the heavier-than-iron elements

Stars with uniform composition evolve by perpetual contraction until the entire star becomes iron and no more nuclear reactions and energy production can take place. If such a star ended its life in an explosion, then the exploding star would only ever enrich the interstellar medium with the end product, namely iron. Where would silicon come from, for example? From stars which end their lives in a state which is unable to ignite silicon? Only if the star reaches the point of mass ejection as an onion is there any chance that various elements can be ejected into space. This is the reason why the non-mixing of stars leading to the stellar onion model is so fundamental. The onion model allows the stellar core to develop the extreme conditions needed for the synthesis of iron without destroying all the lighter elements.

6.42 Are the Derived Abundances Correct?

In 1957, Thomas[231] published his by now classic study. His main point was that the conditions in stellar atmospheres, in the region where the observed spectral lines form, differ substantially from those in the laboratory. The main difference being the fact that, in the hot gas in the laboratory, all photons have the same temperature, while this is not the case in stars. Thomas discussed only the Sun, but if the effect had been true, the implication would have been that all earlier work, starting with Milne and continuing with Unsöld, Chamberlain and Aller, and Baschek, would have

[231] Thomas, R., Astrophys. J. **125**, 260 (1957).

been seriously in error.[232] Thomas' work was extended by Jefferies and Thomas,[233] who tore apart the typical arguments in favor of the simple theory and paved the way for the developments in the coming decades. Astrophysicists were in a deep dilemma. The new procedure was extremely complicated and demanding. How could they analyze thousands of lines without computers? Only in the mid-1960s did the situation change, when sufficiently fast computers became available. In many ways, the effect expresses itself as deviations from the Saha equation, much as Greenstein had suggested.

With such a dramatic gauntlet thrown into the arena, Unsöld[234] rushed to see what could be salvaged from the old theory. He discovered that, for most spectral lines (those that form sufficiently deep in the atmosphere), the deviations are negligible and the demon is not so terrible. Astrophysicists learnt to cope with this problem, and recent abundance determinations even take such small effects into account. The subject is very technical and would carry us too far aside, so will not be discussed here. Suffice it to say that most abundance determinations were not in fact affected.

6.43 Complicated Reality

The clear picture that emerged from the Vatican Meeting, namely, a simple relation between stellar age, composition, and kinematic properties, did not survive for long. Even in 1959, when Father McCarthy[235] summarized the conference and presented the open questions left in its wake, he could not have picked up better questions to demonstrate the complex but extremely interesting reality that was being uncovered.

Even before the Vatican Meeting convened, Morgan[236] and Morgan and Mayall[237] demonstrated that the dominant stellar population in the nuclear bulge of the great Andromeda galaxy contains old, but metal rich stars. Hence the answer to the first question posed by McCarthy: *Has the change in chemical composition been the same in all parts of the galactic system*?, is no! One finds that the three parameters, age, composition, and kinematic properties mix and distort the idealistic picture to which the consensus in the meeting had converged. As a matter of fact, even in 1958,[238] Öort had already realized that the diversity exhibited by the combinations of the three parameters belies the simple picture.

[232] All abundance derivations assumed local thermodynamic equilibrium (LTE). Thomas claimed that the conditions do not justify this assumption and that non-local equilibrium thermodynamics (NLTE) should be assumed. The consequence is an enormous complication in the calculations and completely different relative abundances.

[233] Jeffries, J.T. and Thomas, R.N., Astrophys. J. **127**, 667 (1958).

[234] Unsöld, A., Phys. **171**, 44 (1962).

[235] McCarthy, M.F., JRASC **53**, 211 (1959).

[236] Morgan, W.W., PASP **68**, 509 (1956).

[237] Morgan, W.W. and Mayall, N.U., PASP **69**, 291 (1957).

[238] Öort, J.H., R. A. **5**, 25 (1958).

The observations which showed the need for a 'higher resolution' picture accumulated over the years, but alas the 'final answer' has not been found yet, despite very extensive ongoing research. So we present here the gist of the discoveries, so that the reader can get a general impression and appreciate what to expect.

In 1959, hardly two years after the Vatican Meeting, the American Astronomical Society organized a symposium on *The Differences among Globular Clusters*,[239] and many of the suspicions corroborated by observations were presented at the conference. Morgan repeated the observation of GCs and confirmed[240] the much earlier observation by Mayall[241] regarding the existence of two groups of GCs. Morgan went as far as to define 8 types of globular cluster! He was an expert in astronomical morphology and paid attention to the smallest details. Morgan himself was happy with the discovery of two GCs (NGC 6528 and NGC 6553), the metal abundance of which was significantly higher than in the common GCs, but less than in open clusters. When Sandage heard Morgan's talk at the conference, he immediately mentioned the case of the Andromeda galaxy, where the bulge of the Galaxy resembles globular clusters from the metal abundance point of view, while according to the simple picture one would expect GCs to avoid the core of the Galaxy in their revolution around its center.[242] Why should the nucleus of a galaxy be poor in metals? Was the nucleus the first to form?

Two schools of galaxy formation emerged. The Eastern, or Moscow school, led by Zeldovich and Novikov,[243] who proposed the top–down model. The basic logic was that, as a body contracts, the Jeans' mass[244] decreases, and therefore, as large mass objects become unstable and contract, smaller objects will become successively unstable.

The Western, or Princeton school, led by Peebles[245] pushed for the bottom–up scenario, namely, small objects clustering to form bigger ones. According to this scenario, the globular clusters were the first to form and the galaxies were formed out of the merger of two or more smaller objects.

It is amazing that this controversy was carried out while ignoring all information about the observed variations of abundances. These parallel and disconnected lines of research continue to a large extent even today. But the Galaxy is made of stars, the composition of which is indicative of their history. It seems that those who tried to form the large scale structure and galaxies were just ignoring the various discovered abundances and what they meant, and vice versa.

---

239 Sandage, A.R., Astrophys. J. **64**, 425 (1959).

240 Morgan, W.W., Astrophys. J. **64**, 432 (1959).

241 Mayall, N.U., Astrophys. J. **104**, 290 (1946).

242 Every time a GC passes through the plane of the Galaxy, it loses stars by tidal forces. The effect is strongest near the dense core of a galaxy.

243 See, for example, Zeldovich, Ya.B., and Novikov, I.D., IAUS **29**, 280 (1968); Doroshkevich, A.G., Zeldovich, Ya.B., and Novikov, I.D., Sv. A. **10**, 731 (1967).

244 The Jeans mass is the critical mass for the collapse of a uniform gas under its own gravitation.

245 Peebles, P.J.E., Ast. Space Sci. **31**, 403 (1974).

Coming back to what can be inferred from stellar abundances, according to Hartwick in 1976,[246] the paucity of clusters near the galactic center is caused by removal of gas, say by a series of strong supernovae.

In 1978, Searle and Zinn[247] advocated the opposite scenario to the one suggested by Eggen, Lynden-Bell, and Sandage in 1962, who considered primarily the kinematic properties of GCs. In this picture, the elements were formed during the fast collapse, and no one would expect differences to develop during the fast collapse.

In view of the observed abundance differences between different GCs, Searle and Zinn suggested that the Galaxy was built from small protogalaxies, which subsequently merged to form the present galactic halo. The stars may have already started forming in the protogalaxies, rather then in one giant collapsed galaxy. This is called hierarchical formation from the bottom up. One has to explain why different protogalaxies have different abundances of heavy elements, or why the history of the protogalaxies differed from one to the other. The identification today, in certain galaxies, of stellar populations of mixed abundances, is taken to imply previous galaxy collisions and mergers. The truth is probably in-between these two extremes: Eggen, Lynden-Bell, and Sandage's simple collapse and Searle and Zinn's hierarchical build-up, which may not be identical in all galaxies.

In 1981, Ostriker and Cowie[248] and Ikeuchi[249] adopted what is called the hydrodynamic approach. The basic idea is that a mass of the order of $10^8 M_\odot$ is the seed. The seed collapses and fragments to a stellar system, and massive stars explode as SN about 10^7 years later. The SN causes a blast wave in the interstellar medium, composed mainly of unprocessed material and freshly synthesized material. The fragments collapse under their own gravity. In this way the next generation of stellar-sized objects form. The SNs accelerate the formation of the next generation of stars by exerting pressure via expanding gas. The problem is that the model cannot produce the largest observed voids between galaxies. In addition, this model ignored all information about abundance distributions in the Galaxy.

The problem of the role of dark matter in galaxy formation was not yet appreciated and the recent finding of a massive black hole (BH) at the center of most galaxies was not known either. The situation and emphasis have changed in recent years but information from abundances is still not fully implemented to constrain the models.

All of the above leads to complex models of galaxy formation and evolution based on assumptions of the following kind[250]:

- The gas is chemically homogeneous at all times. It is well mixed and has a single, but time dependent, composition. A newly born star has the same composition as the galactic gas at its birth.

---

[246] Hartwick, F.D.A., Astrophys. J. **209**, 418 (1976).

[247] Searle, L. and Zinn, R., Astrophys. J. **225**, 357 (1978).

[248] Ostriker, J.P. and Cowie, L.L., Astrophys. J. **L127** (1981).

[249] Ikeuchi, S, Pub. Astron. Soc. Jpn **33**, 211 (1981).

[250] Tinsley, B.M., Fundamentals of Cosmic Physics **5**, 287 (1980).

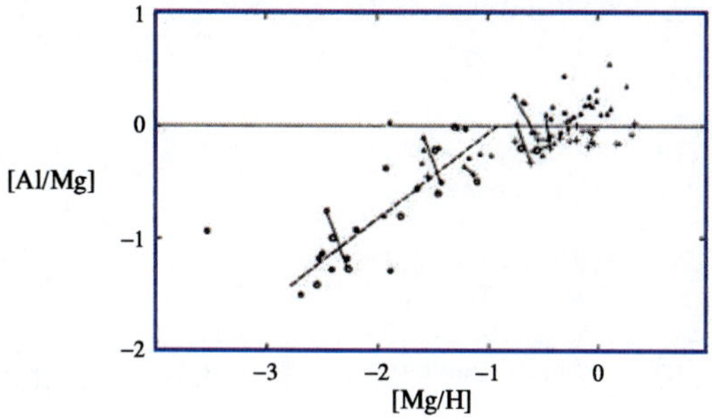

Fig. 6.30 The ratio of aluminum to magnesium abundances as a function of the abundance of magnesium, after Gratton and Sneden (1986). $[X] \equiv \mathrm{Log}(X/X_\odot)$, so $[X] = 0$ means solar abundance

- The Galaxy is closed, with no additional accretion from outside. The disk of the Galaxy accepts mass flow from the halo, but not radially.
- The initial mass function of stars is time-independent (the physics of star formation does not change with composition).
- The gas is recycled instantaneously (fast mixing on the galactic scale).
- SNs spread the synthesized matter throughout the Galaxy.

The observational evidence indicated rather soon that such a model was too simplified to explain the fine details. Models of the Galaxy's formation and chemical evolution are still too simplified to cope with all available observations.

In 1986, Gratton and Sneden[251] discussed the changes in relative abundances as a function of the absolute amount of the elements. A typical result is shown in Fig. 6.30, where the ratio of aluminum to magnesium is given as a function of the relative concentrations of magnesium to hydrogen. Aluminum has atomic number 13 and magnesium 12, and one expects there to be no difference between the ratios when the amount of magnesium changes, unless there are major differences in the nuclear physics and evolution of different stars. But what we see is that the expected result is recovered only for a high absolute abundance of magnesium, or abundances similar to that of the Sun ($[\mathrm{Mg}/H] \sim 0$). At lower abundances the amount of aluminum relative to magnesium decreases as the amount of magnesium decreases.

In 1993, Morrison[252] discovered a significant number of metal-poor stars with low velocity (rather than high velocity). Along the same lines, Sommer-Larsen, Beers, and Alvarez[253] discovered that extremely metal-poor stars in the Galaxy can be described by a non-rotating halo (this is not a new phenomenon), and with a thick disk rotating at about 200 km/s. This is once more an indication that the picture of galactic collapse, and at the same time enrichment with metals, is far from reality. If the enrichment took place during the collapse, one would expect the composition to change gradually with the height above the disk, but observation shows that there is no such change in the composition in the outer halo. The same phenomenon is

[251] Gratton, R.G. and Sneden, C., Astron. Astrophys. **287**, 927 (1986).

[252] Morrison, H.L., Astrophys. J. **106**, 578 (1993).

[253] Sommer-Larsen, J., Beers, T.C., and Alvarez, J., BAAS **24**, 1177 (1992).

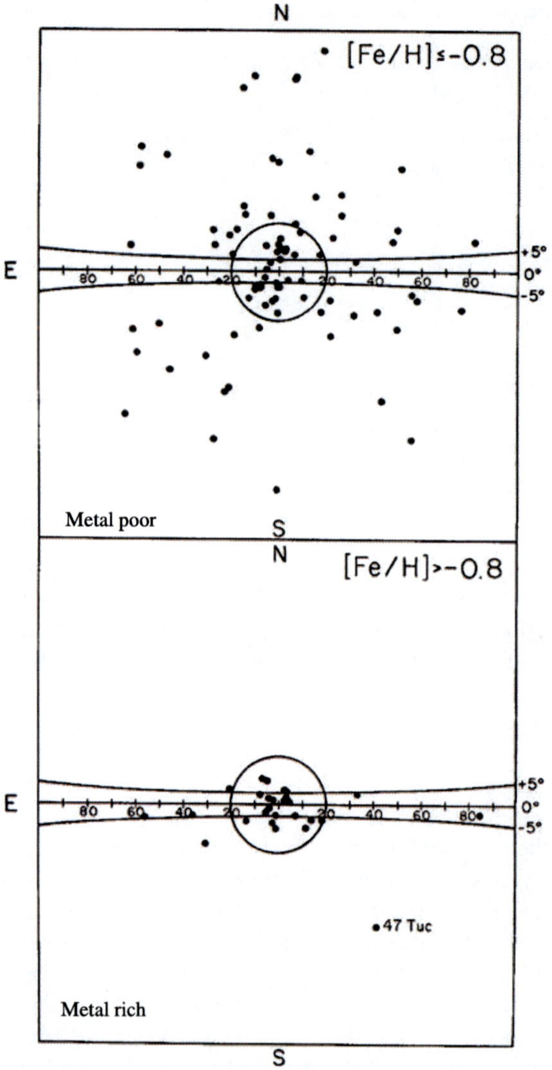

Fig. 6.31 Distribution of globular clusters in the Galaxy. The *upper panel* shows the metal-poor clusters and the *lower panel* the metal-rich clusters. *Continuous lines* are ±5° above the plane of the Galaxy (galactic latitude). The unique cluster 47 Tuc, which has high metal abundance, is marked. After Zinn (1985)

observed in globular clusters, i.e., there is no difference in composition between those globulars which reach the outer parts of the halo.

Three pieces of observational evidence show that the globular clusters are probably composed of two different groups. One is relatively blue, metal-poor, and lives in the outskirts of the halo/Galaxy. The second group is redder, with a higher metal abundance, and the members are more concentrated at the center of the Galaxy and are younger. In 1985, Zinn[254] published Fig. 6.31. The figure shows that there is also

[254] Zinn, R., Astrophys. J. **293**, 424 (1985).

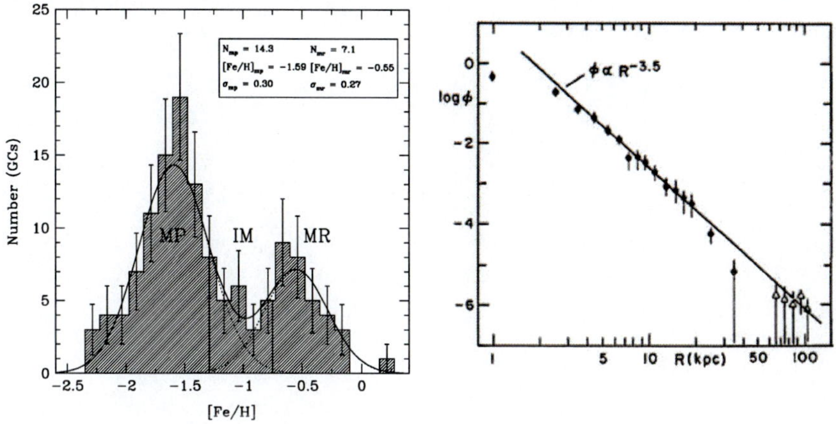

Fig. 6.32 Number of globular clusters as a function of the abundance of iron relative to hydrogen. The distribution shows a double peak (bimodality). After Zinn (1985) and Côté (1999). Number of GCs per unit volume of the Galaxy as a function of the distance from the center of the Galaxy. After Zinn (1985)

a kinematic difference between the two groups of GCs, namely, the closer metal-rich clusters show significant rotation. One may expect from the idea of a simple collapse that the abundance of the heavy elements would show a smooth behavior, but this is not observed.

The number of GCs as a function of the abundance of metals in not monotonic, but shows two peaks (see Fig. 6.32 left, Zinn 1985 and Côté[255]). The number of GCs as a function of the distance from the center of the Galaxy shows a gap (see Fig. 6.32 right). These two results emphasize the existence of two groups of GCs, probably with different origins.

Recently, in 2004, Alves-Brito and Barbuy[256] analyzed 5 giant stars in 47 Tucanae and found high metallicity. Furthermore, the abundances of the α elements O, Mg, Si, and Ti are enhanced relative to iron, as in metal-poor halo stars. On the other hand, the abundances of the α element Ca (relative to iron) and the odd-Z element Na, are similar to those observed in the Sun.

We have seen how the amount of metals is a copious source of information regarding the formation of the Galaxy. The deviations of the relative abundance of one or two species from the universal abundance pattern imply different sources or patterns of nuclear reactions which operated in different stars. On the other hand, we learn about the formation of the complicated and detailed structure of the Galaxy. The problem of element synthesis has been coupled to the problem of Galaxy formation, and consequently both were and still are open questions in astrophysics. The search for element genesis has turned from a goal into a tool!

[255] Côté, P., Astrophys. J. **118**, 406 (1999).

[256] Alves-Brito, A. and Barbuy, B., IAU Symposium # 222, ed. by Storchi-Bergmann, Ho, and Schmitt 549, 2004.

Chapter 7
Big and Little Bangs

7.1 Paving the Way to the Big Bang

Bethe's discoveries explained the source of energy of the main sequence stars but did not elucidate how the elements heavier than helium were formed. Three years later Bethe posed the problem of the synthesis of the elements and suggested, somewhat in despair and for want of a better explanation, that the synthesis of the elements might be independent of the energy source of stars. Chandrasekhar and Henrich[1] wrote similarly that:

> It is now generally agreed that the chemical elements cannot be synthesized under the conditions now believed to exist in stellar interiors. Consequently, the question of the origin of the elements is left open. On the other hand, the striking regularities which the relative abundances of the elements and their isotopes reveal (e.g., Harkin's rule[2]) require some explanation.

But the elements must be formed somehow, unless they are assumed to exist in the initial state of the Universe, i.e., unless they are primordial!

At the same time it was well known that the abundance of the chemical elements is very uniform, that is, all stars exhibit the same composition. Hence, Chandrasekhar and Henrich set out to look for the answers in the pre-stellar stage of the Universe. The basic assumption was therefore that, prior to the formation of stars, the conditions in the Universe were such that the elements could be formed. To allow for the formation of elements like silicon or iron out of whatever state the matter was in, they assumed complete equilibrium between all nuclei, neutrons, electrons, and positrons. This was basically an idea of Weizsäcker's from 1938,[3] to whom the authors gave proper reference, and who in turn gave proper reference to Sterne 1933,[4] but not to any prior

[1] Chandrasekhar, S., and Henrich, L.R., Astrophys. J. **95**, 288 (1942).

[2] The Oddo–Harkins rule states that elements with an even atomic number are more abundant than elements with an odd atomic number, see Sect. 11.1.

[3] von Weizsäcker, C.F., Phys. Zeit. **39**, 633 (1938).

[4] Sterne, T.E., MNRAS **93**, 736 (1933).

G. Shaviv, *The Synthesis of the Elements*, Astrophysics and Space Science Library 387, DOI: 10.1007/978-3-642-28385-7_7,

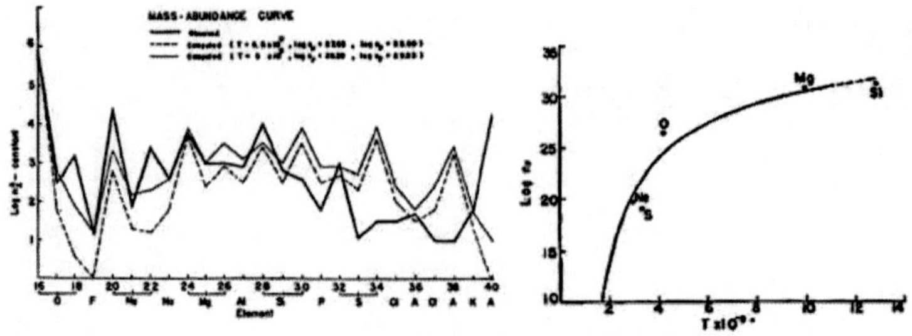

Fig. 7.1 *Left* comparison between theoretical abundances, as calculated by Chandrasekhar and Henrich (1942) assuming equilibrium of nuclei, and observed abundances. The *continuous line* is the observed, and the two other lines are calculations for two temperatures (8×10^9 and 10^7 K). There is agreement for $A \sim 28$, but no agreement for $A \sim 38$. *Right* dependence of the density on temperature needed to obtain agreement between theory and observation of the relative abundance of the isotopes of the same element. The *continuous line* is the neutron density at which the hydrogen to helium ratios agree with observation. Chandrasekhar and Henrich (1942)

attempts. Weizsäcker had shown that the hypothesis that all elements can be formed by assuming a single temperature and thermodynamic equilibrium was probably a bad one, and that several stages had to have taken place. Even a great idea cannot do miracles! The calculations just dealt with equilibrium, assuming fixed conditions of temperature and density. The question as to what happens to the energy, and in particular whether energy was released or absorbed, was ignored.[5]

On the other hand, this was the first calculation with all the known nuclei. Yet, it was not a full calculation. Chandrasekhar and Henrich, lacking a big computer, assumed the density of neutrons and calculated all abundances relative to the number of neutrons. They confirmed Weizsacker's result that it is indeed impossible to find a single temperature and density pair which will result in a good fit between theory and observations of the elements, in particular oxygen and potassium and the elements heavier than potassium, say iron and nickel. The example they provided contradicts observation (see Fig. 7.1 left).

The density needed to obtain a good agreement with the observed H to He abundance ratio is shown in Fig. 7.1 (right) as a function of temperature. For each temperature, there is only one density at which the ratio of the isotopes of the marked element came out close to the observed one. The conditions under which various heavy isotopes are synthesized are marked with their chemical symbol. When the H to He ratio agrees with observation, other abundance ratios do not. They thereby confirmed Weizsacker's claim, although he himself did not actually carry out anycalculation.

[5] In nuclear statistical equilibrium, one assumes that all nuclei compose one system and that the number of particles in a given state, in other words, the number of nuclei of a given kind, is proportional to the Boltzmann factor $\exp(-E_{\mathrm{bind}}/kT)$, where E_{bind} is the binding energy of the nucleus. Thus, more tightly bound nuclei are accordingly more abundant.

7.2 The Synthesis Must Have Been Explosive

As there appeared to be no single case which provided a wide agreement with observation, the authors were led to hypothesize that the matter was in a state of expansion from very dense and hot conditions. At the beginning, the iron group of elements was synthesized. As the matter cooled, part of the matter *must have been frozen* into the mixture. At lower temperatures and densities, the elements with lower atomic weights, like oxygen to potassium, were formed. At still lower temperatures the hydrogen to helium ratio must have been fixed. As the temperature fell to below 4×10^9 K, the assumption of equilibrium ceased to be valid, and the composition of the 'soup of elements' froze into place. In short, Chandrasekhar and Henrich conceived the idea that the formation of the heavy elements occurred in a huge explosion. The calculations they carried out were static, and the authors did not provide any explanation for how the freezing of elements could happen.

The analysis of Fig. 7.1 (right) shows the different conditions under which the various predicted isotopic ratios agreed with the observed values. Also shown is a line combining all conditions that yield the observed H to He ratio. While each species requires a different pair of values of density and temperature, for every temperature there is a density for which the H to He ratio agrees with observation. The fact that the H to He ratio is not very sensitive to temperature and density will become important later.

So what could be the cause of the original expansion and cooling?

> One suggestion is that it may be connected with the beginning of the expansion of the Universe. Another (possibly related) suggestion is that it might have arisen from the loss of energy by neutrino emission in the manner contemplated by Gamow and Shönberg in a different connection.[6]

In other words, Chandrasekhar and Henrich proposed that the giant explosion which formed the elements was the entire Universe, or what we call today the Big Bang. The authors warned that:

> The considerations of this paper should be regarded as of purely exploratory nature and agreements should not be over-stressed.

And indeed, the conditions were very extreme (by the standards of that time). When the density of the Universe was 10^7 g/cm^3, the radius of the Universe was a mere 0.03 light years. The idea that the elements were formed at the beginning of the Universe had very important consequences. In particular, the composition of all stars had to be identical, exactly as was actually observed to high accuracy.[7] But Chandrasekhar and Henrich and their followers overlooked a rather old result by Urey and Bradley[8] from 1931, a result which was confirmed time and again, and which showed that the assumption of equilibrium does not lead to good agreement

---

[6] Gamow, G., and Schönberg, M., Phys. Rev. **59**, 539 (1941).

[7] These were the days before the existence of different stellar populations was discovered.

[8] Urey, H.C., and Bradley, C.A., Phys. Rev. **38**, 718 (1931).

with the details of the abundances. This fact naturally poses a dilemma: if equilibrium does not work, how come the abundances are so uniform? The dilemma was solved when Chandrasekhar and Henrich suggested that (a) the elements were formed prior to the birth of stars, and (b) the synthesis took place in an explosion, where the temperature and density change continuously. These two assumptions 'almost' explain the observed abundance of the elements. An explanation of the word 'almost' will come later.

7.3 Questions About the Validity of Equilibrium

The results of Chandrasekhar and Henrich did not deter different researchers from seeking other conditions or assumptions of equilibrium in an attempt to obtain good agreement with observation. The equilibrium assumption was extremely attractive in its simplicity. However, all reached the same conclusion: the details do not work. Singwi and Rai,[9] Jensen and Suess,[10] and Cherdyncev[11] attempted equilibrium with α particles, but they could only get a rather poor agreement with observation. Various additional assumptions were made, such as an equilibrium between particles composed only of neutrons in a very dense neutron star core, but to no avail: the agreement remained poor.

Ubbelohde[12] considered isotope equilibrium due to absorption and emission of neutrons only. To get any resemblance to the observed abundances, he had to assume a freezing temperature of 10^{10} K, which by all considerations is a bit strange! It was surprising that, in spite of the large scatter between the observed and calculated abundances, he claimed to have obtained a good agreement when all previous researchers had described an agreement of the same quality as poor.

Wataghin[13] examined the physical assumptions underlying the hypothesis that the elements are in equilibrium, and found many reasons why this particular hypothesis should not actually be valid. For example, at the conditions of temperature and density Weizsacker and Chandrasekhar and Henrich had assumed, the energy in the radiation field is so high that the equivalent mass, namely $m = E/c^2$ must be taken into consideration. At a temperature of a billion degrees, the mass of the radiation is just 8.4 g/cm^{-3}, but at 10^{10} K the mass of the radiation is 8.4×10^4 g/cm^3. Wataghin concluded that all previous calculations neglected many such effects and required major corrections

---

9 Singwi, K.S., and Rai, R.N., Proc. Natl. Inst. Sci. India **12**, 291 (1946).

10 Jensen, J.H.D., and Suess, H.E., Naturewiss. **32**, 374 (1944); ibid. **34**, 131 (1947).

11 Cherdyncev, V., Ast. J. Soviet Union, Akad. Nauk CCCR **17**, 1 (1940); ibid. **17**, 12 (1940); ibid. **18**, 1 (1941); Compt. Rend. URSS **33**, 22 (1941).

12 Ubbelohde, A.R., Proc. Phys. Soc. Lond. **59**, 96 (1948).

13 Wataghin, G., Phys. Rev. **66**, 149 (1944).

Unlike previous authors, who assumed static conditions and carried out the calculations as if the external conditions did not change in time, Gamow[14] confirmed Wataghin's conclusions and added even more arguments to show that it was an error to assume static conditions. But no real improvement was obtained regarding the agreeement with observation.

At the same time, Klein, Beskow, and Treffenberg[15] applied the equilibrium assumption as implemented by Chandrasekhar and Heinrich to 'massive stars', and Beskow and Treffenberg[16] discussed the structure of such objects. The as-yet unobserved 'supermassive stars' were introduced as an alternative to the initial phases of the expanding Universe. These objects would be called in to help in later years.

7.4 Hoyle 1946: The Key Is in the Stars

The idea that helium must be synthesized in stars was advocated by Hoyle in 1946.[17] Hoyle pointed out that:

> The usual view that stars contain at least ten per cent by mass in the form of heavy elements is inconsistent with data concerning both the composition of interstellar material and the composition of stellar atmospheres.

Moreover, observations of the interstellar medium implied that:

> At the time of condensation of the stars, at least 99 per cent by mass must be in the form of hydrogen.

However, unless the stars were fully mixed, this does not explain how the amounts of helium and other elements seen on the surface of the stars today were actually formed. Somehow Hoyle overlooked Öpik's 1938 paper about inhomogeneous stars which, on the one hand, aimed to explain the formation of giant stars, but on the other, implied that stars cannot be fully mixed. Indeed, no reference to Öpik was given by Hoyle.

At that time Hoyle was convinced that accretion of gas from the interstellar medium was an important process,[18] and argued that the conversion of hydrogen into helium would be offset, at least in some stars, by accretion. But could that be the case? If stars accrete matter while moving through space, one expects either that all stars were born at the same time or that the amount of heavy elements should be a function of the stellar age. But the general feeling in those days was that stars tended to exhibit a rather uniform composition.

---

[14] Gamow, G., PRL **70**, 490 (1946).

[15] Klein, O., Beskow, G., and Treffenberg, L., Ark. Mat. Astr. Fys. **33B**, 1 (1946).

[16] Beskow, G., and Treffenberg, L., Ark. Mat. Astr. Fys. **34A**, # 13; ibid. # 14. (1947).

[17] Hoyle, F., MNRAS **106**, 255 (1946). In contrast to what Hoyle liked to quote: The fault, dear Brutus, is not in our stars, But in ourselves, that we are underlings. Julius Caesar (I, ii, 140–141).

[18] Hoyle, F., and Lyttleton, R.A., MNRAS **101**, 227 (1941).

A few months later, Hoyle[19] put forward the idea that the heavy elements might be synthesized in supernova explosions. Stars more massive than the Chandrasekhar limit cannot end their life as white dwarfs, and hence, so argued Hoyle, they must eventually contract indefinitely. But since stars rotate, this contraction speeds up the rotation. The process can therefore continue only until the rotation becomes so fast as to eject material from the star. This process should continue until the mass decreases to below the Chandrasekhar limit and the star can subsequently become a white dwarf. Hoyle demonstrated that the maximal temperature (about 5×10^9 K) that the ever-contracting (well-mixed) star must reach is sufficient to establish an equilibrium between the different nuclei.

Hoyle claimed that, in his model, in contrast to Chandrasekhar and Henrich's, where only one part in a million of the mass was converted into heavy elements, a significantly higher fraction of the mass would be processed. Hoyle reasoned that it was not sufficient to show that the correct abundances were obtained at the cooking temperature, but that the relevant abundance was the one obtained after cooling and dispersing the matter into space.

What are the central conditions when rotational instability sets in? Hoyle took a star of $10M_{\odot}$ with an initial rotation velocity of 1 km/s. This is a rather low rate of rotation. For comparison, the Earth rotates at a speed of 1/2 km/s, while the Sun rotates at about 2 km/s (at the equator). Massive stars rotate at equatorial velocities of 100–400 km/s. The initial central density (when on the main sequence) is about 3 g/cm^3. Under these conditions the central density at the onset of the rotational instability is, according to Hoyle, 1.67×10^{17} g/cm^3, and the central temperature is above 4×10^9 K. Hoyle thus justified the assumption of taking this temperature for the calculation.

However, it should be noted that such a high density is above the mean density of nuclear matter. For some reason Hoyle did not pay attention to this high density. It should have been clear that slowly rotating stars must encounter some other fate well before rotational instability sets in, because such densities are above the incompressible nuclear densities (about 10^{15} g/cm^3). If the initial rotation velocity is $\leq$15 km/s, then densities above 10^{10} g/cm^3 are obtained when the instability sets in, and this is a plausible density. In this range, the temperature when the rotational instability sets in is above 4 billion degrees, whence nuclear equilibrium can occur. Hoyle naturally, assumed a homogeneous star and added the unexplained sentence *nuclear reactions prevent the temperature from rising appreciably above 10^{10} C* (yes, C and not K!).

The explosion scenario encountered several problems. The basic idea was that an explosion must be driven by nuclear reactions. However, when compared to the observed rise time of the light in a SN (it takes about two days for the light of the supernova to reach maximum), it was found to be much too short. The explanation was that the star is so big that it needs about two days from the start of the explosion in the core until the luminosity from the surface peaks. No dynamic calculation was carried out, and there was no comparison between the energy released per unit mass

[19] Hoyle, F., MNRAS **106**, 343 (1946).

due to nuclear reactions and the gravitational binding energy, to see to what extent there might be sufficient nuclear energy to drive the matter out of the star. Hoyle must have relied on the kinetic energy in the rotation to provide the energy and drive explosive mass ejection.

7.5 What Do the Numbers Imply?

Consider the rotating SN. The energy of the ejected mass at infinity is $Mv^2/2$, where M is the mass ejected and v the velocity. According to Hoyle, this energy must be of the same order of magnitude or less than the gravitational energy $GM_{\rm ch}/R$, where $M_{\rm ch}$ is the Chandrasekhar limiting mass. Looking at SN 1054AD, Hoyle assumed that $M \approx 15 M_\odot$, $v \approx 10^3$ km/s, and $R_{\rm ch} = 10^9$ cm, whence it follows that $Mv^2/2 \approx 1.5 \times 10^{50}$ erg and $GM_{\rm ch}^2/R_{\rm ch} \approx 6 \times 10^{50}$ erg. So Hoyle concluded that:

> Accordingly, the supernova of 1054AD satisfies the requirement of the theory of rotational instability.

But the argument put forward by Hoyle did not involve the rotation energy at all. All it implied was that the difference in gravitational energy between the initial and final states is greater than the kinetic energy of the remnant expanding gases.

7.6 The Discovery of the Expanding Universe: Theory

As soon as Einstein published his general theory of relativity in a series of papers in 1915, attempts to apply the new theory to the entire Universe were made by Einstein[20] himself and by de Sitter (1872–1934).[21] Both assumed an empty universe without mass. To be more precise, they ignored the pressure inside the Universe. Einstein looked for a static solution, while de Sitter allowed for a dynamic one. It would be found later that, technically speaking, de Sitter's solutions are also static, but this was due to the fact that his models assumed an empty universe, i.e., in his approximation, the mass did not affect the results.

Then de Sitter calculated the gravitational field produced by a hollow spherical shell with total mass M and inner and outer radii R_1 and R_2, respectively. De Sitter compared the units of space and time inside and outside the hollow sphere and found them to change in a way that *cannot be verified by observation of the motions of bodies inside the shell relatively to each other*. De Sitter then proceeded to argue that the difference in the unit of time can, however, be checked by observing light which emanates from a point within the mass, or outside the shell. According to de Sitter's solution, *light coming from distances must appear displaced towards the violet as*

---

[20] Einstein, A., Berl. Ber., 1917, p. 142; Ann. d. Phys. **55**, 241 (1918).

[21] de Sitter, W., Proc. Akad. Wetesch. Amsterdam **19**, 1217 (1917); ibid. **20**, 229 (1917).

compared with light from a source inside the shell, an effect which became known as the de Sitter effect.[22] In other words, the de Sitter effect results from the fact that space geometry is affected by gravitation, and the geometrically induced redshift would appear as an apparent Doppler-induced velocity. According to de Sitter, the result emerges from the fact that we always measure distance and time in certain combinations and never say, distance alone. Moreover, it should be recalled that at this moment in time de Sitter was looking for a steady-state solution.

In the second paper de Sitter derived a formula for the *velocities due to inertia*, and noticed that there is no preference of sign. He then compared three nebulae[23] and argued that the speeds were very large and showed no preference of sign. De Sitter then took the mean of the three (!) velocities to find 600 km/s and assumed an average distance of 10^5 pc to obtain a radius of curvature of 3×10^{11} AU. He admitted that this result had no practical value, but one could say that the first measurement of the distance–velocity relation by de Sitter yielded 6,000 km/s per Mpc in 1917.

De Sitter's solution was found before the Great Debate on the nature of the nebulae was resolved, and de Sitter did not know at the time of writing that extragalactic objects actually existed. Consequently, de Sitter[24] could only speak about distant stars, and estimated that the shift observed in faint and hence remote stars would be equivalent to less than 1/3 km/s. Happy with this result, de Sitter wrote:

> It is well known that stars of all spectra actually show a small systematic displacement towards the red.

It was not the displacement to the violet that he had predicted. Anyway, Einstein did not like the solution that de Sitter had found and criticized it.[25] Assuming that the shift of stars to the blue does not exceed 1 km/s, de Sitter calculated the total mass within the radius of curvature of the Universe.

It was Lanczos (1893–1974) who demonstrated[26] by a simple change of coordinates that the de Sitter static solution could be interpreted as an expansion of the Universe and that the redshift predicted by de Sitter was a genuine recession velocity. However, the direct solutions for the expanding Universe (without manipulations with the coordinates) were found by Alexander Friedmann (1888–1925)[27] in 1924. Friedmann's solutions are known today as the Friedmann world models. Einstein believed in 1923 that he had found an error in Friedmann's 1922 paper

[22] When de Sitter described Lemaitre's solution in 1930 (BAN **5**, 211, 1930), he named the effect *the expanding Universe*.

[23] The nebulae were Andromeda, NGC 1068, and NGC 4594, with velocities −311, 925, and 1,185 km/s, respectively.

[24] de Sitter, W., MNRAS **77**, 155 (1916); ibid. **78**, 3 (1917).

[25] Einstein's criticism had a bad impact on de Sitter who abstained from cosmology for almost a decade and resumed his interest only after Hubble's discovery of the redshift–distance correlation.

[26] Lanczos, C., Phys. Zeit. **23**, 539 (1922).

[27] Friedmann, A.A., Zeit. fur Phys. **10**, 377 (1922); ibid. **21**, 326 (1924).

because it did not agree with his expectations,[28] but Friedmann demonstrated that this was not the case. Consequently, Einstein withdrew his objection to the result.[29] It was difficult to shake Einstein's belief in the static Universe. Friedmann had essentially shown that Einstein's and de Sitter's solutions[30] were the only static ones, and that all other solutions were either expanding or contracting. This dramatic result was confirmed in 1929 by Tolman (1881–1948)[31] and Robertson (1903–1961m).[32] A summary of the static versus non-static solutions to Einstein's equations was published by Robertson in 1933,[33] when it had already become clear that static solutions did not agree with observations.

In 1927, the Jesuit priest Abbé Lemaitre (1894–1966), who was unaware of Friedmann's solutions to Einstein's equation, discovered[34] a solution which included mass and pressure, and effectively rediscovered that Einstein's solution for the Universe is unstable, i.e., it behaves like a giant explosion. It was in this paper that Lemaitre derived the *apparent Doppler effect*, which implied that *the receding velocities of extragalactic nebulae are a cosmical effect of the expansion of the Universe*, with velocity proportional to distance.

In contrast to Friedmann, who was happy simply to derive the theory,[35] Lemaitre then turned to observation in an attempt to verify the prediction. To that end, he used a list of 42 nebulae whose velocities had been found by Strömberg.[36] As for the distance

---

28 *It is a capital mistake to theorize before one has data. Insensibly one begins to twist facts to suit theories, instead of theories to suit facts.* Sherlock Holmes, Adventures of Sherlock Holmes, A Scandal in Bohemia, 1891, p. 189.

29 This story is an example of how even eminent scientists can be captive of their own subjective feelings and have problems freeing themselves from personal biases by means of scientific logic.

30 When de Sitter came up with his solution, Einstein did not like it and criticized it, giving rise to bad mutual feelings. This should not be the result of a scientific criticism if properly expressed. After close to a decade, they made up and agreed upon the solution called the Einstein–de Sitter model. Eddington recounts that, when Einstein visited him and they discussed the model, Einstein remarked that he *did not think the paper was important, but de Sitter was keen on it.* Sometime later, de Sitter wrote to Eddington about a visit to Cambridge and added: *You will have seen the paper by Einstein and myself. I do not myself consider the result of much importance, but Einstein seemed to think it was.* Kerszberg, P., *The Invented Universe: The Einstein–de Sitter Controversy*, Clarendon Press, 1989, p. 403.

31 Tolman, R.C., PNAS **15**, 297 (1929).

32 Robertson, H.P., PNAS **15**, 822 (1929). It was in this publication that Robertson rectified the earlier deductions by Friedmann, which were not entirely satisfactory.

33 Robertson, H.P., Rev. Mod. Phys. **5**, 62 (1933).

34 Lemaitre, G., Annales de la Société Scientifique de Bruxelles **47**, 49 (1927). After writing the paper, Lemaitre met Einstein in Brussels and they discussed its content. Einstein's verdict was: *The calculations are right but your understanding of physics is abominable.* Midbon, M., *A Day Without Yesterday: George Lemaitre and the Big Bang*, Commonwealth, 24 March 2000. At that time Einstein still believed in a static universe, and those who disagreed 'did not understand physics'.

35 Friedmann's interests included meteorology, and in July 1925, he took part in a balloon flight to an altitude of over 7 km. However, he did not try to see if observation confirmed his solution.

36 Strömberg, G., Astrophys. J. **61**, 353 (1925).

of these nebulae, Lemaitre followed Hubble,[37] who had shown that the extragalactic nebulae all had the same absolute magnitude of -15.2. If so, the distance r, in parsecs, of an extragalactic nebula with apparent magnitude m is $\log r = 0.2m + 4.04$. Since the error is expected to increase with the distance, Lemaitre chose to assign a weight of $1/\sqrt{1+r^2}$ to each nebula at a distance r, where r is expressed in Mpc.

What Lemaitre did next was to correct the velocities for the speed of the Sun. He calculated the mean distance and the mean velocity, finding 0.95 Mpc and 600 km/s, respectively, which imply a coefficient of 625 km/s per Mpc. More accurately, if one assumes a linear relation between the velocity and distance, then the coefficient is the one found by Lemaitre. But Lemaitre did not prove that the observations implied a linear relation. However, the theory does. Applying the data provided by Hubble, Lemaitre predicted that light from galaxies at a distance of 0.087 of the present radius of the Universe will never be seen, because their light is completely shifted towards the invisible infrared. Lemaitre could not imagine the space age, with infrared telescopes placed on board satellites moving well outside the Earth's atmosphere. Similarly, he could not imagine that the light from remote but powerful galaxies would be shifted from the ultraviolet into the visible range.

It is interesting to learn what Lemaitre thought about the observational data. In a footnote (p. 36), he stated that (author's translation from French):

> If one does not attribute weights to the observations, one finds 670 km/s per 1.16×10^6 parsecs, or about 575 km/s per Mpc. Certain authors looked for a correlation between v and r, and obtained only a very weak correlation. The error in the distance determinations of the individual objects is of the order covered by the observation interval, and the proper velocity of the nebulae is large (300 km/s according to Strömberg). Accordingly, it seems that the negative results stand neither for nor against the relativistic interpretation of the Doppler effect. All that the inaccuracy in the observations allows us to do is to suppose that v is proportional to r and to try to avoid a systematic error in the determination of v/r.

One may refer to Lundmark for the error estimate, which Lemaitre did not provide.

The theoretical result found by Lemaitre was also obtained in 1928 by Robertson,[38] who, so it seems, did not know about Lemaitre's 1927 paper. Robertson noted in his second paper that he discovered Lemaitre's 1925 paper[39] only after the completion of his own, but apparently he was unaware of Lemaitre's important paper of 1927.

In these early stages of the development of the ideas it was not clear what exactly happened 'at the beginning' of the Universe. Was there a singularity, in other words, did the density and temperature tend to infinity? Or did the Universe start expanding from a static state? And if so, what was the static state? The question of the elements was not considered. According to Lemaitre, the Universe started *from some primeval matter*, the properties of which were ill-defined.

---

[37] Lemaitre stated that the distances were from Hubble, but gave no reference. We suspect that the reference is Hubble, E., Astrophys. J. **64**, 321 (1926).

[38] Robertson, H.P., Philos. Mag. **5**, 835 (1928); ibid. Philos. Mag. Suppl. **5**, 385 (1928).

[39] Lemaitre, G., J. Math. Phys. **4**, 188 (1925).

7.7 The K-Term

As soon as the velocities of stars had been measured, it was discovered that the sum of all stellar velocities relative to the Sun does not vanish. So to force the vanishing of the mean, an extra constant was added artificially. This constant was called the K-term or K-effect.[40] The non-vanishing term is the K-effect. Astronomers soon discovered that different classes of stars have different K-terms. However, the stellar K-term is never larger than 10 km/s.

7.8 Discovering the Recession of the Nebulae

In 1912, Slipher (1875–1969m) reported on the first Doppler measurement of the radial velocity of the Andromeda Nebula (−300 km/s).[41] Slipher noticed that:

> The magnitude of this velocity, which is the greatest hitherto observed, raises the question whether the velocity-like displacement might not be due to some other cause, but I believe we have at present no other interpretation for it.

The equipment Slipher used was amazing. Slipher attached his spectrograph to a 24-inch refractor. In some cases, Slipher applied exposure times as long as 40 h.

In 1917, Slipher[42] published a list of 20 nebulae with their Doppler shifts. This discovery, which was the basis for over a decade of attempts to find a velocity–distance correlation, is considered to be one of the most important discoveries of the Lowell Observatory. The confusing fact was that the 25 nebula were not isotropically distributed in the sky. Most of the positive velocities were close to one direction and the few nebulae in the opposite direction had negative velocities. So it seemed a bit premature to start claiming the expansion of the nebula system. This anisotropy, which eventually turned out to be accidental, was ignored towards the end of the 1920s, and its neglect allowed the discovery of the distance–redshift correlation. However, it did not prevent attempts to find a velocity–direction correlation.

The fact that most nebulae show a shift of their spectral lines toward the red was already known to Campbell[43] as early as 1911 and to Slipher[44] in 1915, long before the location of the nebulae (inside or outside our Milky Way) was known. Slipher measured 14 spiral nebulae, and out of these two showed negative velocities and three showed no velocity at all. All the rest showed positive velocity, meaning that they recede from us. Slipher explained the results by noticing that the

---

[40] The term 'K correction' is probably due to Carl Wilhelm Wirtz (1918), who referred to the correction as a 'Konstante' (or constant, in German).

[41] Slipher, V.M., LowOB **2**, 56 (1913).

[42] Slipher, V.M., PAPhS **56**, 403 (1917).

[43] Campbell, W.W., *Note on Radial Velocities of Nebulae*, Astr. Nachr. **188**, 346 (1911).

[44] Slipher, V.M., *Spectrographic Observations of Nebulae*, Pop. Astron. **23**, 21 (1915).

velocities of the nebulae were about 25 times greater than the average stellar velocity.[45] Campbell and Kapteyn discovered that stars with 'advanced' stellar spectra move faster. Consequently, Campbell hypothesized that the radial velocities are associated with the evolutionary stages of the nebulae. Slipher's explanation of the high velocities was that the nebulae are islands in the so-called theory of 'island universes'. In other words, they are stellar systems like ours, but at great distances. In other words, the spiral nebulae are extragalactic. And this was in 1917, several years before the Great Debate, which took place in 1920.

7.9 First Attempts to Confirm the de Sitter Model

In 1918, the couple Shapley and Shapley[46] analyzed the differences in properties between globular clusters and spiral nebulae. They found that the radial velocities of globular clusters are predominately negative, a fact which led them to the hypothesis that globular clusters are extragalactic objects falling onto the general galactic system.

As for the spiral nebulae:

> The brighter spiral nebulae as a class, apparently regardless of the gravitational attraction of the galactic system, are receding from the Sun and from the galactic plane—a remarkable condition that has been little emphasized heretofore.

The reason for this remarkable conclusion at this point in time was that, although the number of available velocities of spiral nebulae was rather limited, out of 25 spirals, all but three had positive velocities (meaning that they are moving away from us), and the velocities were 150 km/s and above. These velocities are much higher than the (negative) velocities of the globular clusters.

The Shapleys stressed several times in the paper that:

> Globular clusters as a class appear to be rapidly approaching the galactic system; spiral nebulae as a class are receding with high velocities.

They explained the fact that the spirals have receding velocities in terms of repulsive forces which act between the spirals and the Milky Way.

Another conclusion of the Shapleys was that:

> The speed of spiral nebulae is dependent to some extent upon apparent brightness, indicating a relation of speed to distance or, possibly to mass.

Naturally, this fitted the Shapley's idea of a repulsive force acting on the spirals, since they stated that:

---

[45] Reynolds, J.H. (Obs. **40**, 131, 1917) questioned Slipher's results and Slipher replied (Obs. **40**, 304, 1917), commenting on the accuracy of the measurements and the fact that the mean velocity of the spirals was found to be 570 km/s while the mean velocity of stars is about 20 km/s. As for the measurements of the speed of the Great Andromeda Nebula, Slipher got −300 km/s, and he cited Wright who got −304 km/s, Wolf who got −300 to −400 km/s, and Pease who got −329 km/s. It is surprising that nobody asked how our Galaxy could hold such fast-moving nebulae.

[46] Shapley, H., and Shapley, M.B., Astrophys. J. **50**, 107 (1919).

> The hypothesis demands that gravitation be the ruling power of stars and star clusters, and that a repulsive force, radiation pressure or an equivalent, predominate in the resultant behavior of spiral nebulae.

It is interesting to note that de Sitter's 1917 paper was published just a year before the Shapleys completed their work, but in a quite obscure location,[47] not frequently accessed by astronomers. And neither were the papers published a year earlier in the Monthly Notices mentioned by Shapley (although his papers contained other references from this journal, published at roughly the same time). On the other hand, the Annalen der k.k Universitäts-Sternwarte in Wien was consulted. The Shapleys had the first indication that the system of spiral nebulae expands. They did not know what the recession velocity depended on, but the overall expansion was obvious.[48]

Once de Sitter's paper became known, various attempts were made to confirm de Sitter's model. One of the first attempts to discover the spectral shift (or velocity) versus distance relation was carried out by Ludwig Silberstein (1872–1948),[49] who applied de Sitter's result to derive a formula for the shift of the spectral lines emitted by stars. For distant stars, Silberstein found that, in the limit of small velocities, the Doppler shift is

$$\frac{\Delta\lambda}{\lambda} = \pm\frac{r}{R},$$

where R is the radius of curvature of the Universe and $\Delta\lambda$ the shift in wavelength λ of the line. Note that the sign can be positive or negative. Silberstein applied the formula to a list of stellar clusters for which he found mostly negative velocities, but one positive one. He also applied it to the Small and Large Magellanic Clouds, which displayed positive velocities. At that time it was not known that the Magellanic Clouds are outside the Milky Way, or indeed that there were any objects outside the Milky Way.

Application of Silberstein's law to this mix of objects gave a value of $R = 6.0 \times 10^{12}$ astronomical units or about 10^8 light years for the radius of curvature of the Universe. Figure 7.2 gives the data Silberstein used to derive the 'radius of the Universe'. All the objects listed in the table are globular clusters, save the Magellanic Clouds. For example, the implied distance to the globular cluster NGC

---

[47] de Sitter, W., Koninklijke Nederlandsche Akademie van Wetenschappen Proceedings **19**, 367, 527, 1217 (1917).

[48] These fantastic observations did not prevent Shapley from arguing during the Great Debate that the spirals are within the Milky Way. He had the expansion of the Universe in his findings. Was this personal bias?

[49] Silberstein, L., MNRAS **4**, 363 (1924). Silberstein considered himself an expert on relativity. At the Royal Society's 1919 meeting, where Eddington reported his successful trip to see the solar eclipse in Brazil, in which he confirmed the general theory of relativity, he approached Eddington with some degree of skepticism, inquiring about the claim that there were only three men who actually understood the theory. Silberstein, as can be imagined, hoped to hear from Eddington that he belonged to the group, along with Eddington and Einstein. When Eddington refrained from replying, Silberstein pressed Eddington not be *so shy*, whereupon Eddington retorted: *Oh, no! I was wondering who the third one might be!* As told by Chandrasekhar to Isaacson (*Einstein: His Life and Universe*, Simon and Schuster, 2008, p. 262).

	cD. kil./sec.	r. (Astron. units.)	R. (Astron. units.)
N.G.C. 5024	−170	$3{\cdot}8 \times 10^9$	$6{\cdot}7 \times 10^{12}$
5272	−125	2·8	6·7
6205	−300	2·2	2·2
6333	+225	5·0	6·7
6341	−160	2·5	4·7
6934	−350	6·7	5·7
7078	− 95	2·9	9·1
Lesser Magellanic Cloud	+150	4·1	8·2
Greater Magellanic Cloud	+260	3·8	4·4

Fig. 7.2 The original data of Silberstein (1924)

6934 is 6.7×10^9 astronomical units or 100,000 light years, which is about the size of the Milky Way. These results implied that the Small and Large Magellanic Clouds lie outside the Milky Way. Silberstein sent his paper for publication on 18 January 1924, before Hubble's discovery that the nebulae are extragalactic. But it appears that Silberstein was unaware of the astronomical debate, because his results, although highly inaccurate, would have implied the same result that Hubble found shortly afterwards, namely that the spiral nebulae are extragalactic. On the other hand, the possibility of the globular clusters being as far away as he claimed would have implied unacceptably high intrinsic luminosities for the stars in these clusters. In short, a lack of astronomical facts gave rise to unacceptable results.

Eddington[50] attacked Silberstein's proposal for determining the distance of a remote star by observing the displacement of the spectral lines at six months' interval. Silberstein claimed that this method would separate the ordinary Doppler effect of the unknown motion of the star from the distance effect predicted by de Sitter. Eddington argued that Silberstein neglected any effects of the short intervals of time and space between the two observations and of the distortion of the waves by the local gravitational field.

On the other hand Eddington suggested the 'double drift theory of star motions'. This theory refers to two observed stellar systems which differ in their mean velocity. The difference in mean speed, argued Eddington, must have arisen during their formation, because:

> It is difficult to see how gravitation towards the centre of the Universe could separate the motions of the stars into two systems, if they originally formed one system.

Silberstein's results were also criticized by Lundmark (1889–1958m).[51] The criticism shows the extent to which people can be locked up in dogmas. It was in this year that Hubble published his discovery of Cepheids in Andromeda and settled the 'island universe' debate. But Lundmark was still captive of the idea that the nebulae were galactic. First, Lundmark claimed that the data on the Doppler shifts, which were taken from Slipher, were suspiciously large, questioning whether the shifts in spectral lines were not caused by something that had nothing to do with velocity.

---

50 Eddington, A.S., Nature **113**, 746 (1924).

51 Lundmark, K., MNRAS **84**, 747 (1924).

Lundmark could not believe that such high velocities could be real. Next he pointed out that the theory of Weyl[52] and Eddington did not allow for a negative velocity, in contrast to Silberstein's result.[53] As for the globular clusters used by Silberstein, Lundmark claimed that:

> These objects are probably among the most distant celestial objects we know at present, but how do we know that they are so far away that the effect of the curvature on space-time outweighs the effect of the real motions of the clusters themselves?

Lundmark repeated Silberstein's analysis for various groups of celestial objects. Figures 7.3 and 7.5 show the data Lundmark compiled for the globular clusters and the spiral nebulae. From the globular cluster data, Lundmark found a much greater radius of curvature for the Universe than Silberstein. As for the spiral nebula result (Fig. 7.5), Lundmark stated that:

> We find that there may be a relation between the two quantities, albeit not a very definite one.

Lundmark gave a table of distances and velocities of spiral nebulae. He based his distances on a *hypothetical parallax* derived from total magnitude and apparent diameter under the assumption that:

> The apparent angular dimensions and the total magnitude of the spiral nebula are only dependent on the distance.

This amounts to saying that they are standard candles. Lundmark claimed that:

> An inspection of the table (of distances and velocities) will show that a computation of R [the radius of curvature of the Universe] from the individual values of V [the velocity] will give inconsistent values for the radius of curvature.

However, no details were given. Next, Lundmark's calculations showed that:

> Plotting the radial velocities against these relative distances, we find that there may be a relation between the two quantities, although not a very definite one. If this phenomenon were due to the curvature of space-time, we could derive the mean linear distance or determine the scale of our relative distance.

If we take Lundmark's data for the 43 spiral nebulae at face value, ignoring the data for the globular clusters and assuming the existence of a linear relation, we calculate the Hubble constant to be 71.2 ± 55.6 km/s per Mpc, a value extremely close to the present day accepted value (!), but with a large uncertainty. Lundmark, on the other hand, obtained much larger numbers and consequently argued that:

> The simplified formula is not justified in this case.

Lundmark erred by lumping the globular clusters and the spiral nebulae together, although he admitted that this lumping together *certainly is open to objection*. Anyway, Lundmark's method gave much larger distances for the nebulae and consequently a smaller Hubble constant. Evidently, Lundmark had a blind date with destiny and missed it.

---

[52] Weyl, H., Phys. Zeits. **24**, 230 (1923).

[53] It was difficult to believe that such high peculiar velocities could actually exist.

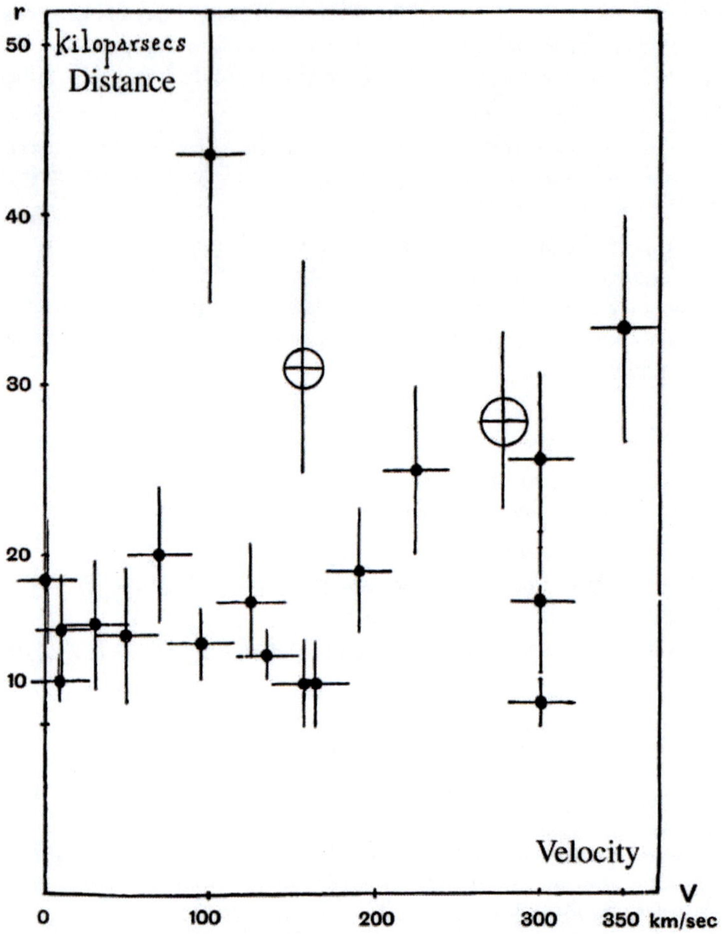

Fig. 7.3 The relation between velocities and distances of globular clusters. *Circles* are the values for the two Magellanic Clouds. Lundmark (1924). Is there any sign of a linear correlation?

Shortly after Lundmark's criticism appeared in press, a similar criticism was published, this time by Strömberg.[54] Strömberg concluded that:

> We have found no sufficient reason to believe that there exists any dependence of radial motion upon distance.

Further attempts were carried out in 1922–1924, when Wirtz (1875–1939)[55] examined the statistics of the radial motions of spiral nebulae. Out of the 29 nebulae he had data for, he found 25 with positive velocities, so that the average speed of the

[54] Strömberg, G., Astrophys. J. **61**, 353 (1925).

[55] Wirtz, C., Astr. Nachr. **215**, 349 (1922).

spiral nebulae was +840 km/s. In 1924, Wirtz[56] tried to relate the velocities to the distances. To this end, he searched for a reliable distance indicator, for example, the apparent diameter, or luminosity of the spiral nebulae, assuming of course that they were all identical and could serve as 'standard candles'. Wirtz took the data for the apparent diameters of spiral nebulae from Curtis[57] and Pease,[58] discovering that the velocity v of the spiral nebula relates to the diameter D according to

$$v \ \text{[km/s]} = 914 - 479 \log D.$$

Here the diameter is given in arc minutes. Wirtz was careful and added to this result an evaluation of the reliability of the correlation just found. Various attempts to confirm the velocity–distance relation which emerged from the de Sitter models were carried out at that time. In 1923, Eddington derived $v \sim R^2$, a year later Weyl obtained $v \sim \tan R$, and we have already mentioned the Silberstein result. Hence, the logarithm was not a surprise. We can say that Wirtz almost confirmed de Sitter's model (we say 'almost' because the expansion law in the de Sitter model does not contain the logarithm).

Wirtz's contribution is hardly ever cited and when it is mentioned, it is only by a few historians of science. In 1936, a short time before his death, Wirtz wrote a half page note[59] reminding the reader of his discoveries back in 1921 and 1924, and before Hubble in 1929, but to no avail.

While the measurement of the redshift is tedious but straightforward, this is not the case with the distances. In 1926, Hubble carried out extensive research on the distance of extragalactic nebulae, measuring the distances of about 400 objects.[60] Hubble deduced an improved distance–apparent luminosity relation, which improved the accuracy of the distances so measured. In addition, the distances to more remote nebulae were measured.

In 1929, Hubble (1898–1953m)[61] decided to resolve the paradox:

> Determinations of the motion of the Sun with respect to the extragalactic nebulae have involved a K-term of several hundred kilometers which appears to be variable. Explanations of this paradox have been sought in a correlation between apparent radial velocities and distances, but so far the results have not been convincing. The present paper is a re-examination of the question, based on only those nebular distances which are believed to be fairly reliable.

---

[56] Wirtz, C., Astr. Nachr. **222**, 21 (1924).

[57] Curtis, H.D., Lick Public. #13 (1918). This was the fellow who argued with Shapley in the Great Debate that the spiral nebulae were extragalactic objects.

[58] Pease, F.G., Mt. Wilson Cont. **132** (1917); ibid. **186**, 1920.

[59] Wirtz, C. Zeit. f. Astrophys. **11**, 261 (1936). *Ein literarischer Hinweis Zur Radialbewegung der Spiral Nebel.*

[60] The objects were taken from a compilation by Hardcastle (Hardcastle, J.A., MNRAS **74**, 699, 1914), which was prepared from a two-hour exposure with a 10 inch telescope!

[61] Hubble, E., PNAS **15**, 168 (1929).

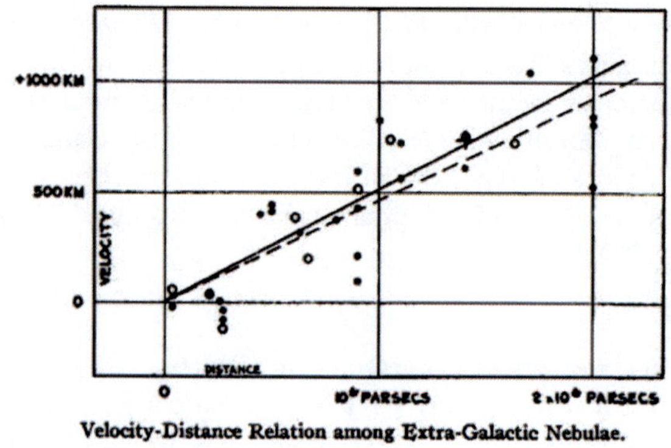

Fig. 7.4 The original velocity–distance relation discovered by Hubble in 1929

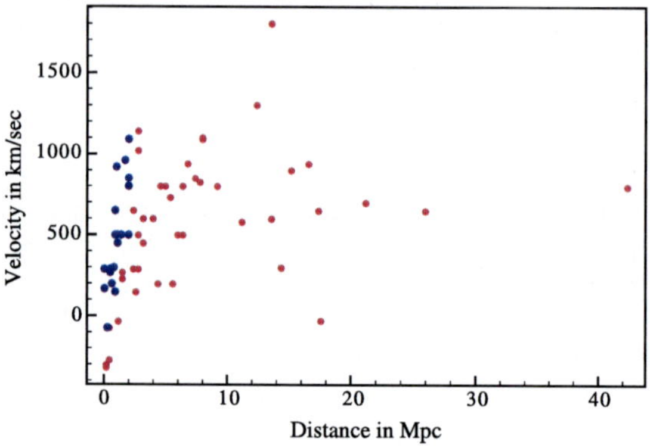

Fig. 7.5 *Red data* is from Lundmark (1925) and includes 43 spiral nebulae, while *blue data* is from Hubble (1929) and includes 23 spiral nebulae

Hubble had 24 pieces of data. According to Hubble, the data indicated a linear correlation between distance and velocity. It is crucial to stress that Hubble did not assume a linear relation, but plotted the data and demonstrated that a linear relation was a good approximation. We have checked the numbers, taking into account only the data from the table, and found the linear correlation $v = (470 \pm 278)r$. Hubble's original correlation is shown in Fig. 7.4. In other words, Hubble succeeded in reducing the scatter in the data and obtaining a convincing correlation. He thus showed that there exists a correlation and it is linear.

Figure 7.5 compares the data used by Hubble and Lundmark. Hubble's distances were significantly smaller than Lundmark's, and consequently the 'Hubble constant' constant is that much larger.

In Fig. 7.6, we can see what happened to the particular data used by Hubble. Red squares are Hubble's data, where distances were measured by Hubble and velocities by Slipher. Green squares are additional data provided by Humason. The velocities of three nebulae originally measured by Slipher are shown in blue. The arrow marks the updated/changed velocities. One can see that the revision of the velocities significantly improved the correlation. Additional data, which is not shown in the figure,

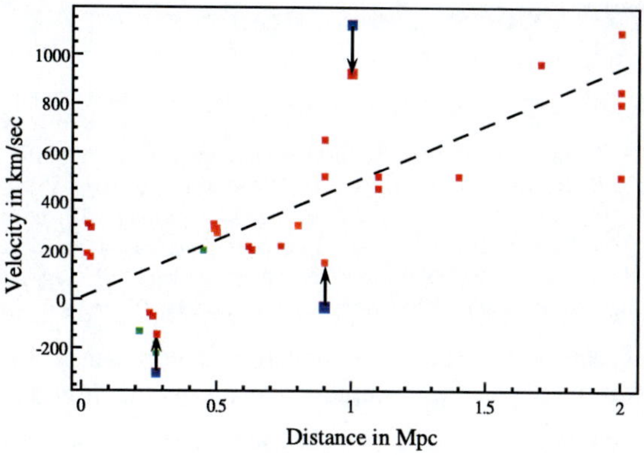

Fig. 7.6 Changes in the data used by Hubble in his velocity–distance correlation, 1929

was provided by Humason, who measured a velocity of 3,779 km/s for NGC 7619 at a distance of 7.8 Mpc.

To appreciate Hubble's results, we mention that in October 1929 Perrine (1867–1951m) published[62] an analysis of the motion of spiral nebulae and globular clusters, still mixing the two, and in particular after establishing the spirals as extragalactic objects. Perrine examined various correlations between the radial velocities and degree of ellipticity, between size, elongation, and galactic latitude, and between distance, diameter, and residual velocities, and concluded that:

- The radial velocities of the spiral and globular (structureless) nebulae vary with apparent size, the smaller ones having the higher velocities.
- The variation is not linear but increases rapidly among the smaller ones, being satisfactorily represented by the inverse square of the diameter.

So far so good, but he continued with the conclusion that:

- The relationship is to size (or mass) and not to distance.

There were additional conclusions, all of them mistaken, probably because of poor statistics. Perrine did not assume that all spirals have the same size, and managed to show that the observations imply that the K-term is given by $K = V\sqrt{d}$, where d is the apparent diameter.

62 Perrine, C.D., Astr. Nachr. **236**, 329 (1929).

7.10 Criticism from Astronomers

Oort[63] criticized Hubble, claiming that:

> It appears impossible to derive even a rough estimate of the distance of a cluster of nebulae as long as there may be such large and still undeterminable differences between the average luminosity of isolated and cluster nebulae as indicated in the first section. We may be surprised at the accuracy with which Hubble and Humason appear to be able to derive the relative distance of various clusters from apparent magnitudes, agreeing beautifully with the distance derived from radial velocities, but these do not help us in getting absolute values.

Hubble and Humason responded to the criticism without naming the criticizer, who was a renowned astronomer, by claiming that it was of minor importance:

> Differences tend to cancel out, and the final numerical result agrees with the Mt. Wilson result.

But it does not agree, as Oort's result (for the Hubble constant) was about half the result due to Hubble and Humason.

On the other hand, Hubble and Humason were very confident in their analysis and estimated, in their 1931 paper, that:

> This number is correct to within 20%.

But estimates have changed by a factor of 7 since then![64]

7.11 Understanding the Recessional Velocities

The correlation found by Hubble was sufficiently accurate to cause people to believe in it. At a meeting of the Royal Astronomical Society early in 1930, de Sitter admitted that neither his nor Einstein's solution could represent the observed Universe. Eddington was stunned and asked *one puzzling question: Why should there be only these two solutions?* Eddington supposed that it was a consequence of the fact that people had only looked for static solutions. The participants of the meeting were unaware of Lemaitre's solution. Indeed, Lemaitre's 1927 paper was published in an almost totally inaccessible journal, never visited by astronomers. Consequently, it did not attract any attention and would have been left to rest in peace, if Lemaitre had missed the commotion in London. As it was, Lemaitre wrote to Eddington, his previous mentor, and drew his attention to the discoveries he had made three years earlier.

Eddington[65] immediately recognized the importance of Lemaitre's paper, and realized that his own solution was not the right one. Eddington's paper is actually a review of Lemaitre's article, because as he said:

---

[63] Oort, J.H., BAN **5**, 105 (1931); ibid. **6**, 155 (1931).

[64] See Trimble, V., H_0: *The Incredible Shrinking Constant 1925–1975*, PASP **108**, 1073 (1996), which discusses the way biases and improvements have substantially reduced the Hubble constant.

[65] Eddington, A.S., MNRAS **90**, 668 (1930).

> My original hope of contributing some definitely new results has been forestalled by Lemaitre's brilliant solution.

Eddington thus presented Lemaitre's solution, giving him the credit for the discovery that *the Einstein world is unstable*. He praised Lemaitre for this solution and showed that Tolman's suggestion was unacceptable. However, when Eddington gave the recession velocity of the nebulae, he just provided a number and did not specify who had obtained it, let alone how it had been obtained. Eddington noted the conflict between stellar ages and the age of the Universe, but did not offer any solution.

Eddington was convinced that Lemaitre's paper should be salvaged from its anonymity, and asked for the paper to be translated into English and published in the MNRAS. In the meantime, Lemaitre published a note explaining the apparently paradoxical behavior of a mass point in an expanding Universe, this time in a journal much more accessible to astrophysicists.[66] But Lemaitre described de Sitter's solution, not his own! The English version of the 1927 paper came out in 1931,[67] in an abridged version. It is not known who decided what to include and what to omit from the translation. However, the first paper includes the formula:

$$\frac{R'}{R} = \frac{v}{cr},$$

and the statement:

> From the discussion of available data, we adopt $R'/R = 0.68 \times 10^{-27}\,\text{cm}^{-1}$.

Here $R'/R = (R_2 - R_1)/R_1$, where R_1 and R_2 are the radii at times t_2 and t_1, respectively. The idea is that this is a universal number. In Lemaitre's words:

> $v/c = (R_2/R_1) - 1$ is the excess radius of the Universe when the light was emitted to the radius when the light is received.

What Lemaitre meant was that the change in radius (=distance) of the galaxy implies the velocity of expansion as given by the above expression. This figure amounts to 629 km/s per Mpc. However, there are no details to say how the number was calculated (see above). Lemaitre noted that the numerical results implied *quite recently for stellar evolution*, which would mean that the Universe is younger than the stars, but did not infer that there might be a problem with that.

In 1930, Lemaitre discussed the expanding Universe[68] and in 1933,[69] he discussed the evolution of the expanding Universe from an initial state. This was a revolution

---

[66] Lemaitre, G., BAN **5**, 273 (1930).

[67] Lemaitre, G., MNRAS **91**, 483 (1931); ibid. 490 (1931). The only information in the paper is that it was translated by permission. No submission date is given, and it is not clear whether Lemaitre saw the translation before publication. The translator of Lemaitre's 1927 paper, as it appeared in the MNRAS, is not specified. In the official site of the George Lemaitre Center for Climate Research in Louvain (http://www.uclouvain.be/en-316446.html), it is claimed that Eddington translated the article and added a long comment. I could not find any confirmation of this claim.

[68] Lemaitre, G., BAN **5**, 273 (1930).

[69] Lemaitre, G., PNAS **20**, 12 (1934).

in thinking. The Universe has not existed for an infinite time, but was created at some moment in time. This sounded to many like creationism, and particularly when it came from a Catholic priest! Consequently, many resented the idea. Lemaitre spoke about the mean density of clusters of nebulae and its connection to *Hubble's ratio of distance to spectroscopic velocities by the approximate relation*

$$\frac{dr}{dt} = r\sqrt{\frac{4\pi}{3}G\rho_0},$$

and there is no mention of his own observational result for the coefficient of correlation. Lemaitre mainly discussed the mean density ρ_0 of the Universe, and tried to infer the cosmic model from this parameter.

Naturally, the discovery of the recessional velocities, even if only apparent, stirred up discussions about the initial state out of which the expansion had started. The questions were:

- What causes the expansion?
- What was the initial state which started the expansion?

One of the first explanations as to why the Universe expands was given by Tolman,[70] who suggested that the expansion might be due to the conversion of matter into radiation. As the radiation accumulates, the radiation pressure increases and causes the expansion. In all these papers, the physics of the initial state was not discussed and remained, therefore, an enigma. This explanation of the expansion was also suggested by Lemaitre.[71]

Immediately after Hubble's discovery, Lemaitre started to look for solutions to Einstein's equations which yield expansion. By 1931, Lemaitre succeeded in constructing two possible models for the Universe. In the first model, Lemaitre presented a Universe with a constant total mass (meaning a closed universe), and increasing radius as time goes by. This paper was the English translation of the 1927 paper published by the Brussels Scientific Society, a translation and publication under Eddington's recommendation. It is followed by a paper by Eddington in which he explains that the expansion results from *formation of condensation, which may be named the 'stagnation' of the Universe*. Later he withdrew this explanation. In this model, the radius of the Universe increases without limit from an initial value R_0 at time $t \to -\infty$, which means that the Universe always existed and will expand forever.[72] This feature of the solution was later a fundamental property of Gamow's model for the Universe, namely that it started out from a state with finite properties.

Lemaitre provided a possible explanation for the expansion of the Universe (compare with Tolman's explanation):

> It seems to suggest that the expansion has been set up by the radiation itself.

[70] Tolman, R.C., PNAS **16**, 320 (1930).

[71] Lemaitre, G., MNRAS **91**, 483 (1931).

[72] The initial radius R_0 is given by $R_0 = rc/v\sqrt{3}$, where r and v are the distance and velocity of a galaxy and c the speed of light.

In a static Universe, light emitted by matter travels around the space, comes back to its starting point, and accumulates indefinitely. This light gives rise to radiation pressure, which causes the Universe to expand. As Lemaitre explained:

> It seems that this may be the origin of the velocity of expansion which Einstein assumed to be zero, and which in our interpretation is observed as the radial velocity of extragalactic nebulae.

In the second model, published together with the first one,[73] Lemaitre proposed that the Universe started from an initial 'stagnation' state characterised by extreme conditions. No further specification was given, in particular, neither the composition nor the temperature and density were specified.

7.12 Why Did Hubble Not Believe in an Expanding Universe?

It seems that at no time did Hubble believe that the spectral wavelength shifts represented real velocities. At the end of the paper, after establishing that the K-term depended only on distance, he pointed out that:

> The outstanding feature, however, is the possibility that the velocity–distance relation may represent the de Sitter effect. [...] The displacements of the spectra arise from two sources, an apparent slowing down of atomic vibrations and a general tendency of material particles to scatter.

The words 'expansion' and 'Universe' were not mentioned at all. As a matter of fact, Hubble remained skeptical about 'the expansion of the Universe', and refrained from using such a term, even when the rest of the community, and in particular the theoreticians, accepted the idea of cosmic expansion.

Objectively, Hubble had good reason at that time to be suspicious of this explanation. His determination of the 'Hubble constant' yielded an age of 2×10^9 years for the Universe, while Jeans estimated that the stars were about a factor 10^3 older *and stellar structure was more advanced.* More specifically, Hubble had several technical objections. In 1936,[74] Hubble thought that:

> It is evident that the observed result (of applying a K-correction $B = 2.94$) is accounted for if the redshifts are not velocity shifts.

And in the discussion of the paper, Hubble expressed his doubts about the expanding Universe as a possible solution. Further, mainly technical points were raised by Hubble over the years.[75]

At the end of their paper, Hubble and Humason reiterated:

---

[73] Lemaitre, G., MNRAS **91**, 490 (1931). The two models were published back to back in the same journal issue.

[74] Hubble, E., Astrophys. J. **84**, 517 (1936).

[75] Hubble, E., *The Observational Approach to Cosmology*, Clarendon Press, Oxford, 1937.

> The interpretation of redshift as actual velocities, however, does not command the same confidence, and the term 'velocity' will be used for the present in the sense of 'apparent' velocity, without prejudice as to its ultimate significance.

The authors did not provide an error estimate because of possible systematic errors in the magnitudes and color indices. But they made the statement:

> It is believed, however, that the uncertainty in the final result is definitely less than 20%.

They trace the difference in this result from *the revision Shapley made in the standard unit of distance*.

In 1935, Hubble and Tolman[76] wrote:

> The most obvious explanation of this finding is to regard it as directly correlated with a recessional motion of the nebulae. [...] Nevertheless, the possibility that the redshift may be due to some other cause [...] should not be prematurely neglected. [...] Until further evidence is available, both the present writers wish to express an open mind with respect to the ultimately most satisfactory explanation.

In 1942, Hubble wrote in an article to Science[77]:

> It may be stated with confidence that redshifts either are velocity shifts or they must be referred to some hitherto unrecognized principle in nature.

Moreover, after analysis, Hubble concluded:

> Thus the assumption that redshifts are not velocity shifts but represent some hitherto unknown principle operating in space between the nebulae leads to a very simple, consistent picture of a Universe so vast that the observable region must be regarded as an insignificant sample.

In 1953, Hubble delivered the George Darwin lecture[78] and discussed the *law of redshifts*. Hubble noted three phases in the history, the first of which was *a discovery phase which ended with a crude formulation in 1928–1929*. There was no mention of Lemaitre. Hubble argued that, *if the redshifts do measure the expansion of the Universe*, then we can trace its history. In the paper, Hubble rejected the term 'apparent velocity' and used just 'velocity', but by this he meant $cd\lambda/\lambda$ or *redshifts expressed on a scale of velocities*. More specifically, Hubble once again raised the conflict between the age of the Earth and the implied age for the Universe, if the expansion solution were assumed. And more technically:

> When no recession factors are included, the law will represent approximately a linear relation between redshift and distance. When recession factors are included, the distance relation (becomes) nonlinear.

A few months after Hubble published his sensational correlation, Zwicky,[79] also at Caltech, found another explanation for the redshift, namely gravitational drag of

---

[76] Hubble, E. and Tolman, R.C., Astrophys. J. **82**, 302 (1935).

[77] Hubble, E., Science **95**, 212 (1942).

[78] Hubble, E.P., MNRAS **113**, 658 (1953). Hubble passed away before he was able to revise the paper before publication, as he intended. The paper was edited by A.R. Sandage.

[79] Zwicky, F., PNAS **15**, 773 (1929); Phys. Rev. **34**, 1623 (1929); ibid. **48**, 802 (1935).

Fig. 7.7 *Left* the galaxy NGC1023, whose recession velocity was measured by Slipher. A massive black hole was discovered in this galaxy (Bower et al., Astrophys. J. **550**, 75, 2001). *Right* the galaxy NGC 4594, known as the Sombrero galaxy. A supermassive black hole was also discovered in this galaxy (Kormendy, Astrophys. J. **335**, 40, 1988)

light. This suggestion explains why the frequency shift is wavelength independent. Zwicky was even able to provide a theoretical estimate of the Hubble constant, given the mean density of the Universe.

In recent years, Burbidge et al.[80] and Arp[81] have consistently discovered very close pairs of quasi-stellar objects (QSO) with separations <5 arcsec and very different redshifts. The estimates that these situations are due to random projection are very small. Consequently, they conclude either that this is evidence that QSOs have significant non-cosmological redshift (like from a black hole) components, or that the pairs must be explained by gravitational lensing (Fig. 7.7).

7.13 The Exploding Universe

The original idea that all the elements were formed in the initial explosion of the Universe is due to Gamow[82] in 1946. In a short Physical Review Letter, Gamow states that:

> It is generally agreed at present that the relative abundances of various chemical elements were determined by physical conditions existing in the Universe during the early stages of its expansion, when the temperature and density were sufficiently high to secure appreciable reaction rates for the light as well as for the heavy nuclei.

---

80 Burbidge, G., Hoyle, F., and Schneider, P., Astron. Astrophys. **320**, 8 (1997).

81 Arp, H., *Current Issues in Cosmology*, ed. by Pecker and Narlikar, Cambridge University Press, 2006.

82 Gamow, G., PRL **70**, 752 (1946).

Gamow cited Weizsacker and also Chandrasekhar and Henrich as responsible for the concept of element synthesis prior to the birth of the first stars in the Universe. However, argued Gamow, the elements could not have been formed in an equilibrium process, as assumed by previous researchers. The way Gamow reached his conclusion is interesting. He noticed that the binding energy of the nuclei increases linearly with the atomic weight, that is to say $E_{\rm bind} \sim A$. Consequently, argued Gamow, any equilibrium calculation would invariably lead to an exponential decrease of abundances. On the other hand, observations show that the exponential decrease takes place only through the first half of the elements, whereas the abundances of the heavier elements remain nearly constant. It was this simple and beautiful physical reasoning that led Gamow to conclude that the elements could not have formed in an equilibrium scenario.

The initial idea was therefore that the Universe started from some substance that was extremely dense and hot. Lemaitre called it 'the primeval atom'.

Gamow was apparently prejudiced by previous calculations, which assumed that the elements were produced at a temperature of 10^{10} K and a density of 10^6 g/cm^3. Consequently, Gamow's second main point was that:

> The conditions necessary for rapid nuclear reaction were existing only for a very short time so that it may be quite dangerous to speak about an equilibrium state which must have been established during this period. [Because] at the epoch when the mean density of the Universe was of the order of 10^6 g/cm^3, the expansion must have been proceeding at such a high rate that this high density was reduced by an order of magnitude in only about one second.

The Universe expanded so fast that there was hardly time to establish any kind of equilibrium, like the one considered by Chandrasekhar and Henrich.

The nice fact about equilibrium is that the initial state of the matter has no effect on the final result: the outcome depends only on the temperature and density. But, if equilibrium does not hold, one has to calculate the rate of all reactions and assume an initial state. Gamow pointed out that the time in the Universe when the conditions were appropriate for the nuclear reactions to take place was shorter than the lifetime of the neutron, which he *estimated* as one hour.[83] Hence, the initial matter, the matter that Lemaitre and Eddington talked about, had to have been very neutron rich. Accordingly, Gamow postulated that the heavy elements were formed by successive neutron captures, so that the Coulomb barrier was not a problem:

> We can anticipate that neutrons forming this comparatively cold cloud were gradually coagulating into larger and larger neutral complexes which later turned into various atomic species by subsequent processes of β emission.

As there is no nucleus composed of two neutrons, the process had to wait for the first neutrons to decay to protons, whence the typical time scale was, according to Gamow, onehour.

[83] Gamow knew that the neutron is more massive than the proton, and hence when left alone it decays to a proton. This was before an accurate half-life was known for the neutron.

7.14 A Rival Theory: The Elements Form in Stars

Almost in parallel, Hoyle,[84] who never accepted the idea of element synthesis prior to the birth of the first stars, adopted the general ideas of Chandrasekhar and Henrich, but applied them to stars. He postulated that, before the existence of stars, the Universe contained only hydrogen. In other words, the first stars were composed of pure hydrogen. At that time it was not known that it is quite difficult to form stars from pure hydrogen. When an interstellar gas collapses and increases its density from about 10^{-24} g/cm^3 to about 1 g/cm^3, the mean density of the Sun, it heats enormously, and if it cannot cool, the gas pressure will increase due to the heating, and very soon this rising pressure will prevent continuation of the collapse. Hence, the ability to cool is a prerequisite for a continuous collapse of the cloud to form a star (recall that the density must increase by 24 orders of magnitude). The cooling of the gas occurs mainly via the heavy elements that radiate through emission in lines, and if there are no heavy elements, the gas has difficulty releasing heat into space. All this implies that pure hydrogen stars have difficulty in forming.

By 1946, the conversion of hydrogen to helium in main sequence stars was known, but how helium converts into carbon was still a conundrum. There was no known way whereby hydrogen could synthesize into helium without the presence of carbon and nitrogen. On the other hand, the synthesis of the elements assuming equilibrium required temperatures in excess of 4 billion degrees, and the question was: how does a star move from hydrogen burning temperatures (a few times 10^7 K) to the high temperatures at which equilibrium can be assumed.

Hoyle separated the discussion into a region where helium is in equilibrium and a region where all the heavy elements are in equilibrium. However, the numbers were such that the helium equilibrium region required higher temperatures than the heavy element equilibrium (because helium is more tightly bound than the heavy elements) and consequently Hoyle's star had to meander in temperature. Since Hoyle assumed that helium conversion to heavy elements was a process in equilibrium, he thereby circumvented the problem that there is no stable nucleus with $A = 5$, the problem which prevented direct synthesis of hydrogen from continuing beyond helium.[85]

Hoyle considered two possibilities: either the temperature rises sufficiently high for the nuclear reactions to become important before the onset of rotational instability (a temperature of about 4×10^9 K is required), or a state of instability is reached before nuclear reactions become important. As the rotating star contracts, it rotates faster. The contraction can go on only until the centrifugal force at the equator overcomes the gravitational force. Hoyle considered the equality of the forces as an approximation to the point of instability. Once the instability sets in, the synthesized elements are ejected into space. The process should continue until all the mass above

[84] Hoyle, F., MNRAS **106**, 343 (1946).

[85] One can write the equation for the equilibrium $4\,^4\text{He} \rightleftharpoons {}^{16}\text{O}$, for example, and in this way transform helium into oxygen and even gain energy. But is the assumption of equilibrium justified? In this way one could avoid the barrier of the non-existence of a stable nucleus with $A = 5$ or $A = 8$ by 'letting statistical mechanics do the job'.

the Chandrasekhar limit is removed. Hoyle did not explain why the process of mass ejection should stop at the Chandrasekhar mass, rather than before or after. We can only guess that he followed Eddington in his antipathy for believing in the existence of black holes (note that the term black hole did not exist at that time, but the notion of a collapse to a singularity was known from the work of Oppenheimer and Snyder[86]).

So according to Hoyle, only after reducing the mass to below the Chandrasekhar limit was the star allowed to cool down and stop the endless run to indefinitely high temperatures and densities. But as the star collapses, we face the same problem as in star formation, namely, collapse cannot proceed without cooling. Hoyle brilliantly calculated that, during the collapse, the internal energy of the gas would be converted into the binding energy of helium. In other words, the difference in the binding energies between the heavy elements and helium is contributed by the kinetic energy of the hot particles. This is in effect almost, but not quite, the first discovery of the trigger which induces the collapse of the star to form a supernova, but it was not yet recognized as such.

Hoyle identified the collapse and the ejection of the heavy elements with a supernova. During the gigantic explosion, matter cools and nuclear reactions cease. The composition is then roughly the one which corresponds to the lowest temperature and density at which reactions still had time to operate. By comparing the energies in the initial and final states, Hoyle found a good agreement with the observed energy in SN 1054 (the Crab Nebula) as reported by Baade[87] and Minkowski.[88]

7.15 Statistical Equilibrium at Zero Temperature

In 1947, van Albada[89] took the idea of statistical equilibrium one step further and proposed to explain the formation of the heavy elements assuming equilibrium of the nuclei. In an attempt to simplify the calculations, he assumed zero temperature. In this case, the equilibrium really means that the only nucleus present is the one for which the total energy per unit mass is minimum. What happens is that, as one increases the density, the electrons, which obey the Fermi–Dirac gas equation, increase their energy due to the compression, which decreases their 'living space'. When the energy of the electrons increases beyond the energies released in β-decays, they penetrate into the nuclei, inducing the conversion of protons into neutrons.

Consider a simple β-decay, in which a neutron is converted into a proton and an electron but the energetic electron leaves the nucleus (despite the attraction between the electron and the remaining protons):

$$^{A}X_{Z-1} \longrightarrow {}^{A}X_{Z} + e^{-} + Q, \qquad \text{low density (laboratory)},$$

[86] Oppenheimer, J.R. and Snyder, H., Phys. Rev. **56**, 455 (1939).

[87] Baade, W., Astrophys. J. **96**, 188 (1942).

[88] Minkowski, R., Astrophys. J. **96**, 199 (1942).

[89] van Albada, G.B., BAN **10**, 394 (1946); Astrophys. J. **105**, 393 (1947).

Table 7.1 The dependence of the most abundant element on density, first found by Albada (1946)

Density (g/cm^3)	5×10^6	4×10^{10}	10^{11}	2×10^{11}
A	60	100	140	180
Z	27	37	45	52

where Q is the total kinetic energy of the particles and ${}^AX_Z$ is any nucleus with atomic weight A and atomic number Z. The equation is written here the way van Albada wrote it, namely without the (anti) neutrinos which should appear on the right-hand side of the equation. These were the days before the existence of the neutrino was confirmed, and many physicists remained skeptical. Under laboratory conditions, namely, low density, space does not contain many electrons, so there are a lot of vacant places for the electron to go into.

However, when the density is very high, all possible locations for the electrons are occupied, and as the density rises, so does the energy of the electrons outside the nuclei. At some point, the energy of the most energetic electron reaches the threshold energy Q to drive the reaction in the other direction, viz.,

$$e^- + {}^AX_Z + Q \longrightarrow {}^AX_{Z-1} \quad \text{high density (very dense stellar cores).}$$

This is called the inverse β process. What van Albada realized, therefore, was that, as the density increases, the electrons are driven into the nuclei, thereby reducing the number of protons and increasing the number of neutrons. The increase in density causes 'neutronization' of the matter, whence van Albada rediscovered what Sterne[90] and Hund[91] had predicted over a decade earlier.

As an example, van Albada found that the atomic weight of the most abundant nucleus increases with density, as can be seen from Table 7.1. But can we increase the density and obtain nuclei heavier than uranium? Van Albada showed that the limit was $A = 250$, and that, as the density continues to increase, the nuclei disintegrate into neutrons. The competition between the nuclear and Coulomb forces does not permit the existence of nuclei heavier than $A = 250$, irrespective of what the density is. The implications are that, as the temperature or density (or of course both) become extreme, matter first converts into a neutron-rich kind of matter which later disintegrates into neutrons and protons. The nuclei cannot survive these extreme conditions. They can exist only as long as the environment is not too energetic. When the energy of the individual particles is of the order of the binding energy of the nuclei, collisions cause a complete break-up of the nuclei. The heavy elements can be formed and survive only within a relatively small range of temperatures and densities.

Where are the elements formed according to van Albada? While in the first paper he reconsidered the original idea of Chandrasekhar and Henrich and required enormous densities of 10^{12} g/cm^3 and relatively low temperatures (to prevent the destruction of the formed elements), in the second paper van Albada considered a

---

[90] Sterne, T.E., MNRAS **93**, 736 (1933).

[91] Hund, F., Ergeb. Exakten Naturwiss. **15**, 189 (1936).

massive rotating star which was too massive to cool and become a white dwarf. As the star cools and contracts, it rotates faster and faster near the equator until it starts to shed its equatorial layers into space. These are not the layers enriched with heavy elements. The remaining core becomes unstable (although it is not clear why) and explodes (so van Albada hypothesized), scattering all the freshly produced heavy elements into space:

> When the central region, where the heavy elements are formed, becomes too large, the star reaches instability. A slight increase in the external pressure will cause the nuclei to expel electrons, and if the weight of the outer layers is not sufficient, the process will give rise to a violent explosion. In this way part of the heavy elements formed in the interior of the star will be distributed over space.

The fast removal of the elements from the star, i.e., by an explosion, is needed in order to freeze the composition. There is no calculation to make this scenario explicit and van Albada did not explain why the process of expelling electrons would give rise to an explosion. However, he should be credited with the idea that there can be a mini-explosion. One does not need a big explosion to form the elements: smaller explosions would do it just as well.

By then, 1947, the criticisms of equilibrium calculations by Wataghin and Gamow's idea of cosmic explosion had been published, but they were not cited by van Albada (Hoyle was not mentioned either). Lastly, when an instability occurs, the star collapses. In an explosion, all layers move outward. Neither Hoyle nor van Albada explained how the inward collapse converts into an outward flow, namely, an explosion. This is the really big problem when attempting to explain a supernova.

7.16 The αβγ Paper

The first ever calculation of Gamow's idea of nucleosynthesis during the initial expansion of the Universe was carried out in 1948 by Alpher (1921–2007), Bethe, and Gamow.[92] These were the good old days when the Bureau of Ordnance, U.S. Navy, supported such far-reaching and imaginative research (Fig. 7.8).

Since the calculation was a rate calculation, they had to assume an initial state, and this was a highly compressed neutron gas, the so called 'overheated neutral nuclear fluid'. The state of the Universe before it reached this state, and indeed the way it reached it, were of no concern to Gamow and his associates. Once it had reached this initial state, the entire Universe expanded together with the gas, and the neutrons decayed into protons and electrons. Some of the newly formed protons captured some of the remaining neutrons to form deuterium. Once deuterium had formed, it captured some of the remaining neutrons to build up heavier and heavier elements. The building-up process had to compete with the expansion. For a slower expansion,

---

[92] Alpher, R.A., Bethe, H., and Gamow, G., PRL **73**, 803 (1948), submitted 18 February, published 1 April. Note the unique date of publication which was not to Alpher's liking. Amazingly, the

Fig. 7.8 A 1949 composite picture with Robert Herman on the *left*, Ralph Alpher on the *right*, and George Gamow in the *center*, like the genie coming out of the bottle of 'Ylem', the initial cosmic mixture of protons, neutrons, and electrons from which the elements were supposedly formed. Copyright uncertain

higher A elements are built (higher abundance) and vice versa. It is clear that the build-up of the elements is linked to the expansion rate of the Universe.

Another issue is the β-decays. Naturally, when flooded with neutrons, all nuclei absorb copious amounts of neutrons and build up neutron-rich nuclei which later disintegrate in a multitude of decays.[93]

(Footnote 92 continued)
review by Alpher and Herman, Rev. Phys. **22**, 153 (1950), entitled *Theory of the origin and relative abundance distribution of the elements*, was also published on 1 April. In the book *The Creation of the Universe*, New York, Viking Press, 1961, rev. edn., Gamow explained how Bethe's name got into the paper. The paper was sent to Bethe with the names of Alpher and Gamow. Bethe's name (in absentia) was added during the preparation for print. Bethe did not object and later joined in writing the discussions. However, the editor was not exultant about Gamow's joke, and removed the words 'in absentia'. Gamow's typical and incessant humor dubbed this the αβγ paper, and it was this joke that dictated the order of the authors. Several years later, when the theory encountered difficulties, rumor had it that Bethe had contemplated changing his name to Zacharais.

Recently, Alpher told the following version of the story (Alpher, J., *Gamow Symposium* ASPC **129**, 127, 1997). Bethe told Alpher that he did not mind having his name added to the paper since he thought that the idea might even be right. So they added 'in absentia'. But the editor of the journal removed it. Gamow told Alpher that he, Gamow, had something on Bethe, so that he could not refuse the offer to join as author. The secret was that Bethe, Beck, and Riezler had published a letter to the editor of Naturewissenschaften which was a beautiful spoof of Eddington's *Fundamental Theory*. The editor had missed the point that the letter was a pure parody and published it. Soon afterwards, Bethe got an invitation to present colloquium talks on the subject. Consequently, the journal added to the master head for Letters that the Editor was not responsible for the content of the published letters. But there was no need to make use of this secret.

[93] A simple observation of the nuclei shows that, for low A, the nuclei have roughly $N \approx Z$, where N is the number of neutrons. For very heavy nuclei, $N \approx 2Z$, and there are no nuclei with $N \gg Z$. Hence, if such nuclei form during massive irradiation with neutrons, they decay soon after into more stable nuclei in which the above relation between N and Z is satisfied. Nuclei with $Z \leq N \leq 2Z$ are in the 'valley of stability' of the nuclei.

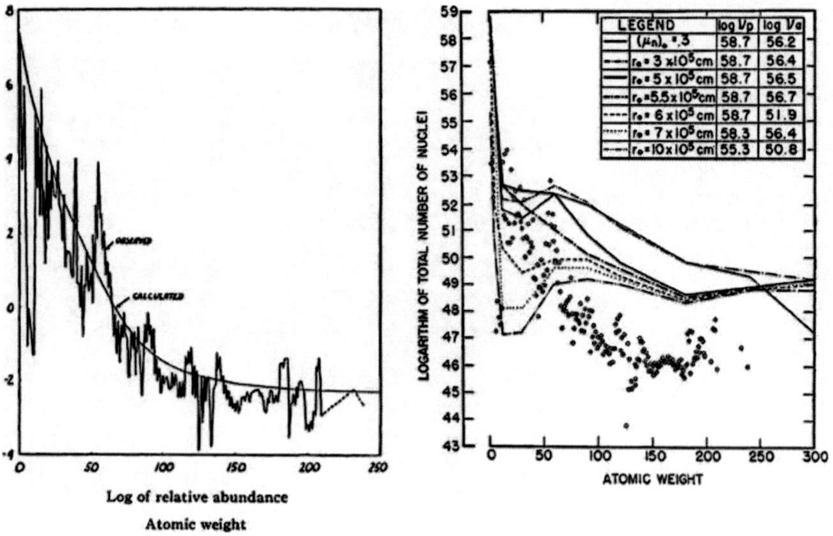

Fig. 7.9 *Left* comparison between the first αβγ nucleosynthesis calculation and the observed abundances. From Alpher et al. (1948). Note the discrepancy in the abundances of the light elements. *Right* the failure of the equilibrium calculation in predicting the abundances of the heavy elements. From Beskow and Treffenberg (1947). *Lines* are predictions for various sets of parameters. *Dots* are observed abundances

The initial matter was a neutron gas, so the probabilities for neutron absorption (in physicists' language, the cross-sections for absorption/capture of neutrons) by all existing nuclei were needed. Such information was unavailable, although it was essential for designing atomic bombs. The 272nd meeting of the American Physical Society was held in Chicago in June 1946. This conference will long be remembered as the great nuclear physics meeting at which a large amount of the work done as part of the famous Manhattan Project was finally brought to light. The American clearance authorities were generous, and 97 papers from the American atomic bomb project were declassified, including some neutron absorption data. Among the invited speakers, we may mention in particular Hughes[94] from the Argonne National Laboratory, who supplied ample neutron absorption data. There were about 43 speakers who presented various pieces of data measured for the Manhattan Project. It was this data that allowed Alpher, Herman, and Gamow to carry out their calculations.

Despite all the uncertainties in the rates of the various nuclear reactions, the results were quite impressive, as can be seen from Fig. 7.9 (left). To obtain this fit to the abundance curve, they had to assume a value for the time integral of the neutron density, which is essentially proportional to the number of neutrons available per unit mass, or in short, the number of neutrons there are available for absorption. Since the initial mass was entirely in the form of neutrons, this value implied that

[94] Hughes, D.J., Phys. Rev. **70**, 106 (1946).

the process started some 20 s after the start of the expansion and continued for no more than 670 s, implying that all the elements were synthesized in just 11 min. The theoretical slope changes from a fast exponential decrease for $A \leq 100$ to a flatter curve for $A > 100$, more or less as the observations go and as predicted by Gamow.

However, the light elements (with $A = 5$–10) showed a marked discrepancy, amounting to a factor of about 10^7, between the observed and the calculated. The total number of neutrons at $t = 1$ s was 10^{21} cm^{-3}, and by the time the synthesis ended, the matter density was $\rho_m = 10^{-6}$ g/cm^3, while the equivalent mass of the radiation was $\rho_{rad} = 1$ g/cm^3 and the temperature 0.6 billion K. Until that time, the Universe was 'radiation dominated', in the sense that by far the most mass/energy in the Universe was in the form of radiation, with the elements contributing a trifling amount.

The amazing comparison with observed cosmic abundances was very reassuring, suggesting that the theory was on the right track. The conclusions were far-reaching. They implied that all stars were formed from the same initial composition. The entire Universe had the same composition. And there is a further important point: the chemical elements which form the Earth need not be formed in stars, removed from the stars, spread all over the Galaxy, and then condensed to form planets like the Earth. The theory was very simple.

The αβγ paper summarized the idea and presented the graph shown in Fig. 7.9 (left). No additional details were given and the reader was referred to a forthcoming paper by Alpher. This was the only paper on the subject which carried Bethe's name. Bethe was never again involved with early Universe element synthesis.

The apparent success of the αβγ theory allowed Gamow[95] to invert the problem, using the nice agreement with observation to infer the conditions in the expanding Universe. In doing so Gamow corrected an error of a factor of 10^8 in the parameters of the expansion.

The publication of the short Physical Review Letter was accompanied with unusual publicity. The editor of Science Service wrote a news column and Alpher's doctorate thesis[96] was presented to the public through the statement that *the Universe was created in about 5 min*, an estimate based on the neutron half-life. The public demonstrated insatiable thirst for such imaginative science, to the point that about 200 journalists and other people, crowded the hall in which Alpher defended his thesis. In reply to a question from the examiners, Alpher remarked that the nucleosynthesis took 300 s. The idea that primordial nucleosynthesis took just 5 min caught the headlines and the imagination of the public. As the reader can imagine, the Sunday supplements of all major newspapers included detailed stories, along with cartoons and an avalanche of mail from religious fundamentalists.

---

[95] Gamow, G., PRL **74**, 505 (June 1948); Rev. Mod. Phys. **21**, 367 (1949).

[96] The creation of the elements at the beginning of the Universe was the second problem Gamow suggested to Alpher. The first problem was the possible growth of condensations in an expanding Universe. However, Yevgeny Lifshitz had just solved this problem, and the resulting paper reached Gamow only after Alpher had already solved the problem, getting identical results. Alpher had to choose a new subject and Gamow suggested the creation of the elements, in the thesis which became so famous.

Gamow,[97] who was excited about the model and its implications, used the obsolete term 'Ylem' to describe the primordial matter out of which all elements formed, namely, the neutron-rich matter. Gamow added a note that[98]:

> According to Webster's New International Dictionary, second edition, the word 'Ylem' is an obsolete noun meaning 'the primordial substance from which the elements were formed'.

7.17 An Opposing Theory: The Cosmological Steady State

In the same year that the $\alpha\beta\gamma$ theory was announced with plenitude of fanfare, in 1948, Bondi (1919–2005) and Gold (1920–2004)[99] came up with the so called steady-state theory of the expanding Universe. To explain how the Universe could expand and yet still appear the same at all times, they postulated the *perfect cosmological principle*, which states that:

> The Universe is homogeneous and stationary in its large scale appearance as well as in its physical laws.

As they claimed:

> It is clear that an expanding universe can only be stationary if matter is continuously created within it.

Bondi and Gold argued that experiments in physics must be reproducible. The Universe around us appears to change continuously so that the environment around us does not repeat. Does the density of the Universe affect the results of laboratory experiments? If so, and we want to keep the basic laws of physics, then the perfect cosmological principle follows. But is it so obvious that the mean density in the Universe does or does not affect the results of experiments? Recall the Mach principle,[100] which states that the inertia of a piece of matter is determined by the entire distribution of matter in the Universe.

Bondi and Gold also investigated the physics of mass creation, while all previous suggestions only alluded to this question. First, how much mass must be created to preserve the appearance of the Universe? The calculation is simple and the result is approximately 10^{-43} g/s in every cubic centimeter of the Universe. This rate corresponds to one hydrogen atom per cubic meter per 3×10^5 years,[101] a rate which is

---

[97] Gamow, G., Rev. Mod. Phys. **21**, 367 (1949).

[98] Gamow, G., *The Creation of the Universe*, Courier Dover, 2004.

[99] Bondi, H., and Gold, T., MNRAS **108**, 252 (1948).

[100] 'Mach's principle', a term coined by Einstein in 1918, states that the source of all inertia lies in the effect of the stars around us. This principle is said to have played an important role in shaping Einstein's general theory of relativity, although it turns out that the final theory does not satisfy the principle. It is not clear what Einstein understood by 'Mach's principle'. And if we examine Mach's own discussion, there is cause for confusion, since there are several candidates for this name. In the case of the steady state theory, it is the idea that the the distant stars are the source of inertia and thereby affect local physics.

[101] The rate of particle creations in the steady state model is given by $3nH$, where n is the observed present day number density and H is the Hubble constant.

many orders of magnitude less than we could detect in the laboratory. And where does the creation of matter take place? Various considerations led the authors to assume that it must take place *in the intergalactic spaces*.

Next came the question of the temperature at which the new matter would be created, and the amount of energy it would have? By considering the amount of radiation coming from intergalactic space (actually none was observed), Bondi and Gold concluded that the temperature of the newly created atoms must be less than 50,000 K.

And is the new matter charged? Are only charged particles created? Again the authors concluded that, if there is any average charge excess of the created matter, the average charge–mass ratio must not exceed a very low limit. However, if a neutral particle is created, can it be a hydrogen atom? That would mean creating a proton and an electron, i.e., two particles. Or does it mean that a neutron is created? If a neutron is created, for example, we should be able to observe the radiation emitted during the decay of the newly formed neutron into a proton and an electron. If the proton and the electron are created separately, then we should observe the radiation emitted when they combine to form neutral hydrogen atoms. After considering the various possibilities, Bondi and Gold reached the conclusion that the creation of a hydrogen atom can be most easily concealed in space because it would emit no radiation, so long as it is created in the ground state. (But if the temperature is anywhere close to 50,000 K, most of the hydrogen atoms would be ionized.) An important conclusion of this discussion was that the newly created matter would be no more massive than hydrogen, and definitely not atoms like helium or more massive than that.

Several scientists contributed to the steady state theory: Bondi and Gold supplied the basic idea and reasoning. Hoyle[102] provided the mathematical formalism and McCrea (1904–1999)[103] proved that the notion of continuous creation was consistent with Einstein's general theory of relativity.[104]

7.18 Inventive but Not Really New

The nature of the expansion of the Universe perplexed researchers, despite the fact that the general theory of relativity predicted no static solution. In 1935, Hubble and Tolman[105] wrote a thorough analysis, aiming to distinguish between the explanation that the Universe expands in the sense that the nebulae move apart with a velocity proportional to their distance, and other alternative explanations. They claimed that:

> If the redshift is not due to recessional motion, its explanation will probably involve some quite new physical principles.

---

[102] Hoyle, F., Nature **163**, 166 (1948); Astrophys. J. **108**, 372 (1948).

[103] McCrea, W.H., Proc. Roy. Soc. **206**, 562 (1951).

[104] A detailed exposition of this interesting modification of the steady state theory is beyond the scope of this book. Hence, we only mention that this formalism, introduced by Hoyle and called the C-field, yields the cosmological constant which was introduced by Einstein.

[105] Hubble, E., and Tolman, R.C., Astrophys. J. **82**, 302 (1935).

Considering recent discoveries in cosmology, it is not clear that their conclusion holds today, since even with the currently accepted assumption that the redshift implies the expansion of the Universe, we still need new physics because of the problem of dark matter.

The idea of mass creation, or alternatively the idea that new mass is poured into our Universe from other unobserved universes, was not new in 1948, when it was put forward by Bondi and Gold. Several scientists were baffled by these astrophysical phenomena and suggested that the law of mass conservation might be violated on cosmic scales.

In 1928, Jeans wrote[106]:

> The type of conjecture which presents itself, somewhat insistently, is that the centers of the nebulae are of the nature of singular points, at which matter is poured into our Universe from some other and entirely extraneous spatial dimension, so that, to a denizen of our Universe, they appear as points at which matter is being continually created.[107]

Years later, Hoyle, Fowler, and the Burbidges[108] applied an almost identical idea to explain the energy output of extremely energetic radio sources. This idea due to Jeans was put forward about a year before Hubble's discovery of the expansion of the Universe. A similar idea was suggested by Nèeman.[109]

More ideas were put forward to 'stop the expansion of the Universe', but to no avail. Sambursky[110] assumed, for example, that the Universe was static, but that the Planck constant $\hbar$ decreased with time. This assumption easily explains the redshift and is equivalent to an apparent expanding Universe with invariable Planck constant. Sambursky even estimated the required rate of change of the Planck constant that would be needed to explain Hubble's discovery.

Nernst claimed in 1937[111] that, for thermodynamic reasons which go beyond our scope here, the Universe must be static. To avoid 'heat death', Nernst postulated continuous creation of energy as a solution. Nernst was completely ignored by the British school, which advocated mass creation. No physics of mass creation was considered, let alone what kind of mass or atoms were supposedly formed.

Jordan[112] also suggested mass creation as a solution to the expanding Universe. The title of Jordan's 1938 paper referred to empirical rather than theoretical cosmology. Yet it was a pure speculation about the behavior of the constants of physics. The basic theory resembled Dirac's (see Chap. 4), in the sense that one of the constants of nature must be assumed to vary in time.

---

106 Jeans, J.H., *Astronomy and Cosmogony* (1928), p. 352.

107 Today we believe, on the contrary, that the large galaxies have a black hole at the center, which swallows matter falling onto it.

108 Hoyle, F., Fowler, W.A., Burbidge, G.R. and Burbidge, E.M., Astrophys. J. **139**, 909 (1964).

109 Nèeman, Y., Astrophys. J. **141**, 1303 (1965).

110 Sambursky, S., Phys. Rev. **52**, 335 (1937).

111 Nernst, W., Zeit. fur Physik **106**, 663 (1937).

112 Jordan, P., Naturewiss. **25**, 513 (1937); ibid. **26**, 27 (1938).

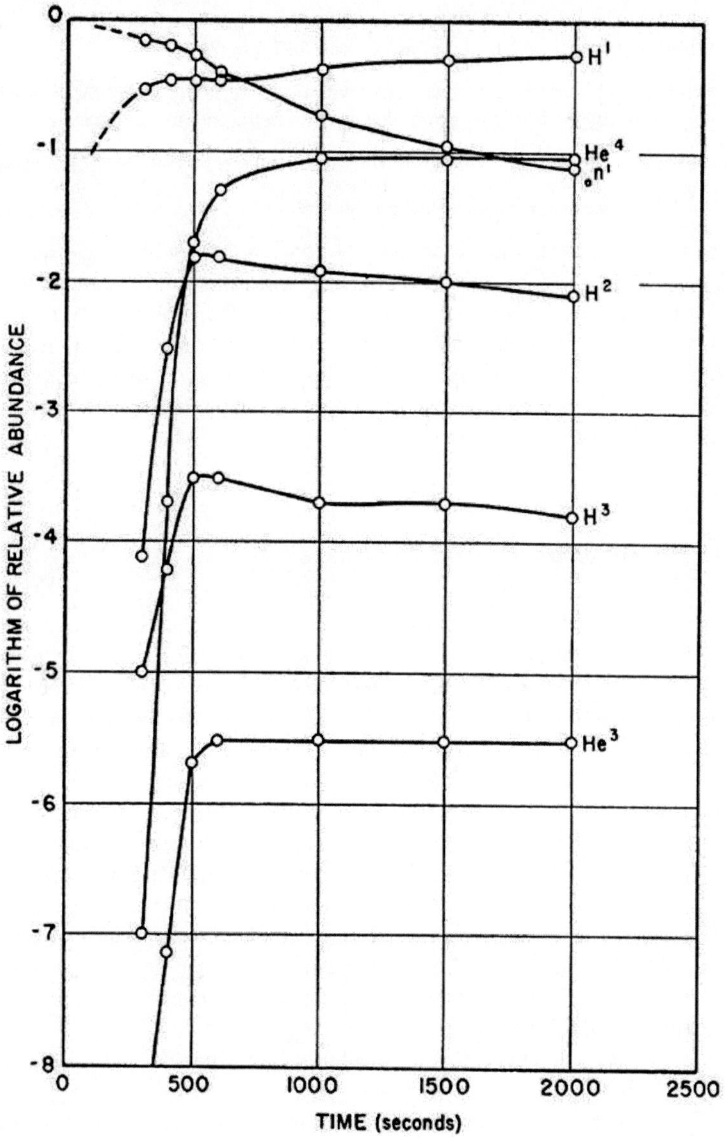

Fig. 7.10 Abundances calculated by Fermi and Turkevich, as reported by Alpher and Herman (1950). The calculation started with a neutron concentration of 10^{21} cm^{-3}, or about 1.7×10^{-3} g/cm^3 at 1 s

In 1937, Tolman reviewed the state of cosmology. Among the possibilities he mentioned as alternatives was Milne's explanation that the nebular recession was a kinematic problem that did not involve any gravitation, or theory of gravitation. Tolman cited Robertson, who had claimed that Milne's theory:

> [...] merely leads to a treatment which agrees with that of relativity except for the greater flexibility but decreased predictive power that result from the omission of dynamics.

Next he mentioned Zwicky's tired light theory,[113] which he described by saying:

> We must be very open-minded as to such possibilities.

However, none of these authors discussed the form of the newly created mass.

One may be puzzled by the need for such a drastic supposition as the non-conservation of mass. Recall that, when faced with the problem of energy conservation in β-decay, the idea of energy (and hence mass) non-conservation had been invoked by Bohr back in the early 1930s. In this context then, energy/mass creation is not such a wild hypothesis, provided it helps to solve a problem. The important fact is that the theory is testable and refutable.

7.19 Problems with the αβγ Theory

As discussed above, the entire theory hinged on neutron absorption data, and fortunately these were declassified in the celebrated 1946 meeting of the American Physical Society in Chicago. The data measured in the laboratory had to be adjusted to the special conditions in the Universe, but this was a relatively small problem.[114] Since not all the required numbers were known and the data varied immensely from one nucleus to the next, Alpher and Herman chose to fit the data with a simple formula and use it in their calculations (see Fig. 7.10). The basic problem they faced was that the data used for the calculations were for heavy nuclei and not for nuclei with $A < 25$. The actual probabilities used for these nuclei were therefore extrapolations to low A by means of a simple fitting formula. What Alpher and Herman apparently did not know was that the fitting formula's predictions for neutron absorption by light nuclei were way off the mark. In particular, the predicted probability for neutron absorption for the $A = 4$ nucleus was very high, whereas it should in reality be vanishingly small, if not zero.

Figure 7.11 shows the original data of Hughes (left), together with the fit (right). Note the lack of data for $A < 20$. Moreover, in 1949, Ross and Story from the Atomic Energy Research Establishment in Harwell (UK) published[115] data under

---

[113] Zwicky, F., PNAS **15**, 773 (1929). The idea was that the photon becomes tired on its way to us, loses energy, and is thus redshifted. This is not the gravitational redshift experienced by a photon coming out of a gravitational potential well. The term 'tired light' was coined by Tolman in 1931. For a review of its status today, see Moore, M., Dunning-Davies, J., arXiv:0707.3351.

[114] The data were averaged over the spectrum of neutrons emitted in fission of uranium in nuclear reactors. The data needed for the astrophysical calculations were averages over the Maxwellian velocity distribution of the nuclei.

[115] Ross, M. and Story, J.S., Rep. Prog. Phys. **12**, 291 (1949).

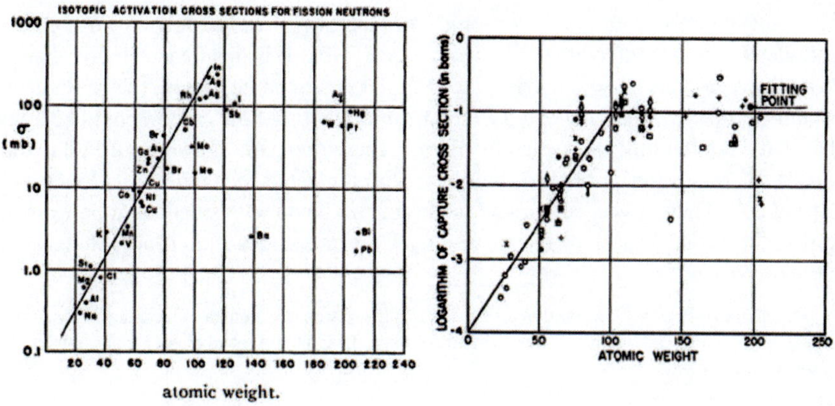

Fig. 7.11 *Left* the data for neutron capture by nuclei. Fit by Hughes et al. (1948). What is shown is the cross-section (called the probability in the text). *Right* the data for neutron capture by nuclei showing the continuation of the fit to $A = 1$. Alpher and Herman (1950)

the title *Slow neutron absorption cross-section of the elements*, and included data on elements with $A < 20$ at more relevant energies than the data published by the Americans. This publication escaped the notice of Alpher and Herman.

Soon after the publication of the αβγ paper, Fermi and Turkevich (1916–2002) repeated the calculations. The results were never published, but they were referred to by Alpher and Herman[116] as Fermi and Turkevich 'private communication'. Fermi and Turkevich's results are shown in Fig. 7.10. In contrast to Alpher and Herman, Fermi and Turkevich only tried to solve for the abundances of the very light elements, and did not attempt to derive them for all the elements. More importantly, Fermi and Turkevich used the actual data for the neutron absorption coefficients, instead of a rather poor fitting formula.[117] We say 'poor' because the neutron absorption coefficients change significantly from one nucleus to the next, and this belies any simple fitting formula, as can be seen from Fig. 7.11 (left). Moreover, it became clear that the neutron absorption probability increases exponentially until $A = 100$ and then becomes essentially constant.[118] What can also be seen is that Gamow's intuition about the general behavior of the abundances was correct. Thus, the prediction of the heavy A part of the abundance is generally correct.

[116] Alpher, R.A., and Herman, R.C., Rev. Mod. Phys. **22**, 153 (1950).

[117] Fermi built the first nuclear reactor in 1942, known as the Chicago Pile #1, and succeeded in running a controlled self-sustained chain reaction. He thus knew all about neutron absorption in nuclei. Turkevich was a chemist who served as Fermi's assistant during the Manhattan Project. He later used radiochemical methods to investigate the composition of the Moon.

[118] Another problem was that Hughes et al.'s data were cross-sections for 1 MeV neutrons, which are called fast neutrons. On the other hand, what Alpher et al. needed was the cross-section for thermal neutrons. The probability of interaction of a neutron rises at low energies, whence the average of a thermal distribution has a significant effect on the total absorption.

Fermi and Turkevich soon discovered a fact that was already known from attempts to understand element synthesis in stars, namely, that it is practically impossible to cross the $A = 5$ and $A = 8$ barriers (see Fig. 7.12). For some reason, this previously established fact, which prevented Bethe from predicting how carbon could be synthesized from helium, was called the Fermi–Turkevich gap. The work of Fermi and Turkevich was never published, but they communicated the results to Alpher and Herman.[119] Thus, it was clear rather soon that no heavy elements could be formed during the initial phases of the expanding Universe. Indeed, in 1950, Alpher and Herman wrote[120]:

> The major difficulty faced by this theory is the non-existence of nuclei at atomic weights 5 and 8, with the consequence that a formation chain through the lightest elements has not yet been constructed.

Regarding the finer details, Alpher and Herman wrote:

> There is at present no quantitative demonstration that such a detailed feature as the iron peak is consistent with the non-equilibrium theory.

The suggestion by Fermi and Turkevich for overcoming the gap, creating elements heavier than ^{4}He, and saving the $\alpha\beta\gamma$ theory went via the reactions

$$^4\mathrm{He} + {}^2\mathrm{H} \longrightarrow {}^6\mathrm{Li} + \gamma, \qquad {}^3\mathrm{H} + {}^3\mathrm{H} \longrightarrow {}^4\mathrm{He} + 2\mathrm{n}, \qquad {}^3\mathrm{H} + {}^4\mathrm{He} \longrightarrow {}^7\mathrm{Li} + \gamma\,.$$

Gamow criticized this idea because, even with this reaction, the amount of heavy elements obtained was too small by a factor of a million. Fermi and Turkevich considered the possibility that there might be a resonance in the last reaction that could create significant amounts of ^{7}Li. The analysis showed that a resonance at an energy of 400 keV would do the job and save the $\alpha\beta\gamma$ theory. But the nearest resonance at the time was at 4 MeV, much too high to have any effect. This time the astrophysical reasoning for having a nuclear resonance at a predicted location did not work. Another way to overcome the *mass 5 crevasse* was by changing the initial conditions and assuming that, at the beginning, a significant fraction of the Ylem was protons. Gamow had no better suggestion.

A few years later, in 1954, Hoyle[121] pointed out that the difficulty the $\alpha\beta\gamma$ theory had in getting past the $A = 8$ nucleus would not exist in stars because of the much higher densities occurring there.[122] It is this reaction that distinguishes between the synthesis in the Big Bang and stellar production. This is the bottleneck which allows the synthesis of the heavy elements in stars, but not in the Big Bang. For details see Chap. 8.

---

[119] In the review, the authors wrote that they got the information as a private communication.

[120] Alpher and Herman, Rev. Mod. Phys. **22**, 153 (1950); ibid. erratum **22**, 406 (1950).

[121] Hoyle, F., Astrophys. J. Suppl. **1**, 121 (1954).

[122] The simple way to overcome the $A = 8$ barrier is through a triple collision of α particles. Such a reaction is extremely sensitive to the density, because three bodies have to collide, not just two. The density in the $\alpha\beta\gamma$ theory was 10^{-7} g/cm^3 some 500 s after the start of the expansion, whereas the density in stars, where this reaction is predicted to occur, is at least 10^5 g/cm^3.

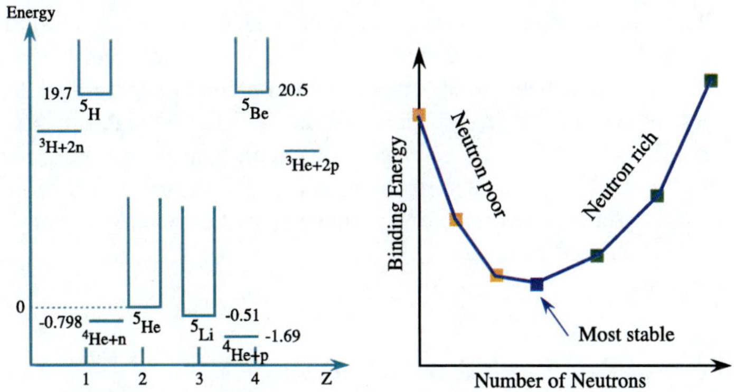

Fig. 7.12 *Left* energy diagram for the $A = 5$ system. The lowest states are either helium with a free neutron or helium with a free proton, but not an $A = 5$ nucleus. There is no stable bound state. The numbers near the chemical symbols are the energies above the ground level energy of ^{5}He. The energies of the free particles are negative and are the lowest. *Right* cut across the valley of stability for a nucleus with some fixed atomic weight Z. As the number of neutrons increases, the nucleus becomes more and more stable, until the most stable nucleus for this Z is reached. As the number of neutrons continues to increase, the nucleus becomes less stable. For every number of protons, there is a preferred number of neutrons which creates the most stable nucleus

And if the inability to cross the $A = 5$ and $A = 8$ barriers was not sufficient, additional problems arose. In 1956, Suess and Urey[123] analyzed the abundance curve and found the following difference between the low and high atomic weight nuclei. In the region $A < 90$, they found that, at mass numbers where two stable isobars exist, the one with the smaller neutron excess has the higher nuclear abundance. The situation reverses in the high A, i.e., nuclei in the range $A > 90$. They concluded, therefore, that for $A > 90$ there must be at least two neutron capture processes. The first reaction must lead to the formation of nuclear species which are neutron rich, while the other must lead to the formation of neutron-deficient nuclei. The assumed neutron capture must be slow enough to ensure that β-unstable nuclides have time to decay before the next neutron is absorbed.

As we know, nuclei cannot have too many neutrons or too few. For a given number of protons, the binding energy is minimal for an optimal number of neutrons. If there are deviations from this optimal (most tightly bound/stable) nucleus, i.e., if there are too few or too many neutrons for a given number of protons, the nuclei are not properly bound/stable. Hence, there is a valley of stability, as shown in Fig. 7.12. The nuclei in the valley of the binding energy are the most stable ones. As one climbs the hill on either side, more and more unstable nuclei are reached. So Suess and Urey drew the conclusion that elements formed on the two sides of the valley of stability must have been formed in two different processes.

In the first process, the rate of neutron captures by the nuclei is extremely high and the nuclei absorb the next neutron before they have time to relax and decay to a more

[123] Suess, H.E., and Urey, H.C., Rev. Mod. Phys. **28**, 53 (1956).

stable nucleus. The second process is the opposite one, i.e., the neutron capture is so slow that the nuclei have ample time to relax between successive captures of neutrons. Hence, it is clear that at best the $\alpha\beta\gamma$ theory would predict only the abundances of the elements formed on the neutron rich side of the valley of stability. The nuclei on the other side of the valley, the neutron-deficient ones, must have been formed in some other process. This was a serious blow to the $\alpha\beta\gamma$ theory, which had claimed exclusivity, in the sense of being able to produce all the elements and their isotopes via a single neutron capture process.

7.20 Why Not Breakup (Top–Down) Rather Than Buildup (Bottom–Up)?

As early as 1934, Lewis (1875–1946m)[124] suggested that:

> […] a great part of the matter in the Universe is composed chiefly of iron and nickel, like the metallic meteors, and such material, which is thermodynamically stable with respect to all spontaneous transmutation.

However, at extremely high temperatures it is bombarded by energetic cosmic rays, so hypothesized Lewis, to produce the material of the Earth's crust. He compared the composition of the two types of meteors, the stony and the metallic, and arrived at the conclusion that they show a genetic relationship.

Lewis was developing the theory of chemical thermodynamics, so he examined the problem as a physical chemist. He essentially discussed equilibrium reactions, although this term did not appear in the paper.[125] Back in 1922, in an address before the Astronomical Society of the Pacific,[126] he examined the temperature at the center of stars. He assumed that the maximum temperature of the Sun is 5×10^6 K, adopting the number from Eddington, and that the maximum temperature in stars is 100×10^6 K. Lewis concluded that such a temperature would be sufficiently high to give rise to equilibrium processes, and in particular that:

> The radioactive nuclei would be forming from substances which under terrestrial conditions are the product of radioactive decay.

---

[124] Lewis, G.N., Phys. Rev. **46**, 897 (1934).

[125] Lewis wrote in 1922:

> The newer methods of thermodynamics enable us to calculate from the heat of a reaction, the temperature at which a certain degree of reversal would be observed, and although it would be unsafe to conclude that methods which were proved satisfactory would retain full validity when the temperature range is 10^8 K, nevertheless, we seem to be justified in accepting in a general way the results of such thermodynamic calculations.

[126] Lewis, G.N., PASP **34**, 309 (1922).

However, the temperature in stars, so he argued, would not be sufficiently high to synthesize most of the radioactive nuclei. As an example he gave radium. Since the stars cannot synthesize the heavy radioactive nuclei, he concluded that:

> We must have recourse to some form of radiation of higher frequency than any radiation known, permeating all space and responsible alike for the formation of the radioactive elements and for the heat of the Sun and stars.

The 1922 paper explains why Lewis came up with the new suggestion that the original elements were iron and nickel, while all the rest were formed by irradiation. Just to give one example, Lewis wrote the following reaction as caused by cosmic radiation:

$$^{30}S_{14} + \text{C.R.} \longrightarrow {}^{26}Mg_{25} + {}^{4}He_{2}.$$

This is what would later be called the α-process. Thus, Lewis wanted to create all the elements via the α-process, starting from iron.

In 1949, Goeppert-Mayer (1906–1972) and Teller (1908–2003)[127] went to the other extreme and suggested that the heavy elements were the result of *the break-up of primordial nuclear fluid*, similar to the fission process. In this picture, the original nucleus is much heavier than uranium. As it breaks, it releases the extra neutrons that fission the residual nuclei, exactly as in a fission bomb, where about 2.5 neutrons are released per fission. The absorption of the neutrons by the debris of the fission gives rise to an abundance distribution that can be compared with the observed one. For this reason, this theory was known as the polyneutron theory. The question as to how the original superheavy nucleus was formed or where it came from was a taboo. The debris, initially rich with neutrons, would let the extra neutrons evaporate from the nuclei until the process stopped. Mayer and Teller tried to work out the stellar conditions (neutron-rich environment like a neutron star) where this process could operate. The idea was essentially Pokrowski's,[128] adapted to the newer nuclear data and the existence of fission.

In the discussion of their theory, Mayer and Teller said this:

> There is conclusive evidence that, at the time of production of the heavy nuclei, the proportion of neutrons considerably exceeded that which is now present in nuclei. Evidence for this neutron excess comes from two sources. First, without such a neutron excess it is not possible to understand that the heavy isotopes of heavy elements are much more abundant than the lightest isotopes. Second, in the absence of a neutron excess, it is very hard to find any method by which the heavy nuclei could be built up at all.

This was the basis of Gamow's theory.

Band[129] extended the Mayer–Teller model to even lighter elements by considering hot neutron stars, in contrast to cold neutron stars as assumed by Mayer and Teller. With hindsight, the fit was amazing and apparently implied that the crucial factor in shaping the abundance curve was rather the nuclear binding energy, and much less

---

127 Mayer, M.G., and Teller, E., Phys. Rev. **76**, 1226 (1949).

128 Pokrowski, G.I., Phys. Zeit. **32**, 374 (1931).

129 Band, W., Phys. Rev. **80**, 813 (1950).

the specific location, process, or initial composition of the matter. According to Band, the primordial neutron-rich fluid first separated into two phases: the first condensed to form stars, which eventually cooled to become the Mayer–Teller polyneutron stars, while the second became the atmosphere of the stars, composed of light elements. Band, as a matter of fact, spoke about neutron stars surrounded by an atmosphere of light elements. The Mayer–Teller theory was checked by Peierls, Singwi, and Wroe,[130] who claimed that:

> At an early stage in the expansion of the Universe, as a homogeneous fluid of nuclear density and low temperature, it is shown that, for reasonable values of the constants, this will, on expansion, leave the matter in the form of droplets of the same properties as those found in the Mayer–Teller 'polyneutron' theory.

The results found were that:

> This model leads necessarily to an abundance curve in which the amount of heavy elements is at least comparable to that of the light elements, contrary to experience.

7.21 The BBC Debate and the Invention of a Sensational Buzzword

At the end of 1949, the BBC radio service organized a transatlantic debate between the proponent of the steady state theory, Hoyle, and the proponent of explosive nucleosynthesis, Gamow. During this debate Hoyle treated the $\alpha\beta\gamma$ theory with contempt, and described it as the 'Big Bang' theory, a name which, to Hoyle's dismay and disappointment, was instantly adopted.[131] Gamow was fast to embrace the new buzzword, and in his next book he referred to:

> [...] the theory of the Expanding Universe, now known as the 'Big Bang' hypothesis.

In a BBC interview in the early 1950s, Hoyle[132] repeated the contagious label the 'Big Bang' with due irony when he referred to explosive nucleosynthesis. The buzzword inflamed the imagination of scientists and non-scientists alike, and has remained with us ever since as the description of the birth of the Universe.

Gamow's feeling of elation can be felt from the motto he chose for his book *The Creation of the Universe*, published in 1952. This motto was taken from Immanuel Kant in *Algemeine Naturegeschichte und Theorie des Himmels*, where he described his theory of the formation of the Solar System: *Give me matter and I will construct a world out of it.*

---

[130] Peierls, R.E., Singwi, K.S., and Wroe, D., Phys. Rev. **87**, 46 (1952).

[131] Gamow, G., *The Creation of the Universe*, Viking Press, 1952. In this book, Gamow used the expression 'the great expansion', and not yet 'the Big Bang'. See also Hoyle, F., *The Nature of the Universe*, a series of broadcast lectures, B. Blackwell, Oxford, 1950.

[132] *The Nature of the Universe*, a series of broadcast lectures by Hoyle, F., B. Blackwell, Oxford, 1950.

While Hoyle's 'steady state' and Lemaitre's 'Big Bang' were the two most popular models used to explain Hubble's observations, other ideas were proposed as well. These alternatives for explaining how the spectral lines can be shifted as a function of the distance included the Milne model,[133] Tolman's oscillatory universe,[134] and Zwicky's tired light[135] hypothesis. However, these ideas did not involve any synthesis of elements, and hence are not discussed here, save to say that they were later proven untenable.

The attempts to find alternative models to avoid the Big Bang, and with it element synthesis, did not die in the 1950s. In 1953, Kapp,[136] who was a professor of electrical engineering but indulged in philosophical astrophysics, summarized the situation as follows:

> The theory of continuous (or spontaneous) origin of matter has been reached by the process of elimination.

Even in 1940,[137] Kapp considered the following possibilities:

1. The matter in the Universe existed for all time.
2. The matter and energy existed for a finite time.
3. The matter and energy are created continuously.

Kapp rejected the first possibility on the grounds that it contradicted the second law of thermodynamics. According to this law, all systems tend towards a limiting condition and must reach it, provided that they have sufficient time. On similar general principles, he reached the conclusion that only continuous creation was an option.

What Kapp actually did was to instate symmetry in time, allowing matter to be created and destroyed and thereby keeping the balance of matter in the world. If this were not the case, argued Kapp, the new matter falling onto galaxies would have conferred upon them an ever-increasing masss, whence the observed masses of galaxies should extend over a much wider range. None of this is regarded as relevant today.

7.22 Hayashi: Fixing the Initial State

In 1950, Hayashi (1920–2010) pointed out that the basic assumption of Gamow and his associates concerning the nature of the Ylem was untenable if the the expansion started from very high temperatures and the physics we know is valid at higher

---

133 The original cosmology paper is: Milne, E.A., MNRAS **94**, 3 (1933); ZA **6**, 1 (1933), but see also Milne, The Obs. **104**, 120 (1944), where Milne lectures on cosmology before the Royal Astronomical Society and expounds his grievances about cosmological models.

134 Tolman, R.C., Phys. Rev. **38**, 797 (1931).

135 Zwicky, F., PNAS **15**, 773 (1929).

136 Kapp, R.O., Obs. **73**, 113 (1953).

137 Kapp, R.O., *Science versus Materialism*, Methuen, 1940, Chap. 24.

temperatures than those assumed for the original explosion. Alpher and Herman discussed temperatures of 4–7 × 10^9 K, whereas Hayashi discussed temperatures of the order of 10^{12} K and below. At such high temperatures one can assume that the neutrons and the protons are in equilibrium, i.e.,

$$n + e^+ \rightleftharpoons p, \qquad p + e^- \rightleftharpoons n.$$

Hence the ratio η = number of neutrons/number of protons is determined largely by the temperature, and is not a free parameter.[138] Consequently, it is not reasonable to start the calculation at a lower temperature with a guess for the ratio η. Let us recapitulate. Gamow objected to the equilibrium assumption and suggested that the rates must be calculated. Hayashi entered the scene and showed that, at the beginning, when the temperature was higher and the expansion faster, an equilibrium necessarily existed before the temperature decreased to such values that equilibrium was no longer valid and the rates had to be included.

Another very important feature of Hayashi's discovery was that the ratio η depends only on cosmological parameters. The state of matter is connected to the state of the Universe, so we can learn about one from the other, and vice versa. In this way, Hayashi established that, at the beginning of element formation, the ratio η was nearly 1:4, and this:

> [...] whatever the physical conditions at higher temperatures, especially at the epoch $t = 0$, when the Universe is singular according to the current theory, may be.

As a consequence, the predicted hydrogen to helium ratio became 6:1, whereas observations at that time gave a ratio between 5:1 and 10:1.

Hayashi remarked that, according to Gamow's original assumption, the Ylem was made up of pure neutrons, and consequently it was impossible to explain how the buildup of the elements jumped over the 'crevasses' at $A = 5$ and 8, as was shown by Fermi and Turkevich. Hayashi argued that, since he had proven that a significant amount of protons existed when the elements first began to be cooked, the problem was thereby alleviated and heavy elements could be built in the modified αβγ theory. He was wrong.

The actual numbers have since changed. Hayashi assumed a neutron lifetime of about 1,000 s, whereas the most recent value is 885.4 s.[139] The lifetime of the neutron determines when the buildup of the elements can take place, and this change in the lifetime of the neutron is equivalent to an almost 30% change in temperature. If the expansion is too slow, all the neutrons decay into protons and then one faces the problem that two protons cannot form a stable nucleus, and vice versa. If the expansion is very fast, the density falls too quickly for the particles to interact and form heavier particles.[140]

---

[138] Again, the neutrinos emitted in the process were originally omitted.

[139] Arzumanov et al., Phys. Lett. B **483**, 15 (2000); Alpher, Follin, and Herman used a neutron half-life of 768 s, as published by Robson, J.M., Phys. Rev. **83**, 349 (1951).

[140] In equilibrium, the ratio of neutrons to protons is given by $n/p = \exp(\Delta mc^2/k_B T)$, where Δm is the mass difference between the neutron and the proton.

Instead of starting with some arbitrary conditions, Hayashi took the calculation to an earlier cosmic time. It is surprising that first-rate scientists like Fermi and Gamow, two of the developers of the theory of β-decay, which is after all about the decay of the neutron, did not dare extrapolating the Universe a bit further in energy (just a factor of 5–10), so as to include the neutron and proton reactions and convert the αβγ theory into a more robust theory by eliminating the need to speculate about the initial state. Moreover, it is amazing that these eminent physicists overlooked the role of neutrinos in the Big Bang, even though they were investigating the problem before the existence of the neutrino was confirmed.

Hayashi returned to the problem six years later.[141] The idea was to find a way to create the heavy elements. They thus studied the buildup of light nuclei, in fact up to $A = 20$, out of initial protons and neutrons in the early stages of the expanding Universe, taking account of the α-capture reactions. Hayashi and Nishida assumed a matter density of $\rho_m = 10^6$–10^7 g/cm^3 at $T = 10^{10}$ K, much higher than the value assumed by Alpher and Herman, and established that the relative abundances of hydrogen, helium, and oxygen were consistent with present observations. To overcome the 'crevasses', or the Fermi–Turkevich gap, Hayashi and Nishida applied the old value due to Salpeter[142] for the $3^4\text{He} \to {}^{12}\text{C}$ reaction, which had since been significantly updated in the upward direction (see Chap. 8). Thus, the heavy elements could be produced by applying stellar densities and Big Bang temperatures. However, at a given temperature, the Big Bang densities are many orders of magnitude lower than stellar densities, so the calculations were not relevant to the Big Bang. On top of this, their results did not agree with the observation that the amounts of C and O were much more than the total amount of all the heavier elements lumped together.

7.23 Cold Big Bang?

Zeldovich (1914–1987)[143] assumed the other extreme, namely, extremely high densities and vanishing temperatures. The original idea, according to Zeldovich, was due to Landau,[144] but the detailed calculation was that of van Albada, who was not cited. It is easy to show that the end product of the suggested process must be pure helium, thus aggravating the problem. Zeldovich circumvented this unacceptable result by assuming that the initial matter was not composed of neutrons, but of protons and electrons. But as we know, under normal conditions this cannot happen, because neutrons decay into protons and electrons. So Zeldovich assumed that the Universe was full of neutrinos to such an extent that the reaction

[141] Hayashi, C., and Nishida, M., Prog. Theor. Phys. **16**, 613 (1956).

[142] Salpeter, E.E., Astrophys. J. **115**, 326 (1952); Ann. Rev. Nucl. Sci. **2**, 4 (1953).

[143] Zeldovich, Ya.B., Soviet Phys. JETP **16**, 1102, 1395 (1963).

[144] Landau, L., Dokl. Akad. Nauk SSSR **17**, 301 (1937).

$$\mathrm{p}+\mathrm{e}^- \longrightarrow \mathrm{n}+\nu \quad \text{or} \quad {}^{A}X_Z+\mathrm{e}^- \longrightarrow {}^{A}X_{Z-1}+\nu, \quad \text{with } A=Z=1,$$

is prevented, because the neutrinos obey the laws of a Fermi–Dirac degenerate gas, and since, as assumed by Zeldovich, all neutrino states are filled, the new neutrino has no vacant location to go to according to the Pauli principle. The reaction is thus suppressed. For this to be the case, the density must be extremely high and the temperature very low. In this case, the end product is molecular hydrogen. Here, the theory shares the premises of the steady state theory, and requires helium, as well as all other elements, to be created in stars.

7.24 The Early 1950s

The results from the αβγ theory depend on the rate of expansion of the Universe and the lifetime of the neutron. New results for the rate of expansion of the Universe were obtained by Behr,[145] who discovered that the implementation of the standard candle method to galaxies may lead to a bias in which distant galaxies appear closer than they really are. The implication of this bias was a new value for the Hubble constant[146] and an age of only 1.71 billion years. This dictated parameters for Big Bang nucleosynthesis that did not yield any helium.

This result for the age of the Universe provided a favourable wind for the proponents of the steady state theory. Moreover, the inability of Big Bang cosmology (irrespective of the value of the Hubble constant and the rate of expansion of the Universe) to account for elements heavier than helium caused the tide of Big Bang enthusiasts to ebb, at least among nuclear astrophysicists. But it was clear that there was a problem with the Hubble constant, since it was also in conflict with the age of the Earth, as measured by means of radioactive dating.

In 1953, Gamow[147] criticized the steady state theory, bringing up two arguments. The first had to do with the discrepancy between the age of the Universe, as measured from recession of the galaxies, and the age of the Earth, which necessarily had to be the smaller of the two but seemed to be bigger. The problem of the age contradiction disturbed astrophysicists, and when the steady state theory was conceived, it provided a solution to this problem. So Gamow pointed to recent data by Baade and Sandage which straightened up the numbers, thereby removing one of the problems with the Big Bang theory and thus reducing the sucesses of the steady state theory.

The second argument was the fact that there are stellar populations. This means that there is evolution of stars, and not all stars are identical. Furthermore, relying on a communication from Baade, Gamow pointed out that the age of nearby galaxies was:

---

[145] Behr, A., Astr. Nachr. **279**, 97 (1951).

[146] This was before the discovery of the error in the Cepheid luminosity. Hubble's value was 580×10^{-6} km/s per parsec, while Behr got a factor 0.45 smaller.

[147] Gamow, G., Astron. J. **59**, 200 (1954).

> [...] remarkably constant, which speaks against the possibility of mixed-age populations predicted by the theory of the steady state universe.

While the steady state theory was never the favorite for many astrophysicists, it now fell out of grace.

7.25 Alpher, Follin, and Herman Correct the Theory

Five years after the idea was presented to the scientific world, the first correct calculation of the Big Bang synthesis was carried out by Alpher, Follin, and Herman.[148] In particular, they extended Hayashi's calculation to still higher energies (about 100 MeV or $\sim 1.2 \times 10^{12}$ K) which correspond to an epoch of 10^{-4} s after the Big Bang. They essentially narrowed the allowed range of the neutron to proton ratio. It was no longer a free parameter, but became connected to the cosmological parameters. In this way a plateau in the helium abundance was obtained, and the plateau agreed much better with observations. The value of the plateau in the Hayashi case was 40% by mass.

7.26 B^2FH 1957: An Attempt at a Global Picture

One of the most famous papers in nuclear astrophysics was published in 1957 by Burbidge, Burbidge, Fowler, and Hoyle (known as the B^2FH paper). This seminal paper is a monumental compendium of nuclear astrophysics, bringing order to the theory of the synthesis of the elements.[149] It will be discussed at length in Chap. 9. We mention it at this point because it essentially claimed that all elements were synthesized in stars. The aphorism of the paper: *It is the stars, The stars above us, govern our conditions*, from King Lear, Act IV, Scene 3, disclosed the basic manifesto the authors aspired to convey.

They pointed to four theories of the origin of the elements. The first three were as follows:

- The non-equilibrium theory of Gamow, Alpher, and Herman (which is essentially neutron capture theory).
- The polyneutron theory of Mayer and Teller.
- The equilibrium theory of Klein, Beshkow, and Treffenberg.

According to the authors:

> Each of these theories possesses some attractive features, but none succeeds in meeting all of the requirements. It is our view that these are mainly satisfied by the fourth theory

[148] Alpher, R.A., Follin, J.W., and Herman, R.C., Phys. Rev. **92**, 803, 1347 (1953).

[149] Burbidge, E.M., Burbidge, G.R., Fowler, W.A., and Hoyle, F., Rev. Mod. Phys. **29**, 547 (1957).

> in which it is proposed that the stars are the seat of origin of the elements. In contrast with the other theories which demand matter in a particular primordial state for which we have no evidence, this latter theory is intimately related to the known facts that nuclear transformations are currently taking place inside stars. This is a strong argument, since the primordial theories depend on very special initial conditions for the Universe.

Moreover, according to the authors, the elements must be spread all over the Universe, and stars and supernovae are well placed to carry out this task. Primordial theories, so argue the authors, certainly distribute matter uniformly in the Universe, but they add that:

> This disagrees with observation. There are certainly differences in composition between stars of different ages.

They also add a word of caution, pointing out that:

> It is not known for certain at the present time whether all of the atomic species heavier than hydrogen have been produced in stars, without the necessity of element synthesis in a primordial explosive stage of the Universe.

While the authors were ready to have all the elements synthesized in stars, they pointed out that the nuclear physics would yield the same results whether the synthesis took place in stars or in a primordial explosion, provided of course that the conditions were alike. Notwithstanding everything that had been discovered regarding stellar populations and even Hoyle's idea of galactic evolution, the authors, or more accurately Hoyle, did not accept it as evidence of evolution.

The authors concluded that stars, in contrast to any theory of primordial synthesis, offer a wide range of conditions. This is important because not all elements can be synthesized under the same conditions. The implication was that the abundances observed in nature are the result of many independent processes of synthesis and ejection from stars, as well as mixing, whence there is no single process that yields all the elements.

7.27 Hoyle and Tayler 1964: This Is no Retreat

At the beginning of the 1960s, cosmologists faced two extreme predictions: the Big Bang theory produced too much helium relative to what was observed in stars, while the steady state theory produced no helium at all.

As more observations and abundance determinations accumulated, cracks started to appear in the 'stars-make-them-all' theory. First, O'Dell[150] analyzed the planetary nebulae observed in the classical globular cluster M15 and found a helium abundance of 0.18 ± 0.03 by number. Planetary nebulae are small mass, dying stars surrounded by an ejected envelope of gases, and M15 is an old population II cluster. So how come this old object contained so much helium? To confirm that his discovery raised

[150] O'Dell, C.R., Astrophys. J. **188**, 1018 (1963).

a real problem, O'Dell measured the ratio of oxygen to hydrogen and checked that it was low, suggesting that:

> The material of this nebula might not have been adulterated by the products of nuclear reactions.

In other words, helium burning might not have taken place. At the same time, measurements of the helium abundance in young B stars[151] gave a helium to hydrogen ratio of 0.16 by number. Thus, the helium to hydrogen ratio in very young and very old objects was found to be almost the same, a result that could be interpreted as a hint that there exists a universal helium abundance irrespective of the age of the star. The stars are not mixed so the surface abundance would be the initial abundance. In the case of a planetary nebula, a star that has ejected its outer layers, it is the composition of these layers that O'Dell measured.

In 1964, in a paper entitled *The mystery of the cosmic helium abundance*,[152] Hoyle and Tayler (1929–1997) discussed the source of helium in very old stars, and revived, somewhat indirectly, the idea of a universal synthesis of helium, or a synthesis in supermassive objects preceding the formation of stars in our Galaxy. They reviewed the observations of helium and reached the conclusion that it ranged from 27% by mass and above, and they suggested that the helium abundance, even in the oldest stars, could not be lower. But Hoyle refused to accept the rationale whereby this fact supports Big Bang nucleosynthesis. At that point Hoyle was still right, because calculations like Smirnov's,[153] based on the Big Bang model, gave too high a helium abundance compared with observed values.

Hoyle and Tayler began their paper with the words:

> It is usually supposed that the original material of the Galaxy was pristine material.

Put another way, no elements were formed before the stars and the galaxies existed:

> However, the presence of helium, in a ratio by mass to hydrogen of about 1:2, shows that this is not strictly the case. [...] if present observations of a uniformly high helium content in our galaxy and its neighbors are correct, it is difficult to suppose that all the helium has been produced in ordinary stars.

The purpose of the paper was to show that:

> Most if not all of the material in our everyday world [...] has been cooked to a temperature in excess of 10^{10} K.

The conclusions reached were:

> (i) The Universe had singular origin or is oscillatory, or (ii) the occurrence of massive objects has been more frequent than has hitherto been supposed.

In principle, Hoyle was almost ready to accept that helium was produced in the much loathed Big Bang, but not quite.

The paper claimed that:

---

[151] Aller, L.H., *The Abundances of the Elements*, Interscience, New York, 1961.

[152] Hoyle, F. and Tayler, R., Nature **203**, 1108 (1964).

[153] Smirnov, Yu.N., Soviet Astr. **8**, 864 (1965).

- The helium cannot be produced in stars. Assume all the stars in the Galaxy convert hydrogen into helium and that this is the source of stellar luminosity. Then from the total luminosity of the Galaxy, Hoyle and Tayler calculated that $10^9 M_\odot$ of helium were produced during the lifetime of the Galaxy (assuming that the total luminosity of the Galaxy did not change). If we include the emission in the ultraviolet and infrared, the number increases by about a factor of 3. But the visible mass of the Galaxy is $\sim 10^{11} M_\odot$, and this means that the total amount of helium in the Galaxy should be about 0.01. This is much too low relative to the observed helium in stars (0.16–0.18). This particular claim was used time and again, even as late as 1998.[154]
- The helium was produced by a big explosion. This is essentially a repetition of the argument due to Alpher, Follin, and Herman, and the result is that $He/H \simeq 0.14$ by number, with this being rather insensitive to the initial conditions.[155] Hoyle and Tayler concluded that *if the Universe originated in a singular way, the He/H ratio cannot be less than about 0.14.* Nonetheless, argued Hoyle and Tayler, if in any object the ratio is found to be appreciably less than 0.14, then clearly, the Universe did not have a singular origin. But measurements and models for the Sun[156] gave $He/H = 0.09$. This was a strong constraint, because a single case was sufficient to ruin the theory.

Finally, the authors asked:

- Could the conditions in other objects be adequate for the synthesis of helium? One possibility is very massive stars (having about $5 \times 10^5 M_\odot$), in which the conditions mimic those of the Big Bang. The discussion was not complete, as Hoyle and Tayler admitted, because it should also have included an explanation of how to eject the material into galactic space. Just creating the elements deep inside stars was not sufficient.

Hoyle and Tayler's paper was instrumental in shifting the general view back towards Big Bang nucleosynthesis, at least for helium, even though this was not Hoyle's original intention.

7.28 The Cosmic Microwave Background Radiation

In 1934, following the theoretical work by Lemaitre on an explosive universe filled with radiation, Tolman showed[157] that the radiation from the primeval fireball would cool adiabatically and maintain its black body spectrum. That is to say, the spectral

---

154 Burbidge, G., and Hoyle, F., Astrophys. J. Lett. **509**, L1 (1998).

155 But it is sensitive to the number of different neutrinos assumed.

156 Sears, R.L., Astrophys. J. **140**, 477 (1964).

157 Tolman, R.C., *Relativity, Thermodynamics, and Cosmology*, Clarendon Press, Oxford 1934. See also Phys. Rev. **38**, 797 (1931); PNAS **17**, 153 (1931).

distribution would be that of a black body, but with decreasing temperature. Moreover, the temperature decreases as the radius of a given amount of mass increases, i.e.,

$$T \propto \frac{1}{R}.$$

A byproduct of the Big Bang hypothesis is therefore the existence of a general background radiation which fills the entire Universe and appears isotropic. In one of the early papers on the synthesis of the elements,[158] Alpher, Herman, and Gamow had already calculated the expected time dependence of the density and temperature of the expanding gases. However, a specific prediction regarding the temperature the radiation should have today was not made.

The first concrete value for the radiation temperature was given by Alpher and Herman.[159] They specified the densities they assumed today and when the nucleosynthesis took place:

$$\rho_{\text{matter, nuc}} = 1.78 \times 10^{-4}\,\text{g/cm}^3 \quad \text{and} \quad \rho_{\text{rad, nuc}} = 1\,\text{g/cm}^3,$$

$$\rho_{\text{matter, now}} = 10^{-30}\,\text{g/cm}^3 \quad \text{and} \quad \rho_{\text{rad, now}} \cong 10^{-35}\,\text{g/cm}^3 \quad (T_{\text{now}} \cong 1\,\text{K}),$$

whence the radiation temperature today had to be about 1 K. The word 'background' was not used. The term adopted was 'cosmic radiation'. In the formulation by Alpher and Herman:

> It is possible to examine a variety of cosmological problems, including for example, the origin of cosmic radiation [...]. If it should prove that cosmic radiation is a universal phenomenon, its characteristics should perhaps be connected with the original element-forming process.

The following footnote appeared in the paper:

> For example, see Lemaitre for some interesting views in this connection. However, with the cosmological model considered in the present paper it seems unlikely that primordial cosmic radiation could have survived through the early epoch of the expanding Universe when mass density was quite high.

In short, they did not believe that cosmic radiation would be discovered one day. What Alpher and Herman claimed, quite correctly, was that the synthesis of the elements became linked with the parameters of the expanding cosmos.

In further papers, all within three years, Alpher and Herman repeated the prediction but changed the numerical value to accord with revisions in the cosmic parameters. The maximum estimated temperature of the background radiation was 28 K.[160] This important conclusion about the temperature of the background radiation was not mentioned in the abstract of the papers. They would probably never have dreamt that this radiation would be discovered about 17 years later.

---

158 Alpher, R.A., Herman, R., and Gamow, G., PRL **74**, 1198 (1948).

159 Alpher, R.A., and Herman, R.C., Rev. Mod. Phys. **22**, 153 (1950).

160 Alpher, R.A., and Herman, R.C., Phys. Rev. **84**, 80 (1951).

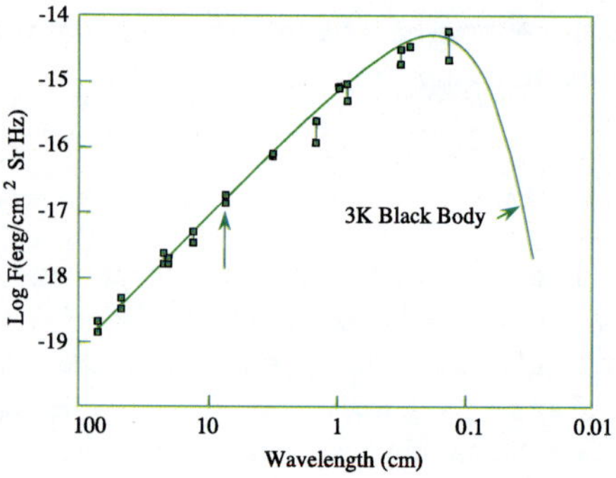

Fig. 7.13 The different measurements of the 3 K background radiation, summarized for the year 1978. The *arrow* along the x axis indicates the wavelength of the original measurement by Penzias and Wilson in 1965. Note that this first measurement was at a wavelength where the intensity of the radiation is almost a factor of a thousand below the peak

A radiative background can be due to an explosion like the Big Bang or due to the thermalization of the energy emitted by certain abundant cosmic sources. Eddington[161] for example, calculated the equivalent of the radiation from all stars and found $T = 3.18\,\mathrm{K}$. But this is just a mind-boggling coincidence.

Some unexplained signs that such a background exists already appeared in the literature as early as 1937, when Dunham,[162] Adams,[163] and McKellar[164] found that certain molecules of CH and CN in interstellar space must have been excited by radiation at a temperature of 2.3 K, from an unknown source. The phenomenon remained a mystery for over 25 years, until the discovery of the background radiation by Penzias and Wilson led the way for a group of scientists working independently to suggest that these molecules were in equilibrium with the background radiation field.[165]

When Penzias and Wilson were employed by Bell Laboratories, one of their tasks was to build a very clean, i.e., noise-free, radio antenna, operating at 4,080 MHz (a wavelength of about 7.3 cm). However, they discovered[166] that, wherever they directed their antenna, it collected noise. The level of the noise they could not get rid of was equivalent to $3.5 \pm 1\,\mathrm{K}$ (see Fig. 7.13). Moreover, they established that:

> This excess temperature is, within the limits of our observations, isotropic, unpolarized, and free from seasonal variations.

During a visit to MIT in Cambridge, Massachusetts, Penzias mentioned to Bernard Burke the unexplained noise detected in the antenna. Burke drew Penzias' attention to the theoretical work by Peebles and Dicke at Princeton, and Penzias and Wilson

[161] Eddington, A.S., *The Internal Constitution of the Stars*, Dover Pub., New York, 1959, p. 371.

[162] Dunham, T., PASP **49**, 277 (1937).

[163] Adams, W.S., Astrophys. J. **93**, 11 (1941); ibid. **97**, 105 (1943).

[164] McKellar, A., Pub. Dominion Astron. Obs. **7**, 251 (1941).

[165] Field, G.B., Herbig, G.H., and Hitchcock, J.L, Astrophys. J. **71**, 161 (1966); Field, G.B., and Hitchcock, J.L, PRL **16**, 17 (1966); Shklovsky, I.S., Anst. Circular No 364, Acad. Sci. USSR (1966); Thaddeus, P. and Clauser, J.F., PRL **16**, 19 (1966).

[166] Penzias, A.A., and Wilson, R.W., Astrophys. J. **142**, 419 (1965).

contacted them soon afterwards. Dicke sent Penzias Peebles' preprint about the radiation of the Universe. It was Dicke who suspected that the background radiation, which Alpher and his associates had written about, might actually be detected. According to Peebles' estimate, the temperature of the background radiation should be 10 K.

A possible explanation of this excess came from the Princeton group headed by Dicke, whose paper[167] appeared, upon agreement between the authors, side by side with Penzias and Wilson's paper. This Nobel prize-winning paper is less than two pages long. It is interesting to point out that, in a note added in proof, Penzias and Wilson mentioned that, in 1962, Pauliny-Toth and Shakeshaft[168] had already measured a minimum sky temperature of 16 K at a frequency of 404 MHz. However, they had not made any reference to the general background radiation. Previous noise measurements by Hogg and Semplak,[169] DeGrasse et al.,[170] and Ohm[171] yielded only upper limits and no detection of any noise. A posteriori analysis of Ohm's results by Wilson[172] showed that his system temperature was 3.3 K higher than predicted. When black body radiation has a temperature of 3.5 K, the maximum of the radiation is at a wavelength of 0.083 cm, some distance from the wavelength used by Penzias and Wilson. It was the extremely high sensitivity of their antenna and their qualifications as superb experimentalists that allowed the discovery.

Finally, Penzias and Wilson reported that the radiation was isotropic but did not specify the degree of isotropy. It is important to stress that, for the background radiation to be the Big Bang relic radiation, it had to be isotropic to a very high degree. In fact, the radiation is not strictly isotropic. It was later discovered that it is not strictly isotropic on two levels. First, as the Earth moves in space along with the Sun and the Galaxy, there is a Doppler shift to the blue in the direction of motion and a shift to the red in the opposite direction. Actually, the Milky Way is falling towards the Virgo cluster of galaxies at a speed of 600 km/s. If one removes this anisotropy from the radiation, one is left with radiation which is isotropic at a level of 10^{-5}. Fluctuations at this level confirm that the basic radiation is a relic radiation, and the shape of these fluctuations discloses many details of the physics of the Big Bang. At the time of the Nobel award, in 1978, the measured anisotropy was less than 0.001 K.[173]

Until the discovery of the cosmic background radiation, the only evidence for the expansion of the Universe was the recession of the galaxies. The cosmic background

---

167 Dicke, R.H., Peebles, P.J.E., Roll, P.G., and Wilkinson, D.T., Astrophys. J. **142**, 414 (1965).

168 Pauliny-Toth, I.I.K., and Shakeshaft, J.R., MNRAS **124**, 61 (1962).

169 Hogg, D.C., and Semplak, R.A., Bell Syst. Tech. J. **40**, 1331 (1961).

170 DeGrasse, R.W., Hogg, D.C., Ohm, E.A., and Scovil, H.E.D., Proc. Natl. Electron. Conf. **15**, 370 (1959).

171 Ohm, E.A., Bell Syst. Tech. J. **40**, 1065 (1961).

172 Wilson, R.W., Nobel address, 1978.

173 An interesting story in this connection is reported by I. Novikov in his book (Naselsky, P.D., Novikov, D.I., and Novikov, I.D., *The Physics of the Cosmic Microwave Background*, Cambridge

radiation presented new and independent evidence that the Universe was expanding long before the galaxies were formed.

In the paper that preceded Penzias and Wilson's, entitled *Cosmic black body radiation*, Dicke and his associates first told how Penzias and Wilson scooped them with the discovery and then continued to analyze the implications. The following solutions were suggested at that time:

- Continuous creation (Bondi, Gold, and Hoyle).
- New matter is intimately related to the existence of singularities.

(Footnote 173 continued)
University Press, 2006). Novikov reports a telephone call he got in 1983 from T. Shmaonov, who directed his attention to a paper he published in 1957 in *Pribory i Teknika Eksperimenta*, which was part of his PhD thesis. In the paper Shmaonov reported that:

> We find that the absolute effective temperature of the radioemission background [is] 4 ± 3 K.

Apparently, Shmaonov did discover the cosmic background, but failed to appreciate the meaning of the discovery. Only some 27 years later did Shmaonov realize what he had found and called Novikov to inform him about the publication in a journal that cosmologists never look at. The conclusion is that it is not sufficient to discover a new phenomenon. It is also crucial to appreciate that this is the case, and then of course if possible to attempt to understand the source of the phenomenon. Unfortunately, one will not be credited for late appreciations of the meaning of one's discoveries (especially if others have done it before).

Similarly, Delannoy, Denisse, Le Roux, and Morlet (Delannoy, J., Denisse, J.F., Le Roux, E., and Morlet, B., An. Astrophys. **20**, 222, 1957) reported the detection and absolute measurement of a weak radiation flux at a frequency of 900 MHz, which is a wavelength of 33 cm. The signal was about one thousandth of the noise in the receptor. They complain that they did not have any reference to measure the power of the signal. So they decided to measure several celestial radio sources for reference at 4 to 5 different frequencies. After a careful set of measurements, they reached the conclusion that *the sky temperature is not higher than twenty degrees kelvin*. This work was based on Le Roux's 1956 thesis, Delannoy's Diplome d'Etude Supérieure in 1955, and Morlet's Diplome d'Etude Supérieure also in 1955. However, the paper did not contain any explanation or guesses as to the nature of the phenomenon.

A year earlier, Seeger, Westerhout, and Van de Hulst (Seeger, Ch.L., Westerhout, G., and van de Hulst, H.C., BAN **13**, 89, 1956) had reported the intensity of the background radiation at 400 MHz. This time the researchers interpreted the result as the galactic background, probably because they found large differences in temperature between different directions. The differences were from 100 K near Cygnus down to 16 K, which was the estimated coldest point.

McGee, Slee, and Stanley (McGee, R.X., Slee, O.B. and Stanley, G.J., Au. J. Ph. **8**, 347, 1955) were confident that they had measured the galactic background, which was not uniform, and the coldest temperature they measured was 10 ± 5 K.

Le Roux and his coauthors followed Seeger and his coauthors and assumed that they had observed the galactic background. However, Seeger et al. accepted the idea of a galactic background from McGee and his colleagues. In this way, an incorrect explanation, which remained unchallenged, prevented the researchers from identifying what they had observed with the relic radiation. Moreover, the prediction by Alpher was there in the literature all the time, but it got ignored. Sometimes it is worth spending an hour in the library before spending months near the radio telescope.

- The singularity results from a mathematical over-idealization, the requirement of strict isotropy or uniformity, and would not occur in the real world (Wheeler,[174] Lifshitz and Khalatnikov[175]).

The physics community did not feel comfortable with the singularity, and the above authors suggested how to avoid it. So Dicke and his associates wrote that, in the third case:

> It is essential to suppose that, at the time of maximum collapse, the temperature of the Universe would exceed 10^{10} K, in order that the ashes of the previous cycle would have been reprocessed back to the hydrogen required for the stars in the next cycle.

Penzias and Wilson identified the newly discovered radiation with:

> [...] the thermal radiation remaining from the fireball, if we can trace the expansion of the Universe back to a time when the temperature was of the order of 10^{10} K.

Irrespective of whether the Universe oscillates, and the question of whether there is or is not a singularity, the temperature would have had to have been above 10^{10} K for none of the systems we know (nuclei of all sorts) to survive. In that case, the Universe must have started from a state in which no bound system of nucleons existed.

The authors credited Alpher, Herman, and Gamow with the idea that the elements built up as the Universe cooled. The previous suggestions by Alpher and Herman, in particular the idea that the Universe should be filled with relic radiation at a temperature of a few degrees, were forgotten.

Penzias and Wilson shared the 1978 Nobel prize for one of the most fantastic discoveries in the history of mankind, which confirmed so nicely that our Universe started from a very dense and hot state. Wilson's address at the Nobel ceremony was entitled: *The Cosmic Microwave Background Radiation*, and Penzias' address was entitled: *The Origin of Elements*, touting the message that he considered to be the most important implication of the background radiation, namely the synthesis of the elements in the Big Bang.

7.29 Peebles (1966): Getting the Helium Abundance Right

In 1966, in a short communication in the Physical Review Letters and accompanied by a long paper in the Astrophysical Journal,[176] Peebles investigated the implications of the discovery of the background radiation for the predicted helium production in the Big Bang. He thus obtained a helium abundance of 26–28% by mass, based on the best cosmological data known in 1966. This was a major step in establishing the

---

174 Wheeler, J.A., 1964, in *Relativity, Groups and Topology*, ed. by DeWitt & DeWitt, Gordon & Breach, New York.

175 Lifshitz, E.M., and Khalatnikov, I.M., Adv. Phys. **12**, 185 (1963).

176 Peebles, P.J.E., PRL **16**, 410 (1966); Astrophys. J. **146**, 542 (1966).

Table 7.2 Comparison between the mass fractions found by Wagoner, Fowler, and Hoyle's Big Bang calculation, with a particular choice for the ratio (sum of neutrons and protons)/(number of photons), and observation in the Solar System

	D	^{3}He	^{4}He	^{7}Li
Solar system	2×10^{-4}	6×10^{-5}	0.27	10^{-8}
WFH theory	3×10^{-7}	1×10^{-5}	0.27	2×10^{-8}

connection between cosmology and the creation of the light elements. Consequently, the prediction was that the oldest stars should contain at least this amount of helium.

Peebles' paper in the Astrophysical Journal began with the words:

> Subsequent to the suggestion by R.H. Dicke (1964, unpublished) that one ought to search for the residual, thermal radiation left over from the early stages of expansion of the Universe (Dicke, Peebles, Roll, and Wilkinson, 1965), a new isotropic microwave background was discovered (Penzias and Wilson 1965) […].

The truth is that (a) Alpher, Herman, and Gamow were the first to predict in published papers the existence of the relic radiation, and (b) Penzias and Wilson worked completely independently, and not 'subsequent' to any suggestion. It is, however, correct that Penzias and Wilson's discovery was a serendipitous one, and it took place while Dicke and his colleagues were intentionally attempting to discover the relic radiation.

7.30 Wagoner, Fowler, and Hoyle: Light Element Indicators

A massive calculation of a large number of nuclear reactions was carried out by Wagoner, Fowler, and Hoyle, hereafter denoted by WFH,[177] in a paper entitled: *On the synthesis of elements at very high temperatures*. This calculation included all the updated physics of elementary particles (including all the known kinds of neutrino), as well as nuclear physics, and set the standard for future calculations.

The first conclusion reached by the authors was this:

> If the recently measured microwave background radiation is due to primeval photons, then significant quantities of only D, ^{3}He, ^{4}He, and ^{7}Li can be produced in the universal fireball.

The abundances so calculated were found to agree quite well with Solar System abundances, as can be seen from Table 7.2. However, no heavy elements were formed, to Hoyle's delight. A glance at the table reveals that the agreement was close but not perfect. To reconcile the cases of ^{3}He and D and bring them even closer, the authors

[177] Wagoner, R.V., Fowler, W.A., and Hoyle, F., Astrophys. J. **148**, 3 (1967). Here is an example of a change in views: before the CMB was found and cosmology became *a decent science*, Willie Fowler argued that it was *a dream of zealots*. Now he was ready to join in the festivities.

found it necessary to argue that isotope separation had taken place within the Solar System and changed the relative isotope abundances. So the authors summed up:

> Unless we appeal to isotopic fractionation it is necessary to conclude that one or more of the following must hold good:
>
> - Our calculation contains appreciable inaccuracies.
> - The measured thermal radiation at 3 K is not primeval.
> - The parameters of the expanding Universe are substantially different.
> - The observed abundances are due to local processes and do not represent universal cosmic abundances.
> - At least one of the original assumptions at the base of the calculation is untrue.

The ^{4}He content of population II stars was not known at that time, so WFH relied on theoretical calculations by Faulkner[178] and Christy,[179] which indicated that the He abundance in RR Lyrae variable stars was in the range 20–30% by mass.

So had Hoyle changed his thinking? Not really. WFH made the following point. The average density of galactic material is $(3\text{–}7) \times 10^{-31}$ g/cm^3. Of this about one third is probably helium, giving an average helium density of $\approx 10^{-31}$ g/cm^3. Since the conversion of 1 g of hydrogen to helium yields 6×10^{18} erg, the average total energy production, if helium had come from hydrogen, would have been $\approx 6 \times 10^{-13}$ erg/cm^3. This energy density, if thermalized, would yield a temperature of just 3 K. The cosmological expansion decreases the baryon density as R^{-3}, while the radiation density decreases as R^{-4}, so the coincidence is an accident if the measured 3 K is a relic of a cosmological fireball. Expansion increased the radius since the fireball by a factor of about 10^9 so that no such coincidence could have been obtained over most of the expansion. It would be accidental if it occurred only in the present epoch. In contrast, this would not be the case if the observed radiation resulted from the thermalization of energy from recent hydrogen to helium conversion in stars.

Years later, in 1998, Burbidge and Hoyle[180] returned to this apparent coincidence and claimed that:

> This result strongly suggests that the ^{4}He was produced by hydrogen burning in stars and not in the early stages of a Big Bang. In addition, we show that there are good arguments for believing that the other light isotopes, D, ^{3}He, ^{6}Li, ^{7}Li, ^{9}Be, ^{10}B, and ^{11}B were also synthesized in processes involving stars. By combining these results with the earlier, much more detailed work of Burbidge et al. and of Cameron, we can finally conclude that all of the chemical elements were synthesized from hydrogen in stars over a time of about 10^{11} years.

This was an attempt to discredit the background radiation as evidence in favor of the Big Bang. Notwithstanding the conclusions in the WFH paper, which Hoyle had signed, he did not relinquish his steady state theory.

The second conclusion reached by WFH was that imploding massive stars can convert the universal D and ^{3}He into C, N, O, Ne, and Mg, together with some

---

178 Faulkner, J., Astrophys. J. **144**, 978 (1966).

179 Christy, R.F., Astrophys. J. **144**, 108 (1966).

180 Burbidge, G., and Hoyle, F., Astrophys. J. Lett. **509**, 1 (1998).

heavier elements, in the amounts observed in the oldest stars. The $A = 5$ barrier was overcome via the following set of reactions:

$$^{3}\mathrm{He} + {}^{4}\mathrm{He} \longrightarrow {}^{7}\mathrm{Be} + \gamma, \quad \text{followed by} \ {}^{7}\mathrm{Be} + {}^{4}\mathrm{He} \longrightarrow {}^{11}\mathrm{C} + \gamma,$$

or

$$^{3}\mathrm{H} + {}^{4}\mathrm{He} \longrightarrow {}^{7}\mathrm{Li} + \gamma, \quad \text{followed by} \ {}^{7}\mathrm{Li} + {}^{4}\mathrm{He} \longrightarrow {}^{11}\mathrm{B} + \gamma.$$

The first reaction operates in the pp chain. The second requires a significantly higher temperature than those found in stars deriving their energy by means of the pp chain, and hence is practically non-existent in stars. It so happens that in supermassive stars with $10^5 M_\odot$ and above, the ^{3}He survives the high temperatures needed for the beryllium to absorb a helium nucleus. If the imploding supermassive stars can reach temperatures higher than about 10^9 K, the $3\,\alpha \to {}^{12}$C of Salpeter (see Chap. 8) can operate and produce metals of the iron group. The most important result now is the prediction of lithium synthesis. Consequently, measuring the amount of lithium in very old stars became a prime target for abundance determinations. We will return to this problem in Chap. 14.

As more cosmological data became available, Wagoner[181] returned in 1973 to the synthesis of the elements in the 'universal fireball', and this time the title of his paper was *Big Bang Nucleosynthesis Revisited*, to let the reader know that the first paper, the 1967 paper with Hoyle, was actually a bona fide Big Bang nucleosynthesis model.

Many calculations of Big Bang nucleosynthesis (today dubbed as BBN) were carried out after the pioneering work of WFH, improving the accuracy and updating the nuclear and cosmological data. By 2007, there were over 2,260 papers on various nuclear aspects of the Big Bang theory.

7.31 Stars Cannot Do it All

The difficulties with the steady state theory led Hoyle and Narlikar,[182] just a year before the WFH paper was published, to propose that matter was created irregularly much along the lines suggested by McCrea.[183] The idea was that mass creation occurred only in certain 'pockets' in space, and that these pockets developed into galaxies. As we know today, the centers of many galaxies are very active and contain black holes, show violent eruptions, and emit copious amounts of energy. In these objects, so claimed Burbidge, Burbidge, and Sandage,[184] pre-stellar matter of unknown form converts into the commonly known form of matter. Similar ideas were

---

181 Wagoner, R.V., Astrophys. J. **179**, 343 (1973).

182 Hoyle, F., and Narlikar, J.V., Proc. Roy. Soc. A **290**, 162 (1966).

183 McCrea, W.H., MNRAS **128**, 335 (1964).

184 Burbidge, G.R., Burbidge, E.M., and Sandage, A.R., Rev. Mod. Phys. **35**, 947 (1963).

put forward in 1967 by Nèeman and Tauber,[185] who suggested that inhomogeneities in the expanding Universe (later called white holes) could provide the energy for the recently discovered mysterious quasi-stellar objects (QSOs or quasars). Since then it has become clear that the engines of these fantastic objects are in fact massive black holes which accrete mass via a disk.

Hoyle continued his uphill fight against the Big Bang until his death in 2001. In a recent book, Hoyle, Burbidge, and Narlikar[186] attempt to provide in a methodical way evidence that mass is created in the cores of active galactic nuclei (AGN),[187] where massive black holes reside. What they have claimed for years is that the astronomical community incorrectly interprets the cosmological data by using the standard Big Bang model. The three argue that this basic presupposition is wrong. They review the defects and inconsistencies that exist within the interpretation of the cosmological data. This is followed by an extensive presentation of the authors' own alternative view of the status of observations and how they should be explained.

As more and more damaging evidence accumulated, Hoyle invented new variants of the theory, the last being the statistically steady state, which claims that creation does not occur everywhere, but statistically. In 1965, he came close to a temporary recapitulation of the steady state theory,[188] only to return later with more vigor. Since the simple steady state model had fallen out of favor, Hoyle invented the quasi-steady state model[189] and the inhomogeneous steady state model. The major creation episode took place when the mean universal density was about 10^{-27} g/cm^3. Creation occurred when certain conservation equations were not satisfied. Since the main death-blow evidence for the steady state was the background radiation, Hoyle suggested various ways to create the background as a combination of the radiation emitted by a collection of distant objects.[190] If the background radiation is due to the contributions of many sources, as these authors claimed, then it should exhibit fluctuations at a level which is not observed. The observed fluctuations in the background radiation are at a level of 10^{-5}, much too low for this explanation.

From Hoyle's obituary by Maddox,[191] we learn about a personal grudge:

> The business of the steady-state Universe has always been more contentious [...]. Perhaps he relished the inevitable row. But he did not enjoy the personal animosity engendered between him and the late Martin Ryle, then one of the big-hitters in British astronomy. Hoyle believed

---

185 Nèeman, Y., and Tauber, G., Astrophys. J. **150**, 755 (1967).

186 Hoyle, F., Burbidge, G., and Narlikar, J.V., *A Different Approach to Cosmology*, Cambridge University Press, 2000.

187 An active galactic nucleus (AGN) is a compact region at the center of a galaxy that has a much higher luminosity than a normal galaxy. The present most widely accepted paradigm for AGNs is that the radiation results from accretion onto a supermassive black hole with mass in the range of a million solar masses and more. AGNs are the most energetic large scale objects in the Universe.

188 Hoyle, F., Nature **208**, 111 (1965).

189 Hoyle, F., Burbidge, G., and Narlikar, J.V., Astrophys. J. **410**, 437 (1993).

190 Narlikar, J.V., Viswakarma, R.G., Hajian, A., Souradeep, T., Burbidge, G., and Hoyle, F., *Inhomogeneities in the microwave background radiation interpreted within the framework of the quasi-steady state cosmology*, Astrophys. J. **585**, 1 (2003).

191 Maddox, J., Nature **413**, 270 (2001).

that the ill-feeling went back to a disagreement at a conference in 1951, in which he had come out on top. Ten years later Ryle[192] got his own back at a media event, which—so general opinion had it—spelled the end of the steady state theory with Hoyle's seeming failure to respond to the evidence marshaled against it.

7.32 Breaking a Symmetry: A Remaining Question

When a particle meets its antiparticle, they annihilate each other and leave just two photons. The process can go in both directions: an electron and a positron for example, annihilate each other leaving two photons, or two photons interact to form an electron–positron pair. In a similar way we have the reactions:

$$\mathrm{p} + \overline{\mathrm{p}} \rightleftharpoons 2\gamma,$$

where $\overline{\mathrm{p}}$ is an antiproton. In our world, the atoms are composed of positive protons in the nucleus and negative electrons moving around them. However, in antimatter, negative antiprotons would be in the nucleus and the positron would move around them. Physics is symmetric with respect to this change.

The creation of the electron–positron pair or a proton–antiproton pair takes place at a temperature of roughly

$$T_{\mathrm{e}\overline{\mathrm{e}}\ \mathrm{pair}} = \frac{m_{\mathrm{e}}c^2}{k_{\mathrm{B}}} \quad \text{and} \quad T_{\mathrm{p}\overline{\mathrm{p}}\ \mathrm{pair}} = \frac{m_{\mathrm{p}}c^2}{k_{\mathrm{B}}},$$

where m_{e} and m_{p} are the masses of the electron and the proton, respectively. Since the mass of the proton is 1,840 times higher than that of the electron, the corresponding temperature is that much higher. An electron–positron pair forms at $T = 5.93 \times 10^9$ K, while the proton–antiproton pair forms at 1.09×10^{13} K, which lies beyond the temperatures discussed in this book.

If the symmetry between protons and antiprotons prevails, like the symmetry between electrons and positrons, then the end product of the Big Bang would have been photons, with neither matter nor antimatter, because the matter would have annihilated all the antimatter and disappeared, leaving only a certain characteristic radiation as testimony to the annihilation processes. So if we see matter today, this means that the symmetry was broken at one stage, leaving some extra matter and annihilating all the rest. For example, if the proton is stable while the antiproton is not, and has some small, yet not vanishing, probability of decaying into other particles, in this case, the consequence would be that matter would remain in the form of

---

[192] Martin Ryle (1918–1984) got the Nobel prize with Antony Hewish for the discovery of the pulsars, which were later identified as very fast-rotating radio sources. The discovery paper had the following names as authors: Hewish, A., Bell, S.J., Pilkington, J.D., Scott, P.F., and Collins, (Nature **217**, 709, 1968). Hoyle expressed misgivings for the omission from the prize of Jocelyn Bell, the PhD student who did the research. See, Judson, H.F., New York Times, 20 October 2003.

protons, while all the antimatter would disappear. Such a case is not hypothetical. It has been observed in the laboratory in the case of particles called kaons.

The probability required for this symmetry breaking is miniscule, just $1/10^9$, and this small probability would lead to a Universe composed only of matter. But if this small deviation from symmetry had not existed, our Universe would have contained just radiation and no matter of the kind we know. So it is the breaking of the symmetry between matter and antimatter which resulted in our material Universe, and the breaking of the symmetry must have taken place when the temperature was at least $\approx 10^{13}$ K, or about one millionth of a second after the beginning, and in any case, before the appearance of the neutron and the proton.

In 1956, Goldhaber[193] was apparently the first to push the temperature of the Big Bang to such high temperatures as to require the equilibrium between matter and antimatter. To explain the fact that today all the antimatter has disappeared, he invoked statistical fluctuations, the consequence of which, it so happened, is matter.

Alpher and Herman responded[194] by showing that the annihilation energy would be so high that the problem must be inverted, i.e., if there was antimatter, how come we do not see the effects of the annihilation energy? If Goldhaber were right, then the initial conditions for the Big Bang as assumed by Alpher and Herman would have to be changed accordingly. In the standard model of the Big Bang, there is no room for antimatter today.

Still, in 1964, McCrea[195] suggested that matter–antimatter separation could be put to work to deliver continuous creation, while Alfvén and Klein[196] assumed the symmetry to go to high temperatures, whence the Universe would have started with a dilute medium made of matter and antimatter. However, these suggestions, which were based on the idea that at very high temperatures matter and antimatter may be in equilibrium, did not explain how come there was a surplus of matter after the matter and the antimatter had annihilated each other.

In 1964, Cronin and Fitch and their assistants[197] discovered the corresponding symmetry breaking in the laboratory and won the Nobel prize in 1980.

7.33 The Present Day Picture: Most of the Universe Is Dark

The temperature in the Universe at an age of a few seconds was in the range of MeV. At this moment, it was already determined that all antimatter would annihilate, leaving some extra matter. This was the period of the nuclear furnace, operating out of equilibrium, in which the light elements were formed. The emerging matter

---

[193] Goldhaber, M., *Speculations on cosmogony*, Science **124**, 218 (1956).

[194] Alpher, R.A., and Herman, R., Science **128**, 904 (1958).

[195] McCrea, W.H., MNRAS **128**, 335 (1964).

[196] Alfvén, H., and Klein, O., Ark. Fys. **23**, 187 (1963).

[197] Christenson, J.H., Cronin, J.W., Fitch, V.L., Turlay, R., PRL **13**, 138 (1964).

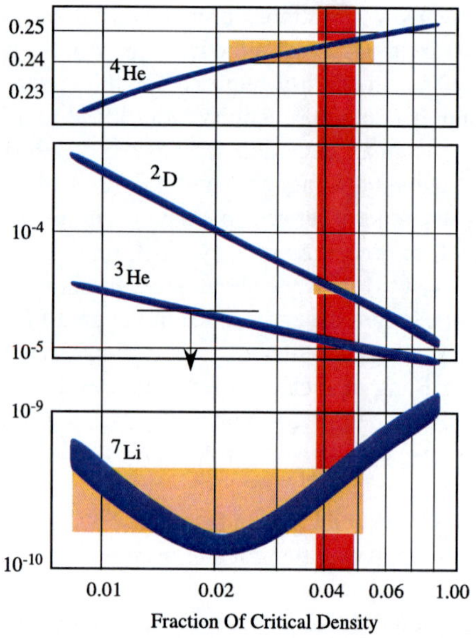

Fig. 7.14 Light element abundances as a function of the cosmological parameters of the Universe, as found by recent calculations. The *top panel* shows the predicted helium abundance. The *second* and the *third panels* show the abundances of ^{2}D and ^{3}He and of ^{7}Li, respectively, relative to the abundance of hydrogen. The observed abundances depicted in *orange*, provide constraints on the cosmological parameters. *Blue stripes* indicate the predicted abundances as a function of the density in the Universe. The *red zone* marks the density range in which the theory predicts all the abundances. Only an upper limit is provided for ^{3}He

composition was hydrogen, about 75% by mass, deuterium, a few parts in 10^{-5} relative to hydrogen, ^{3}He, like deuterium, ^{4}He, about 24% by mass, and ^{7}Li, just a few parts in 10^{-10} relative to hydrogen, and practically nothing heavier. The present accuracy in the cosmological parameters is sufficient to narrow the inaccuracy in the helium mass fraction resulting from the Big Bang to 0.15%, which is fairly low. The interplay between the abundances of the light elements and the density of the Universe as conceived in the present picture is shown in Fig. 7.14.

The discoveries of dark matter and dark energy brought the amount of hadronic matter[198] down to a mere 4%, but did not change the facts that the light elements were synthesized in the Big Bang and all the rest of the elements had to be synthesized in stars.

[198] Hadronic matter is ordinary matter. Hadrons are particles that feel the strong force. The most common hadrons are the proton and neutron. They make up the nucleus and the lion's share of our own mass. The electrons only contribute about 1/1,800th of the mass of ordinary matter. The importance of hadronic matter in cosmology lies in what it cannot do. It cannot account for all the gravitation we infer from the motions of stars in galaxies. Astrophysicists are therefore compelled to invoke the existence of 'dark matter' to provide the majority of the gravity in galaxies. But it has so far escaped direct detection.

Chapter 8
How Nature Overcomes Its Own Barriers

Figure 8.1 shows the light nuclei arranged in the (Z, N) plane, where Z is the number of protons and N the number of neutrons. We see that there are no stable nuclei for $A = Z + N = 5$ and $A = Z + N = 8$. This means that hydrogen cannot fuse with helium to form any $A = 5$ nucleus and helium cannot fuse with another helium to form an $A = 8$ nucleus. These are the $A = 5$ and 8 nuclear barriers. But we know that the nuclei heavier than helium are synthesized. So how does this come about?

Figure 8.2 (left) shows why the nuclide ^{12}C is the most stable nuclide with 6 protons and neutrons. It is simply because there is no lower energy state of the $A = 12$ system. As one moves away from the symmetric nuclide of carbon $(Z = N = 6)$, the ground state energy increases, leaving the symmetric carbon as the most stable nuclide out of all nuclei with $A = 12$. The unstable nuclei decay via β^- (electron emitted by the conversion of a neutron into a proton) or via β^+ (positron emitted by the conversion of a proton into a neutron). The reader may ask why the process does not continue and the proton in the ^{12}C does not convert into a neutron by ejecting an electron. As a matter of fact, it would happen if the system of 5 protons and 7 neutrons had an energy state lower than the energy state of 6 protons and 6 neutrons.

8.1 Unsuccessful Attempts to Overcome the $A = 8$ Barrier

Bethe's work supplied the fundamental theory of the energy sources of stars on the main sequence but did not provide a solution to the question as to how the catalysts carbon and nitrogen were formed and how they found their way into stars. The energy source of the stars after the phase of hydrogen burning and evolution remained a mystery. Scientists were confronted with the problem that the fusion of two helium nuclei does not yield a stable nucleus, and the fusion of 3 particles, say three α particles giving ^{12}C in a triple collision, is very unlikely at the densities known to exist in stars.

G. Shaviv, *The Synthesis of the Elements*, Astrophysics and Space Science Library 387,
DOI: 10.1007/978-3-642-28385-7_8,

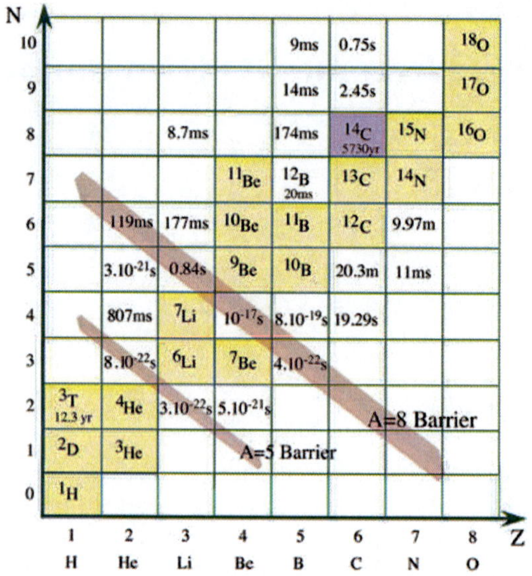

Fig. 8.1 The light elements in the (Z, N) plane. Stable nuclei are *yellow*. Unstable elements with long lifetimes are very *light pink*. Unstable nuclei with very short lifetimes are *white*. The $A = Z + N = 5$ and $A = Z + N = 8$ nuclear barriers are shown as *straight lines*. The nuclear forces do not yield a stable nucleus for these combinations of protons and neutrons. There is only one nucleus with $A = 4$, either stable or unstable. Although it is very stable and has a large binding energy, it has no excited state, only the ground state

The first barrier in the synthesis of helium was the fact that two α particles do not have a bound state. The nuclear levels of the $A = 8$ system are shown in Fig. 8.2 (right). The system of 8 nucleons does not have a stable ground state below which there is no state of free particles. It is rather surprising that, while α type nuclei like ^{12}C, ^{16}O, ^{20}Ne, and so on, show extra stability relative to neighboring nuclei, the simplest of all, the 2α system is unstable.

In his methodical and exhaustive way, Bethe analyzed all nuclear possibilities in his famous paper of 1939, among them the only reactions possible with an α particle, i.e., $^7\mathrm{Li} + \alpha \rightarrow {}^{11}\mathrm{B}$ and $^7\mathrm{Be} + \alpha \rightarrow {}^{11}\mathrm{C}$. These reactions offered a way to overcome the $A = 8$ barrier, although this solution left unanswered the questions as to where the lithium 7 and beryllium 7 came from.

Bethe tried various reactions between stable nuclei which lead to ^{12}C, but these reactions involved relatively rare species and raised the question as to how they themselves would have been synthesized. Consequently, Bethe summed up by saying that *the formation of nuclei heavier than* ^{4}*He can occur only in negligible amounts.*

Finally, Bethe examined the three-body reaction, in which three α particles interact simultaneously with one another to produce the carbon nucleus. Since *collisions of an α particle with one other particle, proton or α, do not lead to stable nuclei*, the alternative was to assume a triple collision. Three types were plausible:

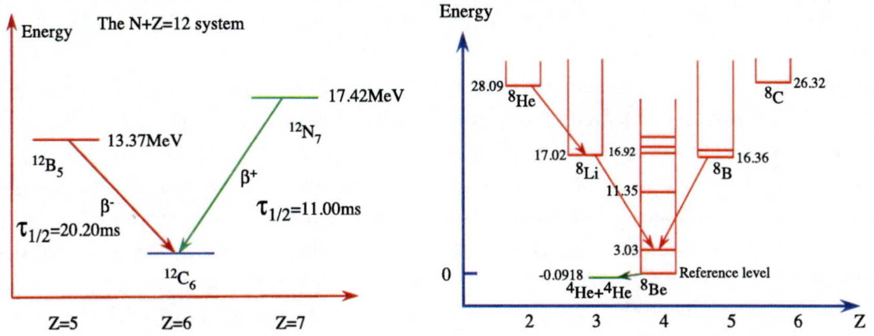

Fig. 8.2 *Left* the $A = 12$ nuclei system. The diagram explains why, of all nuclei with 12 protons and neutrons, the ^{12}C is the most stable. The other nuclei with $A = 12$ are all unstable and decay via β^+ or β^- into ^{12}C. The energies above the ground state of carbon are given in MeV. Also given is the half-life of the unstable nuclei. After Ajzenberg-Selove and Lauritsen (2008). *Right* the energy levels of the nuclear system with $A = 8$ for different atomic numbers Z. The nucleus ^{8}B can decay into an excited state of ^{8}Be, which may decay eventually to the ^{8}Be ground state. But the ground state of ^{8}Be is above the state of two free α particles (*green*), so the fast breakdown of ^{8}Be is unavoidable. Consequently, ^{8}Be does not exist in nature (Color figure online)

- $^4\mathrm{He} + 2\mathrm{p} \rightarrow {}^6\mathrm{Be}$,
- $2^4\mathrm{He} + \mathrm{p} \rightarrow {}^9\mathrm{B}$,
- $3^4\mathrm{He} \rightarrow {}^{12}\mathrm{C}$.

The first two reactions were rejected because they do not lead to a stable nucleus. The third reaction leads directly to ^{12}C (see Fig. 8.2 left). However, Bethe knew from Eddington's models that the temperatures in main sequence stars are around 2×10^7 K, while according to Bethe's estimates, the triple collision of α particles requires temperatures of the order of 10^9 K, which were way beyond the range of temperatures prevailing in main sequence stars according to Eddington. Hence, Bethe dropped the idea of a three-body reaction.

In 1951, von Weizsacker[1] discussed the problem and claimed that the energy source could still be gravitation if only the condensation were extremely large,[2] which is essentially a Milne type model. Weizsacker noted that there were giants among the old population of stars as defined by Baade. So these stars must have had a very long lifetime. Consequently, they must be relatively low mass stars. If giants are main sequence stars which have exhausted their hydrogen fuel in the core and energy is no longer produced in the core, then there need not be any temperature gradient in the core. Weizsacker thought that the fusion of H into He was the only process that could operate in stars because he believed that there was no way to overcome the $A = 5$ and 8 barriers (see Fig. 8.2 right). Consequently, nuclear energy was not an option for these stars, owing to the assumption that they had exhausted their hydrogen and no other fusion could take place. From this point of view, Weizsacker

[1] von Weizsacker, C.F., Astrophys. J. **114**, 165 (1951).

[2] Jordan, P., *Die herkunft der Sterne*, 1947, p. 38.

put the giants and the white dwarfs in the same category of stars. Still, Weizsacker added a reservation that, in view of results known at this time, there was a possibility of hydrogen burning outside the isothermal core.[3] However, the idea Weizsacker put forward was that the difference between white dwarfs and giants should be that giants have masses above the Chandrasekhar limiting mass[4] and hence have enough gravitational energy in the contraction towards extremely high densities, even as high as nuclear densities.[5] However, no detailed calculations were carried out, just speculations. Weizsacker's ideas and attempts to explain the giants were typical of the hopeless situation before the insurmountable $A = 5$ and 8 barriers.

In 1949, Gamow published a paper on *relativistic cosmology*,[6] in which he brought up an idea Wigner had communicated to him in private:

> Another ingenious method for crossing the mass 5 crevasse was proposed by E. Wigner. It is known as the method of the 'nuclear chain bridge'. Wigner's plan is that all that is required for building a chain bridge is an assumption that there was originally one single nucleus on the right-hand side of the crevasse. Such an assumption can easily be granted, since some building up is still maintained across the crevasse by the reaction $^{4}\mathrm{He} + ^{3}\mathrm{T} \rightarrow ^{7}\mathrm{Li} + \gamma$ in spite of the low probability of its occurrence.

Gamow published his rejection of Wigner's idea in a book,[7] just when Salpeter and Hoyle started the work on the triple α process, which ended a few years later with the formulation of the detailed nuclear physics. Gamow gave no reference to where Wigner had published his suggestion.

8.2 Pursuing the Structure of the Carbon Nucleus

The story begins as early as 1917, when Harkins[8] observed that the binding energies of the so-called α nuclei (^{12}C, ^{16}O, ^{20}Ne, ^{24}Mg, and so on) are particularly large. Consequently, he suggested a periodic table for nuclei (see Fig. 8.3 left). The basis for this periodic table was Harkins' idea that the basic building block of nuclei is the

---

[3] Richardson, R.S. and Schwarzschild, M., Astrophys. J. **108**, 373 (1948); Gamow, G. and Teller, E., Phys. Rev. **55**, 791 (1939); Gamow, G. and Keller, G., Rev. Mod. Phys. **17**, 125 (1945); Gamow, G. Phys. Rev. **67**, 120 (1945).

[4] Chandrasekhar, S., MNRAS **95**, 207, 226, 676 (1935).

[5] Nuclear densities are densities at which the nuclei touch each other. The radius of the nucleus is about 10^{-5} times the radius of the atom, taking the latter to be the radius of the outer occupied electronic shells. The atom has an infinite number of electronic shells with radius increasing to infinity. But as the number of electrons in each atom is finite, they occupy only the lowest energy levels and for this reason any particular atom has a finite radius, the radius of the highest occupied level. Normal matter densities on the Earth are $\sim 1\,\mathrm{g/cm^3}$. If the matter is compressed to densities as high as $10^{15}\,\mathrm{g/cm^3}$, the nuclei touch each other and form one gigantic nucleus. This is nuclear matter.

[6] Gamow. G., Rev. Mod. Phys. **21**, 367 (1949).

[7] Gamow, G., *The Creation of the Universe*, Viking press, New York, 1952.

[8] Harkins, W.D., J. Am. Chem. Soc. **39**, 856 (1917); Phys. Rev. **15**, 85 (1920).

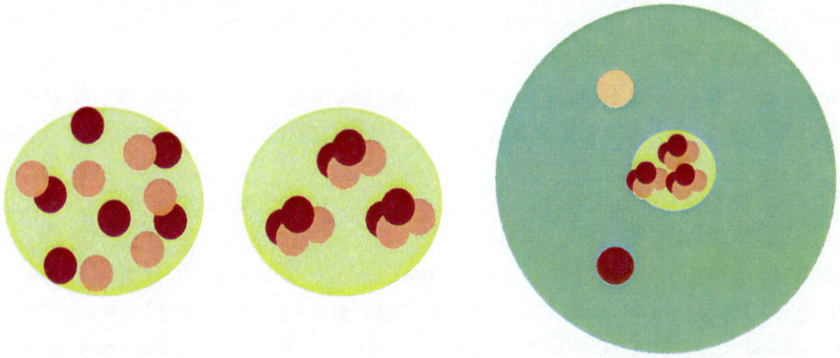

Fig. 8.3 *Left* two possible structures of the carbon nucleus. *On the left*, the nucleons move randomly inside the nucleus. *On the right*, there are three clusters of α particles and the α particles move randomly inside the nucleus. *Right* the structure of ^{14}N according to the revised α-particle nuclear model

α particle. Harkins' original idea was put forward before the neutron was discovered. After the discovery of the neutral particle, the model was modified and the nucleus was considered to be made of a core of α particles with up to 3 nucleons moving around it (see Fig. 8.3 right).

Consequently, the following fundamental question was raised: to what extent would a nucleus of type $4n + 2$ yield a nucleus of type $4n$, where $n = 1, 2, 3, \ldots,$ and an α particle if bombarded with nuclei with atomic weight 2, viz., deuterium? In symbols, the reaction is: $^{4n+2}X_{2n+1} + {}^{2}D_{1} \rightarrow {}^{4n}Y_{2n} + \alpha$, where X is any nucleus and Y is the element that precedes the element X in the periodic table.

The difference between the two models is fundamental. The α particle is fully symmetric and has zero spin, so it must obey Bose–Einstein statistics. If so, then according to these statistics, the three α particles can be at the same energy level. On the other hand, the protons and the neutrons both have spin 1/2 and so obey Fermi–Dirac statistics, which means that no two protons or two neutrons, can be in the same energy level because of the Pauli principle. Obviously the behavior of the particles in the potential well and the way they fill up the levels is completely different in the two cases. The difference should be expressed in the binding energy, which is the energy of the ground state, as well as in the energy levels. If there is a potential well and the particles moving inside it obey Fermi–Dirac statistics, we can expect periodicities in the properties.

On the other hand, if the α particles preserve their zero spin while they move in the nucleus, then they behave like bosons and can all stay in the same nuclear level. Obviously, the binding energy in such a case is larger than in the case of individual protons and neutrons (because in the latter case the nucleons are spread over many energy levels, while in the former case they are all in the lowest, most tightly bound level).

The idea that nuclei are composed of α particles emerged from the fact that the α particle is extremely stable and we do not know whether it has any excited levels, like deuterium, which is composed of a neutron and a proton and does not have any excited state, only a ground state. Hence the helium nucleus is not unique from this point of view. The binding energy of deuterium is 2.225 MeV while that of an α particle amounts to 28.3 MeV, which is significantly more than just two deuterium nuclei put together. It seems that the potential well inside which the α particle resides is very deep and still does not possess any excited state. Lauritsen and Crane[9] investigated this point in 1934, and indeed did not discover any excited state. Feenberg[10] assumed, on the basis of the experiment by Crane et al.,[11] that there might be one excited state, despite the fact that a year earlier two of the authors had rejected this possibility. However, to obtain such a state theoretically, Feenberg had to assume that the range of the nuclear force exceeded 2×10^{-13} cm, which was not acceptable. Margenau[12] carried out the full calculation on the basis of Feenberg's forces, but found that the excited levels were unstable unless the range of the forces was even greater, namely 2.8×10^{-13} cm. But Crane et al. investigated the γ that emerged from the reaction

$$^{7}\mathrm{Li} + \mathrm{p} \rightarrow 2\alpha,$$

and concluded that:

> On the basis of theoretical estimates of the lifetime for radiation and for separation of the two alpha-particles, we must think of the radiation as being emitted predominately after the excited alpha-particle has escaped from the field of the other alpha-particle. The γ-rays are thus characteristic of free alpha-particles, and only the probability of excitation will depend upon the mechanism by which the reaction occurs.

It looks as though they missed the unstable short-lived level in ^{8}Be. This happened in 1935.

The α model was not restricted to light elements. As we know, the heavy radioactive elements emit α particles and not protons or neutrons. Consequently, after Rutherford and his collaborators[13] had sufficiently accurately determined the location of the energy levels in radium C (which is the element bismuth), Gamow[14] attempted to explain the level with a model of a tightly bound core and an α particle moving around it. Wilson[15] showed that Gamow might be right, since the energy levels corresponded very nicely to a single particle moving in the potential well of a harmonic oscillator.

---

9 Lauritsen C.C. and Crane, H.R., PRL **46**, 537 (1934).

10 Feenberg, E., Phys. Rev. **49**, 328 (1936).

11 Crane, H.R., Delsasso, L.A., Fowler, W.A., and Lauritsen, C.C., Phys. Rev. **48**, 100, 102, 125 (1935).

12 Margenau, H., PRL **53**, 198 (1938).

13 Rutherford, E., Proc. Roy. Soc. **A131**, 684 (1931); Rutherford, E., Lewis, W.B., and Bowden, B.V., Proc. Roy. Soc. **A142**, 347 (1933).

14 Gamow, G., Nature **131**, 433 (1933).

15 Wilson, H.A., PRL **44**, 858 (1934).

Recent investigations of the structure of the α particle[16] show that:

- Unlike normal nuclei, where the density is quite constant, the density is much more concentrated towards the center.
- Three-body forces (in contrast with the more common case of forces between two bodies) are very important in this unique nucleus.

These facts explain the extra binding energy of the α particle. If the α particle is so tightly bound, no wonder Harkins reached the conclusion that it was the basic building unit for nuclei.

Before leaving the question of the existence of an α particle in the nucleus, we should mention Schrödinger,[17] who, in the late 1920s, criticized the participants in a Berlin seminar for their lack of imagination. In his spontaneous reaction, he exclaimed:

> Just because you see an alpha particle coming out of the nucleus, you should not necessarily conclude that inside they exist as such.

We should remember this particular example of Schrödinger's wisdom when dealing with the α model for nuclei. Indeed, in 1960, Mang[18] demonstrated how right Schrödinger was: the α particle forms only during escape from the nucleus. Another example is Fermi's theory of β decay. Fermi considered the proton and the neutron as a quantum system with two states. A transition between the two states is induced when the creation of an electron and a neutrino takes place. In this picture, the particles emitted from the nucleus are created at the moment of escape from the nucleus.

8.3 The Structure of the ^{12}C Nucleus

In 1933, in an attempt to solve the original question, Lewis et al.[19] bombarded a nitrogen target with deuterium projectiles. They discovered that the hypothesis was indeed correct and that the products were as expected, namely a $4n$ nucleus plus an α nucleus. One can therefore, consider the nucleus of ^{14}N as made up of a cluster of three α nuclei plus a proton and a neutron. The incoming deuterium just picks up the proton and the neutron, becomes an α particle, and leaves the scene. However, the experimentalists noticed that the energy of the emitted α particle obtained in this disintegration was only about one half of what the difference in mass implied, i.e., the kinetic energy Q_{kin} was only one half of Q as given by

[16] Glöckle, W., and Kamada, H., PRL **71**, 971 (1993). See also Nogga, A., Kamada, H., Glöckle, W., and Barrett, B.R., arXiv:nucl-th/0112026v2, 28 May 2002.

[17] In Jensen, J.H.D., in Nobel Foundation, Nobel Lectures **4**, Physics, 1963–1970, Elsevier, Amsterdam, 1972.

[18] Mang, H.J., Phys. Rev. **119**, 1069 (1960).

[19] Lewis, G.N., Livingston, M.S., and Lawrence, E.O., PRL **44**, 55 (1933).

$$m(^{14}\mathrm{N}) + m(^{2}\mathrm{D}) - m(^{12}\mathrm{C}) - m(^{4}\mathrm{He}) = Q,$$

where $m(\mathrm{X})$ means the mass of the nucleus X. Since they did not have the apparatus required to detect γ rays, they could only claim that:

> The energy of the α particles obtained in this disintegration is only about one-half of that which should be set free in the process $^{14}\mathrm{N} + {}^{2}\mathrm{D} \rightarrow {}^{12}\mathrm{C} + {}^{4}\mathrm{He}$.

No explanation was provided for the missing energy.

In 1935, Lawrence et al.[20] investigated this reaction, but again without any γ-ray detectors. They checked that a ^{12}C nucleus is indeed formed and that two groups of α particles emerged with different energies. Because the α particles had two different energies, they concluded as follows:

> It appears that there are two γ-ray levels in ^{12}C, one of 3.8 MeV and another of 4.7 MeV, though of course the possibility that one (or both) of the levels belong to the alpha-particles cannot be excluded.

In other words, the α particle could have an excited state in contrast to previous findings. The lost energy went into exciting the nucleus, and in this way was lost to the mass–energy conservation equation. Here they pointed to an independent piece of evidence discovered by other researchers: Bothe and Becker,[21] who observed 5 MeV γ rays when they bombarded beryllium with α particles (in the reaction $^{9}\mathrm{Be} + {}^{4}\mathrm{He} \rightarrow {}^{12}\mathrm{C} + \mathrm{n}$). Crane and Lauritsen[22] observed γ rays from boron bombarded by deuterium (in the reaction $^{11}\mathrm{B} + {}^{2}\mathrm{D} \rightarrow {}^{12}\mathrm{C} + \mathrm{p}$). In 1934, Crane et al.[23] identified their 5.6 MeV γ with the 5 MeV γ of Bothe and Becker, and ascribed it to the existence of an excited state in ^{12}C. This was the first direct discovery that ^{12}C has an excited state, and not the α particle. The energies of the states were not accurately, known but their existence was strongly suspected.

8.4 A Digression to ^{8}Be

In 1937, Bethe[24] applied the theory of radioactive α particle decay to estimate the lifetime of the ^{8}Be nucleus. As he did not know the exact energy of the ground level, he gave a table (see Table 8.1), in which the lifetime was given as a function of the uncertain energy. The crossing time has been included in the table. This is the time needed for the α particles to meet each other. As is clear from the table, the lifetime of the ^{8}Be nucleus may be even 10^{6} times longer than the crossing time. The fact

[20] Lawrence, E.O., McMillan, E., and Henderson, M.C., Phys. Rev. **47**, 273 (1935).

[21] Bothe, W., and Becker, H., Zeits. f. Physik **76**, 421 (1932).

[22] Crane, H.R., and Lauritsen, C.C., Phys. Rev. **45**, 497 (1934).

[23] Crane, H.R., Delsasso, L.A., Fowler, and Lauritsen, C.C., Phys. Rev. **46**, 1109 (1934).

[24] Bethe, H.A., Rev. Mod. Phys. **9**, 69 (1937). This extensive review of nuclear physics justifiably earned the title 'Bethe's Bible'.

Table 8.1 Bethe's estimate for the lifetime of the ^{8}Be nucleus as a function of the energy excess above the state of two free α particles (in seconds).

Energy excess (keV)	50	100	200	300	400
$R = 2.5 \times 10^{-13}$ cm	4×10^{-13}	3×10^{-14}	2×10^{-15}	2×10^{-19}	5×10^{-20}
Crossing time [s]	4×10^{-22}	2.8×10^{-22}	2.0×10^{-22}	1.6×10^{-22}	1.4×10^{-22}
$R = 5.0 \times 10^{-13}$ cm	7×10^{-14}	4×10^{-17}	3×10^{-19}	3×10^{-20}	1×10^{-20}
Crossing time [s]	8×10^{-22}	5.6×10^{-22}	4×10^{-22}	3.2×10^{-22}	2.8×10^{-22}

The second row gives the crossing time

that the lifetime of the level is so long relative to the crossing time means that there is a resonance in ^{8}Be.

8.5 Back to New Developments in the Structure of ^{12}C

The correct explanation was put forward five years later, in 1940, by Gaerttner and Pardue,[25] then at the Kellogg Radiation Laboratory in CalTech. Using the same nuclear reaction, $^{14}\text{N} + {}^2\text{D} \rightarrow {}^{12}\text{C} + \alpha$, Gaerttner and Pardue discovered that, in addition to the emerging α particle, γ rays at 1.9, 3.1, 4.0, 5.3, and 7.0 MeV were also produced. They explained their origin as being due to excited levels in ^{12}C at energies of 4.32 and 7.2 ± 0.4 MeV above the ground state. The authors even stated that the level at 7.2 MeV is *weakly coupled with the ground state*, meaning that the probability of a transition from the 7.2 MeV state to the ground state is very small and was not observed. Here we note that the energy of $^8\text{Be} + \alpha$ is 7.367 MeV above the ground state of ^{12}C. Hence only a level above this rest mass energy plus the kinetic energy can be directly reached by the $^8\text{Be} + \alpha$. At stellar temperatures of the order of 10^8–10^9 K, the nuclei have kinetic energies of about 0.01–0.1 MeV, so the kinetic energy is not the dominant part.[26] We conclude that, for the colliding ^{8}Be and α particle to enter into an energy level of ^{12}C, the energy must be above 7.366 MeV, but not by very much. Our knowledge of the structure of the ^{12}C nucleus at that time is shown in Fig. 8.4. The green zone spans the uncertainty in the location of the energy level just above the energy of $^8\text{Be} + \alpha$ or 3 α. Since the reaction rate is very sensitive to the exact value of the energy level, accurate positions are mandatory.

Most nuclear reactions at low energies have two quite independent steps, formation and decay:

- Formation. So far we have discussed the probability of formation. Due to the strong interaction between the particles and the relatively long time the particles stay together, the nucleus 'forgets' how the particular level was reached. The newly

[25] Gaerttner, E.R., and Pardue, L.A., Phys. Rev. **57**, 386 (1940).

[26] The relevant particles are those with energies equal to the Gamow peak, and not those with the average energy.

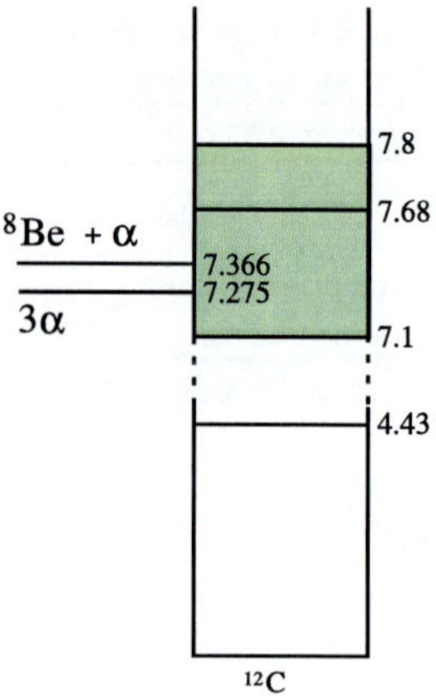

Fig. 8.4 The energy levels of the ^{12}C nucleus relative to the rest mass energy of $^8Be + \alpha$ and 3α. The level at 7.68 MeV can decay into a $^8Be + \alpha$ or 3α. All energies are in MeV

formed excited ^{12}C lives for a while, exactly like the ^8Be nucleus lives for a while, before it decays. Nuclear physicists call the new nucleus the compound nucleus.

- Decay. The new nucleus may have several modes of decay. It can disintegrate back into the original constituents or lose energy in one way or another, and descend into a bound state, provided there is one, even if it is only the ground state. No bound nucleus can be formed without loss of energy. The initial state of free particles has positive energy and the final bound state has negative energy. The difference in energy must be removed from the system. The most common way of losing energy is by emitting a γ. However, other ways exist, but with lower probabilities.

As a rule, nuclear reactions which take place in stars are extremely slow and consequently it is extremely difficult to measure the rates directly in the laboratory. So nuclear physicists look for alternative ways to create the target nucleus and to see how it decays, then use the principle of reversibility: if the new nucleus C decays, say, into $A + B$, then the inverse reaction exists, namely $A + B +$ kinetic energy $\rightarrow C$. If C does not have a stable ground state the reaction takes place, but without a real product. However, if C has a lower bound state which is below the energy of $A + B$, then a stable C nucleus can be formed, provided that the transition is allowed by quantum conservation laws.

In 1940, Holloway and Moore[27] at Cornell, Bethe's home university, repeated the experiment with the reaction $^{14}N + ^2D \rightarrow ^{12}C + \alpha$. This time they carefully analyzed the kinetic energies of the emitted α particles and adopted the explanation given by Gaerttner and Pardue, namely, that the missing energy went into excited states of carbon at energy levels of 7.62 and 4.37 MeV. This appears to be the first reliable identification of the excited level of ^{12}C at 7.62 MeV. The γ emission is then from the decay of the nucleus from the excited states down to a lower state which is not the ground state. Holloway and Moore also suggested that the 7.62 MeV state may be unstable against α emission, i.e., the product ^{12}C further disintegrates into an α and a ^8Be nucleus. According to the principle of reversibility, if ^{12}C decays into

[27] Holloway, M.G., and Moore, B.L., Phys. Rev. **58**, 847 (1940).

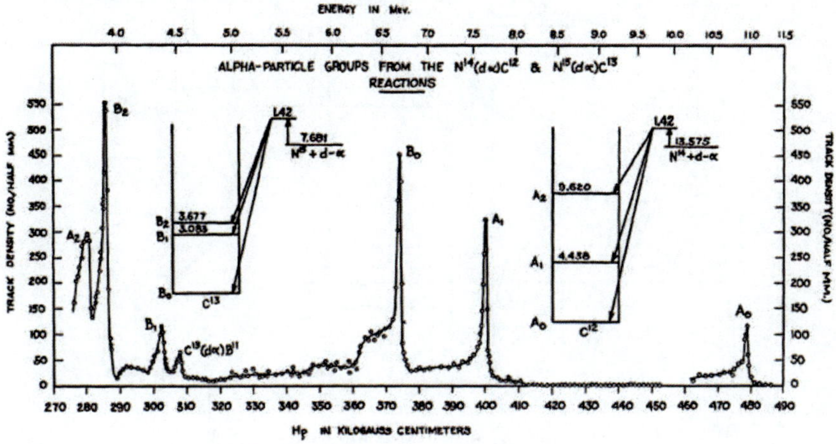

Fig. 8.5 The results of Malm and Buechner (1951). On the *right-hand side*, the implied structure of ^{12}C is shown. No level is seen around 7 MeV

^{8}Be and an α, the inverse reaction must be possible as well. Indeed, they were right, because the energy of this state is above the energy of ^{8}Be + α. This was the first hint of evidence that ^{8}Be + α can fuse into ^{12}C via a resonance level at just the right energy.

It is interesting to note what Holloway and Moore wrote about this energy level:

> The corresponding excited state of ^{12}C would be unstable against alpha-emission, but it is still easily conceivable that such a state could not actually emit alpha particles because of selection rules.[28]

The structure of the energy levels of carbon was, however, still vague. In 1948, Guggenheimer et al.[29] carried out the same experiment (bombarding nitrogen with deuterium) once more and confirmed the existence of a level at 7.3 MeV in ^{12}C. No new information was given about the quantum numbers of the state.

Attempts by Gibson to reach this energy level through the reaction ^{11}B + ^{2}D → ^{12}C + n failed (Fig. 8.5).[30] Malm and Buechner[31] repeated the experiment with the same reaction but got no evidence of a transition to any level around 7 MeV. These difficult experiments yielded contradictory results, but there was a good physical reason why it was so difficult: we shall find later that the probability of decay by emitting a γ is very low (a few parts in 10^{-4}), and this is why some experimentalists missed the decay.

---

[28] Not every transition which is energetically possible actually takes place. As well as energy conservation, the transition between any two energy levels must satisfy all other conservation laws, and in particular momentum and angular momentum conservation. The selection rules tell us which transition is possible and which is forbidden owing to violation of at least one conservation law.

[29] Guggenheimer, K.M., Heitler, H., and Powell, C.F., Proc. Roy. Soc. London **A190**, 196 (1947).

[30] Gibson, W.M., Proc. Phys. Soc. London **A62**, 586 (1949).

[31] Malm, R., and Buechner, W.W., Phys. Rev. **81**, 519 (1951).

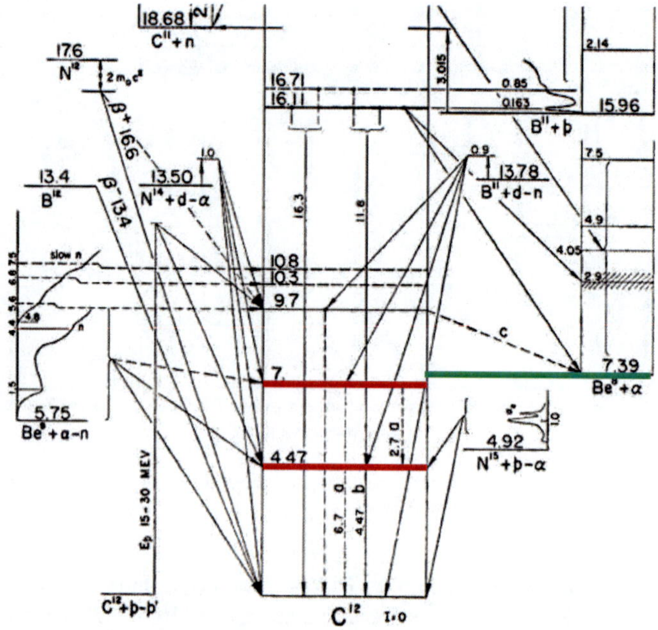

Fig. 8.6 The 1950 compilation of the ^{12}C level by Hornyak et al. Marked in *red* are the 7 MeV and the 4.47 MeV levels. Also marked (in *green*) is the $^{8}\mathrm{Be}+\alpha \rightarrow {}^{12}\mathrm{C}$ reaction. The *broken line* implies that it is not certain that the upper level can be reached by this reaction

In 1950, Terrell[32] bombarded ^{9}Be with α particles with the express purpose of creating the excited ^{12}C nucleus via the reaction $^{9}\mathrm{Be}+\alpha \rightarrow {}^{12}\mathrm{C}+\mathrm{n}$. Only one γ ray appeared to emerge. Once more, no evidence for the existence of 'our' level could be discovered.

In the same year, Johnson[33] investigated the neutrons emitted by the reaction $^{11}\mathrm{B}+{}^{2}\mathrm{D} \rightarrow {}^{12}\mathrm{C}+\mathrm{n}$. He found the ^{12}C levels at 4.4 and 9.6 MeV, but again no indication was found of the level near 7 MeV.

In 1959, Hornyak et al. summarized the information about the energy levels of ^{12}C. On the basis of the experiments $^{14}\mathrm{N}+{}^{2}\mathrm{D} \rightarrow {}^{12}\mathrm{C}+\alpha$ and $^{11}\mathrm{B}+{}^{2}\mathrm{D} \rightarrow {}^{12}\mathrm{C}+\mathrm{n}$, the level at 7 MeV is marked (see Fig. 8.6), while the experiment with the reaction $^{9}\mathrm{Be}+\alpha \rightarrow {}^{12}\mathrm{C}+\mathrm{n}$, which should lead to the same carbon level, is marked with a broken line to indicate that it is not certain. They just gave the three experimental results for the ^{8}Be and did not evaluate them. As for the levels of ^{12}C, based on

[32] Terrell, J., Phys. Rev. **80**, 1076 (1950).

[33] Johnson, V.R., Phys. Rev. **86**, 302 (1952).

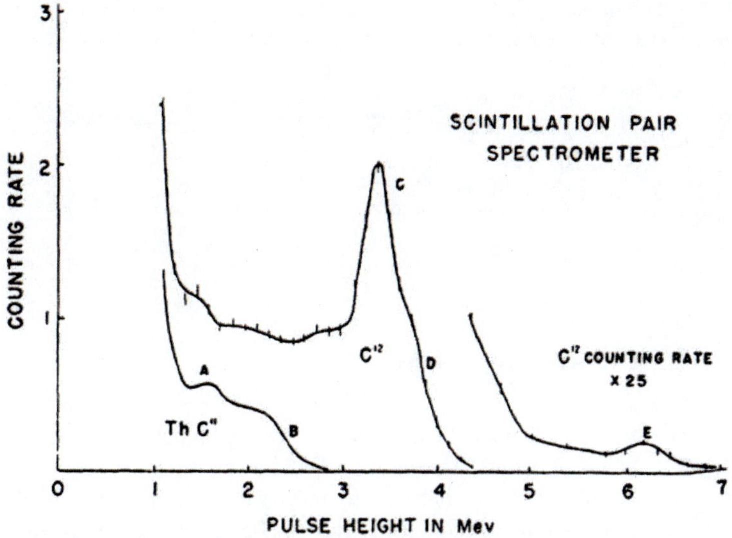

Fig. 8.7 The results of Pringle et al. (1950), where the small peak E is interpreted as *production of an electron–positron pair due to weak* ^{12}C γ *ray at 7.2 MeV*

Tollestrup et al.,[34] they remove the 3.0 or 2.5 MeV level, as there was no evidence to support its existence (Fig. 8.7).[35] They also reported a claim by Pringle et al.[36] that:

> There is evidence for the existence of a very weak one at approximately 7.2 MeV. [...] A detailed study in this region is difficult because of the weakness of the source at this energy and the low relative intensity of the γ ray estimated at about 1 percent of the lower energy component.

Hence, they concluded that there is a strong level at 4.40 MeV and a weak high energy at 7.2 MeV, in agreement with Bradford and Bennett.[37] This explained why the level was there, but exasperated the experimentalists because not all experiments showed its existence.

[34] Tollestrup, A.V., Fowler, W.A., and Lauritsen, C.C., Phys. Rev. **76**, 428 (1949).

[35] The figure in the paper has a small error. The level at 7 should be 7.1 as given in the text.

[36] Pringle, R.W., Roulston, K.I., and Standil, S., Phys. Rev. **78**, 627 (1950).

[37] Bradford, C.E., and Bennett, W.E., Phys. Rev. **77**, 753 (1950).

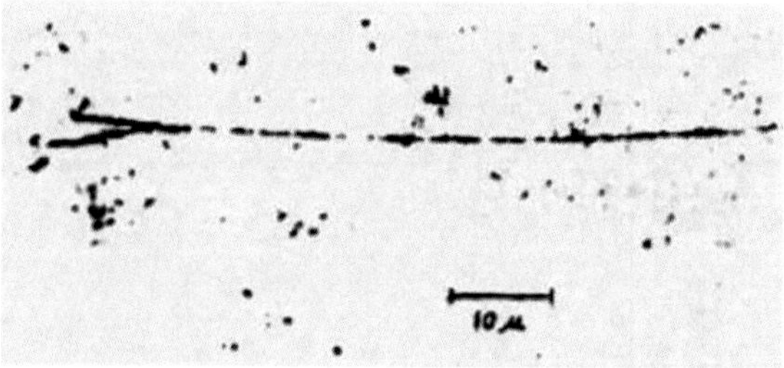

Fig. 8.8 The characteristic appearance of disintegrating ^{8}Be, which is a temporary state in the ^{12}C $\rightarrow$ 3^{4}He decay. The track seen is that of ^{8}Be and the slightly asymmetric V-shaped track is the point where the ^{8}Be decayed into two α particles. From Wilkins and Goward (1950)

8.6 The Last Word: ^{8}Be Is Unstable

We have described Bethe's theoretical result concerning the stability of ^{8}Be. A long list of experimenters attempted to check the system, as discussed in Shaviv.[38]

The most dramatic and convincing experiment was done in 1951 when Miller and Cameron[39] succeeded in measuring the lifetime of ^{8}Be directly by following tracks of ^{8}Be nuclei in emulsion (see Fig. 8.8). At that time, nuclear emulsion[40] constituted one way to detect ionized particles. The half-life thus found was $(5 \pm 1) \times 10^{-14}$ s. At long last, the complete experimental proof that the two α particles do not just cross each other but stay 'bound' for about 10^{-14} s was obtained, and the numbers were close to the maximum duration Bethe had guessed. Further evidence was given by Wilkins and Goward,[41] from whose paper Fig. 8.8 was taken. They also quoted an upper limit for the lifetime of the ^{8}Be nucleus of 9×10^{-14} s. And so the story ended. The system of two α particles is unstable, while the $4n$ nuclei with $n \geq 3$ are the most stable.

8.7 Öpik's Suggested Solution: The Triple Alpha Process

None of the above nuclear data was explored by Öpik in 1951 when he investigated the synthesis of helium into carbon. Öpik[42] assumed what he called somewhat erroneously a triple collision. He assumed that ^{4}He+^{4}He $\rightarrow$ ^{8}Be*, where the star means that it is unstable and hence temporary. Öpik probably knew that ^{8}Be was unstable, but completely overlooked the existence of a resonance (which was very well known

[38] Shaviv, G., *The Life of Stars, The Controversial Inception and Emergence of the Theory of Stellar Structure*, Springer and Magnes Pub., 2009.

[39] Miller, C., and Cameron, A.G.W., Phys. Rev. **81**, 316 (1951).

[40] A nuclear emulsion plate is a photographic plate with a thick emulsion layer. The ionized particles ionize the grains of the emulsion and in this way leave the imprints of their track in the emulsion.

[41] Wilkins, J.J. and Goward, F.K., Proc. Phys. Soc. **63A**, 1173 (1950).

[42] Öpik E. Proc. Roy. Irish Acad. **A54**, 49 (1951). The paper was published over a year after it was read before the Irish Academy.

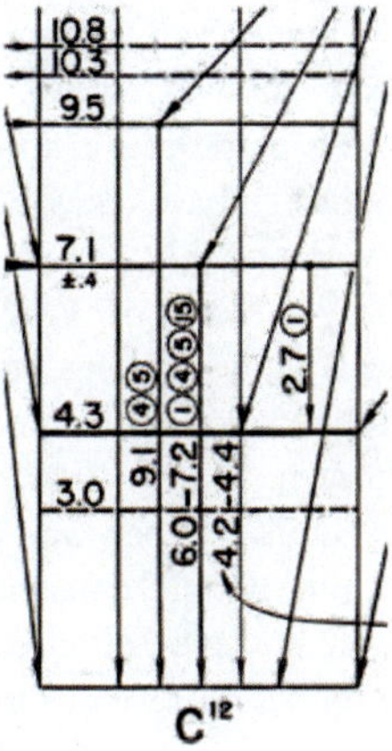

Fig. 8.9 The structure of ^{12}C as it was understood in 1948. For ease of reading the heavy data in the figure, only the relevant part, which describes the structure of ^{12}C, has been copied here. The level at 3.0 MeV was questionable. After Hornyak and Lauritsen (1948)

at that time), so he took the duration of penetration for an ordinary double collision $^4\text{He} + {}^4\text{He} \rightarrow {}^8\text{Be}$ to be $\Delta t = 0.8 \times 10^{-20}$ s and assumed that the third particle had to collide with the other two during this time. At these energies, the speed of the α particle is 2.5×10^9 cm/s or 25,000 km/s. The radius of the α nucleus is about 2×10^{-11} cm. Accordingly, the allotted time he assumed for the collision was the time taken for one α to pass the other.

As can be seen, Öpik's estimate for the lifetime of the ^{8}Be was off by a factor of about 10^6 relative to the published data available in the extensive physical literature and standard tables. The consequence of this underestimate is the same factor in the rate, whence the true rate should have been 10^6 times greater.

What will be the rate of formation of ^{12}C? Öpik wrote that the lifetime during which a helium nucleus can react is proportional to (Fig. 8.9):

(a) the duration of the ^{8}Be nucleus lifetime,
(b) the probability for the two α particles to penetrate each other,
(c) the probability of a third α penetrating the ^{8}Be, and
(d) *the probability of capture*, by which he meant the probability that the excited nucleus would decay to a stable state.

Factors (b) and (c) were given by Gamow's potential penetration theory. For the factor (a), Öpik assumed the time required for the two α particles to fly past one another, so the third particle had to be in the vicinity of the other two α particles when they encountered each other. Factor (d) tells us the probability that, once the excited ^{12}C nucleus is formed, it decays by emitting a γ and descends to the ground state, or any bound state of ^{12}C lower in energy. Öpik gave the final result and did not specify the value of this parameter, which he denoted by q_γ, leaving it up to the reader to guess what value he had adopted and why. As a matter of fact, q_γ was unknown, save for indirect indications from the various experiments that it had to be very small.

Since Öpik ignored the existence of the resonance state, he could not realize how important it was in determining the temperature (and hence the energy) that the α particles must have in order to populate this level. Assuming that the helium burning phase takes 10^{14} s,[43] Öpik found that a temperature of 6×10^8 K was needed to completely consume the helium. Hence, his temperature estimate was off by only a factor of 3–6. The reason why the error in the temperature was so small, relatively speaking, despite the fact that his rate was so far off, is that the temperature dependence of the rate comes mainly from the penetration factors, which were well known.

Öpik's paper was published in a journal which not many astrophysicists visit regularly, and consequently was not implemented in evolutionary calculations. It was, however, cited years later for the idea of the three α particle reaction. From its publication up until mid-2009, Öpik's paper was cited only once, and this was by Salpeter![44] In 1972, Wrubel wrote under the title *Ernst Öpik and the Structure of Red Giants*,[45] and did not mention the idea of the triple alpha reaction. This explains why Salpeter, who worked on the same problem in 1952, did not cite Öpik. He simply did not know about it.

8.8 More Information About ^{12}C

In the meantime, around 1950, Guier and Roberts,[46] who examined the reaction $^9\text{Be} + \alpha \rightarrow {}^{12}\text{C} + \text{n}$, saw the signal of a level in ^{12}C at 7.8 MeV. The signal was not confirmed by the same authors plus Bertini two years later.[47] On the other hand, the energy level they got the second time was 7.5 MeV, in contrast with 7.1 MeV found by Hornyak et al.[48] The uncertainty and confusion was not yet over.

The situation was summarized by Azjenberg and Lauritsen in 1952,[49] in a figure which provided the structure of energy levels in ^{12}C. As can be seen from Fig. 8.10, a level at 4.43 MeV and another at 7.5 MeV appeared with quantum numbers $J = 2^+$

---

[43] If one assumes, as Öpik did, that the stars in a given location of the HR diagram are burning helium, then the number of stars in this location relative to their number along the main sequence is a good estimate of the ratio of the times the stars take to consume their nuclear source.

[44] Salpeter, E.E., when he discussed the history of nuclear astrophysics before 1957, Pub. Astron. Soc. Australia **25**, 1 (2008). The Astrophysical Data System (ADS) is not very accurate with the citations, and I found two citations to Öpik which were not included in the ADS, for example, Dunbar et al., Phys. Rev. **92**, 649 (1953). So ADS results should be treated with caution.

[45] Wrubel, M.H., Irish AJS **10**, 77 (1972).

[46] Guier, W.H., and Roberts, J.H., Phys. Rev. **79**, 719 (1950).

[47] Guier, W.H., Bertini, H.W., and Roberts, J.H., Phys. Rev. **8**, 426 (1952).

[48] Hornyak, W.F., Lauritsen, T., Morrison, P., and Fowler, W.A., Rev. Mod. Phys. **22**, 291 (1950).

[49] Azjenberg, F. and Lauritsen, T., Rev. Mod. Phys. **24**, 321 (1952), published October 1952. The authors' affiliation was with the Kellog Radiation Laboratory, CalTech. Thomas Lauritsen (1915–1973) was the son of Charles Lauritsen (1892–1968m) who established the Kellogg laboratory and was the director from its establishment until his retirement in 1962.

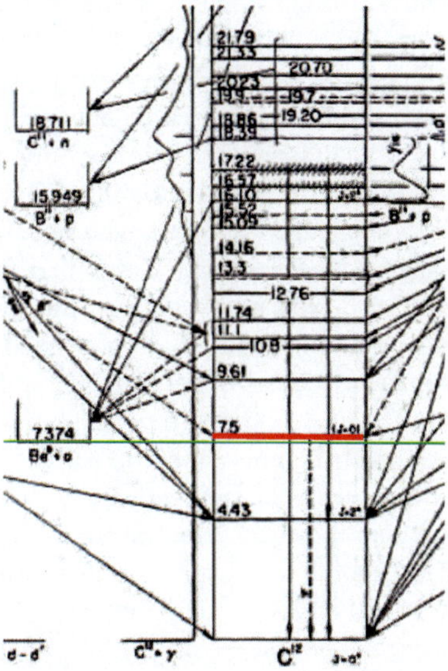

Fig. 8.10 The structure of ^{12}C as understood in 1952. For ease of reading figure, which is heavy with data, the energy level at 7.5 MeV has been marked in *red* and the level of ^{8}Be + α has been marked in *green*. Note that the spin of the level ($J = 0$) is given, indicating that the ^{8}Be + α can create the excited ^{12}C nucleus. It is this coincidence which led later to a philosophical hypothesis. Note also that the spin of the ground state is marked as $J = 0$, so that no direct transition from the 7.5 MeV level to the ground state is possible. From Ajzenberg and Lauritsen (1952)

and $J = 0$, respectively.[50] No parity is assigned to the upper level. The only reference given for this information is *cloud-chamber investigation*, without the details of the paper. However, the only researchers to apply a cloud chamber to a similar problem were Crane, Delsasso, Fowler, and Lauritsen[51] back in 1935.

The data on the quantum numbers of the energy levels is important. It was well known that the ground state of ^{12}C was $J = 0^+$. It was also well known that a transition from a $J = 0^+$ to $J = 0^+$ was strictly forbidden. The spin of the photon is 1. Hence a transition from a zero spin to a zero spin does not leave room for the photon, while nevertheless conserving the angular momentum. Hence, if a bound carbon nucleus is to be the end product, it is not sufficient to have this energy level, because no direct transition is possible from the 7.5 MeV state to the ground state. Another energy level must therefore exist above the ground state and below the level at 7.5 MeV, and it must have the right quantum numbers for the transition to the ground state to go through an intermediate state.

8.9 Salpeter: The Solution Is Two Plus One

In the early 1950s, Willy Fowler, who was already then an insatiable driving force behind the Kellogg laboratory at CalTech, extensively consulted Hans Bethe, who

[50] In 1950, Azjenberg and Lauritsen had the level at 7. MeV and it is not clear why the level 'moved' half an MeV upward.

[51] Crane, H.R., Delsasso, L.A., Fowler, W.A., and Lauritsen, C.C., Phys. Rev. **48**, 100 (1935).

was at Cornell. After a while, Bethe agreed to send Fowler one of his best young men, and Edwin Salpeter (1924–2008) was his choice. Salpeter spent the summer of 1951 at the Kellogg laboratory. It was during this and subsequent stays that Salpeter carried out his research on how helium can be synthesized into carbon.

In October 1952, Salpeter submitted his first short paper to the Astrophysical Journal.[52] In the same month of October, the Physical Review published a paper by Ajzenberg and Lauritsen, who were working at the same Kellogg laboratory, in which the resonance level in ^{12}C was marked at 7.5 MeV with a quantum number assignment of $J = 0$. No parity was given. Salpeter submitted his paper after returning to Cornell, his home university. It so happened that two of the most important results needed for the synthesis of carbon were obtained at the Kellogg laboratory, and published at the same time in different journals, but without reference to each other. Salpeter just published his paper too early.

Salpeter was apparently unaware of the suggestion by Gamow,[53] who wrote about the equilibrium of ^{8}Be, and nor did he know about the work of Öpik, as no reference to these works was given.[54] As for Öpik, it made no difference because his assumption was wrong. As for Gamow, he was investigating an energy source for main sequence stars and not the question of how helium might be synthesized.

Gamow assumed, in 1938, that ^{8}Be is stable and assumed also that, provided the binding energy is small, at the core temperatures of main sequence stars, an equilibrium will be set up, viz.,

$$^{8}\mathrm{Be} \rightleftharpoons 2\,^{4}\mathrm{He}.$$

Even at the relatively low temperatures of main sequence stars, ^{8}Be disintegrates.

When Salpeter attacked the problem, it was already well established that ^{8}Be is unstable, so Gamow's argument, which assumed a stable ^{8}Be, had to be changed in any case. The physical argument was that, if the lifetime of the ^{8}Be is long compared to the crossing time (there is a resonance), then it is permissible to write a dynamic equilibrium equation for the formation of ^{8}Be out of two α particles (see Bethe 1937). Salpeter's important contribution to the problem was to take into account the resonance in the unstable ground state of ^{8}Be, just what Öpik should have known but neglected. Being at the Kellogg laboratory, Salpeter took the 'Kellogg value', namely 95 keV, for the resonance in ^{8}Be. Salpeter gave no reference to indicate where he got the data from.[55]

Salpeter stressed that he assumed no resonances in ^{12}C, ^{16}O, and ^{20}Ne. As for the level of ^{12}C, Salpeter wrote:

> The nuclear γ-ray width for the formation of ^{12}C (but not the one for ^{8}Be) is required. This width has not yet been measured, and the position of resonance levels was not yet known

---

[52] Salpeter, E.E., Astrophys. J. **115**, 326 (1952).

[53] Gamow, G., Phys. Rev. **53**, 595 (1938).

[54] The details of this story were checked personally with Salpeter during my stay at Cornell.

[55] The Kellogg value was from Tollestrup, A.V., Fowler, W.A., and Lauritsen, C.C., Phys. Rev. **76**, 428 (1949).

> accurately enough and an estimate of 0.1 eV was used for this width. Hence the correct production rate could be smaller by a factor of as much as 10 or larger by as much as 1000.[56]

Salpeter hoped that a connection would be found between his rate and carbon stars.[57]

The data available on the energy levels in ^{12}C was not very accurate. For this reason Salpeter had to guess the probability that the excited compound nucleus would decay to the lower bound state, and his guess turned out later to be about a factor of 10 too high. But Salpeter warned that the rate could be in error by a factor of 1,000. However, he pointed out that the effect on the temperature at which helium fuses into carbon would be very small.

After briefly discussing the destruction of ^{12}C by an additional α capture leading to ^{16}O, in the last paragraph of this 3 page paper, Salpeter discussed the subsequent steps in stellar energy generation, namely, how ^{12}C could react with itself to form ^{24}Mg, and how later, the heavy nuclei disintegrate into α particles (by γ emission), on the one hand, and how the heavy nuclei absorb the α particles to form still heavier nuclei, on the other. This would later be called the equilibrium alpha process.

Two years after Salpeter's publication of the triple alpha rate, in 1954, Öpik wrote a paper on the chemical composition of white dwarfs,[58] in which he repeated his calculation of the triple alpha process. He wrote:

> The lifetime of the temporary nucleus ^{8}Be formed is assumed equal to $\sim 8 \times 10^{-21}$ s, being an estimate of the duration of penetration. The lifetime of true ^{8}Be is probably much shorter, about 10^{-22} s.[59] [. . .] Non-resonance capture is assumed in this case also.

Öpik then complained in the paper that Salpeter had not cited him, declaring:

> His method of calculation is not quite clear from his brief note. It seems that the reaction $^4\text{He} + {}^4\text{He} \rightarrow {}^8\text{Be}^*$ he has treated in a manner similar to ours, whereas in $^8\text{Be}^* + {}^4\text{He} \rightarrow {}^{12}\text{C}+\gamma$, he has postulated a resonance process. The outcome is a formula yielding 1.4×10^{13} times higher an energy generation with practically similar temperature dependence as our formula. Of the discrepancy, a factor of 10^3–10^4 seems to refer to the first reaction and is about equivalent to the omission of the probability of penetration. The rest, a factor of 10^9–10^{10}, is about what might be expected for the difference in the rate of a resonance reaction and that of a non-resonance process at $T = 2 \times 10^8$.

To justify this, Öpik pointed to Salpeter's comment that the resonance in ^{8}Be is not known accurately enough.[60] It looks as if Öpik did not even fully comprehend

[56] According to Salpeter in *Nuclear Physics Before 1957*, Fowler considered all previous reports on the existence of such a level as wrong. In this way, he disagreed with Lauritsen and Azjenberg from his own laboratory.

[57] Carbon stars are giant stars containing a higher abundance of carbon than oxygen. In most stars the carbon abundance is lower than that of oxygen.

[58] Öpik, E.J., MSRSL **1**, 131 (1954), *Les processus nucléaires dans les astres*, Communications présentées au cinquième Colloque International d'Astrophysique tenu à Liège les 10–12 septembre.

[59] Here Öpik cited Franzinetti and Payne, Nature **161**, 735 (1948). These researchers investigated high energy cosmic ray induced reactions which led to the emission of α particles from the decay of ^{8}Be, which in turn was the product of the decay of ^{8}Li.

[60] Öpik cited Carlson's result, which was the lowest of all the results. But the idea of a dynamic equilibrium, as suggested by Salpeter, is independent of the exact energy. The exact energy only slightly changes the concentration of ^{8}Be in equilibrium.

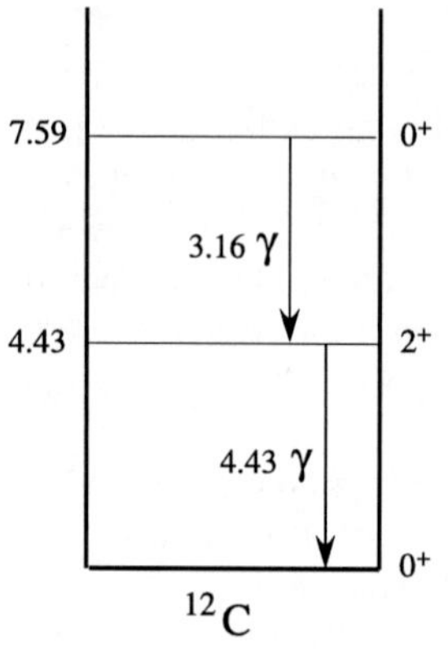

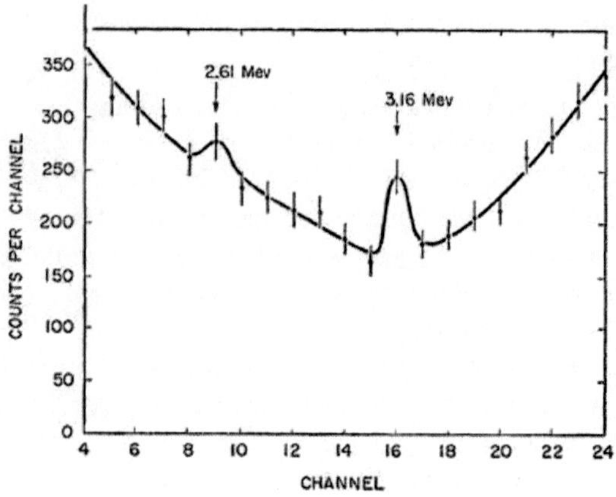

Fig. 8.11 *Left* the status of the level structure of ^{12}C according to Beghian et al. (1953). *Right* the clear evidence for the two groups of γ rays, as discovered by Beghian et al. (1953)

Salpeter's equilibrium idea. Burbidge, Burbidge, Fowler, and Hoyle[61] respected Öpik and cited his erroneous paper. We can trust Willie Fowler to appreciate the difference between Salpeter's and Öpik's treatments.

The young Salpeter was under the influence of the highly articulate and dominant Willie Fowler, who did not trust the floating 'time variable' results for the ^{12}C level, and he carried out the rate calculation without assuming the existence of the energy level. However, he 'smuggled in' a comment at the end of the paper to the effect that a resonance would change the rate by a factor of 1,000.

8.10 Between Salpeter and Hoyle 1952–1954

In 1953, Beghian et al.[62] investigated the reaction $Be^9 + \alpha \rightarrow C^{12} + n$. They detected two groups of γ rays, one at 3.16 ± 0.05 MeV and the other at 4.43 MeV, and concluded that ^{12}C has two levels, one at 4.43 and another at (7.59 ± 0.07) MeV (see Fig. 8.11 left). They also searched for γ rays with energies of up to 7.5 MeV, but found an upper limit of 1/2,500 (relative to previous detections). This implied that the probability that the 7.59 MeV level decays directly into the ground state is less than 1/2,500.

[61] Burbidge, E.M., Burbidge, G.R., Fowler, W.A., and Hoyle, F., Rev. Mod. Phys. **29**, 547 (1957).

[62] Beghian, L.E., Halban, H.H., Husain, T., and Sanders, L.G., PRL **90**, 1129 (1953), submitted April 1953 and published June 1953.

8.11 The Kellogg Laboratory Group Phase I

In the early 1950s, Hoyle was worried by the low rate predicted by Salpeter's calculation for the reaction $3^4\text{He} \rightarrow ^{12}\text{C}$. In 1953, Hoyle came to the Kellogg laboratory and questioned the staff about the possible existence of an excited level of carbon. Hoyle's idea was that a resonance level which corresponds to $T \approx 10^8$ K would accelerate the rate of C production, which otherwise was too low. Ward Whaling together with his visiting associates and a graduate student decided to go to the laboratory and search for this state. They found the energy level exactly where Hoyle had predicted it to be.

It is frequently stated in many articles that Hoyle predicted the existence of the 7.68 MeV level in ^{12}C, and the reference given is Hoyle 1954.[63] When Hoyle wrote his paper, the existence of the level was already known and documented by researchers from Kellogg, as described earlier. However, many of the required details were still missing, and above all Fowler did not trust the experimental results, although some of them were obtained in his own laboratory.

The abstract was presented to the American Physical Society in Albuquerque, New Mexico, 2–7 September 1953. The sensational paper was the last one in the last session. The session itself was on nuclear physics, not astrophysics. It was a victory to astrophysics in general and to Hoyle in particular, under the eyes of the skeptical Fowler,[64] that the location of a nuclear level that wandered so much from one experiment to the other, could be determined by observing the stars. The exact reference is an abstract by Hoyle, Dunbar, et al.[65] with the title: *A state in* ^{12}C *predicted from astrophysical evidence*. The idea was that:

> The observed 'cosmic' ratio of He:C:O can be made to fit the yield calculated for these reactions if the reaction $^8\text{Be} + \alpha \rightarrow {}^{12}\text{C} + \gamma$ has a resonance near 0.31 MeV, corresponding to a level at 7.68 MeV in ^{12}C.

The abstract also claimed, based on Ajzenberg and Lauritsen 1952, that:

> A level had previously been reported at 7.5 MeV.

In the same abstract they report on the experiment which led to the discovery of the level at 7.68 ± 0.03 MeV. The decay mode of the level is not discussed. Shortly afterwards, the two papers were published, one by the experimentalists in the Physical Review and one by the theoreticians in the Astrophysical Journal Supplement. This triumph forged the Fowler–Hoyle collaboration that would contribute enormously to nuclear astrophysics for many years to come.

---

[63] Hoyle, F., Astrophys. J. Suppl. **1**, 121 (1954).

[64] Salpeter, E.E., Publ. Astron. Soc. Australia **25**, 1 (2008).

[65] Hoyle, F., Dunbar, D.N.F., Wenzel, W.A., and Whaling, W., Phys. Rev. **92**, 1095 (1953).

8.12 The New Experiment

Dunbar et al.[66] repeated the $^{14}N + {}^{2}H \rightarrow {}^{12}C + \alpha$ experiment once more, using a special new double focusing magnetic spectrometer, and discovered that the excited level was at 7.68 ± 0.03 MeV, as can be seen from Fig. 8.12. Unfortunately, they did not give the quantum numbers of the levels, and it was impossible to determine whether a stable ^{12}C could be formed by this reaction. This sophisticated and accurate experiment settled once and for all the question about where the level was situated, but could not answer the crucial question as to whether a stable ^{12}C could thus be formed?

8.13 Hoyle: We Must Have Carbon in Our Universe!

In 1954, Hoyle published an extensive investigation into the synthesis of the heavy elements,[67] and discussed the synthesis of carbon as the starting point. This is the paper in which Hoyle suggested a way in which the synthesized elements might spread throughout the Galaxy (see Sect. 6.32). Hoyle followed Salpeter in assuming a dynamic equilibrium for $\alpha + \alpha \rightleftharpoons {}^{8}Be$, and applied statistical mechanics to find the abundance of beryllium. As Salpeter did not provide the equation and only quoted the result, Hoyle cited himself for the equation of equilibrium.[68] In that paper Hoyle discussed the formation of the heavy elements under statistical equilibrium. There was no mention of Gamow's 1938 paper and his equilibrium ^{8}Be reaction.

Once carbon has been synthesized, it can absorb an additional α particle and become oxygen, which in turn can again absorb an α particle and become neon:

$$^{12}C + \alpha \rightarrow {}^{16}O + \gamma,$$

followed by

$$^{16}O + \alpha \rightarrow {}^{20}Ne + \gamma\,.$$

In principle, the ^{20}Ne can capture another α particle to form ^{24}Mg, but the increasing Coulomb barrier reduces the rate so much that α capture can be neglected once past oxygen. The crucial point is therefore that carbon and oxygen are synthesized at the same time, and not one after the other. Thus, on the one hand, carbon is synthesized, and on the other, it is destroyed. At the end of the day, the final amount of carbon depends on the ratio between the rate of formation and the rate of destruction. Hence, Hoyle defined a parameter

[66] Dunbar, D.N.F., Pixley, R.E., Wenzel, W.A., and Whaling, W., Phys. Rev. **92**, 649 (1953).

[67] Hoyle, F., Astrophys. J. Suppl. **1**, 121 (1954).

[68] Hoyle, F., MNRAS **106**, 343 (1946).

Fig. 8.12 Dunbar et al.'s (1953) result. A beautiful manifestation of the predicted 7.68 MeV level

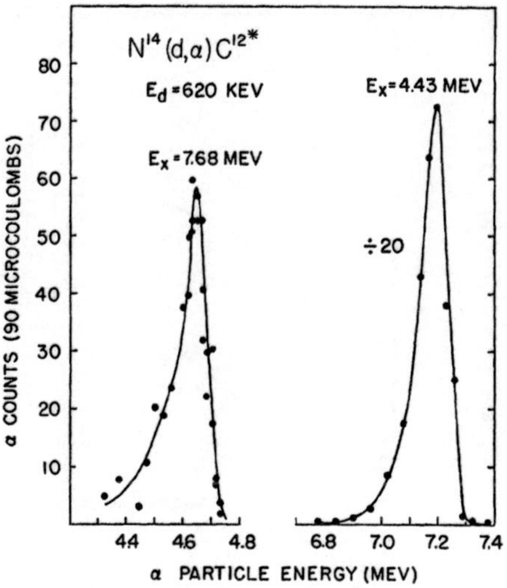

$$\kappa = \frac{A \times \mathrm{Rate}(^{12}\mathrm{C} + \alpha \rightarrow {}^{16}\mathrm{O})}{\mathrm{Rate}(3\alpha \rightarrow {}^{12}\mathrm{C})},$$

where A is a numerical factor, so that relevant values of κ will be of the order of unity. Clearly, κ depends on the temperature as well as the properties of the energy levels in carbon and oxygen. Of all the data which enters into the expression for κ, it is most sensitive to $E_{\mathrm{res}} = E - E(^8\mathrm{Be} + \alpha)$, because

$$\kappa \sim \exp\left(-\frac{E_{\mathrm{res}}}{k_B T}\right) = \exp\left[-\frac{E - E(^8\mathrm{Be} + \alpha)}{k_B T}\right],$$

where E is the energy of the level in the ^{12}C nucleus. All energies are taken above the ground state of ^{12}C. The energy dependence is exponential.

Assuming Salpeter's rate for the triple α reaction and the problematic (see later) subsequent reaction $^{12}\mathrm{C} + \alpha \rightarrow {}^{16}\mathrm{O}$, Hoyle solved for the abundances of carbon, and the results are shown in Fig. 8.13. It can be seen from the figure that, when $\kappa = 1/9$, the amount of carbon produced by the time the helium is exhausted is 1/3, which corresponds to an abundance ratio of C/O = 1/2, the value Hoyle took as the observed cosmic value. Furthermore, one can see that a small variation in κ from 1/3 to 1/15 spans a carbon abundance which covers the observed value. On the other hand, if there had been no resonance at all in ^{12}C, the value of κ would have been orders of magnitude smaller and no carbon would be left at the end of helium burning. If the formation of ^{12}C is much slower than its destruction, the end product is almost pure ^{16}O.

In addition, Hoyle carried out the inverse calculation in a style which is known today as reverse engineering, i.e., he asked what T and E_{res} would have to be in order

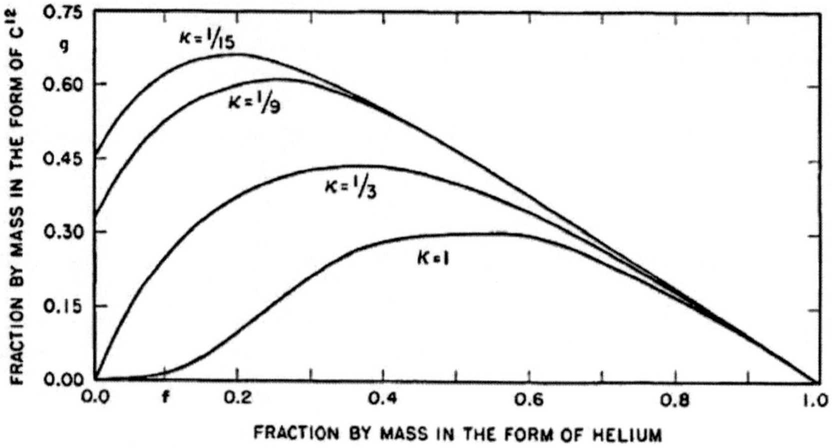

Fig. 8.13 The formation of carbon from helium as a function of Hoyle's parameter κ. This is the figure which convinced Hoyle that ^{12}C must have a resonance

to get the observed abundance, and he found $T = 1.4 \times 10^8$ K and $E_{res} = 0.33$ MeV (above the rest mass of $^8Be + \alpha$), which corresponds to $E = 7.705$ MeV in the ^{12}C nucleus.

In summary, Hoyle was able to apply astrophysical arguments to solve the problem of the carbon energy level, the exact location of which seemed to wander from one experiment to the other. Shortly afterwards, the idea was born that such a resonance must exist, or else the cosmos would have no carbon to support life of the kind we are familiar with. The fantastic story of how astrophysical considerations led to the prediction of a nuclear energy level triggered the invention of the anthropic principle.[69]

8.14 The Kellogg Laboratory Group Phase II

Hoyle's victory was not complete. Cook et al.[70] found it necessary to carry out a new experiment again because:

> Experimental evidence on the character of the 7.7 MeV ^{12}C state is not entirely clear. It seems well established that the state does not radiate directly to the ground state but rather cascades via the 4.43 MeV state.

Furthermore, Cook et al. added that one must:

---

[69] The anthropic principle is a philosophical claim that the Universe must be compatible with the existence of life as it is observed.

[70] Cook, C.W., Fowler, W.A., Lauritsen, C.C., and Lauritsen, T., Phys. Rev. **107**, 508 (1957).

(a) Make sure that the ^{12}C level can be formed by $^{8}Be + \alpha$. The prerequisites for that are the right spin and parity and that there is a non-vanishing probability that it decays by emitting these particles. (b) It must have a finite probability to decay to the lower excited state or the ground state. These questions should be best explored by bombarding ^{8}Be with α particles, but since ^{8}Be is unstable, recourse must be made to study the two possible modes of decay of the excited state of ^{12}C, namely $^{12}C^* \rightarrow {}^{8}Be + \alpha$ and $^{12}C^* \rightarrow {}^{12}C + \gamma$. Then one has to rely on the principle of reversibility of nuclear reactions.

Indeed, Hoyle did not predict the quantum numbers of the level, and without these the prediction was not terribly powerful.

The experiment Cook et al. conducted was innovative, and differed substantially from all previous ones. They created the radioactive isotope ^{12}B, which has an energy of 13.378 MeV above the ground state of ^{12}C and decays via a β decay (in contrast to a γ decay) into all excited levels of ^{12}C with lower energies, and in particular into the two levels relevant for our discussion here (see Fig. 8.14). A special arrangement was set up to detect the emitted α particles using a strongly focusing magnetic spectrometer.

The basic problem that Cook et al. faced with the 7.68 MeV level was as follows. It had been established earlier that there was no direct transition to the ground state. On the other hand, no one had observed this level disintegrating into $^{8}Be + \alpha$.[71] There were merely conflicting estimates of the probability of emitting an α particle. On the one hand, Uebergang[72] and Steffen et al.[73] estimated the probability to be $\ll 50\%$, and on the other hand, Bent et al.[74] and Hornyak[75] estimated it to be $\geq 97\%$.

Cook et al. managed to get only an upper limit for the probability of decay into $^{8}Be + \alpha$. Since this was by far the most probable mode of decay of this level, the exact value was crucial. So here they were forced to rely on Russell, Phillips, and Reich[76] and Heydenburg and Temmer,[77] who had found an upper limit which was about 250 times smaller than the predicted one. With the improved (lower) upper limit they were able to reach their most important result, namely, that the probability for γ decay to the 4.43 MeV level was 0.001–0.0028. It was very small but not zero. Next they found that the probability of decaying through the emission of an electron–positron pair was about 10^{-5}, and hence negligible.

---

[71] Recall that nuclear reactions under astrophysical conditions are usually very slow because the energies are low. So $^{8}Be + \alpha$ yielding ^{12}C and decaying back to the products was hopelessly slow, and hence not observed.

[72] Uebergang, R.G., Aust. J. Phys. **7**, 279 (1954).

[73] Steffen, K.G., Hinrichs, O., and Neuert, H., Zeit. F. Phys. **145**, 156 (1956).

[74] Bent, R.D., Bonner, T.W., McCrary, J.H., Ranken, W.A., Phys. Rev. **100**, 771 (1955).

[75] Hornyak, W.F., Bull. Am. Phys. Soc. Ser. II **1**, 197 (1956).

[76] Russell, J.L., Phillips, G.C., and Reich, C.W., Phys. Rev. **104**, 135, 143 (1956).

[77] Heydenburg, N.P. and Temmer, G.M., Phys. Rev. **104**, 123 (1956).

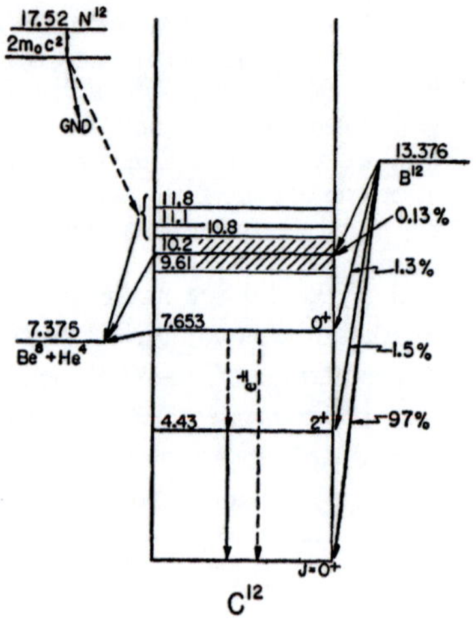

Fig. 8.14 Cook et al. (1957) created the radioactive nucleus ^{12}B which decays into ^{12}C. Most of the decays go into the ground state and only 1.3% go into the investigated level

8.15 Salpeter 1957: Completing the Job

Salpeter's second and updated paper[78] was submitted together with the Cook et al. paper and the two papers were published back to back.

At the beginning of 1957, the spin of the 7.68 MeV level was not yet certain, although very recent experiments seemed to provide the final proof for the quantum numbers of this level. Fowler's distrust in previous results pervaded the community. Fregeau and Hofstadter[79] were interested in using high energy electron beams to investigate the surface of nuclei, and they considered that ^{12}C would be an ideal case. They therefore bombarded ^{12}C with high energy electrons. This and subsequent work of this type won Hofstadter the 1961 Nobel prize.[80]

As early as 1955(!), Fregeau and Hofstadter treated the 4.43 and 7.68 MeV levels of carbon as being so well established that they did not give any reference to prior experiments. Regarding the spin of the 7.68 MeV level, they stated that $J = 0^+$ was *not inconsistent with the experimental results* that they found. The levels emerged very nicely in their electron scattering experiments.

On the other hand, Salpeter argued that the only possibilities were 0^+ or 2^+. If 2^+ was assumed then the probability of electron–positron decay (symbolized in nuclear physics by $\Gamma_{e^\pm}$) could be calculated theoretically, and it was found to be

---

[78] Salpeter, E.E., Phys. Rev. **107**, 516 (1957).

[79] Fregeau, J.H., and Hofstadter, R., Phys. Rev. **99**, 1503 (1955); ibid. **104**, 225 (1956).

[80] The citation of the prize read:

> For his pioneering studies of electron scattering in atomic nuclei and for his thereby achieved discoveries concerning the structure of the nucleons.

1,000 smaller than the one obtained if 0^+ was assumed. This result contradicted the result of the CalTech experiment.[81]

Now came the critical question: what is the probability of decaying back into the incoming channel, ^{8}Be+α? Salpeter resorted to the Wigner theory[82] and calculated an upper limit for the case of 0^+ which was 40% higher than the one adopted by Cook et al. The other possibility was proven by Salpeter to contradict the experimental results. To provide further support for his estimates of the probabilities, Salpeter derived the probabilities by assuming the α model for the ^{8}Be nucleus. The numbers agreed (Fig. 8.15).

Finally came the question: what is the probability Γ_γ of decay into the bound 4.43 MeV level? After all, this was the dominant mode to form a stable carbon nucleus. Again, the estimate depended on the model for the carbon nucleus. If one assumed that the carbon nucleus was made of three α particles, one got a very small probability of decay into the first excited state. So after some lengthy theoretical discussion, Salpeter found that the probability of decay into the ground state plus the probability of decay into the excited state, namely the total probability of decay, was 0.00014. Quite small!

Salpeter made an additional crucial assumption. Since the probability of decay back into the incoming channel (^{8}Be+α) is by far the largest, or in other words, since the leakage to decay modes other than the mode of formation is extremely small, it is reasonable to assume an equilibrium $^8\text{Be} + \alpha \rightleftharpoons {}^{12}\text{C}^*$. The unbound 2$\alpha$ states absorb a third α and become an excited ^{12}C* nucleus. Once this is done and statistical mechanics can be used to calculate the abundance of the excited ^{12}C nuclei, the details of the nuclear reaction become irrelevant! The probability of forming carbon in the ground state is then the number of carbon nuclei in the excited state (as obtained from the equilibrium condition) times the probability that they decay to the ground state of carbon.

At the end of January 1955, the American Physical Society had its annual meeting in New York. In a session on *Reactions of Transmutation and Nuclear Energy Level*, Salpeter presented preliminary results on the effect of the 7.68 MeV level.[83] By now, Salpeter had become aware of the upcoming Ajzenberg and Lauritsen summary (it appeared in his paper as 'to be published') of the energy levels of ^{12}C, and derived the full reaction rate for the first time.

Hayakawa et al.[84] got the information from Salpeter, who also informed them about the probability that the 7.68 MeV excited state decays to the 4.43 MeV level, and submitted their paper to Progress in Theoretical Physics Japan in July 1956, to be published in November 1956. Salpeter's paper, on the other hand, was submitted in March 1957, to be published in July 1957, that is to say, over a year later. So before

---

81 The most recent result, viz., Chemykh, M., and 5 authors, Proc. 13th Intl. Symposium Capture Gamma-Ray Spect., AIP Conf. Proc. 2009 is $\Gamma_{e^\pm} = 6.2 \times 10^{-5}$ eV.

82 Blatt, J., and Weisskopf, V., *Theoretical Nuclear Physics*, Wiley, NY, 1952.

83 Salpeter, E.E., Phys. Rev. **98**, 1183 (1955).

84 Hayakawa, S., Hayashi, C., Imoto, M., and Kikuchi, K., Prog. Theor. Phys. **16**, 507 (1956).

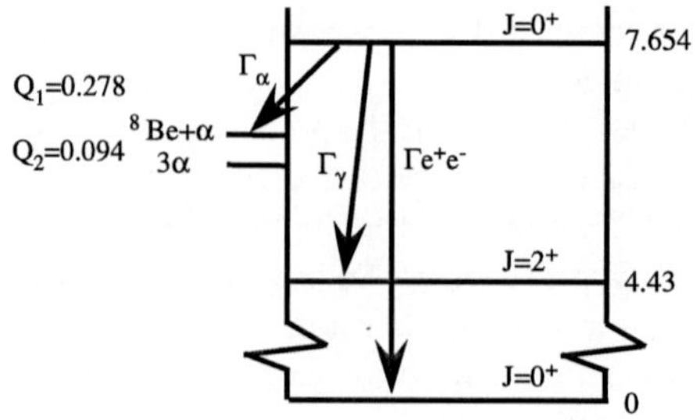

Fig. 8.15 The structure of ^{12}C assumed by Salpeter in 1957. Γ_α is the probability of decay by emitting an α particle, Γ_γ is the probability of decay by emitting a γ photon, and $\Gamma_{e\pm}$ is the probability of decay by emitting an electron–positron pair. This was after Salpeter had obtained the improved results of Cook et al. (1957)

Salpeter managed to publish his results in a journal, Hayakawa et al. published the calculation of the rate including the excited state of ^{12}C, and wrote:

> This rate is about 10^6 times larger than that given by Salpeter.

They meant Salpeter 1952. They went on to state that:

> Such a great difference is due mainly to the fact that he did not take the effect of the recently discovered resonance level in ^{12}C into account. This is one of the most important conclusions obtained in our work.

It may have been the most important result in their paper, but it was Salpeter who told them how he calculated the (correct) rate and what the revised numbers were. For the sake of accuracy, Hayakawa et al. assumed a probability of $(7 \pm 3) \times 10^{-7}$, which is much too low. Fortunately, the American Physical Society published the proceedings, so that the record of priority is easily established.

This time Salpeter cited Öpik as suggesting the $3\alpha \rightarrow {}^{12}$C reaction. But without mentioning Öpik, he wrote further down that:

> If this reaction had to proceed via genuine three-body collision, unassisted by any resonances, its rate would be negligibly small at the relevant temperature of about 10^8 K. The existence of the ^{8}Be ground state level, at 94 keV relative kinetic energy between the two alpha particles, already enhances the reaction rate enormously. A dynamic equilibrium is set up. Even without resonance the rate of carbon formation becomes important on stellar ranges, but since $\alpha + {}^{12}\text{C} \rightarrow {}^{16}\text{O}$ is so much faster, it is clear that, unless the speed of carbon formation starts to compete with carbon destruction, no carbon will be left.

William Alfred Fowler won the Nobel Prize for Physics in 1983 (with Subramanyan Chandrasekhar) and the prize citation said: *For his theoretical and experimental studies of the nuclear reactions of importance in the formation of the chemical elements in the Universe*. For some reason Hoyle's imaginative and extensive contribution was overlooked, and many were dismayed that such a notable contribution to nuclear astrophysics had been missed out. Fowler himself, in an autobiographical sketch, affirmed Hoyle's pioneering contribution.[85]

The exclusion was rectified when, in 1997, Fred Hoyle and Edwin E. Salpeter received the Crafoord prize of the Royal Swedish Academy of Sciences. The prize citation reads: *For their pioneering contributions to the study of nuclear processes in stars and stellar evolution*. The triple alpha reaction is called today the Salpeter reaction, while the level of ^{12}C at 7.68 MeV is called the Hoyle level.

8.16 When Is the Energy Released?

All the steps in the fusion of three ^{4}He into a carbon nucleus consume energy, namely the star has to invest energy as:

$$^4\text{He} + ^4\text{He} \rightarrow ^8\text{Be} - 0.0921\,\text{MeV},$$

followed by

$$^8\text{Be} + ^4\text{He} \rightarrow ^{12}\text{C}^* - 0.286\,\text{MeV},$$

where $^{12}\text{C}^*$ denotes a nucleus in an excited state. Only when the excited nucleus decays into the ground state does it release energy in the process:

$$^{12}\text{C}^* \rightarrow ^{12}\text{C} + \gamma + 7.656\,\text{MeV}.$$

The photon is absorbed by the matter in the star and heats it up.

8.17 More Resonances in ^{12}C

In 1959, Almqvist et al.[86] measured all levels in ^{12}C up to 16.6 MeV. It became clear that the distance between levels is very high for low levels, up to 3–4 MeV, while above about 10 MeV, the density becomes very high. Due to the fact that these levels are rather wide,[87] they all contribute to the rate of the reaction. Consequently, the

---

85 Fowler, W.A., Interviewed by Greenberg and Buge, 1984, CalTech Archives. In his Nobel lecture, Fowler claimed that Hoyle was convinced from his work with Schwarzschild and previous work by Sandage and Schwarzschild, that helium burning should commence in red giants at about 1×10^8 K and not at 2×10^8 K, as the old reaction rate by Salpeter had yielded (see Fowler, 1983 Nobel prize lecture).

86 Almqvist, E., Bromley, D.A., Ferguson, A.J., Gove, H.E., and Litherland, A.E., Phys. Rev. **114**, 1040 (1959).

87 According to the uncertainty principle, the width of the level is connected to the lifetime of the system in that level. From a nuclear reaction point of view, this means that, if the energy of the

locations of the levels up to about 10 MeV are important for the rate of $^8Be + \alpha$ at $T < 10^9$ K. The energy levels higher than 10 MeV correspond to such a high energy range that, by the time the colliding species have enough energy to reach these levels, the temperature is so high that all nuclei disintegrate and hence cannot fuse any further.

In 1994, Freer et al.[88] investigated whether the ^{12}C 7.65 MeV level could decay directly into 3α, rather than always decay to $^8Be + \alpha$ first, whereupon the ^8Be would break into two α particles. The result for this direct decay is less than 4%. This was an additional experimental proof that the triple alpha was not a three-body reaction.

The problems in getting an accurate value for the probability of the transition from the 7.68 MeV level to the 4.43 MeV level continue to create difficulties for physicists. Recent values are: Alburger[89] $(3.3 \pm 0.9) \times 10^{-4}$, Hall and Tanner[90] $(3.5 \pm 1.2) \times 10^{-4}$, Seeger and Kavanagh[91] $(2.8 \pm 0.3) \times 10^{-4}$, and Mak et al.[92] $(4.30 \pm 0.2) \times 10^{-4}$. It is difficult to achieve high accuracy!

8.18 Carbon Structure Is Half the Story: The $^{12}C + \alpha \rightarrow ^{16}O + \gamma$ Reaction

We come to the destruction of carbon. The existence of excited levels in ^{16}O was discovered by Burcham and Freeman,[93] who used the reaction $^{19}F+p \rightarrow ^{16}O+\alpha$ and found the 6.94 and 7.15 MeV levels. However, they did not determine the quantum numbers of the level. In 1950, Millar, Bartholomew, and Kinsey[94] investigated the γ rays from the decay of $^{16}N \rightarrow ^{16}O + e^-$. This radioactive isotope is produced in the cooling water of nuclear reactors via the reaction $^{16}O + n \rightarrow ^{16}N + p$. The half-life of the isotope is 7.35 s. The energy of ^{16}N is so high that it can decay into the first two levels of ^{16}O by emission of an electron. They thus discovered γ rays which corresponded to the levels at 6.133 ± 0.011 and 7.10 ± 0.02 MeV. These results were in good agreement with the values found by Chao, Tollestrup, Fowler, and Lauritsen,[95] viz., 6.136 ± 0.030 and 7.111 ± 0.03 MeV. To complicate the issue, they

colliding particles is within the wide range of the level and not just equal to the prescribed energy, the reaction can proceed.

88 Freer, M., Wuosmaa, A.H., Betts, R.R., Henderson, D.J., Wilt, P., Zurmuhle, R.W., Balamuth, D.P., Barrow, S., Benton, D., Li, Q., Liu, Z., and Miao, Y., Phys. Rev. C **49**, 1751 (1994).

89 Alburger, D.E., Phys. Rev. **124**, 193 (1961), who used the reaction $^{10}B + ^3He \rightarrow ^{12}C + p$.

90 Hall and Tanner, Nucl. Phys. **53**, 673 (1964), who used the reaction $^{10}B + ^3He \rightarrow ^{12}C + p$.

91 Seeger, P.A., and Kavanagh, R.W., Nucl. Phys. **46**, 577 (1963), who used the reaction $^{14}N+^2D \rightarrow ^{12}C + \alpha$.

92 Mak, H.-B., and 4 coauthors, Phys. Rev. C **12**, 1158 (1975).

93 Freeman, J.M., and Baxter, A.S., Nature **162**, 696 (1948); Burcham, W.E., and Freeman, J.M., Phys. Rev. **75**, 1756 (1949).

94 Millar, C.H., Bartholomew, G.A., and Kinsey, B.B., PRL **81**, 150 (1951).

95 Chao, C.Y., Tollestrup, A.V., Fowler, W.A., and Lauritsen, C.C., Phys. Rev. **79**, 108 (1950).

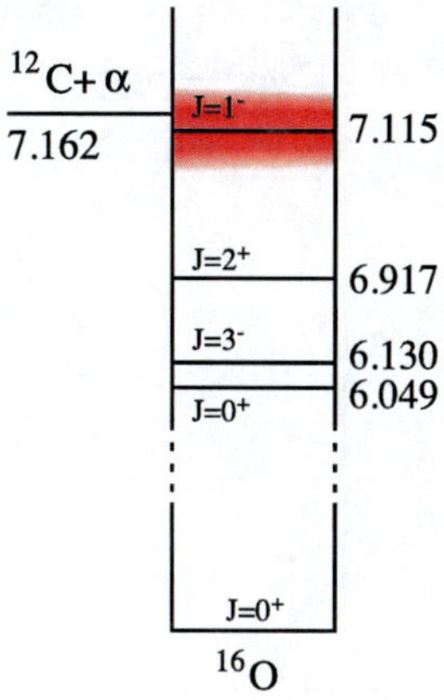

Fig. 8.16 The structure of the oxygen nucleus. Based on Ajzenberg and Lauritsen (1952)

did not discover γ rays that would correspond to the level at 6.91 MeV. They were, however, able to determine the quantum numbers of the level as $J = 1^-$ (Fig. 8.16).

The situation became complicated once more. $^{12}C + \alpha$ has a rest mass energy of 7.162 MeV, while the closest energy level in the ^{16}O nucleus is at 7.115 MeV, that is to say, just below the threshold of the colliding particles. Because all energy levels save the ground state are unstable, they have a certain width in energy. The width ΔE is related to the lifetime Δt through the relation $\Delta E \Delta t \sim \hbar$. The 7.115 MeV level in ^{16}O is sufficiently wide to extend into energies slightly above the $^{12}C + \alpha$ threshold. Thus the carbon plus the alpha combine to form the excited oxygen nucleus via the upper tail of the level. On the other hand, as the width of the level is large, the probability of breakup into $^{12}C + \alpha$ is very small, and as a matter of fact has never been observed.

In 1974, Dyer and Barnes[96] from CalTech attempted to measure the $^{12}C + \alpha$. The next attempt was in 1982, when Kettner et al.[97] from Münster measured the reaction and discovered that it was 3 to 5 times faster then what Dyer and Barnes had found at Caltech. As a consequence, the most abundant species at the end of helium burning was found to be ^{16}O and not ^{20}Ne.

Langanke and Koonin[98] criticized the analysis and the conclusions of the experimenters and repeated the very long theoretical analysis of the experimental data. They could not resolve the discrepancy between the Caltech data (Dyer and Barnes) and the Münster data (Kettner et al.), and fitted each experiment separately (see

96 Dyer, P. and Barnes, C.A., Nucl. Phys. **A233** (1974).

97 Kettner, K.U., and 8 coauthors, Z. Phys. A **308**, 73 (1982).

98 Langanke, K. and Koonin, S.E., Nucl. Phys. **A410**, 334 (1983); ibid., Nucl. Phys. **A439**, 384 (1985). The first paper was published when Langanke was still in Munster. The second paper was published after he joined Koonin in Caltech.

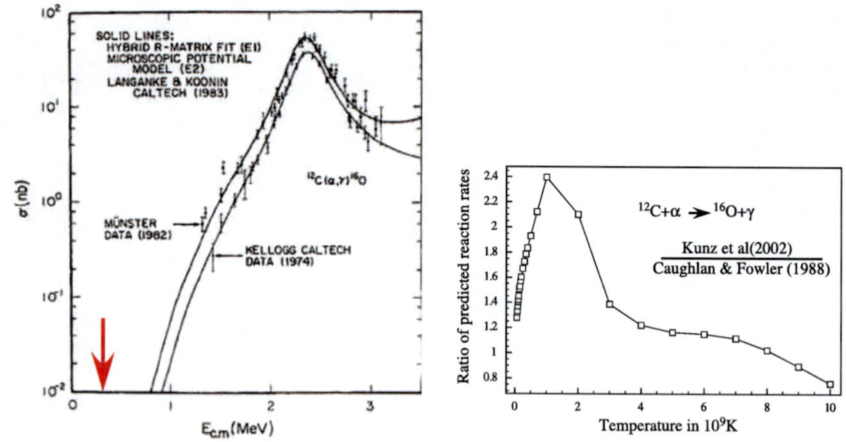

Fig. 8.17 *Left* the capture probability (in units of nanobarns) as measured by the Munster (1982) and the Kellogg (1974) laboratories. The *arrow* marks the energy range at which the reaction takes place in stars. The *continuous lines* are the theoretical fit by Langanke and Koonin (1983), which includes the effect of the resonance. *Right* the ratio between the newest rate calculated by Kunz et al. (2002) and the standard rate compiled by Caughlan and Fowler (1988), used in many calculations of stellar evolution. The ratio is given as a function of the temperature

Fig. 8.17 left). Thus, the Münster data[99] was 1.5 times higher than that of Caltech, which was in turn 3 times higher than what stellar modelers used. As Lankange and Koonin wrote:

> Such higher values [as those found in Münster and Caltech] lead to ^{16}O rather than ^{12}C as the final product of helium burning.

And in this way, they cast a shadow over Hoyle's argument.

The relevance of Hoyle's argument was further questioned by Arnet and Thielemann,[100] who were the first, in 1985, to realize that C burning in the cores of massive stars might not be relevant to nucleosynthesis of the elements in the cosmos because:

> [. . .] the burning product may eventually be locked up in a neutron star remnant after Type II supernova explosion.

In those days the option of a collapse to a black hole, which equally well prevents the products from getting out, was not that popular.

---

[99] The probability for this reaction to take place at thermal energies ($T = 2 \times 10^8$ or 300 keV) was given by $S(300\,\text{keV}) = 350\,\text{keV barn}$ (Münster), 240 keV barn (Caltech), and 80 keV barn in standard nucleosynthesis calculations at that time.

[100] Arnett, W.D., and Thielemann, F-K., Astrophys. J. **295**, 589 (1985).

In 2001, Metcalfe, Winget, and Charbonneau[101] calculated constraints on the $^{12}C+\alpha \rightarrow {}^{16}O+\gamma$ rate from white dwarf oscillations.[102] The value so derived, viz., $S(300) = 290 \pm 15$ keV barn,[103] lies between the values from Caltech and Münster.

In 2002, Kunz et al.[104] carried out an extensive theoretical analysis based primarily on Kunz's experimental Ph.D. thesis.[105] The result of Kunz et al. relative to the standard tables of nuclear reactions compiled by Caughlan and Fowler[106] is shown in Fig. 8.17 (right). We see quite significant changes at the low temperatures which are relevant for quiet helium burning. Clearly, stellar model calculations must be updated.

A major breakthrough took place in 2005 when Schürmann et al.[107] carried out the first ever direct measurement of the $^{12}C+\alpha \rightarrow {}^{16}O+\gamma$ reaction. Direct measurement means in this case that the recoil of the newly formed ^{16}O nucleus was measured. The minimum energy is still high, viz., 1.9 MeV. They nicely confirmed Kunz et al. 2002 at these high energies.

The calculated probability of the reaction $^{12}C + \alpha$ has increased significantly since Hoyle calculated it in 1954 using only the 7.10 MeV level. Consequently, the arguments about the need to synthesize carbon in our world have changed.

8.19 Present Day Perspective: Theory

An obvious question is: how sensitive is the production of C and O in stars to the 3α rate and the energy of the resonance? Schlattl et al.[108] investigated this question

---

[101] Metcalfe, T.S., Winget, D.E., and Charbonneau, P., Astrophys. J. **557**, 1021 (2001).

[102] The oscillations depend on the C/O ratio which is the result of helium burning.

[103] Nuclear astrophysicists write the probability of a nuclear reaction as

$$\sigma(E) = \frac{S(E)}{E} \exp\left[-\left(\frac{E_G}{E}\right)^{1/2}\right],$$

where E_G is the energy of the Gamow peak. The exponent takes care of the tunneling factor and all the nuclear physics resides in the fator $S(E)$. At low energies, what matters is $S(0)$. The unit 'barn' for probability of absorption has its own history. During WWII research on the atomic bomb, American physicists at Purdue University who were bombarding uranium nuclei with neutrons to check the absorption of neutrons, described the uranium nucleus as being 'as big as a barn', i.e., a superb absorber of neutrons. Physicists working on the project adopted the name 'barn' for a unit equal to 10^{-24} cm^2. Initially, they hoped the American slang name would obscure any reference to the study of nuclear structure, but eventually the word became a standard unit in particle physics. Wackeroth and Belkora (eds.), *Cross Section. High Energy Physics Made Painless*, Fermilab Science Education Office. Retrieved 2009-03-13.

[104] Kunz, R., and 7 coauthors, Astrophys. J. **567**, 643 (2002).

[105] Kunz, R. 2002, Ph.D. thesis, Stuttgart.

[106] Caughlan, G.R. and Fowler, W.A., Atomic Data and Nuclear Data Tables **40**, 283 (1988).

[107] Schürmann, D., and 16 coauthors, Eur. Phys. J. **A26**, 301 (2005).

[108] Schlattl, H., and 4 coauthors, A., astro-ph/0307528v2, 30 October 2003; Astrophys. Space Sci. **291**, 27 (2004).

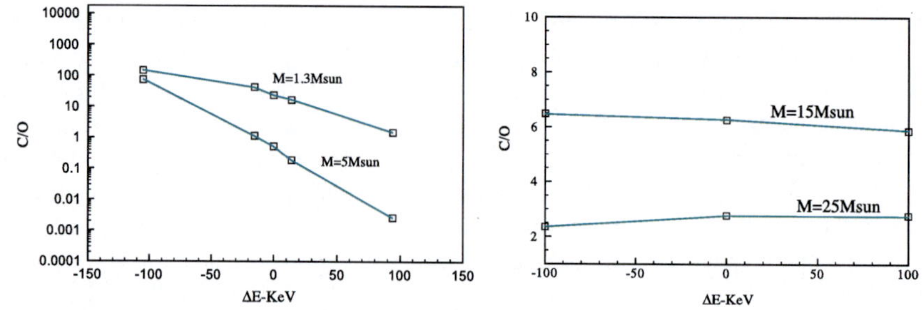

Fig. 8.18 *Left* the C/O ratio obtained in the evolution of low mass stars (1.3 and 5.0$M_\odot$) upon a hypothetical change in the location of the resonance in ^{12}C. *Right* the C/O ratio obtained in the evolution of high mass stars (15 and 25$M_\odot$) upon a hypothetical change in the location of the resonance in ^{12}C

by calculating the evolution of stars when they artificially changed the energy of the level. This is a virtual exercise. If the potential well inside which the nucleons move changes, then all energy levels change their position. Moreover, the binding energy of the fusing particles changes as well, so it is not sufficient to 'change' the location of a single level leaving the rest unchanged. And yet this exercise is very instructive.

Taking their data and plotting the effect of the change in the energy of the level on the C/O abundance ratio, we obtain the results plotted in Fig. 8.18. One can see that, in low mass stars, the C/O abundance ratio changes rapidly with a change in the energy of the resonance. However, these stars do not produce most of the C and O that comes out of stars. These stars become white dwarfs after losing the extra mass, and most of the C and O these stars synthesize ends up buried forever in the white dwarfs.

On the other hand, the C/O abundance ratio hardly changes in massive stars (see Fig. 8.18 right). These are the stars that eject the synthesized elements into space. The complication arises from the fact that Hoyle assumed that $C/O = 1/2$, a ratio obtained in low mass stars, while in massive stars, those which are important to the universal C/O ratio, one finds that the ratio is greater than unity. Actually, the universal abundance is a weighted average of the product of all massive stars.

8.20 The C/O Ratio Today: Observations and What They Imply

Not a single star without oxygen or without carbon has ever been discovered. Detailed observations have shown that the abundance ratio C/O varies from star to star. Gradually, it has become clear that there is no universal value for the C/O abundance ratio which was the basis for Hoyle's analysis, and that different stars expose different ratios. For example, stars with extra high abundance of carbon do not show extra high abundances of oxygen.[109] Examination of the gases ejected from low mass stars shows that, when metals are more abundant by about 50%, the C/O ratio is lower by about 50%.[110] The C/O ratio appears to be higher in stars with low abundance of heavy elements.[111] Finally, the C/O ratio appears to vary from 0.2 to 13 in low mass stars.[112] In short, the picture is more complicated and the last word has not yet been pronounced.

8.21 The Structure of ^{12}C and the α Model: Retrospect

In 1971, Brink[113] concluded from the nuclear cluster model that 'our' state forms a linear chain of three α particles. Further indications from bombardment by electrons imply that this state has an unusually large radius.[114] These properties enhance the probability of decay into three α particles.

This conclusion was supported in 1997 when Pichler, Oberhummer, Csoto, and Moskowski[115] concluded that the 0^+ level was a genuine three-alpha resonance. If so, one would expect the level to decay directly into three α particles, and then Salpeter's equation for the nuclear statistical equilibrium might need to be changed. But the probability of direct decay into three alpha particles appears to be only about 4%. The reason for this is purely statistical. The phase space of ^{8}Be + α is that much bigger than that of the 3 α mode. In layman's terminology, all possible modes of decay have equal probabilities and the number of ways the system can decay into ^{8}Be + α is much higher than the number of different ways it can decay into three α particles.

8.22 Some Recent Developments

The drama is not yet over, and resonances continue to come and go. Examining the structure of ^{12}C as known in 1985,[116] we see levels at 9.641, 10.3, 10.84, and 11.83 MeV. These energy levels turned out to be very wide, and consequently the colliding ^{8}Be and α can enter even these levels provided they have the right quantum numbers. So recent calculations of the rate of the triple alpha have incorporated contributions from these higher levels. Again we find conflicting reports. Angulo

---

109 Bujarrabal, V. and Cernicharo, J., Astron. Astrophys. **288**, 551 (1994).

110 Wang, W. and Liu, X.-W., MNRAS **381**, 669 (2007).

111 Wahlin, R., and 6 coauthors, Memorie della Societa Astronomica Italiana **77**, 955 (2006).

112 Cohen, M. and Barlow, M.J., MNRAS **362**, 1199 (2005).

113 Brink, in *The Alpha Particle Model of Light Nuclei*, Proc. Enrico Fermi school course XXXVII, Varenna, 1966.

114 Takigawa, N., and Arima, A., Nucl. Phys. **A168**, 593 (1971).

115 Pichler, R., Oberhummer, H., Csoto, A., Moskowski, S.A., Nucl. Phys. **A618**, 55 (1997).

116 Ajzenberg-Selove, F., Nucl. Phys. **A506**, 1 (1990).

et al.[117] claim that an important resonance exists at 9.1 MeV, while Fynbo et al.[118] have shown that, in the range 8–11 MeV, there is an elusive 0^+ resonance at 10.3 MeV with a width of 3.0 MeV, and there is no resonance at 9.1 MeV. The important issue is that there is a contribution from several resonances, not only from the 7.68 MeV level. The sad thing is that it will take more time and effort to sort the problem out. Finally, Diget et al. 2005[119] find that a significant contribution to the capture comes from 0^+ in the range of 1–4 MeV above the Hoyle state. These states interfere with the Hoyle state. They discovered that the Hoyle state affects the level even up to 5 MeV away and disturbs a proper analysis of the results. In summary, it appears that, due to the fact that ^{12}C usually has a structure of distinct three α particles, the energy levels are particularly wide, and several energy levels participate in the capture of the α particle, not just one.

8.23 Philosophy: The Anthropic Principle

Hoyle's unique physical reasoning, namely that, since we see carbon in the Universe, there must be a proper energy level in ^{12}C, gave rise to the anthropic philosophical principle: the cosmos must be compatible with the existence of the life that is observed in it. Is it a coincidence? Is it the only one? Many people have found other 'coincidences'. This subject lies in the 'believe it or not' domain. Or should we paraphrase a là René Descartes (1596–1650) *Cogito ergo sum*, and say, we are here, so the Universe allowed us to form. A detailed discussion would carry us too far astray (see Barrow and Tipler[120]).

[117] Angulo, C., et al., Nucl. Phys. **A656**, 3 (1999).

[118] Fynbo, J., et al. Nature **433**, 136 (2005).

[119] Diget and 20 coauthors, Nucl. Phys. **760**, 3 (2005).

[120] Barrow, J.D. and Tipler, F.J., *The Anthropic Cosmological Principle*, Oxford University Press, LC 87-28148, ISBN 9780192821478 (2009).

Chapter 9
Beyond Carbon

9.1 Post-Helium Burning

Once all the helium has been exhausted and energy production by helium fusion into carbon has come to an end, the relentless gravitational force resumes its contraction and the star embarks upon the final steps of energy production by fusion.

The formation of the iron group elements is a very delicate balance between two factors. On the one hand, the charges of the heavy nuclei are very high and consequently the repulsion is so great that even tunneling through the Coulomb barrier, which works for low charge elements, can no longer function. The stars do not become sufficiently hot! As the temperatures reach a few billion degrees, the matter out of which the heavy elements are supposed to form disintegrates. On the other hand, smaller particles like protons and α particles can barely penetrate the Coulomb barrier of the heavier nuclei. Consequently, it is only under a quite restrictive range of temperatures and densities that the buildup of heavy elements can take place, in a process whereby part of the matter disintegrates and at the same time the rest of the matter absorbs the fragments and increases its atomic weight.

9.2 Change of Modus Operandi: Neutrino Losses Take Control

As the temperature increases and reaches $\sim$3–8 $\times$ 10^9 K, a new mechanism for energy loss appears, namely, energy loss via neutrinos. As the star is transparent to neutrinos, such energy loss is immediate. The neutrinos, unlike photons, do not have to diffuse outward in a process of absorption and emission by the intervening matter. Consequently, the time scale for energy losses, and thus also for stellar evolution, shortens enormously.

It is important to stress that a fundamental change takes place here. The star, once driven by photon losses, is now driven by neutrino losses. As the neutrino

G. Shaviv, *The Synthesis of the Elements*, Astrophysics and Space Science Library 387, DOI: 10.1007/978-3-642-28385-7_9, © Springer-Verlag Berlin Heidelberg 2012 losses significantly exceed the photon losses, the time scale of evolution shortens dramatically to become just a few hours, before the final collapse and the supernova explosion.

But there is an additional difference. When the temperature in the core of the star reaches say, 5×10^9 K, the photons in the core of the star have energies of the order of ~0.5 MeV, typical of γ rays. The diffusion of the radiation outward via interaction with the matter causes a gradual decrease in the mean energy of the photons, and the photons which emerge from the surface of the star have energies of a few eV or less. So we observe the energy loss in the form of radiation after its carriers have been enormously downgraded. But the situation with neutrinos is different. As the interaction of the neutrinos with matter is about 10^{20} times weaker, they escape from the star instantly and are not downgraded. The observer gets the neutrinos with the energies they have in the core of the star! Detecting them is like looking directly into the core.

The neutrino losses take place in the core of the star. The star responds by extracting energy from the gravitational field and begins an accelerated contraction with heating which accelerates the neutrino losses. The envelope on the other hand responds very slowly. Thus the core practically separates from the envelope. The core evolves quickly on a time scale of hours and the envelope hardly changes or responds.

9.3 Neutrino Losses

In 1967, Weinberg and independently Salam 1968 proposed a unified theory of weak and electromagnetic interactions. The new theory predicted the existence of a neutral current interaction in the weak force interaction. This interaction had not been observed at the time and some experiments even suggested that it did not exist, or that, even if it did, the strength was one tenth or one hundredth of the usual charged current interaction. In 1973, the neutral current interaction was discovered by an accelerator experiment in FNAL, and the W and Z bosons were found in CERN in 1983. At present, the unified theory of Weinberg and Salam is the standard theory of weak interactions. The first application to stars was carried out by Dicus[1] in 1972, before the Weinberg–Salam theory was confirmed experimentally.

For us here, the neutrinos are like photons in terms of generation and energy, but with two basic differences:

- The interaction constant is different and hence the generation rate is different.
- The neutrino interaction with matter is orders of magnitude smaller, so the neutrinos escape from the stars without any hindrance and induce fast cooling of the core. The control on the evolution of the star is passed from the photons to the neutrinos and the time scale is thereby shortened by orders of magnitude.

[1] Dicus, D.A., Phys. Rev. D **6**, 941 (1972).

The core cools fast, while the envelope, still dominated by photons, hardly notices what is going on in the core.

Fowler and Hoyle (1964) gave a long list of neutrino losses which were expected to operate in stars. We mention here only the dominant three.

Pair Annihilation

The electron–positron pair can annihilate and yield the classical γ photons, but also each of the three kinds of neutrino–antineutrino pair:

$$e^+ + e^- \rightleftharpoons \begin{cases} 2\gamma, \\ \nu_e + \bar{\nu}_e, \\ \nu_\mu + \bar{\nu}_\mu, \\ \nu_\tau + \bar{\nu}_\tau. \end{cases}$$

Photoneutrino

$\gamma + e^- \rightarrow e^- + \nu_e + \bar{\nu}_e$, which is the analogue of Compton scattering of photons by electrons.

Plasma Neutrino

In vacuum, a photon has no mass and cannot decay. In plasma, photons can decay because they move as though they have non-vanishing mass and are called plasmons. The existence of an effective mass implies that the process $2\gamma \rightarrow \nu_e + \bar{\nu}_e$ can take place.[2]

The impact of neutrino losses on stellar evolution can be seen in Figs. 9.1 and 9.2. In the first figure, we see the comparison between the total neutrino losses and the luminosity of the star in the case of a $15M_\odot$ star. Above $\sim 0.6 \times 10^9$ K, the neutrino losses dwarf the photon losses. In the second figure, the comparison is with the various nuclear fuels. Eventually, the neutrino losses overcome all nuclear fuels.

[2] The plasma of stellar matter can oscillate. These oscillations are analogous to oscillations of the electromagnetic field. In the case of the electromagnetic field, quantization of the field yields the classical photon. In a plasma, quantization yields the plasmon. The difference is that the photon has no rest mass while the plasmon does, and so can decay in several ways, e.g., producing neutrinos.

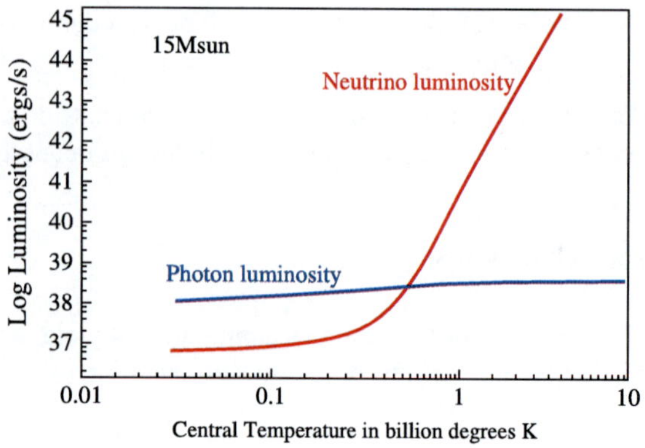

Fig. 9.1 Neutrino luminosity and photon luminosity in a $15M_{\odot}$ star as a function of the central temperature in billions of degrees K. Above 0.5×10^9 K, the neutrino luminosity takes over and controls the evolution of the star

9.4 The Paper B2FH

One of the most fruitful and famous collaborations in nuclear astrophysics started in the fall of 1954, when William (known to his friends as Willy) Fowler spent his sabbatical year in the Cavendish Laboratory, Cambridge, England where he met Fred Hoyle and Margaret and Geoffrey Burbidge.[3] The crossing of the world lines of these four people had a tremendous impact on nuclear astrophysics. The first papers were published while they were still in England.[4] When the time came for Fowler to return to Kellogg, he brought the entire group with him. The precursor to the great work was already published in 1956,[5] and the most famous work was published in 1957.[6] The paper is known today as B2FH[7] and is a compendium of all nuclear reactions in stars and their classification. The terminology used today was invented in this paper. As three of the authors wrote 40 years later (Fig. 9.3):

---

[3] We already mentioned this unique paper when discussing the nuclear reactions of light elements. However, the major breakthrough is in its contribution to the nuclear reactions in the last phases of stellar evolution. Hence we shall now elaborate on this point.

[4] Fowler, W.A., Burbidge, G.R. and Burbidge, E.M., Astrophys. J. Suppl. **2**, 167 (1955); Astrophys. J. **122**, 271 (1955).

[5] Hoyle, F., Fowler, W.A., Burbidge, G.R. and Burbidge, E.M., Science **124**, 1 (1956).

[6] Burbidge, E.M., Burbidge, G.R., Fowler, W.A. and Hoyle, F., Rev. Mod. Phys. **29**, 547 (1957). When completed, the question of where to publish the long paper came up. Chandrasekhar, the editor of the Astrophys. J. began to ask questions about how much of the paper was original and how much a review. Chandrasekhar was unhappy with the answer of 'at least 95%'. The impatient Willy thus called his friend Ed Condon, then the editor of Reviews of Modern Physics, who simply accepted it as-is and published it. See Burbidge, in *Nuclear Astrophysics 1957–2007*, Pasadena, 2007.

As Geoff, who always had grievances about the refereeing processes, said: *If it were fifty years later, it would have been in the hands of referees for months, if not years!*

[7] Some people insist that it should be written as: B^2FH, as in Chap. 7. I compromise and use here B2FH.

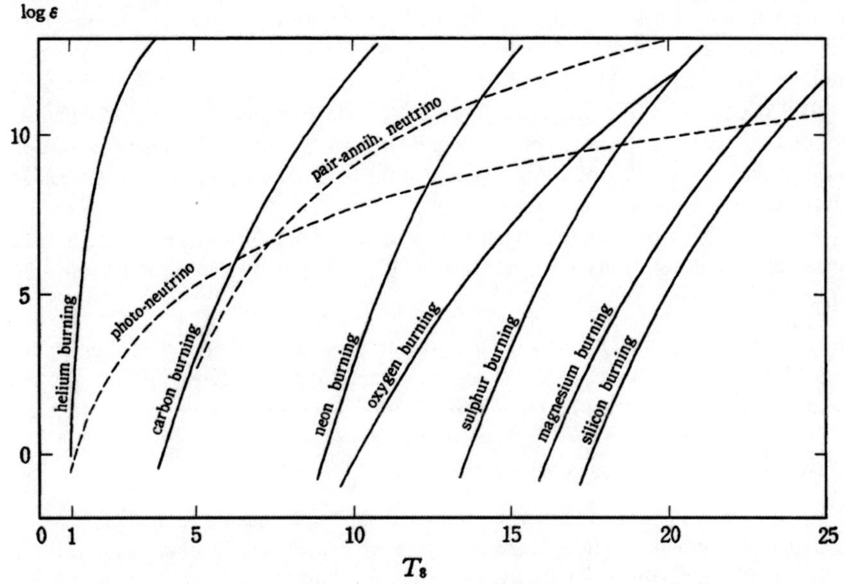

Fig. 9.2 A comparison between the neutrino losses and various nuclear energy generation reactions according to Hayashi et al. (1962). The rates are given in erg/g/s as a function of temperature in units of 10^8 K. The density is 10^4 g/cm^3. Note that several reactions shown in the figure are no longer considered relevant to massive stars. The neutrino rates have been updated since 1962, but the general picture of neutrino losses taking over energy generation remains today

Fig. 9.3 The shortening of the stellar timescale with the rise in temperature according to B2FH. The last nuclear phases take place in seconds. This figure was prepared before the extent of neutrino losses was realized, and hence provides *upper* limits for the timescales

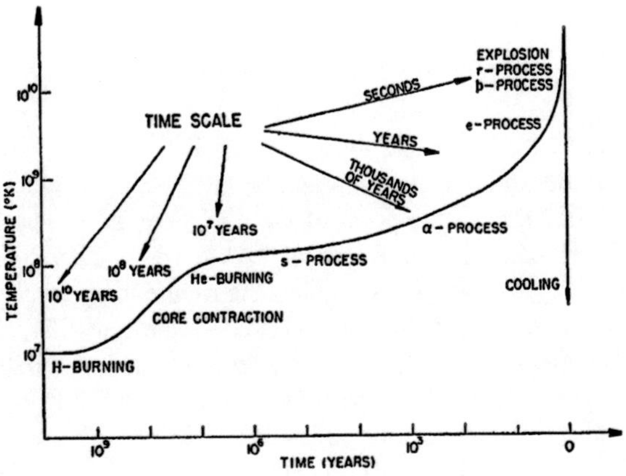

One should not forget the introduction of a lettering notation for the various nuclear processes, α, e, s, r, p, and x, which may have done more for the development of the subject than almost anything else!

The impact and relevance of the paper has still not decayed because it was a blueprint for the theory and evolution of nuclear astrophysics and outlined the various processes in the order they should be described.

The paper by B2FH was to a large extent the response of the astro-nuclear community to the provocation by Suess and Urey,[8] who argued that a single neutron absorption process, like the one suggested in the $\alpha\,\beta\,\gamma$ process, could not synthesize all the elements beyond iron.[9]

By 1957, the fusions of hydrogen into helium and helium into carbon were well understood and not really covered by B2FH. Indeed, B2FH concentrated on two aspects (Fig. 9.4):

- The synthesis of the elements from carbon to iron, at the minimum of the nuclear binding energy. These processes are important for stellar evolution towards the supernova state, where the outer layers contain the processed elements that are ejected into space and eventually enter the next generation of stars.
- The synthesis of elements heavier than iron.

All heavy elements beyond the iron group total 10^{-11} by mass and the B2FH paper is actually about what nuclear reactions may be responsible for the synthesis of these elements. The paper does not discuss what fraction of the star is ejected into space, and consequently what each supernova contributes to the abundance of new elements in the interstellar medium, from which new stars form. Since these processes are very complicated and remain virtually unknown to a large extent even today, only the predicted isotopic composition could be compared with the observed one.

9.5 The Rise and Fall of the α Process

The basic idea is as follows. If ${}^{n_\alpha}X$ is a nucleus composed of n_α α particles, then consider the reaction: ${}^{n_\alpha}X + \alpha \rightleftharpoons {}^{n_\alpha+1}X + \gamma$. Figure 9.5 illustrates the idea of the α process from the point of view of the energy. The energy of ${}^{n_\alpha}X + \alpha$ is always higher than the ground state of ${}^{n_\alpha+1}X$. The rate itself depends critically on the extent to which the target nucleus has the right energy levels for the colliding nuclei to be able to easily form the new excited system. In 1954, Hoyle supposed that the triple α continues until ${}^{20}Ne$, and may even synthesize small amounts of ${}^{24}Mg$.

The initial idea of the α process was the following set of reactions, which may lead from neon to iron:

---

[8] Suess, H.E. and Urey, H.C., Rev. Mod. Phys. **28**, 53 (1956).

[9] A meeting to celebrate the 50th anniversary of the B2FH paper was held in Caltech in 2007. It was at this meeting that Geoff Burbidge told how he had met Maria Mayer at the University of San Diego in 1962, and that it was through discussions with her that he had realized that *stellar nucleosynthesis, as pioneered by Fred (Hoyle), had completely vanquished the earlier ideas of Gamow, Alpher, etc., except for the x-process.*

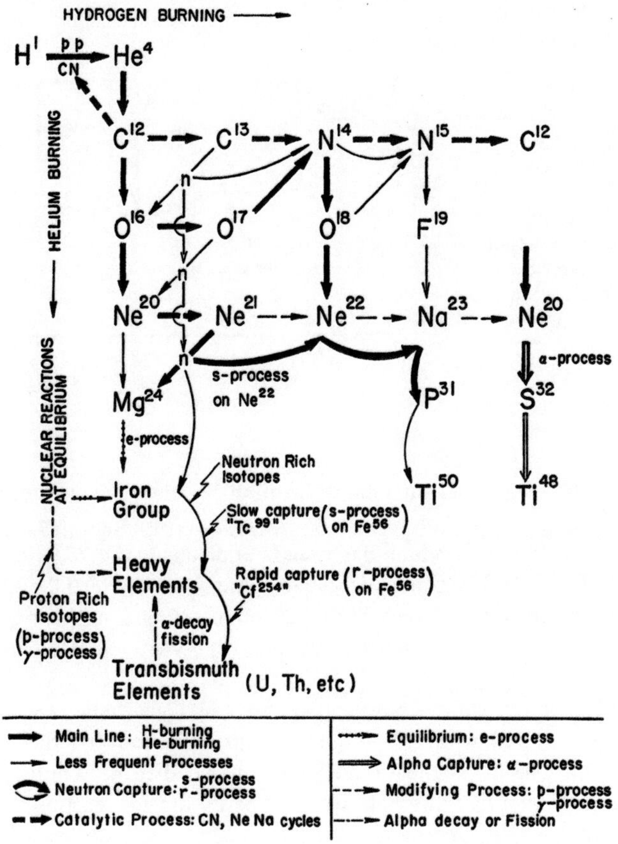

Fig. 9.4 Schematic description of the B2FH grand picture of the synthesis of the heavy elements in stars. There is no trace of carbon and oxygen burning. He burning continues until Ne. The α process goes until Mg. The equilibrium process takes over after Mg. From B2FH 1957

$$
\begin{array}{llll}
{}^{16}\mathrm{O}+\alpha & \rightleftharpoons {}^{20}\mathrm{Ne}+\gamma+4.753\,\mathrm{MeV} & {}^{44}\mathrm{Ti}+\alpha & \rightleftharpoons {}^{48}\mathrm{Cr}+\gamma+7.695\,\mathrm{MeV}\\
{}^{20}\mathrm{Ne}+\alpha & \rightleftharpoons {}^{24}\mathrm{Mg}+\gamma+9.317\,\mathrm{MeV} & {}^{48}\mathrm{Cr}+\alpha & \rightleftharpoons {}^{52}\mathrm{Fe}+\gamma+7.937\,\mathrm{MeV}\\
{}^{24}\mathrm{Mg}+\alpha & \rightleftharpoons {}^{28}\mathrm{Si}+\gamma+9.984\,\mathrm{MeV} & {}^{52}\mathrm{Fe}+\alpha & \rightleftharpoons {}^{56}\mathrm{Ni}+\gamma+7.997\,\mathrm{MeV}\\
{}^{28}\mathrm{Si}+\alpha & \rightleftharpoons {}^{32}\mathrm{S}+\gamma+6.948\,\mathrm{MeV} & {}^{56}\mathrm{Ni}+\alpha & \rightleftharpoons {}^{60}\mathrm{Zn}+\gamma+2.709\,\mathrm{MeV}\\
{}^{32}\mathrm{S}+\alpha & \rightleftharpoons {}^{36}\mathrm{Ar}+\gamma+6.641\,\mathrm{MeV} & {}^{60}\mathrm{Zn}+\alpha & \rightleftharpoons {}^{64}\mathrm{Ge}+\gamma+2.580\,\mathrm{MeV}\\
{}^{36}\mathrm{Ar}+\alpha & \rightleftharpoons {}^{40}\mathrm{Ca}+\gamma+7.040\,\mathrm{MeV} & {}^{64}\mathrm{Ge}+\alpha & \rightleftharpoons {}^{68}\mathrm{Se}+\gamma+2.285\,\mathrm{MeV}\\
{}^{40}\mathrm{Ca}+\alpha & \rightleftharpoons {}^{44}\mathrm{Ti}+\gamma+5.127\,\mathrm{MeV} & {}^{68}\mathrm{Se}+\alpha & \rightleftharpoons {}^{72}\mathrm{Kr}+\gamma+2.153\,\mathrm{MeV}
\end{array}
$$

The process cannot go beyond iron because, as can be seen from the energies of the reactions, the energies beyond iron are significantly smaller than those before, so if the reactions before iron can go, then those after iron will disintegrate the products much faster and prevent any increase in atomic weight.

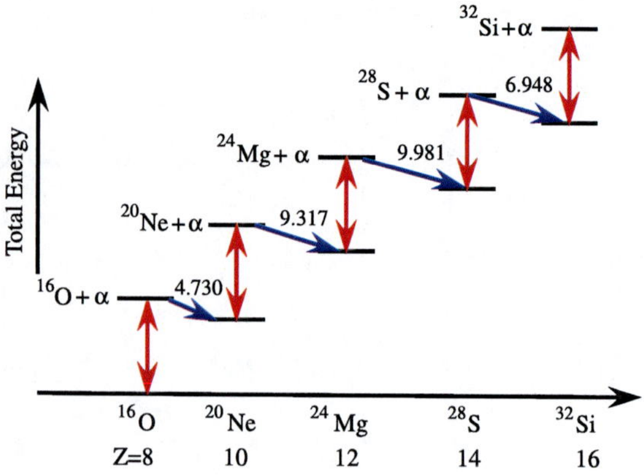

Fig. 9.5 The α process in the (Z, total energy) plane. The absorption of an α by a nucleus $A = 4n$, where n is an integer, yields a system with energy above the one with $A + 4$. Hence it decays into the stable nucleus and is ready to absorb another α. *Red arrows* indicate α absorptions and *blue arrows* are γ decays of the excited nucleus towards the stable ground state

In January 1955, Ajzenberg and Lauritsen published their famous summary on the energy levels of nuclei. A glance at the level diagram from their paper (see Fig. 9.6 right) shows that $^{16}O + \alpha$, which has a mass equivalent of 4.753 MeV above the ground state of ^{20}Ne, corresponds in ^{20}Ne to somewhere between the levels at 4.36 and 5.4 MeV, which were the only ones with unknown quantum numbers. Thus it was impossible to predict whether the reaction $^{16}O + \alpha \rightarrow {}^{20}Ne$ would be slow or fast.

The description of ^{20}Ne was based on the work by Cameron[10] (see Fig. 9.6 left). Cameron experimented with $^{16}O + \alpha$ but could not discover either the level at 4.36 MeV or the one at 5.4 MeV. The implication had to be a red flag, warning that these levels do not have the right spin and parity for this reaction to be relevant to astrophysics.

A month before the paper by Ajzenberg and Lauritsen came out, Freemantle, Prowse, and Rotblat[11] published their result that the state at 5.4 MeV is composed of two states at 4.95 and 5.62 MeV, but no spin and parity assignments were given. Hayakawa et al.[12] realized what the problem was and calculated three rates assuming that the level at 4.95 MeV has three possible spin and parity combinations. As a result of their calculations, they concluded that, at about $T = 1.4 \times 10^8$ K, most of the helium is converted into ^{20}Ne, *in definite disagreement with the observed cosmic abundances*, as they realized. They could have made good use of Hoyle's idea as implemented previously in the case of ^{12}C, suggesting that in this case the energy level in ^{20}Ne has the wrong spin and parity and does not participate in the nuclear reaction.

When B2FH analyzed the options in 1957, they knew about the ^{20}Ne levels at 4.95 and 5.62 MeV and were aware how critical the details of these levels are for the

---

[10] Cameron, J.R., Phys. Rev. **90**, 839 (1953). Not to be confused with Alastaire G.W. Cameron!

[11] Freemantle, R.G., Prowse, D.J. and Rotblat, J., Phys. Rev. **96**, 1270 (1954).

[12] Hayakawa, S., Hayashi, C., Imoto, M., and Kikuchi, K., Prog. Theor. Phys. Jpn **16**, 507 (1956).

α process. So faced with the lack of data, B2FH assumed the theoretical results of Salpeter, although of course he did not have better data either.

In 1956, Hayakawa et al. looked into *helium capturing reactions in stars* and effectively claimed that the process which started with $3\,\alpha \rightarrow {}^{12}\mathrm{C}$ continues to heavier elements and reaches $^{20}\mathrm{Ne}$. They checked the possibility that the process might continue as far as $^{40}\mathrm{Ca}$, but concluded that, at the low temperatures they were considering (less then 4×10^8 K), this would not happen. In essence, they considered what B2FH would later baptize as the α process. They also found that the abundance of $^{16}\mathrm{O}$ turned out to be much less than that of $^{20}\mathrm{Ne}$, in contrast to what Hoyle had found previously, because they had better data. Hoyle's conclusion was a bit premature, since not all options had actually been explored. These authors also found that all heavy ion reactions like $^{12}\mathrm{C} + {}^{12}\mathrm{C}$ would have negligible yield at temperatures at which the α process operates.

In 1959, Cameron[13] investigated various carbon reactions and the possibility of their leading to the synthesis of heavy elements. To this end, he considered the α reactions with carbon as well as the ion–ion reactions with all their complexities (since they yield many nuclei). In the absence of any experiment, Cameron assumed that the $^{12}\mathrm{C} + {}^{12}\mathrm{C}$ reaction could be considered as the inverse of a fission reaction.

Cameron did not calculate the stellar evolution. He assumed that core burning took place at a constant temperature, and did not discuss how the synthesized elements might be ejected from the star. After calculating several cases, Cameron concluded that:

> The carbon burning process will produce overabundances of heavy elements by factors of several thousands relative to a solar abundance.

These were not of course good tidings.

In 1959, Reeves and Salpeter,[14] still without any experimental data, evaluated upper and lower bounds by comparison with other reactions. The difference between the two bounds was a factor of 10^2, and clearly there was no reason to expect a difference between the carbon and oxygen reactions. They ended their paper with the statement that *carbon burning produces plenty of sodium and magnesium*, supporting Cameron's conclusions.

The first experiments on heavy ion–heavy ion reactions were carried out by Almqvist et al. in 1960,[15] and brought a surprise, since they showed that there was a difference between $^{12}\mathrm{C} + {}^{12}\mathrm{C}$ and $^{16}\mathrm{O} + {}^{16}\mathrm{O}$: while the first reaction exhibits

---

[13] Cameron, A.G.W., Astrophys. J. **130**, 429 (1959).

[14] Reeves, H. and Salpeter, E.E., Phys. Rev. **116**, 1505 (1959).

[15] Almqvist, E., Bromley, D.A. and Kuehner, J.A., PRL **4**, 515 (1960); Bromley, D.A., Kuehner, J.A. and Almqvist, E., PRL **4**, 365 (1960).

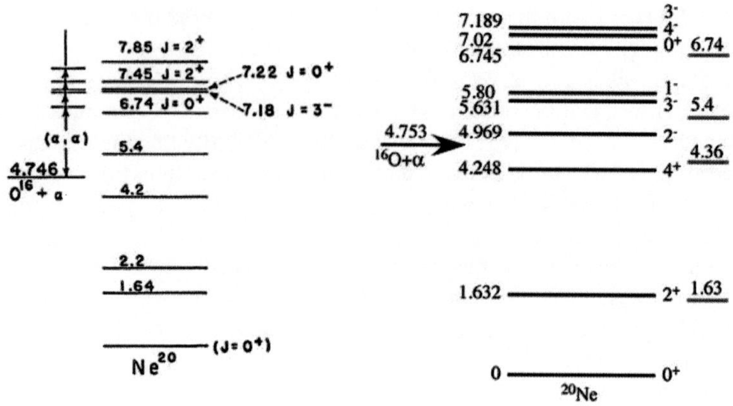

Fig. 9.6 *Left* the structure of ^{20}Ne as determined by Cameron 1953. Note that the energy of $^{16}\mathrm{O}+\alpha$ is not identical in this and the next figure, where the 2.2 MeV level has 'disappeared' and the 5.4 MeV level is found to be a combination of two levels. *Right* the structure of ^{20}Ne as determined by Gove et al. (1963). In *blue* are marked the energy levels as they appeared in Ajzenberg and Lauritsen 1955. The level discovered by Cameron at 2.2 MeV has disappeared

resonances, the second does not.[16] The authors hypothesized that the reactions would eventually get to Si, which could *build up the most stable nuclei near Fe.*

Even before the demise of the α process, Reeves and Salpeter[17] picked up in 1959 Salpeter's suggested but not calculated idea put forward when he had discussed the triple α reaction back in 1952. The fallback set of reactions for the next step were carbon and oxygen burning. Reeves and Salpeter calculated that the lifetime of carbon against destruction by carbon–carbon reactions is 10^5 years at $T = 6 \times 10^8$ K and only 1 year at $T = 8.5 \times 10^8$ K, implying that these reactions would be dominant in this temperature range.

Reeves returned to the problem in 1962[18] and admitted that *theoretical arguments must always be treated with much suspicion*, while the experimenters admitted a possible error of 50%. Reeves introduced a fudge factor which he denoted by $g' = 0.5$ and claimed that:

> The correct physical meaning of this term would be found in a theory of collisions between complex nuclei. Such a theory is still to come.

The conclusion of Reeves after all the corrections was that:

---

[16] Vogt, E. and McManus, H., PRL **4**, 518 (1960) explained the phenomenon as 'molecular states'. But why these states did not arise at no matter what energy in the other reaction was not explained. In ibid., Phys. Rev. **130**, 1140 (1963), they argued that the resonances were due to collective nucleon dynamics in contrast to what you would expect from a nucleus composed of α particles. In other words, the picture of a nucleus composed of α particles is far from ideal.

[17] Reeves, H. and Salpeter, E.E., Phys. Rev. **116**, 1505 (1959).

[18] Reeves, H., Astrophys. J. **135**, 779 (1962).

> The alpha process of B2FH could be responsible for isotopes 36–46 but may be the main source of isotopes 24, 28 and 32. Carbon burning gives rise to a neutron flux which is strong enough to produce large amounts of metals in overabundance.

Only in 1963 were Gove et al.[19] able to conclude from their experiment that the important 4.97 MeV level, which could be the main contributor to the reaction rate, must have an assignment of $J = 2^-$ and was therefore *excluded from participation in helium thermonuclear reactions* (see Fig. 9.6 right). The beautiful α process fell down on details.

In 1964, Fowler and Hoyle,[20] after becoming aware of Gove et al.'s 1961[21] publication (but before they knew about Gove et al.'s (1963) publication), wrote off the oxygen α-capture reaction. Thus, realizing the nuclear facts, Fowler and Hoyle withdrew the B2FH suggestion of the α process.[22] No $J = 0^+$ level was discovered in the energy range which is of interest to thermonuclear reactions in stars. Hence, the total rate is very small.[23] Upon reflection, one could have applied here Hoyle's famous argument in the ^{12}C case, but in reverse. If the reaction had been much faster, no oxygen would have been left at helium exhaustion.

9.6 The Equilibrium *e*-Process

The basic idea postulates that equilibrium between the nuclei is established on a timescale which is significantly shorter than the evolution time of the star at that phase. If the reactions are very fast, then the number of nuclei in a given state will be proportional to the Boltzmann factor $\exp(\Delta E/kT)$, where ΔE is the energy difference between the two nuclei.[24]

The evidence for the operation of this process is based on the observed abundance ratio between isotopes of the same element. As the temperature approaches 4–5 billion degrees, the most abundant element becomes iron, which has the highest binding energy. The simplicity of the *e*-process, in which the abundances depend only on the binding energies of the nuclei, is particularly appealing theoretically. Such an

---

[19] Gove, H.E., Litherland, A.E. and Ferguson, A.J., Phys. Rev. **124**, 1944 (1961); Gove, H.E., Litherland, A.E. and Clark, M.A., Nucl. Phys. **41**, 448 (1963).

[20] Fowler, W.A. and Hoyle, F., Astrophys. J. Suppl. **9**, 201 (1964).

[21] Gove, H.E., Litherland, A.E. and Ferguson, A.J., Phys. Rev. **124** (1961).

[22] There was no 'official' statement in the paper that the α process was defunct, but the possibility of an α process for building up the elements was no longer mentioned there. It should be remarked that Hayashi, Hoshi, and Sugimoto, Suppl. Prog. Theor. Phys. Jpn **22**, 1 (1962) did not mention the α process at all. On the other hand, they included reactions like magnesium and sulphur burning, which are not considered relevant today.

[23] Angulo, C. and 27 other authors, Nucl. Phys. **A656**, 2 (1999).

[24] The ratio between the numbers of nuclei in any given state should contain the number of ways the state can be realized (the statistical weight of the state). However, this important detail for the calculation is not essential for understanding how the *e*-process works.

equilibrium is clearly universal. The equilibrium process is nothing but what Hoyle assumed[25] in 1946 as a way to synthesize the heavy elements up to the iron group.

For the underlying assumption to be valid, a multitude of reactions should be fast relative to the evolution time. When the *e*-process is expected to operate, the timescale is shorter than years.

9.7 Problems with the Grand Picture of B2FH

The great triumph of B2FH element synthesis was in suggesting a process and a plausible site for all the nuclides in existence to be made in at least approximately the right proportions. It was effectively a work plan for generations of nuclear astrophysicists and stellar evolution modelers, as well as astronomers looking for observational evidence to validate or invalidate the various building blocks of the theory.

It was not long before the need for modifications was felt. Already in 1959, Helfer, Wallerstein, and Greenstein[26] had analyzed abundances in several Pop II objects. Their conclusion was:

> The comparison with the prediction of the *e*-process and possibly the *s*-process of B2FH may require some modifications.

In 1964, Suess (1909–1993)[27] criticized the basic assumptions of the two most famous schools of nuclear astrophysics of the day: the Pasadena–Cambridge group (Fowler and Hoyle) and the group at the Institute for Space Studies, New York, led by Cameron. The points of contention were as follows. B2FH did not explain the solar abundance curve and suggested that it was the result of mixing between the products of many stars. Suess objected to this assumption and raised many arguments against it. The most important argument was that the isotopic composition of the elements in meteorites was found to be constant. There was no indication of fluctuations that mixing could have caused.

Suess brought up various characteristic features of the abundance curve which he claimed were not explained by any of the processes suggested by B2FH. As Suess claimed:

> It is shown that, contrary to the belief of many investigators, no evidence exists from nuclear abundances that can be taken as a proof for element synthesis of heavy elements by either one of the processes proposed by B2FH.

Suess remarked that:

> In order to appreciate fully the argument put forward in the preceding sections of this paper, it may be necessary for the reader to spend hours and even days in graphically plotting all

[25] Hoyle, F., MNRAS **106**, 255, 343 (1946).

[26] Helfer, H.L., Wallerstein, G., Greenstein, J.L., Astrophys. J. **129**, 700 (1959).

[27] Suess, H.E., PNAS **52**, 387 (1964). Presented before the Academy, 29 April 1964, by invitation of the Committee on Arrangements for the Annual Meeting.

> existing information on the abundances of the elements and their isotopes to evaluate the various relationships. Anyone who has been doing this will share the opinion of the writer that the present state of our knowledge of the origin of the elements is unsatisfactory.

Maybe, so supposed Suess, the Mayer–Teller polyneutron theory could provide better results. However, his critical paper did not gain much publicity and got little attention from researchers in the field, despite the fact that most observational data on cosmic abundances were taken from Urey and Suess!

The first time Unsöld, the leading theorist on stellar spectroscopy and atmospheres, expressed his criticism[28] of the assumptions of B2FH was in 1969. The occasion was a conference in Pasadena, California, to celebrate the 60th birthday of Jesse Greenstein, a leading observer of stellar abundances. The conference subject was *The Chemical History of the Galaxy*. Unsöld expressed his criticism in the temple of nuclear astrophysics, but nobody listened. He returned to the subject[29] in 1976 without significantly altering his disapproval.

Unsöld posed the following question:

> Did the enrichment of cosmic matter with metals by a factor of about 1000 (from stars like HD122563 to metal strong stars like β Virginis) take place in connection with the nuclear evolution and decay of stars, as we observe them nowadays, or must we invoke a rather different sequence of events?

Can we estimate the nuclear activity of our galaxy? From the age and the luminosity, we calculate that only 3% of the hydrogen has been burnt during the entire lifetime of the Galaxy.

Can we calculate the total mass loss from stars? According to estimates by Reimers,[30] the total mass loss of stars in 10^{10} years is about 3%, a result which makes it impossible to assume that the elements heavier than helium, which constitute about 1.4% of the mass, could have been produced by stars in a galaxy similar to its present constitution. Support for this conclusion also came from the timescale required for the chemical evolution of our galaxy.

In the galactic halo, we observe stars with metal abundances ranging from 1/500 to 1/3 of the solar abundance. At the same time, in the galactic disk which originated from the collapse of the halo, the metal abundance changes from about 1/3 to about twice that of the Sun. Hence, the heavy elements (carbon and above) must have been formed during the quite short time before the collapse of the disk was completed. Only stars more massive than 4–6 solar masses have as short a lifetime as the collapse time of the Galaxy. So did these stars form all the heavy elements? To solve this problem, Truran and Cameron[31] assumed in 1971 that, for some reason, many more massive stars were formed at the beginning of the Galaxy than today. Truran and Cameron even claimed[32] that:

---

[28] Unsöld, A., Science **163**, 1015 (1969).

[29] Unsöld, A., Die Naturwiss. **63**, 443 (1976).

[30] Reimers, D., in *Problems in Stellar Atmospheres and Envelopes*, ed. by Bascheck, Kegel, and Traving, Springer, Heidelberg, 1975, p. 253.

[31] Truran, J.W. and Cameron, A.G.W., Astrophys. Space Sci. **14**, 179 (1971).

[32] Truran, J.W. and Cameron, A.G.W., Nature **225**, 710 (1970).

> In view of the success of the theories of nucleosynthesis in stars, it is very puzzling that the halo stars should have any heavy elements at all.

So how do we solve the problem of the formation of the heavy elements in such a short time? There are many pieces of evidence pointing to a gigantic explosion in the center of our galaxy, similar to the one observed in quasars. This idea is reminiscent of Wagoner, Fowler, and Hoyle's idea of small bangs, as distinct from big bangs. The evidence in support of this idea is the following:

- The metal abundance decreases from the center of the Galaxy outward.
- Giant galaxies seem to exhibit more and stronger explosions in their nuclei, while at the same time they have a higher metal abundance than dwarf galaxies.
- There is sparse data on the composition of matter flowing out of quasars. But the simple fact that one observes only elements with greater solar abundance than iron favors the idea that this matter does indeed contain a more or less solar type mixture of elements.
- From the relative ratio of $^{238}U/^{235}U$, one gets a stellar age of about 8×10^9 years. This age is in good agreement with the age of globular clusters and old galactic clusters. At the time we are discussing, the early 1970s, the Hubble age of the Universe was estimated to be 16×10^9 years. The difference between these two numbers was much too great.

In 1977, in view of the difficulties concerning the production of the elements, apart from hydrogen and helium, Melik-Alaverdyan[33] was ready to follow Suess and abandon the theory of stellar evolution, reverting to Mayer and Teller's idea from 1949 whereby heavy elements were synthesized via fission of ultra-heavy nuclei and subsequent radioactive decay. But where did this process take place? The author resorted to an idea of Ambarzumian, who suggested that the well known phenomenon of stellar flares[34] could be due to the decay of superdense prestellar matter, a hitherto unknown form of matter, capable of remaining in a stable state for a prolonged period and then going over into the ordinary state with the release of an enormous amount of energy.[35]

A decade after the publication of the B2FH paper, two theoretical physicists, Amiet and Zeh[36] wondered whether:

> [...] stellar element synthesis can account for the nuclei found in the Solar System, which should be those of an interstellar gas that existed about 5×10^9 years ago.

Since they had a problem over where the elements were formed, they proposed to:

> [...] draw conclusions on the history of the heavy nuclei solely from their abundance distribution. It will be left to astrophysicists and cosmologists to determine the astrophysical events that provide the physical conditions which according to this analysis must have prevailed.

---

[33] Melik-Alaverdyan, Yu.K., Byurakan Astrophysical Observatory, Astrofizika **14**, 515 (1978).

[34] A flare star is one which experiences non-periodic eruptions with rapid and appreciable increase in luminosity.

[35] Ambarzumian, V.A., Astrofisika **7**, 557 (1949).

[36] Amiet, J.P., and Zeh, H.D., Ziet. f. Physik **217**, 485 (1968); ibid. Phys. Lett. B **25**, 305 (1967).

In short, the idea was to separate the physics from the astrophysics.

The conclusions they reached were as follows. The process of slow neutron capture took place on pre-existing neutron-rich matter (something like Gamow's Ylem). Next, the neutron-rich and proton-rich nuclei were formed by β-decays from progenitors that were produced at a density of 2×10^{10} g/cm^3 and a temperature of 5×10^9 K. No idea was volunteered as to how and where such conditions might exist. Not every difficulty with a theory requires such drastic departures!

9.8 Carbon Burning

When Hoyle discussed the triple alpha reaction and the formation of carbon, he was mainly worried by the destruction of carbon by the reaction $^{12}\mathrm{C} + {^4}\mathrm{He} \to {^{16}}\mathrm{O} + \gamma$. But as the star evolves and the temperature in the core increases, $^{12}\mathrm{C} + {^{12}}\mathrm{C}$ becomes important in destroying the carbon. Hence, Hoyle's original argument must be revised to account for stars in which this destruction does take place and those in which it does not take place, and the carbon is eventually ejected into space. In short, it is far from sufficient to insist that ^{12}C is not destroyed by absorption of an α.

The first to consider how carbon can be further synthesized were Hayashi et al.[37] But this was not the original goal of the authors. B2FH applied the α process starting from ^{20}Ne, or coming after the triple α process, in an attempt to explain the relative cosmic abundances of the α nuclei ^{24}Mg, ^{28}Si, ^{32}S, ^{36}A, and ^{40}Ca. Hayashi et al. realized that when the temperature of the core is $T < 2\times 10^9$ K, the reaction $^{12}\mathrm{C}+{^{12}}\mathrm{C}$ competes with $^{20}\mathrm{Ne}+\gamma \to {^{16}}\mathrm{O}+\alpha$, despite the huge Coulomb barrier, and becomes the main energy source of stars which are 10^2–10^3 times more luminous than the Sun, in contrast with B2FH, who completely ignored the possibility of reactions between the heavy ions like ^{12}C and ^{16}O. Hayashi et al. remarked that, if matter is removed from the star unaltered (never mind how, or whether it is possible), then it still does not reproduce the observed cosmic abundances of the α nuclei.

The carbon-burning phase raises the very interesting question of the fusion of light heavy ions below the Coulomb barrier, and in particular the origin of the very pronounced structures observed in the fusion cross-section at low energies. The devoted experimental and theoretical works do not provide fully satisfactory solutions to explain why six α particles show the resonances and eight α particles do not. However, the detailed analysis would carry us too far astray. Here we note that the heavy ion reactions are extremely temperature sensitive and we find that

$$^{12}\mathrm{C} + {^{12}}\mathrm{C} \propto T^{33}, \quad {^{16}}\mathrm{O} + {^{16}}\mathrm{O} \propto T^{41}, \quad {^{28}}\mathrm{Si} + {^{28}}\mathrm{Si} \propto T^{72}.$$

Hence, an uncertainty of a factor of 10, which is very reasonable in view of the experimental difficulties and existence of resonances, leads to an error of $10^{1/33} = 1.072$ in the burning temperature, and this is at present tolerable.

[37] Hayashi, C., and 3 coauthors, Prog. Theor. Phys. Jpn **20**, 110 (1958).

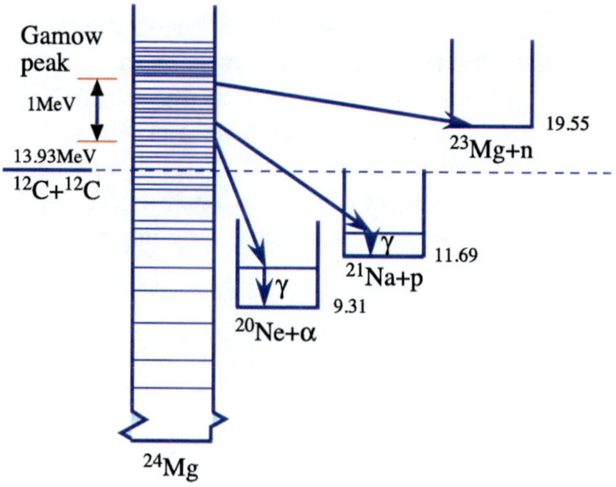

Fig. 9.7 The $^{12}C + ^{12}C$ nuclear reaction and its main products. The Gamow peak is at about 2.4 MeV relative kinetic energy and about 1 MeV wide. It therefore allows for the formation of nuclei like ^{23}Mg which have a ground state above the ground energy of two ^{12}C nuclei. The energy of the colliding ^{12}C nuclei is so high that they combine to form the ^{24}Mg nucleus in a region which contains many nuclear energy levels

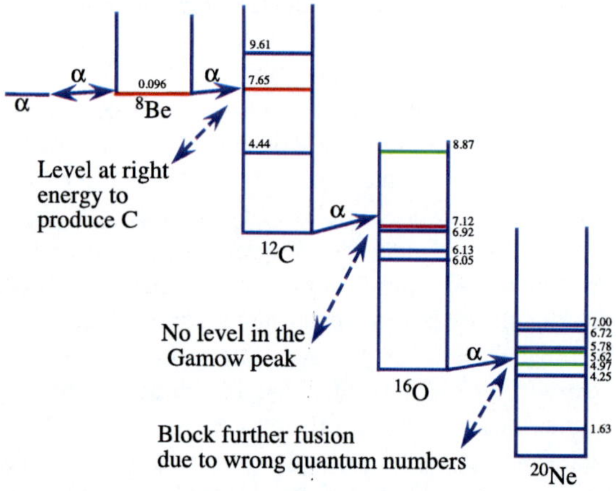

Fig. 9.8 The particular situation of the α nuclei. The resonances in ^{8}Be and ^{12}C which allow the formation of carbon (marked in *red*), the tails of the sub-continuum level (marked in *purple*), and the high level (marked in *green*) which allow a slow buildup of oxygen and prevent complete destruction of carbon, and the levels with the wrong quantum numbers in ^{20}Ne (marked in *green*) which block the conversion of ^{16}O into ^{20}Ne. All this happens at a temperature of $1–4 \times 10^8$ K

The energy diagram for the carbon reaction is shown in Fig. 9.7. The particulars of the energy diagrams of the α nuclei which allow for the formation of carbon are shown in Fig. 9.8. A proper discussion as to whether carbon survives stellar evolution or not must also take into account the carbon–carbon reaction which destroys the synthesized carbon:

$$
\begin{aligned}
{}^{12}\mathrm{C} + {}^{12}\mathrm{C} \longrightarrow {}^{24}\mathrm{Mg}^* &\longrightarrow {}^{23}\mathrm{Mg} + \mathrm{n} - 2.599\,\mathrm{MeV},\\
&\longrightarrow {}^{20}\mathrm{Ne} + \alpha + 4.617\,\mathrm{MeV} \quad \text{(dominant output)},\\
&\longrightarrow {}^{23}\mathrm{Na} + \mathrm{p} + 2.241\,\mathrm{MeV},
\end{aligned}
$$

where ${}^{24}\mathrm{Mg}^*$ is an excited ${}^{24}\mathrm{Mg}$ nucleus. The main burning products are ${}^{20}\mathrm{Ne}$ and ${}^{23}\mathrm{Mg}$ with small amounts of ${}^{21,22}\mathrm{Ne}$, ${}^{24,25,26}\mathrm{Mg}$, and ${}^{26,27}\mathrm{Al}$.

At the same time that carbon burning takes place, a host of 'small reactions' take place, e.g.,

$$
\begin{array}{lll}
{}^{20}\mathrm{Ne} + \alpha \longrightarrow {}^{24}\mathrm{Mg} + \gamma, & {}^{23}\mathrm{Na} + \mathrm{p} \longrightarrow {}^{24}\mathrm{Mg} + \gamma, & {}^{22}\mathrm{Ne} + \alpha \longrightarrow {}^{24}\mathrm{Mg} + \gamma,\\
{}^{25}\mathrm{Mg} + \mathrm{n} \longrightarrow {}^{28}\mathrm{Mg} + \gamma, & {}^{23}\mathrm{Na} + \alpha \longrightarrow {}^{26}\mathrm{Mg} + \gamma, & {}^{23}\mathrm{Na} + \mathrm{p} \longrightarrow {}^{20}\mathrm{Ne} + \alpha,\\
{}^{25}\mathrm{Mg} + \alpha \longrightarrow {}^{28}\mathrm{Si} + \mathrm{n}, & {}^{26}\mathrm{Mg} + \mathrm{p} \longrightarrow {}^{27}\mathrm{Al} + \gamma, & {}^{25}\mathrm{Mg} + \mathrm{p} \longrightarrow {}^{26}\mathrm{Mg} + \gamma\,.
\end{array}
$$

9.9 Neon, Oxygen, and Silicon Burning

The Ne-burning phase is initiated by the dissociation reaction

$$
{}^{20}\mathrm{Ne} + \gamma \rightarrow {}^{16}\mathrm{O} + \alpha - 4.75\,\mathrm{MeV},
$$

followed by the capture of the α particle by another neon nucleus, namely,

$$
{}^{20}\mathrm{Ne} + \alpha \rightarrow {}^{24}\mathrm{Mg} + \gamma + 9.31\,\mathrm{MeV}.
$$

Its rate is evaluated by applying the detailed balance principle to the inverse reaction ${}^{16}\mathrm{O} + \alpha \rightarrow {}^{20}\mathrm{Ne} + \gamma$. At the same time, the buildup reaction ${}^{20}\mathrm{Ne} + \alpha \rightarrow {}^{24}\mathrm{O} + \gamma$ takes place. Consequently, the main products of neon burning are ${}^{16}\mathrm{O}$ and ${}^{24}\mathrm{Mg}$.

Oxygen Burning

$$
\begin{aligned}
{}^{16}\mathrm{O} + {}^{16}\mathrm{O} \longrightarrow {}^{32}\mathrm{S}^* &\longrightarrow {}^{31}\mathrm{S} + \mathrm{n} + 1.500\,\mathrm{MeV},\\
&\longrightarrow {}^{31}\mathrm{P} + \mathrm{p} + 7.678\,\mathrm{MeV},\\
&\longrightarrow {}^{30}\mathrm{P} + {}^{2}\mathrm{D} - 2.409\,\mathrm{MeV},\\
&\longrightarrow {}^{28}\mathrm{S} + \alpha + 9.594\,\mathrm{MeV},\\
&\longrightarrow {}^{30}\mathrm{Si} + 2\mathrm{p} + 0.381\,\mathrm{MeV}.
\end{aligned}
$$

Table 9.1 Summary of the major stellar nuclear energy production phases and element synthesis

Fusing species	Main product	Byproducts	T (10^6 K)	Time scale (year)	Main reaction
H	He	^{14}N	35	8×10^6	$4p \rightarrow He$
He	C, O	^{18}O, ^{22}Ne	190	1.2×10^6	$3\alpha \rightarrow {}^{12}C$ ${}^{12}C + \alpha \rightarrow {}^{16}O$
C	Ne, Mg	Na	870	10^3	${}^{12}C + {}^{12}C$
Ne	O, Mg	Al, P	1,600	0.6	${}^{20}Ne + \gamma \rightarrow {}^{16}O + \alpha$ ${}^{20}Ne + \alpha \rightarrow {}^{24}Mg$
O	Si, S	Cl, Ar, K, Ca	2,000	1.3	${}^{16}O + {}^{16}O$
Si	Fe	Ti, V, Cr, Mn, Co, Ni	3,300	0.03	${}^{28}Si + \gamma \rightarrow {}^{24}Mg + \alpha$ ${}^{28}Si + \alpha \rightarrow {}^{32}S$
Fe	α		6,000	Collapse	${}^{56}Fe + \gamma \rightarrow 13\,\alpha + 4n$

Silicon Burning

It soon became clear that the jurisdiction of the *e*-process had to be limited. If the *e*-process takes over right after carbon and oxygen burning, the result is mostly the iron group nuclei,[38] and very little between silicon and the iron group. Hence, the *e*-process is limited to the last phase, and there should be another phase prior to that.

Truran, Arnett, and Cameron[39] and Bodansky, Clayton, and Fowler[40] realized that the products of carbon and oxygen burning, mostly ^{28}Si, start to photodisintegrate at a temperature of about 3.4×10^9 K, well before the conditions for the *e*-process are satisfied. Hence, the burning of silicon became a possibility. However, the temperature was too low for a direct ${}^{28}Si + {}^{28}Si$ reaction. On the other hand, the high energy photons are powerful enough to disintegrate the silicon into α particles. Thus an effective α process starting from silicon and ending in nickel becomes a reality. Since each reaction is in equilibrium to a good approximation, the entire process of converting silicon into the iron group nuclei was called quasi-equilibrium by Bodansky et al.

One cannot simply write ${}^{28}Si + {}^{28}Si \rightarrow {}^{56}Fe$. Due to the high temperature, part of the silicon is disintegrated by the energetic photons, creating α particles, protons, and neutrons:

$$ {}^{28}Si + \gamma \longrightarrow {}^{24}Mg + \alpha, \quad {}^{24}Mg + \gamma \rightarrow {}^{20}Ne + \alpha, $$

38 The iron group contains isotopes like ${}^{62}Ni_{28}$ to ${}^{58}Fe_{26}$, as well as ${}^{54}Fe_{26}$, which all lie close to the global minimum of the binding energy per nucleon. In other words, while the isotope ${}^{62}Ni_{28}$ has the strict minimum in binding energy per nucleon, other nuclei in the group have binding energies which differ only slightly from it (<0.001 MeV).

39 Truran, J.W., Arnett, W.D. and Cameron, A.G.W., Can. J. Phys. **45**, 2315 (1967).

40 Bodansky, D., Clayton, D.D. and Fowler, W.A., Astrophys. J. Suppl. **16**, 299 (1968).

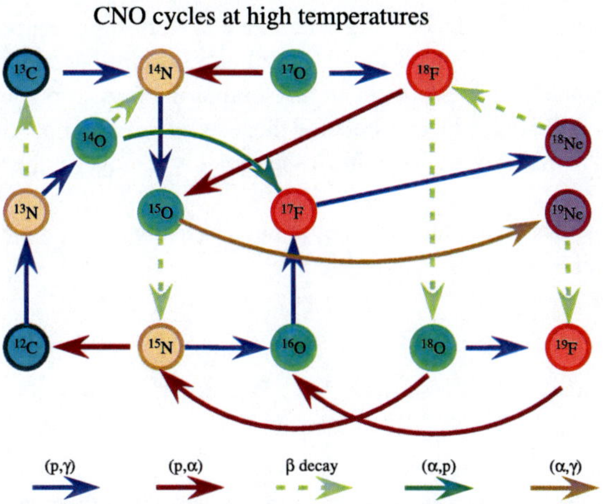

Fig. 9.9 The CNO network of nuclear reactions at high temperatures. *Colored arrows* mark the particular reaction. (p, γ) is proton capture and γ emission. Similarly, (p, α) is proton capture and α emission

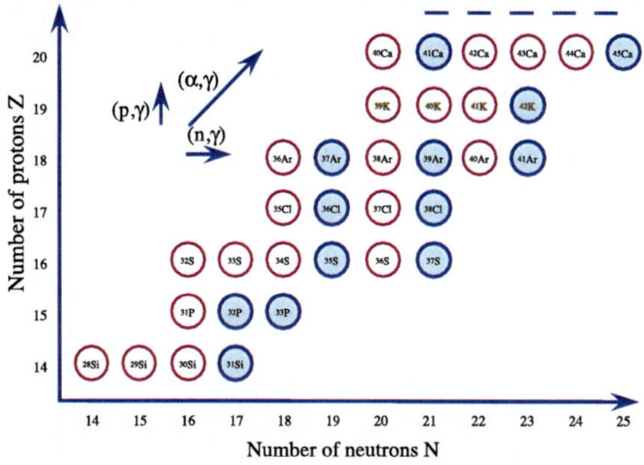

Fig. 9.10 The beginning of the silicon nuclear network. The *upper part* leading to iron is not shown for reasons of size. The reactions which connect between the various isotopes are not shown, to prevent congestion. (p, γ) is proton absorption and emission of a photon, (n, γ) is neutron absorption and γ emission, and (α, γ) is α absorption and γ emission. The inverse reactions exist as well. In some cases proton absorption and particle emission is also possible. *Red circles* enclose stable isotopes, while *blue* enclose unstable isotopes. The *broken line* at the *top right* means that the network continues and only the initial part of it has been shown

all the way down to single α particles, while at the same time the silicon absorbs α particles and builds up. All this takes place in equilibrium. As the temperature rises, the species with maximum abundance shifts gradually towards iron. The energy produced by silicon burning has little effect on the evolution of the star because it is by then completely governed by neutrino losses.

The last phases of silicon burning take place at very high temperatures, and a good approximation is once again a statistical equilibrium based on the rest mass of the nuclei. Since iron is the most tightly bound nucleus, it is clear that, as the temperature rises, the main species become those of the iron group. Although energy is still produced in this way, the neutrino losses quickly remove all the energy thereby generated.

A summary of the main phases in stellar nuclear energy and element synthesis is given in Table 9.1.

9.10 The Modern State of the Art: Nuclear Networks

As the temperature rises to above a few times 10^8 K, many nuclear reactions take place. In the above, we have discussed the most important of them, but this was only the tip of the iceberg. With the progress in computing, it has become possible to include large schemes of nuclear reactions, and those beyond carbon are essentially calculated using nuclear networks. We give two examples. The first is the hot CNO cycle shown in Fig. 9.9 and the second is the first half of the silicon nuclear network shown in Fig. 9.10. Recent nucleosynthesis networks contain about 5,000 nuclei and practically all possible reactions between them.

Chapter 10
Which Star Becomes Which Supernova?

Supernova research is presently in the limelight, and new types of supernova are being discovered, leading to a zoo of types, each with their own particular behavior. As discussed by Shaviv,[1] the theory of supernovas (SN) has not yet produced any 'successful explosion', so we describe here only what the extensive work by various modelers seems to suggest. Masses given relate to the moment just prior to the SN explosion. The initial star on the main sequence would certainly have been more massive and would have lost significant amounts of matter during its evolution toward the critical moment at which it exploded.

Supernovas can be divided into the following major classes: Type Ia, Types Ib and Ic, and Type II.

10.1 SN Type Ia

The major observational features of this type are their rather standard light curve, their lack of hydrogen, and their relative frequency in old type galaxies. Consequently, the widespread thesis is that these SN arise from an old white dwarf that somehow accreted mass, thereby increasing its own mass beyond the Chandrasekhar limit. The collapse of the WD would increase the temperature and ignite the carbon and oxygen, giving rise to a thermonuclear explosion. The lack of hydrogen and appearance in old galaxies hint at small masses, and the standard light curve seems to imply quite a narrow range of masses. The spectra are dominated by lines of iron, which is hypothesized to form in the thermonuclear runaway starting from carbon and oxygen.

The accreted mass must come form a companion star. If the accretion rate is low, the WD will burn it easily and give rise to a nova eruption, whereupon all the accreted mass, and even part of the WD, will be ejected into space. However, if the rate of accretion is prohibitively high, the assumption is that the burning will be stable and

[1] Shaviv, G., *The Life of Stars, The Controversial Inception and Emergence of the Theory of Stellar Structure*, Springer and Magnes, 2009.

G. Shaviv, *The Synthesis of the Elements*, Astrophysics and Space Science Library 387,
DOI: 10.1007/978-3-642-28385-7_10,

Fig. 10.1 *Left* the NGC 2818 planetary nebula. The nebula is about 3.25 light years across. The Hubble image is a composite of exposures through narrow-band filters, presenting emission from nitrogen, hydrogen, and oxygen in the nebula as red, green, and blue hues, respectively. The theory of stellar evolution predicts that this is the fate of stars with masses up to $8M_\odot$ when on the main sequence. The mass of the remnant is about $0.6M_\odot$ and it cools to become a *white* dwarf. Most of the mass is ejected as stellar wind, while the last part forms the nebula seen in the picture when the remaining envelope is thrown off. The various colors are due to different species excited by UV light from the cooling remnant. Credit: NASA, ESA, Hubble Heritage Team (STScI/AURA). *Right* the Cat's Eye Nebula (NGC 6543) was one of the first to be discovered, by William Herschel in 1786, and was the first planetary nebula investigated spectroscopically, by Huggins in 1864. The shape of the gases indicates the complexity of the mass ejection mechanism. The nebula is once again illuminated by UV radiation from the hot central star. Typical temperatures of the gas are $10–15 \times 10^3$ K, while densities are below 5×10^3 particles/cm^3. Credit: NASA, ESA, Hubble Heritage Team (STScI/AURA)

the accreted mass will generate an envelope that increases gradually until the mass goes beyond the Chandrasekhar mass and the star collapses. So far this has not been confirmed in detailed calculations.

10.2 SN Types Ib and Ic, and II

These three types belong to what are called core-collapse SNe, i.e., massive stars or cores stripped of the envelope, which collapse to form a neutron star (NS) or a black hole. Type Ib and Ic are assumed to be those stars which have for some reason lost their envelope. Hence, these two types are frequently referred to as stripped core-collapse SNe. Spectroscopically, they are distinct in their lack of silicon lines ($\lambda = 6{,}355$ Å). They show lines of oxygen, calcium, and magnesium. Type Ic, in contrast with type Ib, show no helium lines at 5,874 Å.

~8–10$M_\odot$ Stars

Stars with masses below $8M_\odot$ lose mass in extensive winds, their mass eventually going below the Chandrasekhar limiting mass so that they can end their lives as white dwarfs (see Fig. 10.1 left).

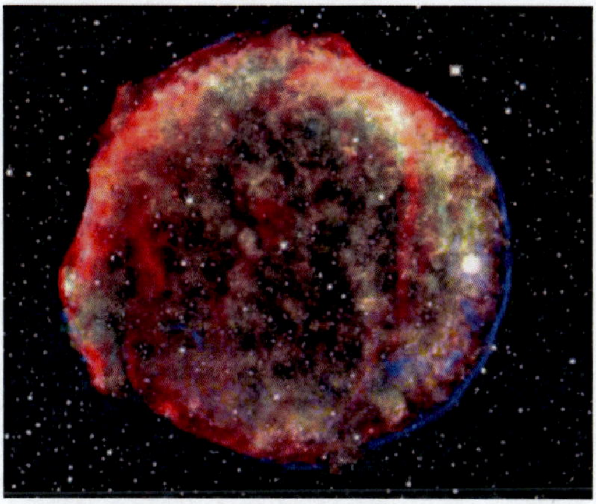

Fig. 10.2 On 11 November 1572, Tycho Brahe noticed a new star in the constellation of Cassiopeia. It was as bright as Jupiter. Legend has it that Tycho was so impressed by the appearance of a new star that he decided to devote his whole professional life to astronomy. However, Tycho must already have been quite knowledgeable about the stars to notice the change at all. He published his discovery a year later in a small book called *De nova stella*, and invented the terms ‘nova’ and ‘supernova’ used today. The object is about 7,500 light years away. The gaseous shells are expanding today at about 9,000 km/s. A subgiant star with heavy-element abundance close to solar was suggested as the likely survivor companion of an exploding white dwarf that produced the type Ia SN 1572 (Ruiz-Lapuente and 10 coauthors, Nature **431**, 1069, 2004.). Credit: NASA, ESA, Hubble Heritage Team (STScI/AURA)

Stars in the mass range ~ 8–$10 M_\odot$ become faint supernovas. Stars which have lost their envelope expose an O + Ne + Mg core that is more massive than the Chandrasekhar limit, and this becomes very dense due to cooling. The electrons become highly degenerate and hence energetic, resulting in inverse β-decays, i.e., conversion of a proton into a neutron. This process reduces the Chandrasekhar limiting mass and, if the star is not too massive, leads to an implosion followed by an explosion.

10–12$M_\odot$ Stars

Mass loss does not succeed in reducing the mass sufficiently, or in building a core in which nuclear reactions can produce iron. As the temperature continues to rise, the strong radiation field disintegrates the iron ($^{56}\mathrm{Fe} + \gamma \rightleftharpoons 13\,\alpha + 4\mathrm{n}$). The internal energy absorbed in this process reduces the gas pressure and causes a collapse. The Fe core, which is relatively small, collapses to form a neutron star, and the resulting SN tends to be faint.

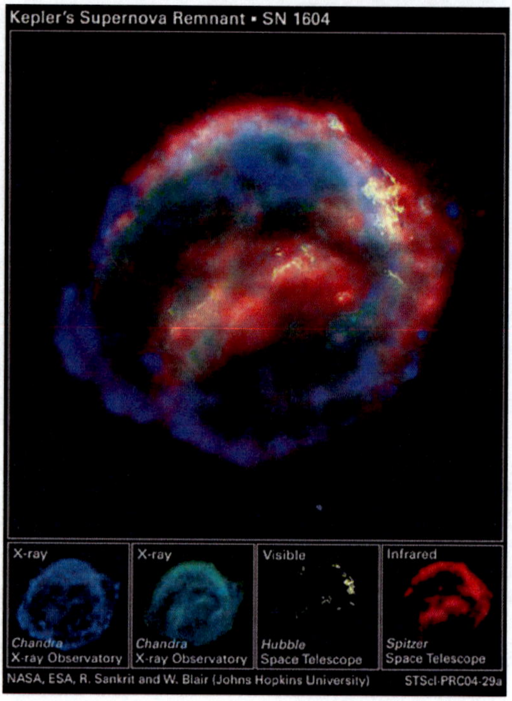

Fig. 10.3 The image of the Kepler 1604 type Ia supernova remnant as observed by the Hubble (visible), Chandra (X rays), and Spitzer (infrared) satellite observatories about four centuries after the explosion. This was the third supernova observed in our galaxy (after the Crab in 1054, and Tycho in 1572). No other well established supernova has since been observed in our galaxy, although statistics from other galaxies imply that there should have been about 4–8. For more information on Kepler's supernova, check the NASA Press Release. Credit: NASA, ESA, Hubble Heritage Team (STScI/AURA)

12–25$M_\odot$ Stars

These stars undergo Fe-core collapse to form a NS, and produce a significant amount of heavy elements. The upper limit is only approximate due to several uncertainties in the theory of stellar evolution of massive stars, regarding rotation and semi-convection (Figs. 10.2 and 10.3).

25–90$M_\odot$ Stars

These stars undergo Fe-core collapse to form a black hole (BH). The SNe seem to bifurcate into two branches: hypernovae (the correct name should be hyper supernovae) and faint SNe. The parameter which distinguishes between the two extremes is probably the angular momentum of the black hole or remnant. The collapsing rotating core may give rise to jets. Large amounts of heavy elements can be expected, produced by neutron capture by the iron group elements.

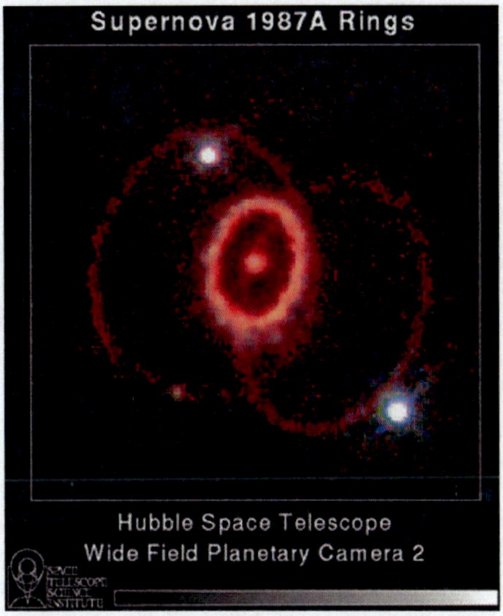

Fig. 10.4 The type II 1987A SN which exploded in 1987 in the nearby galaxy known as the Large Magellanic Cloud (at a distance of 170,000 light years). About three years later, the expanding light from the SN reached the two rings seen in the picture. The exact mechanism which gave rise to the rings is still not clear. It could be due to jets, superposition of stellar winds, a binary star effect, etc. The rings were ejected by the star before it became a SN, during the phase of extensive mass loss. They were not seen at first because they were in the dark, but the strong SN flash subsequently illuminated them. The luminosity of a star is not usually sufficiently powerful to light up its ejecta when it is so far away. Credit: NASA, ESA, Hubble Heritage Team (STScI/AURA)

90–140$M_\odot$ Stars

Stars in this range undergo nuclear instabilities and associated pulsations. Eventually, these stars undergo Fe-core collapse. The outcome is sensitive to the degree of rotation. It may lead to a hypernova-like energetic SN which produces large amounts of ^{56}Ni with subsequent neutron capture (on the seed ^{56}Ni nuclei or other iron group nuclei), thereby synthesizing the heavier-than-iron elements (Fig. 10.4).

140–300$M_\odot$ Stars

The fate of these stars depends on mass loss and rotation. They become pair-instability supernovae.[2] The star is apparently completely disrupted, leaving no remnant. All the heavy elements thereby produced are ejected into interstellar space. The expanding

[2] An instability induced by the strong radiation field losing energy by creating electron–positron pairs. The conversion of radiation energy, which supplies the pressure against gravitation, into the rest mass energy of the electron–positron pair reduces the support against gravitation and triggers collapse.

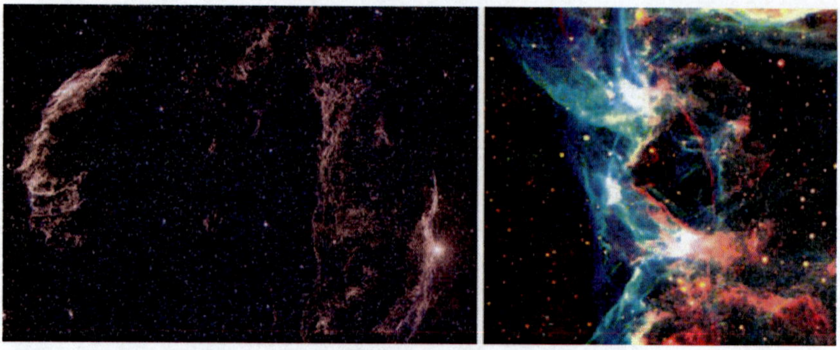

Fig. 10.5 *Left* the Cygnus Loop, also sometimes called the Veil Nebula. Here we see a supernova remnant several thousand years after the explosion. The distance is about 1,500 light years. The Cygnus Loop extends over 3.5°, or about 7 full Moons! Credit: NASA, ESA, Hubble Heritage Team (STScI/AURA). *Right* A magnified small section of the Cygnus Nebula. Different colors show emission from various kinds of UV or optical light, and the white regions are the densest, brightest regions that are cooling most rapidly. The gases shine because they are expanding at high velocity (about 180 km/s) and collide with interstellar matter, thus causing powerful shocks. These shocks heat the gases, which then cool by emitting line radiation. Credit: NASA, ESA, Hubble Heritage Team (STScI/AURA)

gaseous shells are powered by radioactive decays of the multitude of radioactive species produced by neutron capture.

$M \geq 300 M_\odot$ Stars

The evolution is now so fast that an iron core develops before the radiation in a sufficiently large volume of the star can create electron–positron pairs. The collapse should produce a massive black hole (Fig. 10.5).

Chapter 11
Between Two Extreme Nuclear Models

The nuclear reactions responsible for the synthesis of the elements are sensitive to the properties of the nuclei. The resulting abundances of the elements are a direct consequence of nuclear properties. The question as to whether a certain element can be synthesized in a star depends on the temperature of the star, but the relative abundance of its isotopes depends on the nuclear properties. So let us investigate those nuclear properties which control the synthesis.

Unlike the atom or the Solar System, the nucleus does not contain a central body which controls the motion of the nucleons in the space of the nucleus. In the atom, the Coulomb charge of the nucleus governs the motion of the electrons; in the Solar System, it is the gravitational force of the Sun which governs the motion of the planets. In the nucleus, it is the mutual attractive force acting between all particles which governs the motion, and hence there is no a priori reason to expect any parallelism between the structure of the atom and that of the nucleus. While in the Solar System and in the atom one can assume that the interactions between the planets, or between the electrons, are small, whence they may be ignored as a first step (e.g., when deriving Kepler's laws of planetary motion), this is not expected to be the case in the nucleus.

Yet against all physical logic, the history of nuclear theory is a story of wandering between two extremes: the shell model and the collective model. All the physical arguments are in favor of the collective model, i.e., a model in which each nucleon feels all the others, in contrast with the shell model, where one assumes that there exists 'an effective central force' and the nucleons move under the effect of this 'fictitious' force (Fig. 11.1). The collective model predicts that the properties of the nuclei should be a smooth function of the number of particles in the nucleus. However, the abundances of the elements, as well as many other properties of the nucleus, are monotonic neither as a function of the atomic number Z, nor as a function of the atomic weight A, nor as a function of the number of neutrons N. They thus provide strong evidence in favor of the simple picture of individual particles moving under a central force, even though it goes completely against straight-thinking physical intuition.

G. Shaviv, *The Synthesis of the Elements*, Astrophysics and Space Science Library 387, DOI: 10.1007/978-3-642-28385-7_11,

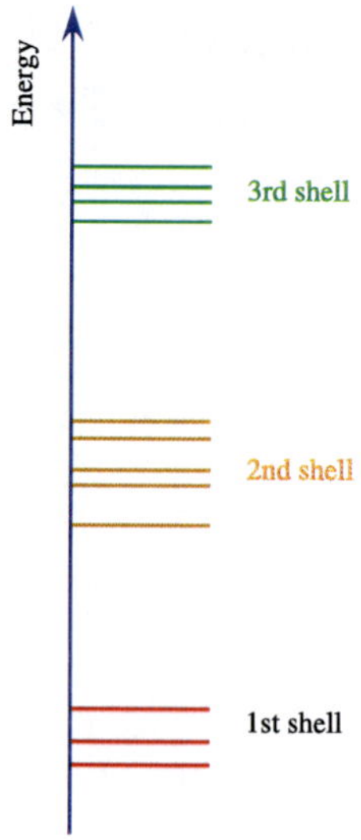

Fig. 11.1 Shells are groups of closely spaced energy levels (closely spaced in energy). The energy difference between shells is greater than the energy difference within a shell

The search for peculiarities and special numbers as indications of what the theory should predict began right from the very moment the existence of the nucleus had become clear. But nuclear theory did not really exist before the discovery of the neutron in 1932, and consequently, whatever theory there was had to tackle two problems:

- The problem of the stability of the nucleus, i.e., explaining how the positive charges hold together and do not dissolve the nucleus into its constituents.
- The difference between the number of positive charges (protons) and the total mass of the nucleus.

The existence of the proton was known, along with the fact that it carries a unit of charge when it is outside the nucleus. But does it have the same charge when it is inside the nucleus? The simple solution was to assume that the nucleus was full of protons and invent 'nuclear electrons' which neutralize the extra protons leaving only Z non-neutralized protons. This of course does not explain why the nucleus does not split into its constituents. The idea of a nuclear force between the particles in the nucleus was not yet born.

11.1 Nuclear Theory Before Nuclear Physics

In 1919, Wolff[1] considered the structure of the nucleus. To overcome the fact that the atomic weight was greater than the atomic charge, he assumed that there were *nuclear electrons* in the nucleus. With this assumption, he essentially followed Stewart,[2]

[1] Wolff, H.Th., Annal. Phys. **60**, 45 (1919).

[2] Stewart, A.W., Phil. Mag. **36**, 326 (1918).

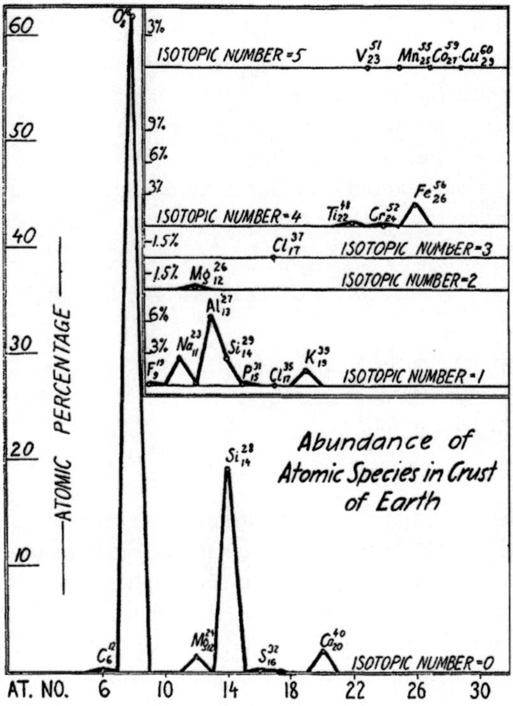

Fig. 11.2 In 1920, Harkins claimed that this graph *shows the variation in the abundance of the atomic species of each isotopic number*

Kohlweiler,[3] and Haas.[4] Still, this fact was not sufficient to stabilize the nucleus, a problem that bothered him in his paper. So Wolff hypothesized that the Coulomb law changes over distances as small as the radius of the nucleus, writing the Coulomb force as $F = 1/R^2 - a_4/R^4 + a_6/R^6$ and fitting the two parameters a_4 and a_6. However, the scattering of α particles showed that the Coulomb law was strictly valid down to the shortest distances. No deviations from this law were discovered.

In 1920, the chemist Harkins, probably influenced by the existence of a periodic table of the chemical elements, published a paper under the title *The nuclei of atoms and the new periodic system*,[5] in which he suggested that the nucleus was composed of α particles (Fig. 11.2). Nuclei with atomic weights which are not $4n$, where n is an integer, have extra particles moving in orbits around a core composed solely of α particles. The fact that the binding energy of nuclei is a small fraction, less than 0.7% of the rest mass, and only α particles are emitted in natural radioactivity, led many to conclude that the nuclei were in fact composed of α particles and protons. Support

[3] Kohlweiler, E., Zeit. f. Phys. Chem. **92**, 685 (1918).

[4] Haas, A., Physik. Zeit. **18**, 400 (1917).

[5] Harkins., W.D., Phys. Rev., **15**, 73 (1920).

for this idea was provided by Rutherford in his Bakerian lecture,[6] delivered in June 1920.[7] The difference, however, as noted by Rutherford, was that Harkins assumed that the mass 3 had one unit of charge and hence was an isotope of hydrogen, while Rutherford had assumed that the particle with mass three had two units of charge and hence was an isotope of helium. Not many years later it turned out that both possibilities exist in nature.

Harkins provided a simple expression for the mass of a nucleus, which was actually the first formula for the binding energy of the nucleus, namely

$$W = 2(Z+n) + \frac{1}{2} + \frac{1}{2}(-1)^{Z-1},$$

where Z is the atomic number, W the atomic weight, and n the number of cementing electrons used to attach helium nuclei. For nuclei with even atomic number, the formula simplifies to

$$W = 2(N+n).$$

Here we already see the difference between nuclei with even and odd numbers of charges Z, breaking the monotonicity in properties.

It is interesting to note that, although he did make a connection between the nuclear properties and the abundance, Harkins did not mention the Oddo–Harkins rule in this paper. It had been discovered a couple of years earlier. It is to Harkins' credit that he noticed that, as a function of atomic number, the abundance of the elements is not a smooth function of the atomic number, whereupon he established the Oddo–Harkins rule stating that even Z nuclei are more abundant than odd Z nuclei.

Harkins gave a table with the masses of the nuclei and classified them according to his model (see Table 11.1), i.e., in terms of how many α particles and how many additional protons (or protons plus nuclear electrons) there were. He then gave the 'deviation from a whole number' and found them to be both positive and negative. It is striking that, 15 years after the publication of the theory of relativity, which clearly predicted that the difference between the mass of the nucleus and 'the whole number of particles in the nucleus' must be negative, in contrast to his findings, Harkins missed the role of special relativity and did not realize that his numbers were wrong and made no sense, even though he was aware of the formula $E = mc^2$. The only unacceptable excuse by Harkins and the referee of the Physical Review is that he published his paper before Aston discovered that the chemically measured atomic weights are due to a mixture of different isotopes. Hence, he had to work with non-integer numbers.

[6] The Bakerian Lecture is a prize-winning lecture of the Royal Society. The series of lectures was established in 1775 by Henry Baker, who donated 100 pounds for a scientific lecture by a Fellow elected by the Royal Society.

[7] Rutherford, E., Bakerian Lecture: *Nuclear Constitution of Atoms*, Proc. Roy. Soc. **A 97**, 374 (June 1920). Rutherford notes that, in November 1919, Fulcher from the National Research Council in the USA sent him *a suggestion that nitrogen and oxygen might prove to be α particles*.

Table 11.1 Harkins' 1920 table of the first nuclei

Atomic number	Atomic weight divided by 4	Deviations from a whole number	Atomic number	Atomic weight divided by 4	Deviations from a whole number
2	1.00	0.00	28	14.40	+0.50
6	3.00	0.00	30	16.34	+0.34
10	5.00	0.00	32	18.12	+0.12
12	6.08	+0.08	34	19.80	−0.20
14	7.07	+0.07	36	20.73	−0.27
16	8.01	+0.01	38	21.91	−0.09
18	9.97	−0.03	40	22.65	−0.35
20	10.02	+0.02	42	24.00	0.00
22	12.02	+0.02	44	25.42	+0.42
24	13.00	0.00	46	26.67	−0.32
26	13.96	−0.04			

Note that Harkins found positive and negative deviations from a whole number. After Harkins 1920.

11.2 Rutherford's Nuclear Model

Rutherford discovered the nucleus[8] in the famous 'Rutherford scattering experiment', an experiment which was actually carried out by Geiger and Marsden,[9] but little else could be done about modeling the nucleus before quantum theory became available in the mid-1920s. In 1920, Rutherford delivered the Bakerian lecture,[10] in which he reviewed the state of the art in nuclear physics. As a matter of fact, all there was to report was what was the assumed building block of nuclei. Once this was settled (unsatisfactorily), the stability of the model could be analyzed. Rutherford assumed that the nuclei were composed of hydrogen and α nuclei and/or particles of mass 3. The problem that the mass of helium is less than the mass of four hydrogen atoms was explained by him, following Sommerfeld,[11] by saying that the binding energy (the conversion of the mass difference into energy) implied that the helium 4 nucleus was very stable. All the experiments he carried out, which disintegrated nitrogen and oxygen by bombarding them with α particles, failed to disintegrate the α particle itself. In addition, Rutherford cited Aston,[12] who had discovered experimentally that:

[8] Rutherford, E., Phil. Mag. **6**, 21 (1909); ibid. **21**, 669 (1911). It is interesting to note that Rutherford calculated the scattering of the α particles assuming classical physics, whereas it should have been calculated using quantum mechanics. However, in this particular case, the coincidence is that the quantum mechanical result is identical to the classical one, and the Planck constant, already known by then but in another connection, cancels from the quantum mechanical calculation.

[9] Geiger, H. and Marsden, E., *On a diffuse reflection of the α particles*, Proc. Roy. Soc. London A **82**, 495 (1909). The experiment was carried out by Geiger and Marsden, but the interpretation was made by Rutherford, and this is the reason for the name.

[10] Rutherford, E., Bakerian Lecture: *Nuclear Constitution of Atoms*, Proc. Roy. Soc. London A **97**, 374 (1920).

[11] Sommerfeld, A., *Atombau und Spektrallinien*, Vieweg und Son, 1919, p. 538.

[12] Aston, F.W., Nature **104**, 393 (1919).

> A fact of the greatest theoretical interest appears to underlie these results, namely, that of more than forty different values of atomic and molecular masses so far measured, all, without a single exception, fall on the whole numbers.[13]

This reminds us of Harkins' hypothesis.

Rutherford claimed that:

> [They] have shown that atoms of mass about 3 carrying two positive charges are liberated by α particles from both atoms of nitrogen and oxygen, and it is natural to suppose that these atoms are independent units in the structure of both gases.

This was long before ^{3}He was discovered in 1934.[14]

Rutherford hypothesized that one nuclear electron could bind two H nuclei (to form what we call today deuterium). Rutherford even hypothesized the existence of an atom of mass 1 which has zero nuclear charge:

> Such an atomic structure seems by no means impossible.

Furthermore, continued Rutherford, such neutral nuclei might have masses 2, 3, 4, or more. It is amazing that Rutherford, who carried out the first nuclear reaction, did not mention that on no occasion had he ever observed a negative nuclear electron coming out of the reaction when he bombarded, say, a nitrogen nucleus with hydrogen nuclei or α particles. Likewise, Rutherford argued that:

> A helium nucleus of mass 4 may be substituted in the complex structure for the corresponding nucleus of mass 3, without seriously interfering with the stability of the system.

The way Rutherford imagined the structure of lithium, carbon, nitrogen, and oxygen is shown in Fig. 11.3. From this picture Rutherford concluded that:

> The chance of a direct collision with one of the four atoms of mass 3 in nitrogen is much greater than the chance of removing an H atom, for it is to be anticipated that the main nucleus would screen the H atom from a direct collision, except in restricted regions facing the H atoms.

Thus Rutherford assumed that the building blocks of the nucleus were fixed in place, i.e., static.

On 10 June 1921, Rutherford delivered a lecture before the Royal Society in which he discussed the stability of atoms.[15] Rutherford discussed natural radioactivity and did not raise the question as to why the nucleus does not disintegrate if it is composed of several positive charges.

---

[13] Strictly speaking, this result implies zero or at best vanishingly small binding energy, and should have prompted him to ask 'how come?' Rutherford noted that the only exception to this rule was hydrogen with mass 1.008. Rutherford did not know about the existence of deuterium, and the deviation of the hydrogen mass from unity led him to exclude the possibility of its being a fundamental building block for more massive nuclei.

[14] Oliphant, M.L.E., Harteck, P., and Rutherford, E., Proc. Roy. Soc. London A **144**, 692 (1934).

[15] Rutherford, E., *The Stability of Atoms*, abridged report of a lecture delivered before the Society on 10 June 1921, but published as Proc. Phys. Soc. London **33**, 389 (1920). Note the date reversal.

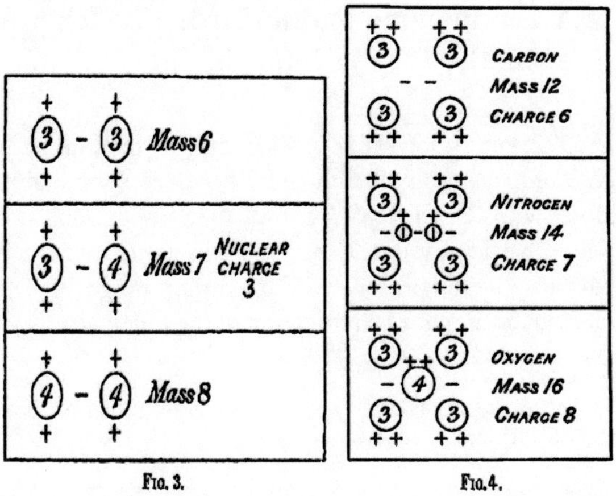

Fig. 11.3 Rutherford's models of lithium, carbon, nitrogen, and oxygen. According to Rutherford: *The carbon nucleus is taken to consist of four atoms of mass 3 and charge 2 and two binding electrons. The change to nitrogen is represented by the addition of two H atoms with a binding electron and an oxygen nucleus by the substitution of a helium nucleus in place of the two H atoms.* From Rutherford's Bakerian lecture 1920

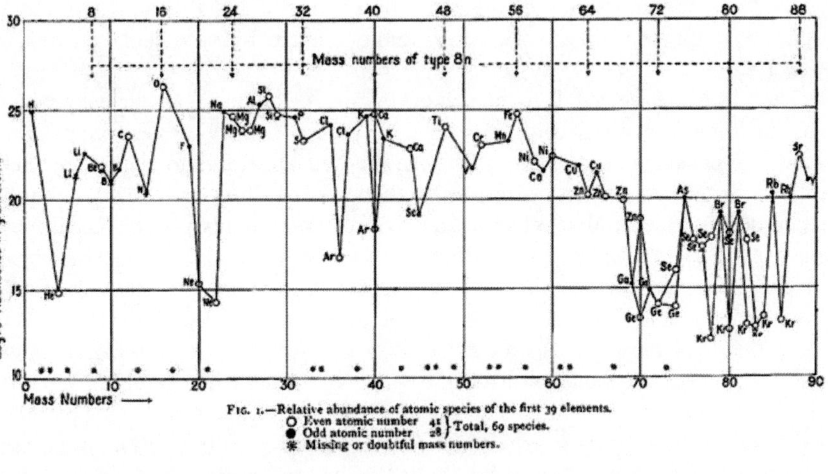

Fig. 11.4 Abundances of the nuclides by mass from Aston 1924. *Black circles* at the bottom indicate masses (2, 3, 5, 8, 13, etc.) for which Aston did not have information

11.3 Isotopic Abundances

Once Aston had discovered that almost all elements have isotopes, he found that the relative abundances of the different isotopes of the same chemical element varied.[16] The curve is not smooth but shows the Oddo–Harkins odd–even effect. The most important conclusion to draw was that the process which controls the relative abundances is not chemical but physical, and involves the nuclei directly (Fig. 11.4).

[16] Aston, F.W., Nature **113**, 393 (1924).

11.4 Challenging Rutherford: The Right Explanation for the Wrong Experiment

In 1922, Pettersson (1888–1966) came to the Radium Institute in Vienna with the purpose of measuring the radioactivity of deep-sea sediments. During the investigation, along with Kirsch (1890–1956), they bombarded lithium, magnesium, and silicon with α particles emitted by ^{210}Po. As a result of the bombardment, they claimed to have discovered protons with maximum range 10–18 cm in air.[17] The distance travelled in the air gave an indication of the kinetic energy of the particles. Consequently, they concluded that:

> The hydrogen nucleus is a more common constituent of the light atoms than one had hitherto been inclined to believe.

This result contradicted similar experiments carried out in Rutherford's group in Cambridge, and triggered two students in Rutherford's laboratory, Bates and Rogers[18] to write a rebuttal of the claim by Pettersson and Kirsch. The students argued that they had never observed protons emitted by polonium, but only α particles.

The debate was heated. It concerned the basic issue of the day in nuclear physics: was the nucleus composed of α particles or protons? In 1923, Pettersson[19] presented an opposing nuclear model to the one advocated by Rutherford and Chadwick, a model that could explain his results. Rutherford and Chadwick[20] interpreted the experiment in which a nucleus bombarded by an α particle emits a proton (and lithium is not such a case) in what is called a direct reaction in the parlance of modern nuclear physics, i.e., the α particle hits a proton which revolves around a core. The single proton, called a satellite particle, moves around a core composed of α particles. The alternative hypothesis presented by Pettersson, equivalent to what became known later as the collective model, was that the energy of the incoming particle was *transferred from the α particle to the nucleus as a whole*, and was shared by all particles in the nucleus, leading to *an explosion which is supposed to have only a limited stability in the case of each of the elements*. Pettersson even carried out simple classical energy considerations which indicated the plausibility of his claim. Rutherford and Chadwick just explained their model and calculated nothing.

Rutherford in turn claimed that he had repeated Pettersson's experiment and did not see protons coming out, whereupon he considered all the work by Pettersson to be wrong. The controversy ended when Chadwick paid a visit to Vienna and

---

[17] Kirsch, G. and Pettersson, H., Sitzunsberichten der Akademie der Wissenschaften in Wien, p. 418, 1924; ibid. Die Naturwissen. **12**, 464 (1924). The full paper is, however: *Über die Verwandlung der Elemente durch Atomzertrümmerung. I*, Die Naturwissen. **12**, 495 (1924). This came out after Pettersson published his theoretical nuclear model.

[18] Bates, L.F. and Rogers, J.S., Nature, September 1923. Pettersson claimed in his paper that Bates and Rogers had *apparently detected H particles*, while they claimed to have observed α particles.

[19] Pettersson, H., Proc. Phys. Soc. London **36**, 194 (1923).

[20] Rutherford, E. and Chadwick, J., Phil. Mag. **42**, 809 (1921); ibid. **44**, 417 (1922).

realized the problem with distinguishing between protons and α particles by means of scintillation counters. Chadwick discussed his findings in Pettersson's laboratory and convinced Pettersson and Meyer (his boss) to admit that *subjective bias was at work*,[21] and had led the researchers astray. Under the suggestion of Chadwick, it was agreed to drop the entire matter and avoid public announcements.

The Rutherford model was not completely free of problems. According to Petterson:

> A satellite of positive charge revolving round the nucleus at a certain distance from it necessarily involves a change in the sign of Coulomb's law for distance of the same order.

This conclusion was also recognized by Rutherford and Chadwick when they said that:

> They implicitly assumed that positively charged bodies attract one another at the very small distance involved. Such attractive forces must exist in order to hold the ordinary composite nucleus in equilibrium, and it seems likely that these attractive forces will extend some distance from the nucleus.

They did not imagine the existence of a nuclear force, but assumed that the Coulomb force changes sign. However, the careful α scattering experiments from various nuclei, in which particles would have experienced very close approaches, indicated that the Coulomb law remained valid down to the smallest available distances, leaving no room for a change from repulsion to attraction. In short, none of the models could explain why the nuclei did not disintegrate into their fundamental units.

Another argument against a change in sign of the Coulomb force at short distances was that, in this case, we could expect the existence of nuclei composed only of positive charges, a phenomenon we do not see in nature. Pettersson, on the other hand, argued, but did not prove, that the nuclear electrons provide the cement which holds the positively charged nucleus together.

The collapse of Pettersson's nuclear experiment pushed the idea of interaction of the incoming particle with the entire nucleus into oblivion. The suggestion of a collective interaction vanished for several years, but when it resurfaced, Pettersson's suggestion was forgotten, probably because of the second part of the hypothesis, namely, the explosion.[22]

---

[21] The Vienna group used scintillation counters to distinguish by eye between scintillations caused by protons and those caused by α particles. This was mission impossible, and the women who had carried out the measurements had become victims of what their boss wanted to discover.

[22] In the wake of the experimental fiasco, incorrectly identifying α particles as protons, even though he had had a good idea to explain these wrong experiments (for details see Rentetzi, M., *Trafficking Materials and Gendered Experimental Practices*: *Radium Research in Early 20th Century Vienna*, Gutenberg, Columbia University Press, 2007), Pettersson was unable to get a position as a physicist and moved back to oceanography. Here he excelled and made several very important contributions, to the point that the Royal Swedish Academy of Science, in commemoration of his achievements, organized a conference in August 1997 with the title: *In the Wake of the Swedish Deep Sea Expedition*: *Development of Paleoceanography as a New Field of Science*, and minted a medal in honor of Hans Pettersson. The medal is awarded for achievements in oceanography.

Rutherford's nuclear model became the 'standard accepted theory' to deal with all attempts at element transmutations. For example, Nagaoka[23] spoke about *slightly detached portions of the nucleus*, namely, a core and particles outside it. It is appropriate to mention that Davies and Horton[24] suggested that hydrogen and helium nuclei might be found after the bombardment of heavy nuclei by α particles, which is exactly what Pettersson claimed but did not find experimentally.

In 1928, Harkins and Kay[25] followed through the logic of the hypothesis about the existence of nuclear electrons and attempted to add an electron to the nucleus of mercury so as to convert it into gold.[26] Since no gold was found in the *exceedingly sensitive tests*, they concluded that either less than one in a billion of the electrons attached itself to a nucleus, or else all or part of the nuclei produced were not sufficiently stable and disintegrated before they could be detected.

The idea of modifying the repulsive Coulomb force was replaced in 1928 by changing the assumed distribution of electric charges on the interacting particles. Thus, Deming (1900–1993)[27] considered the problem of how to predict the binding energy with negative electrons and protons, and found that, by assuming that the charge of the electron and the proton was on the surface of a sphere, he could derive sensible numbers for the binding energy of nuclei.

Not much more was done concerning the physics of the nucleus during the early and mid-1920s, because the main interest was in developing quantum theory. The first application of the revolutionary new quantum theory to the nucleus was carried out by Gamow in 1928.

11.5 Gamow Enters the Game

In 1928, Gamow was still a graduate student in Leningrad, and upon a special recommendation of senior Russian physicists was allowed by Moscow to spend the summer at the Max Born Institute in Göttingen. During his stay he developed the α decay theory, as well as the theory of tunneling.[28]

In his paper Gamow declared that Rutherford's model for the nucleus was all wrong, assuming as he did that the α particle was a fundamental building block of the nucleus and existed in the nucleus before escaping, in contrast to the idea of 4 protons and two negative electrons which combine to form an α particle just before escaping from the nucleus. In 1929, Gamow took part in a discussion at the Royal Society[29] about the physics of the nucleus, and proposed a simple model, which he

---

[23] Nagaoka, H., Nature **116**, 95 (1925).

[24] Davies, A.C. and Horton, F., Nature **117**, 152 (1926).

[25] Harkins, W.D. and Kay, W.B., Phys. Rev. **31**, 940 (1928).

[26] The idea was suggested by Rutherford in his Bakerian address in 1920, p. 394.

[27] Deming, W.E., Phys. Rev. **31**, 453 (1928).

[28] See Gamow's authorbiography: *My World Line*, 1970.

[29] Gamow, G., Proc. Roy. Soc. A **123**, 386 (1929); ibid. Phys. Zeit. **30**, 717 (1929).

called the 'water drop' model. As for the basic building block, Gamow argued that a certain number of protons (not more than three) and electrons could be bound to an α aggregate without forming a new α particle. Gamow claimed that such a unit would be less tightly bound than a nucleus comprising only α particles. In other words, the fundamental particles of the liquid were α particles.

The first model Gamow treated was made up only of α particles. He assumed that there was a mutually attractive force between any two α particles, and he assumed that this nuclear force had a very short range (Rutherford's experiment showed this), so the interaction distance was some very small r^*:

> The sphere of radius r^* is well known in the theory of capillarity as the sphere of molecular action. We can say that the particle inside the liquid has no resultant force acting on it if the distance from the boundary is greater than r^*.

A nucleon inside feels a force from all directions, while a particle near the surface feels a force only from inside. The result is an effective force which acts like surface tension:

> Such a collection of α particles will be very like a minute drop of water where the inside pressure, due to the kinetic energy of quantized motion, is in equilibrium with the forces of surface tension trying to diminish the drop radius. [...] The important point for the nuclear drop model is the question of the quantum number to be ascribed to the different α particles in the drop.

Gamow's solution was this:

> All α particles in the nucleus must be considered to be in the same state with quantum number unity.

This is so because the α particles are bosons and hence can all reside in the same energy level.

Suppose there are n α particles in the nucleus. The total energy is the sum of the potential and kinetic energies. Using Heisenberg's uncertainty principle, he obtained the kinetic energy and then applied the virial theorem (see Sect. 5.2). Hence the total internal energy of the α particles is $E = n(T + V) = c_1 n^{1/3}$, where c_1 is a numerical constant. On the other hand, the repulsive Coulomb energy at the surface is given by the number of particles squared divided by the radius of the nucleus. Since the nucleus is liquid and hence non-compressible, the radius is proportional to $n^{1/3}$. The Coulomb repulsion term is therefore given by $E_c = c_2 n^{5/3}$. The term proportional to $n^{1/3}$ is positive and the term proportional to the repulsion is proportional to $Nn^{5/3}$ and negative. Hence, as N increases there exists an n for which the binding energy becomes negative, and no such nucleus can exist. This was the first prediction of an end to the periodic table, assuming all nuclei to be spherical.

Gamow managed in this paper to get the first hints about the general shape of the valley of stability and an explanation as to why certain combinations of protons and neutrons form nuclei while others do not.

In the wake of Gamow's theory of α-decay, he was convinced that the nucleus was composed of α particles,[30] in other words, Harkins' basic theory. Actually, this

---

[30] Gamow, G., Proc. Roy. Soc. A **126**, 632 (1930).

was one of the main topics in the book on nuclear physics that Gamow published in 1931,[31] before the neutron was discovered.

The assumed 'nuclear electrons' must have had peculiar properties, as Gamow wrote:

> For some unknown reason, although the electrons in the nucleus behave in a peculiar and obscure way, this does not affect very much the laws governing the motion of the nuclear α particles and protons; we can treat nuclear processes involving only α particles and protons independently of the nuclear electrons.

The situation was exacerbated when it was realized that the nuclear electrons had to move inside the nucleus with relativistic velocities. Klein[32] discovered that such electrons, if they move freely in the nucleus, convert into positrons when they hit the nuclear potential well, and in this way can escape from the nucleus and cause its disintegration. This became known as the Klein paradox. In other words, it was impossible to assume the existence of 'free' nuclear electrons in the nucleus. So Gamow began to doubt the existence of electrons in the nucleus.

Gamow's idea of the liquid-drop model evolved gradually from 1928 to 1930. In 1928, Gamow went to visit Bohr in Copenhagen and discussed the nucleus[33] as a collection of interacting α particles. Gamow thanked Bohr for the discussion and his stay in Copenhagen.

The water drop metaphor also included the assumption of non-compressibility of the nucleus, which in turn implied that the radius of the nucleus varied as the 1/3 root of the number of nucleons. This was by then a known result from α particle scattering. Together with the assumption of surface tension, Gamow was able to predict the existence of what later became known as the valley of stability. Effectively, he was the first to obtain the first two terms in the known phenomenological formula for the binding energy of nuclei (see Fig. 11.5).[34]

Gamow's water drop model was reconceived in Copenhagen under the name 'liquid drop'. It is interesting to note that Bohr's first paper in physics was about the surface tension of water drops,[35] and had nothing to do with nuclear physics, although it was written in the year the existence of the nucleus was discovered, i.e., 1909.

Shortly after the historic paper by Houtermans and Atkinson, Breit wrote a very short notice entitled *On the possibility of nuclear disintegration by artificially accelerated protons*.[36] While Houtermans and Atkinson discussed the possible transmutations by charged particles in stars, Breit asked whether this process could be imitated in the laboratory. He hypothesized that this could indeed be done, but added

---

[31] Gamow, G., *Constitution of Atomic Nuclei and Radioactivity*, Oxford University Press, 1931.

[32] Klein, O., Zeit. f. Phys. **53**, 157 (1929).

[33] Gamow, G., *Zur Quantentheorie des Atomkernes*, Zeit. f. Phys. **51**, 204 (1928).

[34] Gamow, G., *Mass defect and nuclear constitution*, Proc. Roy. Soc. A **126**, 632 (1930).

[35] Bohr, N., *Determination of the surface tension of water by the method of jet vibration*, Phil. Trans. Roy. Soc. London A **209**, 281 (1909).

[36] Breit, G., Phys. Rev. **34**, 817 (1929). Breit also criticized Gamow, pointing out that his assumptions and expressions regarding the way the reaction takes place were not clear.

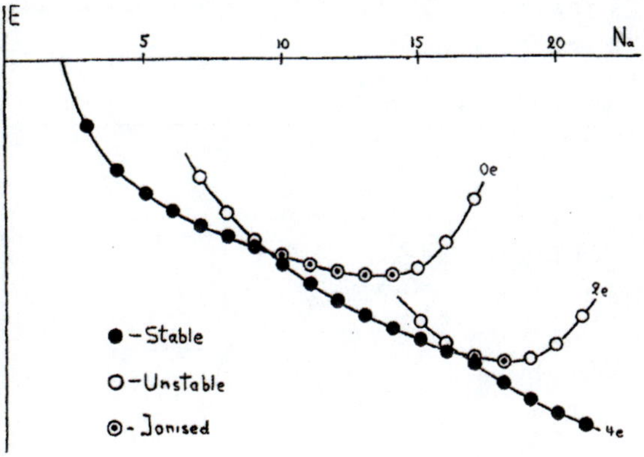

Fig. 11.5 Gamow's first prediction of the shape of the stability valley. N_α is the number of α particles in the nucleus. The valley is a consequence of the interplay between the Coulomb repulsion and the effective nuclear attraction due to the surface tension (paper in Roy. Soc. London)

that, if it failed, this would mean that *deviations from Rutherford's inverse square law of scattering* should be expected. We may say, therefore, that element transmutations confirmed the validity of the Coulomb force down to distances as small as the radius of the nucleus, and that Breit gave the starting signal for a glorious history of laboratory nuclear astrophysics in which stellar nuclear reactions were imitated in the laboratory.

11.6 Signs of Non-Smoothness in Nuclear Properties

In 1930, in the days before the discovery of the neutron, Barton[37] discovered *A new regularity in the list of existing nuclei* when he plotted the number of protons in the nucleus, P, against the number of electrons in the nucleus, E (see Fig. 11.6). The atomic number is $Z = P - E$. Barton noticed that the nuclear data plotted in this way does not produce a smooth curve, but rather three distinct 'clusters' of nuclei. Moreover, the clusters appeared to be symmetrical with respect to their center. If so, suggested Barton, an analogous symmetry should be exhibited by the relative abundances of the nuclei. He was thus among the first to indicate a connection between nuclear properties and abundancc.

Bartlett[38] attempted to explain Barton's clusters by assuming that there exist closed shells in the nucleus:

> The center of the cluster seems to lie about where the shells would be half-completed provided that the closed shells correspond to the masses 36, 64, 100, 144, etc.

It was Bartlett who introduced the idea of nuclear shells.

---

[37] Barton, H.A., Phys. Rev. **35**, 408 (1930); ibid. **34** (1929).

[38] Bartlett, Jr., J.H., Nature **130**, 165 (1932); ibid. Phys. Rev. **41**, 370 (1932); ibid. **42**, 145 (1932).

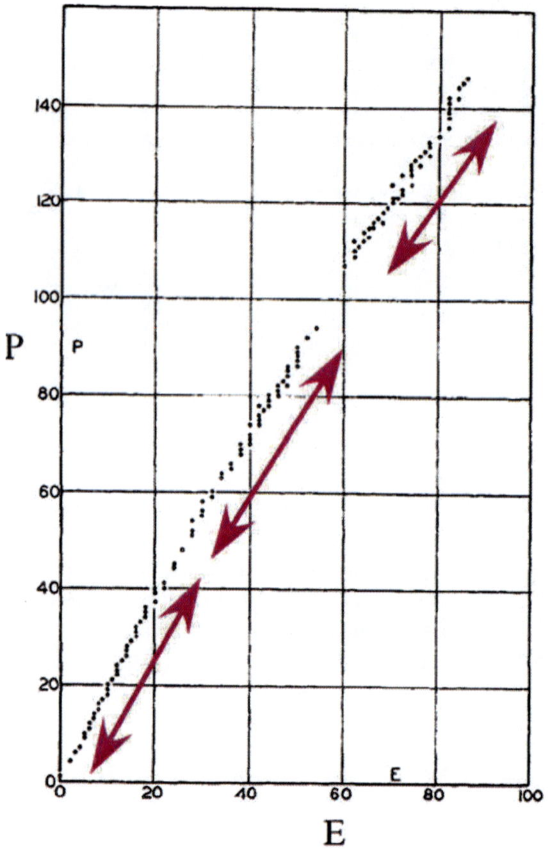

Fig. 11.6 Barton's nuclear clusters. P represents the number of protons and E the number of electrons. E stands for all the nuclear electrons, whether constituting α particles or not. The atomic number is given by $Z = P - E$. *Pink arrows* mark the clusters of nuclei as observed by Barton 1930

11.7 The Elimination of the Nuclear Electrons

When Rutherford delivered his Bakerian lecture in 1920, he hypothesized about the nuclear electrons which neutralize the protons. He went one step further and hypothesized that:

> Under some conditions, however, it may be possible for an electron to combine much more closely with the H nucleus, forming a kind of neutral doublet. Such an atom would have very novel properties. Its external field would be practically zero, except very close to the nucleus, and in consequence it should be able to move freely through matter. [...] It may be impossible to contain it in a sealed vessel. On the other hand it should enter readily the structure of atoms, and may either unite with the nucleus or be disintegrated by its intense field, resulting possibly in the escape of a charged H atom or an electron or both. [...] They

> may be produced in the electric discharges through hydrogen, where both electrons and H nuclei are present in considerable numbers.[39]

Two students in Rutherford's Cavendish laboratory tried to follow up Rutherford's idea[40] and looked for the predicted particle in an electric discharge. No signs of such a particle were discovered.

In 1932, Chadwick discovered[41] an extremely penetrating radiation, as he wrote:

> It was shown by Bothe and Becker[42] that some light elements when bombarded by α particles of polonium emit radiation which appears to be of the γ-ray type. The element beryllium (more accurately ^{9}Be) gave a particularly marked effect of this kind, and later observations by Bothe, by Mme. Curie-Joliot,[43] and by Webster[44] showed that the radiation [. . .] possessed a penetrating power distinctly greater than that of any γ radiation yet found from the radioactive elements.

Moreover, he adds:

> Mme. Curie-Joliot and M. Joliot made the very striking observation that these radiations from beryllium and from boron were able to eject protons with considerable velocities from matter containing hydrogen.

The researchers were in a dilemma: how to discover and verify the existence of radiation which is invisible and carries no charge? For this reason Chadwick discussed *The neutron hypothesis* in his third paper on the neutron, and started the discussion by saying:

> It is evident that we must either relinquish the application of the conservation of energy and momentum in these collisions or adopt another hypothesis about the nature of the radiation.

He thus supposed that the radiation consisted of neutral particles. This situation is reminiscent of the one faced later by Pauli, when he hypothesized the existence of a neutral particle emitted by β-decays. The discovery of the neutron provided the solution to the problem of nuclear stability.

Two consequences of the neutron discovery were (a) the death of the nuclear electron hypothesis, which in any case, did not work, and (b) the introduction of a nuclear force which differed substantially from the classical Coulomb law. In 1935, the Nobel Prize in Physics was awarded to Chadwick in recognition for his discovery of the neutron. Simultaneously, the Nobel Prize in chemistry was awarded to the Joliot couple, *in recognition of their synthesis of new radioactive elements*. Bothe won the Nobel Prize in 1954 for inventing the coincidence method for detecting elementary particles.

---

[39] Many credit Rutherford with naming the 'neutron', and cite the Bakerian lecture. I could not find the name 'neutron' in this paper.

[40] Glasson, J.L., Phil. Mag. **42**, 596 (1921) and Roberts, J.K., Proc. Roy. Soc. A **102**, 72 (1922). Since 1920, many attempts to discover neutral particles had been carried out in this laboratory.

[41] Chadwick, J., *Possible existence of a neutron*, Nature **129**, 312 (1932); ibid. *The existence of a neutron*, Proc. Roy. Soc. London a **136**, 692 (1932).

[42] Bothe, W. and Becker, H., *Eine Polonium-?-Strahlung*, Die Naturwiss. **18**, 894 (1930); ibid. Zeit. f. Phys. **66**, 289 (1930); Zeit. f. Phys. **76**, 421 (1932); ibid. Die Naturwiss. **20**, 757 (1932).

[43] Curie, I., Compt. Rend. **193**, 1412 (1931).

[44] Webster, H.C., Proc. Roy. Soc. A **136**, 428 (1932).

11.8 More Signs of the Individual Particle Behavior of Nuclei

In 1934, in a paper entitled *The genesis of the elements*,[45] Lewis made the following suggestion:

> [...] that a great part of the matter in the Universe is composed chiefly of iron and nickel, like the metallic meteors, and that such material which is thermodynamically stable with respect to all spontaneous transmutations, except at extremely high temperatures, is superficially attacked by cosmic radiation[46] to produce the material represented by the Earth's crust and by stony meteors.

Lewis discussed possible nuclear transmutations and spallation of iron leading to elements with smaller atomic weights. However, Lewis did not calculate the yield to show that he could predict abundances, and furthermore, he ignored all the elements heavier than iron. If indeed iron is the most abundant element in the Universe, then the abundance curve cannot decrease towards iron. Unfortunately, Lewis was misled by the observed abundances on Earth. He ignored the already available observations of stellar abundances.

Gapon[47] tried to calculate the masses of the nuclei of various isotopes in an attempt to show periodicity. The first problem was insufficiently accurate data, so that the 'special numbers' did not appear in the right places. The paper came out shortly after the discovery of the neutron, and the second problem was that his series of papers was effectively written without knowing about the neutron.

In 1933, Landé[48] demonstrated that, in odd-Z nuclei, the unpaired proton contributes the magnetic moment of the nucleus. Consequently, Schmidt and Schüler[49] discovered nuclei with large deviations from spherical symmetry. However, the liquid drop model does not predict such large deviations. A particular success came when they discovered that the nucleus of europium is not spherically symmetric. The model introduced by Landé to explain why particular nuclei were not spherical, and used by Schmidt, was remarkably simple. Landé assumed that:

> One particle only, one proton or one neutron, is responsible for the total spin and magnetic properties of the whole nucleus.

In this way, some special properties of the nucleus, in this case the total angular momentum, were explained as the property of the last nucleon,[50] which surrounds a spherical core. Schüler and Schmidt did not discuss shells, let alone closed ones.

---

[45] Lewis, G.N., Phys. Rev. **46**, 897 (1934).

[46] The cosmic rays were identified by Hess in 1912, a discovery that won him the Nobel Prize in 1936.

[47] Gapon, E.N., Zeit. f. Phys. **79**, 676 (1932); ibid. **81**, 419 (1933); ibid. **82**, 404 (1933); ibid. **84**, 509 (1933); ibid. **84**, 520 (1933).

[48] Landé, A., Phys. Rev. **44**, 1028 (1933).

[49] Schmidt, T. and Schüler, H., Zeit. f. Phys. **94**, 457 (1935).

[50] Schüler, H.J.J. and Schmidt, T., Zeit. f. Phys. **98**, 430 (1936).

However, they did discuss single particles moving outside a closed spherical core, and the issue was that a property of a single nucleon was associated with the property of the entire nucleus.

Returning to the α model, the problem was to determine how the particles move inside the nucleus. It was clear by then that quantum mechanics would not allow such particles to remain at rest, so they had to move. The question was, could they move inside a potential well, never mind how such a potential might be generated? In 1934, Margenau (1901–1997)[51] assumed a simple square well potential. Margenau asked what form the potential should take in order to produce the energy levels proposed by Gamow[52] or Rutherford, Lewis, and Bowden[53] to explain the observed binding energies of the nuclei. What Margenau discovered was that the situation was *hopelessly irreducible* with Gamow's scheme and *no less unsatisfactory* with the scheme of Rutherford et al.:

> The α particle cannot be said to move essentially in a potential field due to the other nuclear constituents.

In the late 1920s, Schrödinger regularly blamed the participants in Berlin seminars for their lack of imagination, claiming that:

> Just because you see alpha particles coming out of the nucleus, you should not necessarily conclude that inside they exist as such.[54]

The unavoidable conclusion was that there are no α particles in the nucleus, and that they form just before escaping.

Still, in 1935, Beck (1903–1988)[55] claimed that:

> There exist regularities in the occurrence of stable isotopes which have to be regarded as an expression of the unknown laws governing the structure of atomic nuclei. [...] The alpha particle plays an important role as an intermediary subnuclear unit in the nuclear structure.

But Beck's claim was largely ignored as it was not backed by any detailed experimental data.

11.9 Elsasser and Guggenheimer

Elsasser (1904–1991) and Guggenheimer were both in Paris, refugees from Nazi Germany, but interested in nuclear physics from different points of view. Progress came when Elsasser[56] and Guggenheimer,[57] motivated by the idea that the nucleus

---

51 Margenau, H., Phys. Rev. **46**, 613 (1934).

52 Gamow, G., Nature **131**, 433 (1933).

53 Rutherford, E., Lewis, W.B. and Bowden, B.V., Proc. Roy. Soc. A **142**, 347 (1933).

54 Jensen, J.H.D., *Glimpses at the history of the nuclear structure theory*, Nobel lecture, 1963.

55 Beck, G., Phys. Rev. **48**, 47 (1935); ibid. Zeit. f. Phys. **47**, 407 (1928).

56 Elsasser, W.M., J. de Phys. **4**, 370 (1933); ibid. **5**, 389 (1934); ibid. **6**, 473 (1935).

57 Guggenheimer, K., J. de Physique **5**, 253 (1934).

contains alpha particles, and the surrounding controversy, determined to look into the problem. Guggenheimer, who was a chemist, searched for nuclear analogy with the atomic periodic table. Elsasser, on the other hand, noticed the existence of 'special numbers' of neutrons and protons which endowed the corresponding nuclei with a particularly stable configuration. By analogy with atomic electrons, he correlated these numbers with closed shells in a model of non-interacting nucleons obeying Pauli's exclusion principle and occupying energy levels generated by a potential well. In parallel, Guggenheimer tried to classify the nuclei according to the number of protons and neutrons, and to show that the alpha model does not work. The periodicities so found were interpreted as reflecting full nuclear shells. Evidence for the magic numbers at $N = 50$ and $N = 82$ soon became evident. But the case of $N = 28$ was less clear. Surprisingly, Guggenheimer considered the fact that the numbers 50 and 82 were identical for protons and neutrons as a coincidence, although there was not supposed to be any difference between a proton special number and a neutron special number.

These hypotheses were not pursued any further at that time, partly because of the apparent paradox that strong inter-nucleon forces would average out in such a simple way, and partly because of the paucity of experimental data in favor of a single-particle description. The lack of rigorous derivation, and what looked to many like blatant numerology, in Bartlett's and Elsasser's physical arguments meant that the community did not take their findings sufficiently seriously. Robert Oppenheimer, as reported by Willy Fowler,[58] was very skeptical of Elsasser's work because Elsasser could not fit the data beyond 20 nucleons. The next special number that Elsasser got was 40. But according to experiment, the next special properties actually occur at 50. Oppenheimer also expressed his doubt directly by telling Jensen that:

> Maria and you are trying to explain magic by miracles.[59]

The discoveries by Elsasser and Guggenheimer were rejected by Bethe and Bacher in 1936, in Bethe's Bible of nuclear physics.[60] They claimed that, although the order of single nucleon orbits proposed by Elsasser and others reproduced the 'special numbers', their model lacked a *theoretical foundation.* A deeper argument they presented concerns the effect of nucleon–nucleon interactions on the single nucleon picture of the shell model. It is fair to admit that this problem has remained with us until now. It is in this respect that the shell model still *lacks theoretical foundation.*

In a way, Elsasser used all the data that was known at the time. Hence, until the late 1940s, when much more data on various nuclear properties, in particular β-decays and nuclear moments like the magnetic moment, had accumulated, nothing happened that could help with the solution of this problem. This is part of the reason for the timing of thebreakthrough.

---

[58] Fowler, W.A., Archives interview by Greenberg, 1983.

[59] Jensen, Nobel address, 1963. The term 'magic numbers' was coined by Wigner to express his contempt for the idea (at the time), but the name caught on among nuclear physicists.

[60] Bethe, H.A. and Bacher, R.F., Rev. Mod. Phys. **8**, 1892 (1936).

11.10 Fermi

On 15 January 1934, Curie and Joliot[61] succeeded for the first time in observing the long sought phenomenon of induced or artificial radioactivity. In the experiment, Curie and Joliot bombarded aluminum with α particles emitted by polonium and created phosphorus. They suggested calling the new isotope thereby created radio-phosphorus. The triumph of Curie and Joliot caused Fermi to change the nature of his research dramatically, and after his great successes with the theory known today as Fermi–Dirac statistics and the β-decay theory,[62] he started to experiment by bombarding all the elements with neutrons rather than α particles. The aim was to demonstrate eventually that nuclear transformations take place in almost every element irradiated by neutrons. The advantages of the neutral neutron over the charged α particle were obvious.

Various elements were irradiated by the newly discovered neutrons emitted by the polonium–beryllium source.[63] The first attempts were negative: none of the bombarded samples showed any appreciable activity, and Fermi realized that to induce artificial radioactivity with neutrons he should use the more intense radon–beryllium source. Another advantage of this source is that the γ radiation emitted by it does not interfere with the observation of a delayed neutron effect. The first applications of the more powerful radon–beryllium source did not yield better results. However, as Fermi irradiated successively more and more elements and reached fluorine and aluminum, he finally succeeded in 1934 in observing a small but significant level of β activity.[64] Fermi's success attracted a host of researchers who followed suit, and volumes 134 and 135 of Nature from 1935 will be found to contain an unusually large number of papers about neutron-induced radioactivity.

The accidental covering of the neutron source by the target led to the discovery of neutron albedo, that is, the reflection of neutrons by matter. When the cover was replaced by matter composed of light elements, like paraffin wax, it was discovered that (a) the neutrons lost energy, i.e., they were slowed down, and (b) the probability

---

[61] Joliot, F., Curie, I., *Artificial production of a new kind of radio element*, Nature **133**, 201 (1934).

[62] Fermi, E., *Versuch einer Theorie der β-Strahlen. I*, Zeit. f. Phys. **88**, 161 (1934).

[63] In those days this was one of the best sources of neutrons. For every 10^6 α particles emitted by polonium, about 30 neutrons are emitted from beryllium. Today, sources like plutonium–beryllium are used because they are more powerful. The reaction is interesting for another reason. It is a two step reaction:

$$^{210}\text{Po} \longrightarrow {}^{208}\text{Pb} + \alpha, \qquad {}^{9}\text{Be} + \alpha \longrightarrow 2\,{}^{4}\text{He} + \text{n},$$

so the polonium emits an α particle, which hits a ^{9}Be nucleus. The beryllium emits a neutron and the leftover ^{8}Be does not stay, but disintegrates into two α particles. ^{210}Po is very radioactive and dangerous to handle. Absorption of the energy of the alpha particle by living tissue causes severe damage. Who knows if carelessness in these experiments were not the cause of Fermi's untimely death.

[64] Fermi, E., La Ric. Scientifica **5**, 283 (1934).

of interaction increased tremendously for the slow neutron. Thus the term 'slow neutrons'[65] was invented for these particles, and their unique properties investigated.

By 1938, Fermi was without doubt the world's leading expert on neutrons, and neutron irradiation became very popular. Fermi continued his work on this topic upon his arrival in the United States. His expertise in neutron capture should be remembered when we recall that he developed the first nuclear reactor, which applied neutron physics, and inflicted a death blow to Gamow's $\alpha\,\beta\,\gamma$ theory (see Sect. 7.19).

The 1938, the Nobel Prize for Physics was awarded to Fermi for his work on *Artificial radioactivity produced by neutron bombardment*[66] and for nuclear reactions brought about by slow neutrons.[67] A summary can be found in the following papers by himself and various collaborators: *Artificial radioactivity produced by neutron bombardment*, Proc. Roy. Soc., 1934 and 1935; *On the absorption and diffusion of slow neutrons*, Phys. Rev., 1936. The theoretical problems connected with slow neutron irradiation are discussed by Fermi in the paper *Sul moto dei neutroni lenti*, Ricerca Scientfica, 1936. The Nobel prize was not awarded for the ingenious β-decay theory, although this theory is mentioned in the Nobel publication, or for the idea of Fermi–Dirac statistics. The importance of either is definitely commensurate with the neutron capture work, or surpasses it.

Returning to our discussion here, in his Nobel address, Fermi stressed that the capture probability of neutrons by various nuclei *varies with no apparent regularity for different elements* by a factor of thousands. Fermi offered no explanation. On the other hand, the sharp and closely spaced resonances discovered in the neutron capture were the trigger for Bohr's theory of the compound nucleus (see next section). The large variations in the neutron capture rate just hinted towards the independent particle model and the existence of closed neutron shells, but nobody at that time actually deciphered the hint. The particular importance of these experiments is that they directed physicists in the following years toward two extreme modes, the compound and the shell modes, as will soon become evident.

It is interesting to note that D'Agostino, a chemist in Fermi's group who participated in the neutron capture experiments, was convinced that nuclei could be organized according to a periodic system like atoms. The group disagreed and ridiculed him.[68] There was thus a non-believer in Fermi's camp who turned out to be right.

The great oversight by Fermi and his group is expressed simply by Meitner and Frisch in the abstract to their identification of the fission process:

> In making chemical assignments, it was always assumed that these radioactive bodies had atomic numbers near that of the element bombarded, since only particles with one or two

[65] The reader should not confuse 'slow neutrons', meaning low energy neutrons, with the s(low)-process, which means slow in time. This process will be discussed in Chap. 12.

[66] Fermi, E., Amaldi, E., D'Agostino, O., Rasetti, F. and Segre, E., *Artificial radioactivity produced by neutron bombardment*, Proc. Roy. Soc. London Ser. A **146**, 483 (1934); Amaldi, E. and Fermi, E., *On the absorption and the diffusion of slow neutrons*, Phys. Rev. **50**, 899 (1936).

[67] Fermi, E., *Radioattività indotta dal bombardamento di neutroni*, La Ric. Scientifica **II**, 15 (1934).

[68] Segre, E., *Autobiographies by Nobel Laureates in Physics*, from the Ann. Rev. Nuc. and Particle Sci. **31**, 1 (1981).

> charges were known to be emitted from nuclei. A body, for example, with similar properties to those of osmium was assumed to be eka-osmium ($Z = 94$) rather than osmium ($Z = 76$) or ruthenium ($Z = 44$).

Hence, they missed the unimaginable result that neutrons can fission uranium, a discovery left for Meitner (1878–1968m), Hahn (1879–1968m),[69] and Strassmann (1902–1980).[70]

Meitner, Hahn, and Strassmann collaborated and were among those who followed Fermi and his group. The year was 1938 and, being Jewish, Meitner had to flee the Nazis, escaping to Sweden.[71] It so happened that Meitner's nephew Otto Frisch, also a refugee, was in Sweden when Meitner received a letter from Hahn describing his chemical proof that one of the products of the bombardment of uranium by neutrons was barium ($Z = 54$), an element with much lower atomic weight than uranium. Meitner guessed correctly that the barium was formed when the neutrons fissioned the uranium. In December 1938, hardly a month after Fermi's Nobel lecture, Hahn and Strassmann sent the paper to Naturwissenschaften, reporting that they had detected the element barium after bombarding uranium with neutrons,[72] and informed Meitner in Sweden. Meitner and Frisch published the correct explanation and its relevance to the drop model.[73] Meitner and Frisch also estimated the exceptionally high energy output to be expected from the fission of uranium (about 200 MeV) and coined the expression 'nuclear fission'. This was a long shot, but a correct one. Explaining fission became the crowning achievement of the 'water drop model' by Bohr and Wheeler (see later), which served to convince physicists of the correctness of the model.

Hahn received the 1944 Nobel Prize for Chemistry for the discovery of nuclear fission.[74] It is deplorable that Meitner's brilliant physical intuition was not recognized by the 1944 Nobel committee as worthy of at least a share in the prize for

---

[69] The lunar crater is shared by Otto Hahn the chemist and by Friedrich II Graf von Hahn (1742–1805), the astronomer who discovered the central star in the Ring Nebula.

[70] Strassmann was recognized by the Yad Vashem Holocaust Memorial in Jerusalem as Righteous Among the Nations for what he did during the war to conceal Jews persecuted by the Nazis. When in 1933 the Society of German Chemists became part of a Nazi-controlled public corporation, he resigned from it in protest. He was consequently blacklisted and lost his job. Hahn and Meitner (before she had to flee) found an assistantship for him at half pay. During the war, the Strassmanns concealed a Jewish friend in their apartment for a long time, putting themselves and their three year old son at great risk. Hahn, Meitner (then in Britain), and Strassmann won the 1966 Enrico Fermi Award. The International Astronomical Union named an asteroid after Strassmann: asteroid 19136 Strassmann. There is also asteroid 6999 Meitner and asteroid 3676 Hahn.

[71] Meitner stayed in Germany for 5 years after the rise to power of the Nazis, despite the fact that she saw how Strassmann lost his job for speaking out against them. She even helped him get some financial support, although not a job, but later found it appropriate to blame all those physicists who stayed in Germany and helped the Nazis.

[72] Hahn, O. and Strassman, F., Naturwiss. **27**, 11 (1939).

[73] Meitner, L. and Frisch, O.R., Nature **143**, 239 (1939).

[74] After the war, Hahn was not allowed by his British captors to go to Stockholm for the Nobel ceremony.

being the person who recognized the phenomenon and gave the explanation.[75] Hahn himself was abundantly clear in his Nobel address about Meitner's crucial role in this discovery.

11.11 Bohr's Declaration Against the Independent Particle Model

On 27 January 1936, Bohr delivered a lecture about *Neutron capture and nuclear constitution* before the Copenhagen Academy. The lecture, which was later published in Nature,[76] was a crusade against the independent particle model, and propaganda for the collective model. No doubt, from a strictly physical point of view, Bohr's brilliant idea was right, but nature sometimes havoc plays with physicists. Bohr observed that the neutron capture resonances, as discovered by Fermi, were narrow in energy. Using the Heisenberg uncertainty principle, Bohr calculated that the time involved was several orders of magnitude longer than 10^{-21} s, which is the time needed for the slow neutron to cross the nucleus. Hence, the nucleus, into which a large amount of energy was added by the penetrating neutron, must remain for what is relatively speaking a very long time in an excited state before it finally decays and yields the products of the reaction.

Bohr and Kalckar[77] claimed that:

> Every nuclear transmutation will involve an intermediate stage in which the energy is temporarily stored in some closely coupled motion of all particles of the compound nucleus.

In other words, Bohr envisioned the nuclear reaction as

$$\text{initial state} \longrightarrow (\text{compound state}) \longrightarrow \text{final state.}$$

Bohr claimed that:

> It is, at any rate, clear that the nuclear models hitherto treated in detail are unsuited to account for the typical properties of nuclei for which, as we have seen, energy exchanges between the individual nuclear particles are a decisive factor. In fact, in these models it is, for the sake of simplicity, assumed that the state of motion of each particle in the nucleus can, in the first approximation, be treated as taking place in a conservative field of force, and can therefore be characterized by quantum numbers in a similar way to the motion of an electron in an ordinary atom.

The fundamental assumption of the shell model is that each nucleon moves in a potential well produced by all nucleons, and except for this mean smooth force, the

[75] See also *Revelations Concerning Lisa Meitner and the Nobel Prize*, Science Week, 17 October 1997.

[76] Bohr, N., Nature **137**, 344 (1936).

[77] Bohr,N. and Kalckar, F., *On the transmutation of atomic nuclei by impact of material particles. I.*, Det Kongelige Danske Videnskabernes Selskab. **24**, 10 (1937).

nucleon is only marginally affected by the rest of the nucleons. This supposition apparently contradicts a basic premise of nuclear physics, claimed Bohr, namely that the interaction between the nucleons is very strong, implying that fast energy exchange takes place between them. Bohr continued with a carefully formulated statement, that was intended to bring a calamitous blow to the shell model:

> In the atom and in the nucleus we have indeed to do with two extreme cases of mechanical many-body problems for which a procedure of approximation resting on a combination of one-body problems, so effective in the former case, loses any validity in the latter.

In short, the structure of the atom is one thing and the structure of the nucleus is another thing. These conflicting descriptions are not to be mixed.

11.12 Support in Question: The Breit–Wigner Formula

Completely independently and even slightly before the appearance of Bohr's Nature paper, Breit (1899–1981) and Wigner (1902–1995)[78] attacked the same problem and provided a beautiful mathematical formulation.[79] As a matter of fact, Bethe,[80] Fermi and his group,[81] Perrin and Elsasser,[82] and Beck and Horsley,[83] all proposed alternative explanations. But Breit and Wigner rejected all these explanations, claiming that:

> The combined evidence of experimental results and theoretical expectation is thus against a literal acceptance of the current theories.

They thus attacked the problem of the resonances observed in slow neutron capture, coming up with the famous Breit–Wigner formula, which was later generally reinterpreted as the mathematical expression for, and full support for, Bohr's compound nucleus idea. But one has to read Breit and Wigner carefully. They wrote the following in the abstract to their paper:

> These facts can be accounted for by supposing that, in addition to the usual effect, there exist transitions to virtual excitation states of the nucleus in which not only the captured neutron but, in addition to this, one of the particles of the original nucleus is in an excited state.

In other words, Breit and Wigner did not take it that all the nucleons shared the energy of the incoming neutron, but rather that only one of them did. Moreover, the

---

[78] Breit, G. and Wigner, E., *Capture of slow neutrons*, Phys. Rev. **49** (1936).

[79] Bohr's paper in Nature consisted mainly of qualitative arguments, typical of the deep thinker Bohr, since the attempts at a mathematical formulation by Bohr and Kalckar had been delayed. In fact, the paper by Bohr and Kalckar came out significantly later.

[80] Bethe, H.A., Phys. Rev. **47**, 747 (1935).

[81] Amaldi, E., d'Agostino, O., Fermi, E., Pontecorvo, B., Rasetti, F. and Segrè, E., Proc. Roy. Soc. A **149**, 522 (1935).

[82] Perrin, F. and Elsasser, W., Compt. Rend. **200**, 450 (1935).

[83] Beck, G. and Horsley, L.H., Phys. Rev. **47**, 510 (1935).

single excited nucleon in the nucleus was assumed to stay inside a potential well in a precisely defined quantum state.[84] Apparently, the two phases assumed in the Breit–Wigner formalism, namely the entrance and the exit channel, were interpreted as support for Bohr's 'intermediate state' idea.

11.13 Domination of the Compound Model

Soon after the lecture, which was more of an address than a lecture, in Copenhagen, Bohr went on a 6 month international sales tour to promote his model, even before he had a concrete mathematical scheme for it.[85] Bohr and Kalckar explicitly criticized Schmidt and Schlüter's independent particle model. However, there was nothing in Bohr and Kalckar's theory to explain why the properties of the nuclei are not smooth as a function of the atomic weight.

A major factor in the widespread belief in the dogma was the massive support Bohr's compound model obtained from Bethe and Bacher in Bethe's Bible. This review had an all-pervading influence on physicists and had won its nickname justifiably. Bethe extended Bohr's theory, assuming that the nuclei are perfect many-body systems. It should be stated that Bethe and Bacher provided a very extensive review of Elsasser's model. Bethe even contributed some additional evidence in favor of the model. But Bethe and Bacher concluded with a final negative verdict:

> In conclusion, we want to emphasize again that reliable conclusions about the shell structure of nuclei can only be drawn when atomic weight determinations will be available which are guaranteed to at least three decimals, i.e., 1 part in 100,000 for atomic weights of the order of 100.[86]

In other words, the masses of the nuclei, so claimed the authors, were not sufficiently accurately known to determine whether those masses were a smooth function of the atomic weight or not. Moreover, strong arguments were presented in the Bible to the effect that spin–orbit coupling should be very weak (but wait and see how the situation changed later).

---

[84] The Breit–Wigner formula is

$$\sigma_{a\to b} = \frac{1}{p^2}\frac{\Gamma_a\Gamma_b}{(E-E_{\mathrm{comp}})^2+(\Gamma/2)^2},$$

where p is the momentum of the relative motion, E is the total energy of the system, E_{comp} is the energy of the resonance in the compound nucleus, Γ_a/Γ and $\Gamma_b\Gamma$ are the probabilities for the corresponding processes, and Γ is the total width (in energy) of the resonance, determining the lifetime of the compound nucleus according to $\tau \sim \hbar/\Gamma$.

[85] Bohr, N. and Kalckar, F., Kgl. Dansk Vid. Selsk. Skr. **14**, 1 (1937).

[86] Bethe, H.A., Rev. Mod. Phys. **9**, 69 (1937), p. 179. This is the second volume of the Bible and is devoted to theoretical nuclear physics. It was written by Bethe alone.

The compound model's greatest success was no doubt the Bohr and Wheeler theory of nuclear fission.[87] But the compound model has its shortcomings. According to this model, the valley of stability should be a smooth valley without small cliffs. But we see steep slopes at, for example, $N = 128$, i.e., very large changes in the binding energy upon the addition of one neutron or one proton. This cliff was called by Gamow 'Heisen-Berg',[88] but it did not adulterate his full support for Bohr's theory. On the other hand, the problem of how to accommodate the entire list of magic numbers with a single potential or theory remained unsolved. The widespread belief in Bohr's theory treated the magic numbers as an irrelevant curiosity, mere numerology, for which there was no room in physics.

The impact of Bohr's lecture and Nature paper on theoretical nuclear physics was to suppress practically all research based on the independent particle model or any variation of it, and to deter young physicists from pursuing 'non-physical' research. Only those who were deeply involved and already eminent, like Wigner and Hund, were not discouraged and continued their exploration into nuclear structure using the independent particle model, whenever the compound model had no appropriate answer. Their work, however, was mostly limited to nuclei lighter than $N, Z \leq 20$, for which the order of single nucleon orbits could be simply understood.[89]

As an example, consider the young Victor Weisskopf,[90] who stated that:

> Under Bohr's influence I was thinking about the compound nucleus.

Another young physicist who subscribed to the compound model was Robert Oppenheimer[91] who, with Bohr's collaborator Kalckar (1910–1938), immediately applied the idea of the compound nucleus to another nuclear phenomenon.[92]

The pressure on supporters of the shell model was not trivial. Gamow met Elsasser in Paris and explained to him that he risked his prospects for getting a job in physics

---

[87] Bohr, N. and Wheeler, J.A., Phys. Rev. **56**, 426 (1939). In their 1939 paper, Bohr and Wheeler did not mention Gamow's priority for the idea of a liquid drop from 1928–1930, although Bohr was undoubtedly aware of it, because Gamow was in Copenhagen at the time the idea was born. The injustice was corrected only for history, when Wheeler in his *Oral History Transcript Interview*, by K.W. Ford, Princeton University, 12 January 1994, said:

> That thought [of Gamow, the author] had fallen in abeyance in the meantime.

Better late than never. Bethe and Bacher 1936 did not mention Gamow's priority on the liquid drop models, and Gamow and Bethe were good friends!

[88] The Heisenberg mountain.

[89] Wigner admitted that he did not really believe in the independent particle model of Elsasser, but changed his mind when Mayer and Teller told him about it after WWII. Wigner then claimed that Hund's results were similar to those of Elsasser and nobody had understood them. Interview of Wigner by C. Weiner and J. Mehra in 1966, Niels Bohr Library and Archives, American Institute of Physics, College Park, MD USA, www.aip.org/history/ohilist/LINK.

[90] Weisskopf, V.F., *Fifty-five years of my life in nuclear physics*, Nucl. Phys. A **527**, 331 (1991).

[91] Kalckar, F., Oppenheimer, J.R. and Serber, R., *Note on resonances in transmutations of light nuclei*, Phys. Rev. **52**, 279 (1937).

[92] The authors referred to a paper *N. Bohr, Science, to be published.*

if he chose to continue exploring the individual particle model.[93] Indeed, Elsasser moved to geophysics and is known for the dynamo theory of the Earth's magnetic field.

Also in 1937, Rose and Bethe[94] took up Bethe and Bacher's suggestion in Bethe's Bible, namely:

> The individual particle model affords one the opportunity to construct a rational theory of nuclear spin and magnetic moments for light nuclei.

This is exactly what Rose and Bethe tried to do in their paper. The authors, who applied the basic results of Feenberg (1906–1977) and Wigner,[95] used essentially spectroscopic methods first developed to handle problems with many electrons in the atom.[96] They assumed spin–total angular momentum and orbital–total angular momentum coupling and discussed the motion of particles in well defined quantum levels. The compound model did not allow for single particles with well defined quantum numbers. However, they added that this was an approximation and Elsasser was not mentioned.

In 1938, Gamow published his book *Structure of Atomic Nuclei and Nuclear Transformation*.[97] In a nutshell, here is what Gamow thought about the individual particle model:

> One may hope that further investigation along these lines will add considerably to our understanding of more detailed problems of structure. Much has already been done with rather overlapping results by Barlett, Gapon, Ivanenko, Elsasser, Guggenheimer, and others; it is not referred to in detail here because the author was never able in studying these articles to remember the beginning when he was reading the end.[98]

In 1940, Peierls[99] wrote a long review of Bohr's theory of nuclear reactions without mentioning the problems of magic numbers at all. One had to be a genuine iconoclast to work on the independent particle model in those days, and ready to risk one's career. Research on the shell model was largely suppressed until the late 1940s when more problematic data for the compound model was uncovered, and when Mayer, Jansen, Suess (1909–1993), and Haxel violated the boycott on the independent particle idea.

---

[93] Johnson, K.E., Am. J. Phys. **60**, 164 (1992).

[94] Rose, M.E. and Bethe, H.A., Phys. Rev. **51**, 205 (1937).

[95] Feenberg, E. and Wigner, E., Phys. Rev. **51**, 95 (1937).

[96] They assumed the Russell–Saunders coupling.

[97] Gamow, G., *Structure of Atomic Nuclei and Nuclear Transformation*, Clarendon Press, Oxford, 1938.

[98] The irony is that the type of states defined by Gamow in 1928, namely states which are unstable and decay (the energy has an imaginary part) are called Gamow states, so you can find papers discussing the Gamow states in the shell model. See, for example, Michel, N., Nazarewicz, W., Ploszajczak, M. and Okolowicz, J., *Gamow shell model description of weakly bound nuclei and unbound nuclear states*, Phys. Rev. C **67**, Issue 5, id. 054311, 05/2003.

[99] Peierls, R., *The Bohr theory of nuclear reactions*, Rep. Prog. Phys. **7**, 87 (1940).

11.14 Elsasser Does Not Back Down

In a short publication Elsasser,[100] who was already in the USA, was quick to respond to Bohr by pointing out how Bohr's ideas concerning the compound nucleus conflicted with observation. Indeed, Bohr's model predicts no differences between a nucleus and a nucleus plus an additional nucleon. Similarly, the model predicts no difference between nuclei with an even or an odd number of nucleons. The experimental facts are, as Elsasser pointed out, that there are certain empirical features, usually referred to as shell formation in nuclei, which are difficult to explain by such a model. An example is the even–odd effect of Harkins and Oddo.[101] This was Elsasser's last paper in nuclear physics.

11.15 From Cosmic Abundances to Nuclear Structure

The evidence for a unique cosmic abundance curve for the elements was already conclusive by the late 1940s. Consequently, Mayer and Teller[102] attempted to explain the abundances by means of the polyneutron theory. This theory will be discussed in Sect. 12.3 and it is mentioned here only because it served as the trigger for the shell model. In her 1963 Nobel lecture, Mayer recounted that it was during her work on the abundances of the elements that she stumbled upon the magic numbers:

> We found that there were a few nuclei which had a greater isotopic as well as cosmic abundance than our theory [...] could possibly explain.[103]

It soon turned out that the polyneutron theory was wrong on many counts, and it was its failure which led Mayer to propose the nuclear shell model.[104]

The first paper by Mayer[105] summarized experimental data showing that nuclei with 20, 50, 82, and 126 neutrons or protons are particularly stable. The data began with Elsasser's suggestion in 1934 and had accumulated since then, and there was a lot of new data. Interestingly for the question of element synthesis, Mayer pointed to the particularly low neutron capture probability by nuclei possessing a magic number of neutrons. It is these probabilities that control the synthesis of the elements

---

[100] Elsasser, W.M., PRL **51**, 55, December (1937).

[101] More specifically, Elsasser claimed that *such features follow immediately from any model of the Hartree–Fock type*, namely a model where the force of all nucleons is averaged. Elsasser tried to explain why the already accepted nuclear force would prefer an even number of nucleons and produce a more tightly bound system in such a case.

[102] Mayer, M.G. and Teller, E., Phys. Rev. **76**, 1226 (1949), submitted June 1949. Note that, although the results of this research were the trigger to Mayer's shell model papers, they were submitted to the journal and published only after Mayer's first papers on the shell model.

[103] Goeppert Mayer, M., *The shell model*, in Nobel Lectures, Physics, 1963–1970, Amsterdam, 1972.

[104] The introduction of the shell model did not rescue the polyneutron model. It remained wrong.

[105] Mayer, M.G., Phys. Rev. **74**, 235 (1948), submitted 16 April 1948 and published 1 August 1948.

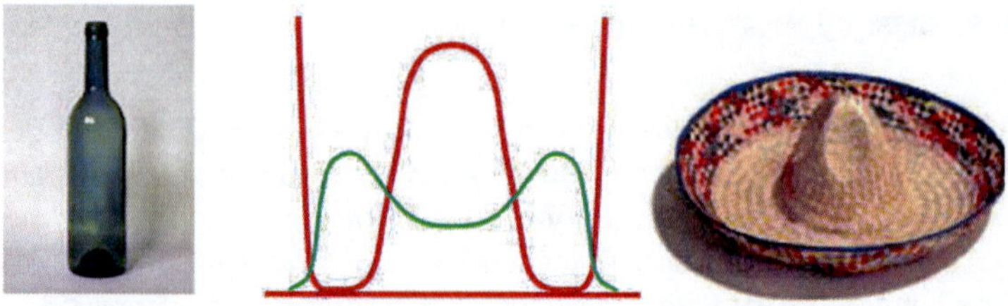

Fig. 11.7 Due to the mutual repulsion of the protons, the density of protons (the green curve) tends to a maximum near the surface and causes a depression at the center. The consequence is a potential which has the shape of a Wine Bottle or a Mexican hat

by the rapid neutron capture process. Mayer mentioned the fact that shell closure at $N = 20$ was understood, but she did not speculate about the other numbers. Mayer also explained why the identification of the atomic electronic closed shells was so easy while the identification of the nuclear closed shell was so difficult. The ionization potential for a closed shell, namely, the energy by which the last electron is bound to the atom, varies by several hundred percent, while the binding energy of the last nucleon in a nuclear closed shell varies by not more than 30%. Could this 30% be the cause of so much ado? Apparently, it could!

Mayer's paper triggered two very interesting and important papers by Feenberg and Hammack[106] and by Nordheim[107] which were published several months after Mayer's paper. The papers arrived on the editor's desk on the same day and were printed back to back. In addition, copies of both papers were sent to Mayer before publication. The papers were important in encouraging Mayer that she was on the right track.

What was important for Mayer was that Feenberg and Hammack, as well as Nordheim, assumed that the nucleon outside the closed shell could be handled as a single particle with well defined quantum numbers. The problem then was to identify the force which ordered the energy levels in such a way that the magic numbers corresponded to full shells. More accurately, the energy levels are not evenly spaced. A group of different energy levels, the energies of which differ from one another by a relatively small amount, is called a shell. The difference in energy between the shells is relatively large compared to the differences in energy between levels within the same shell (see Fig. 11.1). It is important to stress that the shells need not be arranged according to the ordering of the quantum numbers of the energy levels. In other words, an energy level with principal quantum number $n = 4$ can belong, due to whatever effect, to the third shell which mainly contains levels with $n = 3$.

To get an agreement with the magic numbers, Feenberg and Hammack assumed a simple rectangular potential well and succeeded[108] for $N, Z \leq 20$. To obtain a good agreement for more massive nuclei, they had to adopt Elsasser's idea of a central

[106] Feenberg, E. and Hammack, K.C., Phys. Rev. **75**, 1877 (1949), submitted 27 December 1948.

[107] Nordheim, L., Phys. Rev. **75**, 1894–1901 (1949), submitted 27 December 1948.

[108] Actually, a wide range of potential wells yield a good agreement for less than 20 nucleons.

Table I. Proposed schemes for nuclear shells.

No. of particles in nucleus	8	20	50	82
No. of particles in shell	2+6	12	30	32
Feenberg and Hammack	$(1s)^2(2p)^6$ $(1s)^2(2p)^6$	$(2s)^2(3d)^{10}$ $(3d)^{10}$	 $(4f)^{14}(5g)^{18}$	 $(6h)^{22}(4d)^{10}$
Nordheim	$(1s)^2(2p)^6$	$(2s)^2(3d)^{10}$	$(4f)^{14}(3p)^6(4d)^{10}$	$(5g)^{18}(5f)^{14}$
Mayer	$(1s)^2(2p)^6$	$(2s)^2(3d)^{10}$	$(4f)^{14}(3p)^6(5g_{9/2})^{10}$	$(5g_{7/2})^8(4d)^{10}(3s)^2(6h_{11/2})^{12}$
Order of levels in potential well	$1s$, $2p$, $3d$, $2s$, $4f$, $3p$, $5g$, $4d$, $3s$, $6h$, $5f$, $4p$, $7i$			

Fig. 11.8 The comparison between the shell models of Feenberg and Hammack, Nordheim, and Mayer. After Feenberg and Hammack and Nordheim 1949

depression to the well, thereby forming what is known as the wine bottle potential or Mexican hat potential (see Fig. 11.7). The data regarding the total angular momentum of nuclei (called the spin of the nucleus to distinguish it from the spin of a single particle) with a small number of nucleons outside closed shells tended to corroborate the shell model. Consider a core of nucleons with one nucleon outside the core. According to the shell picture, the core is spherical and does not rotate, so it has vanishing angular momentum. The total angular momentum of the nucleus would then be the total (spin plus orbital) angular momentum of the single nucleon moving outside the core.

Feenberg and Hammack realized that a potential with the shape of a wine bottle has its energy levels arranged in the right way. But what would cause a wine bottle potential? The authors looked to the Coulomb repulsion between the protons as the source of the depression. Because of the Coulomb repulsion, they hypothesized, the density of the protons is minimal near the center and maximal near the boundary of the nucleus. The distorted proton distribution would then induce a distortion in the neutron distribution. The authors mentioned that this effect had been investigated over a decade earlier by Wigner[109] and Feenberg,[110] and more recently by Nordheim.

Nordheim, on the other hand, made explicit use of the results of Schmidt[111] for nuclear magnetic moments. Landé's model,[112] in which *a single particle, one proton or one neutron, is responsible for the total spin and magnetic properties of the whole nucleus*, was taken up by Schmidt, who had by then more experimental data than

[109] Wigner, E., Bicentennial Symposium, University of Pennsylvania, 1940.

[110] Feenberg, E., Phys. Rev. **59**, 593 (1941).

[111] Schmidt, T., Zeit. f. Physik **106**, 358 (1937).

[112] Landé, A., Phys. Rev. **46**, 477 (1934).

were known to Elsasser. The level schemes of both Nordheim and Feenberg and Hammack list the various orbits by their orbital angular momentum and characterize the states by the total, L, and the total intrinsic spin, S, as in the Russell–Saunders coupling of atomic electrons (LS coupling).[113]

A few months after the appearance of the two papers by Feenberg and Hammack and by Nordheim, Mayer's breakthrough appeared,[114] when she assumed the spin–orbit coupling[115] and considered what it does to the energy levels. Mayer had the good fortune of discussing the paper with Fermi,[116] who asked her whether there was any indication of strong spin–orbit coupling. She instantly replied: *Yes*.[117] Mayer thanked Fermi for *this remark which was at the origin of this paper*. Note, however, that the wonderfully imaginative Fermi was skeptical at the beginning, but later changed his mind and accepted Mayer's model. Mayer's paper to the PRL was held by the editor who asked Feenberg, Hammack, and Nordheim[118] to compare the three proposed schemes. The two letters, that of Mayer and that of the referees, were published back to back. The authors prepared a comparison table, reproduced here as Fig. 11.8.

---

[113] In a system with many particles, the total angular momentum of the system is the sum of the angular momenta of the particles. The problem is that there are many ways to add the angular momentum, and they do not all yield the same result. One can sum the angular momentum of each particle, say the spin plus the orbital angular momentum, to yield the total angular momentum of a single particle, j, and then sum the j momenta of all particles. This is called the jj coupling. Another way is to sum the orbital momentum of all particles and to sum the spin angular momentum of all particles separately and only then sum the total orbit angular momentum with the total spin momentum to get the total angular momentum of the system. This is called the LS coupling. The final results are different. What scheme to use depends on the interaction between the spin and the orbit.

[114] Goeppert Mayer, M., Phys. Rev. **78**, 16 (1949).

[115] Consider first an atom with not too many electrons. Each electron has a spin and an angular momentum, that is, the charge of the electron moves in the orbit and rotates. Consequently, each moving charge creates a corresponding magnetic field. The magnetic fields thereby produced, i.e., the magnetic field of the spin and the magnetic field of the orbital motion, interact with one another and affect the energy of the electron. This is called the spin–orbit interaction of a single electron. It was found that the spin–orbit interaction of each electron and their mutual interactions among themselves are small for a small number of electrons so to a very good approximation the electronic states are determined mostly by the principal quantum number n. The occupancy of the atomic levels is determined solely by n here, and the higher n is, the higher the energy is (see Fig. 2.15 in which the simplest case of hydrogen atom levels is shown). As the number of electrons increases, the spin–orbit interaction increases. The main effect is that certain levels with $n+1$ (and given l) have a lower energy than the level with n and l. When the spin–orbit interaction is very strong, as in nucleons, the effect increases and distorts the order of the levels on the energy scale, so that a level with a higher n and some $l_1 < n$ may have a lower energy then a level with $n-1$ and $l_2 < n-1$.

The energy levels of a single particle moving in a potential well depend on the shape of the potential. In the simple case of an harmonic potential well ($U \sim r^2$ and $U \sim r^{-2}$), the energies of the levels do not depend on the angular momentum. But in more complicated shapes, this is not the case. So the question is: What shape of potential would yield an energy-level system which reproduces the observed levels in the nucleus?

[116] Mayer, M., in Nobel Lectures, Physics 1963–1970, Elsevier Pub. Co., Amsterdam, 1972.

[117] There are places where it is claimed that her reply was: *Yes, and it explains everything*.

[118] Feenberg, E., Hammack, K.C. and Nordheim, L.W., Phys. Rev. **75**, 1968 (1949).

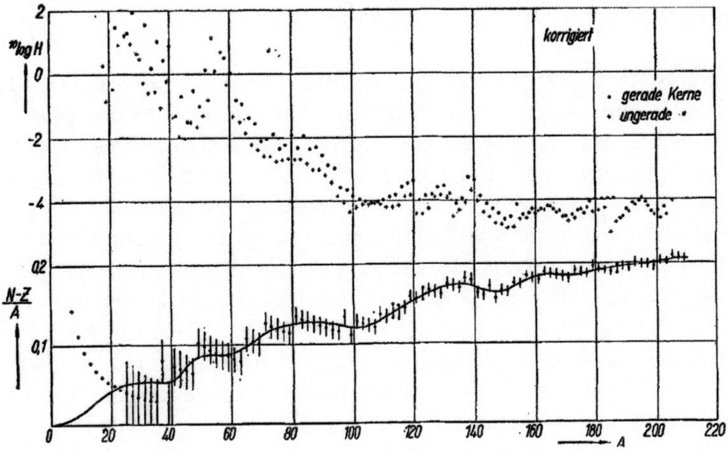

Fig. 11.9 The Jensen and Suess correlation between the cosmic abundance curve of Goldschmidt and $(N - Z)/A$, namely the excess of neutrons over protons divided by the sum of neutrons and protons

All three models successfully explained the arrangement of up to 20 nucleons. The problem was with larger numbers of nucleons, so they claimed that:

> These facts suggest, however, that a rearrangement of levels may be successful.

Consequently, three different rearrangements were suggested. Feenberg and Hammack suggested a change in the Coulomb force, Nordheim suggested a discrimination against levels with high angular momentum, and Mayer's scheme followed the order of levels in a potential well and achieved the breaks at the right locations by assuming an arbitrary and very strong spin–orbit coupling:

> All three schemes give, of course, the empirical shell numbers and a statistical correlation with observed spins and moments.

In particular they wrote:

> The shell structure in nuclei is, however, so pronounced an effect that one may hope to obtain an interpretation even on the basis of such crude approximation as the individual model.

The comparison table according to Feenberg, Hammack, and Nordheim shows that the only differences are in the third shell and above (see Fig. 11.8). However, these differences do not concern the number of particles in each shell, but rather the angular momentum of the levels composing the shells. The principle is the same in all three models, viz., add an agent which destroys the spherical symmetry and induces a situation where, by the lifted degeneracy of the simple potential, the energy levels are rearranged into a new order under the condition of the existence of the magic number of nucleons in closed shells. It is now relatively simple to decide which model holds: simply check the angular momentum.

11.16 Enlightenment Occurs Twice

Victor Goldschmidt (1888–1947) was the first to notice the peculiar curve of abundances and its unique peaks.[119] Goldschmidt[120] also plotted the data as a function of Z, since he knew the isotopic composition from measurements by Aston. He found that certain elements have significantly higher abundances than neighboring elements. These elements did not show any chemical similarities and he thus inferred the fundamental conclusion that the difference in abundance was due to nuclear structure and not chemical reactions. He therefore classified the elements according to Z and N, and not according to $A = Z + N$. The abundance peaks he discovered were at: $Z = 28, 40, 50, 74, 82$, and 90, and at $N = 30, 50, 82$, and 108. Inaccuracies in the data prevented Goldschmidt from concluding that the numbers were identical for protons and neutrons. Today we know that the numbers are the same. Goldschmidt did not live to hear the explanation for his peculiar nuclear numbers,[121] but his colleague Suess[122] discussed the abundance curve with its 'peculiar numbers', or in German *ausgezeichneten Zahlen.* The numbers Suess got were as follows: for neutrons 20, 28, 50 (58), 82, and for protons (20), 26 or 28, 50, 74, and 82. Note that the magic numbers for neutrons and protons are not identical as given in this list. The same 'special numbers' were discovered by the nuclear experimentalist Haxel.

Two months after Mayer's second paper appeared, the Physical Review published the paper by Haxel, Jensen, and Suess[123] with the title *On the 'magic numbers' in*

---

[119] Goldschmidt, V., Naturwiss. **18**, 999 (1930).

[120] It was Goldschmidt who, in his 1937 paper, suggested examining meteorites for primordial abundances. J. Chem. Soc. 655 (1937), the 7th Hugo Müller Lecture delivered before the Chemical Society on 17 March 1937.

[121] As a Jew, Goldschmidt resigned from his Göttingen professorship in 1935, in protest against the treatment of 'non-Aryans', and returned to Oslo (Norway). Suess on the other hand, was involved in the German uranium project and suggested the use of heavy water as a moderator to slow down the neutrons in uranium fission reactors and thereby increase the absorption of neutrons by the uranium. This suggestion brought him to Norway, where the Germans used the hydroelectric power plant to produce heavy water (deuterated water), for which he was a consultant. Being in Norway he met Goldschmidt, and it was during these discussions that Suess realized the meaning of the special numbers discovered by Goldschmidt. A few days after Suess visited Goldschmidt, whose property had been confiscated, he was caught by the Gestapo and put on a freighter to be deported. Just before leaving the harbor, a Norwegian police officer came to claim his release with the excuse that he was important for Norwegian industry. The next day he was smuggled to Sweden. Since we noted the importance of the neutron capture data for construction of the A-bomb (as well as for the synthesis of the elements in stars and nuclear physics), it is interesting to mention at this point that, on the basis of Haxel's measurements and data collection regarding the magnitude of neutron capture by carbon (a magic nucleus), Heisenberg and Weizsäcker concluded that it would not be practical to build a nuclear reactor to produce fissionable material using carbon (or more accurately graphite) as moderator. Haxel's error turned the graphite into a poor moderator. Consequently, the use of heavy water as a moderator became extremely important for the German atomic energy program.

[122] Suess, H.E., Zeit. f. Naturforschung **2A**, 311 (1947).

[123] Haxel, D., Jensen, J.H.D. and Suess, H.E., Phys. Rev. A **137**, 129 (1949). After WWII, Suess managed to get the nuclear theoretician Jensen interested in the problem. Goldschmidt died in England in 1947, just before the shell model festival started.

nuclear structure. The paper was preceded by a short paper[124] in German. This was an independent discovery of the strong spin–orbit effect in nuclear structure.

The history of how Jensen and his collaborators reached their shell model is interesting. A few years earlier, in 1946, Hans, Jensen, and Steinwedel[125] noticed that the masses of the nuclei were not a smooth function of the proton or neutron numbers, and they claimed that the data provided support for the α-particle model.

In 1947, after the death of Goldschmidt, Jensen and Suess wrote a paper in memory of Goldschmidt.[126] The paper contained a comparison between the abundances and the differences between neutron number and proton number divided by their sum (see Fig. 11.9). Next, they tried to explain the observed curve as a consequence of thermodynamic equilibrium between the different species, exactly as Chandrasekhar and Henrich and Lattes and Wataghin[127] had done. There was no attempt to understand why these elements were more stable and possessed this particular property, and no attempt to explain how the hot soup was dispersed in space.

Suess and Haxel independently reached the conclusion that nuclei must exhibit closed shell phenomena, but from different data. Suess examined the abundances and Haxel analyzed nuclear data. Jensen did not know what to do with the magic numbers, and being under the influence of Bohr's 1937 paper, on the one hand, and pressed by his two colleagues, on the other hand, he remained hesitant. But during a visit to Copenhagen, fate arranged for Jensen to stumble upon Mayer's first paper in the Physical Review Letters, and he took the risk of discussing it in a seminar where he also presented his results. Bohr was present and heard Jensen, but this time took the matter seriously. He asked many questions and in particular refrained from attacking Jensen. Bohr's non-negative attitude encouraged Jensen to listen more carefully to Haxel and Suess and to take their findings seriously. Thus, Jensen and his colleagues were driven to look for an agent that would affect the order of the energy levels so as to fit the nuclear levels, and they immediately hit upon the spin–orbit interaction.

However, the dramatic results were apparently too incredible for some people to digest, and the first version of the paper was rejected by the editor *of a serious journal*,[128] who asserted that:

> It is not really physics but rather playing with numbers.

Jensen then sent the paper to Weisskopf, who forwarded it to the Physical Review, and it was published two weeks before Mayer's paper itself was published. It should be

---

[124] Jensen, J.H.D., Suess, H.E., and Haxel, D., Naturwiss. **36**, 155 (1949); ibid. 153 (1949).

[125] Hans, J., Jensen, D., and Steinwedel, H., Die Naturwissenschaften **33**, 249 (1946).

[126] Jensen, J.H.D. and Suess, H.E., Die Naturwissenschaften **34**, 131 (1947). The Geochemical Society, which considers Goldschmidt as the founder of modern geochemistry and crystal chemistry, established the Goldschmidt medal. This paper was the address Suess delivered when he was awarded the Goldschmidt medal.

[127] It is strange that Jensen and Suess cited the American Physical Review Journal as well as Klein, O., Beskow, G. and Treffenberg, L., Ark. f. Math. Astr. och Fysik Stockholm **33**, 1 (1946), and overlooked the extensive work by Chandrasekhar and Henrich (see Sect. 7.1) published in the American Astrophysical Journal.

[128] The journal was Nature.

recalled that Weisskopf was by then a leading authority on nuclear physics who had published many results within the framework of the compound nucleus. However, Weisskopf was quick to recognize the discovery by Jensen and his colleagues. Yet, Jensen was not confident in the physicality of his results until he presented them in front of Bohr in Copenhagen. This time Bohr agreed that:

> The shell model explains why many nuclei do not show energy levels of rotation.

11.17 The Final Vindication: From Three Remains One

The revival of the shell model and the agreement with observation began to convince physicists of the viability of the shell model. It was still far from perfect and further approximations started to appear, in particular with respect to finer details such as the quadrupole moment of the nuclei. For example, Townes, Foley, and Low[129] pointed out that, in order to explain the quadrupole moments, the single nucleon state must be mixed with some nucleon states from the core. In other words, it was not a compound or a shell model, but something in between.

With three models on the table, comparisons with experimental data were very quickly carried out by several groups. Hughes and Couteur[130] showed that ^{5}He agreed with Mayer's scheme and ordering of the levels, but did not agree with the schemes of Feenberg and Hammack and Nordheim. On the other hand, Bethe and Butler[131] were not yet convinced that the shell model was correct and proposed a critical test that was intended to provide information about the accuracy of the picture in which nucleons move individually outside the core. Similarly, Umezawa and collaborators[132] found a nice agreement between Mayer's theory and the Fermi theory of β-decay, although they pointed out some problems for $N > 40$.

Mayer herself added[133] empirical evidence for her model. She calculated the energy levels of a single particle *in a potential between that of a three-dimensional harmonic oscillator and a square well* (see Fig. 11.10). Next, Mayer[134] added more theoretical considerations when she demonstrated that the spin–orbit coupling model agrees well with the idea of a very short-range nuclear force. Béné[135] quickly showed that Mayer's scheme provides a nice agreement with observed nuclear magnetic moments.

In May 1953, a conference on *Nuclear Spectroscopy and the Shell Model* took place in Indiana University and the different schemes were compared. Many technical

---

129 Townes, C.H., Foley, H.M. and Low, W., PRL **76**, 1415 (1949).

130 Hughes, J. and Le Couteur, K.J., Proc. Phys. Soc. A **63**, 1219 (1950).

131 Bethe, H.A., Butler, S.T., PRL **85**, 1045 (1952).

132 Umezawa, M., Nakamura, S., Yamaguchi, Y. and Taketani, M., Prog. Theoret. Phys. **6**, 408 (1951).

133 Mayer, M.G., Phys. Rev. **78**, 16 (1950).

134 Mayer, M.G., Phys. Rev. **78**, 22 (1950).

135 Béné, G.J., J. Phys. Radium **13**, 161 (1952).

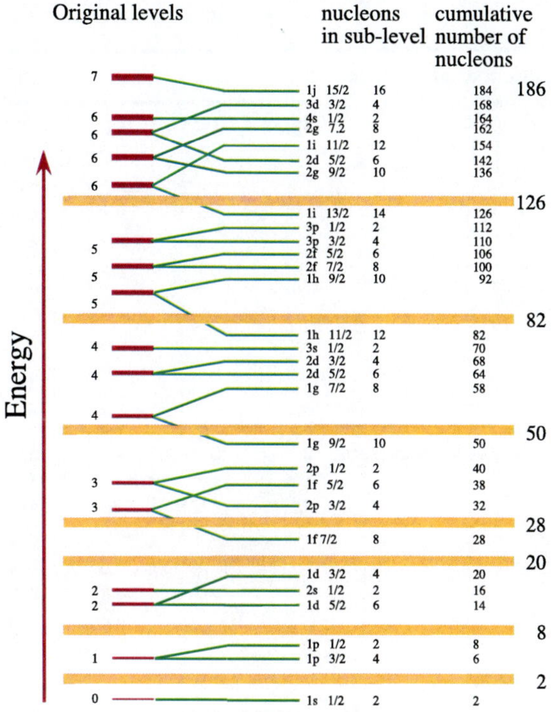

Fig. 11.10 The splitting of the energy levels due to the spin–orbit coupling, leading to the magic numbers. The levels on the left are the energy levels without the effect of the spin–orbit coupling. The width of the level is indicative of the total number of protons/neutrons that can occupy it without violating the Pauli principle. On the right is the total number of 'quantum locations' in the sublevel after the splitting, and further to the right is the cumulative number of nucleons

details were discussed, but no final verdict was published. Somehow each side had an answer to all problems. Note that by now the shell model had gained unofficial recognition as bona fide physics.

The 1963 Nobel Prize in Physics was awarded half to Wigner *for contributions to nuclear physics, elementary particles and symmetry principles in physics* and *for the discoveries concerning nuclear shell structure*, a quarter to Mayer (1906–1972),[136]

[136] It is hard to believe, but here was a Nobel prizewinner without a permanent position. Maria Mayer, born Goeppert, married Joseph Mayer in 1930 in Germany and moved to the New World, where the chemist Joseph got a position in Johns Hopkins University. Maria, however, did not get a paid position because of anti-nepotism rules. She became a 'volunteer associate'. In this position, she could do research.

From Johns Hopkins, the Mayers moved to Columbia, New York, where Maria got a part-time teaching job in Sarah Lawrence College, a liberal arts college for women. From New York, the Mayers moved to the University of Chicago, and Mayer was still without a paid job. It was there that she met Teller and collaborated with him on his polyneutron idea for the formation of the elements, an idea which simply did not work. On the other hand, she was exposed to the real element abundances and started to work alone on the shell model. Her success is now history.

In 1960, three years before she was recognized by the Nobel committee, she was offered a full professorship at the new campus of the University of California at San Diego. Joseph Mayer was appointed a professor in the Chemistry Department. In California, the university authorities were sufficiently clever and flexible to overcome the anti-nepotism rule for discriminating against

Magic number nuclides

Number of protons	2	8	20	28	50	82	126
	^{4}He	^{16}O	^{40}Ca	^{58}Ni	^{112}Sn	^{204}Pb	
		^{17}O	^{42}Ca	^{60}Ni	^{114}Sn	^{206}Pb	
		^{18}O	^{43}Ca	^{61}Ni	^{115}Sn	^{207}Pb	
			^{44}Ca	^{62}Ni	^{116}Sn	^{208}Pb	
			^{46}Ca	^{64}Ni	^{117}Sn		
			^{48}Ca		^{118}Sn		
					^{119}Sn		
					^{120}Sn		
					^{122}Sn		
					^{124}Sn		
Number of neutrons	**2**	**8**	**20**	**28**	**50**	**82**	**126**
	^{4}He	^{15}N	^{36}S	^{48}Ca	^{86}Kr	^{136}Xe	^{208}Pb
		^{16}O	^{37}Cl	^{50}Ti	^{87}Rb	^{138}Ba	^{209}Bi
			^{38}A	^{51}V	^{88}Sr	^{139}La	
			^{39}K	^{52}Cr	^{89}Y	^{140}Ce	
			^{40}Ca	^{54}Fe	^{90}Zr	^{141}Pr	
					^{92}Mo	^{142}Nd	
						^{144}Sm	

Fig. 1. The magic numbers.

Fig. 11.11 The magic nuclei as given in Mayer's Nobel address of 1963. A notable omission from the magic nuclei is the extremely important and abundant nucleus ^{12}C

and a quarter to Jensen (1907–1973). One can say that the Nobel prize was given to those who expanded on the initial work of Elsasser and stayed in the field of nuclear physics. The irony is twofold: the Nobel prize was first awarded for the 'shell model', which Niels Bohr detested and vehemently preached against, before it was awarded in 1975 to Niels Bohr's son Aage Bohr for the 'collective model' his father had initiated and encouraged so many to follow.

Aage Bohr (1922–2009), Ben Mottelson, and Leo Rainwater (1917–1986) were jointly awarded the 1975 Nobel Prize in Physics in equal shares *for the discovery of the connection between collective motion and particle motion in atomic nuclei and the development of the theory of the structure of the atomic nucleus based on this connection,*[137] i.e., for allowing the compound model to take account of individual

(Footnote 136 continued)
women, and offered both Maria and her husband positions. They re-enacted this procedure when astrophysicist Geoff and astronomer Margaret Burbidge were attracted to San Diego. The astronomer Margaret got a position in chemistry.

[137] Bohr, A. and Mottelson, B.R., Physica **18**, 1066 (1952); ibid. Phys. Rev. **89**, 316 (1953); ibid. Phys. Rev. **90**, 717 (1953).

nucleons. Bohr and Mottelson developed the theory, and Rainwater determined the asymmetrical shapes of certain atomic nuclei. In a sense, the Nobel was awarded for reconciling the compound and shell models.

By the time Mayer delivered the Nobel address, there was no longer any question about the identity of the magic nuclei, and she summarized them in the table shown in Fig. 11.11.

It was mentioned earlier that the success of the liquid drop model in describing the fission of uranium was considered as the greatest victory of Bohr's compound model. In 1950, after the publication of the shell model, Meitner[138] published a paper under the title *Fission and nuclear shell model*, showing how the shell model explains the observed features of the fission process, i.e., how the magic numbers 50 and 82 affect the debris of the fission. Thus the liquid drop explains why and how the nucleus fissions, while the shell model explain what it fragments into.

In 1950, thirty five years after his first paper in which he proposed the α-particle model, Harkins wrote[139]:

> In the years 1915 to 1923, the following concept was introduced into nuclear science in about 20 papers. The stability and abundances of nuclear species are determined largely by the relation of special numbers. This concept was received by Rutherford, and later by Goldschmidt, with much approval, but did not meet with so much favor from certain theorists.

Even after claiming his priority in public, and he was quite right to do so, he was still not credited as the originator of the idea.

The apparently fortuitous shell model was imposed upon physicists by observations and against good physical arguments to the contrary. We are here interested in the effects the shell model has on nuclear energy levels and stability of nuclei, because of their importance for the synthesis of the elements. Other aspects are described by Talmi[140] in his review.

The shell model has been discussed at length here. However, the collective or unified model is very successful in describing dynamic properties which result from collective phenomena like rotation of nuclei. In 1953, Bohr and Mottelson[141] developed the theory of the collective model, in particular the rotation of the nuclei. In 1953, Temmer and Heydenburg[142] discovered such nuclear levels, and this finding signaled a victory for the collective model. But the observational evidence of the magic numbers was not washed away by this discovery. So attempts began to incorporate single particle levels (into a 'unified' model). In this way Nilsson and Mottelson[143] managed to predict the magic numbers from the unified model.

---

138 Meitner, L., Nature **165**, 561 (1950).

139 Harkins, W.D., PRL **79**, 724 (1950). The paper was the summary of a talk Harkins gave in 1950 at the Chicago meeting on nuclear shells.

140 Talmi, Y., *Fifty years of the shell model*: *The quest for the effective interaction*, in Nuclear Shells: 50 Years, ed. by Oganessian and Kalpakchieva, World Scientific, 2000.

141 Bohr, A. and Mottelson, B.R., Mat-Fys. Medd. Dan. Vid. **29**, 16 (1953).

142 Temmer, G.M. and Heydenburg, N.P., Phys. Rev. **94**, 1399 (1954).

143 Nilsson, S.G. and Mottelson, B.R., Phys. Rev. **99**, 1615 (1955).

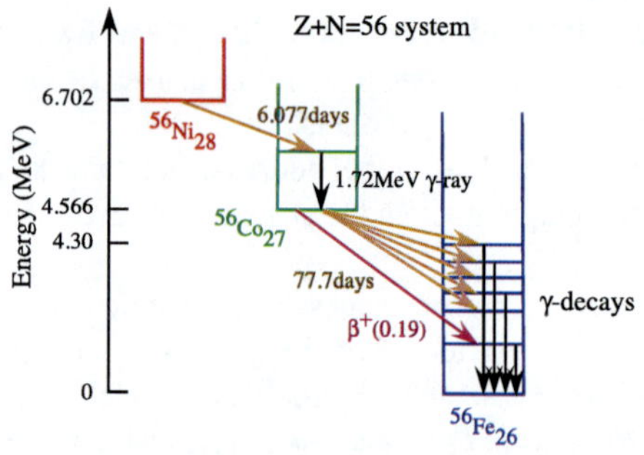

Fig. 11.12 The surprising decay of the doubly magic nucleus $^{56}Ni_{28}$ to the most stable nucleus $^{56}Fe_{26}$. $^{56}Ni_{28}$ decays via electron capture (*brown arrow*) in 6.077 days to give an excited state of $^{56}Co_{27}$. This excited state decays via γ emission (*black arrow*) to the ground state of $^{56}Co_{27}$, which decays 19% of the time via positron emission (*pink arrow*) and the rest of the time via electron capture into excited states of $^{56}Fe_{26}$. The excited iron decays via γ emission to the ground state

11.18 How Nature Teases Physicists and Astrophysicists

The nucleus ^{56}Ni is a doubly magic nucleus ($N = 28$ and $Z = 28$), but against all expectations of any nuclear model it is unstable and decays through electron capture to ^{56}Co and then to ^{56}Fe, as shown in Fig. 11.12. We mention this case here because the energy released in the decay, amounting to 6.702 MeV per nucleus of $^{56}Ni_{28}$, heats the exploding supernova and causes it to shine. It is this unexpected anomaly which provides the energy for the Type I supernovae and allows us to observe them. Consequently, the total light energy emitted by the supernova is proportional to the total mass of ^{56}Ni formed in the explosion. It is worth recalling here that Type I supernovae result from a very small range of stellar masses, and this in turn implies that the range of possible masses of the nickel core is quite narrow. Such supernovae can thus serve as standard candles and allow the determination of their distance by comparing apparent and absolute luminosity (energy output). Since 1998, this type of supernova has been used to infer the accelerated expansion of the Universe. One may therefore say that the above nuclear surprise has helped us to reveal the amazing cosmic acceleration.

Goode and Zamick[144] even asked why the decay was so slow. Clearly, if the decay were faster, the rate of energy release in these supernovae would have been greater, and the supernovae would have been even brighter. The reason, as Goode and Zamick explained, is that the nickel is doubly magic: it comprises a closed core, while the excited state into which it decays is mostly a two-particle two-hole state of cobalt. If this were not the case, the rate of decay would have been about a factor of a hundred higher, and likewise the brightness of the supernova.[145] Additional but rare examples of magic numbers and elements with no stable isotopes are given by

[144] Goode, P. and Zamick, L., PRL **22**, 958 (1969).

[145] The conversion of a proton into a neutron in the nucleus is due to a Gamow–Teller transition.

Fig. 11.13 Spheron model for the nucleus, Pauling 1965. In this realization 45 spheres are arranged in icosahedral closest packing. On the left is a single sphere which constitutes the inner core. Next there is a layer of 12 spheres at the corners of a regular icosahedron, and so on

Kowarski,[146] and attempts to explain the two elements missing from the periodic table are presented by Suess.[147]

If extrapolation can be trusted, the next predicted magic number is $Z = 120$ and $N = 184$. Such nuclei are not found in nature, or at least have not yet been found. But could they be synthesized in the laboratory or in stars? It is not clear, but present indications are that such nuclei would decay quickly via fission.

11.19 A Chemist's Attempt: Linus Pauling

In 1965, the chemist Linus Pauling (1901–1994)[148] revived the old α particle theory under the name of the close-packed spheron theory, illustrated in Fig. 11.13. Pauling claimed that his model explained nuclear properties as well as asymmetric fission. Indeed, fission is asymmetric in the sense that the two large pieces of debris are not identical. According to Pauling, the nucleus is not a sphere of uniform density because there is clustering inside. Pauling assumed that, inside the nuclei, the nucleons may to a first approximation be described as occupying localized orbitals to form small clusters. These small clusters, called spherons by Pauling, are usually helions (α particles), tirons (^{3}H), and dineutrons (^{2}H);[149] in nuclei containing an odd number of neutrons, a ^{3}He cluster or a deuteron may serve as spherons. According to Pauling:

> The close-packed spheron model differs from the conventional liquid drop model of the nucleus in having spherons rather than nucleons as units.

Maximum stability is reached when each spheron ligates about itself the maximum number of neighbors, to produce a nucleus with closest-packed structure. No quantum calculations followed the model. Pauling presented his model as refinement of the shell model:

---

146 Kowarski, L., PRL **78**, 477 (1950).

147 Suess, H.E., PRL **81**, 1071 (1951).

148 Pauling, L., Science **150**, 297 (1965).

149 A frequently used trick is to invent attractive names. So instead of an α model, one invents the name polyhelion.

The spherons are assigned to concentric layers (mantle, outer core, inner core, innermost core) with use of a packing equation.

And what about the structure of the basic spherons? Inspired by the molecular bond, Pauling asserted that core repulsion suggested a triangular structure for ^3H and ^3He and a tetrahedral structure for ^4He. The resemblance of the nucleus to a molecule looked strange at first glance, but the incredible reality has surprises. Pauling published about 25 papers about the spherons, the last one shortly before his death, but they attracted little attention from physicists and could not be found in textbooks.

11.20 What Nuclei Can Exist: The Binding Energy Formula

Crucial for astrophysics and life is the question of what nuclei can exist? Is any combination of protons and neutrons possible? Can we have nuclei with, say, 200 protons and 300 neutrons? While we are still unable to predict the answer from first principles, we can give a temporary answer based on what is called the phenomenological Bethe–Weizsäcker formula. This formula gives the mass, or equivalently the energy, of a nucleus made of Z protons and N neutrons. When compared with the mass of Z protons plus N neutrons, we get the binding energy of the nucleus, namely the energy needed to separate the nucleus into its $N + Z$ nucleons. The binding energy can be expressed in energy units or in atomic masses. In the latter case, it is called the mass defect.

In 1932–1933, Heisenberg (1901–1976)[150] published a series of three papers about the structure of the atomic nucleus and also derived the beginning of the mass formula, without mentioning Gamow's earlier attempts. It is correct that Gamow did not then know about the neutron, but the idea of nuclear surface tension was his, and appeared in the paper. Heisenberg confirmed Gamow's basic result, namely that, as A increases, the mass defect passes through a minimum. This is the minimum which, in 1935, was reconfirmed by Heisenberg's student von Weizsäcker.

Heisenberg derived the following condition for the stability of the nucleus:

$$N/Z = c_1 + c_2 \frac{Z}{A^{1/3}},$$

where $A = N + Z$ and the constants c_1 and c_2 must be found from experiment. In present day terminology, we would say that Heisenberg predicted the neutron drip line. For every Z there exists a maximal N for which there is a stable nucleus.

A year later, Landé,[151] assuming the α-particle model, reached the conclusion (without really doing the calculation) that the binding energy per neutron is constant

---

150 Heisenberg, W., Zeit. f. Phys. Hadrons & Nuclei **77**, 1 (1932); ibid. **78**, 156 (1932); ibid. **80**, 587 (1932).

151 Landé, A., Phys. Rev. **43** (1933).

throughout the periodic system, and is about 0.009 in mass units for even elements. Landé argued as follows:

> If one plots the mass defects according to the old theory, one obtains for the potential energy of the nuclei a distribution having a minimum in the range of element number 50. This minimum is a serious difficulty, because it means that the elements above 50 should be unstable, while in fact they are unstable above 80.

So in effect Landé realized that a minimum in the binding energy per nucleon curve implies instability of elements. This feature was eliminated in Landé's scheme, where he found (wrongly, as it would turn out) *a general decrease of the potential energy with increasing atomic number.*

Eastman[152] took Heisenberg's expressions and found the value of the constants empirically, plotting Fig. 11.14, where N/Z is given as a function of $Z/(N+Z)^{1/3}$ for the different stability conditions.

In 1934, Wick[153] realized that the analogy between a nucleus and a liquid drop was not complete and that there should be a surface term, because nucleons near the surface do not feel the same force as nucleons deep inside the nucleus and surrounded by other nucleons. The idea was picked up a year later by Weizsäcker.[154] Weizsäcker therefore had three terms which were based on the liquid model, viz., the volume attraction by all nucleons, the Coulomb repulsion by the protons, and the surface term (ignoring Gamow's earlier treatment of the problem).

The liquid drop model in this form did not provide satisfactory results for the binding energies of nuclei. Weizsäcker thus proposed an empirical way to calculate the nuclear binding energies. What he actually did was to take the terms dictated by the liquid drop and add corrections due to single nucleon effects, as demonstrated by Elsasser. A general formula was derived, containing three unknown numerical constants. The constants were found by fitting the formula to the data. The formula was:

$$M = (Nm_{\mathrm{n}} + Zm_{\mathrm{p}}) - \alpha A + \beta \frac{(N-Z)^2}{A} + \gamma A^{2/3} + \frac{3}{5}\left(\frac{e^2}{r_0}\right)\frac{Z^2}{A^{1/3}},$$

where $r_0 A^{1/3}$ is the radius of the nucleus, and m_{n} and m_{p} are the masses of the neutron and the proton, respectively. The formula contains four adjustable parameters: α, β, γ, and r_0. The first term is the rest mass of the individual particles when they are separated. The third term is the main binding energy which is assumed to be proportional to the number of nucleons. The fourth term expresses the decrease in the binding energy when the number of neutrons is very different from the number of protons. The fifth term expresses the surface energy, i.e., the fact that a nucleon close to the surface feels attraction from the inside and not from the outside. The last term is the Coulomb repulsion.

---

[152] Eastman, E.D., Phys. Rev. **46**, 1 (1934).

[153] Wick, G.C., Nuovo Cimento **11**, 227 (1934).

[154] von Weizsäcker, C.F., *Zur Theorie der Kernmassen*, Zeit. f. Phys. **96**, 431 (1935).

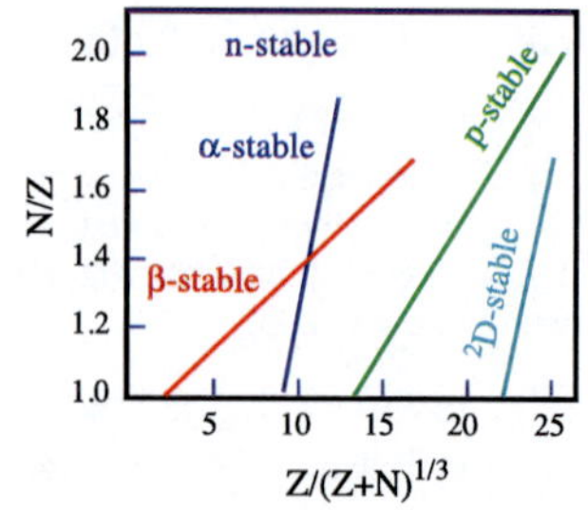

Fig. 11.14 The limit of the ratio of neutrons to protons for the stability of nuclei against emission of various particles, as found by Eastman in 1934 on the basis of Heisenberg's 1932 expression for the binding energy. Nuclei to the left of the line are stable against emission of a particle. For low N/Z, a β-decay appears first, while for high N/Z, an α emission appears first. For $N/Z > 2.1$, the nucleus is unstable for all Z and emits a neutron

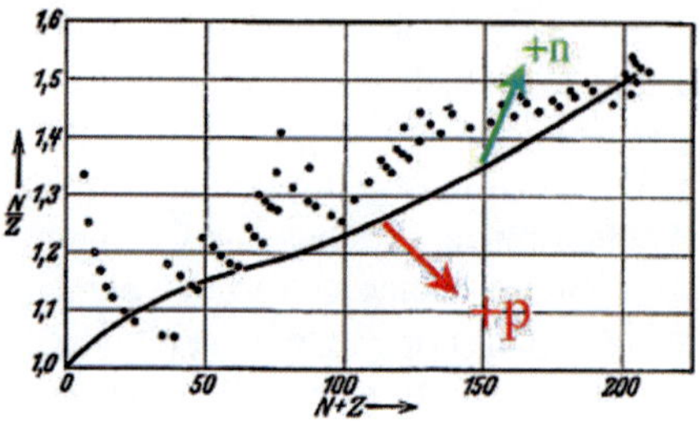

Fig. 11.15 Weizsäcker's result for the maximum number of protons in a nucleus, known today as the proton drip line. The *coloured arrows* mark the direction for adding a proton or a neutron

Weizsäcker realized that this formula applies only to nuclei with even numbers of neutrons or protons. Nuclei containing an odd number of either neutrons or protons have higher mass (less binding energy). So Weizsäcker suggested that, for odd neutron or proton numbers, the mean between the adjacent nuclei should be taken, viz.,

$$M(Z, N+1) = \frac{1}{2}\big[M(Z, N) + M(Z, N+1)\big],$$

$$M(Z+1, N) = \frac{1}{2}\big[M(Z, N) + M(Z+2, N)\big],$$

$$M(Z+1, N+1) = \frac{1}{4}\big[M(Z, N) + M(Z, N+2) + M(Z+2, N) + M(Z+2, N+2)\big].$$

Having obtained this expression, Weizsäcker also calculated the condition on the ratio N/Z for the stability of nuclei. The results are shown in Fig. 11.15. As can be seen from the figure, Weizsäcker calculated the proton drip line, that is, the limiting number of protons that can reside in a stable nucleus for a given number of neutrons.

Bethe and Bacher tried to improve upon Weizsäcker's formula and derived the condition for stability against emission of an α particle. What they found was

$A = N + Z > 147$. The procedure is simple. The most stable nucleus for a given A is given according to Bethe and Bacher's revised mass formula by

$$E_{\text{min}} \times 10^3 = -6.65A + 14.2A^{2/3} + 0.156A^{5/3}\frac{135}{134 + A^{2/3}}.$$

The condition for stability against emission of an α particle is that the lighter nucleus plus the free α particle have lower energy:

$$E_{\text{min}}(A) - E_{\text{min}}(A - 4) > E_{\text{min}}(\alpha) = 3.35,$$

where 3.35/1,000 is the mass excess of the helium atom. A simple manipulation yields the condition for instability as $A > 147$.

But beware! This is not a genuine prediction or explanation of the stability of nuclei, or why there are such nuclei, because the formula used to derive the conditions is phenomenological. Hence, we may say that the conditions are what nature says about the range of stable nuclei.

Many researchers contributed to improve the original Bethe–Weizsäcker total mass formula, and today there are several hundred such formulae (Fig. 11.16). The best provide an error of 1.6–3.5 MeV per nucleus. While the relative error is quite small, for example for a nucleus with $A = 100$, the relative error is $\sim 2 \times 10^{-5}$. This is far from being sufficient for stellar neutron capture processes. The formulae combine features of the collective and shell models.

It is amazing that, despite the exponential increase in computer power, the fundamental problem, i.e., derivation of nuclear structure in general and the shell model in particular from first principles, has remained uncracked. We still cannot assume a nuclear force and derive the binding energies of the nuclei from it. The problem is so complicated that we do not even know how sensitive the overall properties of the shell model are to the fine details of the assumed effective interactions. It might very well be that physicists have not yet discovered the right approach to attack the problem of a large number of nucleons, that is, large, but not sufficiently large for statistical approaches to become valid.

As a final surprising observation, consider the Bethe–Weizsäcker formula for matter composed of neutrons alone. Let the body be large enough for self gravity to become important. Then on the one hand we ignore the protons and on the other hand we should include self-gravity energy in the form GM^2/R. The attractive gravitational force, although it has infinite range, is usually so weak that it cannot hold a large number of neutrons together. However, if $A > 3.4 \times 10^{56}$, or equivalently a mass of $\sim 0.3 M_{\odot}$, we get the minimal mass for a neutron star composed solely of neutrons. More accurately, the matter in a neutron star is 8/9 neutrons and 1/9 protons, but this does not significantly modify the result.

In 1949, in the wake of Gamow and Alpher's theory of element synthesis in the Big Bang, Smart published a paper entitled *The effects of nuclear stability on the*

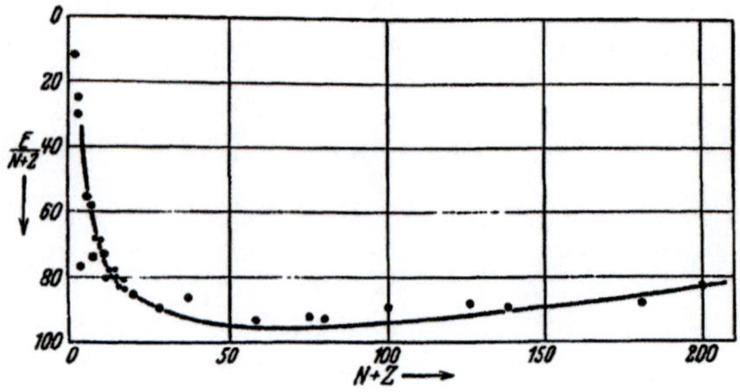

Fig. 11.16 Weizsäcker mass defect formula 1935. Here the mass-defect energy is divided by the atomic weight, so it is binding energy per unit atomic weight. The *continuous curve* is the phenomenological formula obtained by fitting the numerical constants to the available data and the *points* are the data

formation of the chemical elements.[155] He applied nuclear stability considerations to the neutron capture theory for the formation of the chemical elements and reached the conclusion that the nuclei were probably formed with some neutron excess so that both neutron capture and β-decay were effective in determining relative abundances. He claimed that formation followed a path such that the β-decay and capture probabilities were approximately equal. The abundance of nuclei of even atomic weight over those of odd atomic weight could be explained from the general features of the nuclear energy, regardless of the neutron excess at the instant of formation.

11.21 Unstable Elements

We distinguish between two categories of unstable elements. Elements with atomic numbers less than lead ($Z = 82$) and those above it. The unstable elements below lead are further divided into those with low atomic number and those with high atomic number. As a rule, radioactive elements above lead form series of decays, while those below lead involve single decays.

Chemists have a conceptual problem with an 'unstable element'. For example, is Be^8, which lives for 10^{-16} seconds, an element or a bare nucleus? To solve these problems, the International Union of Pure and Applied Chemistry in conjunction with the International Union of Pure and Applied Physics set up a committee to fix *Criteria that must be satisfied for the discovery of a new chemical element to be recognized.*[156] Their basic concern was the transfermium elements ($A \geq 257$ and $Z \geq 100$), which live for a really very short time, so short that their electronic structure cannot be established, whence no chemistry can be done with them. Hence, the first criterion required by the committee is as follows:

155 Smart, J.S., Phys. Rev. **75**, 1379 (1949).

156 Wilkinson, D.H., and 8 authors, Pure Appl. Chem. **63**, 879 (1991).

Table 11.2 Isotopes of promethium and technetium, with their half-lives

Promethium Isotope	Mass	Half-life	Technetium Isotope	Mass	Half-life
^{143}Pm	142.91	265 days	^{93}Tc	92.91	2.73 h
^{144}Pm	143.91	360 days	^{94}Tc	93.91	4.88 h
^{145}Pm	144.91	17.7 years	^{95}Tc	94.91	20.0 h
^{146}Pm	145.91	5.53 years	^{96}Tc	95.91	4.3 days
^{147}Pm	146.92	2.6234 years	^{97}Tc	96.91	2.6×10^6 years
^{148}Pm	147.92	5.37 days	^{98}Tc	97.91	4.2×10^6 years
^{149}Pm	148.92	2.212 days	^{99}Tc	98.91	2.13×10^5 years
^{150}Pm	149.92	2.68 h			
^{151}Pm	150.92	1.183 days			

Discovery of a chemical element is the experimental demonstration, beyond reasonable doubt, of the existence of a nuclide with atomic number Z not identified before, existing for at least 10^{-14} s.

According to this definition, Be^8 is not a 'chemical element', although it is a bona fide nucleus.

A few comments are in order. The committee explains the rather peculiar requirement of 'beyond reasonable doubt' as caused by the fact that reports about the discovery of new elements had occasionally been found to be incorrect. The committee suggests that:

It is often considered desirable to wait with the assignment of priority until the reported results have been confirmed by later work.

This echoes several embarrassments in this connection. Finally, the committee thinks that the highest Z element that one can hope to produce through interaction between stable nuclei or long-lived nuclei is in the most optimistic case around $Z = 190$. We are still a long way off this predicted upper limit.

11.22 Unstable Elements Below Lead

The first unstable element to be found below lead was technetium (see Sect. 6.40). Promethium (Pm), with atomic number $Z = 61$, is another chemical element that is not found on Earth. It was discovered in 1945 by Marinsky, Glendenin, and Coryell in the nuclear waste of a reactor.[157] All isotopes of promethium are unstable, and the longest lived isotope is ^{145}Pm, with a half-life of 17.7 years. All known isotopes of technetium and promethium are listed in Table 11.2 along with their half-lives. For some particular reason, there cannot be a stable nucleus with either 43 protons

[157] Marinsky, J.A., The search for element 61, in *Episodes from the History of the Rare Earth Elements*, ed. by Evans, Boston, Kluwer Pub. 1996.

or 61 protons, irrespective of how many neutrons it may contain. The nuclear forces which keep the protons and the neutrons together in the nucleus do not allow these numbers of protons, and if formed artificially, they decay into other, stable elements.

When Mendeleev attempted to organize the elements according to their properties, he did not recognize any element with atomic number[158] $Z = 43$ or with $Z = 61$, and predicted many of the chemical properties of elements that were only synthesized artificially many decades later.

11.23 The Elements Above Lead

Lead is the most massive stable element. All elements with atomic number higher than bismuth ($Z = 83$) and lead ($Z = 82$) are unstable and decay. Elements with half-lives as long as 4.5 billion years are found on the Earth, but those with much shorter decay times are not found on our planet. The periodic table does not end abruptly at $Z = 82$ or 83. The transition from stable to non-existent nuclei takes place through a series of unstable nuclei with shorter and shorter decay times.

We note that no single proton or neutron is emitted in natural radioactivity, but always a whole helium nucleus, as if the building block of the heavy nucleus was helium. Indeed, the helium nucleus is relatively very stable, in the sense that the two protons and two neutrons are held together very strongly. The binding energy of the helium nucleus divided by its mass, which is the energy needed to break the nucleus into its 4 components divided by its mass (or the energy gained when the four components are brought together) is the highest of all nuclei.

The artificially produced transuranic elements and their decay times are given in Table 11.3. Physicists have succeeded in manufacturing a few elements beyond uranium in the laboratory. The first identification of a transuranium element was by McMillan and Abelson[159] in 1940. In 1999, the group at the Joint Institute for Nuclear Research in Dubna, Russia, reported their success in producing elements with $Z = 114$, as well as $Z = 115$, 116, and 117, by shooting calcium nuclei at a plutonium target. Only 10 years later, in 2009, a team at Lawrence Berkeley National Laboratory, led by Heino Nitsche and Ken Gregorich, confirmed the Russian discovery.[160]

[158] The notion of atomic number did not exist in Mendeleev's day, only the atomic weight. Hence the missing element could not be identified by the atomic number. However, Mendeleev realized that the chemical properties suggested leaving a vacant location in the table for an element to be discovered later.

[159] McMillan, E. and Abelson, P.H., Phys. Rev. **57**, 1185 (1940).

[160] Physics Today, September 2002, p. 15.

Table 11.3 The artificial transuranium elements

93	Neptunium	239	Np	2.4 days
94	Plutonium	244	Pu	8.2×10^7 years
95	Americium	243	Am	7400 years
96	Curium	247	Cm	1.56×10^7 years
97	Berkelium	247	Bk	1380 years
98	Californium	251	Cf	898 years
99	Einsteinium	252	Es	471.7 days
100	Fermium	257	Fm	100.5 days
101	Mendelevium	258	Md	51.5 days
102	Nobelium	259	No	58 min
103	Lawrencium	262	Lr	3.6 h
104	Rutherfordium	263	Rf	10 min
105	Dubnium	268	Db	16 h
106	Seaborgium	266	Sg	21 s
107	Bohrium	272	Bh	9.8 s
108	Hassium	277	Hs	12 min
109	Meitnerium	276	Mt	0.72 s
110	Darmstadtium	281	Ds	1.1 min
111	Roentgenium	280	Rg	3.6 s
112	Ununbium	285	Uub*	10 min
113	Ununtrium	284	Uut*	0.48 s
114	Ununquadium	289	Uuq*	21 s
115	Ununpentium	288	Uup*	87 millisec
116	Ununhexium	292	Uuh*	0.6 millisec

The names reflects great scientists or the laboratories where the elements were produced.

11.24 Why Are There Only 92 Chemical Elements?

Why are there only 92 chemical elements? Why can there not be elements with, say, atomic number $Z = 200$ and atomic weight $A = 1,000$? Alternatively, why are there so many chemical elements? Can there be a situation in which there would be no elements at all and hence no chemistry whatsoever? In short, what determines the grand chemical picture of our Universe?

In fact, it is the interplay between two forces that determines the structure of the nucleus: the Coulomb repulsion between any two protons, and the nuclear attraction between any two nucleons. Consider Fig. 11.17, where the two forces are shown.

The two forces differ in their properties. The nuclear force is very strong but acts only over a short distance, of the order of 10^{-13} cm. The Coulomb repulsion is rather weak relative to the nuclear force, but acts over much greater distances. So, over the short distance the attractive nuclear force wins and preserves the potential well, while over large distances the repulsive Coulomb force wins and destroys the potential well. The figure comes out nicely because the potentials are not drawn to scale.

Fig. 11.17 Formation of the nuclear potential well

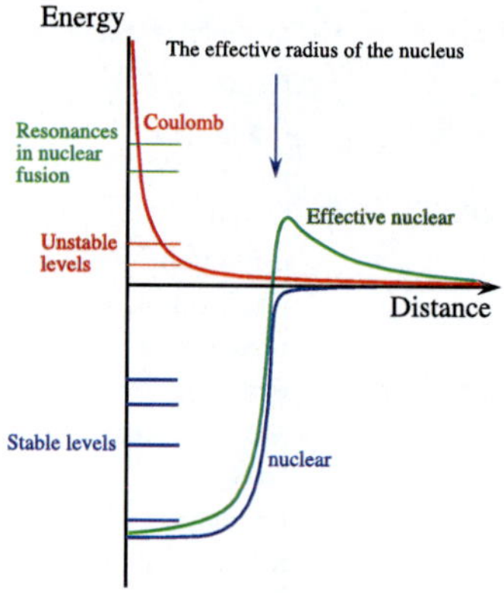

In a given nucleus with protons and neutrons, the glue which holds the protons and neutrons together is supplied by the nuclear attraction which overcomes the Coulomb repulsion between the protons. However, because of the short range, the nuclear attraction operates only between adjacent protons or neutrons while the Coulomb repulsion acts between all protons. When the number of protons is small, say 6 or 8, the most stable nucleus has an equal number of neutrons and protons. But as the number of protons increases so does the Coulomb repulsion and consequently the number of neutrons needed to supply the nuclear attraction to overcome the Coulomb repulsion increases rapidly. In lead, with 82 protons, we find that the number of neutrons is 126. As the number of protons increases to beyond 83 no number of neutrons can stabilize the nucleus and no stable nucleus exists. When Z increases beyond 94, the decay times become shorter than the age of the Earth, and hence no such elements are found on the Earth.

If the nuclear force had not been much stronger than the Coulomb force over short distances, no nuclear potential well could be formed. In a world in which the Coulomb force were stronger than the nuclear one, there would be no nuclei, let alone atoms and chemistry. This discussion has been based on the tacit assumption that the nuclei are spherical or close to spherical. But one can think of other configurations in which the balance between the forces may yield stable configurations with higher atomic weights.

Chapter 12
Synthesis of the Heavier-than-Iron Elements

12.1 Introduction

The famous curve of the binding energy per nucleon tells us that the most stable nuclei are in the iron group. Stellar evolution is therefore the formation of more and more stable nuclei, accompanied by the extraction of the nuclear binding energy which is then radiated away. The nuclei beyond the iron group are less and less stable, and their formation requires energy.

Understanding the synthesis of the heavier-than-iron nuclei (hereafter HTI nuclei) requires us to answer the following questions:

- Which stars synthesize these elements?
- At what stages in the evolution of these stars are such elements synthesized?
- How are the freshly synthesized elements returned to the interstellar medium?
- How do the nuclear properties of the nuclei accord with the physical conditions in the stars in which they are synthesized? Alternatively, do we know the required properties of all nuclei under the physical conditions in which they are formed?

Since the process of HTI synthesis consumes energy, it is plausible that it is not a major factor in the evolution of stars, but a process which produces some trace elements and has little or no effect on the evolution of the star.

While practitioners agree to a large extent on the nuclear steps leading to the synthesis of iron, so that there is a general consensus among scientists on these issues, this is far from being the case for the HTI nuclei. The general idea that the synthesis should be carried out via neutron capture reactions is quite compelling, but the details are still unclear. The basic idea is extremely simple, but myriad details and plenty of open questions remain. In short, the framework of the theory may already be available, but the finer points still need to be worked out.

G. Shaviv, *The Synthesis of the Elements*, Astrophysics and Space Science Library 387,
DOI: 10.1007/978-3-642-28385-7_12, © Springer-Verlag Berlin Heidelberg 2012

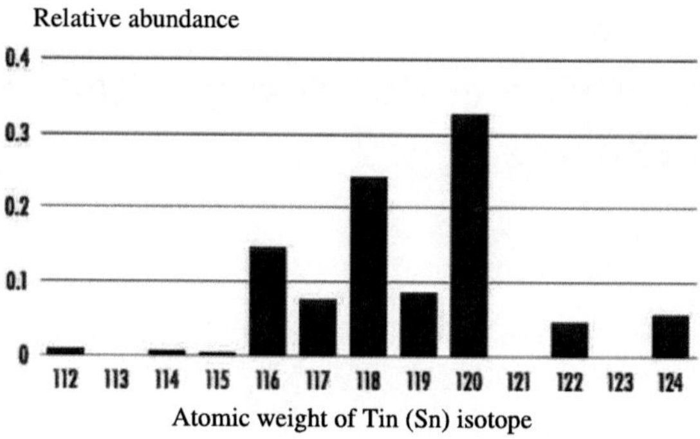

Fig. 12.1 The relative abundance of the stable tin (Sn, $Z = 50$) isotopes. Note that the $N = 50$ isotope does not appear

12.2 Salient Features of the Abundance Curve for HTI Nuclei

The main features of the abundance curve were obvious from the very beginning and have remained unchanged since their discovery:

- An exponential decrease for atomic weight $A > 100$. There seems to be an abrupt change around $A \sim 100$.
- On top of the decrease, an odd–even effect.
- Existence of peaks at $A = 80, 130, 138, 195$, and 208. More accurately, there are double peaks (80 and 90), (130 and 138), and (196 and 208).
- No signs of proton shell closures have been found in the abundance distribution curve. Proton-rich nuclei are extremely rare.
- The distribution of the relative abundances of different isotopes of the same element. One of the classical examples is the isotopes of tin. Tin has the largest number of stable isotopes (ten), with the relative mass fractions given in Fig. 12.1. It is, however, interesting to note that the doubly magic isotope ${}^{100}\mathrm{Sn}_{50}^{50}$ is unstable and decays in 1.1 s. Our question is therefore, what determines the relative isotopic abundance of Sn?

12.3 Overcoming the Coulomb Barrier: Neutron Processes

The synthesis of the elements up to the iron group releases energy, but the continuation of the synthesis beyond the iron group elements extracts energy from the star. The total amount of such elements is extremely small, about 170 atoms beyond the iron group (namely nuclei with $Z > 30$) per 2.8×10^{10} hydrogen nuclei or per 10^6 silicon nuclei. The solar abundances (in numbers of atoms) of the iron group elements are: $\mathrm{Fe} = 9.0\times10^5$, $\mathrm{Co} = 2.3\times10^3$, $\mathrm{Ni} = 5.0\times10^4$, $\mathrm{Cu} = 4.5\times10^2$, and $\mathrm{Zn} = 1.1\times10^3$, according to Grevesse and Anders (1988).[1] So we are talking here about the formation of 170 nuclei out of about 9.54×10^5, or about 1.8×10^{-4} of the iron group which is converted into HTI nuclei. If these nuclei are formed by means of neutron capture, we need at least 1.5×10^4 neutrons,[2] which is almost one neutron per iron nucleus. This is a small amount. Indeed, they are practically trace elements. Consequently, the demand on the star from the point of view of energy invested in the synthesis is negligible.

The abundance of HTI nuclei in cosmic rays,[3] i.e., after acceleration to very high energies, is the sum of all nuclei with $Z > 30 = 4.4 \times 10^{-4}$[Fe], where [Fe] denotes the absolute or relative abundance of iron.

The problem is therefore not the energy balance. This is a major change in the role of nuclear synthesis. Up to iron, the synthesis was central for the energy balance of the star, while beyond iron, it is an unimportant process from the point of view of stellar evolution. We are dealing now with a side process which synthesizes trace elements. However, these elements provide insight into the internal structure of stars and the evolution of galaxies.

Even with the tunneling effect operating, the yield of nuclear fusion reactions (under stellar conditions) beyond iron is totally negligible. The reactions have to be between two heavy ions, both with a high charge. There are no protons or even α particles left. Recall the onion-like structure of advanced stars. The composition changes from high atomic number to a lower one as a function of the radius, and light nuclei and protons will not be found mixed with iron, except in the primordial material out of which the star formed. But even the proton reactions of the heavy elements are very slow at temperatures as high at 10^{10} K. Supposing we have protons and iron nuclei at a temperature of 10 MeV (corresponding to 7.25×10^{10} K), then the peak of the Coulomb barrier for p + p fusion is 0.489 MeV, while for p + Fe it is 650.24 MeV and for Fe + Fe it is 1.25×10^7 MeV, which is truly astronomical. Consequently, the relative rates of penetration of the barrier are 0.83, 1.3×10^{-104}, and $2 \times 10^{-2,001,499}$, which are astronomically small numbers. It is therefore inconceivable that two iron nuclei could fuse in stars. Moreover, even proton absorption reactions cannot take place by the time iron has been synthesized.

The star cannot increase its temperature indefinitely without disintegrating all the nuclei previously formed. A close check of the binding energy as a function of the atomic weight reveals that it is roughly 8 MeV per nucleon. Consequently, at a temperature of about 5.8×10^{10} K, when the mean energy of the photons is 8 MeV, all nuclei disintegrate due to energetic photons or collision with other nuclei.

---

[1] Grevesse, N. and Anders, E., *Cosmic Abundances of Matter*, Am. Inst. Phys., ed. by Waddington, New York, 1, 1988. Recently, Asplund, M., Grevesse, N., and Sauval, J., in ASP Conf. Ser. 336, *Cosmic Abundances as Records of Stellar Evolution and Nucleosynthesis*, ed. by T.G. Barnes, III and F.N. Bash (San Francisco, CA: ASP), 25 (2005) discovered that new models for the solar photosphere yield a total abundance of C, N, O, and Ne which are about half the old values.

[2] The calculation is $170 \times (240 - 56)/2 = 1.5 \times 10^4$ where $(240 - 56)/2$ is the mean number of neutrons needed to synthesize the elements between $A = 56$ and $A = 240$, and 170 is the total number of nuclei. 100% efficiency is assumed.

[3] Prince, P.B., Rajan, R.S. and Tamhane, A.S., Astrophys. J. **151**, L109 (1968).

The nuclei, which in everyday life astound us by their huge amounts of binding energy, are relatively loosely bound at such a high temperatures! Nuclei cannot survive at temperatures of the order of 10^{10} K! As a matter of fact, the maximum temperature for nuclei to survive is significantly lower because, at any temperature, there are a sufficient number of photons with energies ten times the mean energy, and these photons disintegrate atoms and nuclei at a temperature which is $\sim 1/10$ of the equivalent binding energy. Hence, the process of heavy element synthesis must take place at lower temperatures, and the products must be removed from the furnace before the star reaches temperatures of the order of a few MeV. In contrast to popular opinion, stars are not sufficiently hot to allow for such reactions to take place in the core, and on the other hand, they become hot enough to cause the disintegration of iron. Production should take place before the iron can no longer sustain the heat inside the core and disintegrates.

The alternative ideas that the heavy elements result from disintegration, fission, and/or spallation of a superheavy element (like the idea of Mayer and Teller,[4] which they called the break-up hypothesis and the polyneutron model) are not solutions. They just shift the problem elsewhere. The only way we know of today to overcome this barrier is by neutrons, since they do not experience the Coulomb barrier. About $238 - 56 = 182$ neutrons are needed to transform an ^{56}Fe nucleus into a ^{238}U nucleus, assuming no losses. But where does the star find them?

The idea that the elements were synthesized bottom-up by successive neutron absorption is actually due to Gamow and his two collaborators, Alpher and Bethe.[5] These authors noted as early as their first publication that, while the buildup of nuclei by successive neutron capture fails in a big way for light elements, it is quite successful for $A \geq 100$. In addition, they stressed that the timescale of the process should be 10^3–10^4 s. This is a very short timescale by stellar evolutionary standards and hints at an explosive event. The timescale was inferred by looking at β-decays with known decay times, which did or did not take place in the process.

In January 1956, Suess and Urey[6] were the first to realize that no single theory could explain the entire abundances of the elements, and that the neutron absorption theory was a good candidate for explaining the (bottom-up) synthesis of nuclei with $A \geq 70$. Suess and Urey distinguished between two types of reactions:

- A reaction leading to the formation of nuclear species on the neutron-rich side of the energy valley.
- A reaction leading to nuclei on the neutron-deficient side of the energy valley.

They identified the reaction of the first type with the $\alpha\beta\gamma$ theory of Alpher et al.[7] but had no candidate for the second type of reaction.

---

[4] Mayer, M.G. and Teller, E., Phys. Rev. **76**, 1226 (1949).

[5] Alpher, R.A., Bethe, H.A. and Gamow, G., Phys. Rev. **73**, 803 (1948).

[6] Suess, H.E. and Urey, H.C., Rev. Mod. Phys. **28**, 53 (1956).

[7] Suess and Urey knew that Fermi and Turkevitch had shown that the $\alpha\beta\gamma$ theory did not explain the abundance of the light elements and referred only to the high A species.

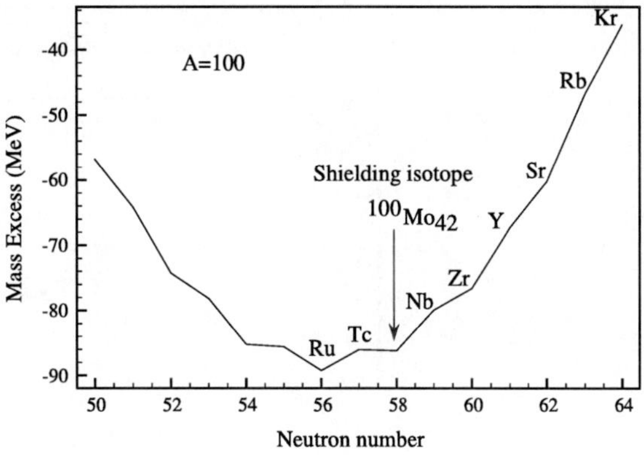

Fig. 12.2 Example of a shielded nucleus. In this case $^{100}Tc_{43}$ is a shielded nucleus because the sequence of β-decays starting from ^{100}Kr stops in the more tightly bound $^{100}Mo_{44}$, and does not reach $^{100}Tc_{43}$ or $^{100}Ru_{44}$

A shielded nucleus is one which cannot be reached in a certain nuclear process because there is another nucleus in front of it which prevents the progress of the process. An example of a shielded isotope is shown in Fig. 12.2. In this case $^{100}Tc_{43}$ and $^{100}Ru_{44}$ are shielded from the r-process nuclei because the sequence of β-decays starting from ^{100}Kr stops in the more tightly bound and stable $^{100}Mo_{44}$ and does not reach $^{100}Tc_{43}$ or $^{100}Ru_{44}$. Neither of these elements can be produced by fast neutron capture (exact definition to follow) with subsequent decay at constant A to the bottom of the β-stability valley. $^{100}Tc_{43}$ is highly unstable and β-decays into $^{100}Ru_{44}$ in 15.8 s.

The reason for the binding energy difference near the bottom of the stability valley is the even–odd effect on the binding energies of the nuclei. When the difference in the binding energy of an extra neutron becomes small, near the bottom of the stability valley, the even–odd effect, which is independent of the number of neutrons, becomes important and accounts for the deviation of the binding energy from a perfect parabola.

Suess and Urey pointed out that the neutron capture theory does not account for the existence of the type of shielded nuclei that have a lower binding energy than their shielding isobars and are on the β^+ side of the energy valley. The abundances of these elements are generally ten times lower than those of the energetically favored type of shielded relatives. Clearly, no neutron capture process can synthesize these heavy elements, and a new, though rarer process must be invoked.

Suess and Urey offered no suggestion as to where the neutrons needed for the process could come from. In the $\alpha\beta\gamma$ theory, the authors postulated that the Ylem, the primordial matter, was essentially made up of neutrons.

In October 1956, Hoyle, Fowler, Burbidge, and Burbidge[8] suggested that the neutrons might be the product of the reaction

$$^{21}\mathrm{Ne} + \alpha \longrightarrow {}^{24}\mathrm{Mg} + \mathrm{n},$$

[8] Hoyle, F., Fowler, W.A., Burbidge, G.R. and Burbidge, E.M., Science **124**, 611 (1956).

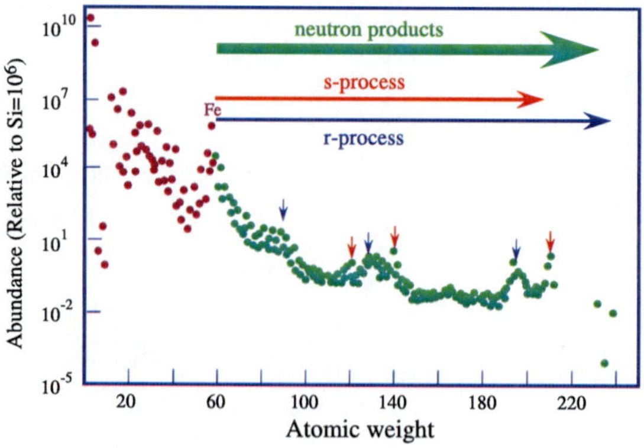

Fig. 12.3 The abundance curve with the abundance peaks caused by the two main neutron capture processes: the s- and the r-processes. The *blue arrow* indicates the r-process peaks, while the *pink* one indicates s-process peaks. *Green points* show the elements synthesized by neutron capture. *Pink* ones are elements synthesized as part of the major evolution process or energy source of the star. Very rare elements heavier than iron but not produced by neutron absorption are not shown. The s-process is predicted to synthesize the isotopes up to $A = 209$ and not beyond, while the r-process does continue to higher A values

which we will call the 21Ne reaction. They realized the problem that nuclei like ^{5}He are neutron poisons[9] for the building of HTI nuclei. Thus, they searched for conditions under which α particles are abundant and were happy to note that ^{5}He is unstable. They were captivated by the idea that it might be possible to synthesize ^{254}Cf via fast neutron capture, as was discovered to be the case in the Bikini H-bomb test.

Figure 9.4 shows the 1957 road map for stellar element synthesis as portrayed by B2FH. Neutrons formed from ^{21}Ne are captured by the iron group and form the HTI nuclei. The arrows in the figure represent the 'flux of matter', and as can be seen, the neutron capture processes are side processes which synthesize trace elements.

Figure 12.3 shows the share of the neutron capture processes in synthesizing elements. All nuclei heavier than the iron group, save extremely rare nuclei which are synthesized by proton capture (to be discussed in Sect. 13.19), are synthesized by neutron capture. The figure also shows the ranges of slow neutron capture, i.e., nuclear synthesis during secular processes (like the s-process), and rapid neutron capture (like the r-process) which takes place during explosive events.

[9] In the slang of nuclear reactor theory, neutron poisons are nuclei that are good absorbers of neutrons and consequently impede any process based on neutron capture.

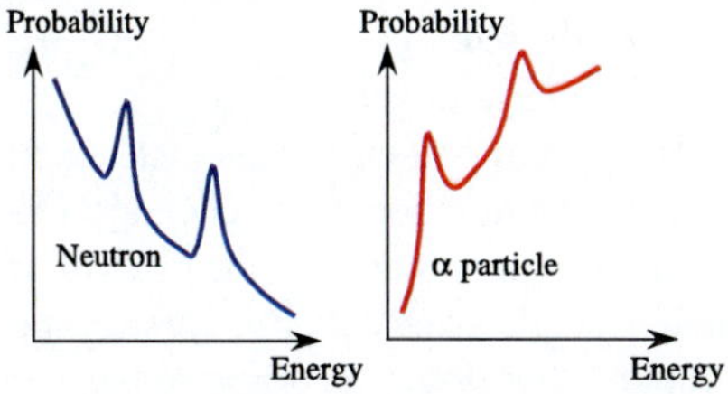

Fig. 12.4 The *left panel* shows schematically the reaction probability of neutrons, i.e., a general decline going as $1/\sqrt{E}$, on top of which there are resonances due to energy states in the compound nucleus. The *right panel* shows schematically the probability for α-particle reactions, which increases with energy as $\exp(-b/E)/E$, on top of which there are resonances. Due to this energy dependence, the energy at which the probability for reaction is measured in the laboratory is very important

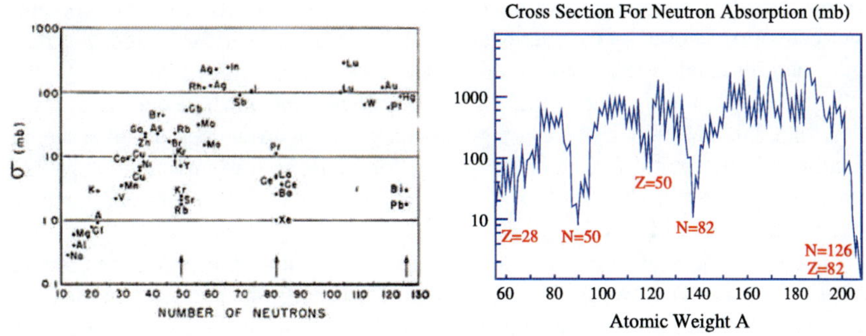

Fig. 12.5 *Left* the first results of Hughes and Sherman (1950), indicating the very low absorption of 1 MeV neutrons by nuclei with magic numbers of neutrons. The data given are for the chemical element which contains several isotopes, not pure isotopes. *Arrows* mark the magic numbers. *Right* the neutron absorption probability (at an energy of 25 keV) as a function of the atomic weight. Nuclei with magic numbers of neutrons are poor neutron absorbers

12.4 Neutron Absorption by Heavy Elements

The probability of neutron capture by various nuclei, or as physicists call it, the cross-section, is the basic data required, apart from the stellar model, to understand the synthesis of the HTI nuclei (Fig. 12.4).

The absorption of neutrons by various elements was needed for the Manhattan project to build an atomic bomb, so it was no wonder that Hughes, Spatz, and Goldstein 1949[10] and Sherman 1950[11] from the Argonne National Laboratory were the first to show that the neutron capture probability of nuclei containing a magic number of neutrons is exceptionally small (see Fig. 12.5 left). It is interesting to note that, in January 1949, six months before Hughes et al. published their paper, Ross and

[10] Hughes, D.J., Spatz, W.D. and Goldstein, N., Phys. Rev. **75**, 1781 (1949), published June 1949.
[11] Hughes, D.J. and Sherman, D., PRL **78**, 632 (1950).

Story[12] in England published the data on neutron capture probabilities measured in Harwell, England, in the form of a table (in contrast to a plot), which masked the general behavior of the data and in particular the existence of minima. Hughes and collaborators plotted the data as a function of the chemical elements, and in this way exposed the effect of magic proton numbers on the capture of neutrons.

The experimental data by Hughes and collaborators was taken as strong evidence in favor of the neutron capture theory for the origin of the elements. Suess and Urey spoke about the 'neutron capture theory' without any further specifications, such as timescales, neutron fluxes, neutron absorptions, and so on. The message was that the local peaks in relative abundances of the elements arise from the magic numbers.

One has to be careful with the touted correlation between neutron captures and abundances. The correlation exists for neutrons with energies in the range of 20–30 keV or temperatures of the order of 10^8–10^9 K, and becomes less pronounced for other neutron energies. Thus, the temperature at which neutron capture operates is effectively disclosed by the energy at which the correlation is discovered. The data published by Hughes and collaborators were for neutrons with energy 1 MeV (the title of the paper was *Capture cross-section for fast neutrons*), where one can still see the effect, but 1 MeV is much too high an energy for neutron capture under stellar conditions. The data of Ross and Story were for neutrons at room temperature (the title of the paper was *Slow neutron absorption cross-section of the elements*), which is too low an energy for stellar reactions.

Neutron capture data relevant to neutron reactions in stars (energies of the order of 25 keV) were rather scarce until the early 1950s, when physicists started to measure the 'relevant data', also needed for the design of fast nuclear reactors.[13] Data on neutron capture has since been collected by various individuals, but a major drive in filling the information gaps was played by the Oak Ridge National Laboratory, Oak Ridge, Tennessee, under Macklin.[14] Once the large liquid scintillator (about 1,000 l) had been completed at the Oak Ridge laboratories (see Fig. 12.6) in the late 1950s, a large amount of new and better data became available. Today, the Oak Ridge National Laboratory houses a modern neutron source facility called ORELA (see Fig. 12.7).

To appreciate the numbers, it is instructive to compare the absorption probability with the projected area of the nucleus. At the minimum (probability of 1 millibarn or cross-section of 10^{-27} cm^2) the neutron 'sees' the nucleus as a circle with area $\sim 1/250$ of its actual size, that is, very small indeed. At the maximum, the neutron 'sees' the nucleus at its real size or slightly bigger.

Progress in the accuracy of absorption probability data obtained by the Karlsruhe group of Käppeler[15] has significantly improved the situation. Typical present day

---

[12] Ross, M. and Story, J.S., Rep. Prog. Phys. **12**, 291 (1949).

[13] Fast nuclear reactors are ones in which the neutrons released in the fission (up to 14 MeV) have no time to slow down.

[14] Gibbons, J.H. and Macklin, R.L., Science **156**, 1039 (1967).

[15] Käppeler, F., Beer, H. and Wisshak, K., Rep. Prog. Phys. **52**, 945 (1989); Beer, H., Voss, F. and Winters, R.R., Astrophys. J. Suppl. **80**, 403 (1992).

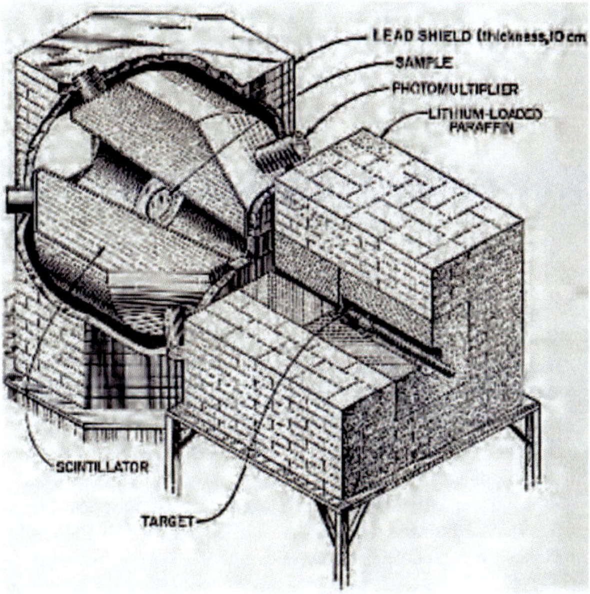

Fig. 12.6 The experimental setup at Oak Ridge National Laboratory used to measure the capture of neutrons by heavy elements. A large liquid scintillator detector was used. The neutron source was the reaction $^7\text{Li} + \text{p} \to {}^7\text{Be} + \text{n}$ and the energy of the neutron was measured by the time of flight from the target to the sample. Stellar nuclear reactions were simulated from the beginning in such Earth-based devices

absorption probabilities of neutrons with 25 keV are shown in Fig. 12.5 (right). The minima in the absorption are clearly evident. Also evident is the odd–even effect.

The effect is even more clearly visible when the split to individual isotopes is made, as in Fig. 12.8, which shows absorption by isotopes with neutron numbers close to the magic neutron number 82. In all nuclei with a given Z, i.e., fixed chemical elements, the one with 82 neutrons has the lowest absorption.

There is much to improve, extend, and discover at the present time in the uncharted territories of the (N, Z) plane. New facilities, mainly in the USA and Germany, are already operating or planned to further improve neutron capture data, in particular those pertaining to unstable nuclei. The new data will bring a major advancement in the drive to unfold the story of HTI synthesis[16] and write the next chapter in this saga.

New nuclear physics facilities are under development. The Radioactive Ion Beam factory (RIB) at RIKEN in Japan is already operational. Furthermore, the Facility for Antiproton and Ion Research (FAIR) will soon be built in Darmstadt, and the Facility for Rare Isotope Beams (FRIB) will be built in East Lansing (Michigan, USA). We may thus expect major changes in our understanding of nuclear structure far from the

[16] Käppeler, F., J. Neutron Res. **13**, 33 (2005).

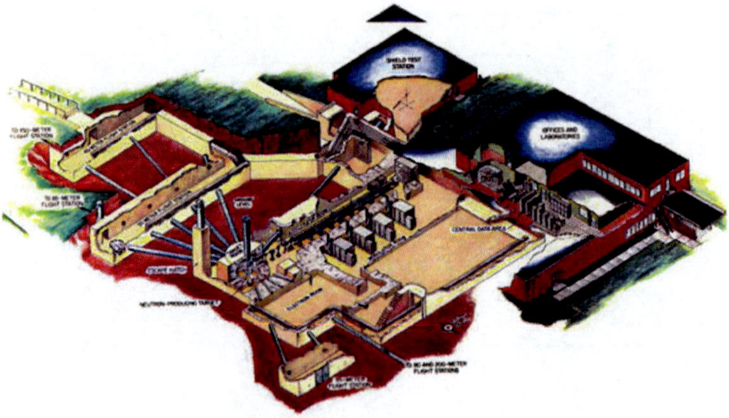

Fig. 12.7 The modern ORELA neutron source facility located in the Physics Division of Oak Ridge National Laboratory. It produces intense, nanosecond bursts of neutrons, with energies from 10^{-3} to 100 MeV. ORELA's present mission is research in nuclear astrophysics. Most of the neutron capture measurements required for the *s*-process are carried out with ORELA, which is typical of the present day machines used in nuclear astrophysics

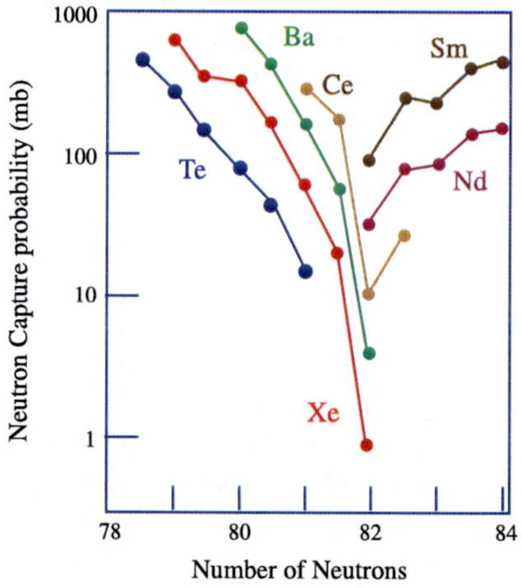

Fig. 12.8 The fast decline of the neutron absorption probability as a function of the number of neutrons and for several elements, i.e., different values of Z, around $N = 82$. The curve pertaining to each element is colored differently. The units for the capture probability are millibarn, or 10^{-27} cm^2. As one approaches the $N = 82$ case from below or from above, the absorption decreases for all Z values

stability region, the way stars synthesize the heavier-than-iron elements, and what can be learnt about the structure of stars and the cosmos by observing synthesized matter.

About 7,500 papers have been published on neutron capture and element synthesis in astrophysics. It is impossible to cover all of them here, so we shall just describe the gist.

12.4.1 The Available Time, Timescales, and Capture Processes

Stars have several timescales. The dynamic timescale, which reflects the response of the star to changes in density for example, is given by

$$t_{\rm dyn} \approx \left(\frac{2R^3}{GM}\right)^{1/2},$$

where R and M are the radius and mass of the star, respectively. The formula may be applied to the entire star or just to the core. When we substitute the relevant numbers for, say, a $25M_\odot$ star, we find that $t_{\rm dyn} \sim 10^3$–10^4 s. This is the typical timescale for an explosion.

If the star has no other energy sources save the gravitational one, then the characteristic timescale is given by

$$t_{\rm therm} \approx \frac{GM^2}{LR},$$

where L is the luminosity of the star. If we substitute in the relevant numbers, we find $t_{\rm therm} \sim 10^4$ years, i.e., about 10^7 times longer than the dynamic timescale. This is the time it takes the energy to flow out as radiation from the core to the surface and be radiated away.

The nuclear timescale is defined as

$$t_{\rm nuc} \approx \frac{E_{\rm nuc}}{L},$$

where $E_{\rm nuc}$ is the total available nuclear energy. Along the main sequence, we find that

$$t_{\rm nuc} \gg t_{\rm therm} \gg t_{\rm dyn}.$$

As the star evolves, the luminosity rises (and to a lesser extent the radius also increases), and consequently all three timescales change, but not at the same rate. The nuclear evolutionary timescales for a $25M_\odot$ star are shown in Table 12.1. The first time a neutron-releasing reaction is available is during helium burning. From the table, we see that the evolution time for core helium burning is 5×10^5 years. The lifetimes of more massive stars are even shorter, and conversely. Thus, secular neutron capture reactions have at most 10^4–10^5 years to operate. The table is taken from B2FH. They hypothesized that the relevant star for the s-process was a $25M_\odot$ star. Today we know that the slow neutron capture process takes place in stars less massive than $8M_\odot$, where the timescales are longer than those for a $25M_\odot$ star.

Let us take all β-unstable nuclei and look for the longest and shortest half-life values, $\tau_{\beta,\rm max}$ and $\tau_{\beta,\rm min}$, respectively. Consider first a process in which the rate of neutron capture is so slow that all unstable nuclei have ample time to decay after

Table 12.1 Available time during nuclear burning in a $25M_{\odot}$ star. After B2FH 1957

Nuclear source	Temperature (K)	Burning lifetime
Hydrogen (CNO cycle)	4×10^7	7×10^6 years
Helium (3α)	2×10^8	5×10^5 years
Kelvin–Helmholtz–Ritter timescale		7.5×10^4 years
Carbon (C+C)	6×10^8	600 years
Neon (α process)	1.2×10^9	1 year
Oxygen (O+O)	1.5×10^9	0.5 year
Silicon (α process)	2.9×10^9	1 day
SN explosion	$(5–10) \times 10^{10}$	1–100 s

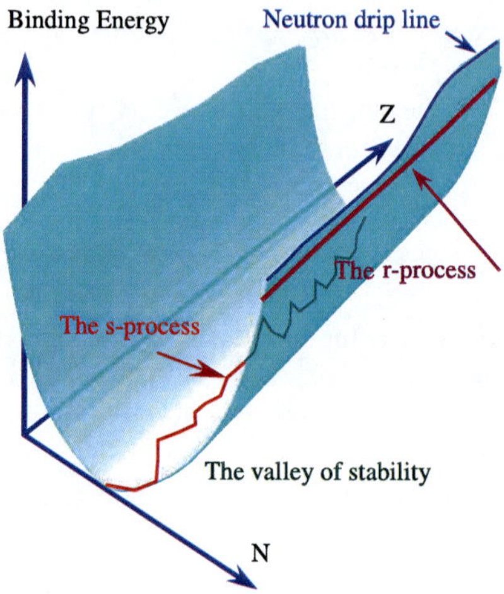

Fig. 12.9 The difference between the paths of the *s*- and the *r*-process

the absorption of a neutron. Clearly, this process will move very close to the bottom of the stability valley (see Fig. 12.9). At a temperature of about 10^8 K, the speed of the thermal neutrons[17] is about 10^8 cm/s. Let us search for a process that lasts 10^5 years. Finally, let us assume that the nuclei that participate in the process expose a 'black body' to the neutrons, in the sense that they absorb every neutron that hits them. A simple calculation[18] then shows that the number density of neutrons

[17] All neutron-releasing reactions produce neutrons at very high energies. It is assumed that the fast neutrons scatter off different nuclei and lose energy in order to adjust their energies to those of the ambient nuclei before they are captured by a nucleus. Neutrons with such relaxed energies are called thermal neutrons. For this to be the case, the scattering probability must be much greater than the capture probability.

[18] The estimate is as follows. The radius of the nucleus is given by $r_{\rm nuc} = 1.3 \times 10^{-13} A^{1/3}$ cm, where A is the atomic weight, for which we assume an average value of 100. The number density of neutrons is then given by $N_{\rm n} = 200/v\sigma T$, where $\sigma = \pi r_{\rm nuc}^2$, $T = 10^4$ years is the duration of the process, and v is the mean velocity of the neutron. The number 200 is the number of neutrons needed to build the entire list of HTI nuclei.

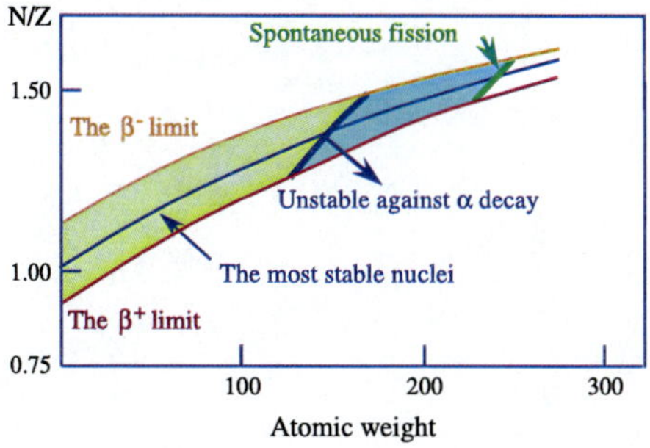

Fig. 12.10 The nuclear stability valley in the plane of the N/Z ratio and the atomic weight A. Marked are the minimal and maximal values of N/Z for which there exist nuclei, as well as the limit for stability against emission of an α particle. Finally, the limit above which all nuclei undergo spontaneous fission is marked in *green*. This is frequently called the β stability valley, since its borders are determined by $\beta^{\pm}$-decays. Nuclei in the *yellow* and *blue* regions have finite decay times. Nuclei beyond the borders decay upon formation. All nuclei beyond the border for α stability decay via α-decay. Nuclei beyond the fission stability limit decay by fission

should be $\sim 10^{8\pm 1}$/cm^3. Different nuclei absorb neutrons with different probabilities, but the global estimate is quite good. This number density dictates the possible sources of neutrons. Such a process was dubbed the slow process or just the s-process by B2FH.[19]

A number density of neutrons equal to 10^8 seems like a very large number, but this is not so. At this stage in the evolution, the density of the matter is above 100 g/cm^3. If all the matter is composed of helium nuclei, the number of nuclei is $100 \times N_{\rm A}/4$, where $N_{\rm A} = 6.023 \times 10^{23}$ is the Avogadro number and 4 the atomic weight of helium. Hence, the neutrons comprise only a part equal to 6.6×10^{-18} of the matter.

An overall view of the valley of stability is shown in Fig. 12.10, where the blue line describes the stable nuclei and the yellow region marks the unstable nuclei. The s-process therefore advances along the blue line and gradually changes the N/Z ratio. This ratio has limits beyond which the process cannot go. The s-process continues until the nuclei become unstable against α-decay in the $A = 204$–210 region. Thus, elements like ^{209}Bi, ^{204}Pb, and ^{206}Pb are synthesized at the end of the s-process. In particular, we find that

$$\mathrm{Bi}^{209} + \mathrm{n} \longrightarrow {}^{206}\mathrm{Pb} + \alpha .$$

Thus, no heavier nuclei can be synthesized by slow neutron capture.

Let us consider two detailed examples of the finer details. Figure 12.11 shows neutron absorption by an iron seed nucleus. The isotopes are arranged according to increasing numbers of neutrons. Thus, ${}^{56}\mathrm{Co}_{27}^{29}$ is unstable and decays into ${}^{56}\mathrm{Fe}_{26}^{30}$,

[19] B2FH assumed two extreme cases based on the possible flux of neutrons. The first was a small neutron flux which leads to a long timescale for neutron capture, ranging from $\sim$100 to $\sim 10^5$ years. This case was baptized the s-process, while the other extreme was baptized the r-process. There was nothing in between.

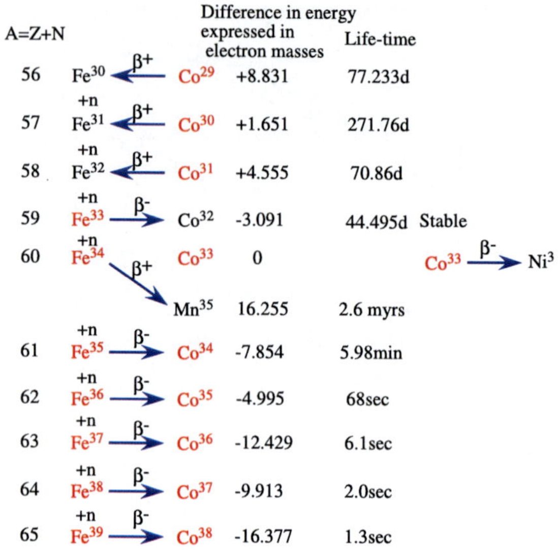

Fig. 12.11 Slow neutron capture by iron isotopes. Stable isotopes are shown in *black* and unstable ones in *red*. Consider the first reaction. The iron with 30 neutrons absorbs an additional neutron and becomes Fe^{31}. This is a stable nucleus, because it has a lower energy than Co^{29}, which decays into it. The neutron absorption process continues as far as Fe^{33}, which decays into Co^{32} in 44 days. When the neutron flux is high, the Fe_{26}^{33} succeeds in capturing another neutron to become Fe^{34}, which decays into Mn^{35} in 2.6 million years. Hence, the Fe^{34} has time to absorb another neutron to become Fe^{35}, and this decays into Co^{34} in about 6 min. This is still a long time for the *r*-process, and Fe^{35} has time to absorb another neutron to become Fe^{36}, which decays into cobalt in 68 s. The process continues until Fe^{39} is produced. The result of the next step, Fe^{40}, disintegrates upon formation. Recall that $Z = 25$ is Mn, $Z = 26$ is Fe, and $Z = 27$ is Co

which subsequently absorbs a neutron to become ${}^{57}Fe_{26}^{31}$ and sets off along the chain of neutron-rich iron nuclei until an unstable nucleus is encountered, i.e., one whose half-life is comparable with the mean time between neutron absorptions.

A second detailed view of the path of the *s*-process is shown in Fig. 12.12 starting with ${}^{158}Cd_{48}^{110}$. As a result of neutron capture by the stable nucleus, the number of neutrons increases up to $N = 115$, where the unstable Cd_{48}^{115} nucleus is reached. The problem is that, by then, the process has gone past any stable isotope of In. In_{49}^{115} is unstable, but with a lifetime of 4.4×10^{14} years. It so happens that this is, on the *s*-process timescale, a nucleus that is to all intents and purposes stable, sitting between two In isotopes that are unstable.

The effectively stable In absorbs a neutron and decays into Sn_{50}^{116}. Now the proton magic number allows for a long list of stable nuclei, so the number of neutrons increases to 121. The stable nucleus Sn_{50}^{122} cannot be reached by the *s*-process (neutron absorption on Sn_{50}^{121}) unless the neutron reaches the nucleus in less than 27 days after its formation. But as the flux of neutrons in the *s*-process is relatively low, the probability for this to happen is extremely small.

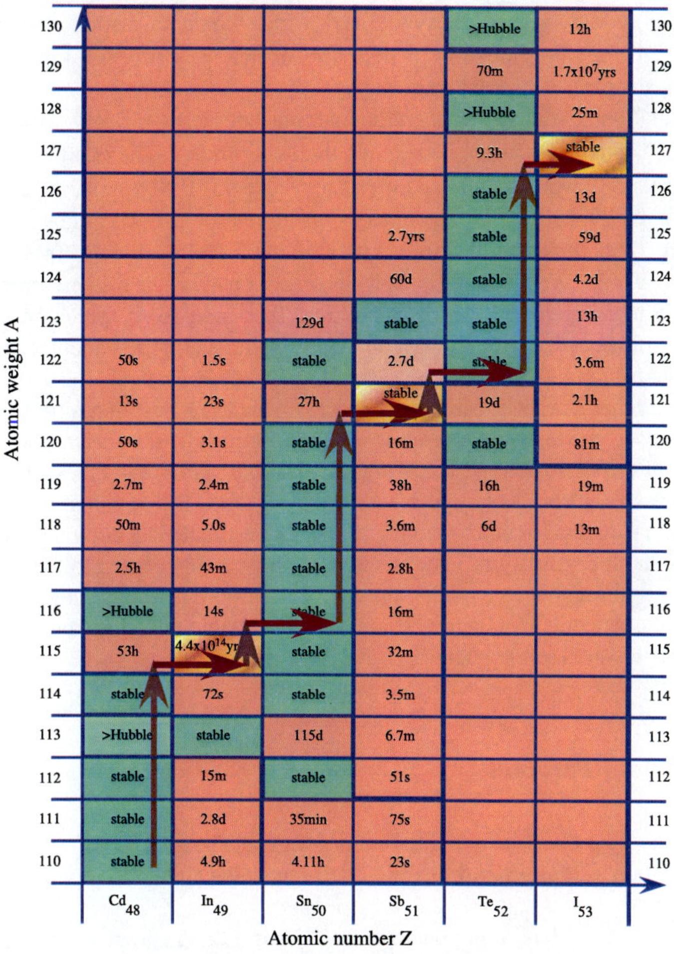

Fig. 12.12 The path of the s-process in the (Z, N) plane. *Yellow vertical arrows* are neutron captures and *horizontal pink arrows* are β-decays. Each (Z, N) box represents an isotope. If the isotope is unstable, the decay time is given, otherwise it is marked as stable. Neutron absorption creates nuclei rich in neutrons. Stable elements are marked by *green* and unstable elements by *light red*. The process operates along the transition from stable to unstable nuclei. The upward climb ends when it encounters a nucleus whose β-decay is sufficiently short for it to decay before being able to absorb another neutron. We indicate the lifetimes of the nuclei close to the path of the s-process. Further away, the lifetimes are much shorter and irrelevant to the s-process. The entry 'Hubble' indicates a half-life longer than the Hubble time, i.e., the age of the Universe. *Yellow squares* mean transition to a higher Z

It seems to be a collusion of nature that, at the very moment that Cd_{48}^{115} becomes unstable, In_{49}^{115} turns out to be stable among unstable neighbors. However, if this were not so, and if the lifetime of In_{49}^{115} had been much shorter, then the s-process would have moved forward via a β-decay to Sn_{50}^{115}, which would have continued to absorb neutrons. The long lifetime of In_{49}^{115} only implies that the s-process skips Sn_{50}^{115} and this nucleus is formed in a different way.

The situation repeats itself when the path reaches Sb_{51}^{121} and I_{53}^{127}. Similarly, the nuclei In_{49}^{113} and Te_{52}^{120} cannot be reached by the s-process. The relative abundance of In_{49}^{113} is 0.0429 while the abundance of In_{49}^{115} is 0.9571. Similarly, the abundance of Te_{52}^{120} is 0.0009, while the abundances of the rest of the Te isotopes are Te^{122}: 0.0255, Te^{123}: 0.0089, Te^{124}: 0.0474, Te^{125}: 0.0707, Te^{126}: 0.1884, Te^{128}: 0.3174, and Te^{130}: 0.3408. We see that Te^{120}, which is not on the route of the s-process, has a relatively low abundance. At the same time, Te^{128} and Te^{130}, which are shielded from the s-process, have the highest abundances, indicating that another process is responsible for the bulk of the Te nuclei.

If on the other hand we search for a process which is so fast that most β-unstable nuclei have no time to decay, a process which must go at least as fast as the dynamic timescale of the explosion, we can simply look at the ratio of the timescales and see that we need a neutron density of at least $\sim 10^{15\pm 1}/\text{cm}^3$. B2FH called this the rapid process, or r-process for short (see Fig. 12.9). The estimate is a minimal one, and in reality the number density of neutrons is much higher. In contrast to the s-process, which passes through stable nuclei, the r-process proceeds through highly unstable nuclei. As soon as the irradiation ceases, all the unstable nuclei go into a sequence of β-decays towards stable nuclei.

12.4.2 The β Parabola

Let us look at the phenomenological mass formula. The general expression can be seen to contain a term proportional to Z^2, because the Coulomb repulsion goes as the number of protons squared, and a term which goes as Z. Thus, from the charge point of view, the behavior is parabolic in Z. Figure 12.13 shows a cut through the valley of stability at constant atomic weight A. Close to the minimum of the parabola, we find one and sometimes several stable nuclei (marked in green). To the right, in orange, we see the unstable nuclei with more protons than neutrons, while to the left, in blue, we see unstable nuclei with more neutrons than protons. At the edges, there exists an as yet unexplored extreme range of unstable nuclei. The right-hand axis shows the behavior of the lifetimes of the nuclei: the further from the minimum of the parabola, the shorter the lifetime.

A typical example of the effect of the mass parabola is visible in the neutron absorption by the first nucleus, namely Fe_{26}, as shown in Fig. 12.11. The isotopes of cobalt with $N \geq 29$ decay into Fe, while Fe isotopes with $N \geq 35$ decay into cobalt. The exceptions are Co^{33}, which is stable, and Fe^{34}, which decays into Mn_{25}^{35}. The decay times are also given to demonstrate how they shorten when N is verylarge.

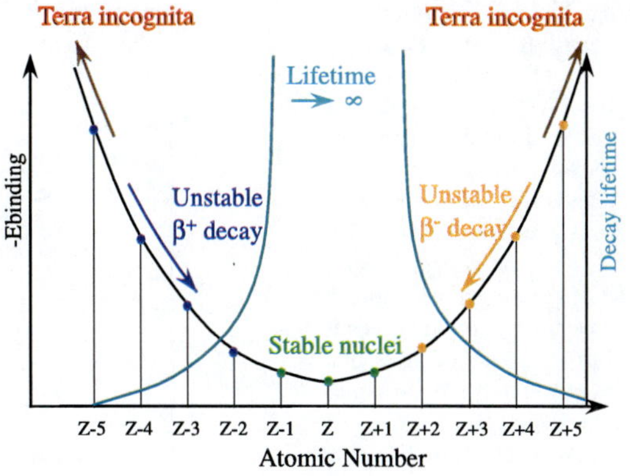

Fig. 12.13 The β parabola. *Green circles* denote stable nuclei, while *orange* are unstable neutron-deficient nuclei, and *blue* are unstable neutron-excess nuclei. The *bluish curve* is a schematic lifetime of the nuclei

12.5 Steady State Versus Transient Neutron Capture

B2FH did not carry out any calculation of the *s*-process. They simply assumed a steady flow. Clayton et al. 1961[20] were the first to carry out a time-dependent calculation for irradiation by neutrons. Soon the basic idea was expanded to series of irradiations and in this way additional parameters, the extent of the irradiation and its time dependence, were introduced into the problem.

12.5.1 *The Steady State Assumption*

The great advantages of the steady state assumption are:

- One can easily solve for the complete abundance curve.
- The results depend only on the temperature, density, and nuclear physics, and not at all on the astrophysical properties of the site where the synthesis takes place or the initial conditions prevailing when the synthesis started.

If N_i is the number density of a species i and σ_i is the probability for the absorption of a neutron by this species, then in steady state, the product $N_i\sigma_i$ is a constant and does not depend on the element. The expectation is that poor neutron absorbers should have high abundances because they are not easily destroyed by a neutron, and conversely. This is indeed the case.

A typical example is shown in Fig. 12.14. Given the far-reaching assumptions of steady state, the agreement with observation is quite amazing. Three cases are shown, calculated in 1957, 1965, and 30 years later. Figure 12.14 shows the product of abundance and absorption probability for low A values. One can see how unimpressive the 'constancy' is. In Fig. 12.15, the abundances predicted by the steady

[20] Clayton, D.D., Fowler, W.A., Hull, T.E. and Zimmerman, B.A., Anal. Phys. **12**, 331 (1961).

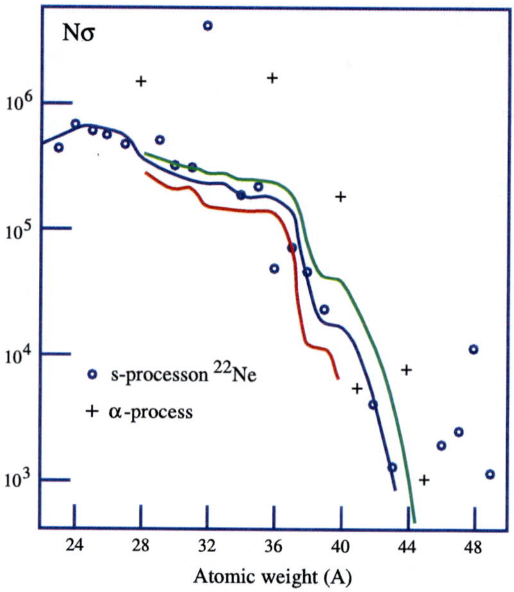

Fig. 12.14 The product (according to B2FH) of abundance and absorption probability, which should be constant in steady state, for relatively low atomic weights. Abundances are based on Suess and Urey (1956). *Circles* are *s*-process isotopes while the *plus signs*, which lie above the other points because of the large abundances of these isotopes, are attributed to the α process. Curves were calculated by Tuttle for different numbers of neutrons made available per ^{22}Ne nucleus. The numbers are, from the bottom curve up, 1.8, 2.1, and 2.4. These are very small numbers and imply a low yield

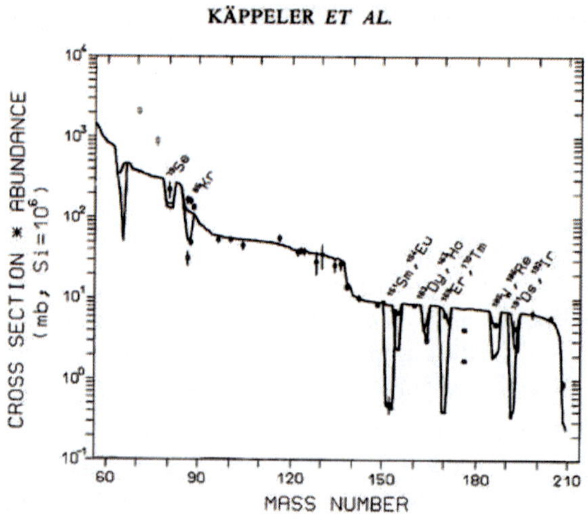

Fig. 12.15 Over thirty years later, improvements in the nuclear data had an important impact on the quality of the agreement with observations. However, $N_i\sigma_i$ is still not a constant, indicating the need for transients to do the synthesis job. After Käppeler et al. (1990)

state assumption are shown for a higher *A* and the fit is reasonable. In view of the results expressed in the right-hand panel, B2FH suggested that:

> The departure of the calculated curve from the plotted points beyond $A \approx 45$ suggests that other processes must be responsible for synthesis in this region.

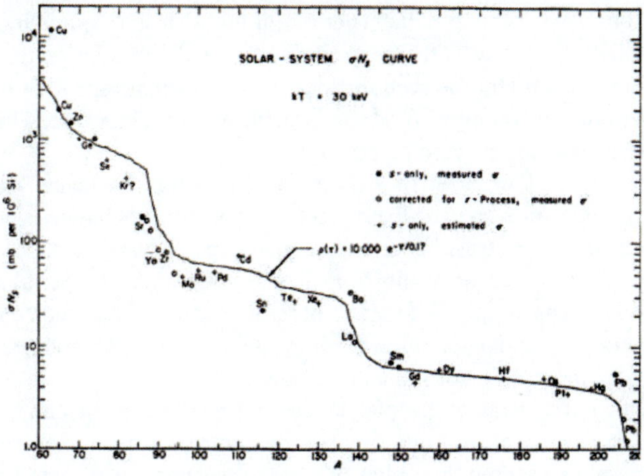

Fig. 12.16 The *solid line* is a calculated curve corresponding to an exponential distribution of integrated neutron flux. After Seeger et al. (1965)

But they had no suggestion as to what it could be. In 1965, Seeger, Fowler, and Clayton[21] produced the curve shown in Fig. 12.16, and the authors commented that:

> A surprisingly good fit has been obtained with a two-parameter curve, although such smooth distribution functions require a much more sudden drop from Sr to Zr than is apparent in the observations.

A closer check reveals that $N_i\sigma_i$ is constant for all species i far from magic numbers, but deviates significantly for magic numbers and values near them, precisely where we most need it.[22]

12.5.2 Non-Steady State Options

In view of the comment by B2FH reported in the previous section, Clayton et al. (1961) suggested that the s-process might be composed of several episodes of neutron irradiation rather than long-time steady state irradiation.[23] The parameters of the irradiation are the duration and the time dependence, which is usually assumed to be an exponential function. There was no a priori reason why the irradiation should be

---

[21] Seeger, P.A., Fowler, W.A. and Clayton D.D., Astrophys. J. Suppl. **11**, 121 (1965).

[22] In recent years this approximation has been called the 'local approximation', since it is meaningful only for a few nuclei and away from magic numbers.

[23] Clayton and coauthors did not specify the implications of the idea since it was a phenomenological approach. It could imply (a) that the irradiation in each star is intermittent and may change with time, (b) that the s-process in different stars takes place under different conditions, or (c) both of the above, and the mixing in the interstellar medium is far from perfect.

exponential, but it is easy to find the general solution with such a function. The new parameter, called fluence, was the neutron flux times the irradiation time, i.e., the number of neutrons that hit the seed nucleus. It soon became apparent that several irradiation periods were required to get a reasonable fit to the abundances, in particular near the bottlenecks of the magic numbers.

After Clayton et al. introduced the exponential irradiation, it became very popular, although it was never proven to be the correct function. Such behavior is not a decree of Nature. In some cases, trust in the ad hoc time dependent s-process was so great that researchers considered the results from their s-process calculations as constraints on stellar models. The simple fact that no other alternatives have been conceived is not a convincing argument for inferring how stellar models should behave when heavy numerical models do not show them what to do.

Once the idea of an irradiation episode was in the air, the extension to several or many episodes was trivial. But in doing so, the number of parameters which enter the theory becomes prohibitively high relative to the number of data known. The difficulties with a single irradiation event led to proposals for several irradiation episodes which can be summarized as follows:

- The weak component, which produces the $A = 56$ to $A \sim 90$ nuclei.
- The main component, which produces the $A \geq 90$ nuclei.
- A strong component, which produces half or more of the $A = 204$–209 nuclei.

Each of the three components has a different exponential distribution of neutron exposures. The three processes must somehow take place in the same star. If the process takes place in different stars and then the products are mixed in the interstellar medium, one has to explain how the three processes produce a continuous abundance curve without noticeable jumps.

It soon became apparent that three irradiation events would be insufficient to produce an acceptable agreement between theory and observation, and the idea of multi-events was suggested by Goriely.[24] In this approach one assumes a large number, up to 500, of s-process events with different irradiation periods but the same temperature and neutron flux. If this is not sufficient, the irradiation times are optimized so as to provide the best fit to observations of abundances in the Solar System. Even then, only about 35 nuclei are fitted. It is conceivable that such a situation reflects some fundamental deficiency in the theory. Is it the exponential assumption, the assumption that the theory can progress without reference to a particular stellar environment? We do not know. It seems that the most plausible way to proceed is to get the basic nuclear data right and then carry out 2D or 3D stellar modeling of the detailed hydrodynamics which takes place in the star. The idea that the s-process can be modeled in a way that is disconnected from the stellar environment is probably wrong, not because it affects the evolution, but because it is very sensitive to the detailed stellar environment.

---

[24] Goriely, S., Astron. Astrophys. **342**, 881 (1999).

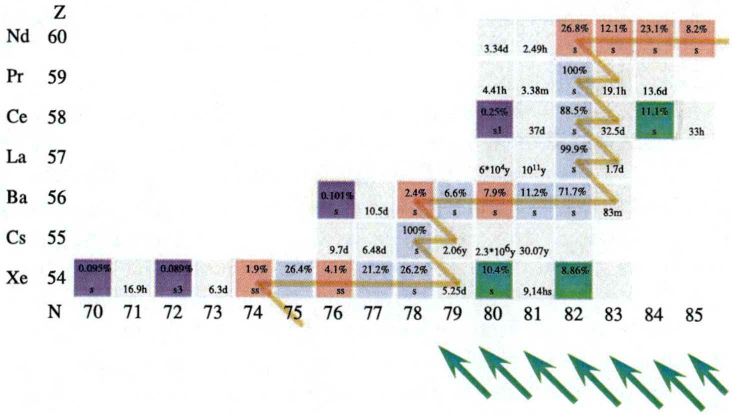

Fig. 12.17 The s-process in the (N, Z) plane around the magic number $N = 82$. Stable elements are marked with an 's' and the half-life is given for the unstable nuclei. The percentage given for each stable species is the cosmic relative abundance of the element. The path of the s-process is marked by an *orange line*, while the direction of the decay of neutron-rich nuclei, products of the r-process, is denoted by *green arrows*. Nuclei marked in *orange* are produced only by the s-process because they are shielded from the r-process by the nuclei marked in *green*. The decay of neutron-rich nuclei toward the minimum of the β-stability valley stops at the *green* nuclei. The stable *pink* nuclei cannot be produced by any kind of neutron capture, and are assumed by default to be produced by the p-process. The *grey* nuclei are produced by both the s- and the r-process

12.6 Classification of Heavy Nuclei

The existence of shielding nuclei implies that not all nuclei are synthesized by all the neutron capture processes. We have pure s-process nuclei, pure r-process nuclei, and nuclei which are produced by both mechanisms. Thus, out of the isotopes of a given element, some may be s-process only while others may be the product of more than one process. The fact that different isotopes of the same element are produced by different processes provides a good tool for assessing the particular role of each process.

When one attempts, for example, to compare the theory of the s-process with observed abundances, it is crucial to know the relative contribution of each process in cases where both processes contribute to the abundance. This is very complicated, and in many cases the procedure is not uniquely defined (see Fig. 12.17).

In general, a phenomenological model of the s-process is used to fit the abundances of the s-only nuclides. Once the parameters of this model have been determined, it is used to predict the s-process contributions to the other s-nuclides. The subtraction of these s-process contributions from the observed solar abundances, for example, leaves for each isotope a residual abundance that represents the contribution of the r-process (if the isotope is neutron-rich) or p-process (if the isotope is neutron-deficient). These nuclides of mixed origins are called sr or sp nuclides.

In the solar case, about half of the isotopes are synthesized in the s-process and about half in the r-process. The p-process contributes something between 0.01 and 0.001 of the HTI sum of isotopes.

Exact observational separation of the various contributions to a given isotope requires measurement of the relative abundances of the different isotopes. Unfortunately, for $Z \geq 30$, the isotopic shift in the spectrum, i.e., the effect on the optical spectrum of the fact that different isotopes have different masses, decreases to such an extent that it is practically impossible to measure the abundances of different isotopes using present day spectroscopic equipment. Unique in this respect are the elements Ba, Eu, Hg, Pt, and Tl. For all other elements, the only measurements available are of meteorite abundances with isotopic breakdowns. Consequently, the separation into the contribution of the two processes is frequently problematic. Caution must be exercised when the contributions of the s- and r-processes to the composition of a given star are customarily evaluated from elements whose Solar System abundances are predicted to be dominated by one or the other of these neutron capture processes.

The traditional probe of the s-process has long been Ba with 70–100% contribution to the Solar System abundance from the s-process. The difficulty with Ba, which has several stable isotopes each with different s- and r-contributions and whose precise abundance determinations raise specific problems, is circumvented by using the essentially mono-isotopic lanthanum.[25] It is traditionally considered that about 75% of the La in the Solar System originates from the s-process. In summary, the disentangling of the observed contributions by the different processes may contain non-trivial systematic errors.

Figure 12.18 shows a typical breakdown of the contribution by the different processes to the abundances of elements as a function of atomic weight. The different shares are no indication of the absolute abundance. In addition, note that different isotopes of the same element may involve different r/s contributions. The classical approach tells us that about half of the HTI nuclei are synthesized by means of the s-process, which takes place in asymptotic giant branch[26] stars with masses lower than $8M_{\odot}$.

12.7 Viable Stellar Neutron Sources

The basic idea behind the neutron sources is as follows: we look for a nucleus with atomic weight $m\,\alpha + \mathrm{n}$ (where m is an integer and n a neutron), i.e., a nucleus made of m α particles plus a neutron. This neutron moves outside a closed and compact shell, and hence should be relatively easy to knock out.

[25] Lanthanum has one stable isotope, ^{139}La, with natural abundance 0.9991. Regarding the rest of its 38 unstable isotopes, ^{138}La has the longest lifetime at 1.05×10^{11} years but a natural abundance of 0.0009. Thus for all practical purposes lanthanum has only one stable isotope.

[26] Asymptotic giant branch stars, or AGBs for short, are stars in an evolutionary phase beyond the ignition of helium in the core.

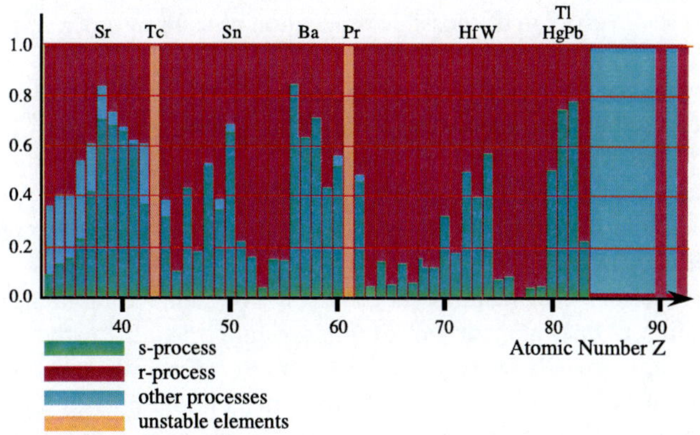

Fig. 12.18 The estimated share of the *s*- and *r*-process in the synthesis of the elements heavier than iron. The *s*-process part is shown in *green* and the *r*-process part in *red*. Other processes are shown in *blue* and unstable elements are shown in *yellow*

One of the most abundant products of helium synthesis through the CNO cycle is ^{13}N, which decays into ^{13}C. Towards the end of hydrogen burning, the matter would contain α particles (the original and the fresh products) which now constitute the majority of the nuclei. Greenstein,[27] who was an astronomer and not a nuclear physicist, and independently Cameron,[28] realized that the exothermic reaction

$$^{13}\text{C} + \alpha \longrightarrow {}^{16}\text{O} + \text{n} + 2.2\,\text{MeV}$$

should be a good stellar neutron source. Indeed, this reaction is the most important and famous neutron source for synthesizing the nuclei in the *s*-process. As this reaction will appear time and time again, let us call it the 13C reaction for short.

B2FH added the following exothermic reactions:

$$^{17}\text{O} + \alpha \longrightarrow {}^{20}\text{Ne} + \text{n} + 0.60\,\text{MeV}, \quad {}^{21}\text{Ne} + \alpha \longrightarrow {}^{24}\text{Mg} + \text{n} + 2.58\,\text{MeV},$$

$$^{25}\text{Mg} + \alpha \longrightarrow {}^{28}\text{Si} + \text{n} + 2.67\,\text{MeV}, \quad {}^{26}\text{Mg} + \alpha \longrightarrow {}^{29}\text{Si} + \text{n} + 0.04\,\text{MeV}.$$

The next reactions in this series are endothermic, that is, they require energy to proceed:

$$^{29}\text{Si} + \alpha + 1.53\,\text{MeV} \longrightarrow {}^{32}\text{S} + \text{n}, \quad {}^{33}\text{S} + \alpha + 2.0\,\text{MeV} \longrightarrow {}^{36}\text{A} + \text{n}.$$

27 Greenstein, J.L., *Modern Physics for Engineers*, ed. Ridenour, McGraw-Hill Book Company, New York, p. 267 (1954). This was really a strange place to suggest such an idea. Greenstein wrote the astrophysics part of the book under the heading *Man's Physical Environment*.

28 Cameron, A.G.W., Phys. Rev. **93**, 932 (1954); ibid. Astrophys. J. **121**, 144 (1955).

All the isotopes involved in the possible production of neutrons are rare species. For example, the most abundant isotope of oxygen is ^{16}O (0.9976 in relative isotopic abundance) and not ^{17}O, which has a relative abundance of only 0.00038. Another example is Ne. The relative abundance of the three stable Ne isotopes is ^{20}Ne: 0.9048, ^{21}Ne: 0.0027, and ^{22}Ne: 0.0925. It is the same story again: the most tightly bound and abundant elements are the α nuclei. The loosely bound nuclei of type $4m + 1$ are significantly less abundant. The relatively rare isotopes are used to produce the even rarer HTI elements.

The relevant source of neutrons at each stage of stellar evolution depends on the composition at that moment. For example, while pristine[29] carbon, which has not participated in the CN cycle, has a 90:1 ratio of ^{12}C to ^{13}C, carbon which has participated in the CN cycle has a 4.6/1 ratio by number in equilibrium. Consequently, less than a quarter of the carbon is available for absorption of α particles. Since the cosmic abundance ratio of carbon to iron is $[^{12}C]/[^{56}Fe] = 6.4$, this means that only $6.4/5.6 \approx 1$ neutron will become available per iron nucleus, and this is clearly too few neutrons for the expected job of synthesizing the HTI elements. This is the first problem.

While the scarcity of neutrons may not be sufficient to worry the theoreticians, another problem is the existence of large amounts of ^{14}N which are, for our purposes here, a poison for neutrons via the reaction[30]

$$^{14}N + n \longrightarrow {}^{14}C + p \longrightarrow {}^{14}N + \bar{\beta}\,.$$

So how do we square this circle? FB2 (as early as 1955[31]) and later B2FH (in 1957) reiterated the supposition, ingeniously hypothesizing that, during the critical phase in which neutrons are released, the product ^{12}C of the 3 α reaction, which should now be abundant, mixes with the envelope and that the mixing brings fresh hydrogen-rich material (from the envelope) into the burning zone (in the core).[32] As a consequence, the hydrogen interacts with the ^{12}C and converts it to ^{13}N. This subsequently decays

[29] Measurements of the $^{12}C/^{13}C$ ratio in different stars yield different values, which are interpreted as indicators of a sporadic and unexplained mixing on top of complete mixing by convection. In a similar but not necessarily identical way, extra mixing will be called upon to rescue the *s*-process. This unexplained but apparently necessary mixing is one of the unsolved problems in stellar evolution. See Charbonnel, C., Brown, J.A. and Wallerstein, G., Astron. Astrophys. **332**, 204 (1998).

[30] When a neutron created by cosmic rays in the Earth's atmosphere collides with the abundant ^{14}N in the atmosphere, it generates exactly the same reaction. The ^{14}C so formed is radioactive with half-life 5,730 years. This radiocarbon is the basis for the carbon dating method discovered by Libby in 1949. Segrè proclaims in his autobiography (Segrè, E., *A Mind Always in Motion: The Autobiography of Emilio Segrè*, Univ. California Press) that it was Enrico Fermi who suggested the (Footnote 30 continued)
idea to Libby in a Chicago University seminar in 1949. In any case, in 1960, Libby was awarded the Nobel Prize in Chemistry for the development of the technique.

[31] Fowler, W.A., Burbidge, G.R. and Burbidge, E.M., Astrophys. J. **122**, 271 (1955).

[32] Here is the exact formulation by FB2: *The essential point made here is that neutron production follows any building of heavier nuclei by helium as long as hydrogen can be introduced into the core by some mixing process or by the growth of the core before the hydrogen is completely consumed.*

into ^{13}C, which is then available for an additional absorption of an α and the release of neutrons. Simple mixing is out of the question as it completely ignores the fact that, outside the region in which He has burnt into carbon, there is a region in which hydrogen burns into helium, and if the outside envelope mixed with the helium burnt material, it would wash away the entire structure of the red giant.

B2FH therefore hypothesized that sporadic mixing between the burning zone and the unburnt envelope leads to a continuous supply, not too much, but just in the right measure, of raw materials needed for the production of neutrons. Moreover, at the same time, ^{12}C is dredged-up from the burning zone into the envelope by the same hypothesized mixing, and creates a star with a surface rich in carbon. Thus, the known 'carbon stars'[33] should according to B2FH be the location where neutrons are released to build the HTI nuclei. B2FH did not do the calculations for a stellar model in which the mixing mechanism they conjectured was at work, and nor did they propose any specific sporadic mixing mechanism. It was just a scenario. Evidence that 'some mixing' does in fact take place between the internal furnace and the envelope was provided by Merrill's discovery in 1952 (see Sect. 6.40). This hypothesis and the unique combination of properties (sporadic and not complete mixing, etc.) still awaits substantiation. Finally, recall that as early as 1954, Cameron, while suggesting the ^{13}C reaction, already realized that there might be a problem with the supply of neutrons. Cameron therefore stated that the ^{13}C reaction was *particularly important in those stars with appreciable internal circulation.* The circulation, timescales, mechanism, and so on, were not specified, however.

The worry about ^{14}N caused mixed feelings due to the following sequence of reactions:

$$^{14}\mathrm{N} + \alpha \longrightarrow {}^{18}\mathrm{F}, \quad {}^{18}\mathrm{F} \longrightarrow {}^{18}\mathrm{O} + \mathrm{e}^+ + \nu, \quad {}^{18}\mathrm{O} + \alpha \longrightarrow {}^{22}\mathrm{Ne},$$

$$^{22}\mathrm{Ne} + \alpha \longrightarrow {}^{25}\mathrm{Mg} + \mathrm{n},$$

which presents another source of neutrons (to be discussed shortly).

FB2 were apparently not so self-assured about the hypothesis and looked for additional neutron sources. They suggested the reaction $^{21}\mathrm{Ne} + \alpha \rightarrow {}^{24}\mathrm{Mg} + \mathrm{n}$ as an alternative source of neutrons which could circumvent the difficulty of the hypothesized mixing between the envelope and the interior. The basic idea was that the ^{20}Ne, which was previously produced during helium burning at an earlier stage of galactic evolution, is converted into ^{21}Ne in the hydrogen-burning shell, at a temperature of 3–5 $\times$ 10^7 K. Since the initial relative abundance is estimated to be $[^{21}\mathrm{Ne}]/[^{56}\mathrm{Fe}] = 14$, the process can start, although the abundance of ^{21}Ne is not sufficient to form nuclei like lead (Pb). Moreover, this suggestion requires the production of ^{21}Ne to be faster than its destruction by, for example, $^{21}\mathrm{Ne} + \mathrm{p} \rightarrow {}^{22}\mathrm{Na} + \gamma$, and it also requires the neutron poison ^{14}N to be removed from the system

---

[33] Any of a class of stars with high carbon-to-hydrogen ratios and primarily low surface temperatures is defined as a carbon star. These stars will be discussed later.

by absorption of an α particle before the neutrons are formed. Marion and Fowler[34] examined the above questions and concluded that the $^{21}Ne + \alpha \rightarrow {}^{24}Mg$ reaction could be *considered as a possible adequate source of neutrons for HTI synthesis in red giant stars*. This statement was later found to be untenable.

B2FH could not decide which was the relevant process because (a) stellar evolution calculations had not yet reached this phase of the evolution at the time the paper was written, and (b) the rates of the competing nuclear reactions had not yet been measured. But it was obvious that the default suggestions could not produce all the HTI elements due to the shortage of neutrons. As the source of neutrons operates before any iron is synthesized by the star, it is clear that iron must exist in the star at its formation. This expectation is not born out by present day observations of extremely old stars. Does it imply that the 'first stars' have not yet been discovered? Or do not exist any longer? The idea of a population III that preceded the population II stars had not yet been born at the time B2FH wrote their paper.

12.8 The Nuclear Physics of the ^{13}C Neutron Source

The reaction $^{13}C + \alpha \rightarrow \left(^{17}O\right)^{*} \rightarrow {}^{16}O + n$, which Greenstein and Cameron suggested as a source of neutrons in stars, was discovered by Jones and Wilkinson in 1951–1953.[35] The interest in the neutron-emitting reaction was considerable since the ^{13}C nucleus was considered to be a core composed of three α particles plus an outside neutron. If this had been so, it should have been easy to knock the neutron out. Jones and Wilkinson demonstrated the feasibility of this idea and 'apologized' that, due to improper equipment, they were unable to estimate the probability of the reaction.

Soon after, in 1952, the neutron reactions appeared in a compilation of nuclear reactions by Ajzenberg and Lauritsen[36] (see Fig. 12.19), but no details were given and the level diagram only tells us that the reaction is energetically possible, while saying nothing about the probability of its occurrence. Ajzenberg and Lauritsen based their compilation on other reactions which led to the same compound nucleus, namely ^{17}O.

When Cameron[37] attempted to calculate the reaction in 1954, he based his data on the sparse information given by Ajzenberg and Lauritsen 1952. Cameron, probably aware of the paper by Jones and Wilkinson since it is cited in Ajzenberg and Lauritsen, although not cited by him, argued that:

> Precision laboratory measurements of the cross-section of the $^{13}C+\alpha \rightarrow {}^{16}O+n$ [...] have not yet been reported.

---

[34] Marion, J.B. and Fowler, W.A., Astrophys. J. **125**, 221 (1957).

[35] Jones, G.A. and Wilkinson, D.H., Proc. Phys. Soc. (London) **64A**, 756 (1951); ibid. **66A**, 1176 (1953).

[36] Ajzenberg, F. and Lauritsen, T., Rev. Mod. Phys. **24**, 321 (1952).

[37] Cameron, A.G.W., Phys. Rev. **93**, 932 (1954); ibid. Astrophys. J. **121**, 144 (1955).

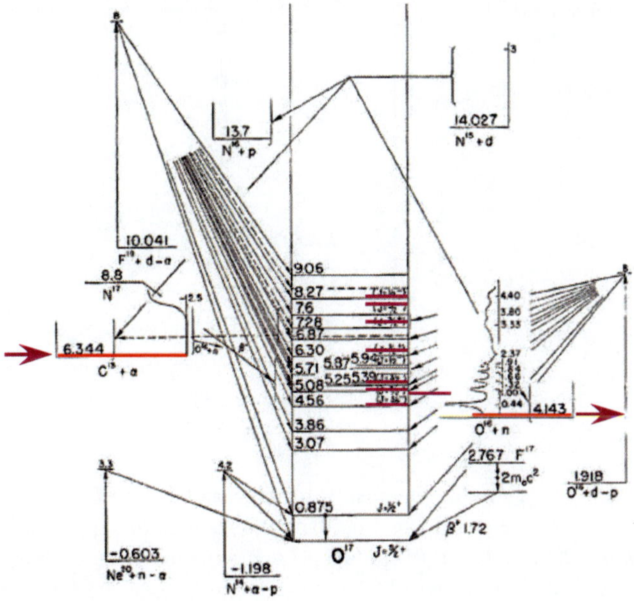

Fig. 12.19 The energetics of the reaction $^{13}C + \alpha \rightarrow (^{17}O)^* \rightarrow {}^{16}O + n$ as shown in Ajzenberg and Lauritsen (1952), based on Jones and Wilkinson's (1951) experiment and many others, all leading to the same compound nucleus ^{17}O. We have added the energy levels used by Cameron in his theoretical calculation of the rate (shown in *pink* and placed *on the right*). The level Cameron used which does not appear in Ajzenberg and Lauritsen is plotted *to the right* of the energy level diagrams. The location of the 13C reaction and its products are marked in *red*. We note that several of the levels used by Cameron cannot be identified with levels in Ajzenberg and Lauritsen, and vice versa. In particular, the level just below the energy of the $^{13}C + \alpha$ does appear in both cases while the level just above, at 6.87 MeV, appears in Ajzenberg and Lauritsen but not on Cameron's list

The structure of the energy levels of ^{17}O was of interest to nuclear physicists in 1957, and several papers were published.[38] The structure of the levels is crucial. But the above-mentioned papers discussed high energy levels which have no effect on the stellar case. A typical result is shown in Fig. 12.20. The resonances are clearly seen and it is obvious that the measurement is at too high an energy. The possible importance of the reaction as an astrophysical neutron source was apparently unknown to the nuclear physics community.

In 1967, Sekharan et al.[39] investigated the probability of the reaction. By then, the B2FH paper was at least 10 years old and the idea of the *s*-process was well known in the nuclear astrophysics community. However, the concerns of Sekharan et al. were not astrophysical, but stemmed from the nuclear physics problem regarding the

[38] Walton, R.B., Clement, J.D. and Boreli, F., Phys. Rev., **107**, 1065 (1957); Rusbridge, M.G., Proc. Phys. Soc. A **69**, 830 (1956); Schiffer, J.P., Kraus Jr., A.A. and Risser, J.R., Phys. Rev. **105**, 1811 (1957).

[39] Sekharan, K.K., and 4 coauthors, Phys. Rev. **156**, 1187 (1967).

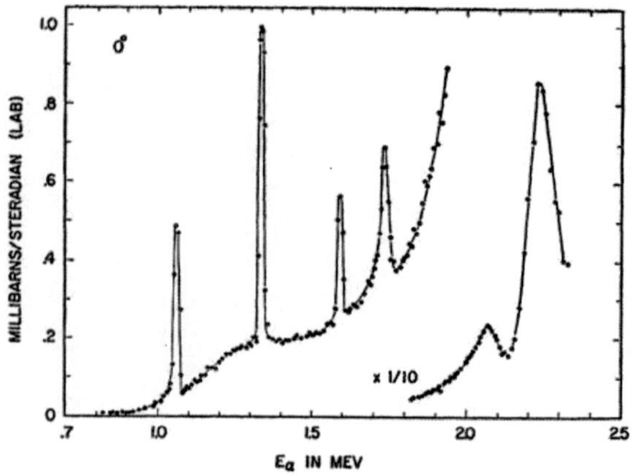

Fig. 12.20 The neutron yield of the ^{13}C reaction at an angle of 0°. The relevant astrophysical range is beyond the edges of the figure

structure of the ^{17}O nucleus. It is interesting to note that it was the first measurement of the probability for the reaction. Previous experiments with this reaction[40] measured only the angular distribution of the products but not the total rate. It should be noted that the inverse reaction, namely $^{16}O + n \rightarrow {}^{13}C + \alpha$ became important for the structure of nuclear reactors, and hence was measured by several authors.[41] No mention was made of its importance for the *s*-process and the physical fact that one can learn about the forward reaction from the reverse reaction. In those days, nuclear physics and astrophysics moved along separate paths.

In 1968, the reaction was considered to be the best source of neutrons in the laboratory,[42] and one would have expected plenty of information to be available. However, the astrophysically relevant information remained meagre. The first experimental attempt to get the astrophysical value for the probability of the reaction was carried out by Davids in 1968,[43] and the results are shown in Fig. 12.21. Two points are immediately clear:

- The fast decline of the neutron yield as the energy decreases. By the time the relative energy between the colliding particles is as low as 0.46 MeV, an energy significantly higher than the expected energy for the reaction to take place in stars, the reaction becomes so slow as to prevent any measurement. Because of the low rate of the reaction, Davids was forced to apply non-standard methods, such as a thick target instead of a thin one.
- Davids did not observe any resonances in the range reported in the figure. Finally, Davids thought that the 13C neutron source operated during the CN cycle, and

[40] Becker, R.L. and Barschall, H.H., Phys. Rev. **102**, 1384 (1956); Walton, R.B., Clement, J.D., and Boreli, F., Phys. Rev. **107**, 1065 (1957); Bonner, T.W., and 3 coauthors, Phys. Rev. **102**, 1348 (1956).

[41] Seitz, J. and Huber, P., Helv. Phys. Acta **28**, 227 (1955); Davis, E.A. and 3 coauthors, Nucl. Phys. **48**, 169 (1963); Lister, D. and Sayres, A. Phys. Rev. **143**, 745 (1966).

[42] Lehman, R.L., Int. J. Appl. Radiat. Isot. **19**, 841 (1968).

[43] Davids, C.N., Nucl. Phys. A **110**, 619 (1968); ibid. Astrophys. J. **151**, 775 (1968).

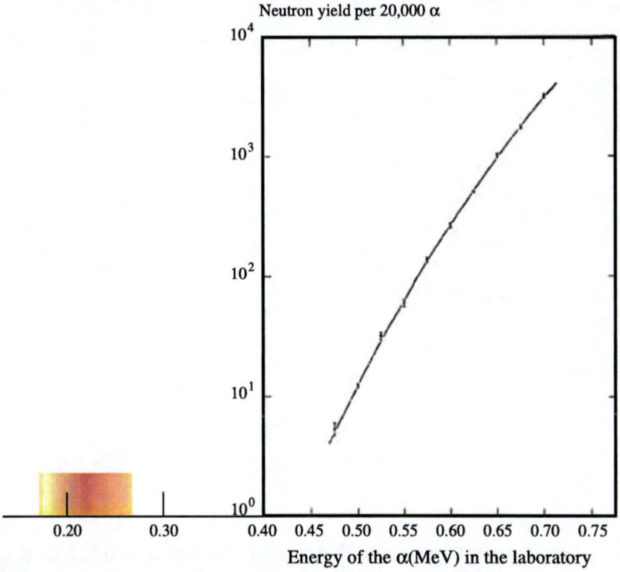

Fig. 12.21 The neutron yield from the reaction $^{13}C + \alpha \rightarrow {}^{16}O + n$ as a function of the relative energy. We have added in *yellow*, the stellar range of energies over which the reaction is expected to operate in stars (about 200 keV). Note that no resonances were observed. After Davids (1968)

dismissed it as a viable stellar source because of the large amounts of ^{14}N poison that would be present. This was not what B2FH had in mind.

The total yield of neutrons from the reaction was measured again only in 1973 by Bair and Haas,[44] but for energies well above those needed for the stellar case (see Fig. 12.22). Bair and Haas saw many resonances, none of which were observed by Davids. So in an attempt to compare their result with his, they averaged the resonances over energy, and to their dismay, their result was just 5% higher at 700 keV and 33% higher at 475 keV. In view of the fact that Davids saw no resonances while they did, they expressed their fluster by stating:

> Such excellent agreement must be to some large extent fortuitous.

Before Kubono et al.[45] carried out their experiment in 2003, the neutron yield rate was measured only down to an energy of 270 keV.[46] Kubono et al. succeeded in measuring[47] the energy levels of ^{17}O in the energy range relevant to the *s*-process.

---

44 Bair, K.J. and Haas, F.X., Phys. Rev. C **7**, 1357 (1973).

45 Kubono, S., and 15 coauthors, PRF **90**, 062501-1 (2003).

46 Drotleff, H.W., et al., Astrophys. J. **414**, 735 (1993) and the references therein for the previous experiments.

47 The experiment was carried out via the reaction $^{13}C + {}^{6}Li \rightarrow {}^{17}O + d$, and not directly, then calculated theoretically.

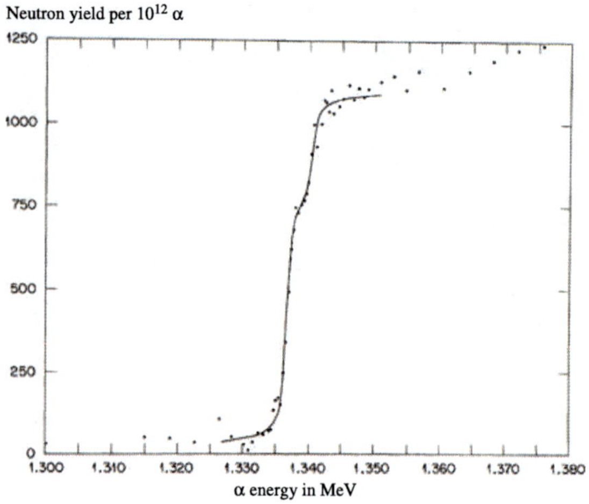

Fig. 12.22 As for Fig.12.21. The experiment shows the existence of a resonance at higher energies than the Gamow peak. A low yield at the Gamow peak energies precludes any measurements at these energies. It was found by other means that ^{17}O has an energy level just below the rest mass energy of ^{13}C + α, and its effect on the rate can only be calculated. The Gamow peak is off range to the left. After Bair and Haas (1973)

Taking all possible errors into account, they reached the conclusion that the maximum value of the rate, as expressed traditionally[48] by the function $S(E)$, is 0.011 (in units of 10^6 MeV barn).

However, a theoretical reanalysis of the experimental data of Kubono et al. by Keeley et al.[49] yielded $S(E = 0) = 0.4$, which is a factor of 40 greater. In other words, just a theoretical analysis using the same data, but improved nuclear theory, changed the result by a factor of forty!

The exact value is important and no wonder the experimentalists tried hard to narrow down the uncertainty. A small value becomes a problem of sufficient neutron supply. A large number implies that more s-process elements are formed by more stars, and consequently should be more abundant than actually observed in the Galaxy.

In 2006, Johnson et al.[50] claimed that, since the cross-section depends on the 6.356 MeV level, which is below the threshold, so cannot be reached directly and is consequently poorly known, the cross-section is uncertain by a factor of ten. Johnson et al. remeasured the reaction $^{13}\text{C} + ^{6}\text{Li} \rightarrow ^{17}\text{O} + \text{d}$ and recalculated the rate. The value they found was a factor of 3 below what appeared in the paper entitled

[48] The rate of nuclear reactions between charged colliding particles with relative kinetic energy E is expressed as $\sigma(E) = S(E)\exp\left[-2\pi(E_G/E)^{1/2}\right]/E$, where E_G is the Gamow energy, which for two colliding species depends only on the atomic numbers (charges) and weights. The function (or constant) S contains all the nuclear physics lumped into a single number or function. S is traditionally expressed as a constant plus a small correction. Hence, the value of the constant (Footnote 48 continued)
$S(E = 0)$ is the most important factor determining the rate of the reaction. All the astrophysical nuclear physics of a given nuclear reaction boils down to the value of $S(E = 0)$, which is known as the astrophysical $S(E)$ factor.

[49] Keeley, N., et al., Nucl. Phys. A **726**, 159 (2003).

[50] Johnson, E.D., and 17 coauthors, PRL **97**, 192701 (2006).

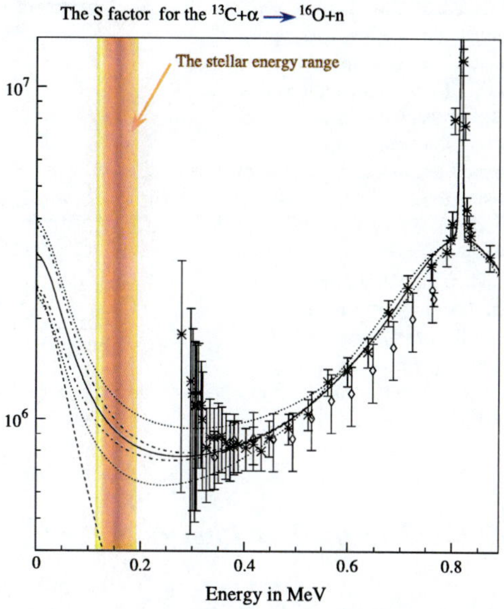

Fig. 12.23 The rate of the reaction $^{13}C + \alpha \rightarrow {}^{16}O + n$ as measured by several experiments. *Lines* show the fit using various approximations for the nuclear levels, and demonstrate the uncertainty in the rate under stellar conditions as well as the uncertainty in nuclear theory. The narrow resonance seen above 0.8 MeV hardly affects the rate. After Johnson et al. (2006)

Compilation of charged particle induced thermonuclear reaction rates by Angulo et al.[51] which was 0.23 (in the above units). In 2008, Hammache[52] remeasured the value (via the reaction $^{13}C + {}^{7}Li \rightarrow {}^{17}O + t$) and recalculated, this time obtaining $S(E = 0) = 0.29$.

It appears from theoretical calculations that a value of $S(E) \sim 0.3$ is sufficient to ensure the neutron supply if mixing exists, while a smaller number would pose a severe problem. We stress that the 'right' number was obtained from another reaction, which creates the same intermediate nucleus ^{17}O, not by a direct measurement, and with the help of a different nuclear theory. In summary, it is plausible to say that the S factor for the main neutron generating reaction is certain to within a factor of about five.

The present situation for this reaction is summarized in Fig. 12.23, where the S factor is shown with the measured rates and the energy range at which the reaction takes place in stars. Because of the energy level at 6.356 MeV (not shown in the figure), the theoretical rate is expected to grow for energies below ~0.2 MeV. It is not implausible that the rate actually increases and the inaccuracy is even around a factor of ten.

About six decades after the *s*-process was invented/discovered, one of the most important parameters was at long last found experimentally, with reasonable but insufficient accuracy.

[51] Angulo, C., and 27 coauthors, Nucl. Phys. A **656**, 3 (1999).

[52] Hammache, F., and 11 coauthors, 10th Symposium on Nuclei in the Cosmos 27 July–1 August 2008.

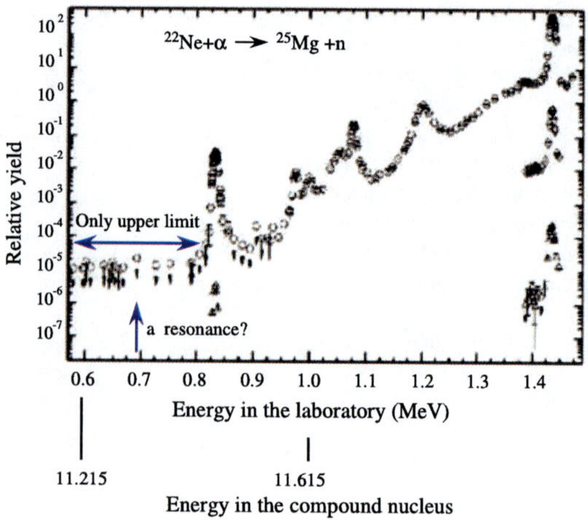

Fig. 12.24 Yield of the reaction $^{22}Ne + \alpha \to {}^{25}Mg + n$. There are only upper limits in the relevant region for astrophysics. The *arrow* marks a possible resonance (unobserved because of a low yield) due to the 11.319 MeV level in ^{26}Mg. The *upper energy scale* gives the energy of the α particle in the laboratory, while the *lower energy scale* gives the energy in the system of the compound nucleus shown in Fig. 12.25. After Jaeger et al. (2001)

12.8.1 The Nuclear Physics of the ^{22}Ne Neutron Source

In 1960, Cameron[53] found that $^{22}Ne + \alpha$ can go to $^{25}Mg + n$, a reaction which we will denote by 22Ne, and also to $^{25}Mg + \gamma$, in equal probabilities, and estimated that up to 40 neutrons could be released for each silicon nucleus, or about one neutron per 2×10^4 iron nuclei. Cameron realized that ^{22}Ne is destroyed only at temperatures above 2×10^8 K and hence, if ^{22}Ne is available at all and if the temperature is below 2×10^8 K, the reaction 22Ne can produce the much needed neutrons.[54] Moreover, before this temperature is reached, the poisons ^{14}N and ^{18}O (operating via the reaction $^{18}O+\alpha \to {}^{22}Ne+\gamma$) are completely consumed by conversion into ^{22}Ne (Fig. 12.24).

In 1966, Reeves[55] analyzed all the possible α-capture reactions as given by B2FH, singling out the reaction 22Ne as a neutron source.[56] Reeves' estimate was that the reaction could produce a neutron density of $10^5/cm^3$ and he stated that:

> Larger stars will transform their ^{14}N abundance into ^{22}Ne and burn it by the ^{22}Ne reaction at the end of the helium burning stage ($T > 2 \times 10^8$) if enough ^{4}He is left then. If the presence of Mg as a neutron poison is neglected, we find that about twenty to thirty neutrons will be captured by the metal group. Here one would expect enhancement of heavier elements accompanied by both carbon and oxygen enhancement.

---

[53] Cameron, A.G.W., Astron. J. **65**, 485 (1960).

[54] Cameron published the idea in a conference and the paper does not contain any additional information. For example, he does not say why the temperature should be below 2×10^8 K to have ^{22}Ne. The reaction $^{22}Ne + \alpha \to {}^{25}Mg + n$ is endothermic with $Q = -0.482$ MeV. Hence the temperature must be at least 2×10^8 K for the reaction to take place. But at these energies the reaction $^{22}Ne + \alpha \to {}^{26}Mg + \gamma$ (with $Q = 10.612$ MeV) also takes place.

[55] Reeves, H., in *Stellar Evolution*, ed. by Stein and Cameron, Plenum Press, New York, 1966, p. 83; ibid. Astrophys. J. **146**, 447 (1966).

[56] Reeves did not mention that Cameron had suggested this reaction as a possible neutron source. On the other hand, Reeves credited Cameron with the idea of the need for mixing in the 13C reaction (Cameron, A.G.W., Chalk River Rep. C.R.L. 41, 1958) which he had stated a year before Fowler,

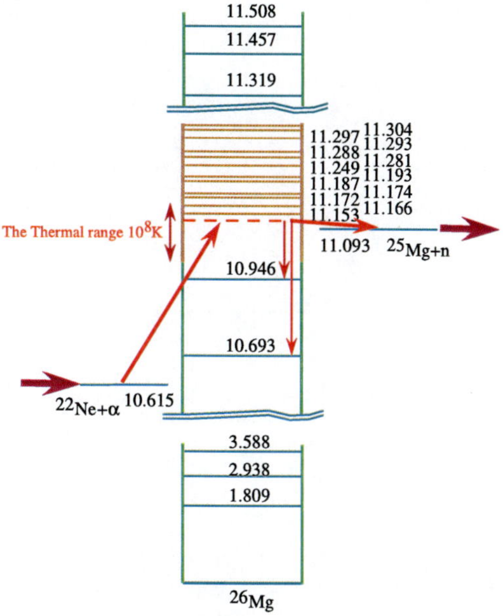

Fig. 12.25 The reaction $^{22}Ne + \alpha \rightarrow {}^{25}Mg + n$ and the structure of the intermediate compound nucleus ^{26}Mg. The relevant levels for the reaction are shown in *orange*. The level at 11.319 MeV might have been observed in the direct experiment

Thus, Reeves established that the 13C reaction should operate in low mass stars at relatively low temperatures, while the 22Ne reaction should operate in massive stars at higher temperatures. But without detailed stellar models, Reeves was unable to establish that his predicted number density of neutrons was much too low. The dividing line between low and high masses was not specified.

The philosophy of this reaction is a bit different from that of the 13C reaction. The ^{22}Ne has two neutrons outside the core of 5α particles, and the incoming α particle is expected to eject one of them and create a new core with 6α particles and one neutron. For this reason it is possible to knock a neutron out from ^{21}Ne, ^{22}Ne, or even ^{23}Ne, while it is very difficult to knock a proton out from these nuclei. Our knowledge of these reactions is still problematic (Koehler 2002[57]), and so therefore are any predictions sensitive to the number of neutrons. As we are aware today, the rate is very sensitive to the structure of the levels in the compound nucleus ^{26}Mg (see Fig. 12.25).

(Footnote 56 continued)
Burbidge, and Burbidge pointed out, without crediting Cameron, that the 13C neutron source would be insufficient if there were no additional sporadic mixing. In 1957, in a short conference paper (AJ **62**, 138, 1957), Cameron discussed the 'admixture of hydrogen'. In his 1954 Astrophys. J. paper, Cameron wrote:

> This is particularly important in those stars with appreciable internal circulation. The currents can carry ^{12}C to the core boundary and into the hydrogen-burning shell.

According to Cameron, the process would only work in the 'proper stars'.

[57] Koehler, P., Phys. Rev. C **66**, 055805 (2002).

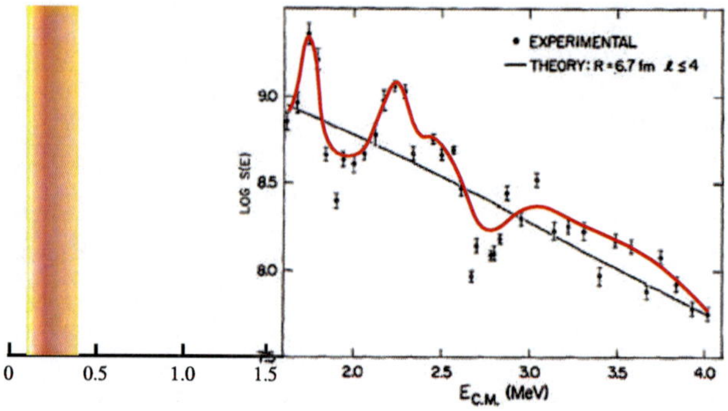

Fig. 12.26 The experimental results and theoretical prediction for the $S(E)$ function for the 22Ne reaction as measured by Ashery in 1969. The *red line* has been added to guide the eye. Note that this experiment went down to an energy of 2 MeV, which is still way above the thermal energy shown in *yellow*. After Ashery (1969)

In 1969, Ashery[58] studied the 22Ne reaction and his results for the $S(E)$ function are given in Fig. 12.26. Ashery was able to measure the rate of the reaction down to an α particle energy of 1.6 MeV, which is still much higher than the energy of the α particles at a temperature of 10^8 K. The red line which connects the data points has been added to the figure to guide the eye and demonstrate the uncertainty compounded in such an extrapolation. Actually, Ashery established that the reaction proceeds mainly through resonances in the compound system formed by the target and the projectile, so that knowledge of the energy levels in the compound nucleus is mandatory. As Ashery stressed, the separation between resonances in this energy range is about 300 keV or less, just the energy range where the reaction takes place in stars. However, he concluded that the relatively small separation between the resonances implies that *a meaningful extrapolation of the S(E) function to lower energies can be made*. But we contend that, just because the separation is of the order of the thermal energy range, the interpolation is not reliable, since a resonance may fall just 'in the wrong place'.

In 1973, Haas and Bair[59] measured the yield of the 22Ne reaction. Being careful with the resonances and not knowing how many of them were really discovered, they plotted the number of levels observed with spacing greater than some spacing D. Their results are shown in Fig. 12.27. As one can see, the observed spacing falls short of the extrapolated values by a factor of about two. If the extrapolation is indeed valid and meaningful, and this is a big 'if', it implies that Haas and Bair managed to observe only about half of the levels (out of 81 expected levels, they saw only 39, and the true average spacing was 52 keV).

Haas and Bair found that the rate of $^{21}\mathrm{Ne} + \alpha$ is just 20% higher than the rate of the competing reaction $^{22}\mathrm{Ne} + \alpha$. While Ashery got $S = 0.037 \pm 0.020$ for the

[58] Ashery, D., Nucl. Phys. A **136**, 481 (1969).

[59] Haas, F.X. and Bair, J.K., Phys. Rev. C **7**, 2432 (1973).

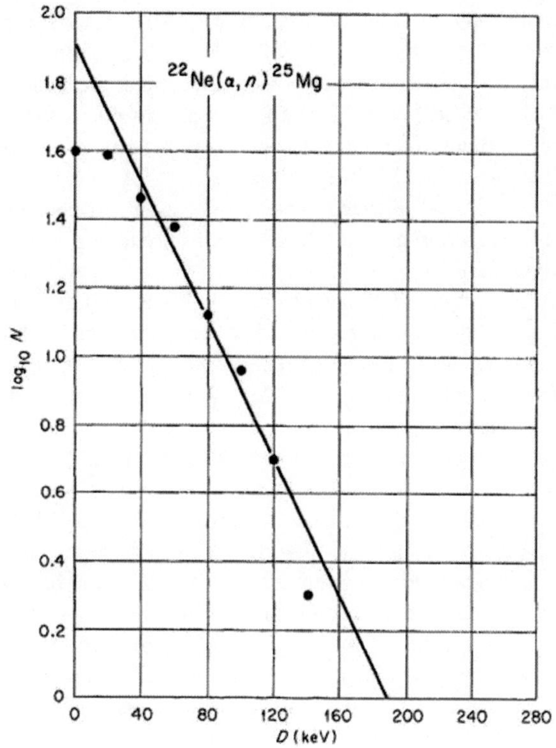

Fig. 12.27 Number of energy levels in the reaction $^{22}Ne + \alpha \to (^{26}Mg)^* \to {}^{25}Mg + \alpha$ with spacing greater than some spacing D (in the laboratory system). The *straight line* is an extrapolation from greater spacing where one expects reliable data. After Haas and Bair (1973)

$^{22}Ne + \alpha$ reaction, Haas and Bair found $S = 0.026 \pm 0.012$. Haas and Bair noted that, where Tanner[60] saw two levels, they did not see any level at all, and on the other hand, where Tanner did not see any level, they did see a level (at 2.27 MeV). The source of the discrepancy was not found. It just goes to show how tricky these experiments are, at the very limit of the possible. Finally, one should note that the results of Hass and Bair involved an error of about 1/3 of the result. Nevertheless, their results were about 40% lower than those of Ashery.

A year later, Mak, Ashery, and Barnes[61] compared the rates of the reactions $^{21}Ne + \alpha \to {}^{24}Mg + n$ and $^{22}Ne + \alpha \to {}^{25}Mg + n$. This was the old suggestion by Fowler and his associates. The results obtained were, that for the same concentrations of ^{21}Ne and ^{22}Ne, the reaction of ^{21}Ne is about 20% faster than the reaction of ^{22}Ne. The problem was thereby reduced to identifying which isotope was more abundant, and in this way they revived the old idea and essentially confirmed the result by Haas and Bair.

---

60 Tanner, N.W., Nucl. Phys. **61**, 297 (1965).

61 Mak, H.B., Ashery, D. and Barnes, C.A., Nucl. Phys. A **226**, 493 (1974).

In 1989, Wolke et al.[62] discovered 15 new resonances and in particular a resonance for the (α, γ) reaction at 0.828 MeV, while Drotleff et al.[63] discovered a resonance in the (α, n) reaction at 0.838 MeV. But was it the same level with two decay options, or two levels? Also, Drotleff et al. proposed a resonance at 0.626 MeV which was not verified. Such a level would significantly affect the results! Wolke et al. found that the rates of 22Ne and $^{22}Ne + \alpha \rightarrow {}^{26}Mg + \gamma$ were at least a factor of 60 lower at $T = 6.5 \times 10^8$ K than the previously adopted value, whence they concluded that the 22Ne reaction played a lesser role in producing neutrons than had previously been thought.

The situation as it stood in 1990 was summarized by Endt[64] and is shown in Fig. 12.28. Endt merely remarked that there were 150 levels in the important range for our reaction. As can be seen, the $^{22}Ne + \alpha$ reaction reaches an energy of 10.612 MeV in the compound nucleus ^{26}Mg, so the structure of the levels is needed at a rather high energy.

As a consequence, the specific contribution of each level had to be added on top of the smooth $S(E)$ function. They concluded that two resonances, one at $E_\alpha = 0.097$ MeV (which corresponds to $E = 10.69$ MeV in the compound nucleus) and the other at $E_\alpha = 0.400$ MeV (which corresponds to $E = 10.95$ MeV in the compound nucleus), were important and all the rest would not affect the reaction. The levels which were discovered via photoneutron and neutron capture were not observed in the $^{22}Ne + {}^{6}Li \rightarrow {}^{26}Mg + d$ reaction, indicating that these levels have a very low probability of being fed via the α-capture reaction.

Figure 12.29 (left) shows the ratio of the $^{22}Ne + \alpha \rightarrow {}^{25}Mg + n$ reaction rate to the $^{22}Ne + \alpha \rightarrow {}^{26}Mg + \gamma$ reaction rate as a function of temperature in units of 10^8 K. The solid line shows the result by Giesen et al. and the short-dashed line is the result of Caughlan and Fowler,[65] which was for many years the standard data in nuclear astrophysics. We see that the $^{22}Ne + \alpha \rightarrow {}^{26}Mg + \gamma$ reaction dominates at the lowest temperatures, killing the 22Ne for such low temperatures. However, the purpose of the figure is to demonstrate the effect of the two additional resonances found by Giesen et al. on the ratio of the two reactions. There is also an effect on the absolute value. It was only after the work of Giesen et al. that the importance of the 22Ne reaction as a neutron source at high temperatures was finally established.

---

[62] Wolke, K., and 9 coauthors, Z. Phys. A **334**, 491 (1989).

[63] Drotleff, H.W., and 7 coauthors, Z. Phys. A **338**, 367 (1991). Two of the collaborators on this paper also worked together on the previous one.

[64] Endt, P.M., Nucl. Phys. A **521**, 1 (1990).

[65] Caughlan, G.R. and Fowler, W.A., At. Data Nucl. Data Tables **40**, 291 (1988). Mainly in collaboration with Caughlan and Zimmerman, but sometimes including others, Fowler used to publish periodical compilations of relevant astrophysical nuclear data. These data soon became the 'bible' for nuclear astrophysics and stellar evolution. With the death of Fowler, different groups adopted the important role of compiling consistent nuclear data for astrophysics. See, for example, the US Nuclear Reaction Data Network (USNRDN), or Nuclear Data for Nuclear Astrophysics, Oak Ridge National Laboratory, or NACRE, the European Compilation of Reaction Rates for Astrophysics.

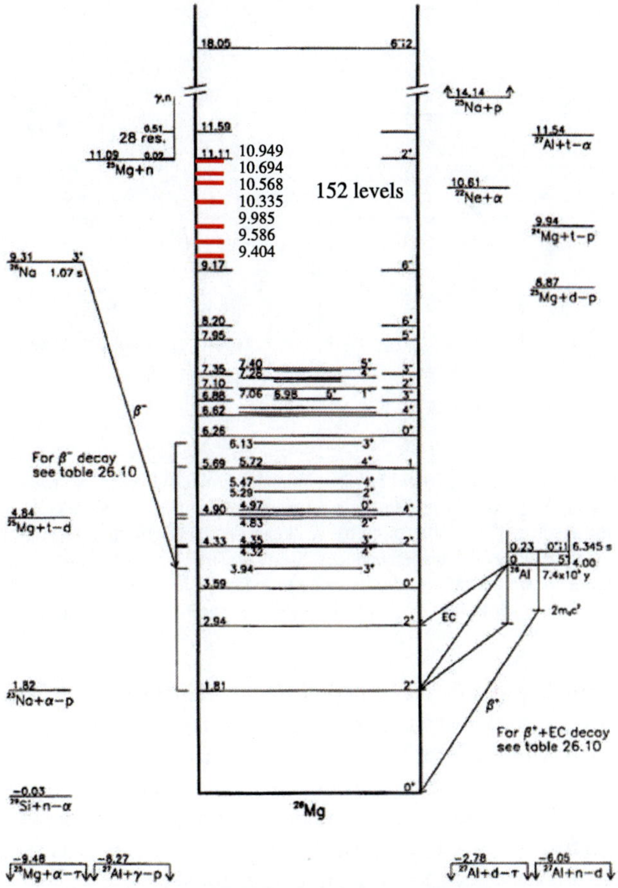

Fig. 26.3. Energy levels of ^{26}Mg; for more detail, see table 26.4.

Fig. 12.28 Knowledge in 1990 of the compound nucleus ^{26}Mg, as summarized by Endt (1990). In the important range for the ^{22}Ne reaction, Endt noted that there are 150 levels. The specific levels which were later found to be important are marked in *red* and the energies are given on the *far left*. Note the direction of the *pink arrows* which mark the entrance and exit channels of the reaction. The exit channel is higher than the entrance channel. Hence the energy threshold for the reaction is 0.48 MeV, or a temperature of 3.9×10^8 K

The 22Ne reaction continued to provide surprises when in 1991 Harms, Kratz, and Wiescher[66] discovered several more resonances, despite the quite detailed work described above. The effect on the rate is shown in Fig. 12.29 (right) in comparison

[66] Harms, V., Kratz, K.L., and Wiescher, M. Phys. Rev. C **43**, 2849 (1991).

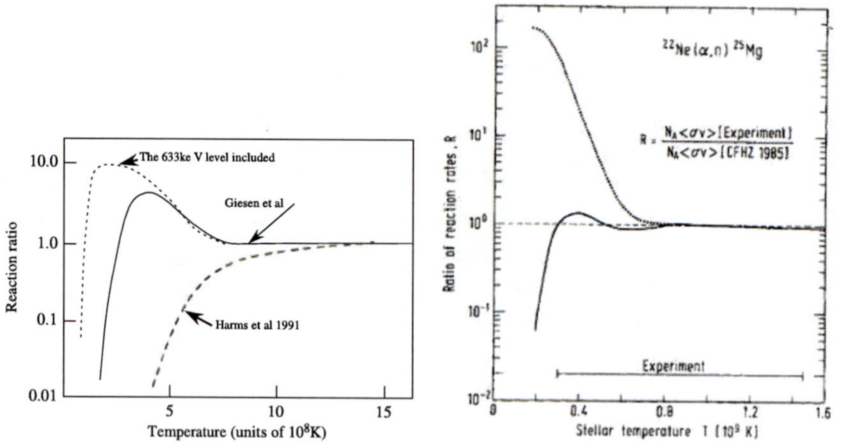

Fig. 12.29 *Left* ratio of the rates $\left({}^{22}\mathrm{Ne}+\alpha \rightarrow {}^{25}\mathrm{Mg}+\mathrm{n}\right) / \left({}^{22}\mathrm{Ne}+\alpha \rightarrow {}^{26}\mathrm{Mg}+\gamma\right)$ found by Giesen et al. for the same ratio as found by Caughlan and Fowler (1988). The *short-dashed curve* demonstrates the effect of including Giesen's proposed resonance at 633 keV. The *long-dashed curve* is the comparison with the rate found by Wolke et al. (1989) and Harms et al. (1991). After Giesen (1993). *Right* effect of the new resonances discovered by Harms et al. on the rate of the 22Ne reaction. The figure shows the ratio of the new rate to the old rate as calculated by Caughlan et al. (1985). The *dotted curve* shows the effect of the resonance at 634 keV. After Harms et al. (1991)

with the standard data as compiled and prepared by Caughlan et al.[67] Harms, Kratz, and Wiescher concluded that their result showed that the 22Ne reaction was less efficient as a neutron source for the *s*-process than had been assumed by Caughlan et al. However, the final verdict must wait until the resonances below 0.8 MeV have been sorted out. Note the experimental range in the figure. This shows how nature may tease scientists, placing the largest change just outside the experimental range and just where it is expected to be most important for stars.

Once the results of Harms et al. had been published, Bazan and Truran[68] examined the impact that the lack of knowledge of what goes on at lower energies, in particular at 635 keV, has on the *s*-process. Bazan and Truran realized that:

> Under no circumstances can the M-MS-S-C star sequence[69] be achieved at any mass capable of undergoing dredge-up with the low Ne rate.

Hence, they continued:

> A second model involving enhanced diffusion of ^{13}C throughout the convective shell is proposed to give the correct observed properties of AGB star abundances.

---

[67] Caughlan, G.R., Fowler, W.A., Harris, M.J. and Zimmerman, B.A., At. Data Nucl. Data Tables **32**, 197 (1985).

[68] Bazan, G. and Truran, J.W., BAAS 23Q1411B, 1991. The 179 AAS Meeting, Atlanta, USA.

[69] Borderline cases between M class and S class stars are called MS stars. The sequence M→MS→S→C is supposed to be a sequence of increasing carbon abundance (with age) for carbon stars, i.e., stars rich in carbon, in the AGB.

We will explain the meaning of the relevant stars shortly. In the meantime, note how delicate the model is, when somewhat better data can cause a shift in concept.

The experimental work is very difficult and frequently leads to puzzles. In 1991, Drotleff et al.[70] and Harms, Kratz, and Wiescher[71] argued that the experimental data suggests a resonance at an α-particle energy of 0.623 MeV. The effect of such a resonance on the rate of the 22Ne reaction could be critical. In 1993, Drotleff et al.[72] returned to the problem and succeeded in showing that the 0.623 MeV 'resonance' is actually a ghost resonance which creeps in due to the reaction $^{11}B + \alpha \to {}^{14}N + n$, where boron (B) is just pollution in the sample/target used in the experiment! This reaction has a strong resonance at 0.606 MeV. But this is not 0.623 MeV. However, the α particles slow down a bit even in the thin target and reach the energy of the resonance. The problem was therefore the boron contamination at concentrations below one part per million!

The rate Drotleff et al. got was in good agreement with Caughlan et al. 1985 in the extrapolation, but a factor of 3 higher at $T = 3.5 \times 10^8$ K compared with the revised Caughlan and Fowler 1988.[73] In summary, Drotleff et al. argued that their new techniques allowed the investigation of all relevant resonances in 22Ne, and they reached the conclusion that 22Ne can produce the neutrons needed for the s-process. However, they added a reservation:

> The situation will only change if a resonance exists in the competing (α, γ), near or below the (α, n) threshold, that is strong enough to diminish the abundance of ^{22}Ne in stellar nucleosynthesis scenarios.

The low-lying resonances, which are extremely difficult to reach directly in experiments, continue to bother nuclear physicists. In 1993, Giesen et al.[74] implemented a classic trick of nuclear physics, using the reaction $^{22}Ne + {}^6Li \to {}^{26}Mg + d$ to probe the α-unbound levels of ^{26}Mg. They obtained the total rate from theoretical models for which they fitted the parameters based on their experiment. Giesen et al. proposed a possible resonance at 0.633 MeV which may be important for low temperatures (see Fig. 12.30).

The story did not end there. In 1994, Davis and Abegg[75] returned to the energy levels in ^{26}Mg and discovered 92 new resonances and 21 possible resonances in the energy range $3.8 \leq E_\alpha \leq 11$ MeV with intervals of 15 keV. The situation is so complicated that the authors end their paper with the hope that:

> [...] the reader will accept the identification as a guide to the structure of ^{26}Mg and not as a complete explanation of the natural parity levels which is not accessible from the data.

---

70 Drotleff, H.W., and 7 coauthors, Zeit. f. Phys. A Hadrons and Nuclei **338**, 367 (1991); ibid. Nucl. Astrophys. Conf. 23, 1991.

71 Harms, V., Kratz, K.L. and Wiescher, M., Phys. Rev. C **43**, 2849 (1991).

72 Drotleff, H.W. and 9 coauthors, Astrophys. J. **414**, 735 (1993). The group of authors includes several of those who made the suspected discovery.

73 The source of this claim is not clear, because the difference between the two compilations of Caughlan et al. is just a few percent or less.

74 Giesen, U., and 10 coauthors, Nucl. Phys. A **A561**, 95 (1993).

75 Davis, C.A. and Abegg, R., Nucl. Phys. A **571**, 265 (1994).

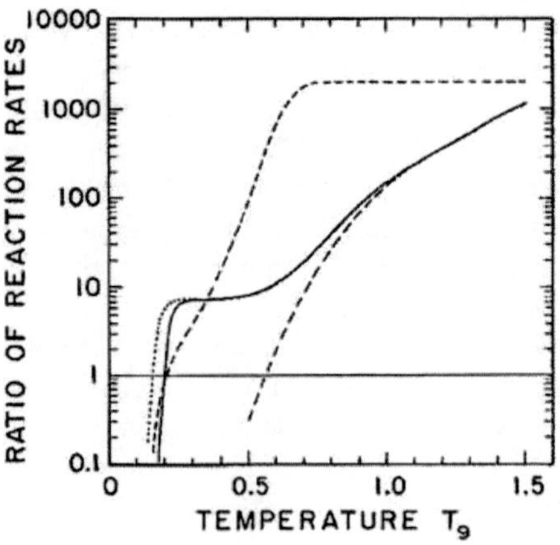

Fig. 12.30 The ratio of the $^{22}Ne + \alpha \rightarrow {}^{25}Mg + n$ reaction rate to the $^{22}Ne + \alpha \rightarrow {}^{26}Mg + \gamma$ reaction rate as a function of temperature in units of 10^9 K. The *solid line* shows the result of Giesen et al. and the *short-dashed line* is the result of Caughlan and Fowler (1988). The *long-dashed lines* show the ratio calculated on the strength of the data before Giesen et al. got their result. The effect of the suggested 0.633 MeV resonance is shown by the *dotted line*. After Giesen et al. (1993)

Fig. 12.31 The effect of the better measurement by Fey et al. A comparison with the compilation of nuclear data known as NACRE and the results of Käppeler et al. (1994). The rates are from NACRE. After Jaeger et al. (2001)

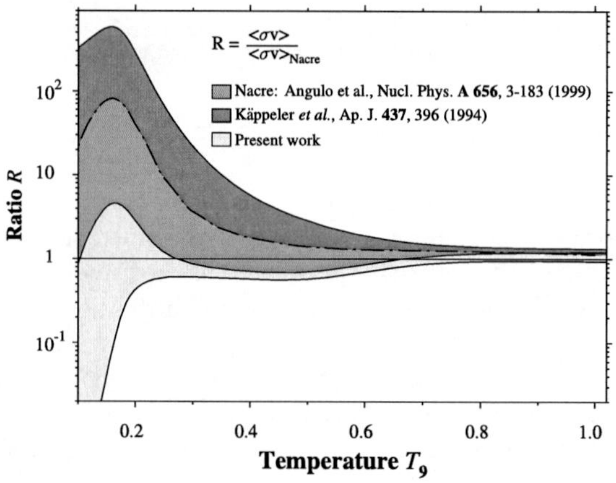

Jaeger et al.[76] revisited the reaction and measured it with an unequalled accuracy. They redetermined the parameters of all resonances in the energy range from 0.57 MeV up to 1.47 MeV. As can be seen from Fig. 12.31, the change in the rate at $T = 2 \times 10^8$ K is about a factor of 100. The effect on the star may not be so large because 22Ne assumes the dominant role of neutron production at somewhat higher temperatures. For all resonances in this energy range, new resonance parameters have been measured. For a possible resonance at about 635 keV (not to be confused

[76] Jaeger, M., and 6 coauthors, PRL **87**, 202501-1 (2001).

with the resonance at 0.623 MeV), a new upper limit for its strength was obtained. Is it the resonance observed by Endt at 623 keV and not observed by Drotleff et al.?

On the basis of the new data, improved reaction rates were calculated as a function of temperature. The new uncertainty limits were considerably smaller than in previous determinations, ruling out the large enhancement factors, up to 500, assumed too early on by too many optimistic astrophysicists in some stellar model calculations. The main reason for such a large change can be traced to the NACRE assumption of constant S factor for energies below 820 keV. This demonstrates once more the danger of extrapolation from the higher energies where the measurements are carried out to the lower energies where the rates are terribly slow. By the way, the 'recommended' value includes the hypothetical resonances taken at 10% of the upper limit of their strength.

In 2003, a large group of authors, Fey et al.[77] which includes Jaeger's coauthors, summarized as follows:

> The present experimental means have been exhausted. Only a new experiment in an underground laboratory will improve the sensitivity because there the limiting background will be reduced considerably.

Pignatari et al.[78] checked the effect of the uncertainties in the 22Ne and 13C reactions on the s-process. They concluded, as might have been guessed, that:

> More precise extrapolations or measures of the 22Ne rate are necessary.

On the other hand, the uncertainties in the 13C reaction cut no ice because ^{13}C burns in a convective zone. However, recently, the possibility was raised of ^{13}C burning in a radiative zone. If this is justified, the conclusion is the opposite.

A year later, Karakas et al.[79] also examined the inaccuracy in the rate of neutron production and reached the opposite conclusion, namely, that the astrophysical uncertainties (to be discussed later) are so great that their effect on the accuracy of the final answer is much greater still.

This discussion of neutron sources may have been somewhat too detailed, but we have tried to convey the problematics and the uncertainties so that the reader can appreciate the state of the art this part of nuclear astrophysics finds itself in. Table 12.2 summarizes what has been measured so far.

12.9 Where Does the Reaction Take Place?

Based on Lugaro and van Raai,[80] Table 12.3 summarizes the accepted view today with respect to the operation of the two most important neutron sources.

---

[77] Fey, M., and 14 coauthors, Nucl. Phys. A **718**, 131 (2003).

[78] Pignatari, M., Gallino, R., Käpper, F. and Wiescher, M., Nucl. Phys. A **758**, 541 (2005).

[79] Karakas, A.I., and 4 coauthors, astro-ph/0601645v1 27 Jan 2006; ibid. Astrophys. J. **643**, 471 (2006); Karakas, A. and Lugarno, M., Nucl. Phys. A **758**, 489 (2005).

[80] Lugaro, M. and van Raai, M., J. Phys. G: Nucl. Part. Phys. **35**, 014007 (2008).

Table 12.2 Properties of the main stellar neutron sources

Reaction	Energy release (MeV)	Relevant stellar energy range Gamow energy range (keV)	Lowest energy measured in the laboratory so far (keV)
^{13}C	2.21	170–250	270
^{22}Ne	−0.47	470–700	850

Table 12.3 Summary of the most important properties of stellar neutron sources

Neutron source	$^{13}C + \alpha \rightarrow {}^{16}O + n$	$^{22}Ne + \alpha \rightarrow {}^{25}Mg + n$
Mass of relevant star	$M < 4M_\odot$	$M > 4M_\odot$
Maximum T (K)	3×10^8	3.8×10^8
Intershell mass ($M_\odot$)	10^{-2}	10^{-3}
Timescale (years)	10^4	10
Neutron density (n/cm^3)	10^8	10^{11}
Neutron exposure (mbarn^{-1}, $Z = Z_\odot$)	0.3	0.02

12.10 Carbon Stars: Where the HTI Nuclei are Synthesized

The s-process and the carbon stars are unique. Carbon stars are the only place where we see the product of the synthesis which takes place in the deep interiors of stars, dredged-up to the surface where they can be observed. This does not happen with the fusion of any nucleus lighter than iron. For example, the iron seen in the outer layers of any star was synthesized in an earlier generation of stars. The situation is quite paradoxical. We are highly confident about the reactions whose products we do not see on the surface, and we still have problems with the reactions whose products we do see on the surface.

The so-called carbon stars are red giants in which there is an excess of carbon in the atmosphere. The word 'excess' means that the amount of carbon relative to hydrogen is significantly greater than is commonly observed in stars. Carbon stars were discovered as early as 1867, when Secchi[81] carried out the first attempts to classify the spectra of stars. The classical carbon stars are traditionally classified as types N and R, and are giant stars with low surface temperatures, whose spectra exhibit abnormally strong lines of carbon molecules like C_2, CH, and CN.[82] On the other hand, they do not show signs of molecules like TiO, which are generally observed in stars with the same luminosity and surface temperature. In the HR diagram, these stars are found in a region which is otherwise the location of stars with 'normal surface composition' (see Fig. 12.32). Thus, in the same region in the HR digram, we find both types of stars. An additional unique feature of the carbon stars is the

[81] Secchi, A., *Catalogo Della Stelle*, Paris, Gauthier-Villars, 1867.

[82] The subgroup of carbon stars which exhibit molecular bands of CH are called CH stars. Similarly, those that exhibit CN bands are called CN stars. The R stars show no enhanced barium lines. The N type shows many lines of s-process elements.

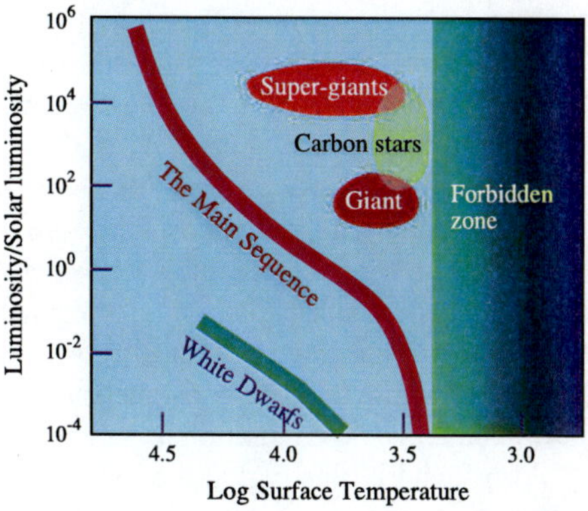

Fig. 12.32 HR diagram showing the location of the carbon stars. These are stars inside which the *s*-process operates. The process responsible for the synthesis may also bring the freshly synthesized elements to the surface of the star, so that they can be observed. The *s*-process elements thus observed where not synthesized in a previous generation of stars

existence of many HTI elements believed to be produced in the *s*-process. On the other hand, no spectral lines of HTI elements are seen in the spectra of R type stars (which have even cooler surfaces).

In 1940, McKellar[83] discovered lithium in carbon stars. This finding, which preceded the discovery of technetium by Merrill, did not impress astronomers as much as the discovery of the short-lifetime radioactive technetium because too little was known about the nuclear reactions of lithium, the formation of lithium in the Big Bang, and the consumption of lithium along the main sequence. But today we know that lithium burns at the relatively low temperature of 2×10^6 K, so whenever a convective zone extends from the surface down to temperatures higher than the above, all the lithium is destroyed and we see main sequence stars depleted of lithium. The reappearance of lithium in carbon stars may mean that, either the star was so massive on the main sequence that its envelope convective zone was very shallow, or that lithium was resynthesized after the star left the main sequence. Thus, lithium, like technetium, signaled to astronomers that the synthesized matter reached the surface, but too little was known at the time for them to appreciate what that observation was telling them. What is clear is that the convective zone of the carbon stars cannot extend for too long into a region as hot as 2×10^6 K, or else all the lithium would be destroyed. But *s*-process elements require higher temperatures for their synthesis, hence the problem. The clues provided by lithium are not yet fully understood.

In addition, this domain in the HR diagram hosts stars with other peculiarities of surface composition. For example, some of the stars show very weak hydrogen lines. As hydrogen is so abundant in stars, its spectral lines are usually very strong. Consequently, weak hydrogen lines are a unique phenomenon and may imply that these stars have lost their entire envelope. Recall that SN types Ib and Ic are supposed to originate from progenitors which have lost their envelope. Where are the progenitors in this part of the HR diagram?

---

[83] McKellar, A., PASP **52**, 407 (1940).

Most of the stars under discussion here are variable stars known as R Coronae Borealis stars.[84] These stars fade dramatically and unpredictably by a factor of up to one thousand within a few weeks, before recovering their original brightness in several months. The spectacular fading is usually hypothesized to be caused by the formation of sooty dust clouds above the surface of the star which scatter the emerging radiation. The atmospheres are, as a rule, poor in hydrogen and rich in carbon and nitrogen. Many questions regarding the evolutionary state of these stars and the physical mechanism of dust formation remain unanswered. For example, if the radiation is absorbed by dust and consequently diminished, it must come out when the dust reradiates at different wavelengths. So far it has not been confirmed that the decline in the visible radiation is accompanied by a rise in the infrared radiation.

Also found in this region are the barium stars. These show an enhancement of carbon-like 'normal' carbon stars, but in addition show abnormally strong lines of barium, strontium, and other s-process elements.

A crucial piece of evidence was supplied by the 1952 discovery by Merrill of Tc lines in S stars (see Sect. 6.40). Merrill's discovery provided direct evidence that (a) nuclear reactions take place in stars and (b) there are mixing processes which bring the products to the surface. Along the same lines, the enhanced abundance of the $A \geq 75$ elements in S stars[85] and the BaII in carbon stars led several researchers[86] to discover that the chemically peculiar S stars, enriched in elements heavier than iron, contain the unstable isotope ^{99}Tc ($\tau = 2 \times 10^5$ years) in their spectra. It became evident that ongoing nucleosynthesis takes place somewhere in the interior of these stars and that the products are brought to the surface. More accurate and quantitative spectroscopy confirmed that Tc is very abundant in S stars and also in the more evolved C stars.[87]

For many years, carbon stars were considered a mystery.[88] The mystery remained until the mid-1970s, when detailed computer models of stars with two burning shells were calculated and 'almost proved' the imaginative hypothesis of B2FH about the need for mixing between the envelope and the burning zone.

It is important for us here to note that these stars belong to an old disk population of stars. In other words, the lower the amount of metals (i.e., the older the star is), the

---

84 The R Coronae Borealis stars were amongst the first variable stars to be discovered, as early as 1797, by Pigott.

85 S is a low surface temperature giant star, similar to class K5 or M, whose spectrum displays bands from zirconium oxide, in addition to titanium oxide which is characteristically exhibited by K and M class giant stars. Other s-process elements, for example yttrium oxide and technetium, are also enhanced, hinting at the existence of HTI elements.

86 Buscombe, W. and Merrill, P.W., Astrophys. J. **116**, 525 (1952); Bidelmann, W.P., Astrophys. J. **117**, 377 (1953); Merrill, P.W. and Greenstein, J.L., Astrophys. J. Suppl. **2**, 225 (1956); Teske, R.G., PASP **68**, 520 (1956).

87 Smith, V.V., Astrophys. J. **273**, 742 (1983); Dominy, J.F. and Wallerstein, G., Astrophys. J. **310**, 371 (1986); Little-Marenin, I.R. and Little, S.J., AJ **93**, 1539 (1987); Wallerstein, G. and Dominy, J.F., Astrophys. J. **326**, 292, (1988); Kipper, T., IAUS **145**, 317 (1991).

88 In 1956, Bidelman, W.P., reviewed our understanding of the carbon stars and the title of the review was *The Carbon Stars: An Astrophysical Enigma*, in *Vistas in Astronomy*, ed. by Beer **2**, 1428.

higher the number of carbon stars. Thus, despite the larger amount of seed nuclei, the stellar evolution is such as to reduce the formation of carbon stars (and s-process elements?). Does this have something to do with the effect of metals on the internal workings of stars?

A new twist was discovered in 1984 by McClure,[89] who discovered that many of the CH stars are binaries. In 1980, McClure, Fletcher, and Nemec[90] showed that the BaII stars are probably all members of binary systems. McClure thus asked whether the Pop II CH stars which also show high abundances of HTI elements might also be binaries. This is not a trivial question. Several researchers[91] concluded that there are fewer binary systems in Pop II stars than in Pop I. McClure investigated this question and concluded:

> It is quite possible that all CH stars are binaries.

This option immediately raises a completely different possibility as to how the s-process elements reached the surface. The close binaries are more difficult to destroy by stellar collisions than the widely separated ones. For this reason they live longer as a binary than the widely separated binaries. Close binaries can transfer mass from one star to the other. As the stars which compose the binary system seldom have the same mass, the more massive of the two stars evolves faster, expands and reaches the point of mass transfer to the companion. The idea is that the mass transfer continues until the layer rich in synthesized s-process elements is exposed or the envelope enriched with s-process elements is transferred to the surface of the companion.

12.10.1 *The Rosetta Stone?*

The fact that FG Sagittae is a unique star was discovered in 1944 by Hoffmeister[92] while searching for variable stars, but it did not attract the attention of astronomers. In 1960, Richter[93] published the long term behavior of FG Sagittae, indicating a change in the luminosity by a factor as large as 50 since the beginning of the twentieth century, on top of which smaller luminosity variations were observed (see Fig. 12.33). Richter also carried out a spectral analysis, but gave no information about anything unusual regarding s-process elements.

During the years 1960–1967, the spectrum of the star was investigated by Herbig and Boyarchuck,[94] who found that the spectrum changed during the 7 years of obser-

[89] McClure, R.D., Astrophys. J. Lett. **280**, 31 (1984).

[90] McClude, R.D., Fletcher, J.M. and Nemec, J.M., Astrophys. J. Lett. **238**, L35 (1980).

[91] Abt, H.A. and Levy, S.G., AJ **74**, 908 (1969); Crampton, D. and Hartwick, F.D.A., AJ **77**, 590 (1972); Barry, D.C., Nature **268**, 510 (1977); Gunn, J.E., and Griffin, R.F., AJ **84**, 752 (1979).

[92] Hoffmeister, C., Astr. Nachr. **274**, 176 (1944).

[93] Richter, G., Astr. Nachr. **285**, 274 (1960).

[94] Herbig, G.H. and Boyarchuck, A.A., Astron. J. **153**, 509 (1974).

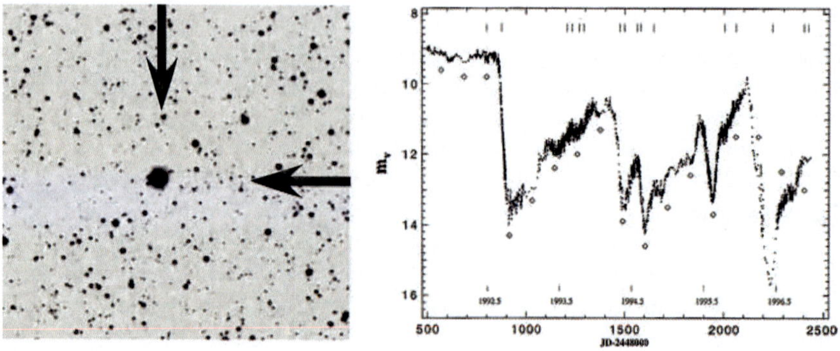

Fig. 12.33 *Left* FG Sagittae as photographed with the 48″ Schmidt telescope on 21 September 1951. The star appears not as a point but surrounded by a nebula. *Right* the time variability of FG Sagittae as found by the AAVSO (American Association of Variable Star Observers). Most of the data was collected in the visual range and not fully reduced to total luminosity. Astronomers measure time in Julian days. Julian day 2,400,000 corresponds to 12:00, 16 November 1858. The figure extends over 50,005 days and hence ends in 1995

vations from B8 Ia to A5 Ia, which implies cooling. By 1975, the star had continued to cool and exposed a G2 spectrum. However, a dramatic change took place when Langer et al.[95] discovered that the spectral lines of several *s*-process elements had began to appear in the spectrum of FG Sagittae sometime in 1967, and since then had increased their strength to the point that present day values are about 25 times the solar value. Since such a change is extremely unusual, the authors were especially careful with their spectral analysis. In the 1970s, the impression was that the rate of enrichment was beginning to decline. Kraft,[96] taken aback by the phenomenon, called the star the Rosetta stone of nucleosynthesis. Further observations revealed that FG Sagittae had ejected a planetary nebula some 6,000 years earlier.

In 1977, Kipper and Kipper[97] found that, while the abundances of the *s*-process elements changed, those of the iron peak elements remained unchanged. In 1996, Richter[98] claimed that:

> Observations in the years 1992 and 1993 show that FG Sge might have evolved to an R Coronae Borealis star.

This time the information was more quantitative. Over the past 100 years, it had brightened by a factor of more than 70 and then cooled off at a rate of 340 K/year between 1955 and 1965, and at a rate of 250 K/year between 1969 and 1974. Apparently, the thermodynamic changes had induced the changes in the spectrum.

---

[95] Langer, G.E., Kraft, R.P. and Anderson, K.S., Astron. J. **189**, 509 (1974).

[96] Kraft, R.P., S&T **48**, 18 (1974); Langer, G.E., Kraft, R.P. and Anderson, K.S., Astrophys. J. **189**, 509 (1974).

[97] Kipper, T.A. and Kipper, M.A., Sov. Astron. Lett. **3**, 220 (1977).

[98] Richter, H., Sterne **72**, 73 (1996).

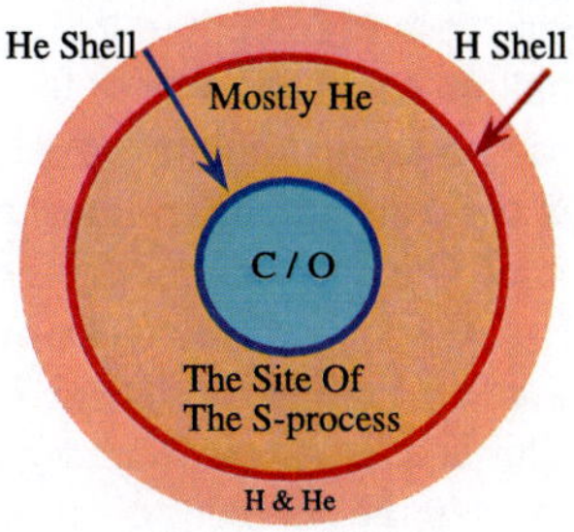

Fig. 12.34 Schematic structure of an AGB star where the *s*-process take place

The full story of FG Sagittae has not yet been told, but the presently accepted view[99] is that FG Sagittae experienced the last episode of *s*-process element formation during the 'last thermal pulse' (see Sect. 12.15) and ejected the rest of the envelope as a planetary nebula.

12.11 The Stellar Asymptotic Giant Branch Phase

The asymptotic giant branch, or AGB for short, is the phase in which the star becomes a supergiant (Fig. 12.34). All stars above the main sequence in the HR diagram are called giants because their radius is greater than the radius on the main sequence. The supergiants are the stars with the largest radius and furthest away from the main sequence. Figure 12.35 (left) shows the location of the asymptotic giant branch in the HR diagram.

12.11.1 Towards the AGB Phase of Stellar Evolution

The conversion of hydrogen to helium in the core of the star causes the contraction of the core and expansion of the envelope, so that the star leaves the main sequence by increasing its radius. The gas pressure in the core is proportional to the number of particles. The conversion of hydrogen into helium takes four protons and gives back one helium nucleus. As a consequence, the gas pressure decreases and the core contracts under the pressure of the envelope. But the contraction of the core releases gravitational energy. Part of the gravitational energy so released is radiated away (this is the Kelvin–Helmholtz–Ritter process), but the rest is absorbed by the envelope and as a consequence the envelope expands.

The expansion of the envelope continues until it almost reaches the forbidden zone on the right of the HR diagram at low surface temperatures. The surface temperature is now so low that hydrogen and helium, which are fully ionized inside the

[99] Schönberner, D., In *Hydrogen-Deficient Stars*, ASP Conference Series, ed. by Werner and Rauch, **391**, 139 (2008).

star, recombine in the low temperature envelope. The ionizations of hydrogen and helium cause the star to be dynamically unstable.[100] Consequently, the expansion can proceed along the edge of the forbidden zone, that is, increasing the luminosity at constant surface temperature, but never enter into this zone. A star which gets into the forbidden zone collapses and moves out of it, whence the name. Figure 12.35 (right) shows three tracks of stars of different masses as they converge along the border of the forbidden zone. Thus, asymptotically, the tracks of all stars converge. This is a fundamental change in the way stars evolve. Along the main sequence, the mass of the star controls its position and the main sequence is the sum (in geometry one would call it the envelope) of the positions of all stars that burn hydrogen. Every star has its own track as it gradually moves out of the main sequence. If we know the mass of the star, we know its position along the main sequence. There are no stellar tracks along the main sequence.[101] The opposite is true along the red giant branch. All stars evolve along the same track. The envelope of stellar positions is the actual track of the stars.

12.12 The Helium Flash: Does it or Does it Not Mix

Once the hydrogen in the core is exhausted, the devoid-of-hydrogen core contracts and heats up until helium ignites. The core consequently expands. The envelope contracts during the expansion of the core. For many years it was suspected that

[100] In equilibrium, the pressure of the gas equals the gravitational pressure. A prerequisite for the stability of a star is that, if the star contracts for some reason, the gas pressure, which acts outward, should grow faster than the gravitational pressure which acts inward. If this is not the case, the increase in the inward pull of the gravitational pressure will win and the star will collapse, and vice versa. So where does ionization come into this? The gas pressure due to the motion of the particles is equivalent to energy per unit volume. The neutral ion has a binding energy which does not contribute to the pressure. To ionize, the binding energy must be supplied somehow, and this is taken from the kinetic energy of the gas. Upon compression of a neutral gas, the gas is ionized and consequently takes energy from the kinetic motion of its particles (the traditional gas pressure) and uses it to pay for the binding energy. Thus, two things happen in ionization: energy which contributes to the pressure is 'lost' and the number of particles changes by a factor of two. Analysis shows that the entire story can be formulated by the following condition: if γ is equal to the logarithmic change in pressure due to a logarithmic change in density under adiabatic conditions (no heat exchange), then γ must be greater than 4/3 for stability. In ionization of hydrogen, $\gamma \approx 1.1$, and hence hydrogen ionization may induce an instability. The number 4/3 appears as a magic fraction, but it is not. One can show that, upon compression, the gravitational pressure increases as $1/r^{4/3}$ and the gas pressure must change faster for stability. Similarly, at high temperatures, iron disintegrates as follows: $^{56}\mathrm{Fe} + \gamma \leftrightarrow 13^4\mathrm{He} + 4\mathrm{n} - 124.9\,\mathrm{MeV}$ and internal energy is paid for by pressure/kinetic energy. When this happens a collapse follows and triggers a supernova. The interplay between binding energy and kinetic energy of the particles may give rise to oscillations. Indeed, many of the stars which are near the forbidden zone exhibit oscillations.

[101] In the past, before nuclear energy was discovered, it was believed that mass annihilation (Jeans) could be the energy source. According to this hypothesis, stars are born massive at the top of the main sequence and then lose mass by converting it into energy, and gradually 'slide' along the main sequence. Had this been the case, the main sequence would have been the track of the stars and not the envelope.

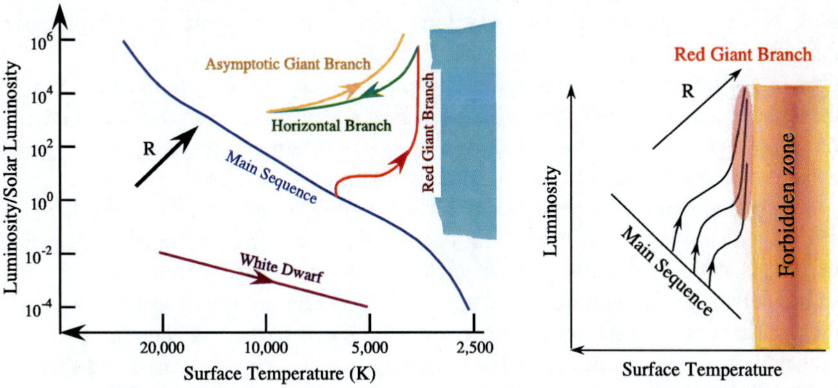

Fig. 12.35 *Left* the asymptotic giant branch (AGB), the red giant branch (RGB), and the horizontal branch (HB) in the HR diagram. The main sequence is the location of hydrogen burning stars with different masses. For each stellar mass, there is a single well defined location along the curve shown. On the other hand, the other marked 'locations' on the diagram are the tracks of all stars with different masses in the HR diagram, and for this reason there are *arrows* which mark the direction. The *light-blue zone* on the right is the forbidden region. The direction in which the radius of the star increases is marked with a *black arrow* and R. *Right* the red giant branch in the HR diagram. The changes in the core cause the expansion of the star. However, the expansion is limited on the right by the forbidden zone. As a consequence the tracks of all models, irrespective of mass, converge near the limit, where they increase their luminosity at almost constant surface temperature. So asymptotically, all tracks converge. An expansion with constant luminosity is limited by the border of the forbidden zone. The red giant branch is marked by a *red region*

helium ignition in low mass stars might be a traumatic event for the star and disrupt it. The reason given was as follows. The electrons in the core of low mass stars are highly degenerate because the temperature is relatively low and the density high. Owing to this degeneracy, which is a purely quantum effect and has no analogue in classical physics, the electrons contribute the majority of the gas pressure which holds up the upper layers. The contribution of the ions to the total pressure is negligible. But a degenerate electron gas has the unique property that it is insensitive to temperature.[102] As helium ignites and releases nuclear energy, it heats the core, but the pressure of the core does not change. The ions become very hot but their contribution to the pressure remains negligible relative to the contribution of the electrons. It is an amazing phenomenon. The ions are very hot, but the electrons around them behave as if the temperature is zero and does not affect them. The rate of the nuclear reactions depends solely on the temperature of the ions. Consequently, the temperature rises and with it the rate of the nuclear reactions and the rate of energy release. Since the total pressure does not depend on the temperature, it does not change when the temperature rises.

[102] The pressure of a highly degenerate gas is given by $P(\rho) = K\rho^{5/3}$, where K is a constant which does not depend on the temperature. If the density is very high, the electrons move with speeds close to the speed of light and the pressure is given by $P(\rho) = K_0\rho^{4/3}$. In this respect the electrons behave like radiation.

Thus the ions experience a runaway temperature and energy release. Eventually, the temperature rises so much that the degeneracy of the electrons is lifted and the pressure of the electrons once again becomes sensitive to the temperature. But by then a huge amount of energy has been released and we have a very hot core that acts like a powerful piston on the envelope. The question is therefore: what is the fate of the star after the release of the large amount of energy inside?

The phenomenon was predicted/described by Mestel in 1952.[103] Given the lack of detailed simulations of the helium flash, Hoyle and Schwarzschild[104] suggested in 1955 that the horizontal branch was the post-giant phase, containing survivors of the helium flash. With no calculated models to rely on, the hypothesis was that the remnants of the flash would move to the blue part of the HR diagram, i.e., with higher surface temperatures. In 1958, Haselgrove and Hoyle[105] assumed that the helium flash causes no mixing of the core and envelope (the mild flash hypothesis) and found that, with this assumption, they failed to explain the horizontal branch.

The failure of the classical assumption of 'no mixing at all' pushed the pendulum to the other extreme, and people supposed that the consequence of the helium flash would be complete mixing (the very strong flash hypothesis). A simple change in the degree of mixing was also raised as a possibility. However, this assumption led to worse results: the theoretical model moved all the way to the blue zone of the HR diagram, where the theoretical main sequence based on helium burning would have been, had such a main sequence existed.

The first calculations of the helium flash were carried out by Schwarzschild and Selberg[106] and Schwarzschild and Härm[107] and encountered many difficulties which required major simplifications. Yet they were inconclusive vis-à-vis the question of disruption or dynamic effects, and above all the question of whether there was mixing, and if so how much, was left open. The nuclear energy generation in Schwarzschild and Härm's calculation reached the astronomical rate of 10^{14} erg/g/sec and the luminosity of the core exceeded the incredible value of $10^{12}L_{\odot}$. However, practically nothing of this luminosity appeared on the surface of the star, and only a tiny amount of energy reached the hydrogen layer at the edge of the helium core. The net effect seemed to be lifting the degeneracy and turning the core into a convective one in which helium burns peacefully. The mountain of the helium flash appeared in the end to be a mere molehill!

---

[103] Mestel, L., MNRAS **112**, 583 (1952). Mestel investigated the energy source of white dwarfs and concluded that any nuclear energy source would be unstable:

> [...] because the pressure of degenerate gas depends almost entirely on the density and is almost independent of temperature. This means that a white dwarf whose energy sources are insufficient to supply its surface radiation cannot restore the balance by contracting.

No mathematical formulation was given. Mestel's argument as to why white dwarfs cannot have a nuclear energy source is applied to the question of why the helium flash is expected to be explosive.

[104] Hoyle, F. and Schwarzschild, M., Astrophys. J. Suppl. **2**, 1 (1955).

[105] Haselgrove, C.B. and Hoyle, F., MNRAS **118**, 519 (1958); ibid. **119**, 112 (1959).

[106] Schwarzschild, M. and Selberg, H., Astrophys. J. **136**, 150 (1962).

[107] Schwarzschild, M. and Härm, R., Astrophys. J. **136**, 158 (1962).

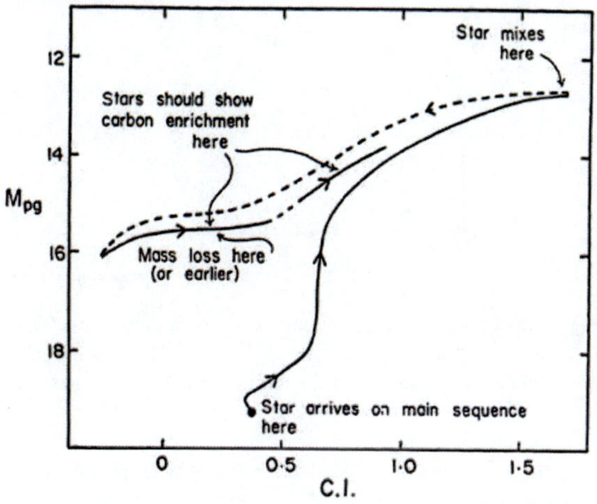

Fig. 12.36 The schematic evolution through the helium flash according to Woolf (1964). M_{pg} is the photographic magnitude, which for our purposes here is equivalent to the logarithm of the luminosity. C.I. is the color index which is roughly equivalent to the surface temperature. A high C.I. means low surface temperature

In 1964, Woolf[108] suggested that the evolution through the helium flash might proceed as shown in Fig. 12.36. Woolf assumed mixing, but apparently not enough mixing to bring the star all the way to the helium main sequence. Woolf also predicted the locations in the HR diagram at which carbon-rich stars should be visible. Thus the mixing should be 'incomplete', 'sporadic', or 'selective', or anything else but complete.

A similar approach, namely 'jumping' over the helium flash (in professional terms it is called 'parametrizing the unknown') and examining the possible structure of the 'survivors' was adopted by Faulkner[109] in 1966. Faulkner's initial models were post helium flash models in which no mixing took place. These were models with a helium burning core surrounded by a hydrogen-rich envelope. Faulkner discovered a new phenomenon, namely that the luminosity of the star was determined by the mass of the helium core and the envelope was to a large extent independent of the core. Stars with the same core mass but different envelope masses had practically the same luminosity. The total mass ceased to control the total luminosity.

However, Faulkner's models had:

> [...] a fairly uniform helium concentration in the envelope of globular clusters at a value ≈ 0.35.

Globular clusters are Pop II, and as such, a high helium content is problematic, unless the helium synthesized in the core was mixed into the envelope during the helium flash and in retrospect invalidated Faulkner's initial supposition.

More complicated models of post-helium stars were calculated by Nishida and Sugimoto,[110] who assumed three different zones in the star rather than two. Later

[108] Woolf, N.J., Astrophys. J. **139**, 1081 (1964). Woolf was in Princeton and discussed his hypothesis with Schwarzschild.

[109] Faulkner, J., Astrophys. J. **144**, 978 (1966).

[110] Nishida, M. and Sugimoto, D., Prog. Theor. Phys. Jpn **27**, 145 (1962).

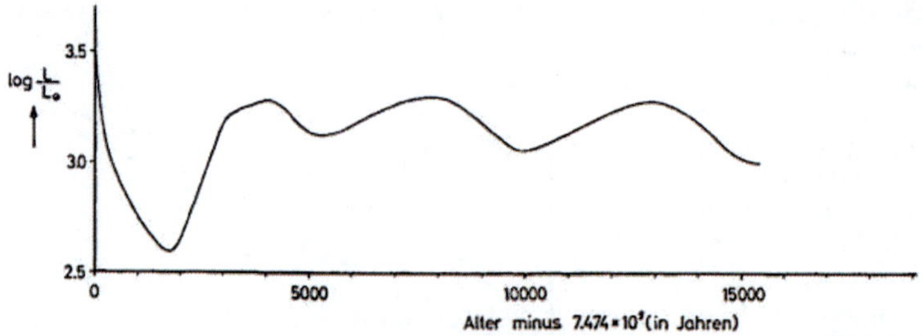

Fig. 12.37 Thomas' 'thermal pulses' of the helium shell. The *horizontal axis* shows years minus 7.474×10^9. Apparently, the mass divisions used in the difficult calculations were not small enough to capture the fast changes in the luminosity. After Thomas (1967)

Sugimoto[111] demonstrated that the power of the helium flash depends on what you assume in the treatment of convection, and when the development of convection is made more difficult, the core explodes, a result that tells us that the outcome is very sensitive to the detailed physics.

In 1967, Thomas[112] calculated the evolution of a $1.3M_{\odot}$ star as it traverses the helium flash. For the first time, Thomas included the energy losses via neutrinos[113] in the calculation, and consequently found that the center of the helium core was cooler than its outskirts, leading to a helium flash starting at the surface of the core and not at its center. Thomas found that no overall mixing happens after the flash, but that partial mixing does occur. This mixing brings carbon to the surface of the star and increases its surface abundance by a factor of 4. In addition, a helium shell forms and exhibits thermal pulses.

Meanwhile, Schwarzschild and Härm,[114] unaware of Thomas' results, proceeded with their calculations and reached practically the same results as Thomas with respect to mixing, although the details were slightly different. A comparison between Thomas' results and those of Schwarzschild and Härm is shown in Figs. 12.37 and 12.38. Note that the details of the luminosity time variation are quite different in the two calculations. These results were a confirmation of what Weigert[115] and Rose[116] had found previously. However, Schwarzschild and Härm went further by calculating 13 successive flashes. Weigert found that the first few pulses were not sufficiently powerful to induce any mixing. On the other hand, the models by Rose did not include any hydrogen envelope. The calculations were extremely time-consuming, as Schwarzschild and Härm pointed out:

---

[111] Sugimoto, D., Prog. Theor. Phys. Jpn **32**, 703 (1964).

[112] Thomas, H.C., Zeit. f. Astrophys. **67**, 420 (1967), submitted August.

[113] The neutrino losses mainly cool the center of the star. For the nature of neutrino losses, see Shaviv, G., *The Life of Stars*, Springer-Magnes, 2009.

[114] Schwarzschild, M. and Härm, R., Astrophys. J. **150**, 961 (1967), submitted 1 May, but published December 1967.

[115] Weigert, A., Zeit. f. Astrophys. **64**, 395 (1966).

[116] Rose, W.K., Astrophys. J. **146**, 838 (1966); ibid. **150**, 193 (1967).

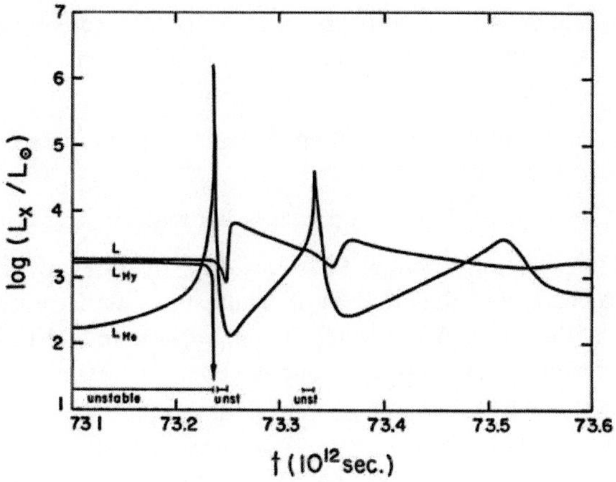

Fig. 12.38 The 'thermal pulses' of the helium shell as found by Schwarzschild and Härm (1967). L is the surface luminosity, L_{Hy} is the luminosity of the hydrogen shell, and L_{He} is the luminosity of the helium shell. The time axis relates to some arbitrary fiducial time. After Schwarzschild and Härm (1967)

> In the present investigation, brute force was applied and the star's evolution was followed in detail through thirteen cycles.

Schwarzschild and Härm did not explain the difference between this calculation and the previous one, and why the thermal pulses had not been seen before. In particular it was not clear why the pulse did not relax to a stable state. However, a close examination shows that there is secular evolution. For example, the peak flash luminosity increased by a factor of 10 in 13 thermal pulses. Perseverance and brute force sometimes pay.

Each flash causes convection and mixing. As the flashes come one after the other, the convective zone extends further and further and approaches the hydrogen burning shell. By the ninth flash, it reaches the hydrogen shell and mixes hydrogen with helium. Since the hydrogen shell does not correspond to a sharp discontinuity in the abundance of hydrogen, the penetrating convection gets further and further into the hydrogen shell from one cycle to the next. However, the amount of hydrogen mixed with helium is very small.[117]

Sanders,[118] who was in Princeton with Schwarzschild at that time, immediately understood that Schwarzschild and Härm had discovered where the old mixing idea of B2FH might actually take place. He noted from Schwarzschild and Härm's calculations that:

- The fraction of stellar mass in the convective zones of the pulses remains essentially constant at around 7%.
- Of the matter in the convective layer of any given pulse, 80–90% was also in the convective layer of the previous pulse.

[117] We do not go into the details at this point, because it will be found that it does not work. We will discuss the details where the chances of a successful mechanism are better.

[118] Sanders, R.H., Astrophys. J. **150**, 971 (1967).

- The determination of the edge of the convective zone was very inaccurate, and hence the available numbers of protons and consequently neutrons were also highly inaccurate.

Sanders did not carry out any calculations but discussed plausibilities and as such concluded that:

> Helium shell flashes may present the likely astrophysical site for *s*-process nucleosynthesis.

The pendulum continued to swing. In 1969, Edwards[119] carried out a hydrodynamic calculation of the problem, in contrast to a stellar evolution calculation, and found that the convective core reaches a temperature of 10^9 K. Moreover, the core cannot expand fast enough to subdue the energy released in such an event, and explodes. According to Edwards:

> The future evolution of the star seems very questionable. The helium flash might leave the core in a highly expanded state. There is then the possibility that the core becomes convective throughout (much more like after an explosion), mixing into the envelope, as it slowly returns to a stable state. Any conclusion to date concerning horizontal branch stars may have to be radically altered.

The high temperature lasting about 4.1 seconds caused Molnar[120] to suggest that even *r*-process elements might be produced through the reaction $^{18}O + \alpha \rightarrow {}^{21}Ne + n$ and subsequently $^{21}Ne + \alpha \rightarrow {}^{24}Mg + n$, reviving the old B2FH idea for a neutron source. Similarly, Cameron and Fowler[121] suggested that the helium flash might be the location of the *s*-process synthesis, while Truran and Cameron[122] even suggested that *p*-process elements might be produced in the helium flash. This was close, but too early to call it a hit.

The situation appeared explosive, and in 1970, when Rose and Smith[123] set out to check the possibility that a helium flash in a $0.85M_\odot$ star might be so powerful as to eject part of the envelope and become a planetary nebula, the conclusion was that, despite the huge luminosity of $2.6 \times 10^7 L_\odot$ in the second flash, the star would remain in hydrostatic equilibrium. Indeed no signs of dynamic effects were observed in the calculations. It is amazing how resilient stars are. As for the mixing, the authors did not discuss this question. The results were supported by Tomasko.[124] Since the full problem was still beyond the reach of even the most powerful computers, different researchers had to make a variety of assumptions to make the calculation tractable, and hence it became important that different treatments should yield similar results.

In the same year, Rood[125] found that he could explain the AGB stars as post helium flash stars with a little bit of mixing. On the other hand, Demarque and Heasley[126]

---

119 Edwards, A.C., MNRAS **146**, 445 (1969).

120 Molnar, M.R., Astrophys. J. **163**, 203 (1971).

121 Cameron, A.G.W. and Fowler, W.A., Astrophys. J. **164**, 111 (1971).

122 Truran, J.W., Cameron, A.G.W., Astrophys. J. **171**, 89 (1972).

123 Rose, W.K. and Smith, R.L., Astrophys. J. **159**, 903 (1970).

124 Tomasko, M.G., Astrophys. J. **162**, 125 (1970).

125 Rood, R.T., Astrophys. J. **162**, 939 (1970).

126 Demarque, P. and Heasley, J.N., MNRAS **155**, 85 (1971).

found no mixing whatsoever in the helium flash experienced by a $1.19M_\odot$ star rich in metals ($Z = 0.06$). A similar approach was adopted by Sugimoto[127] and Giannone and Refsdal.[128]

The pendulum swung again. In 1974, Sweigart[129] found no hydrogen mixing in the flash. Similar 'non-violent and non-mixing' results were found by Paczynski,[130] although the first hints were found by Christy-Sackman and Paczynski[131] who realized that:

> The bottom of the hydrogen layers has been approached very closely by both the flash-driven intershell convective region and by the outer convective envelope, the former approach [...] being within $2 \times 10^{-4}M_\odot$ in mass and a factor 1.8 in density.

The latter was a barrier which prevented mixing. As can be seen from Fig. 12.39 (left), the very steep rise in hydrogen abundance at the shell creates a very steep density fall, while the pressure and temperature have a slower rate of change.[132] The pressure, however, changes much less than all other quantities, so can be considered as barely modified. Suppose now that a blob of gas rises in the convective zone and penetrates into the non-convective zone above it, in which the density is decreasing very fast. If the blob rises slowly, the pressure inside equalizes to the pressure outside and the internal density changes only due to the change in the outer pressure. Since the pressure hardly changes, the blob, which was buoyant in the dense region and rose, will hardly expand at all and will feel denser than the surroundings as soon as it enters the region with smaller molecular weight, whence it will sink back. This is like light oil placed over heavy water, a stable situation. The opposite, water over oil, is unstable. Hence convection has a problem in penetrating a very steep density distribution like the one which appears in burning shells.

In 1976, Despain and Scalo[133] managed to derive a criterion according to which the long-awaited conclusion was finally proven: there is no reason why the inter-shell convective region cannot make contact with the hydrogen-rich layer, if the flash is

---

127 Sugimoto, D., Prog. Theor. Phys. Jpn **45**, 761 (1971).

128 Giannone, P. and Refsdal, S., Astron. Astrophys. **12**, 28 (1971).

129 Sweigart, A.V., Astrophys. J. **189**, 289 (1974).

130 Paczynski, B., Astrophys. J. **192**, 483 (1974).

131 Christy-Sackman, I.J. and Paczynski, B., Mem. Soc. R. Sci. Liege **8**, 335 (1975). The first indication of the core-mass–luminosity relations is due to Paczynski in Acta Astr. **20**, 47 (1970), and there is no mention of Faulkner who discovered and stated it, or Milne, E.A., MNRAS **9**, 4 (1930), who discovered it but did not tout it.

132 Suppose the extreme case that there is a discontinuity in the composition. The hydrogen has one molecular weight and the helium another. The equation for an ideal gas is $P = R_{\rm gas}\rho T/\mu$. The pressure P and the temperature T must be continuous through the discontinuity or else there would be an infinite pressure force or infinite energy flux, respectively. Hence, the entire abrupt change in molecular weight is compensated for by a change in the density, which in our example becomes discontinuous. Under less extreme conditions, when the abundances change quickly, we observe the same phenomenon, namely a large variation in density and small changes in pressure and temperature. This fast change in density creates a barrier for the convection: light matter over heavy. (The opposite, heavy over light, is unstable.)

133 Despain, K.H. and Scalo, J.M., Astrophys. J. **208**, 789 (1976).

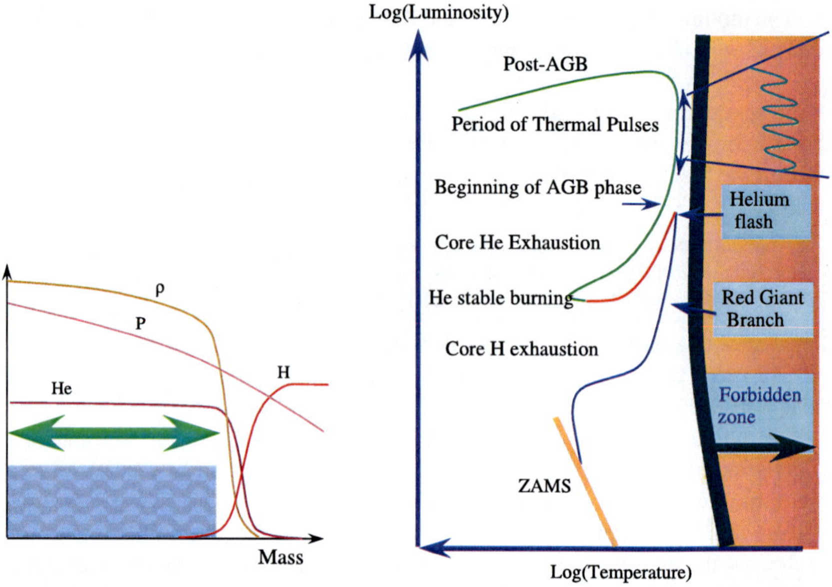

Fig. 12.39 *Left* the sharp fall in density at the location of the hydrogen shell is too steep to allow for the convective currents to penetrate into it. The helium has a higher molecular weight than the hydrogen. When a helium blob rises out of the helium convective zone, it finds itself surrounded by hydrogen which is lighter, and consequently sinks, turning it into a barrier for convection. *Right* the evolution of a $1M_{\odot}$ stellar model in the Log (total luminosity)–Log (surface temperature) plane. The star leaves the main sequence upon the contraction of the core and the expansion of the envelope and its cooling. The star cannot cross into the forbidden zone, so moves upward, almost touching the forbidden zone until the He ignites in the core, causing fast expansion (marked in *red*) and ends with the steady burning of He. When the He in the core is exhausted, the phase of two shell burnings (He and H) starts, and with it a phase of many thermal pulses and the AGB phase

sufficiently powerful. No mixing was found in their calculation. Thus, the energy released during the thermal pulses, although quite powerful, was nevertheless not sufficiently powerful to induce the mixing that had been so long hoped for. Yet the problem was not over, and in 1977, Paczynski and Tremaine[134] assumed that the unknown degree of induced mixing could be treated as a free parameter. With this assumption, they were able to find a value for their mixing parameter which explained the appearance of certain carbon molecules on the surface of post helium flash stars.

By then the idea of synthesizing the HTI elements in the helium flash had been practically squeezed dry. Theoreticians on the one hand abandoned their hope that the helium flash would save the s-process and on the other hand looked for the case of two burning shells as a new and more promising potential site for the operation of the s-process.

[134] Paczynski, B. and Tremaine, S., Astrophys. J. **216**, 57 (1977).

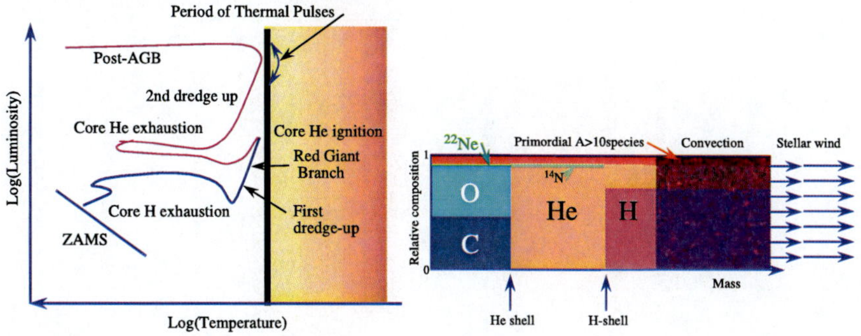

Fig. 12.40 *Left* outline of the evolution of a $5M_{\odot}$ stellar model. The ignition of He causes a fast expansion (marked in *red*) and ends with the steady burning of He. When the He in the core is exhausted, the phase of two shell burnings (He and H) begins, and with it the phase of many thermal pulses and the AGB phase. *Right* the structure of a star at the beginning of the AGB phase (not to scale). The figure shows the distribution of the various isotopes. *Blue arrows* mark the location where the hydrogen and helium burning shells will develop. The envelope has a deep convective zone and the star experiences extensive mass loss

In retrospect, this is an example of how nature can play havoc with observational results. All stars with extended envelopes appear in the HR diagram close to the forbidden region. Thus, when a star approaches the helium flash and when a star has two burning shells which proceed outward, their locations in the HR diagram will be very close. It is thus no trivial matter to distinguish the phase a star is in when it lies in this region.

12.13 Core Helium Burning

At the ignition of helium in the core, the moment of the helium flash, the star is at the tip of the red giant branch. The rapid expansion of the core and the development of convection in the core drive the star down from the tip, as shown in Fig. 12.39 (right) for the case of a $1M_{\odot}$ star and in Fig. 12.40 (left) for the case of a $5M_{\odot}$ star. Helium burns in the core with no surprise to report.

12.14 Two Burning Shells

The contraction/expansion repeats during the consumption of helium, at the end of which the star has a carbon/oxygen core (plus trace amounts of byproducts), a helium layer (plus trace amounts of hydrogen burning byproducts), and what is left of the original envelope. The composition of such a star is shown in Fig. 12.40 (right). Now devoid of an energy source, the core contracts, and the star ignites the helium shell

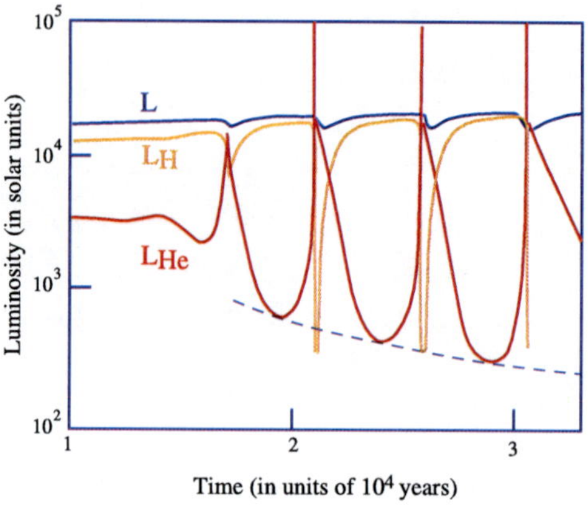

Fig. 12.41 Luminosity behavior during the thermal pulses in a $5M_\odot$ star. The hydrogen shell (*yellow line*) and the helium shell (*red line*) erupt one after the other and never simultaneously. The enormous rise in luminosity of the helium shell causes the development of a convective zone between the helium and hydrogen shells. As a result of the large L_{He}, the region between the two shells becomes convective and expands, and the luminosity of the helium shell decreases. The *blue line* is the luminosity emitted from the surface of the star. The luminosity of the star does not reveal the storms that the core has to cross

at the boundary of what was previously the helium core. In this way the star has two burning shells, which is an unstable situation, but now the burning switches between the two shells. The burning shells alternate, so either the hydrogen shell or the helium shell is at its maximum power, but never both at once, as can be seen from Fig. 12.41.

12.15 Why Cigar Burning is Unstable in Stars

In 1965, Schwarzschild and Härm were investigating the helium flash[135] when they discovered that nuclear burning in a thin shell is unstable. Thermal pulses were discovered numerically by Schwarzschild and Härm in 1967[136] and explained by

135 Schwarzschild, M. and Härm, R., Astrophys. J. **142**, 855 (1965). The analysis of Schwarzschild and Härm gave the following result: if the rate of nuclear energy generation is given by $\epsilon \propto T^\nu$ and ΔT is the change in temperature, and if $\Delta T/T > 4/\nu$, the burning is stable:

> This condition states that the perturbation of the energy source will outweigh the perturbation of the flux divergence only if the layer is thick enough to contain a temperature drop fulfilling the above condition.

Since the pp-chain depends on the temperature as $\nu \approx 4$, they predicted that hydrogen shells burning via the pp-chain would be stable and all other nuclear burning fuels would be unstable.

136 Schwarzschild, M. and Härm, R., Astrophys. J. **150**, 961 (1967).

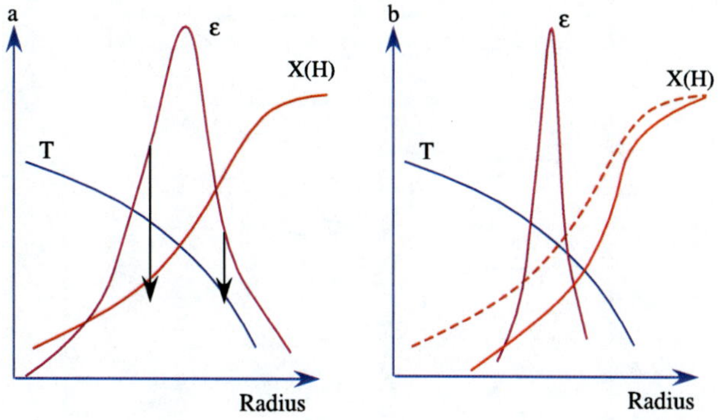

Fig. 12.42 *Left* the temperature distribution as a function of the radius and the mass fraction of hydrogen $X(H)$. The nuclear rate ε is very sensitive to the temperature and much less to the hydrogen abundance. Consequently, the product of the energy generation per unit mass and the hydrogen abundance is very large at the high temperature behind the burning shell and much lower in front of it. As the exhaustion of hydrogen is greater at higher temperatures (as marked by the *large arrow* behind the peak) and much less in front of the shell (the *smaller arrow*), $X(H)$ becomes steeper and the burning zone narrower. *Right* as the hydrogen is consumed more behind the maximum and less in front of it, the shell narrows as it moves outward. The *dashed line* is the earlier hydrogen abundance. Only microscopic dissipative processes, not usually considered in stellar models, would prevent the shell from approaching almost zero width. There is no need to consider dissipative processes because the shell becomes unstable long before such processes become important

Rakavy and Shaviv[137] as follows. The basic physics of the burning shell instability is the same as the physics of the helium flash: for some reason, the connection between temperature and pressure is disrupted.

Consider Fig. 12.42 (left) where the initial state of a burning shell is shown. The abundance of the fuel changes quickly (in fact, rises as we move out), while the temperature decreases slowly throughout the region where the abundance increases. The nuclear energy generation $\varepsilon(T, X)$ is proportional to a high power of the temperature and to the abundance (or the abundance squared). In any case, the sensitivity to the temperature is much greater than the sensitivity to the abundance, and hence the rate of nuclear energy release decreases quickly with temperature. Consequently, the fuel burns faster on the rear side (shown by a long arrow) and slower at the front (a short arrow). As a result, the shell narrows with time (see Fig. 12.42 right). So let us suppose for the sake of simplicity that the shell width reduces to zero.

---

[137] Rakavy, G. and Shaviv, G., Astrophys. Space Sci. **1**, 347 (1968). Rakavy and Shaviv provided a general stability analysis, but a full exposition would carry us outside the scope of this book. An explicit expression was derived which spelt out the dependence on the sensitivity of the temperature to density changes, the sensitivity of the pressure to changes in entropy, and the sensitivity of the pressure to changes in density. The sensitivity of the nuclear reactions to temperature enters only as a weighting function which gives higher weights to the value of the above parameters where the density is highest. Consequently, the dependence on the particular fuel is significantly reduced.

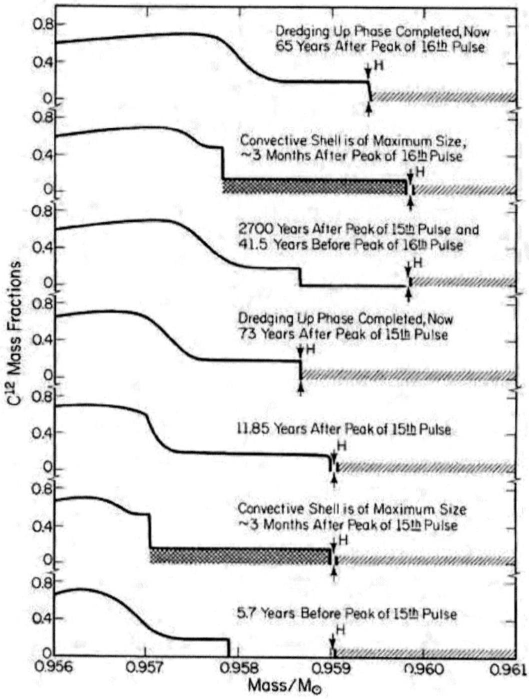

Fig. 12.43 Details of the behavior of the intershell and envelope convective zones occurring in a $7M_{\odot}$ star between the 15th and 16th thermal pulses. The figure explains how the dredge-up takes place, according to the calculation by Iben. The panels show the state at successive times with time running upward. The *continuous curve* is the mass-fraction of ^{12}C. The inner convective zone is depicted *cross-hatched*, while the convective zone is shown *singly hatched*. The times at which the convective zones are shown are the moments when they have maximum extent. *Arrows* indicate the hydrogen–helium interface. The two convective zones never meet, but cover alternatively a small region between the two shells. After Iben (1976)

The stability of the star requires that, if for some reason the temperature rises a bit and increases the rate of energy production, there will be some kind of feedback which causes the temperature to return to the original value. Such feedback might be expansion, for example. Upon expansion the temperature and density decrease and with them the rate of energy generation. However, if the shell has a vanishingly small width, any expansion is not really meaningful, and so does not lead to a decrease in temperature, whence there is no negative feedback to stabilize the star. For this reason, stars do not burn stably like a cigar, i.e., from one side to the other at a constant stable rate. When the shell narrows sufficiently, it becomes unstable and this leads to thermal pulses.

12.16 Two Types of Mixing, but Do They Mix?

The fast rise in the luminosity of the helium shell leads to a breakdown in the energy transfer by radiation, so that radiation alone is incapable of transferring so much energy. Consequently, convection develops between the helium and hydrogen burning shells. As before, the large density drop caused by the hydrogen abundance distribution constitutes a barrier for convection. A typical situation is shown in Fig. 12.43 after Iben.[138] Before discussing the details, we have to turn to the second convective zone which develops in the envelope. The envelope of stars with low surface temperatures is always convective. When the outer convective zone extends inwards into the region where the nuclear burning takes place and brings products of the nuclear burning up to the surface, we call this dredge-up. The question is, however, how deep does the outer convective zone penetrate?

The figure shows how the two convective zones, the inner and the outer convective zones, expand and contract but never meet. The steep gradients prevent the convective zones from extending all the way and mixing the entire star. The steep gradient barriers allow only for trace amounts of ^{12}C to be dredged-up, so the signature of the effect should be an enhanced surface abundance of carbon. According to the $^{12}\mathrm{C}/^{13}\mathrm{C}$ ratio, it should then be possible to determine the degree of burning that the carbon has undergone. The details are sensitive to the assumptions made in the treatment of convection, for which we still do not have a comprehensive theory.[139] Since present day physics is not sufficient to fully describe the phenomenon, the mixing is parametrized and presents one of the major unknowns/uncertainties in the problem of the s-process.

Figure 12.44 (left) shows the details of the optimistic classical picture. We start at the moment at which the helium shell is dormant. The H shell is active and above it there is a convective envelope which does not reach as deep as the H shell. The difference in mass between the He and H shells is only $10^{-2}M_{\odot}$. As the He shell begins its runaway, its nuclear energy generation rises, and with it the radiative flux. The breakdown of radiative energy transfer causes a convective zone which expands rather quickly outwardly from the He shell. The convective zone expands out to mass m_2. The He shell energy generation rises, then eventually decreases, and as it decreases, the helium convective zone retreats and disappears. Now the outer convective zone expands inwards and reaches mass m_1, which is beyond mass m_2. In this case, there is mixing of matter lying between m_2 and m_1, and s-process elements synthesized deeper than the H shell are dredged-up to the surface. The mixing is partial because at one time the inner zone reaches out to this mass and at another time the outer zone reaches in to it. At no time do the two zones touch each other. The alternative, pessimistic view is shown in Fig. 12.44 (right), where m_1 is external to m_2. The two convective zones do expand as before, but they never reach

[138] Iben, I., Jr., Astrophys. J. **208**, 165 (1976).

[139] As the full theory of convection is still elusive, parameters are introduced and the values of the parameters are adjusted so as to agree with the observations. Unfortunately, different values of the parameters are found for different situations, providing us only with ad hoc theories.

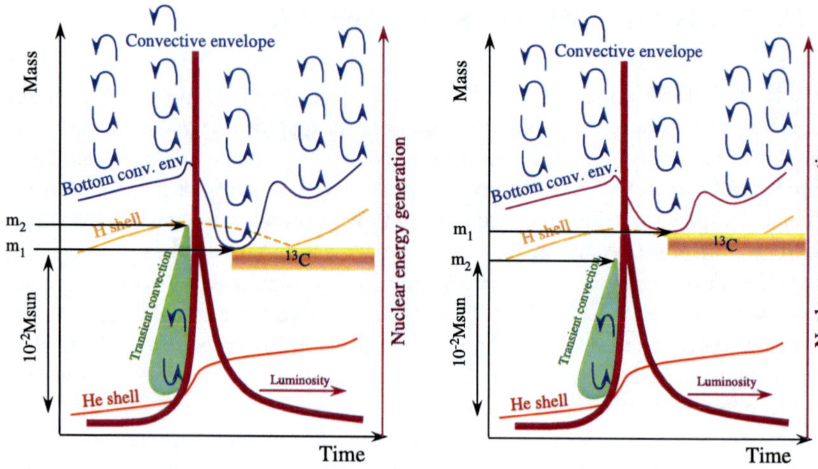

Fig. 12.44 *Left* the structure of the thermal pulse as a function of time. Only one pulse is shown but the repetition is quite accurate, except for the slow outward propagation and secular changes. We also show the rise in He shell nuclear energy generation (*pink*) as the cause for the appearance of the transient convective zone between the two shells (marked in *green*). *Right* the same as on the left, but with the difference that the helium convective zone does not extend so far out and the outer convective zone does not extend far enough in to cover the same mass region. In this way the α particles in the helium convective zone do not mix with the ^{13}C and the protons

the zone visited by the other convective zone. No mixing takes place in this case. Does the case shown in Fig. 12.44 (left) represent the mixing envisaged by B2FH?

12.17 The End of the AGB Phase and Planetary Nebulae

Long before the hydrogen and helium burning shells have time to reach the surface, extensive mass loss reduces the mass of the star. The mass loss continues until it exposes the core of the star, which may be covered by only a very thin layer of hydrogen and helium, too small or too cool (or both) to burn, and nuclear energy production is extinguished.

12.17.1 Mass Loss

Almost all stars experience mass loss. The Sun, for example, loses mass at a rate of $9.13 \times 10^{-14} M_\odot$/year. In the case of the Sun, this mass loss is slightly greater than the mass lost from the conversion of hydrogen into helium ($6.3 \times 10^{-14} M_\odot$/year), and has practically no effect on the Sun. The mass loss increases along the main sequence as the stellar mass increases, and luminosity increases even faster.

As the star leaves the main sequence and its envelope expands, the radius increases, and consequently the gravitational attraction at the surface decreases. On the other hand, the luminosity increases and with it the radiation pressure. The result is that, as the star becomes a red giant and until it ends the AGB phase, mass loss is a dominant controller of the evolution. Moreover, the peculiarities of the AGB phase, like thermal pulses, s-process synthesis, etc., all cease with the erosion of the envelope, not by nuclear burning and conversion of hydrogen into helium, but with the dispersion of the envelope via the wind and exposure of an almost 'naked' star. Mass loss represents a significant part of the stellar material which is returned by the stars to the interstellar medium, and it is therefore very important for enriching the interstellar medium with synthesized elements and passing the new elements to the next generation of stars to be formed from this matter.

A self-consistent theory of stellar mass loss is not yet available. However, there are dimensional arguments without a physical mechanism. As such, these are very simple and only constitute a rough guide. As the mass loss takes place on an evolutionary timescale, it is plausible that the relevant timescale is the Kelvin–Helmholtz–Ritter timescale. If so, then the rate of mass loss should be given by

$$\text{rate of mass loss} = \dot{m} = \frac{M}{\tau_{\mathrm{KHR}}} = \frac{M}{GM^2/RL} = \frac{1}{G}\frac{RL}{M}.$$

Since this is a dimensional plausibility argument, one multiplies the equation by a fudge factor and writes it as

$$\dot{m} = \eta \frac{R_* L_*}{M_*},$$

where asterisks indicate that quantities are measured in solar units and

$$\eta = (8 \pm 1) \times 10^{-14} \mathrm{M}_\odot/\mathrm{year}$$

is a numerical constant usually found by attempting to fit observations. Note that this η, which was derived from a fit for red giants, agrees quite well with the solar mass loss.

The outcome of such a rate of mass loss is that, in 10^4 to 10^5 years, the envelope is removed and brings the AGB phase to an end. The nuclear timescale, that is, the time it takes the helium shell to convert the entire envelope into C and O, is at least an order of magnitude longer, and hence the removed envelope largely consists of the original hydrogen and helium enriched by what the various dredge-ups bring to the surface. It is the mass loss that affects the number of thermal pulses and with it the amount of s-process elements and peculiarities.

In absolute numbers, a $1\mathrm{M}_\odot$ star may lose about $0.2\mathrm{M}_\odot$, and an $8\mathrm{M}_\odot$ star may lose all the mass above $\sim\mathrm{M}_\odot$, so as to allow the remnant to become a white dwarf, and end its active life as a cooling star.

How reliable are these numbers? There is single measurement of the mass loss from population II stars, and recall that carbon stars are in fact Pop II stars. As for Pop I stars, reliable measurements of mass loss were carried out for about half a dozen stars and the information about the dependence on the stellar parameters is very meagre.

Probably the best investigated system is α Scorpio A+B, better known as Antares,[140] which is a binary system and not a single star. By analyzing the visible light, Kudritzki and Reimers[141] found that the rate of mass loss is $7 \times 10^{-7} M_{\odot}$/year and its mass is $\sim 7M_{\odot}$. In 1983, Hjellming and Newell[142] observed radio emission from the clouds that surround Antares, and measured a mass loss rate of $2 \times 10^{-6} M_{\odot}$/year. In contrast to Kudritzki and Reimers, who analyzed the stellar visible light, Hjellming and Newell analyzed the nebulosity around Antares, a cloud of gas observed only via its radio emission. In 1987, Hagen, Hempe, and Reimers[143] determined the rate of mass loss by examining the ultraviolet radiation emitted from the circumstellar matter and found a mass loss rate of $\sim 10^{-6}$, in reasonable agreement with the radio result. In 2007, Baade and Reimers[144] analyzed the spectrum of the companion star and found that there is probably a continuous outflow at a rate of $3 \times 10^{-7} M_{\odot}$/year, accompanied by sporadic events in which an increase in mass loss rate is observed that is up to 30 times the 'average', whence they conclude that *former mass-loss determinations should be interpreted with care*. In other words, it is highly uncertain. It also explains the extra high results found by Van der Hucht et al.[145] and Bernat.[146]

Eventually, in 2001, Marsh et al. succeeded[147] in observing Antares in the infrared, thereby discovering the dust around it. More importantly, these observations revealed the clumpiness and irregularity of the dust distribution and mass loss (see Fig. 12.45) and consequently posed a major theoretical problem for calculations regarding the evolution of AGB stars.

---

[140] Antares, the brightest star in the Scorpion constellation, is a supergiant. Its radius is 565 times the solar radius, the surface temperature is about 3,500 K, and the mean density is about 10^{-6} g/cm^3. It would easily float if placed in the Earth's atmosphere. However, the Earth's orbit around the Sun comfortably enters more then twice inside this supergiant star.

[141] Kudritzki, R.P. and Reimers, D., Astron. Astrophys. **70**, 227 (1978).

[142] Hjellming, R.M. and Newell, R.T., Astrophys. J. **275**, 704 (1983).

[143] Hagen, H.J., Hempe, K. and Reimers, D., Astron. Astrophys. **184**, 256 (1987). The authors explain some of the difficulties which led several previous determinations astray, such as blending of spectral lines.

[144] Baade, R. and Reimers, D., Astron. Astrophys. **474**, 229 (2007).

[145] van der Hucht, K.A., Bernat, A.P. and Kondo, Y., Astron. Astrophys. **82**, 14 (1980).

[146] Bernat, A.P., Astrophys. J. **252**, 644 (1982).

[147] Marsh, K.A., Bloemhof, E.E., Koerner, D.W. and Ressler, M.E., Astrophys. J. **548**, 861 (2001). Strictly speaking, in 1937, O.C. Wilson and R.F. Sanford (PASP **49**, 221, 1937) discovered forbidden spectral lines of iron in the spectrum of the companion of α Sco which indicated expansion. Wilson and Sanford did not have a good explanation for how such lines could occur in the spectrum of a star, since they normally appear only in tenuous nebulae. Thus, Wilson and Sanford actually observed signs of a nebula, but did not recognize it.

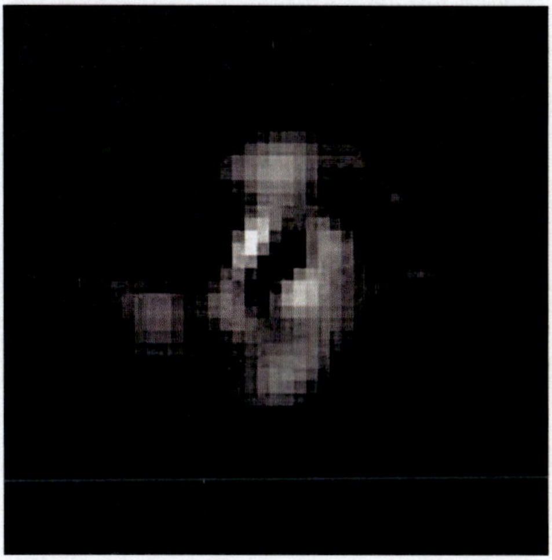

Fig. 12.45 The dust clouds ejected by Antares. The dust is clumpy, indicating non-spherical mass ejection. After Marsh et al. (2001)

We have discussed the mass loss determination from Antares at some length since this is the best analyzed case, the aim being to demonstrate the problems and uncertainties involved, and the possible impact these uncertainties may have regarding the evolution of AGB stars and the *s*-process.

Observations of planetary nebula allow estimates of the rate of mass loss, and the findings are a rate as high as $\sim 10^{-5} M_{\odot}$/year, which is about a factor of 10 higher than the value predicted by the above formula. In one case, the nebula PN K648, the estimate is as high as a factor of 300 greater than what the 'classical wind' mass loss formula would predict. Consequently, Renzini[148] called the last phase, the one which ejects the observed nebula, a super-wind. The nature of the super-wind or fast wind is still unknown, despite a few unsuccessful attempts to penetrate the mystery.[149]

12.18 What Becomes of a Star After the AGB Phase?

The remnants, devoid of nuclear energy, contract and heat up, as shown in Fig. 12.46 (right). Stars less massive than $1 M_{\odot}$ never reach the ignition temperature of the $^{12}C + ^{12}C$ reaction and eventually cool to become white dwarfs. But if the mass is somewhat higher than $1 M_{\odot}$, the star ignites carbon, generally in an explosive way, and does not become a white dwarf.

---

148 Renzini, A., *Effects of Mass Loss on Stellar Evolution*, ed. by Chiosi and Stalo, Dordrecht, Reidel, 1981.

149 Wood, P.R., Astrophys. J. **190**, 609 (1974); Tuchman, Y., Sack, N. and Barkat, Z., Astrophys. J. **234**, 217 (1979).

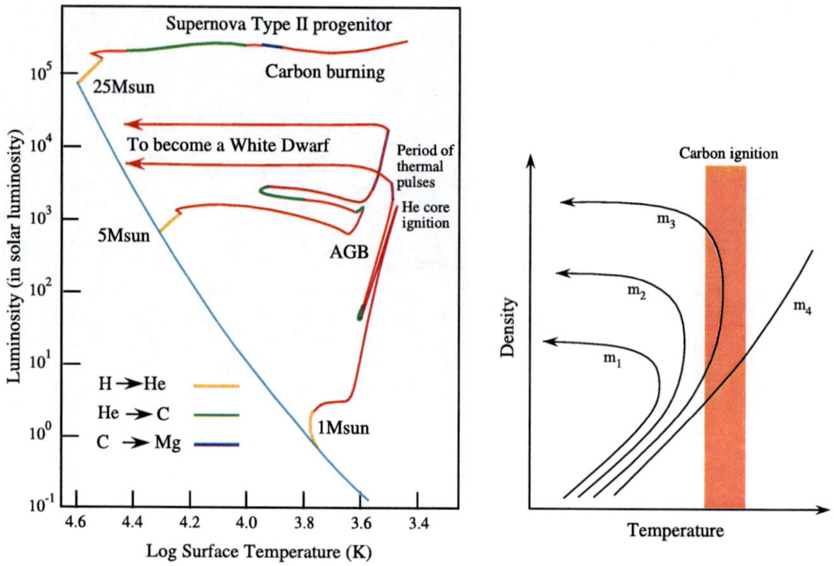

Fig. 12.46 *Left* the evolution of 1, 5, and 25M$_\odot$ models. The phase of hydrogen burning is shown in *orange*, the phase of core helium burning in *green*, and the phase of carbon burning in *blue*. The period of thermal pulses is shown in *pink*. We see that the phase of thermal pulses is limited in mass, since only stars in the mass range 1 to ~8M$_\odot$ undergo this phase. These stars lose mass via a wind which contains the *s*-process elements, and end their lives as white dwarfs. More massive stars end their lives as supernovae. Such stars synthesize only *r*-process elements. *Right* the cooling path of the central stars of planetary nebulae in the temperature–density plane. The masses of the stars are $m_1 < m_2 < m_3 < m_4$ and $m_3 > 1M_\odot$. The more massive the star, the higher its maximum temperature. The density always rises. Consequently, all stars more massive than 1M$_\odot$ ignite carbon. The lower mass stars become white dwarfs. In this way the most massive white dwarf has a mass slightly less than 1M$_\odot$. This unique evolution of stars is due to quantum effects (see Shaviv, G., *The Life of Stars*, Springer, 2009)

Our interest here is in the extensive mass loss which appears as a slowly expanding planetary nebula. The nebula provides an excellent signature for a dying AGB star, and we expect to see products of the *s*-process in the nebula.

As the *s*-process elements are trace elements, their identification in nebular spectra occurred rather late, when Péquignot and Baluteau[150] discovered emission lines of Se, Br, Kr, Xe, and possibly other HTI elements in the optical spectrum of PN NGC 7027, as was confirmed by Zhang et al.[151] and Sharpee et al.[152] Additional investigations by Dinerstein[153] and Sterling et al.[154] can be summarized as follows:

150 Péquignot, D. and Baluteau, J.P., Astron. Astrophys. **283**, 593 (1994).

151 Zhang, Y. and 4 coauthors, Astron. Astrophys. **442**, 249 (2005).

152 Sharpee, B., and 8 coauthors, Astrophys. J. **59**, 1265 (2007).

153 Dinerstein, H.L. Astrophys. J. **550**, L233 (2001).

154 Sterling, N.C., Dinerstein, H.L., and Bowers, C.W., Astrophys. J. **578**, L55 (2002).

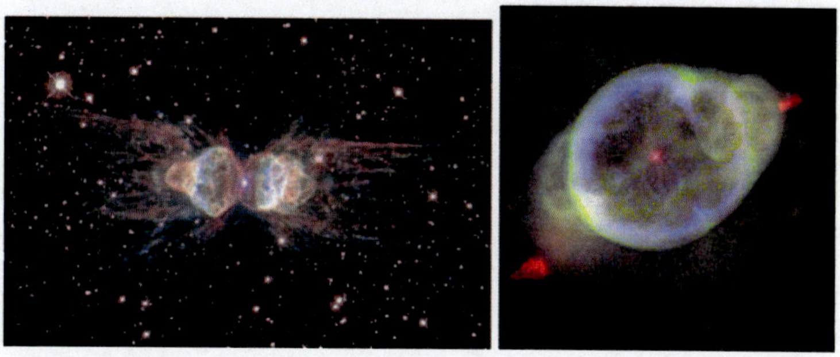

Fig. 12.47 *Left* Menzel 3 (the Ant Nebula) is a bipolar planetary nebula composed of a bright core and four distinct high-velocity outflows. The unique structure of the nebula demonstrates the intricacy of the mass outflow. The Hubble Heritage Team (STScI/AURA). *Right* the planetary nebula NGC3242, the progenitor of which requires no extra mixing. Credit HST/NASA/ESA

abundances of n-capture elements were found in only 11 planetary nebulae,[155] which represents too small a sample compared with the expectation that the progenitors of PNe are AGB stars. In several of the nebulae in which s-process elements were discovered, the relative abundances were found to be up to a factor of 10 above the solar value.

Many planetary nebulae appear in stellar binary systems. We have discussed in some detail the evolution of single AGB stars that end their lives as planetary nebulae. Clearly, if a significant fraction of the planetary nebulae are ejected from binary systems, this should affect the morphology, as claimed by the proponents of this view and as can easily be verified in the example of the planetary nebulae shown in Fig. 12.47. But then we should analyze the action of the s-process in binary systems, and this may be quite different from what we have argued to be the case in single AGB stars.

12.19 The Mass Limit

Somewhere between $8M_{\odot}$ and $10M_{\odot}$, the nature of the evolution changes. The helium core does not become degenerate and the evolution is more peaceful, at least for a while. Figure 12.46 (left) shows the basic difference between the low mass and high mass stars. While the low mass stars ignite helium in a flash, the massive stars ignite helium under 'peaceful' conditions and evolve to the right, where they develop, on a relatively short timescale, an iron core surrounded by shells of different isotopes with decreasing atomic weight. Eventually the iron core collapses and triggers a supernova collapse, the product of which is a compact neutron star, or a black hole, or nothing, i.e., it may be that no remnant is left. The massive stars end their lives as

[155] In the interactive *Catalogue of Galactic Planetary Nebulae* by Köppen, there are 1185 objects.

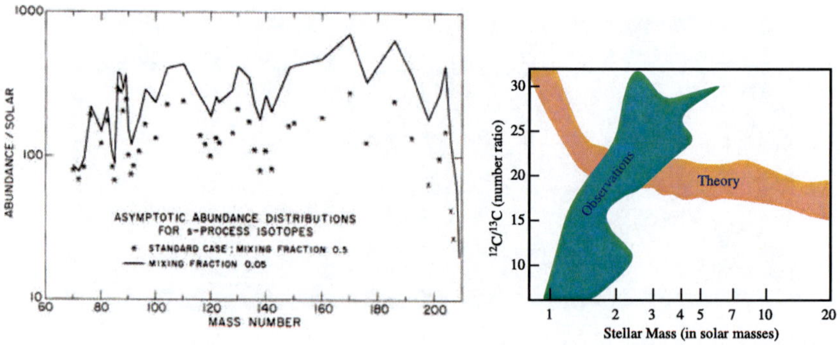

Fig. 12.48 *Left* calculated abundance/solar abundance as found by Truran and Iben (1977), assuming two extreme degrees of mixing. As can be seen, the ratio is far from constant. *Right* observed $^{12}C/^{13}C$ number ratio in red giant stars as a function of the mass (after Busso et al. 1999). The *green area* covers the observed values, while the *brown area* covers the spread in various theoretical models

supergiants and do not move to the left. The core causes the stellar explosion before mass loss removes the entire envelope.[156]

12.20 Comparison with Observation: But What to Compare?

In 1977 Truran and Iben[157] examined the *s*-process induced by the 22Ne source under different possible degrees of mixing in the thermal pulses. The results are shown in Fig. 12.48 (left). They formed the basis for the statement by Truran and Iben that:

> In the mass range $70 \leq A \leq 204$, the asymptotic *s*-process abundance distributions are found to fit extremely well the solar abundance pattern, remarkably independent both of the degree of ^{22}Ne exhaustion per pulse and of the degree of overlap of the helium convective zones during successive pulses. However, these distributions are very sensitive to the neutron capture cross-section of those light nuclei which are progenies of ^{22}Ne.

The positive statement by Truran and Iben was subsequently contested by Mathews et al.[158] and indeed problems surfaced shortly afterwards. The 22Ne reaction yields a neutron density above $10^{11}/cm^3$, which, as shown by Despain,[159] suited neither the *s*- nor the *r*-process and led to a complicated situation concerning the identification

---

[156] But beware, the progenitor of SN1987A was a blue star, to the general surprise of the theoreticians!

[157] Truran, J. and Iben, I. Jr, Astrophys. J. **216**, 797 (1977).

[158] Mathews, G.J., Takahashi, K., and Ward, R.A., Astrophys. J. **302**, 410 (1986).

[159] Despain, K.H., Astrophys. J. **236**, 648 (1980).

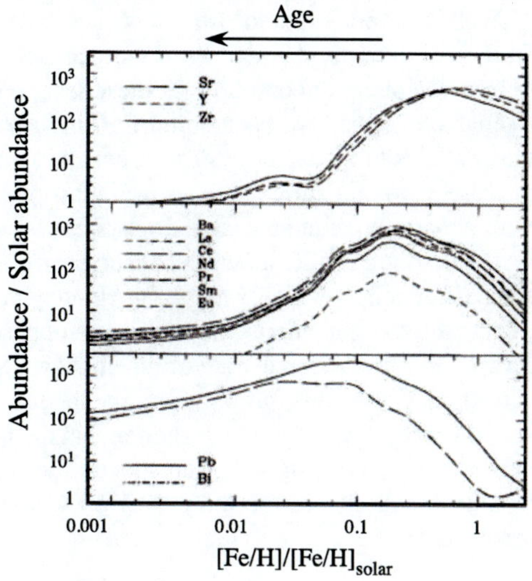

Fig. 12.49 The product of the *s*-process in stars with different amounts of heavy elements. The *horizontal axis* shows the iron to hydrogen abundance relative to the solar value. The lower the iron to hydrogen abundance, the older the star. Hence stellar age goes *from right to left*. The *vertical axis* gives the abundance relative to the solar abundance. After Busso et al. (1999) and Gallino et al. (1998)

of *s*-only and/or *r*-only elements, as well as creating an element distribution that deviates significantly from the solar one. With this result the hopes for an overall fit between a purely theoretical approach and observation evaporated.

Yet a detailed comparison between the stellar conditions suitable for 13C burning and the results obtained by adopting a phenomenological approach was presented by Käppeler et al.[160] This is presented in Fig. 12.15. For the first time a remarkable general agreement is seen, in sharp contrast with the many problems previously encountered with the 22Ne source. However, as stated by the authors, the averaging of stellar contributions from different generations of stars was essential. Despite this success, the idea of 13C burning in the convective pulses was also shown to have some drawbacks. In general, it provided values of $N_n = 4 \times 10^8$–10×10^8/cm^{-3}, higher than previously expected, unless very slow ingestion rates were assumed. Moreover, Bazan and Lattanzio[161] pointed out that there could be a risk of producing excessive energy by ^{13}C burning, thus modifying the pulse structure in a direction that would increase N_n up to values incompatible with the *s*-process.

The general picture that emerges is the following. The *s*-process takes place under many circumstances (Fig. 12.49). Different stars carry the process differently and the parameters of the process change in each star. The *n* th thermal pulse is slightly different from the $(n + 1)$th. Next, in different pulses, the stars eject the synthesized elements into interstellar space, where the elements sometimes mix very effectively and sometimes less effectively. So the chances of a single *s*-process taking place in a given star explaining all the relevant data are minute. Is this good?

---

[160] Käppeler, F. and 4 coauthors, Astrophys. J. **354**, 630 (1990).

[161] Bazan, G. and Lattanzio, J.C., Astrophys. J. **409**, 726 (1993).

Well, astronomers and astrophysicists have turned the problem upside down. Observations of the abundances of s-process elements in different locations are used to infer the physical conditions at these locations. This procedure would have been a reliable one, had we been sure of all the necessary nuclear data, leaving all the uncertainties to the astrophysical or cosmic setup. But unfortunately, this is not yet the case. However, progress in securing this procedure is encouraging, and it constitutes a major drive in nuclear astrophysics.

Figure 12.48 (right) shows the observed $^{12}C/^{13}C$ ratio on the surface of red giant stars as a function of stellar mass.[162] Along with a huge spread in values, the general trends of the observations and theory are opposed. It is difficult to believe that this reveals an observational bias of an unknown sort, and it seems more likely that something essential is missing from the theory.

However, it should not be forgotten that the best and least shakable observational evidence that AGB stars do actually do the job is the fact that Tc is widespread in S stars and also in the more evolved C stars, as has been confirmed by many scientists.[163]

12.20.1 Is There Mixing or Not?

The observational results and their interpretations are still not clear. For example, Charbonnel and do Nascimento[164] attempted to determine the fraction of low mass stars that underwent 'extra mixing' on the red giant branch using statistical means. They found that:

> [...] 96% of evolved stars show a $^{12}C/^{13}C$ ratio in disagreement with the standard predictions.

They concluded that:

> [...] 96% of low-mass stars do experience an extra mixing process on the RGB.

On the other hand, Palla et al.[165] examined the planetary nebula NGC 3242 and concluded that:

> The spectrum indicates that the progenitor star did not undergo a phase of deep mixing during the last stages of its evolution.

---

[162] Busso, M., Gallino, R. and Wasserburg, G.J., Annu. Rev. Astron. Astrophys. **37**, 239 (1999); Lambert, D.L. and Ries, L.M., Astrophys. J. **248**, 228 (1981); Luck, R.E. and Lambert, D.L., Astrophys. J. **256**, 189 (1982); Kraft, R.P. and 3 coauthors, AJ **106**, 1490 (1993); Kraft, R.P., PASP **106**, 553 (1994); Gilroy, K.K. and Krishnaswamy, K., Astrophys. J. **347**, 835 (1989); Gilroy, K.K. and Brown, J.A., Astrophys. J. **371**, 578 (1991); Charbonnel, C., Astrophys. J. **453**, L41 (1995).

[163] Smith, V.V. and Wallerstein, G., Astrophys. J. **273**, 742 (1983); Dominy, J.F. and Wallerstein, G., Astrophys. J. **310** (1986); Wallerstein, G. and Dominy, J.F., Astrophys. J. **330**, 937 (1988); Kipper, T., in *Evolution of Stars: The Photospheric Abundance Connection*, ed. by Michaud and Tutukov, IAU Symp. 145 (1991).

[164] Charbonnel, C. and do Nascimento, J.D., Astron. Astrophys. **336**, 915 (1988).

[165] Palla, F. and 4 coauthors, Astrophys. J. **568**, L57 (2002).

They thus left the issue unresolved.

Pavlenko et al.[166] examined the $^{12}C/^{13}C$ ratio in giant stars in globular clusters, which the reader will remember are population II stars. They concluded that:

> This suggests complete mixing on the ascent of the red giant branch.

In other words, prior to the helium flash. This accordingly is:

> [...] in contrast to the values predicted by current models.

So there may already be problems with the models leading to the helium flash.

And again, Lebzelter et al.[167] examined the evolution of the C/O and the $^{12}C/^{13}C$ ratios along the AGB observationally. They observed an increase in both C/O and $^{12}C/^{13}C$ as compared with the prediction of the 'standard' theory and argued that:

> The low carbon isotopic ratios of the two C-stars in our sample indicate the late occurrence of moderate extra mixing. The extra mixing affects the most luminous AGB stars and is capable of increasing the abundance of ^{13}C, while leaving unchanged the C/O ratio, which has been fixed by the cumulative action of several third dredge-up episodes.

12.21 Consequences of Complexity: Each Computer Code has its Solution

The *s*-process depends on very fine details in the structure of stars and their time evolution. The computer codes used to calculate the evolution have become very complicated, to the point that a comparison between the results obtained by different codes is sometimes embarrassing.

Straniero et al.[168] chose a set of parameters and obtained a self-consistent model for a $3M_\odot$ star exhibiting regular dredge-up above $M_H = 0.63M_\odot$ and evolving into a C star, while Frost and Lattanzio,[169] Herwig et al.[170] and Mowlavi et al.[171] could not reproduce the same results. It is hard to believe that the synthesis of the elements could depend on a computer code!

However, one should not be surprised. The problem is extremely complicated, the numerical codes are computing monsters, and the physics input is not always identical. Small changes in the input, in particular when carried over a long evolution time, may grow into meaningfuldifferences.

---

166 Pavlenko, Y.V., Jones, H.R. and Longmore, A.J. MNRAS **345**, 311 (2003).

167 Lebzelter, T., and 5 coauthors, Astron. Astrophys. **486**, 511 (2008).

168 Straniero, O., and 6 coauthors, Astrophys. J. **440**, 85 (1995); ibid. and 5 coauthors **478**, 332 (1997).

169 Frost, C.A. and Lattanzio, J.C., Astrophys. J. **473**, 383 (1996).

170 Herwig, F., Schönberner, D., and Blöcker, T., Astron. Astrophys. **340**, L43 (1998).

171 Mowlavi, N., Jorissen, A., and Arnould, M., Astron. Astrophys. **334**, 153 (1998); Qian, Y-Z, Vogel, P. and Wasserburg, G.J., Astrophys. J. Astron. Astrophys. **334**, 153 (1998).

12.22 Variations with the Metallicity

The metallicity refers to the level of all elements with atomic weights $A \geq 12$. A glance at the abundances will show that the main contributors are C, N, O, Ne, and Fe. It is generally assumed that the relative proportions of the different species is universal, but it is not. Still, the sum total of all these species, called the metallicity, is an indicator of the age of a star and the amount of iron in it. Hence, one expects a variation in the metallicity with age. Indeed, metallicity is considered as a proxy for age. The more 'metals' there are, the younger the star, and vice versa.

We have seen that the s-process is sensitive to the details of the unknown mixing mechanisms. Gallino et al.[172] assumed that all stars behave in the same way, and in particular, that the details of the mixing do not vary with the amount of metals the star contains. Under this condition, they calculated how the amount of HTI elements changes with the time the star was born. One would expect the relative amount of heavy elements formed to be independent of the age or the amount of metals, but apparently this is not the case. The most metal-poor stars, those that were first to be born, produce the largest amount of lead and bismuth. On the other hand, the younger stars with the largest amount of seed nuclei predominantly produce the lighter HTI elements. Attempts to explain this amazing outcome were made by Clayton[173] and by Mathews et al.[174] However, the entire process is non-linear and complicated, and defies simple reasoning.

We expect the central stars of planetary nebulae to show s-process elements. Pereira et al.[175] analyzed the central star of the planetary nebula HD 149427 and found that it was slightly metal poor. Barium and strontium, two key elements for s-process action, are not enriched relative to solar abundances. Carbon and oxygen are underabundant.

Ziegler[176] analyzed the central stars of the PNe K1-16, NGC7094, and Abell 78. They found iron deficiency of up to a factor of 10 and explained it by postulating a *very efficient s-process*, so that all the iron was transformed into HTI elements. Similarly they reported nickel deficiency by a factor of about 10. However, there is no measurement of any s-process element to corroborate this explanation.

In 2008, Sterling and Dinerstein[177] carried out a survey of planetary nebulae, in search of s-process elements. They used the lines of Kr and Se as a proxy for these elements. Their findings were as follows. Kr tends to be more strongly enriched than Se. A large variation in the ratio $X(\mathrm{Kr})/X(\mathrm{Se})$ was found along with a positive correlation between s-process enrichments and the C/O ratio. Only 44% of the PNe in their sample (41 objects out of 94) displayed significant s-process enrichments,

---

[172] Gallino, R., and 7 coauthors, Astrophys. J. **497**, 388 (1998).

[173] Clayton, D., MNRAS **234**, 1 (1988). This and the following attempts were carried out before the pubication by Gallino et al.

[174] Mathews, G.J., Bazan, G. and Cowan, J.J., Astrophys. J. **391**, 719 (1992).

[175] Pereira, C.B., Baella, N.O., Daflon, S. and Miranda, L.F., Astron. Astrophys. **509**, 13 (2010).

[176] Ziegler, M., and 4 coauthors, J. Phys. Conf. Series **172**, 012032 (2009).

[177] Sterling, N.C. and Dinerstein, H.L., Astrophys. J. Suppl. **174**, 158 (2008).

which they translated to the conclusion that a total of at least 20% of all existing PNe display signs of s-process action. The authors conjectured that the enrichment was self-enrichment of the star and not an enriched primordial gas out of which the stars formed. The reason for this assumption is that, in 1999, Travaglio et al.[178] found that the scatter in the abundance of other light n-capture elements (Sr, Y, and Zr, or $Z = 38$–40) was roughly a factor of 1.5, which is much less than the scatter in the observed PNe. The analysis is complicated once more by the fact that many PNe are binary stars in which the two stars interact with one another. Furthermore, on the basis of CNO abundances, distribution in the Galaxy, etc., they concluded that the PNe that do show s-process elements are the outcome of AGB stars with masses greater than 3–4$M_{\odot}$, which experience a weaker s-process than AGB stars with masses of less than 3$M_{\odot}$.

It is not clear why only about 1/5 to 1/2 of the PNe show signs of the s-process. If it is a mass effect, why is it a sudden effect and not a smooth transition from rich to poorly enriched stars?

12.23 Astrophysical Problems Affecting the s-Process Theory

Semi-Convection

The name semi-convection may be misleading. Semi-convection is in fact convection, but not due to thermal conditions (very hot matter below cold matter and breakdown of heat conduction by means of radiation).

Massive stars fuse hydrogen via the CNO cycle, which is highly temperature dependent. Consequently, the temperature gradient is significant and the energy generation in the star is very intense. The energy flux is so intense that the radiative flux fails to carry it, and the radiative transport of energy becomes unstable against mass motions. The instability creates mass motions in which the energy is carried outward by the internal energy of the gas, a much more powerful process than energy transfer by radiation. Thus, when convection sets in, it is more effective than radiation in carrying the energy outward.

As time goes on, hydrogen is depleted in the core, and consequently the core shrinks, gradually leaving behind a zone in which the composition (and the molecular weight) varies gradually. The composition at each point is the one that existed in the core when the core extended to that point.

Schwarzschild and Härm (1958), who were the first to encounter the problem, treated this zone as a 'semi-convection', which means to say that the chemical composition was adjusted by partial mixing so as to maintain convective neutrality after the mixing. Thus, in Schwarzschild and Härm's treatment, the condition is applied

178 Travaglio, C., and 5 coauthors, Astrophys. J. **521**, 691 (1999).

after mixing sets in. Sakashita and Hayashi (1959, 1961) applied the condition before mixing took place, including the effect of the change in molecular composition with location.

The confusion over what criterion to use and the extent to which mixing is partial or complete continues to this day.[179] In any case, if there is a molecular gradient at the beginning, it is now a question of how fast the mixing currents will wash it away and 'solve the problem of semiconvection'. As soon as the region is homogeneous, the confusion disappears.

The condition for stability against convection can be formulated in two ways. When a blob of a slightly lower density than the surroundings rises, we have convection. This condition, as can be imagined, is formulated as a relation between the run of the density and pressure throughout the region. This is known as the Ledoux criterion.[180] The run of the density compounds the variation in the composition because, in the virtual experiment, the blob does not mix with the surroundings. This condition does not consider the radiative transfer problem at all.

The second possibility is to consider a blob which has a slightly higher temperature than its surroundings and ask when this will rise. This condition is formulated as a relation between the run of the temperature and the run of the pressure throughout the star. This is known as the Schwarzschild criterion.[181] Since the radiative transfer depends on the run of the temperature, this condition compounds the radiative conductivity of the matter, and through it the composition.

One soon realizes that the Schwarzschild criterion contains the radiative conductivity, while the Ledoux criterion is free of it, so there is no reason to expect the two criteria to be identical and yield identical results. And sure enough, they do not. Only when the medium is homogeneous and the contribution of the radiation to the total pressure can be neglected are the two conditions identical.

Semi-convection is the term applied to the 'slow' mixing that goes on in a region that is stable according to the strict Ledoux criterion, but unstable according to the Schwarzschild criterion. However, note carefully that we start with the Ledoux condition and approach the Schwarzschild one.

---

[179] Kato, S., PASJ **18**, 374 (1966); Gabriel, M., Astron. Astrophys. **1**, 321 (1969); ibid. **6**, 124 (1970); Stothers, R., MNRAS **151**, 65 (1970); Simpson, E.E., Astrophys. J. **165**, 295 (1971); Stothers, R. and Chin, C.-W., Astrophys. J. **198**, 407 (1975); Sreenivasan, D.J. and Wilson, W.J.F., Astrophys. Space. Sci. **53**, 193 (1978); Stevenson, D.J., MNRAS **187**, 129 (1979); Langer, N., Sugimoto, D., and Fricke, K.J., Astron. Astrophys. **126**, 207 (1985); Spruit, H.C., Astron. Astrophys. **253**, 131 (1992); Merryfield, W.J., Astrophys. J. **444**, 318 (1995); Canuto, V.M., Astrophys. J. **534**, L1 (2000); Shibahashi, H., Comm. Asteroseismol. **158**, 38th LIAC/HELAS-ESTA/BAG, 2008.

[180] Ledoux, P., Astrophys. J. **105**, 305 (1947).

[181] Schwarzschild, K., Gött. Nach., **1**, 41 (1906). The condition is named after Karl Schwarzschild (1893–1916m) the father of Martin Schwarzschild, whose research we have discussed in detail in this book. Karl discovered the criterion when he discussed the stability of the solar atmosphere. In his book, *Structure and Evolution of the Stars*, the son Martin discusses the stability named after his father, but refrains from calling it the Schwarzschild criterion and does not even give reference to his father's discovery paper (Schwarzschild, K., Nachrichten Königlichen Gesell. Wiss. **1**, 41 1906). I find this modesty out of all proportion.

It is postulated that mixing does not contribute appreciably to energy transport, and radiation remains the main carrier. The effect, however, is slow mixing. No final solution to the problem of semi-convection has been found. Uncertainty regarding semi-convection affects our knowledge of the evolution of massive stars as soon as they leave the main sequence.

Mixing by Convective Overshoot and Undershoot

An obvious possible mechanism for mixing beyond a formal convective boundary is overshooting. This refers to the situation when blobs of matter manage to leave the unstable region and penetrate some distance into the stable region.[182]

Stimulated by the 2D hydrodynamic convective models of Freytag et al.[183] which showed such partially mixed zones, Herwig et al.[184] introduced an ad hoc exponential decay in the convective velocity. This produces the partial mixing required by the 13C source. Unfortunately, it also introduces parameters associated with overshoot, which determine the size of the ^{13}C pocket.

Mixing by Gravity Waves

A recent suggestion by Denissenkov and Tout[185] is that gravity waves at the bottom of the convective envelope can produce partial mixing beyond the convective boundary. The convective region contains mass motion and turbulence, which give rise to acoustic as well as gravity waves that can propagate at least a certain distance into the non-convective region. In a way, it is like overshoot. For reasonable assumptions, the resultant 13C pocket is about the size required to match observed abundances.

Rotation and Mixing by Rotation

As all stars rotate, this is a major problem for current models! One should also consider the fact that magnetic fields are generated by rotation, but these have not been included in the computations of AGB stars so far, and could represent a way to 'save the s-process'. Magnetic fields could in fact enhance the coupling between the core

---

182 Shaviv, G. and Salpeter, E.E., Astrophys. J. **184**, 191 (1973).

183 Freytag, B., Ludwig, H.-G., and Steffen, M., Astron. Astrophys. **313**, 497 (1996).

184 Herwig, F., Blöcker, T., Schönberner, D., and El Eid, M., Astron. Astrophys. **324**, L81 (1997).

185 Denissenkov, D. and Tout, C.A., MNRAS **340**, 722 (2003).

and the envelope, thereby decelerating the core and reducing the strength of rotational shear mixing.[186] The problem remains open and requires further exploration.

Mass Loss: Observation and Theory

We have seen the scant observations and determination of mass loss on the one hand and the possible complication of clumped, sporadic, eruptive, and other forms of mass loss on the other. The theory of stellar structure is still struggling with these problems.

[186] Spruit, H.C., Astron. Astrophys. **333**, 603 (1998).

Chapter 13
A Process in Search of an Environment: The *r*-Process

13.1 The *r*-Process: A High Neutron Flux for a Short Time

While the *s*-process can be described as a low neutron flux occurring over a long period of time, the *r*-process can be described as the opposite, namely, irradiation by a high neutron flux for a very short time. Hence, while the *s*-process could be described as a quiet, slow, hydrostatic evolution, the *r*-process is usually associated with the most violent phases in stellar life—the explosion that puts an end to the 'normal' life of the star. It seems that no in-between process exists, only the two extremes. Notwithstanding, problems with the explosion scenario have led astrophysicists to ponder on less violent circumstances where a milder variant of the *r*-process might take place. But whatever the situation, we are now looking for an extreme source of neutrons in an expanding environment. The adjective 'extreme' refers to the nuclear properties while the requirement of expansion refers to the astrophysical site, so that the synthesized isotopes can be ejected into space. Alpher, Bethe, and Gamow considered the Big Bang to be just such an explosive situation, but in this specific case we are supposedly immersed in the products.

As we saw in Chap. 7, the αβγ theory cannot synthesize anything beyond the light elements if it starts from protons or neutrons. But the experimental evidence from the heavier-than-iron elements, and in particular the existence of the actinides,[1] strongly hints that these elements were formed by neutron irradiation. Particularly for Hoyle, leading advocate for stellar synthesis of the elements, the failure of the αβγ theory was a victory on the one hand and a challenge on the other, since it raised the problem of how to synthesize the actinides in stars.

[1] The actinide group includes the heaviest elements that terminate the periodic table. The group contains 14 chemical elements with atomic numbers from 90 to 103, thorium to lawrencium. Only thorium and uranium appear in nature in significant amounts. The other actinides are either man-made elements or decay products with very short decay times relative to the age of the Earth. All actinides are radioactive, with uranium and thorium having half-lives commensurate with the age of the Earth. More massive elements do not exist in nature, and when formed artificially, decay very quickly.

G. Shaviv, *The Synthesis of the Elements*, Astrophysics and Space Science Library 387, DOI: 10.1007/978-3-642-28385-7_13, © Springer-Verlag Berlin Heidelberg 2012

Two events/experiments ignited the imagination of Hoyle and his colleagues[2] and led to the idea that neutrons can form the transuranium elements along with the actinides up to the transuranium elements. The first was the detection of ^{254}Cf in the Bikini H-bomb[3] test, which implied that the massive neutron flux converted the uranium and metal remnants into transuranium elements during the fast explosion. The second was the discovery that the light intensity of Type I supernovae decays exponentially with a time constant of 55 ± 1 days, as discovered by Baade et al.[4] Consequently, Burbidge et al. identified the decay time of the SNe (see Fig. 13.1) with the decay of californium 254. The fundamental question was therefore how ^{254}Cf formed during or shortly prior to the SN explosion (Fig. 13.2).

BHBCF suggested the following four-step process for producing the HTIs:

- Collapse of the core of the star as a trigger of the supernova.
- Proton absorption reactions in the outer collapsing envelope, causing matter to reach a temperature of 6×10^9 K and a density of 10^5 g/cm^3.
- α-capture reactions release neutrons in reactions like $^{21}\mathrm{Ne} + \alpha \rightarrow {}^{24}\mathrm{Mg} + \mathrm{n}$ at this high temperature.
- Neutrons captured by the iron group elements produce the HTI elements. All in all, BHBCF expected the capture of about 200 neutrons converting ^{56}Fe into ^{254}Cf in about 200 s, which is a very long time by supernova collapse timescales.[5]

---

[2] Burbidge, G.R., Hoyle, F., Burbidge, E.M., Christy, R. and Fowler, W.A., Phys. Rev. **103**, 1145 (1956), hereafter referred to as BHBCF. According to Geoff Burbidge, *An Accidental Career*, Ann. Rev. Astron. Astrophys. A **5**, 1 (2007), it was he who discovered the paper in the Phys. Rev. which described what was found in the Bikini experiment.

[3] Fields, P.R., and 8 coauthors, Phys. Rev. **102**, 180 (1956). The authors described the discovery of transuranium elements including californium, plutonium, americium, curium, berkelium, einsteinium, and fermium in the *resulting debris of the H-bomb*. As for californium, the following isotopes were identified in the debris: ^{249}Cf, ^{252}Cf, ^{253}Cf, and ^{254}Cf. No ^{250}Cf was detected, probably because it is stable against β-decay. Moreover, ^{251}Cf, the longest living isotope of californium, was not mentioned, and no explanation was given. Californium was discovered by Thompson, Street, Ghiorso, and Seaborg in 1950 by bombarding ^{242}Cm by α particles. The last author got the credit for the discovery. The New Yorker magazine teased the discoverers by claiming that they should have named californium ($Z = 97$) and berkelium ($Z = 98$) 'universitium' and 'ofium', respectively, reserving 'berkelium' and 'californium' for $Z = 99$ and 100. Seaborg's reply was that they *named it first for 97 and 98, berkelium and californium, thus forestalling the possibility of some New Yorker finding the next two elements and naming them newium and yorkium*. The unpublished reply of the magazine was that they were working on it. The irony is that californium, which was synthesized by an intense neutron source, is used today as an efficient neutron source.

[4] Baade, W., and 5 coauthors, PASP **68**, 296 (1956). The collaborators on this observational paper were Baade and the group that wrote the previous theoretical paper BHBCF. The observations were carried out solely by Baade.

[5] The subsequent expansion of the SN takes months and years, and it may take up to hundreds of thousands of years before the last sign of an SN explosion disappears. The short timescale is restricted to the collapse and fast expansion, and the existence of the proper conditions for the r-process.

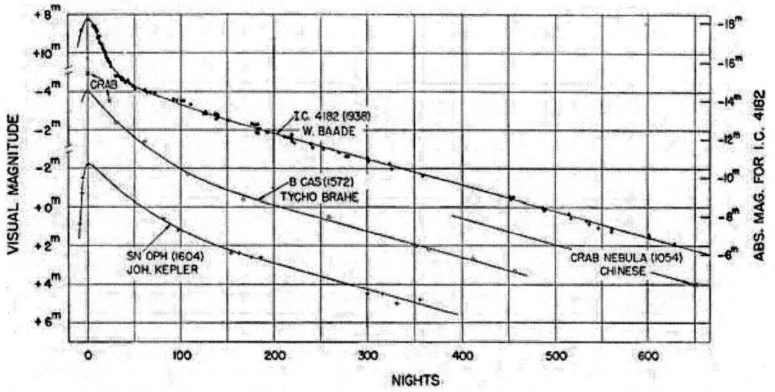

Fig. 13.1 The light-curve of SN in the IC 4182 galaxy, which rekindled the idea of fast neutron synthesis of the elements and in particular californium, in supernova explosions. The *horizontal axis* represents days and the *vertical axis* represents the visual magnitude, which is equivalent to minus the logarithm of the energy emitted in the visible range (greater magnitude means fainter). The straight-line decline implies an exponential light intensity decay with a half-life of 55 ± 1 days. Three further SN light curves have been added. All data reduction was carried out by Baade. After B2FH 1957

A few comments are in order here:

- First, note that the suggested neutron source is the same as the one suggested for the s-process, and later rejected.
- The suggested supernova was a Type I SN, which is not expected to create a neutron star.
- The formation of Cf is supposed to take place in a region composed mainly of ^{4}He, ^{12}C, ^{16}O, and ^{20}Ne, with about 10^{-3} of the mass in the form of the seed nucleus ^{56}Fe. More specifically, the synthesis should not take place in the collapsing core with an extremely short timescale.
- The authors estimated that, since Ne was so abundant in their assumed initial composition, the reaction $^{21}\text{Ne} + \alpha \rightarrow {}^{24}\text{Mg} + \text{n}$ would yield several hundred neutrons for each iron nucleus.
- Recent measurements of the half-life of ^{254}Cf yield 60.5 days, in clear contrast with the measured timescale for SN luminosity decay. Another point is that californium has several long-lived isotopes, the longest of which is ^{251}Cf, with a half-life of 898 years. So why would we observe the formation of the particular isotope which has a half-life similar to the relevant SN timescale and no others, such as ^{252}Cf, which has a half-life of 2.645 years and one would therefore expect to find powering the SN remnants?
- The authors were unable to find a neutron source for SN Type II.
- The energy supply for the SN explosion was assumed to be from typical proton capture reactions which operate in the CN cycle, namely explosion of nuclear fuel.

The idea that ^{254}Cf decay might be the energy source which powers the SN remnants was soon abandoned, but the concept of a large neutron flux operating for a few

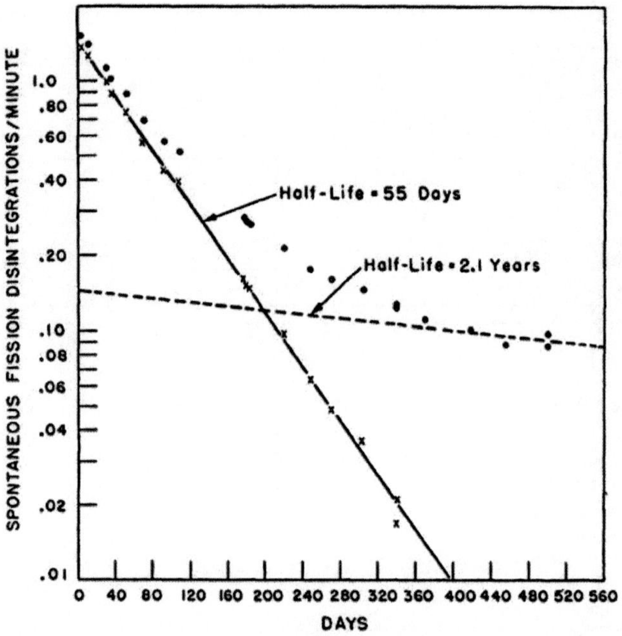

Fig. 13.2 The spontaneous fission decay of the californium fraction from the thermonuclear debris. Burbidge et al. hypothesized that the decay curve observed by Fields et al. in the debris of the Bikini H-bomb test was due to the same phenomenon as the decay of the supernova remnants. The californium was synthesized during the explosion in a tiny fraction of a second by the very large neutron flux absorbed by the metal structure of the bomb and uranium. After Fields et al. 1956

seconds at most and creating all the HTI elements up to and beyond uranium remained to be worked out in detail by B2FH shortly afterwards.

B2FH were the first to classify the HTI elements according to the process responsible for their synthesis, as s-only, r-only, $r + s$ isotopes, and p-process. In this way they obtained the schematic abundance curve given in Fig. 13.3 (left). Each abundance peak is split into peaks of almost equal height. The s-peak is always at a higher atomic weight than the r-peak. Note also that, while for lighter-than-iron elements the abundance decreases exponentially, the abundances appear to level off for $A > 100$. Moreover, if the abundance curve results from two independent processes, it is surprising that the s-peaks should have roughly the same height as the r-peaks. Why should the total yield of two independent processes be of the same order of magnitude? These are isotopes which are produced in almost equal quantities by the completely independent s- and r-processes. How could that be?

In their 1957 paper, B2FH discussed the r-process at length. First, they assumed equilibrium, namely, if $X(A, Z)$ is any nucleus, then

$$\text{formation} \equiv X(A, Z) + \text{n} \rightleftharpoons X(A+1, Z) + \gamma + Q(A, Z) \equiv \text{destruction} .$$

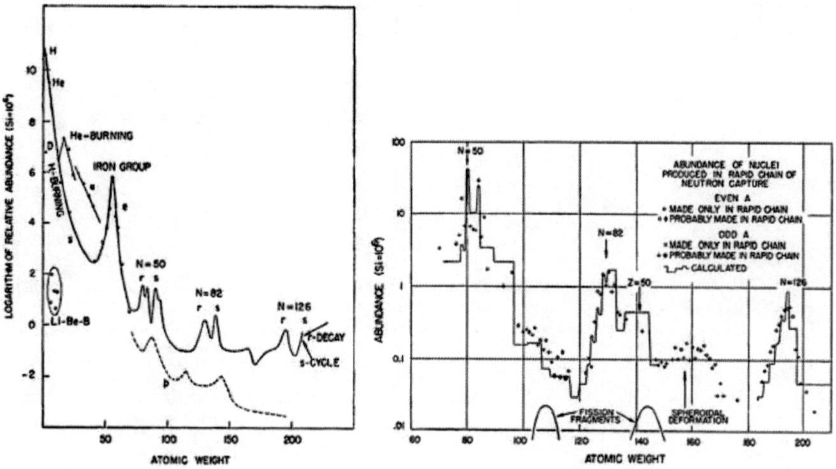

Fig. 13.3 *Left* the schematic curve of atomic abundances as a function of atomic weight, as depicted by B2FH on the basis of the data of Suess and Urey 1956, and which became the framework for the *s*- and the *r*-processes. After B2FH 1957. *Right* the B2FH comparison of the *r*-process abundances with observations. The observed abundances are from Suess and Urey 1956, which are essentially the Solar System abundances. Note that the three peaks are nicely reproduced

The equation implies that if the external conditions like temperature, density, and neutron flux do not vary, then the abundances, and in particular the ratio $N_{X(A,Z)}/N_{X(A+1,Z)}$, where $N_{X(A,Z)}$ is the concentration of isotopes with atomic weight A and atomic number Z, are constant and given by well known expressions in terms of the external conditions. $Q(A, Z)$ is the energy difference for any specific reaction. Missing nuclear data forced the authors to extrapolate, most importantly in a smooth way, and apply certain approximations to solve the large number of equations the *r*-process involves. The value of A (or Z) at which the number of neutrons reaches the magic number was kept as a free parameter. In this way, B2FH succeeded in fitting the three known peaks of the *r*-process element distribution as shown in Fig. 13.3 (right).

In 1960, Hoyle and Fowler[6] discussed the role of supernovae in the synthesis of the elements. In the paper, which is basically discursive, they concluded that the *r*-process takes place in SN Type I, in which the luminosity (or energy supply) is caused by transuranium element decay *with half-lives ranging from 40 to 70 days but in many cases close to 55 days*. The scenario envisaged by Hoyle and Fowler is exhibited pictorially in Fig. 13.4 (left). The *r*-process operates for a very short time during the fast expansion of the SN debris.

Hoyle and Fowler returned to the californium hypothesis arguing that:

> The hypothesis proved fruitful in suggesting the existence of a process in which neutrons are added very rapidly to nuclei of medium atomic weight.

[6] Hoyle, F. and Fowler, W.A., Astrophys. J. **132**, 565 (1960).

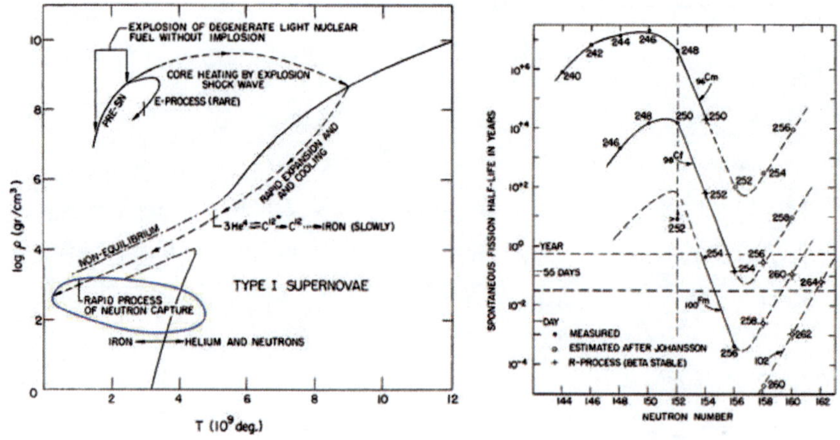

Fig. 13.4 *Left* the 1960 conceptual model of Hoyle & Fowler for SN Type I explosions. The core contracts and heats. At $T \sim (1.5\text{–}2.5) \times 10^9$ K, the nuclear fuel composed of ^{12}C, ^{16}O, and ^{20}Ne ignites and explodes. Subsequent cooling freezes the composition that existed at $\rho = 10^8$ g/cm^3 and $T = 3 \times 10^9$ K. In this state, the matter is composed of iron peak isotopes. The shock wave from the explosion moves rapidly outward and enters the helium–neutron zone. The expansion and cooling are very rapid and the slow 3α reaction has no effect. The high ratio of neutrons to seed nuclei results in the *r*-process formation of HTI elements by neutron capture at $\rho = 10^3$ g/cm^3 and $T = 10^9$ K (marked by a *blue circle*). *Right* the connection between the number of neutrons in transuranium isotopes and the half-life for spontaneous fission as depicted by Hoyle and Fowler, which limits the range of isotopes required to power the expanding SN envelope, provided this is indeed the correct mechanism. As can be seen, there are several candidates with roughly the same decay times. The nuclear data is, however, outdated. For example, ^{252}Cf decays 96% of the time via α decay and only 3.09% of the time via spontaneous fission, and the half-life is 2.645 years. But this was not the reason why the californium idea was abandoned

Furthermore, they claimed that resulting abundance calculations were *in good agreement* for isotopes with $A > 80$. This time Hoyle and Fowler realized that, besides ^{254}Cf, there might be additional transuranium elements whose decay could power the expanding gases of the SN, e.g., ^{260}Fm with an expected decay time of 10–100 days, or ^{258}Cf and ^{258}Fm. Hoyle and Fowler provided a figure (see Fig. 13.4 right) from which it could be inferred which of the transuranium elements was a possible candidate to power the SN expanding envelopes. It is mind-boggling to think that transuranium elements are formed in a fraction of a second in the expanding envelopes of gigantic explosions and later power the exquisite images of SN remnants for many days and years, eventually yielding the uranium and thorium which allow cosmological dating.

Californium was never observed in the spectrum of SN remnants. It was just the coincidence between the then known half-life of ^{254}Cf and the decay time of the supernovae, together with the total energy release in spontaneous fission (which is amply sufficient), that led Burbidge and his collaborators to this idea.[7]

[7] The first measurement of the spectral lines of any Cf isotope, in this case ^{249}Cf, only took place in 1962, when 14 lines were observed by Conway, Hulet, and Morrow (Conway, J.G. and Hulet, E.K.

In 1961, Coryell,[8] who measured many of the half-lives of heavy elements, published a review of the *r*-process[9] in the Journal of Chemical Education, a journal which is not frequently visited by astrophysicists, to say the least. Consequently, some of the new ideas expressed in this paper never made their way into the astrophysical literature, and the paper was to a large extent ignored. Although the title of the paper was *The chemistry of creation* and it was published in the Journal of Chemical Education, there is no chemistry there, only nuclear physics and a little bit of astrophysics. But maybe the paper was not appreciated because Coryell ended his article with *sincere hope that the new alliance (cosmochemistry) will bring new perceptions for the chemistry of Creation*, and the reference he gave was Moshe.[10]

Coryell was strongly influenced by chemistry, where the electronic shell structure completely controls the chemical properties of the atoms, and he assumed that the nuclear shells would be equally dominant. He thus assumed that the magic numbers were universally valid. Yet, the nuclear odd–even effect is of the order of 1.5–2.5 MeV and the shell structure effect is of the order of 2 MeV relative to the mean energy per

---

(Footnote 7 continued)
and Morrow, R.J., J. Opt. Soc. Am. **52**, 222, 1962). The sample contained a mixture of Cf isotopes. Hence, even if any Cf isotope got into the remnants of the SN, it would have appeared as unknown spectral lines. The only case known to the author in which a discovery of Cf in the spectrum of a star was claimed is by Gopka et al. (Gopka, V.F., and 4 coauthors, *Kinematics and physics of celestial bodies* **24**, 89, 2008). The original Russian text was published in Kinematika i Fizika Nebesnykh Tel **24**, 125 (2008) in the case of Przybylski's star. However, no confirmation has been forthcoming so far.

8 Charles DuBois Coryell (1912–1971) was an American chemist. In 1942, he was Chief of the Fission Products Section, at both the University of Chicago and the Clinton Laboratories (now Oak Ridge National Laboratory) in Oak Ridge, Tennessee. His group was responsible for finding the properties of radioactive isotopes created by the fission of uranium and for developing a process for chemical separation of plutonium.

In 1945, along with Marinsky and Glendenin, Coryell was one of the discoverers of the element promethium (see Sect. 11.22). Marinsky and Glendenin distilled the new element by extraction from fission products and by bombarding neodymium with neutrons. They managed to isolate it using ion-exchange chromatography. The discovery was only announced at the meeting of the American Chemical Society in September 1947. It was upon the suggestion of Coryell's wife Grace Mary that the team named the new element after the mythological god Prometheus (*Promethium Unbound: A New Element*, ORNL Review **35**, # 3,4, 2002). An earlier version has it differently. On 29 September 1947, Time magazine wrote the following in an article entitled *Nervous elements* (http://www.time.com/time/magazine/article/0,9171,804273,00.html):

> Element 61 will have no official name. Dr. Hopkins has called it illinium. Mr. Glendenin wants to call it prometheum after the Greek god Prometheus, giver of fire. One convention wag suggested grovesium, after loud-mouthed Major General Leslie R. Groves, military chief of the atom bomb project.

The name prometheum was eventually adopted by the International Union of Chemistry in 1949.

The Charles D. Coryell Award of the Division of Nuclear Chemistry and Technology of the American Chemical Society, awarded annually to undergraduate students doing research projects in nuclear-related areas, is named in his honor. Along with Felbeck, Coryell discovered how oxygen molecules attach to the iron atoms of hemoglobin.

9 Coryell, C.D., J. Ch. Ed. **38**, 67 (1961).

10 Moshe Ben Amran (Moses), Genesis, Chapter I, Verses 1–5 (ca. 1250 R.C.E.).

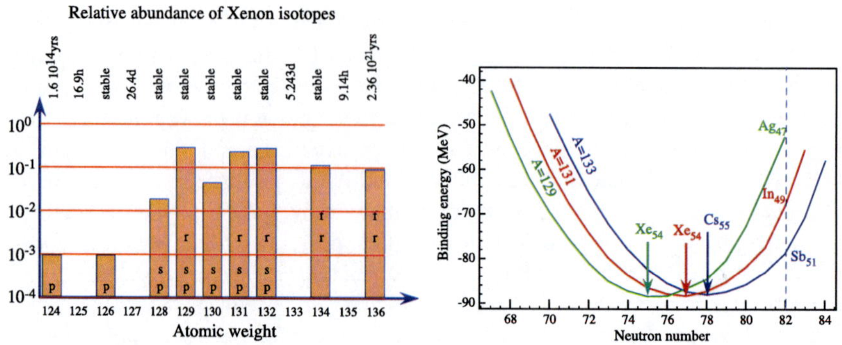

Fig. 13.5 *Left* relative abundances of the xenon isotopes. *Letters* denote the synthesis process and 'f' stands for fission product. The isotopes with atomic weights 125, 127, 133, and 135 do not exist in nature. The excess in the p-process-only isotopes and the excess in the r-process-only isotopes are always simultaneously observed. *Right* three cuts across the β-stability valley for $A = 129$, 131, and 133. If the r-process accumulates at $N = 82$, then depending on the particular nucleus (Ag_{47}, In_{49}, or Sb_{51}), the decay series ends at different stable isotopes. It is tempting to attribute the r-process peak at Xe ($Z = 54$) to decay of the isotopes with $N = 82$. The nuclear binding energy is given in MeV

Table 13.1 Half-lives for r-process decay towards stable xenon isotopes

Symbol	Z	$A = 129$	$A = 131$	$A = 133$
Ag	47	44 ms		
Cd	48	27 s	?	
In	49	61 s	0.282 s	180 ms
Sn	50	2.23 m	56.0 s	1.45 s
Sb	51	4.40 h	23.03 m	2.5 m
Te	52	69.6 m	25.0 m	12.5 m
I	53	1.57×10^7 years	8.0207 days	20.8 h
Xe	54	Stable	Stable	5.243 days
Cs	55	32.06 h	9.689 days	Stable

nucleon of about 8 MeV. Consequently, the nuclear shell structure is not expected a priori to have the same impact on the nucleus as the electronic shells on the atom.

Consider, for example, the formation of xenon. The relative abundances of the xenon isotopes, as well as the particular process responsible for their formation, are shown in Fig. 13.5 (left). The isotopes with $A = 125$, 127, 133, and 135 (all with odd A values) have very short lifetimes. They decay shortly after formation, and do not exist in nature. So how is the r-process contribution obtained, according to Coryell? Consider Fig. 13.5 (right), which shows three cuts across the valley of stability at constant A. The r-process advances along the neutron-rich side of the stability valley until it encounters a nucleus with a magic neutron number, in this case $N = 82$. As the probability for neutron absorption is small at magic numbers, the isotope accumulates. Once the irradiation ceases, the products decay quickly (see Table 13.1, Fig. 13.6) towards the stable isotope with minimum binding

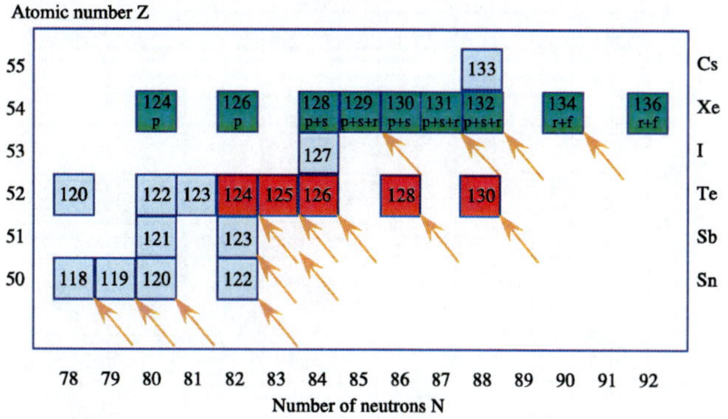

Fig. 13.6 Chart of the elements and processes which shows how Xe isotopes are shielded from the r-process. The Xe isotopes are shown in *green* and the shielding isotopes in *red*

energy, as shown in Fig. 13.5 (right). Note the relatively long lifetime of $^{129}I_{53}$ before the chain of decays ends in a stable Xe isotope. Since the life of stars is much shorter in the phase when the r-process operates, deviations from the universal Xe abundance are expected. In other words, we may see Xe and I in stars before they reach the final stable state. According to this simple picture, the r-process elements should have peaks about 5 atomic weights off the peaks of the s-process. Indeed, Coryell took the data from Suess and Urey, plotting Fig. 13.7 and attributing the peaks around bromine, xenon, and platinum to r-process pile-up at $N = 50$, 82, and 126, respectively.

B2FH, and independently Cameron,[11] were less impressed by the assumption that the neutron magic numbers were also valid far away from the stability valley, and argued that the neutron capture process would continue until the neutron binding energy fell below 2 MeV, at which point the energetic photons that exist at the high temperatures at which the process takes place would dissociate the nuclei faster than they were formed. Nonetheless, the two models are not really independent, because the magic numbers affect the binding energy of the neutrons and their resistance to disintegration by photons.

Frequently, however, the sequence of isotopes ends before the binding energy of the last neutron decreases below 2 MeV. An example is shown in Fig. 13.8, where the binding energy of the last neutron in indium ($Z = 49$) isotopes is shown as a function

[11] Cameron, A.G.W., Ann. Rev. Nucl. Science **8**, 299 (1958); Unclass. Rep. Atomic Energy of Canada, CRL-41 (1957). Cameron remarks that:

> It must be emphasized that the straitness of the exponential decay (Fig. 13.1) in the last half of the curve is very uncertain because of possible error in the normalization and calibration of the photographic magnitude scale for faint magnitudes.

In short, beware of rushing to dramatic conclusions! In addition, Cameron argued with B2FH about the source of neutrons. While B2FH suggested the α reactions on ^{21}Ne and ^{25}Mg, Cameron proposed the $^{13}C + \alpha$ reaction.

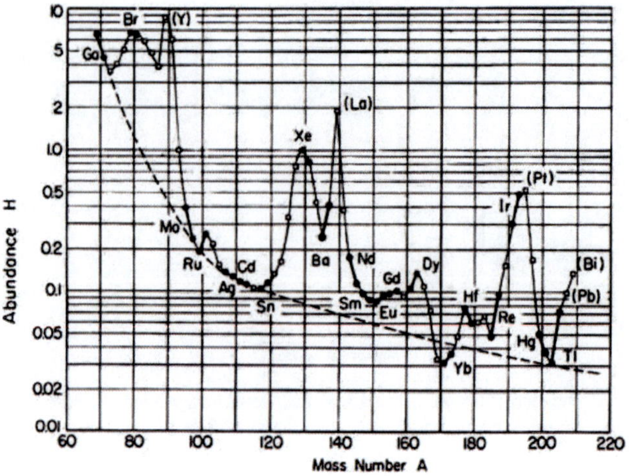

Fig. 13.7 Abundance of odd-A elements in the range $A = 69$–209 relative to silicon ($=10^6$). (The *vertical axis* is marked with a confusing H.) The abundances marked with *filled circles* connected by a *thick line* denote chemical elements having more than one stable isotope with odd A. The original data was from Suess and Urey, and Coryell had the idea of looking at the odd-A and even-A abundances separately

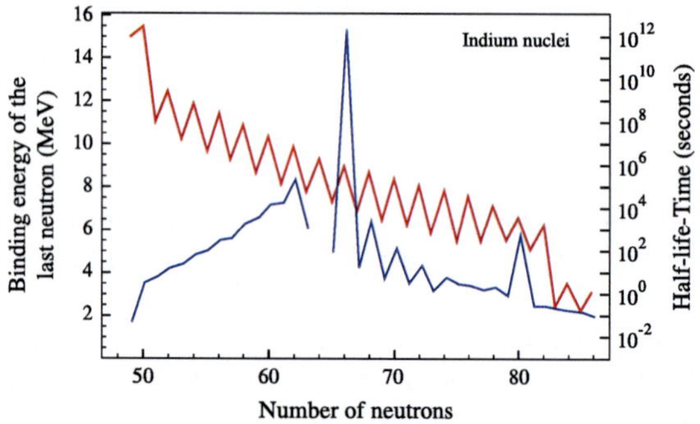

Fig. 13.8 The energy (in MeV) of the last neutron (*red*, on the left) and the lifetime of the nucleus (*blue*, on the right in seconds) as a function of the number of neutrons for the case of indium ($Z = 49$). The gap in the lifetime curve is due to the only stable isotope $^{113}\mathrm{In}_{49}^{64}$ with a relative abundance of 4.32%. The lifetime of the unstable isotope $^{115}\mathrm{In}_{49}^{66}$, which has a relative abundance of 95.72%, is 4.41×10^{14} years, longer than the age of the Universe. The unstable isotope is a factor of 21 more abundant than the stable isotope. The zigzag in the binding energy curve is due to the odd–even effect of the nuclear force. The large drop at $N = 50$ is the magic number effect. Data from Audi, G., Nucl. Phys. A **729** (2003)

of the number of neutrons in the nucleus. The gradual decrease is clearly visible as the number of neutrons increases. The greatest jump between neighboring isotopes takes place at $N = 50$. There is no indium isotope with fewer than 49 neutrons. The neutron in the last unstable nucleus has a binding energy of 3 MeV and a half-life of a second.

Coryell's contribution, which was not recognized at all by the community, was to point to the continuity in the abundance curve, which implies universality. Had there been a sharp discontinuity in the abundance curve, it would have been natural to assume two or more distinct processes with independent yields. Since Coryell did not carry out any calculation, he did not know that there was a problem in predicting the peaks. In the case of B2FH, the various approximations they made resulted in three peaks which were adjusted with three parameters, leaving the impression that there was no problem whatsoever.

Coryell asked how far the r-process would go, and answered that it would continue until *fission will intervene*. But he observed that the abundance curve does not show any hump that could easily be correlated with fission products. Coryell's explanation was:

> It may be the result of averaging out of the many kinds of fission that may occur, with recycling of the fast-neutron fission products.

In other words, Coryell tacitly assumed many different neutron irradiation events taking place under a variety of conditions.

In 1963, Hoyle and Fowler[12] changed their minds about the astrophysical site where the r-process takes place. In a paper which suggests supermassive stars with masses above $10^4 M_\odot$ as an explanation for the newly discovered and completely enigmatic quasi-stellar objects, they conceded that the original idea for the site of the r-process, namely the explosive outburst of relatively low mass stars in the range 1.2–1.5$M_\odot$, was untenable because there was no way that matter could pass through states in which the conditions would be favorable for the r-process. As an alternative, Hoyle and Fowler now advocated the operation of the r-process in supermassive stars associated with the recently discovered quasi-stellar radio sources,[13] whose nature was under vociferous and emotional debate at the time. Hoyle and Fowler argued that:

> The materials ejected have an important bearing on nucleosynthesis, particularly in their relation to what we have called the r-process.

The idea of supermassive stars was meant to kill two problems with a single solution.

---

[12] Hoyle, F. and Fowler, W.A., Nature **197**, 533 (1963).

[13] Quasi-stellar radio sources, or QSOs, were discovered in the early 1950s as point radio sources, and all attempts to identify them with visible objects failed until Hazard, Mackey, and Shimmins succeeded in using lunar occultation to obtain a sufficiently accurate location in the sky to allow Schmidt (1963) to discover a faint stellar-like object with a redshift of 0.158, which was considered enormous at that time. The discovery of the redshift implied right away a distance indicating that the object was extremely powerful. As can be imagined, various hypotheses about the nature of these QSOs were tossed into the air before it became clear that the QSOs were at the heart of certain types of galaxies. One of the hypotheses about the nature of the QSOs, advocated by Hoyle and Fowler, was that they were supermassive stars of 10^4–$10^7 M_\odot$. Having made this suggestion to solve the problem of the QSOs, Hoyle and Fowler conjectured that QSOs might also harbor the r-process.

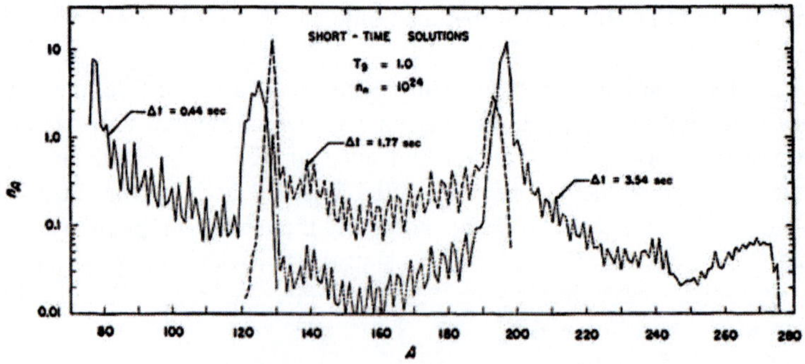

Fig. 13.9 The short time results for the r-process, assuming $T = 10^9$ K and $n = 10^{24}/\text{cm}^3$ or about 1.66 g/cm^3 of neutrons. After Seeger, Fowler, and Clayton 1965

The next major step in the r-process theory came with the publication of Seeger, Fowler, and Clayton,[14] a publication the authors described as *Handbuch der sr-Prozess*. Seeger et al. adopted the idea of exponential neutron irradiation along the same lines as discussed before for the s-process. With this assumption it became possible to solve the problem and obtain explicit expressions, in terms of the nuclear data, for the relative abundances. It became evident once more that the duration of the irradiation, which was a free parameter, would be crucial.

The calculation of Seeger and collaborators followed B2FH and assumed equilibrium, i.e., that the photodissociation[15] of the nuclei would be fast, and with it also the neutron capture. By fast we mean faster than the changing thermodynamic conditions so that the state of equilibrium is established 'instantaneously' relative to the astrophysical evolution. While the assumption does indeed tremendously simplify the calculation, it requires information about the number of possible nuclear states available for each nucleus. Unfortunately, this information is unavailable and must be extrapolated from the vicinity of the stability valley.

The most important conclusion drawn by Seeger et al. was:

> It is impossible in the r-process as described here to have any material reaching the third peak while it remains in the first peak. Thus we must say that solar r-process material was produced in two distinct types of events: one type lasting about 4 s to produce the first peak and most of the second, and another type lasting for long enough times that cycling occurs.

In other words, just the nuclear physics implies that at least two independent and not necessarily similar astrophysical sites must operate. The results of Seeger et al. which demonstrate the need for more than one irradiation event are shown in Figs. 13.9 and 13.10.

---

[14] Seeger, P.A., Fowler, W.A. and Clayton, D.D., Astrophys. J. Suppl. **11**, 121 (1965).

[15] Two reasons make photodissociation important in the r-process and unimportant in the s-process. Firstly, the s-process path is along tightly bound nuclei with mean binding energies per particle of 8 MeV. Secondly, the temperature at which the r-process takes place is about a factor of 10 higher, so the photons are more energetic.

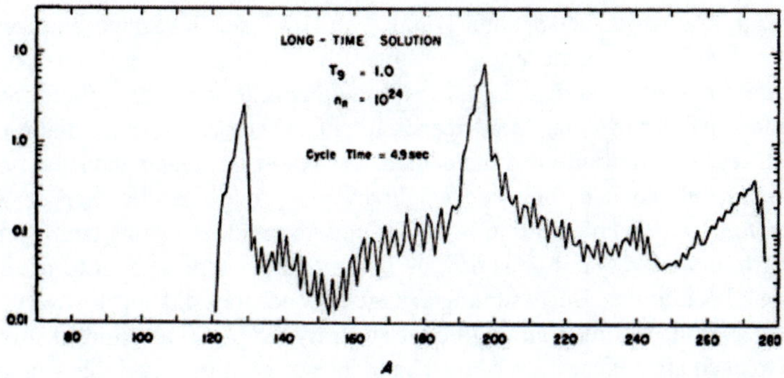

Fig. 13.10 The long time solution for the *r*-process abundances. The relative abundances become constant in time, but the absolute scale doubles every 4.9 s. After Seeger, Fowler, and Clayton 1965

As for the astrophysical site, Seeger, Fowler, and Clayton adopted Fowler's idea of supermassive stars and argued that the conditions for the *r*-process appear in massive stars with masses of $\sim 10^{5\pm1} M_{\odot}$. Since the more massive the star, the faster its evolution, the supermassive stars evolve extremely fast and *are thus capable of producing r-process elements early in the history of the Galaxy*, argued Seeger et al. Iben[16] and Fowler[17] showed that such stars were unstable and, in the wake of the collapse, would ignite the nuclear fuel, causing it to explode and thereby creating, so contended Seeger et al. the proper expanding environment for the *r*-process.

The aphorism of the paper, which certainly reflects Willie Fowler's artistry, was this:

> Some things are hurrying into existence, and others are hurrying out of it; and of that which is coming into existence, part is already extinguished. [Quoted from Marcus Aurelius.[18]]

Supermassive stars were among the highlights in the 1971 meeting on *Supermassive Objects in Astrophysics*, celebrating Willie Fowler's 60th birthday,[19] but they soon fell out of favor when it became evident that even more massive objects are formed in the Universe but soon fragment into smaller objects, whence the coherent supermassive star is not observed.

So far, the modeling was carried out assuming constant temperature and density. The first to improve upon this assumption was Schramm,[20] when he assumed that the *r*-process takes place dynamically, in the sense that the temperature and density are likely to change very quickly, as they would in an explosion. Schramm was actually

16 Iben, I. Jr., Astrophys. J. **138**, 1090 (1963).

17 Fowler, W.A., Astrophys. J. Suppl. **9**, 1 (1964).

18 Marcus Aurelius Antonius (121–180), *Meditations*, Harvard Classics 1909–1914, The Sixth book, 15.

19 Trimble, V., Nature **232**, 607 (1971).

20 Schramm, D.N., Astrophys. J. **185**, 293 (1973).

preceded by Cameron, Delano, and Truran[21] and by Sato, Nakazawa, and Ikeuchi.[22] However, Schramm's predecessors assumed that the rate of neutron capture was equal to the rate of photodisintegration (namely equilibrium) throughout the entire calculation, and in particular after 'freeze-out', i.e., the cessation of irradiation. When this incorrect assumption is implemented, the result is greater than the observed difference in abundances between neighboring r-process nuclei. Seeger, Fowler, and Clayton circumvented the problem by adding together results from somewhat different temperatures and densities, as if the reality were a mixture of different background conditions. But assuming a range of conditions did not cure the problem, and Seeger et al. assumed an additional arbitrary (unphysical) kind of mixing by adding to each abundance 15% of the abundance of each of its neighboring nuclei.

Schramm assumed that the temperature and density varied in an exponential way, and took the expansion time as a free parameter. He then succeeded in showing that the conditions assumed by Seeger et al. in which they got only two abundance peaks, actually yielded the three peaks.

In 1971, Cowan and Hantel[23] took the idea of fission-powered supernovae one step further. They estimated that the decay of the r-products could give rise to a high concentration of fissionable nuclides at the bottom of the valley of stability. Provided the expansion of the remnant is not too fast, they estimated that the matter could become supercritical and multiply the number of neutrons, leading to a gigantic atomic bomb powering the supernova even further. Arnett[24] analyzed this idea and concluded that, despite the non-existence of any better model:

> The site of the r-process is presently so poorly known that this idea cannot be completely ruled out.

In view of the current state of affairs of the r-process, Arnett's statement is still valid.

In 1974, Sato[25] repeated the dynamic calculation, but this time included the effect of the energy release by β-decays, and found that the three peaks could not be reproduced simultaneously. The only way to obtain the three peaks was by averaging over several different events. This is the present situation.

13.2 *r*-Process Signatures in the Solar System Abundance Curve

The prominent inability of the s-process to synthesize isotopes beyond barium, because the products undergo α-decay, is an unequivocal signal of the need for

21 Cameron, A.G.W., Delano, M.D. and Truran, J.W., *Conference of properties of nuclei far from the region of β-stability*, Leysin, Switzerland (1970).

22 Sato, K., Nakazawa, K. and Ikeuchi, S., Prog. Theo. Phys. **49**, 1166 (1973).

23 Cowan, G.A. and Hantel, E.G., Astrophys. J. **167**, 101 (1971).

24 Arnett, W.D., Ann. Rev. Astron. Astrophys. **11**, 73 (1973).

25 Sato, K., Prog. Theor. Phys. **51**, 726 (1974).

a special process to synthesize these elements. Equally, there are several isotopes that cannot be synthesized by the s-process and are on the neutron-rich side of the β-stability valley.

The most important observational data that the r-process theory has to explain is the Solar System abundance shown in Fig. 13.11, as derived by Seeger et al. after fitting the s-process contribution.[26] A more modern result is shown in Fig. 13.12. The major difference between the two lies in the treatment of the s-process elements. Once the fit to the s-process elements is complete, the 'leftover' abundances are attributed to the r-process. The conspicuous point is the existence of 'leftover' peaks which do not coincide with the peaks of the s-process. Next, there are four peaks with different widths. Unlike the peaks of the s-process, which are associated with the magic numbers, it does not appear at first sight to be the case with the r-process abundances. Indeed, given that the r-process path goes through the very neutron-rich side of the valley of stability, it is possible that the magic numbers are not universal, but depend on the height in the stability valley where the particular r-process passed.

13.3 *r*-Process Elements in the Sun

Obtaining the abundances of the r-process elements in the Sun is crucial for understanding the operation of this process in the cosmos. The solar abundances provide the ruler according to which any deviation from 'normal' can be identified, stated, and eventually explained. Any error in this fundamental measuring stick translates automatically into an 'anomaly' in need of explanation. By 'normal', we mean solar abundance. However, if in fact the the r-process elements were synthesized in several events occurring under different conditions and consequently exhibited a spectrum of abundances, then the solar composition would not be universal and we should not be surprised to find large deviations in the relative abundances throughout the Universe.

Goriely[27] investigated how far one should be able to trust the derived solar r-abundances, taking into account the uncertainties in the basic nuclear data and the missing but confirmed r-process. He concluded that:

> When considered simultaneously, the remaining uncertainties appear to be such that all the solar abundances of the s-only nuclei can be explained satisfactorily. The resulting uncertainties in the s-component of the solar abundances give rise to large errors in the solar r-abundances of the s-dominant isotopes.

---

[26] The fit is first made only to about thirty s-process elements. Once this is done, the resulting s-process contribution to the s- and r-process elements is estimated. Even today the predictions are based on a parametric model, referred to as the canonical exponential model, initially developed by Clayton et al. (Clayton D.D., Fowler W.A., Hull T.E., Zimmerman B.A., Ann. Phys. **12**, 331, 1961) and perfected by Käppeler et al. (Käppeler F., Beer, H., and Wisshak, K., Rep. Prog. Phys. **52**, 945, 1989). Only iron is assumed as a seed isotope in this model and the neutron densities and temperatures remain constant during neutron irradiation.

[27] Goriely, S., Astron. Astrophys. **342**, 881 (1999).

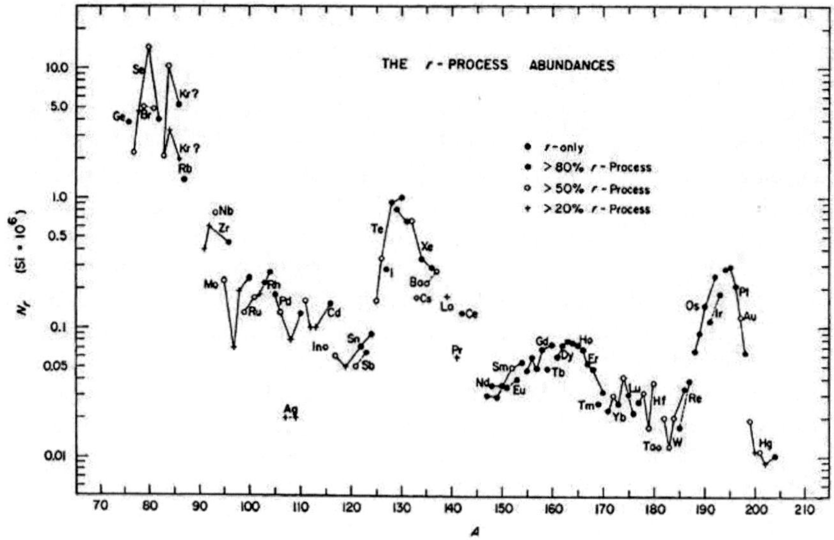

Fig. 13.11 Abundances of r-process elements in the Sun. Abundances of isotopes which belong to the same element are connected by a *continuous line* if Z is even and by a *broken line* if Z is odd. The gaps contain isotopes to which the contribution of the r-process is estimated at less than 20%. After Seeger, Fowler, and Clayton 1965

Goriely based his analysis on the multi-event, effectively multi-parameter, fit to the s-process elements, and ignored the possibility of many seed processes and time-variable neutron fluxes.

Another annoying problem in determining the r-abundances is that the fit for the s-process contribution is a general one which is designed to fit as many isotopes as possible. It is not a fit to each isotope. When a fit is an overall fit, the details are usually compromised.

Some r-process residuals are seen to suffer from remarkably large uncertainties, which quite clearly cannot be ignored when discussing the r-process and the virtues of one or another model for this process. This concerns in particular the elements Rb, Sr, Y, Zr, Ba, La, Ce, and Pb. Some of them, and in particular Ba or La, are often used as tracers for the levels of s- or r-processing during galactic history. Note that some researchers rush ahead and use these elements in other contexts as if the data were highly certain. Obviously, conclusions drawn in such cases are questionable. The uncertainties blur any picture one might try to draw from spectroscopic observations and from simplistic theoretical considerations.

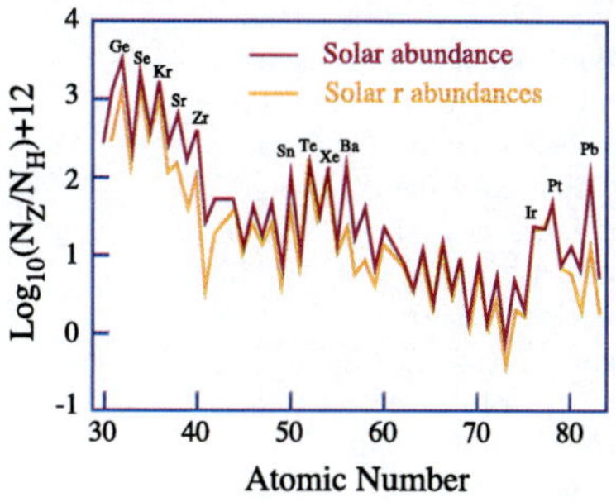

Fig. 13.12 Recent derivation of the solar r-process abundances. After Goriely, S., Astron. Astrophys. **342**, 881 (1999)

Just recently the community was taken aback when Aspund, Grevesse, and Sauval[28] discovered that the solar abundance of iron and a few major elements lighter than Ne must be corrected by a factor close to two.

13.4 Some Characteristics of the Process

The typical β-decay time is ~0.1 s, independently of A. If this is so, the time needed to convert the seed nucleus Fe into Cf is shorter than about 7 s. The initial phase of the r-process yields only unstable nuclei, while in the second phase, the nuclei decay towards the more stable ones. Consequently, there is a shift of about 10 mass units in the abundance maxima relative to the magic numbers.

The r-process builds up elements that are more and more unstable. Put another way, it climbs the walls of the valley of stability. This process continues until the newly formed nucleus disintegrates by a photoreaction. The temperature at which the r-process is assumed to take place is of the order of 10^9 K. But at such temperatures, photons with energies of about 2 MeV become abundant, so when the energy difference $B(A+1, Z) - B(A, Z)$ between the binding energies of neighboring nuclei amounts to about 2 MeV, any capture of a neutron is followed by a quick disintegration. The process thus effectively stops increasing the number of neutrons and waits for the β-decay to shift Z by one unit (see Fig. 13.13).

The problem in calculating the nuclei at which the r-process stops building a heavier nucleus and starts a new sequence is that the binding energies of nuclei far away from the stability valley are poorly known. It is simply difficult to measure them. As an escape route, one applies theoretical estimates, all based on the Bethe–von Weizsacker phenomenological formula, or one of the multitude of available variations. Two problems with such a procedure emerge right away:

[28] Asplund, M., Grevesse, N. and Sauval, A.J., in *Cosmic Abundances as Records of Stellar Evolution and Nucleosynthesis*, ed. by Barnes and Bash, ASP Conf. Series **336**, 25 (2005); Grevesse, N., Asplund, M., Sauval, A.J. and Scott, P., Astrophys. Space Science **328**, 179 (2010).

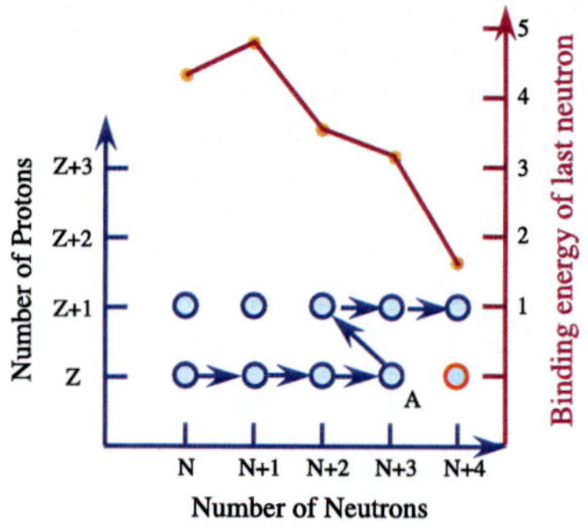

Fig. 13.13 Neutron absorption continues until the binding energy of the last neutron is below about 2 MeV, at which point the the energetic photons eject the additional neutron immediately and the nucleus has time to β-decay and increase Z. The binding energy of the last neutron is shown *on the right*

- The numerical coefficients used in the formula are found from a fit to the known nuclei, namely the stable nuclei near the bottom of the stability valley, but we know already that the nuclei far from the bottom may, and effectively do, behave differently.
- The fit of any phenomenological formula is to many nuclei with the logic that the more nuclei you attempt to fit to, the more accurate the formula is expected to become. This may be true for the average, but it is far from the truth in any particular case.

Add to this the discontinuous changes in the binding energies that take place at magic numbers and we begin to fully appreciate the hot water we are getting into! Yet, even this is not the entire story. At very high atomic weights, the nuclei become non-spherical, and it is to be expected that the further a nucleus is from the bottom of the valley of stability, the greater will be its non-sphericity. Naturally, this too will add to the uncertainty in the estimate of the binding energy, not to mention the probability of absorbing a neutron.

Estimates of the time nuclei spend at the end of the r-process chain before they decay yield about 1/2 s, when the nuclei are near magic numbers. Add to this about 50 more β-unstable nuclei which the process must encounter and we reach the conclusion that the entire r-process should take no more than a few seconds. There is simply no more time available, because the buildup of more and more unstable nuclei requires faster and faster rates. Classical estimates for $T \sim 10^9$ K lead to neutron densities of $N \sim 10^{24}$ n/cm^3. In other words, about one gram per centimeter cube is neutrons. (The total density of matter is several orders of magnitude higher.) The same estimates produce times of the order of 2–3 s for the entire synthesis, from the seed to the heaviest isotope. In the Big Bang, the original idea was that all the elements would be synthesized in just three minutes. In the r-process, all the HTI elements are synthesized in a few seconds!

When the atomic weight reaches 230–270, a new phenomenon takes place: fission. The buildup disintegrates into two large fragment nuclei and 2–3 neutrons. The fragments absorb neutrons and start the entire process again. This is called cycling, because the neutrons are cycled between the fission fragments and the heaviest nuclei which undergo fission.

At the neutron magic numbers, there will be a sequence of waiting points because the binding energy of the next neutron is very low, and as soon as the nucleus absorbs a neutron, a photon will kick it out. With no way to increase the number of neutrons any further, the nuclei will wait for β-decays to increase Z and move the nucleus into a region in which a significant neutron binding energy is reached. Only then can capture and buildup resume. From the time estimate given above, it can be expected that the resulting pileup of nuclei (with large numbers of neutrons) will lead to peaks in the abundance curve. Once the irradiation stops, the sequence of β-decays will bring the nuclei to the bottom of the valley of stability. This decay takes place at constant A.

Bearing in mind the above caveats, let us discuss possible sites for the r-process.

13.5 Site-Free High-Temperature *r*-Process Models

The reader will soon realize that we do not know where in the cosmos the r-process operates. Facing the problems of identifying appropriate astrophysical sites for the r-process to occur, the natural question that arises is the extent to which general properties of the r-process can be ascertained irrespective of the detailed characteristics and conditions of one site or another.

The r-process takes place when the neutron concentration is so high that neutron capture by nucleus A is faster than β-decay of the resulting nucleus $A + 1$, at least for a substantial number of neutron-rich nuclides that are not located too far from the valley of nuclear stability. Such conditions provide by definition a natural way to transform any pre-existing material into highly neutron-rich species. By high neutron concentration we mean concentrations as high as $10^{25}/\text{cm}^3$ or more. In other words, at least a few grams per cubic centimeter of neutrons. The hypothesized high neutron flux is crucial in all attempted realizations of the process, so one assumes a given neutron flux, fixed for some time or an exponential[29] function of time, and one disregards possible sources of such a flux. This is the buildup part.

The destruction of the heavy products, according to the prevailing assumption, is by photons. We know that, at a sufficiently high temperature, the photons are amply energetic to dissociate all nuclei. Hence, the photons determine the maximal height A for a given Z. Clearly, the higher the neutron flux, the greater the value of A before the photons are able to stop the growth of A. The tacit assumption is that the temperature is sufficiently high to supply enough photons with energy of more than, say, 2 MeV, so that photodisintegrations can counteract the neutron flux. However, the temperature is assumed to be not so high that all nuclei disintegrate, i.e., only the

[29] If the neutron flux and the physical conditions are constant, it is called the canonical model. If the thermodynamic conditions and neutron flux are allowed to vary in time, it is called the dynamic model.

very neutron-rich ones do, those in which the binding energy of the last neutron is about 2 MeV or less. This model therefore requires fine-tuning!

Then we need to expel the products from the furnace! Hence, the process cannot take place in matter that will undergo dramatic changes after irradiation, like falling into a black hole or a neutron star. The optimal situation is if the process takes place in matter doomed to be expelled from the star and dispersed in space, hopefully without additional reactions.

The above three sub-processes combine to form the r-process, against a background of changes in the nuclear properties for different numbers of neutrons and protons. The most important property is the closed shell and the magic number phenomenon. The binding energy of the neutron just beyond a magic number is particularly small, so any added neutron is quickly expelled, giving rise to an accumulation of nuclei at this point. Provided, of course, that magic numbers exist for very neutron-rich nuclei and keep their properties as we know them at the bottom of the valley of stability. However, other nuclear properties like deviation from spherical symmetry play a role in changing the properties of the nuclei.

As discussed above, it was soon found that a single irradiation event could not simulate the entire abundance curve, and more than one event is needed to save the model. This property of the r-process prevails today, and the popular way to circumvent it is by assuming what is referred to as a multi-event r-process model.[30] The r-process elements observed in a given star were, in all probability, synthesized in a previous generation of stars, expelled into space in a violent event, mixed, and then incorporated in the observed star. Hence, if indeed many events with a spectrum of properties (neutron fluxes, temperature, density, and time variability) contributed to the observed abundances, then the multi-event assumption does seem appropriate. However, this is a Pyrrhic victory! The mixing of a multitude, i.e., hundreds or even millions of events[31] would wash all unique signatures out of the synthesis. If the number of events contributing to the r-process elements in a given star is small, one should expect stars to expose a variety of abundance curves, and indeed this seems to be the case.

Once one assumes a site-free model, one can add additional free parameters, for example, the seed nuclei or the ratio between the protons (or electrons) and the total number of neutrons and protons, which of course embraces even unrealistic situations.

13.6 A Site-Free High-Density r-Process Scenario

Can we say something about the r-process which is environment-independent, that would prove the existence of the process? As early as 1965, an alternative to the

[30] Goriely, S. and Arnould, M., Astron. Astrophys. **312**, 327 (1996).

[31] Arnould, M., Goriely, S. and Takahashi, K., Phys. Rep. **450**, 97 (2007), p. 176.

high-T r-process canonical model was proposed by Tsuruta and Cameron.[32] The suggestion hinges on the fact that, as a result of the operation of free-electron captures by nuclei, very high densities (say $\rho > 10^{10}$ g/cm^3) can drive material deep into the neutron-rich side of the valley of nuclear stability. This so-called neutronization of the material can operate even at vanishingly low temperatures. The astrophysical plausibility of this scenario in accounting for the production of the r-nuclides has been questioned since its birth and consequently remained practically unexplored until the composition of the outer and inner crusts of neutron stars[33] and the decompression of cold neutronized matter resulting from tidal effects of a black hole on a neutron-star companion[34] were investigated. The decompression of cold neutron star matter has recently been the subject of further study.

The idea that collapse to high densities might lead to neutronization of the matter (electrons are captured by the protons and converted into neutrons) was proposed by Hund as early as 1936.[35] The original idea that neutrons could do the synthesis job can be traced to Flügge,[36] who published it in an observatory bulletin. But those who read such bulletins are apparently not the same as those interested in the collapse of stars! Consequently, none of these ideas were implemented in the theory of the synthesis of HTI nuclei before Tsuruta and Cameron brought them back from oblivion.

It is important, even crucial for the standard r-process that neutron absorption–γ emission reactions should be faster than neutron absorption followed by a β-decay. This situation may not be valid under all conditions. Goriely and Arnould[37] investigated this point and produced Fig. 13.14 (left), which indicates the domain of validity of the equilibrium assumption. The r-process can in principle take place when this assumption is not valid, but then its features will differ from those exhibited by the classical hot process.

13.7 The Multi-Event *r*-Process Model

Goriely and Arnould[38] and Bouquelle et al.[39] showed that, by proper mixing of the outcome of a large number of astrophysical events, the solar abundance can be quite well reproduced. In its formulation, it is identical to the multi-event model used for the s-process in the decomposition between the Solar System s- and r-abundances (see Sect. 12.5.2).

---

32 Tsuruta, S., Cameron, A.G.W., Can. J. Phys. **43**, 2056 (1965).

33 Baym, G., Bethe, H.A. and Pethick, C.J., Nucl. Phys. A **175**, 225 (1971).

34 Lattimer, J.M., Mackie, F., Ravenhall, D.G. and Schramm, D.N., Astrophys. J. **213**, 225 (1977).

35 Hund, F., Erg. d. exacten Naturwis. **15**, 189 (1936).

36 Flügge, S., Veroeffentlichungen der Universitaets-Sternwarte zu Goettingen **3**, 2 (1933).

37 Goriely, S. and Arnould, M., Astron. Astrophys. **312**, 327 (1996).

38 Goriely, S., and Arnould, M., Astron. Astrophys. **312**, 327 (1996).

39 Bouquelle, V., and 4 coauthors, Astron. Astrophys. **305**, 1005 (1995).

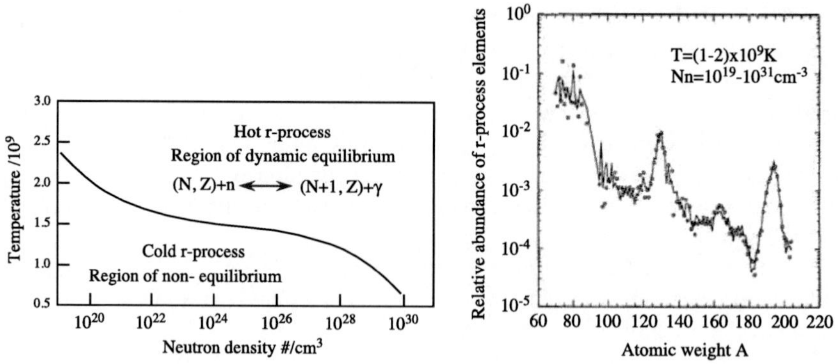

Fig. 13.14 *Left* the boundary between the hot and the cold *r*-processes is determined by the condition for dynamic equilibrium between neutron absorption and γ absorption. The curve is calculated according to the Hauser–Fershbach nuclear theory. See Goriely, S. and Arnould, M., Astron. Astrophys. **312**, 327 (1996). *Right* the multi-event fit of the *r*-process products to the solar composition by Goriely and Arnould 1996

The multi-event *r*-process model is a mixture of many events. But how many? From attempts to reproduce the Solar System abundance curve, one may conclude that about as many as 10^7 supernovae were needed for it to get its present shape. However, when such a large number of events is needed, one expects fluctuations to be small from one star to another, and this is not observed. On the other hand, if the *r*-process elements are a consequence of many events, this diminishes the potential to make inferences about the relevant parameters or where such isotopes are synthesized.

13.8 The Basic Problem: Lack of Nuclear Data—An Example

Since the path of the *r*-process lies along highly unstable nuclei, a typical problem today is the difficulty in measuring the nuclear data for isotopes which are frequently produced in a non-pure way (several isotopes are produced at the same time) and for a minute fraction of a second.

As an example, let us examine the nuclear properties of the actinides as given in Tables 13.2 and 13.3. These tables show the data for the actinides (nuclei with $Z \geq 90$) as they stood in the year 2000.[40] Table 13.2 gives the half-lives of the actinides along the valley of stability. The abundances of the stable nuclei and those with exceptionally long half-lives are also given. A notable omission from the table is ^{235}U, with half-life 7.038×10^8 years and natural abundance 0.72% of ^{238}U.

[40] The data is from the *Berkeley Laboratory Isotopes Project*, LBNL, 2000.

Table 13.2 The last elements

Atomic number Z	Symbol	Name	Longest stable isotope, decay product	Lifetime	Atomic abundance
82	Pb	Lead	Stable 206, 207, 208	0.05–0.28	
83	Bi	Bismuth	Stable 209		0.0016–0.005
84	Po	Polonium	^{209}Po	102 years	
85	At	Astatine	^{211}At	7.214 h	
86	Rn	Radon	^{222}Rn	3.8235 days	
87	Fr	Francium	^{223}Fr	21.8 m	
88	Ra	Radium	^{226}Ra	1600 years	
89	Ac	Actinium	^{227}Ac	21.773 years	
90	Th	Thorium	^{232}Th	1.405×10^{10} years	0.026
91	Pa	Protactinium	^{231}Pa	32,760 years	
92	U	Uranium	^{238}U	4.468×10^{9} years	0.0072–0.0079
93	Np	Neptunium	^{237}Np	2.144×10^{6} years	
94	Pu	Plutonium	^{239}Pu	24,110 years	
95	Am	Americium	^{243}Am	7,370 years	
96	Cm	Curium	^{247}Cm	1.56×10^{7} years	
97	Bk	Berkelium	^{247}Bk	1,380 years	
98	Cf	Californium	^{251}Cf	898 years	
99	Es	Einsteinium	^{252}Es	471.7 days	
100	Fm	Fermium	^{257}Fm	100.5 days	
101	Md	Mendelevium	^{258}Md	51.5 days	
102	No	Nobelium	^{257}No	25 s	
103	Lr	Lawrencium	^{262}Lr	3.6 h	
104	Rf	Rutherfordium	^{263}Rf	10 m	
105	Db	Dubnium	^{268}Db	16 h	
106	Sg	Seaborgium	^{266}Sg	21 s	
107	Bh	Bohrium	^{272}Bh	9.8 s	
108	Hs	Hassium	^{277}Hs	12 m	
109	Mt	Meitnerium	^{276}Mt	0.72 s	
110	Ds	Darmstadtium	^{281}Ds	1.1 m	
111	Rg	Roengenium	^{280}Rg	3.6 s	
112	Uub	Ununbium	^{285}Uub	10 m	
113	Uut	Ununtrium	^{284}Uut	0.48 s	
114	Uuq	Ununquadium	^{289}Uuq	21 s	
115	Uup	Ununpentium	^{288}Uup	87 ms	
116	Uuh	Ununhexium	^{292}Uuh	0.6 ms	

The abundance is calibrated to the abundance of Si $= 10^6$. Values taken from Ehmann, W.D., J. Chem. Ed. **38**, 53 (1961)

Now consider Tables 13.4 and 13.5. Green letters indicate that the decay time is of the order of the r-process timescale, while black indicates that the decay time is longer. Thus, isotopes in black accumulate neutrons. Underlined isotopes decay via α-decay, and hence, if the process gets stuck at such a nucleus, it does not

Table 13.3 Known decay times for the actinides: the proton-rich side

A	Z = 82	Z = 83	Z = 84	Z = 85	Z = 86	Z = 87	Z = 88	Z = 89	Z = 90	Z = 91	Z = 92	Z = 93
181	45 ms											
182	55 ms											
183	300 ms											
184	0.55 s											
185	4.1 s	44 μs										
186	4.83 s	15.0 ms										
187	**18.3 s**	35 ms										
188	**24 s**	0.21 s										
189	**51 s**	680 ms										
190	**1.2 m**	6.3 s	2.0 ms									
191	**1.33 m**	12 s	15.5 ms									
192	**3.5 m**	**37 s**	0.0332 s									
193	**2 m**	**67 s**	0.42 s	40 ms								
194	**12.0 m**	**95 s**	0.392 s	40 ms								
195	**15 m**	**183 s**	4.64 s	0.63 s								
196	**37 m**	**308 s**	5.8 s	0.253 s	3 ms							
197	**8 m**	**9.33 m**	53.6 s	0.35 s	65 ms							
198	**2.40 h**	**10.3 m**	1.77 m	4.2 s	64 ms							
199	**90 m**	**27 m**	**5.48 m**	7.2 s	0.62 s							
200	**21.5 h**	**36.4 m**	**11.5 m**	43 s	0.96 s	19 ms						
201	**9.33 h**	**108 m**	**15.3 m**	89 s	7.0 s	48 ms						

Empty locations belong either to unstable nuclei for which we have no data or to nuclei that do not exist. *Underlined* isotopes decay via α-decay, isotopes written in *bold face* decay via electron capture, and a *double underline* means decay via emission of a proton progress towards isotopes with higher Z, but on the contrary decays to elements with $Z - 2$. Such isotopes do not allow the progress of the process to higher Z elements. Similarly, isotopes written in bold face decay via electron capture (an electron that passes through the nucleus is captured by a proton and converted to a neutron). Electron capture[41] reduces Z, and hence, like an α-decay, stops the advance of the process to higher Z elements. Isotopes which decay via the emission of an electron (the inverse of the previous process) increase the atomic weight and are fine from the point of view of the r-process. Thus, the r-process requires the isotopes which appear black (decay time longer than, say, 0.1 s) but are neither underlined nor boldfaced. A double underline means decay via emission of a proton, which again reduces Z. This is a rare mode of decay and it also cuts short the r-process. Finally, isotopes which undergo spontaneous fission are shown in blue and clearly prevent the advancement of the process.

Note that, for all elements, the isotope with the highest number of neutrons decays with half-life longer than the timescale of the assumed r-process. That is to say that the data for the nucleus with the maximum number of neutrons, which β-decays into the nucleus with higher Z, is not known and must be calculated from theory.

In the case of polonium, all established isotope decay times are for either α-decay or electron capture, and there is no experimental information about neutron-rich polonium isotopes which decay via inverse β and thus allow the progress of the r-process towards higher Z values.

Finding the universal abundances of the actinides is problematic, and at best we have ratios like U/Th measured from meteorites.[42] Some heavy elements (with atomic number $A > 69$) are produced by the r-process, where lighter elements serve as seeds. Although this is characteristic of supernovae and neutron star mergers, uncertainties persist about where the r-process occurs, because stellar models are too crude to allow the precise quantification of this phenomenon. As a result, there are many uncertainties and assumptions in the models used to calculate the production ratios of actinides. Current estimates of the U/Th production ratio range from ~0.4–0.7.

13.8.1 The Nuclear Binding Energy

[41] The electron in the lowest energy state spends a small but finite amount of time moving in the nucleus, leading to the reaction $(Z, N) + e^- \rightarrow (Z-1, N+1) + \gamma$, i.e., while moving through the nucleus, it can be trapped by a proton and converted into a neutron. This happens in the laboratory. The situation in stars is different. Only the elements with very high Z keep the most strongly bound electrons that move through the nucleus. However, the electrons released by all ionized elements create a sea of electrons which move around freely, and in particular can pass through the nucleus. These electrons can be captured by the nuclei and convert a proton into a neutron. Thus, dense stellar environments enhance the rate of electron capture.

[42] Dauphas, N., Nature **435**, 1203 (2005).

Table 13.4 Known decay times for the actinides: the neutron-rich side

A	*Z* = 82	*Z* = 83	*Z* = 84	*Z* = 85	*Z* = 86	*Z* = 87	*Z* = 88	*Z* = 89	*Z* = 90	*Z* = 91	*Z* = 92	*Z* = 93	*Z* = 94
202	**5.25e4 years**	**1.72 h**	**44.7 m**	**184 s**	10.0 s	0.34 s	0.7 ms						
203	**51.873 h**	**11.76 h**	**36.7 m**	**7.4 m**	45 s	0.55 s	1.0 ms						
204	1.4e17 years	**11.22 h**	**3.53 h**	**9.2 m**	1.24 m	1.7 s	59 ms						
205	**1.53e7 years**	**15.31 days**	**1.66 h**	**26.2m**	2.8 m	3.85 s	210 ms						
206	stable	**6.243 days**	**8.8 days**	**30.0 m**	5.67 m	15.9 s	0.24 s	22 ms					
207	stable	**31.55 years**	**5.80 h**	**1.80 h**	**9.25 m**	14.8 s	1.3 s	22 ms					
208	stable	**3.68e5 years**	2.898 years	**1.63 h**	24.35 m	59.1 s	1.3 s	95 ms					
209	3.253 h	stable	102 years	**5.41 h**	**28.5 m**	50.0 s	4.6 s	0.10s					
210	22.3 years	5.013 days	138.3 days	**8.1 h**	2.4 h	3.18 m	3.7 s	0.35 s	9 ms				
211	36.1 m	2.14 m	0.516 s	**7.214 h**	**14.6 h**	3.10 m	13 s	0.25 s	37 ms				
212	10.64 h	60.55 m	0.299 μs	0.314 s	23.9 m	**20.0m**	13.0 s	0.93 s	30 ms	5.1 ms			
213	10.2 m	45.59 m	4.2 μs	125 ns	25.0 ms	34.6 s	2.74 m	0.80s	140 ms	5.3 ms			
214	26.8 m	19.9 m	164.3 μs	558 ns	0.27 μs	5.0 ms	2.46 s	8.2 s	100 ms	17 ms			
215	36 s	7.6 m	1.781 ms	0.10 ms	2.30 μs	86 ns	1.59 ms	0.17 s	1.2 s	14 ms			
216		3.6 m	0.145 s	0.30 ms	45 μs	0.70 μs	182 ns	0.33 ms	0.028 s	0.20 s			
217		97 s	10 s	32.3 ms	0.54 ms	22 μs	1.6 μs	69 ns	0.252 ms	4.9 ms			
218			3.10 m	1.5 s	35 ms	1.0 ms	25.6 μs	1.08 μs	109 ns	0.12 ms	1.5 ms		
219				56 s	3.96 s	20 ms	10 ms	11.8 μs	1.05 μs	53 ns	42 μs		
220				3.71 m	55.6 s	27.4 s	18 ms	26.4 ms	9.7 μs	0.78 μs			
221				2.3 m	25 m	4.9 m	28 s	52 ms	1.68 ms	5.9 μs			
222				54 s	3.823 days	14.2 m	38.0 s	5.0 s	2.8 ms	2.9 ms	1.0 μs		
223				50 s	23.2 m	21.8 m	11.435 days	2.10 m	0.60 s	6.5 ms	18 μs		
224					107 m	3.33 m	3.66 days	**2.78 h**	1.05 s	0.79 s	0.9 ms		
225					4.5 m	4.0 m	14.9 days	10.0 days	8.72 m	1.7 s	6 ms		
226					7.4 m	49 s	1600 years	29.37 h	30.57 m	1.8m	0.35 s	35 ms	

(continued)

Table 13.4 (continued)

A	*Z* = 82	*Z* = 83	*Z* = 84	*Z* = 85	*Z* = 86	*Z* = 87	*Z* = 88	*Z* = 89	*Z* = 90	*Z* = 91	*Z* = 92	*Z* = 93	*Z* = 94
227					22.5 s	2.47 m	42.2 m	21.773 years	18.72 days	38.3 m	1.1m	0.51 s	
228					65 s	38 s	5.75 years	6.15 h	1.9116 years	22 h	9.1 m	**61.4s**	4 ms
229						50 s	4.0 m	62.7 m	7340 years	**1.50 days**	**58m**	4.0 m	
230						19.1 s	93 m	122 s	7.538e4 years	**17.4 days**	20.8 days	**4.6 m**	
231						17.5 s	103 s	7.5 m	25.52 h	32,760 years	**4.2 days**	**48.8 m**	
232						5 s	250 s	119 s	1.405e10 years	1.31 days	68.9y	**14.7 m**	**34.1 m**
233							30 s	145 s	22.3 m	26.967 days	1.592e5 years	36.2 m	**20.9 m**
234							30 s	44 s	24.10 days	6.70 h	2.455e5 years	**4.4 days**	8.8 h
235									7.1 m	24.5 m	7.038e8 years	396.1 days	**25.3 m**
236									37.5 m	9.1 m	2.342e7 years	**1.54e5 years**	2.858 years
237									5.0 m	8.7 m	6.75 days	2.144e6 years	45.2 days
238										2.3 m	4.468e9 years	2.117 days	87.7 years
239											23.45 m	2.3565 days	24,110 years
240											14.1 h	61.9 m	6563 years
241												13.9 m	14.35 years
242											16.8 m	5.5 m	3.733e5 years
243												1.8 m	4.956 h
244												2.29 m	8.08e7 years
245													10.5 h
246													10.84 days
247													2.27 days

Empty locations belong either to unstable nuclei for which we have no data or to nuclei that do not exist. *Underlined* isotopes decay via α-decay, isotopes written in *bold face* decay via electron capture, and a *double underline* means decay via emission of a proton. *Green* letters indicate that the decay time is of the order of the *r*-process timescale, while *black* indicates that the decay time is longer.

Table 13.5 Known decay times for the actinides: the last elements

A	$Z = 95$	$Z = 96$	$Z = 97$	$Z = 98$	$Z = 99$	$Z = 100$	$Z = 101$	$Z = 102$	$Z = 103$
232	**79 s**								
233									
234	2.32 m								
235	15 m								
236									
237	**73.0 m**	2.1 s							
238	**98 m**	**2.4 h**	**144s**	21 ms					
239	**11.9 h**	**2.9h**	39 s						
240	**50.8 h**	27 days	**4.8m**	1.06m					
241	432.2 years	**32.8 days**	**3.78 m**	9 s					
242	16.02 h	162.8 days	**7.0 m**	3.49 m	40 s	0.8 ms			
243	7,370 years	29.1 years	**7.0 m**	**10.7 m**	**21 s**	0.18 ms			
244	10.1 h	18.10 years	**4.35 h**	19.4 m	**37 s**	3.3 ms			
245	2.05 h	8,500 years	**4.94 days**	**45.0 m**	**1.1 m**	4.2 s	0.35 s		
246	39 m	4,730 years	**1.80 days**	35.7 h	**7.7 m**	1.1 s	1.0 s		
247	23.0 m	1.56e+7 years	1,380 years	**3.11 h**	**4.55 m**	35 s	35s		
248		3.40e+5 years	9 years	333.5 days	**27 m**	36 s	**7 s**		
249		64.15 m	320 days	351 years	**102.2 m**	**2.6 m**	**24 s**		
250		9,000 years	3.217 h	13.08 years	**8.6 h**	30 m	**52 s**	0.25 ms	
251		16.8 m	55.6 m	898 years	**33 h**	**5.30 h**	**4.0 m**	0.8 s	
252		2 days		2.645 years	471.7 days	25.39 h	**2.3 m**	2.30 s	
253				**17.81 days**	20.47 days	**3.00 days**	**6 m**	1.7 m	1.3 s
254				60.5 days	275.7 days	3.240 h	**10 m**	55 s	13 s
255				85 m	39.8 days	20.07 h	**27 m**	3.1m	22 s

(continued)

Table 13.5 (Continued)

A	$Z = 95$	$Z = 96$	$Z = 97$	$Z = 98$	$Z = 99$	$Z = 100$	$Z = 101$	$Z = 102$	$Z = 103$
256				12.3 m	25.4 m	157.6 m	**78.1 m**	2.91 s	28 s
257						100.5 days	**5.52 h**	25 s	0.646 s
258						370 μs	51.5 days	1.2 ms	3.9 s
259						1.5 s	96 m	58 m	6.3 s
260							31.8 days	106 ms	180 s
261									39 m
262									216 m

Empty locations belong either to unstable nuclei for which we have no data or to nuclei that do not exist. *Underlined* isotopes decay via α-decay, isotopes written in *bold face* decay via electron capture, and a *double underline* means decay via emission of a proton. Isotopes which undergo spontaneous fission are shown in *blue*.

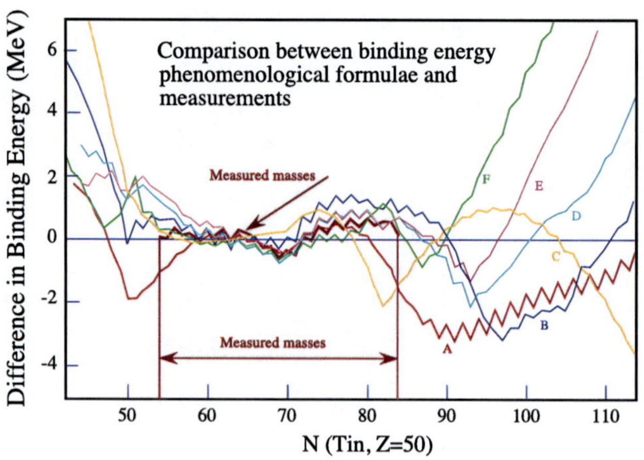

Fig. 13.15 Comparison between several phenomenological formulae for the binding energy and the measured binding energies of tin isotopes. After NuPECC (Associated Committee of the European Science Foundation) Report, *Nuclear Physics in Europe: Highlights and Opportunities*, ed. by Vervier, J., and 7 coeditors, December 1997

Figure 13.15 shows how a small randomly selected group of phenomenological formulae describe the nuclear binding energies. The formulae are obtained from a general fit to all of the isotopes with known binding energies, over 2,000 in total. The figure shows the quality of the fit to the different isotopes of tin. The effect on tin has been chosen here because tin is the element with the most stable isotopes (ten), and hence with the best data for a fixed Z.[43] The large number of stable isotopes is associated with the fact that $Z = 50$ is a proton magic number. As can be seen, the fit to the binding energies of the stable isotopes is reasonable and the different formulae provide similar and close answers. However, when one considers isotopes outside the range of stability, the different formulae deviate from each other and even diverge. So which formula should we trust? In effect, this kind of behavior is common to all fitting formulae, not just those for binding energies, i.e., fitting formulae are not to be trusted outside the range where they have been fitted.

13.9 Neutron Absorption Data

The situation with neutron absorption probabilities is not much better. Figure 13.16 shows, with reference to Brown and Loyola,[44] a summary of the data available for neutron absorption by the actinides. Observing the figure, one soon realizes that the

[43] Against all odds, the doubly magic nucleus $^{100}Sn_{50}$ is highly unstable and its half-life is 1.1 s.

[44] Brown, D.A. and Loyola, B., UCRL-WEB-210095, 2005-01-31.

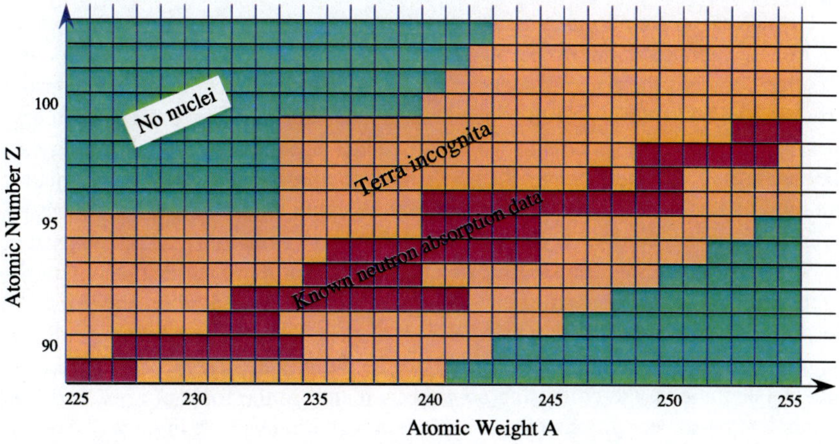

Fig. 13.16 A summary of the available neutron absorption data for the actinides. *Pink* implies reliable data. *Orange* means there is no data, and *green* means there are no nuclei with that particular number of protons and neutrons. After Brown and Loyola, *Big actinide scorecard*, 2005

situation concerning neutron absorption data is no better than for decay times to the ground state.

13.10 Where Could the *r*-Process Operate? Suggested Sites

The *r*-process requires extreme conditions, such as prohibitively large concentrations of neutrons, high (though not too high) temperatures, etc., so it is no wonder one has to look for esoteric astrophysical phenomena as possible sites for the *r*-process to have its few seconds of grace. Let us consider here a few scenarios, suggested on the basis of the above missing knowledge in nuclear physics.

13.10.1 Neutron Star Mergers

Even though a merger of two neutron stars, both products of supernovae, may sound fictitious, we note that such a merger is the best possible source of gravitational waves.[45] Ongoing experiments attempting to detect gravitational waves may soon tell us how frequent the phenomenon is.

---

[45] Gravitational waves represent one of the great challenges of present-day fundamental physics. So far, no direct detection of gravitational waves has been confirmed, but the chances of doing so

It is natural and tempting to assume that the surroundings of a neutron star, where neutrons are abundant, might be a place where the r-process occurs. The fundamental question concerns the circumstances under which neutrons become available for the process. The neutrons in a neutron star exist at a density well above the neutron drip line. Then we need a process in which the matter decompresses while preserving the neutrons. Processes that might result in decompression of neutronized matter are tidal forces of collisions between a black hole and a neutron star, as suggested by Lattimer and Schramm,[46] or a collision between two neutron stars, as suggested by Symbalisty and Schramm.[47] But apparently, none of these authors were terribly happy with the above suggestions, for in 1985, Symbalisty, Schramm, and Wilson[48] described the above two ideas as *exotic*, and suggested that the fast-rotating collapsing model of Symbalisty[49] is more likely to be the proper site. According to their model calculations, the collapsing core ejects matter in the form of a jet, and it is in this jet that the conditions for the r-process would prevail. In Fig. 13.17, we see the basis for the author's conclusion about the formation of a jet. Unfortunately, further calculations regarding jets, and jets are a frequent phenomenon, have so far failed to actually produce one. The formation of jets has thus become one of the most pressing problems in astrophysics. Many different kinds of jets are seen, but none is predictable from scratch. Last but not least, the authors did not carry out a full calculation, and were content with a global estimate. According to their estimate, if every supernova generated such a jet, there would be an overproduction of r-process elements. They thus concluded that, within the uncertainties inherent in the assumptions, the problem of overproduction should not arise.

The decompression of cold neutron matter was investigated by Lattimer et al. who showed that the process is more complicated than the hot r-process, because it starts with neutrons and protons and not with iron, and it advances through even more neutron-rich isotopes. In a way, this is a modern variation of the $\alpha\,\beta\,\gamma$ theory. If the neutron magic numbers are valid for such neutron rich isotopes, than the subsequent β-decays should end at the bottom of the valley of stability at a lower Z value

(Footnote 45 continued)
improved recently with the discovery by Burgay and colleagues (Burgay, M., and 12 coauthors, Nature **426**, 504, 2003) of two neutron stars that orbit around each other in just 2.4 h. This orbital period is three times shorter than that of the Hulse–Taylor binary pulsar, hitherto the closest known double neutron star. Double neutron star systems are considered to be one of the best sources of gravitational waves. The energy loss via gravitational waves causes the neutron stars to spiral in until the two stars collide and merge. During the last minute of their lives, there is an enormous release of gravitational radiation that is likely to be detectable on Earth. Several instruments have been built that would be capable of picking up such gravitational waves, including VIRGO in Italy (http://www.virgo.infn.it/central.html), GEO600 in Germany (http://www.geo600.org/), TAMA in Japan (http://tamago.mtk.nao.ac.jp/), and LIGO, the Laser Interferometer Gravitational Wave Observatory, based on sites in Washington and Louisiana, USA (http://www.ligo.caltech.edu).

46 Lattimer, J.M. and Schramm, D.N., Astrophys. J. Lett. **192**, L145 (1974).

47 Symbalisty, E., and Schramm, D.N., Astrophys. J. Lett. **22**, 143 (1982).

48 Symbalisty, E.M.D., Schramm, D.N., and Wilson, J.R., Astrophys. J. Lett. **291**, 11 (1985).

49 Symbalisty, E.M.D., Astrophys. J. **285**, 729 (1984).

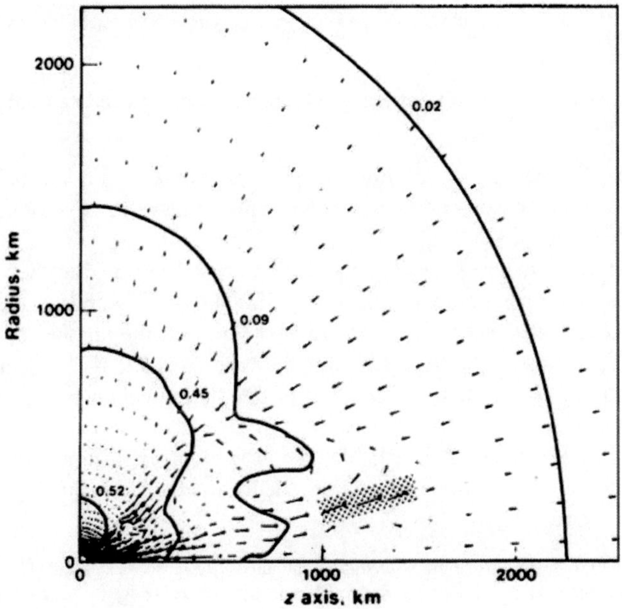

Fig. 13.17 The state at the end of the calculation. *Curves* are isodensity curves in units of 10^8 g/cm^3. The neutron star is shrouded by an imploding envelope extending over 2,000 km. One can see the beginning of a jet close to the *horizontal axis*. The *dotted rectangle* denotes a region where the velocity is greater than the speed of sound, and it is the existence of such a region which the authors interpreted as a jet with a total mass of $10M_\odot$. After Symbalisty, Schramm, and Wilson 1985

than the one where the classical r-process ends, which is an additional problem. No comparison with the observed abundances was carried out.

In 1989, Meyer[50] returned to the decompression problem for neutron matter, this time allowing for temperature changes due to energy released in the process. In principle, Meyer demonstrated that the decompression of the cold neutron-rich matter would release enough energy to heat the matter to r-process temperatures. Meyer concluded that:

> One advantage for decompressing initially cold neutron star matter as the mechanism for the r-process is that it tends always to give rise to r-process conditions, regardless of the hydrodynamics.

A kind of plausibility argument which is raised from time to time[51] is the following. The overall fraction of mass in the form of r-process nuclei is $\sim 10^{-7}$, which means that, out of the $10^{11}M_\odot$ of material in the Milky Way, about $\sim 10^4M_\odot$ consists of r-process elements. The rate of supernovae in the Galaxy is about one per century and the age of the Solar System is roughly $\sim 10^{10}$ years. Hence, the demand on SNe

---

[50] Meyer, B.S., Astrophys. J. **343**, 254 (1989).

[51] Mathews, G.J. and Cowan, J.J., Nature **345**, 491 (1990).

is to convert $\sim 10^{-4}$ of their mass into r-elements. Mathews and Cowan thus make the following inference:

> The r-process either occurs in a small subset of supernovae, or it occurs in a small region of the supernova.

In fact, a smaller region than the minimal size that is taken into account in supernova models. In other words, the process takes place in such a small volume of the supernova that it is usually neglected.

In 1999, Freiburghaus, Rosswog, and Thielemann[52] returned to the problem of neutron star mergers and found that the process might indeed synthesize the r-process elements, but only in a very narrow range of the ratio of the number of protons to the number of neutrons. On the other hand, the outer layers in the neutron star have a variable ratio of protons to neutrons, so very fine tuning would be required for the process to operate.

In 2001, Sumiyoshi et al.[53] calculated the prompt explosion[54] of an $11 M_\odot$ supernova model. They compared their result to a calculation by Käpper et al.[55] and concluded that:

> The results are in good agreement with solar abundances of r-process elements for $A > 100$ [...] such events may be responsible for the abundances of the heaviest r-process nuclei.

One may check Fig. 13.18 to try to convince oneself.

In 2003, Wanajo et al.[56] repeated the basic idea of Sumiyoshi et al. but for an 8–$10 M_\odot$ model with a core composed of O-Ne-Mg. The first attempts yielded no r-process elements whatsoever. They then heated the matter artificially and found that, for a narrow range of the proton to neutron number ratio, they could reproduce the observed r-process abundances. However, the total amount of r-elements thereby produced exceeds the galactic requirement of $10^4 M_\odot$ by two orders of magnitude. They solved the problem by stipulating that only a small fraction of the elements thus synthesized escapes from the neutron star, i.e., most of the ejected matter falls back onto the core and only a small fraction, which was not estimated, escapes.

Hillebrandt, Kodama, and Takahashi[57] calculated the r-process in a SN explosion and assumed the r-process to take place in $0.43 M_\odot$ of the mass. Such a high mass is way off the estimate of $10^{-4} M_\odot$ by Mathews and Cowan. If correct, it means flooding the Galaxy with r-process elements. In another calculation, Kitaura, Janka,

52 Freiburghaus, C., Rosswog, S. and Thielemann, F.-K., Astrophys. J. **525**, 121 (1999).

53 Sumiyoshi, K., and 5 authors, Astrophys. J. **562**, 880 (2001).

54 When the iron core of a massive star collapses gravitationally, a shock wave forms and rebounds as the density in the core reaches nuclear densities. The rebounding shock stalls as a result of energy losses to nuclear dissociation and neutrinos. This is the 'prompt' phase of the supernova mechanism. The subsequent phase is a 'delayed' mechanism, whereby the shock is reenergized by heating of material by the intense neutrino flux (Bethe, H.A. and Wilson, J.R., Astrophys. J. **295**, 14 1985).

55 Käpper, F., Beer, H. and Wisshak, K., Rep. Prog. Phys. **52**, 945 (1989).

56 Wanajo, S., and 6 coauthors, Astrophys. J. **593**, 968 (2003).

57 Hillebrandt, W., Kodama, T. and Takahashi, K., Astron. Astrophys. **52**, 63 (1976).

Fig. 13.18 The abundances found by Sumiyoshi et al. 2000 in the neutrino-driven wind. *Dots* are abundances from Käppeler et al. 1989

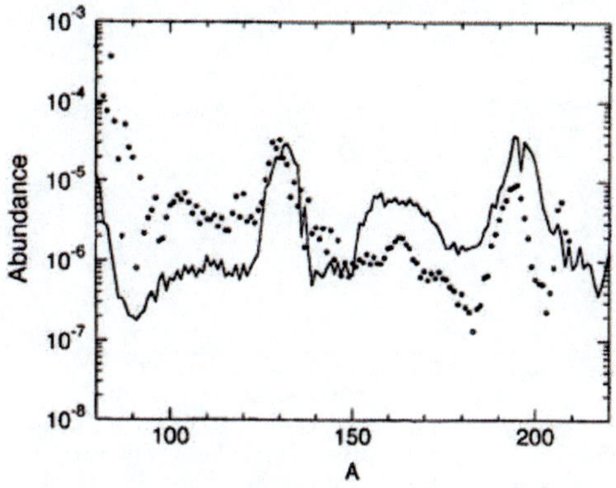

and Hillebrandt[58] calculated the collapse of an 8–10$M_\odot$ core of a star[59] composed of O-Ne-Mg, just the model considered by B2FH. They found that the matter ejected during the first second has proton to (proton plus neutron) ratio ≥ 0.46, and no matter with proton to (proton plus neutron) ratio $\ll 0.46$ is ejected. But as the authors concluded, the conditions under which the matter is ejected exclude the operation of the *r*-process.

13.10.2 r-Process in Neutrino Winds

The problem of ejecting the products led to the idea that, if the *r*-process could take place in a neutron-rich wind, then part of the problem could be solved. Meyer et al.[60] had the idea of a neutrino-driven wind and this was elaborated by Woosley et al.[61] The basic idea was that the wind should start where matter is characterized by nuclei with an excess of neutrons. The matter disintegrates upon expansion into nucleons, and later mostly reassembles into α particles plus free neutrons. The α particles reassemble to form the heavy seed nuclei on which the surplus of neutrons can be absorbed to form the *r*-process elements. If on top of this condition the wind has the correct density and timescale, and operates to yield the proper total ejected mass, then it looks as though we have a viable model/site for the *r*-process.

But how do we create the 'correct' conditions? Woosley et al. assumed a wind heated by neutrinos that emerge from the collapsing core. We recall that, if one

---

58 Kitaura, F.S., Janka, H.-Th. and Hillebrandt, W., Astron. Astrophys. **450**, 345 (2006).

59 The timescale of the envelope is much longer than the timescale of the core, so it is sensible to assume that the envelope is frozen during the collapse of the core. Astrophysicists use the term 'frozen' to imply that the composition does not change any further, a kind of abeyance. There is no change of phase as the term might be taken to imply.

60 Meyer, B.S., and 4 coauthors, Astrophys. J. **399**, 656 (1992).

61 Woosley, S.E., and 4 coauthors, Astrophys. J. **433**, 229 (1994).

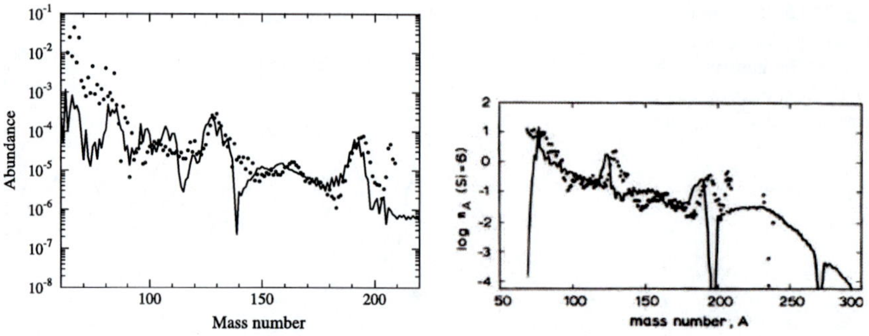

Fig. 13.19 *Left* comparison (*solid line*) between the calculation of Sumiyoshi et al. and the solar r-element abundances (*filled circles*) as calculated by Käpper, Beer, and Wisshak 1989. After Sumiyoshi et al. 2001. *Right* the *continuous line* is the theoretical model of Hillebrandt et al. 1976 and the *full circles* are r-process abundances according to Allen, Gibbons, and Macklin, in *Advances in Nuclear Physics* **4**, ed. by Baranger and Vogt, Plenum Press, 1971. After Hillebrandt et al. 1976

assumes that the wind is shock heated, the problem is that the model is too successful, in the sense that too much r-process matter is produced. Woosley et al. showed that, in the neutrino-driven explosion, only a small amount of matter is heated to the required high temperatures, thus circumventing the overproduction problem.

But shortly afterwards, Hoffman, Woosley, and Qian[62] stated that:

> We find that the standard wind models derived in that paper are inadequate to make the r-process, though they do produce some rare species above the iron group.

Indeed, r-process elements were produced in the model, but they did not resemble the observed solar abundances. On the other hand, Otsuki and collaborators[63] claimed that including the effects of general relativity would change the situation, and they managed to find a set of parameters for the wind under which the r-process advances up to $A \sim 200$. In another publication by the same group, Sumiyoshi et al.[64] derived the predicted abundances of the r-process as found in this model. The results are shown in Fig. 13.19 (left). It seems again that several winds with different parameters are needed to obtain a reasonable agreement. Wanajo[65] expanded the model and argued that:

> Our results have confirmed that the neutrino-driven wind scenario is still a promising site in which to form the solar r-process abundances. However, our best results seem to imply both a rather soft neutron-star equation of state and a massive proto-neutron star that is difficult to achieve with standard core-collapse models. We propose that the most favorable conditions perhaps require that a massive supernova progenitor forms a massive proto-neutron star by accretion after a failed initial neutrino burst.

---

[62] Hoffman, R.D., Woosley, S.E. and Qian, Y.-Z., Astrophys. J. **482**, 951 (1997).

[63] Otsuki, K., Tagoshi, H., Kajino, T., and Wanajo, S., Astrophys. J. **533**, 424 (2000).

[64] Sumiyoshi, K., and 4 coauthors, PASJ **52**, 601 (2000).

[65] Wanajo, S., and 3 coauthors, Astrophys. J. **554**, 578 (2001).

In other words, it is not clear that such a site exists, and if it exists, that it allows the process to do what it is expected to do.

In parallel, Thompson, Burrows, and Meyer[66] carried out an exhaustive investigation of neutrino-driven proto-neutron star winds and reached the conclusion:

> The potential for the *r*-process in proto-neutron star winds remains an open question.

And this despite the failure to find a proper set of parameters and a scenario for the *r*-process.

Recently, Roberts, Woosley, and Hoffman[67] included time variability, improved physical assumptions, and generally better physics, with the following conclusion:

> It thus seems very unlikely that the simplest model of the neutrino-driven wind can produce the *r*-process. At most, it contributes to the production of the $N = 50$ closed shell elements and some light *p*-nuclei. [...] Similar to most of the work on the neutrino-driven wind after Woosley et al. (1994), we find that it is unlikely that the *r*-process occurs in the neutrino-driven wind unless there is something that causes significant deviation from a purely neutrino-driven wind.

There is no shortage of ideas, but the solution is not yet here. For example, Suzuki and Nagataki[68] proposed magnetic proto-neutron star winds driven by Alfvén waves.[69] They showed that such a wind, if it existed, should create the right conditions for the *r*-process. However, their calculations were restricted to dynamics alone, and did not include the resulting nuclear synthesis.

Moreover, Suzuki and Nagataki ignored the work by Ito et al.[70] who had killed the idea three years earlier. Ito and collaborators calculated stellar winds from neutron stars with magnetic fields as strong as 10^{12}–5×10^{15} gauss, which is an incredibly high magnetic field, and found that:

> Even with a magnetar-class field strength,[71] $\sim 10^{15}$ gauss, the features of the wind dynamics differ only little from those of non-magnetic winds and the conditions for an effective *r*-process are not realized.

---

66 Thompson, T.A., Burrows, A. and Meyer, B.S., Astrophys. J. **562**, 887 (2001).

67 Roberts, L.F., Woosley, S.E. and Hoffman, R.D., Astrophys. J. **722**, 954 (2010).

68 Suzuki, T.K. and Nagataki, S., Astrophys. J. **628**, 914 (2008). Published May 2008.

69 An Alfvén wave in a plasma is a low-frequency (compared to the ion cyclotron frequency) traveling oscillation of the ions and the magnetic field. The ion mass density provides the inertia and the magnetic field line tension the restoring force.

70 Ito, H., Yamada, S., Sumiyoshi, K. and Nagataki, S., Prog. Theor. Phys. **114**, 995 (2005). Submitted April 2005, published November 2005.

71 A magnetar is a neutron star with an imagined strong magnetic field of up to 10^{15} gauss. At such strong fields, the vacuum is polarized and atoms deform into long macaroni shapes. In short, the physics as we know it on Earth changes completely. Magnetars are observed as soft gamma repeaters, sporadically emitting bright bursts of energy over an estimated period of the order of 10^4 years. It is hypothesized that the crust and the penetrating magnetic field become unstable for some reason, causing a flare which ejects up to 5×10^{-9}–$5 \times 10^{-7} M_\odot$ (Gelfand, J.D., and 9 coauthors, Astrophys. J. **634**, L89, 2005). The frequency of such events and the total amount of matter they might expel is so far unknown.

Discoveries in recent years have shown that jets are a ubiquitous phenomenon, appearing in a wide range of cosmic objects from star formation and accretion processes to supernovae and galaxies. The possibility of the r-process taking place in certain types of jets emerging from compact objects has not yet been investigated with sufficient detail to allow for a safe conclusion as to whether these are the so desperately sought-after sites for the r-process. One of the troubles here, as with supernova explosions, is that we have not yet reached the stage where jets can be predicted from first principles. In all present day jet calculations, the jet is assumed to exist to begin with.[72]

Maybe features like magnetic fields, rotation, binary star systems, additional modes of energy transfer such as waves of different kinds, should be included in the calculation. Could it be that these parameters, which were usually considered of secondary importance, are so crucial for supernova collapse and the creation of the r-process elements? Or could the process be inverted and a successful r-process scenario be made to tell us about supernova mechanisms?

13.10.3 Neutron Captures in Exploding He- or C-Rich Layers

It has long been recognized that neutrons could be released in the He- or C-rich layers of SN type II heated by an outward-moving shock wave. This has raised the hope that an r-process could develop in such locations. This is exactly the idea Hoyle and Fowler advocated for some time, but later abandoned.

A neutron-capture episode could be encountered in explosive He burning as a result of the neutrons produced by (α, n) reactions on pre-existing Ne or Mg isotopes. This neutron production could be augmented by inelastic scattering on ^{4}He nuclei of neutrinos streaming out of the forming neutron star at the centre of the exploding star.[73] Parametric studies have demonstrated, however, that too little neutron supply could be achieved to allow for the production of a Solar System type of r-nuclide abundance pattern. Only a limited redistribution of the pre-explosion heavy nuclide abundances is possible. This conclusion is confirmed by the use of recent stellar models,[74] which do indeed predict far too low neutron densities to generate a well-developed r-process. This shortcoming has been shown to be circumvented only if totally ad hoc and implausible astrophysical assumptions are made, particularly concerning the initial amount of the relevant Ne or Mg isotopes and/or of the heavy seeds for radiative neutron captures (especially Ba).

---

[72] See Nishimura, S., and 6 coauthors, Astrophys. J. **642**, 410 (2006). The calculations without neutrino transport show that one does get synthesis of the r-process which is non-isotropic. But when an approximate treatment of the neutrino transport is introduced, alas, the r-process disappears.

[73] Epstein, R.I., Colgate, S.A. and Haxton, W.C., PRL **61**, 2038 (1988).

[74] Rauscher, T., Heger, A., Hoffman, R.D. and Woosley, S.E., Astrophys. J. **576**, 323 (2002); Meyer, B.S., The, L.-S., Clayton, D.D. and El Eid, M.F., Lunar Planet. Sci. **35**, 1908 (2004).

Parametric studies of explosive carbon burning along lines similar to those followed for explosive He burning have also been conducted. In particular, the production of the elements up to Zr by neutron captures has been computed by Wefel and collaborators,[75] who concluded that only some contribution to the ultra-heavy cosmic rays or to some isotopic anomalies can be expected, whereas the bulk of the Solar System r-nuclides just beyond iron cannot be produced in such a way. A very limited production of some light ($A \leq 88$) r-nuclides is predicted in the C- and Ne-rich shells.

13.10.4 The r-Process in the Decompression of Cold Neutron Star Matter

Suppose that by some event a piece of a neutron star is ejected. The coalescence of two neutron stars and of certain neutron star–black hole binaries may be accompanied by the ejection of decompressed cold neutron star material. This matter can be highly neutron rich to serve our purpose. Following the early work of Lattimer et al. and also Symbalisty and Schramm,[76] the first detailed calculation of the r-process that accompanies the decompression of neutron star material was performed by Meyer.[77] The expansion was only followed down to densities around the neutron-drip density ($\rho_{\text{drip}} \approx 3 \times 10^{11}$ g/cm^3), however. In contrast, only densities below this value were considered by Freiburghaus et al.[78] who assumed the initial composition of the material to result from a statistical equilibrium at high temperatures ($T \approx 6 \times 10^9$ K) and with the neutron-to-proton ratio taken as a free parameter. However, these selected high temperatures cannot plausibly be reached in the unshocked crust material of a neutron star that eventually gets dynamically stripped and expands very quickly from the surface during a merger.

An improved model calculation of the composition of the dynamically ejected material from cold neutron stars has been performed[79] making use of a more detailed treatment of the physics during decompression. The evolution of the matter density is modeled by considering the pressure-driven expansion of a self-gravitating clump of NS matter under the influence of tidal forces on an escape trajectory. The expansion is characterized by the expansion timescale τ_{exp}, defined as the time needed for the initial density to drop by three orders of magnitude. In summary, despite a series of suggested esoteric models, we do not have an established theory for the r-process elements.

---

[75] Wefel, J.P., Schramm, D.N., Blake, J.B. and Pridmore-Brown, D., Astrophys. J. Suppl. **45**, 565 (1981).

[76] Symbalisty, E., and Scramm, D.N., Astrophys. Lett. **22**, L143 (1982).

[77] Meyer, B.S., Astrophys. J. **343**, 254 (1989).

[78] Freiburghaus, C., Rosswog, S. and Thielemann, F.K., Astrophys. J. **525**, L121 (1999).

[79] Goriely, S., and 4 coauthors, Nucl. Phys. A **758**, 587c (2005).

Table 13.6 Isotopes of Molybdenum found in nature

$N+Z$	N	Lifetime	Relative abundance on Earth
92	50	$>190\times10^{18}$ years	0.1477
93	51	4.0×10^{3} years	
94	52	Stable	0.0923
95	53	Stable	0.1590
96	54	Stable	0.1668
97	55	Stable	0.0956
98	56	$>100\times10^{12}$ years	0.2419
99	57	2.7489 days	
100	58	8.5×10^{18} years	0.0967

13.11 What Can We Learn from the *r*-Process Abundances

Despite the uncertainties, there is a lot to be learnt from the observed abundances of the elements and the isotopic distribution, in particular, when the abundances and the isotopic ratios differ from the 'usual' one, whatever 'usual' means in this context. An exclusive role is played by the radioactive isotopes found in meteorites.

13.12 The So-Called Mo Anomaly

There are 35 known isotopes of molybdenum (Mo) with $A = 83$ to $A = 117$. Seven isotopes, with atomic masses 92, 94, 95, 96, 97, 98, and 100, occur naturally. Of these, five are stable, with atomic masses from 94 to 98. The unstable isotopes of molybdenum decay into isotopes of niobium, technetium, and ruthenium. The isotopes found in nature are given in Table 13.6. The isotopes ^{92}Mo and ^{100}Mo are unstable but occur in nature. The reason is their extra long half-lives, much longer than the age of the Universe. Consequently, if they are formed by any process, we should find them in nature.

Isotopic analysis of chondritic meteorites[80] shows a difference in isotopic composition relative to the common molybdenum found on Earth. Such differences are called 'anomalies' in the literature, but provide us with very valuable information about the origin and history of the elements and meteorites.

We define

$$\varepsilon^i = \left[\frac{{}^i\mathrm{Mo}/{}^{96}\mathrm{Mo}}{\left({}^i\mathrm{Mo}/{}^{96}\mathrm{Mo}\right)_{\mathrm{solar}}}\right]\times 10^4$$

[80] Dauphas, N., Marty, B. and Reisberg, L., Astrophys. J. **565**, 640 (2002); Yin, Q.Z., Jacobsen, S.B. and Yamashita K., Nature **415**, 881 (2002).

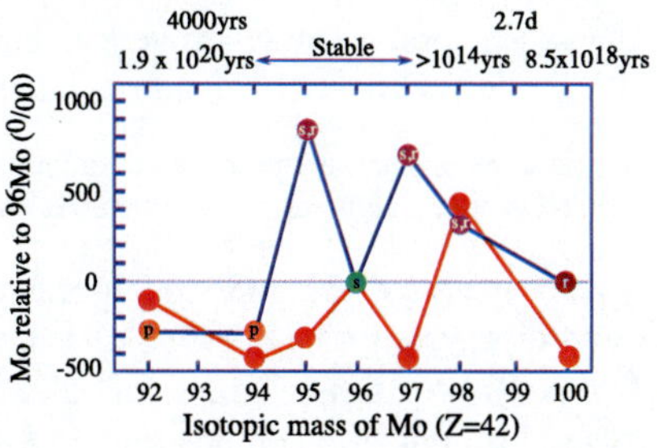

Fig. 13.20 The molybdenum relative isotopic composition as measured in type X grains found in meteorites and showing an excess of nuclei formed by the s- and r-processes (joined by a *purple line*). *Green* marks an isotope formed by the s-process alone, *orange* denotes isotopes formed by the p-process, and *brown* stands for isotopes formed by the r-process. *Full red circles* are the solar relative abundances (joined by a *red line*). Half-lives are given *at the top*. Note that the effects are actually under 10%. After Yin et al. 2002

as a measure of the deviation from the reference. The so-called Mo anomaly results are displayed in Fig. 13.20. They exhibit abundance enhancements of the sr-isotopes ^{95}Mo, ^{97}Mo, and of the r-only isotope ^{100}Mo. There is a claim that excesses of the p-isotopes ^{92}Mo and ^{94}Mo are also found in bulk material and that the p- and r-nuclide anomalies are, rather strangely, strongly correlated.

On the other hand, the only resolvable excesses found by Chen and collaborators[81] were ^{92}Mo and ^{95}Mo in their analysis of bulk iron and carbonaceous chondrite meteorites,[82] and some Allende calcium–aluminum-rich inclusions. These results convinced Chen et al. that the p- and r-processes are essentially decoupled.[83] Another problem concerns the positive ^{97}Mo anomaly. It naturally raises the prospect of a contribution to this nuclide from the in situ decay of ^{97}Tc, which has a half-life of 2.6×10^6 years, and which was present in the primitive material from which the

---

[81] Chen, J.H., Papanastassiou, D.A., Wasserburg, G.J., and Ngo, H.H., Lunar Planet Sci. **35**, 1431 (2004).

[82] Chondrites are stony meteorites that have not been modified due to melting or differentiation of the parent body. They formed when various types of dust and small grains that were present in the early Solar System accreted to form primitive asteroids. Embedded in these meteorites, one finds pre-solar grains, which predate the formation of our Solar System and crystalized elsewhere in the Galaxy. The carbonaceous chondrites or C chondrites are a class of chondritic meteorites comprising at least 7 known subgroups. These are among the most primitive known meteorites. Several groups of carbonaceous chondrites contain relatively high percentages (3–22%) of water, as well as organic compounds. These meteorites are composed mainly of silicates, oxides, and sulfides, and minerals such as olivine and serpentinite. Some famous carbonaceous chondrites are Allende, Murchison, Orgueil, Ivuna, Murray, and Tagish Lake.

[83] Chen et al. suggested that the claimed correlation between the p-process enhancement and the r-process enhancement might be due to *possible technical issues*.

Solar System formed. The explanation is still controversial, and Dauphas et al.[84] for example, do not accept it. Similarly, Yin et al.[85] contested that:

> The multiple r- and p-process components were produced in different sources, transported in different forms into interstellar space, and preserved in the early solar nebula.

In 1998, Nicolussi et al.[86] analyzed 32 individual graphite grains from the Murchison[87] meteorite and discovered a huge range of isotopic abundances. The ratio $^{96}Zr/^{94}Zr$ varied from 0.074 to 10 times the solar value. Most of the graphite grains show close to terrestrial isotopic Mo composition. In 8 grains, they measured both Zr and Mo and found s-process isotopic characteristics. On the other hand, the Mo in four grains was nearly normal, while the $^{96}Zr/^{94}Zr$ in those grains was 10.4 ± 1.3 and 2.5 ± 0.3 times the solar value, hinting at an r-process signature. Thus, this meteor was apparently assembled from matter that had different nucleosynthesis histories.

Recently, Burkhardt et al.[88] measured the molybdenum isotopes in a series of meteorites and plotted a sequence of ε^iMo versus ε^{92}Mo, i.e., the deviation from solar abundance of a given isotope as a function of the deviation of ε^{92}Mo. The results are shown in Fig. 13.21. The authors argued that:

> Given the observed patterns the anomalies can only be explained in terms of a nucleosynthetic origin.

This implies that there were *regions with a variable mix of nucleosynthetic components* in the nebula out of which the Solar System formed. Similar results were found by Nicolussi et al.[89] who analyzed pre-solar SiC grains, although all the grains analyzed were from the Murchison meteorite.

The question can be expressed in different ways. If the theory is correct, we expect the data collected from various meteor species to represent a straight line with the same slope (all species were enhanced by the same factor). But clearly, this is not the case. The explanation may lie either in the astrophysical site or in the basics of the r-process. The authors preferred an astrophysical resolution[90]:

> The large-scale isotope heterogeneity of the Solar System observed for molybdenum must have been inherited from the interstellar environment where the Sun was born, illustrating the concept of 'cosmic chemical memory'.

---

84 Dauphas, N., Marty, B., Reisberg, L., Astrophys. J. **565**, 640 (2002).

85 Yin, Q.Z., Jacobsen, S.B. and Yamashita, K., Nature **415**, 881 (2002).

86 Nicolussi, G.K., and 5 coauthors, Astrophys. J. **504**, 492 (1998).

87 The Murchison meteorite is named after Murchison, Victoria, in Australia where it fell in September 1969. It is one of the best studied meteorites due to its large mass (over 100 kg). The meteor is rich in organic compounds.

88 Burkhardt, C., Kleine, T., Oberli, F. and Bourdon, 41st Lunar and Planetary Science Conference, 2010, 2131.

89 Nicolussi, G.K., and 5 coauthors, Geo. et Cos. Acta **62**, 1093 (1998).

90 Dauphas, N., Marty, B., and Reisberg, L., Astrophys. J. **565**, 640 (2002).

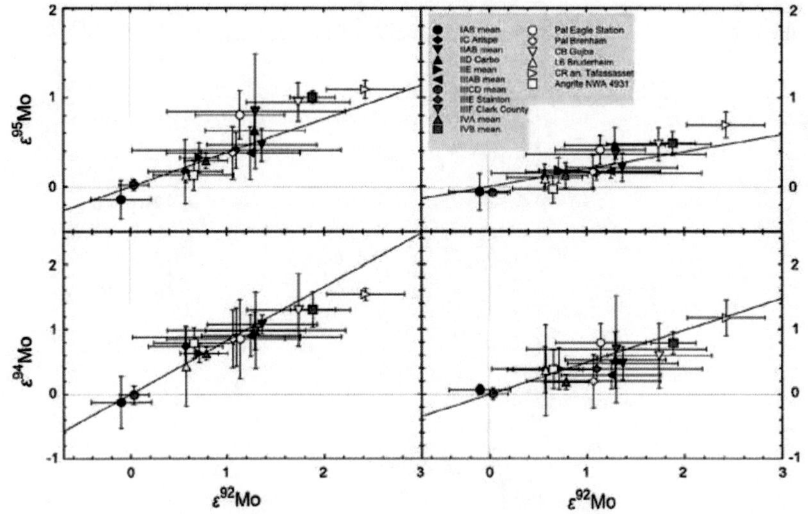

Fig. 13.21 The Mo isotopic composition as found in iron meteorites. Shown are the deviations ε^{i}Mo as a function of the deviations ε^{92}Mo. If the molybdenum was from the same source, all lines would be expected to have the same slope. After Burkhardt 2010

We remark that the experimentalists trust the theoreticians but not other experimentalists, while the theorists distrust other theorists but put their faith in the experimentalists.

13.13 The Xe Anomaly and a Revolution

The xenon isotopes in meteorites played a key role in our understanding of how the Solar System interacts with the rest of the Universe. The known xenon isotopes are given in Table 13.7.

A revolution in the science of meteorites came in 1931 when Paneth and Koeck[91] applied radioactive methods to measure the age of meteorites and discovered that all ages fall below the age of the Solar System and the Earth. Paneth and Koeck concluded that all these meteorites must be of solar origin, and hence questioned the idea of hyperbolic orbits inferred for the majority of meteoritic fireballs.[92] Öpik, on the

[91] Paneth, F. and Koeck, W., Zeits. f. Phys. Chem., Festband, 145 (1931); ibid. Nature **125**, 490 (1930).

[92] The von Niessl–Hoffmeister cataloque: von Niessl, G. and Hoffmeister, C., Denksch. d. Akad. d. Wiss., Wien, Math. Naturw. I00 (1925), is a catalogue of meteors that showed fireballs observed from two distant points. Hyperbolic orbits imply very high velocities and non-solar origin. The issue is controversial even today. See Woodworth, S.C. and Hawkes, R.L., Proc. 150th Coll. of the IAU, Gainesville, Florida, USA, 1995, Gustafson and Hanner, eds.

Table 13.7 Xenon isotopes on Earth

A	*N*	Lifetime	Relative abundance
124	70	$>48 \times 10^{15}$ years	0.000,952
125	71	16.9 h	
126	72	Stable	0.000,890
127	73	36.345 days	
128	74	Stable	0.019,102
129	75	Stable	0.264,006
130	76	Stable	0.040,710
131	77	Stable	0.212,324
132	78	Stable	0.269,086
133	79	5.2475(5) days	
134	80	$> 11 \times 10^{15}$ years	0.104,357
135	81	9.14(2) h	
136	82	$> 10^{21}$ years	0.088,573

other hand, argued against the idea of Paneth and Koeck in 1933,[93] claiming that the meteors could indeed be younger than the Solar System, having formed somewhere in interstellar space after the Solar System had formed, and hence entering the Solar System after its formation. The problem was that, beyond the radioactive dating, little else could be done to show who was right.

The first attempts to measure relative isotopic abundances did not lead to any meaningful discovery as the methods for isotopic analysis were not sufficiently accurate. A piece of circumstantial evidence favoring the possibility of an extra-solar origin came with the B2FH paper, which advocated stellar synthesis and later ejection into space. The first experimental results confirming this hypothesis came in 1960, when Reynolds[94] and Rowe and Kuroda 1965[95] discovered the enrichment of ^{129}Xe and $^{131-136\mathrm{Xe}}$ from the decay of ^{129}I (half-life 10 Million years) and ^{244}Pu (half-life 82 Million years). Since the half-life of the mother isotope is very short relative to the age of the Earth, it is clear that these isotopes were formed long after the formation of the Earth.

In 1987, Lewis et al.[96] isolated nanodiamonds[97] with an average size of ~2.6 nm from meteorites, and showed that they were pre-solar and originated outside the Solar

[93] Öpik, E., Pop. Astron. **41**, 71 (1933).

[94] Reynolds, J.H., PRL **4**, 351 (1960).

[95] Rowe, M.W. and Kuroda, P.K., J. Geophys. Res. **70**, 709 (1965).

[96] Lewis, R.S., and 4 coauthors, Nature **326**, 117 (1987).

[97] Observations of interstellar extinction imply that up to 10% of interstellar carbon could be bound up in interstellar diamond (Lewis, R.S., et al. Nature **339**, 117, 1989).

System (see Ott[98]), thereby vindicating Öpik's claim. Isotopic analysis indicated the noble gas xenon-HL which exhibits both the heavy and the light Xe isotopes relative to solar abundance (HL indicates just that, i.e., it contains both heavy and light isotopes). Since the Xe found in meteoritic diamonds contains both p- and r-process products, the natural thing was to assume that they were synthesized in a supernova.[99]

Recent results established the findings of Lewis and collaborators. A typical Xe-HL isotopic pattern is displayed in Fig. 13.22, where we see that the p-process elements and the r-process isotopes are more abundant in the pre-solar diamonds relative to ${}^{130}\mathrm{Xe}^{76}_{54}$ than in the Sun. The case has become even more exciting with the discovery of a Te component, referred to as Te-H, made solely of the r-isotopes ${}^{128}\mathrm{Te}$ and ${}^{130}\mathrm{Te}$, accompanying Xe-HL in pre-solar diamonds.[100]

The idea that interstellar dust may contain diamonds is due to Saslaw and Gaustad,[101] who discovered that diamonds can provide the required interstellar extinction observed in the far ultraviolet. Pure graphite grains, which were till then considered to be the predominant component of interstellar dust, just do not yield the right extinction. Consequently, the prevailing theory for the origin of pre-solar diamonds is that they condense out of hot carbon vapors in atmospheric outflows of red giant stars, in which C is more abundant than oxygen, and with masses of 8–10$M_\odot$, which are on their way to become supernovae.

Excesses in the p-process-only isotopes and in the r-process-only isotopes are observed to occur simultaneously (recall the dilemma with molybdenum). This correlation has been observed in all diamond separates from all kinds of meteorites. The close association of the two different nucleosynthetic processes is indeed unexpected and intriguing.

Several conclusions can be drawn here. Firstly, there is no universal abundance of Xe. Secondly, since some of the meteorites are younger than the age of the Solar System, the contribution of each of the three neutron capture processes may have changed in time, or the Xe of the Sun and the Xe of the meteorites might have been synthesized in at least two different sites with different properties. Attempts to relate the production of the younger Xe to one site or another (like SNe or shocks) are probably premature, as no site has so far been confirmed.

(Footnote 97 continued)
Nanodiamonds with a similar lognormal size distribution to meteoritic nanodiamonds are observed in the circumstellar disks of the Herbig emission stars HD97048 and Elias 1 (Van Kerckhoven, C., et al. Astron. Astrophys. **384**, 568, 2002), in carbon-enriched proto-planetary nebulae (Hill, H.G., et al. Astron. Astrophys. **336**, 41, 1998), and even in interplanetary dust (Dai, Z.R., et al. Nature **418**, 157, 2002). The process of nanodiamond formation is not understood, but their ubiquitous spread throughout the cosmos probably implies that there is a low temperature, low pressure way for them to form.

98 Ott, U., Nature **364**, 25 (1993).

99 Heymann, D. and Dziczkaniec, M., Proc. Lunar Planet. Sci. Conf. **10**, 1943 (1979); Clayton, D.D., Astrophys. J. **340**, 613 (1989).

100 Richter, S., Ott, U, Begemann, F., Nature **391**, 261 (1998); Huss, G.R. and Lewis, R.S., Geochim. Cosmochim. Acta **59**, 115 (1995); Pepin, R.O., Becker, R.H. and Rider, P.E., Geochim. Cosmochim. Acta **59**, 4997 (1995).

101 Saslaw, W.C. and Gaustad, J.E., Nature **221**, 160 (1969).

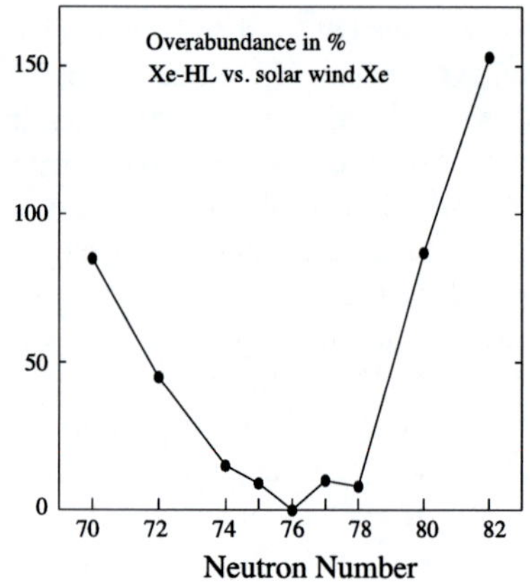

Fig. 13.22 Typical isotopic pattern of the Xe-HL component carried by pre-solar diamonds. The overabundances (in %) are relative to solar Xe. The two lightest and two heaviest isotopes are of pure p- and r-origins, respectively. Normalization is to the s-only isotope ^{130}Xe. After Lewis et al. 1993

13.14 Determining the Abundance of HTI Elements in Stars

Unlike the case of the Solar System, where we have meteorites to analyze, only spectroscopic studies can provide information about abundances of the various elements in stars. Unfortunately, the isotopic wavelength shifts in HTI elements are in general too small to be detectable by present day instruments and telescopes. The only known exceptions for elements with $Z > 30$ are Ba, Eu, Hg, Pt, Tl, and, tentatively, Pb. Apart from these, we cannot distinguish spectroscopically between the isotopes. Consequently, the contributions of the s- and r-processes to the composition of a given star are customarily evaluated from elements whose Solar System abundances are predicted to be dominated by one or the other of these neutron capture processes. In other words, one uses proxies to represent all the HTI elements produced by the process.

The classical proxy for the s-process has for a long time been Ba, as it is estimated that 70 to 100% of it is produced by the s-process. To avoid the difficulties with Ba, which has several stable isotopes with different s- and r-contributions, and whose precise abundance determinations raise specific problems,[102] the essentially mono-isotopic La has become popular in recent investigations.[103] It is classically considered that about 70% of the La in the Solar System originates from the s-process. Nonetheless, in one of the best data compilations of astrophysical data, Palme and Beer[104] give an estimate that the r-process contribution to ^{139}La is 0.0183–0.2480 relative to Si $= 10^6$, which is an embarrassingly wide range for a proxy.

[102] Simmerer, J., and 5 coauthors, Astrophys. J. **617**, 1091 (2004). Additional problems with the spectroscopy of Ba are discussed in the paper.

[103] Winckler, N., and 5 coauthors, Astrophys. J. **647**, 685 (2006).

[104] Palme H. and Beer H., in Landolt Börnstein, New Series, Group VI, Astron. Astrophys. **3**, Subvol. a (Berlin, Springer), p. 196.

Table 13.8 Measures for the isotopic mix

Species	Pure-r	Pure-s	Measure	Reference
Ba	0.46	0.11	$\frac{N(^{135}\text{Ba}) + N(^{137}\text{Ba})}{N(\text{Ba})}$	Lambert, D.L. and Allende, P.C., MNRAS **335**, 325 (2002)
Sm	0.36	0.09	$\frac{N(^{147}\text{Sm}) + N(^{149}\text{Sm})}{N(\text{Sm})}$	Lawler, J.E., Den Hartog, E.A., Sneden, C., and Cowan, J.J., Astrophys. J. S. **162**, 227 (2006)
Eu	0.47	0.53	$\frac{N(^{151}\text{Eu})}{N(\text{Eu})}$	Aoki, W., and 8 coauthors, Astrophys. J. **592**, 67L (2003)

The proxy for the r-process is usually Eu ($Z = 63$), with a relative abundance of 0.0460–0.0526 relative to Si $= 10^6$. Europium has one stable isotope ^{153}Eu with a relative abundance 0.522 and one isotope ^{151}Eu with a relative abundance of 0.482 and a half-life of $5^{+11}_{-3} \times 10^{18}$ years.[105] Europium, barium, and samarium are ideal elements for isotopic analysis because relatively large energy shifts exist between the stable isotopes, and they have specific measures of the r- and s-process contributions to the isotopic mix, as given in Table 13.8.

The use of a single proxy for a process is fine under the assumption of a single event in which the elements formed. But if this assumption has to be abandoned, then additional proxies should be used, each representing the relevant process, or a different mix of processes, otherwise the result may be meaningless.

While metallicity in stars means the sum of all isotopes heavier than helium, it is usually also represented by a proxy, the abundance of iron. Iron has many spectral lines in the visible range, and hence is easy to measure. The deep reason for lumping the elements from C to Fe under the heading of 'metals' is that the structure of stellar atmospheres, where the spectral lines form, is sensitive to the number of free electrons. The elements heavier than helium and hydrogen have relatively low ionization energies, whence these are the ones that supply most of the electrons and not the much more abundant hydrogen and helium. Consequently, the structure of the atmosphere is most sensitive to the total amount of elements heavier than He. If now the relative abundances of these elements are constant, one can use, for example, the abundance of iron to compare the 'metallicity' of a star with that of the Sun, defining

$$\left[\frac{\text{Fe}}{\text{H}}\right] = \text{Log}_{10} \left.\frac{N(\text{Fe})}{N(\text{H})}\right|_{\text{star}} - \text{Log}_{10} \left.\frac{N(\text{Fe})}{N(\text{H})}\right|_{\text{Sun}},$$

where N is the number of atoms per cubic centimeter. In words, [Fe/H] (square brackets) is the logarithm of the ratio in number density of Fe to H in the star divided by the same ratio in the Sun. When the relative abundance of the lighter-than-iron

105 Belli, P., and 14 coauthors, Nucl. Phys. A **789**, 15 (2007).

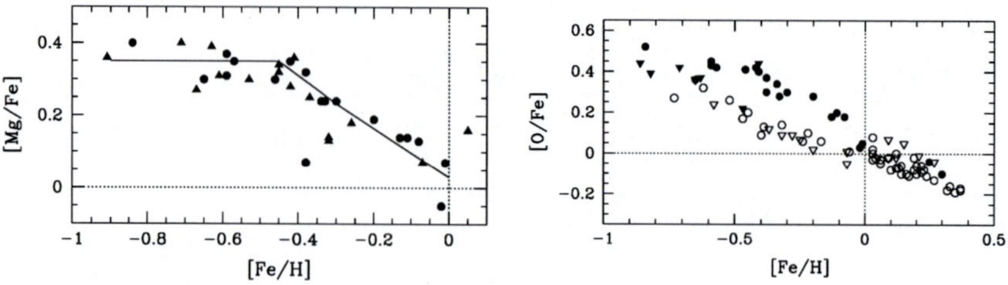

Fig. 13.23 *Left* the ratio (Mg/Fe) versus the ratio (Fe/H). Data from Bensby et al. 2004. *Right* the ratio (O/Fe) versus the ratio (Fe/H). Data from Bensby et al. 2004

species is not constant, for example, in carbon stars where the amount of carbon is much higher than in the Sun, iron loses its meaning as a proxy for 'metallicity'.[106]

Now back to the sky. Observations show that the older the star, the higher its [Mg/Fe] ratio, and this is true for all elements lighter than Fe. This means that, in older stars, there is more Mg with respect to Fe than there is in young stars. It is as though the synthesis of the elements stopped before getting to Fe, forming enough Mg and not sufficient Fe. Similarly for oxygen. Figure 13.23 shows the ratios [Mg/Fe] and [O/Fe] as a function of [Fe/H]. Magnesium is an α-nucleus, like carbon and oxygen. On the one hand we see that [O/Fe] is nicely proportional to [Fe/H], but on the other we see that [Mg/Fe] saturates for very old stars, implying that the early stars did not generate as much Mg as the Fe would have implied.

Hence, the relative abundance of magnesium [Mg/Fe] is not always proportional to the ratio [Fe/H], which in this case is a proxy for age. The absolute, as well as the relative, abundances change with stellar population age. The use of a single proxy washes away such details.

13.15 Neutron Capture During the Early Evolutionary Phases of the Galaxy

In order to assess the synthesis of the r-process elements during the evolution of the Galaxy, one must look at metal-poor stars. Christlieb and colleagues[107] divided the metal-poor stars into two categories:

- r-I metal-poor stars with $2 \leq \dfrac{X[\mathrm{Eu}]/X[\mathrm{Fe}]}{(X[\mathrm{Eu}]/X[\mathrm{Fe}])_\odot} \leq 10$ and $\dfrac{X[\mathrm{Ba}]/X[\mathrm{Eu}]}{(X[\mathrm{Ba}]/X[\mathrm{Eu}])_\odot} \leq 1.$
- r-II metal-poor stars with $\dfrac{X[\mathrm{Eu}]/X[\mathrm{Fe}]}{(X[\mathrm{Eu}]/X[\mathrm{Fe}])_\odot} > 10$ and $\dfrac{X[\mathrm{Ba}]/X[\mathrm{Eu}]}{(X[\mathrm{Ba}]/X[\mathrm{Eu}])_\odot} \leq 1.$

106 The 'metals' are also important for the interior of the star, where they control the radiative opacity, or how much radiation energy is absorbed per unit length. Thus, all the heavier-than-He elements are lumped into a single variable Z.

107 Christlieb, N., and 17 coauthors, Astron. Astrophys. **428**, 1027 (2004).

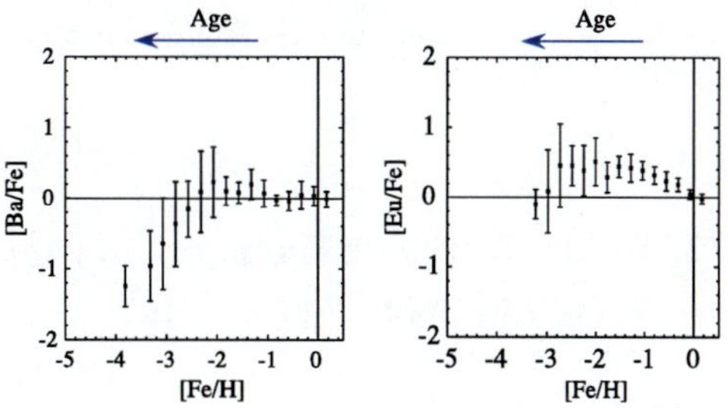

Fig. 13.24 The ratios [Ba/Fe] and [Eu/Fe] versus the ratio [Fe/H]. Due to the large scatter from star to star, Cescutti et al. averaged the data into bins. The data are from various sources, the details of which can be found in Cescutti et al. 2006

The reason for the condition

$$\frac{X[\mathrm{Ba}]/X[\mathrm{Eu}]}{(X[\mathrm{Ba}]/X[\mathrm{Eu}])_{\odot}} \leq 1$$

is that Eu can be produced by the s-process as well, and this is a way to distinguish between 'pure' r-process-enhanced stars and stars that were enriched by material produced in the r- and the s-process.[108]

The chemical evolution of the proxies barium and europium in the Milky Way was discussed by Cescutti and collaborators,[109] and the observational data is shown in Fig. 13.24. As the data has a large scatter, the authors averaged it over small bins, and the error is expressed in the range of the data. We find that the s-process, as represented by Ba, was less powerful at the beginning of the Galaxy. The variation from star to star, as represented by the large errors, is quite significant. The r-process seems to hint towards a similar behavior, but the lack of data for very old stars precludes the drawing of a definite conclusion.

It is also interesting to analyze individual stars. Christlieb and coauthors[110] analyzed ultralow metallicity stars like HE 0107-5240,[111] which has [Fe/H] ~ -5.3 or an iron abundance which is 1/200,000 that of the Sun. Frebel and coauthors[112] analyzed the star HE 1327-2326, with [Fe/H] ~ -5.5, or an iron abundance of 1/315,000 that of the Sun, and Norris[113] analyzed the star HE 0557-4840, which has [Fe/H] ~ -4.8, or 1/63,000 the solar iron abundance. These two stars are strange because the only element produced by neutron capture and observed in these stars

---

[108] Astrophysicists usually write

$$[\mathrm{Eu/Fe}] \equiv \mathrm{Log}_{10}\frac{X(\mathrm{Eu})/X(\mathrm{Fe})}{(X(\mathrm{Eu})/X(\mathrm{Fe}))_{\odot}},$$

which is a shorter notation. We adhere to the more cumbersome notation to stress that, if the solar abundance changes, so will all the absolute results.

[109] Cescutti, G., and 5 coauthors, Astron. Astrophys. **448**, 557 (2006).

[110] Christlieb, N., and 8 coauthors, Nature **419**, 904 (2002).

[111] HE indicates a star from the Hamburg/ESO Survey of metal-poor stars.

[112] Frebel, A., and 20 coauthors, Nature **434**, 871 (2005).

[113] Norris, J.E., Astrophys. J. **670**, 774 (2007).

is strontium (Sr), with an overabundance by a factor of 10 relative to the Sun, i.e., $[Sr/Fe]/[Sr/Fe]_\odot \sim 10$.

13.16 Are r-Process Elements in Metal-Poor Stars Solar Abundance Compatible?

In 1996, Sneden et al.[114] analyzed the ultra-metal-poor star CS 22892-052[115] with $[Fe/H] \sim -3.1$ or an iron abundance that is 1/1,250 the solar value, and were able to determine the detailed abundance distribution of elements with $Z > 30$. They found that:

> The elements from $56 \leq Z \leq 76$ in CS 22892-052 are very well matched by a scaled Solar System r-process abundance distribution.

This statement could be made because the authors detected five elements which were never before observed in such metal-poor stars. On the other hand, they added that:

> A scaled solar s-process cannot be plausibly fitted to the observed data.

The age estimate for the star is $15 \pm 3 \times 10^9$ years, so the first conclusion was that the r-process operated before these old stars were formed. If the r-process elements observed in these stars were synthesized earlier, in other stars, this implies that r-process elements are synthesized in very massive stars which have a very short life, because only very massive stars can do the job so quickly. Are we going back to the idea of Hoyle and Fowler about supermassive stars? To complicate the problem, the star is found to be rich in carbon (by a factor of 10 relative to the Sun).

The results of Sneden and coauthors led researchers back to the old claim that the r-process is 'universal'. However, Goriely and Arnould[116] analyzed the possibilities and concluded that there were two:

- The abundances in CS 22892-052 demonstrate that the r-process yields are almost perfectly solar, irrespective of the properties of the production site. This universality would imply that the relative r-element abundances are the product of global galactic processes like mixing, which completely washes away specific site signatures.
- The whole $A \geq 70$ r-process abundance pattern is a result of the invariance of nuclear properties in the $56 \leq Z \leq 76$ range.

---

114 Sneden, C., and 5 coauthors, Astrophys. J. **467**, 819 (1996).

115 This is also known as Sneden's star.

116 Goriely, S. and Arnould, M., Astron. Astrophys. **322**, L29 (1997).

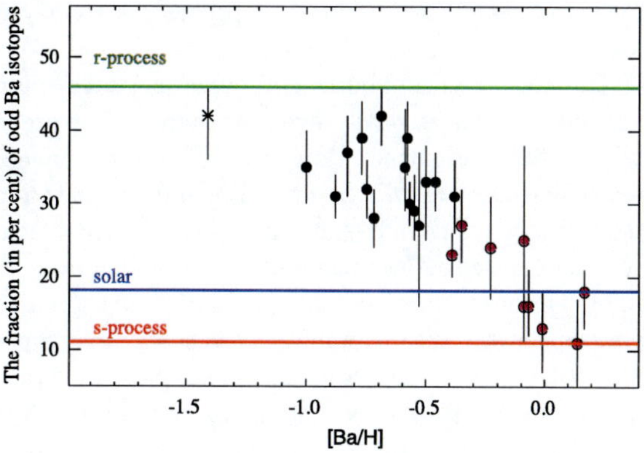

Fig. 13.25 Fractional abundance of the odd A barium isotopes versus [Ba/H], as found by Mashonkina and Zhao (2006) for 25 cool dwarf stars with $-1.35 \leq$ [Fe/H] ≤ 0.25. *Solid circles* represent stars found at relatively high altitudes above the disk of the Galaxy, while *open circles* represent a younger population, found at lower altitudes above the disk of the Galaxy. *Horizontal lines* indicate the solar fraction of the odd isotopes (18%) and the predicted values for a pure r-process (46%) and a pure s-process (11%) production of barium

13.17 The Galactic History of Neutron-Capture Elements

The giant star HE 1523-0901[117] is an extremely metal-poor star with $[\mathrm{Fe/H}] = -3.0$, or 1/1,,000 of the solar iron, and hence very old. At the present time, it has the strongest enhancement of neutron-capture elements associated with the r-process. The abundance of the r-process elements relative to iron is 1.8 times the solar ratio. The spectrum of HE 1523-0901 shows numerous strong lines of about 25 neutron-capture elements, such as those of Sr, Ba, Eu, Os, and Ir. This result is counterintuitive because the simple expectation would be that, the more iron there is, the greater should be the ratio of r-process to iron abundances. But we see that things are the other way round here, i.e., the less Fe, the more r-process elements there are (Fig. 13.25).

As early as 1963, Wallerstein et al.[118] observed old stars in which the iron abundance was about a factor of 100 less than the solar value, and discovered relatively large amounts of europium in the star HD165195, but not in two other old stars with similar iron abundance. Considering various possibilities for the way Eu was formed, they advocated the idea that Eu might be made in supernovae of large mass if iron or s-process nuclei were initially present, but with very low yields if iron were absent.

The two other stars examined by Wallerstein et al. did not exhibit the same deficiency of heavy elements as shown by HD122563, and consequently they assumed that the matter in these stars, although quite old as can be inferred from the very low iron abundance, was processed by other stars.

The stars HD122563, HD165195, and HD221170 have metal-to-hydrogen ratios of 800, 500, and 500 compared with the Sun. However, HD122563 is different because in its case all elements heavier than zinc are deficient compared with iron

[117] Frebel, A., arXiv:0812.1227v1 [astro-ph] 2008, *10th Symposium on Nuclei in the Cosmos*, USA.

[118] Wallerstein, G., and 4 coauthors, Astrophys. J. **137**, 280 (1963).

by a factor of 50. Wallerstein et al. interpreted this as follows: when the flux of free neutrons became available, the progenitors of the light s-process were present, but the iron peak elements were rare or absent. The europium seen in HD122563 may have been produced by the s-process, or so they advocated, which would be an unexpected result.

Despite the uncertainties and open questions enumerated above, we can establish that the Galaxy was first enriched with r-process elements, and only later with s-process elements. However, this general statement requires caution. The time dependent data of La/Eu is not sufficiently accurate to pinpoint the moment at which the s-process came into operation. This is partly related to the quite large abundance scatter observed at low metallicities. The situation appears to be very different for the light neutron-capture nuclides (Sr in particular), which appear to have had a different nucleosynthesis history. The easiest way to overcome this problem is to assume different r-process components, mainly producing the light or the heavy r-nuclides, thus in some respects reviving the original idea of Seeger et al.[119] that there might have been two or more components. The possibility of the s-process producing nuclei like Sr early in galactic history cannot be disproved at the moment.

The fluctuations in the abundance of the proxy Eu relative to Fe increase with decreasing [Fe/H]. The fluctuations are probably due to improper mixing of matter ejected by different supernovae or other r-process sources giving rise to large variations.

13.18 Solar System Nucleo-Cosmochronology

In their seminal work in 1957, B2FH and Cameron already suggested that the long-lived ^{232}Th, ^{235}U, and ^{238}U could be used to estimate the age of the r-nuclides from the present meteoritic content of these nuclides. The first attempt was carried out by Fowler and Hoyle,[120] using two models. The first assumed that the Galaxy had been an *autonomous system*, for which they got an age estimate of $1.5^{+0.5}_{-0.3} \times 10^{10}$ years. The second model, which assumed the *steady state cosmology and galactic–intergalactic exchange of matter*, yielded a Hubble constant for the expansion of the Universe of $1.1 \pm 0.6 \times 10^{10}$ years. Cameron[121] worked the other way round. Assuming that the age of the Solar System was 4.55×10^9 years, he concluded that the initial ratio $^{235}\mathrm{U}/^{238}\mathrm{U}$ was 0.301 (while today the value is 0.0072).

In 1964, Clayton[122] suggested using the ratio $^{187}\mathrm{Re}/^{187}\mathrm{Os}$ as a chronometer. At that time the value of the half-life of ^{187}Re was poorly known, and estimates varied

[119] Seeger, P.A., Fowler, W.A. and Clayton, D.D., Astrophys. J. Suppl. **11**, 121 (1965).

[120] Fowler, W.A. and Hoyle, F., Astrophys. J. **70**, 345 (1960).

[121] Cameron, A.G.W., Icarus **1**, 13 (1962).

[122] Clayton, D.D., Astrophys. J. **139**, 637 (1964).

from 16×10^{10} to 3×10^{10} years.[123] The advantage/disadvantage of the suggestion is that initial amounts depend on the *s*-process theory for lighter-than-actinide nuclei. Clayton proposed several more methods based on unstable isotopes with proper decay times, but in all these methods the contribution of the *s*- and the *r*-processes must be reliably separated. Clayton reached the conclusion that galactic nucleosynthesis started at least 5×10^9 years before the formation of the Solar System, *and perhaps considerably earlier.*[124]

Then came a period in which the chronology of the Galaxy was combined with detailed models of chemical evolution of the Galaxy (Truran and Cameron,[125] Reeves,[126] Talbot and Arnett,[127] Tinsley[128]). But soon the complexity on the one hand and the lack of fundamental data on the other pushed Schramm[129] towards *a model independent nucleochronology*. With such a simple model, Schramm found that the age T_G of the Galaxy was 6×10^9 years $\leq T_G < 10 \times 10^9$ years which agreed well with contemporary estimates of the Hubble constant (55 km/s/Mpc). However, Schramm summarized by saying:

> Detailed conclusions about galactic evolution cannot at present be given because of the large uncertainties in the input parameters.

---

[123] The accurate half-life is 4.12×10^{10} years. Strangely enough, rhenium has a stable isotope ^{185}Re with relative abundance 0.374, while the unstable isotope ^{187}Re has a relative abundance of 0.626, so that the unstable isotope is more abundant than the stable isotope. The β-decay of rhenium to osmium is one of the lowest in energy of all nuclei which β-decay (just 2.5 keV).

[124] It is instructive to read Clayton's introduction:

> One such calculation has been made several times, most recently by Fowler and Hoyle (1960), Cameron (1962), and Dicke (1962). Those calculations are based on the fact that the relative abundances of ^{235}U, ^{238}U, and ^{232}Th at the time of the Solar System formation are known. [...] Unfortunately, the relative production rates of those nuclei in supernovae are not known, but must be calculated from theories of nucleosynthesis. Therein lies the uncertainty of that method, for the calculation must proceed in a phenomenological way by counting the number of progenitors of each nucleus and making an estimated correction for the odd–even abundance effect. Fowler and Hoyle conclude that the $^{235}U/^{238}U$ production ratio is 1.65 ± 0.15, as is the $^{232}Th/^{238}U$ production ratio. On the other hand, Cameron calculates a value 1.45 ± 0.15 for the $^{235}U/^{238}U$ production ratio and states that 1.65 is a minimum for the $^{232}Th/^{238}U$ production ratio, which may be considerably larger.

Fowler and Hoyle got an age of 11.6×10^9 years while Cameron assumed an increase in the rate of nucleosynthesis shortly before the formation of the Solar System and got an age of 14×10^9 years for uniform synthesis and 6.5×10^9 years for sudden synthesis. In view of such a gap, Clayton suggested the ^{187}Re–^{187}Os method. As Clayton implied, repeating the same calculation many times does not make it more accurate or correct.

[125] Truran, J.W. and Cameron, A.G.W., Astrophys. Space Science **14**, 179 (1971).

[126] Reeves, H., Phys. Rev. D **6**, 3363 (1972).

[127] Talbot, R.J., Jr. and Arnett, W.D., Astrophys. J. **197**, 551 (1975).

[128] Tinsley, B.M., New York Academy of Sciences, Annals **262**, 436 (1975).

[129] Schramm, D.N., *The age of the elements*, Scientific American **230**, 69 (1971).

Almost a decade later, Yokoi, Takahashi, and Arnould[130] examined the chronology of ^{187}Re–^{187}Os and reconfirmed Schramm's summary by stating that:

> ^{187}Re–^{187}Os cannot be regarded yet as a truly reliable chronometer.

Two decades later, in 2001, Arnould and Goriely[131] asserted that:

> Nucleocosmochronology has the virtue of always being able to provide numbers one can interpret as ages. The real challenge is to evaluate the reliability of these predictions. [...] We are much more pessimistic concerning the classical ^{232}Th–^{238}U and ^{235}U–^{238}U pairs.

The stakes are high, so reducing the errors and uncertainties would provide a method to decipher many cosmic conundrums. The general push towards a resolution of the problems continues, even at an accelerating rate.

13.19 The p-Process

The p-process was postulated to produce about 35 stable, but rare, isotopes between ^{74}Se and ^{196}Hg on the proton-rich side of the valley of stability. As B2FH found no way to produce these isotopes by neutron capture, they suggested a special process which they termed the p-process. The solar abundances of the p-process elements are one to two orders of magnitude lower than values for the respective s- and r-process nuclei.[132] However, so far it seems to be impossible to reproduce the solar abundances of all p-isotopes by a single event.

After B2FH had referred to the p-process as the one that produced the elements the s- and the r-processes could not synthesize, little was done to reveal the details. In 1964, Fowler and Hoyle[133] faced unsurmountable problems in synthesizing the proton-rich nuclei ^{50}Cr and ^{58}Ni, and consequently hypothesized that the iron group captures protons, neutrons, and α particles, but added right away that:

> The equilibrium abundance of these nuclei is too small for this process to be effective.

In 1969, Clayton and Woosley[134] attacked the problem of the formation of these nuclei. First they mentioned the result due to Bodansky, Clayton, and Fowler[135] that the abundances they got during silicon burning nicely reproduced the most abundant elements in the range $28 \leq A \leq 57$. The agreement was so good that Bodansky

---

[130] Yokoi, K., Takahashi, K. and Arnould, M., Astron. Astrophys. **117**, 65 (1983).

[131] Arnould M. and Goriely S., in *Astrophysical Ages and Time Scales*, ed. by T. von Hippel et al. ASP Conference Series **245**, 252 (2001).

[132] Anders, E. and Grevesse, N., Geochim. Cosmochim. Acta **53**, 197 (1989); Rosman, K. and Taylor, P., Pure and Appl. Chem. **70**, 217 (1998).

[133] Fowler, W.A. and Hoyle, F., Astrophys. J. Suppl. **9**, 201 (1964).

[134] Clayton, D.D. and Woosley, S.E., Astrophys. J. **157**, 1381 (1969).

[135] Bodansky, D., Clayton, D.D. and Fowler, W.A., PRL **20**, 161 (1968); ibid. Astrophys. J. Suppl. **16**, 299 (1968).

et al. claimed that this was *probably the major source of nucleosynthesis in this mass range*. This supposition was strengthened when Clayton, Colgate, and Fishman[136] claimed that, if Bodansky et al. were right, then young SNe would exhibit well defined γ lines from the decay of ^{56}Ni. On the other hand, the γ lines were not expected if Fowler and Hoyle were right. The detection of such γ lines when SN 1987 exploded eighteen years later[137] was a great victory to Bodanky et al. although not a complete vindication, because at the same time Bodansky et al. found that silicon burning did not produce ^{58}Ni and ^{58}Fe in the observed amounts, since these nuclei could not be synthesized by neutron irradiation. Indeed, Clayton and Woosley claimed that:

> The position of ^{58}Ni in the chart of nuclides warrants a *p*-designation, although its source may not be the same *p*-process that is responsible for the heavier proton-rich nuclei.

When Clayton and Woosley analyzed the situation they concluded that *both conjectures fail*, and suggested that the ^{58}Ni nucleus might be produced in an equilibrium process dominated by ^{56}Ni. But what about the *outstanding success of Bodansky et al.* in fitting the observed abundances? Clayton and Woosley argued that it was *not necessarily relegated to a fortuitous coincidence*, and it was plausible that the conditions were appropriate in two or more locations in the exploding star. Thus, two *p*-type processes are not needed.

In 1972, Truran and Cameron[138] argued that:

> As supernova events provide the most likely site for the *p*-process synthesis, we suggest that the nucleosynthesis conditions examined by Howard, Arnett, and Clayton[139] might be realized in the helium zones of stars which experience the phenomenon of helium-burning shell flashing described by Schwarzschild and Härm.

In the same year, Audouze and Schramm[140] suggested using ^{146}Sm with half-life 7×10^7 years for dating. The idea exploits the fact that both the stable isotope ^{144}Sm and the radioactive one are products of the *p*-processes alone. Thus, it should determine the time between the cessation of the *p*-process and the formation of solid bodies in the Solar System.

Audouze and Truran[141] assumed that a shock wave was responsible for the *p*-elements. The main nuclear reaction in this model is proton capture followed by emission of a γ line. They found that there was a restricted range of temperatures and densities ($T \sim 2 \times 10^9$ K and $\rho \sim 10^4$ g/cm^3) for which enhanced production of *p*-elements would take place *with relative concentrations consistent with Solar System abundances*. However, they found that the results were very sensitive to the temperature. A small deviation in the temperature changes the resulting abundances and destroys the similarity to Solar System abundances. Indeed, one of the annoying problems in these attempts to 'fit the observed Solar System abundances' is the

---

136 Clayton, D.D., Colgate, S.A. and Fishman, G.J., Astrophys. J. **155**, 75 (1969).

137 See details in Shaviv, G., *The Life of Stars*, Springer and Magnes Pub. 2009.

138 Truran, J.W. and Cameron, A.G.W., Astrophys. J. **171**, 89 (1972).

139 Howard, W.M., Arnett, W.D. and Clayton, D.D., Astrophys. J. **165**, 495 (1971).

140 Audouze, J. and Schramm, D.N., Nature **237**, 447A (1972).

141 Audouze, J., and Truran, J.W., Astrophys. J. **202**, 204 (1975).

extreme sensitivity to initial conditions. A trifling deviation and the compatibility between theory and observation evaporates. However, since they assumed proton absorption reactions, it was not at all clear to what extent protons, even at extremely low concentrations, would survive stellar evolution up to this temperature and not be completely consumed long before.

At the same time, Woosley and Howard[142] adopted the inverse model, i.e., since there are no protons, one has to create them. Thus, γ absorption followed by the release of protons, neutrons, and α particles would do the job. In 1981, Ward and Beer[143] showed that:

> The two models differ sharply in their prediction for the relative production of ^{114}Sn and ^{115}Sn. Audouze and Truran predict a roughly comparable production of ^{115}Sn while the calculation of Woosley and Howard yields essentially no ^{115}Sn. However, both models concur in their prediction of a negligible production of ^{113}In.

Consequently, the authors favored the Audouze–Truran model despite its failure with regard to ^{114}Sn. Evidence to the contrary was given by de Laeter, Rosman, and Loss,[144] who supported the Woosley–Howard model.

The pendulum moved once more, in 1989, when Prinzhofer, Papanastassiou, and Wasserburg[145] analyzed the presence of ^{146}Sm in two meteorites and found that the production rate of this nucleus was not compatible with the photodisintegration model for the production of *p*-process nuclides.

In 1990, Rayet, Prantzos, and Arnould[146] decided to parametrize the explosion and applied newly obtained nuclear data. Moreover, they assumed as initial conditions that a previous *s*-process had left remnant seed nuclei which were to absorb the protons. They managed to obtain a fit to the solar *p*-nuclei in which the difference between observed and predicted values ranged from about 3–60%. However, problems found in previous attempts, such as underproduction of ^{92}Mo, ^{94}Mo, and ^{96}Ru, as well as a few other nuclei, did not disappear. The authors concluded that:

> In spite of the improvements reported above with respect to Woosley and Howard's calculations, we feel it premature to elaborate upon detailed comparison between our predicted *p*-process yields and the Solar System data.

In 1995, Rayet and coauthors[147] presented what they called *the first quantitative calculation of p-process yields from SN II explosion of* $Z = Z_\odot$ *stars*. Since they used a range of masses, they averaged the results weighted by the frequency of formation of the stars. The good news was that the results were found to be insensitive to the mass of the star, and in general within a factor of 4 of the observed values. The bad news was that the previous standing problems with ^{92}Mo, ^{94}Mo, and ^{96}Ru were not cured. And neither were the Xe *p*-isotopes correctly reproduced.

---

[142] Woosley, S.E. and Howard, W.M., Astrophys. J. Suppl. **36**, 285 (1978).

[143] Ward, R.A. and Beer, H., Astron. Astrophys. **103**, 189 (1981).

[144] de Laeter, J.R., Rosman, K.J.R. and Loss, R.D., Astrophys. J. **279**, 814 (1984).

[145] Prinzhofer, A., Papanastassiou, D.A. and Wasserburg, G.J., Astrophys. J. **344**, L81 (1989).

[146] Rayet, M., Prantzos, N. and Arnould, M., Astron. Astrophys. **227**, 271 (1990).

[147] Rayet, M., and 4 coauthors, Astron. Astrophys. **298**, 517 (1995).

If there is a problem with ^{92}Mo, ^{94}Mo, and ^{96}Ru, perhaps they are produced in different processes? Schatz and coauthors[148] suggested that ^{92}Mo and ^{96}Ru could be synthesized in X-ray bursts in low mass binary systems.[149] The problem with ^{94}Mo was not mentioned, however. The catch is that thermonuclear bursts on the surface of neutron stars do not eject mass from the star. Consequently, the authors guessed *an escape factor of about 1%* to get their result.

In 1997, Sauter and Käppeler[150] measured the probability of proton absorptions on the ^{92}Mo, ^{94}Mo, ^{95}Mo, and ^{98}Mo and found them to be a factor of 2 to 4 greater than what the nuclear theory had given and the values that had been used previously. But the situation did not change.

In 2000, Costa and coauthors[151] suggested that, if the rate of the reaction $^{22}\mathrm{Ne} + \alpha \rightarrow {}^{25}\mathrm{Mg} + \mathrm{n}$ which generates the neutrons, as it appears in compilations of nuclear data, was wrong by a factor of 50, then the problem of the Mo and Ru isotopes would be solved without the need for exotic solutions. The authors mentioned additional nice features that such a correction would procure, but warned that *this array of pleasing features has of course not to be viewed as a proof of the validity* that such an error actually exists.

As if there were not enough problems, in 2001, Davis and coauthors[152] measured Mo isotopes in SiC grains found in primitive meteorites, which are believed to condense out of the expanding carbon-rich atmosphere of a star. Since the abundance of ^{92}Mo was significantly lower than commonly found and this isotope is easily destroyed by a neutron flux, they suggested that they had identified:

> [...] a new isotopic signature not previously recognized, which is neither *s*-, *r*-, nor *p*-process nucleosynthesis, but rather appears to be the result of a short neutron burst.

We thus realize that, if the synthesized matter inside the star passes extreme conditions like a shock on its way out, the relative abundances may change and give us additional headaches.

Another attempt was carried out in 2003 by Fujimoto and coauthors[153] when they calculated the *p*-process in supernovae that are hypothesized to form when the central star accretes heavily from a disk around it. The synthesis of the *p*-process is assumed to take place in the accretion disk. Again, a nice overall agreement is found, but the problems of Mo and Ru are not alleviated. The question of how the products come out was not discussed.

---

148 Schatz, H., and 12 coauthors, Nucl. Phys. A **621**, 417 (1997).

149 Low mass X-ray binaries are systems in which one star radiates in the X-ray range. The X-ray radiator is a neutron star or black hole. The objects are quite bright in the X rays, but faint in visible light.

150 Sauter, T. and Käppeler, F., Phys. Rev. C **55**, 3127 (1997).

151 Costa, V., and 4 coauthors, Astron. Astrophys. **385**, L67 (2000). See also Rayet, M., Costa, V., Arnould, M., Nucl. Phys. A **688**, 64 (2001); Goriely, S., and 3 coauthors, Astron. Astrophys. **375**, L35 (2001).

152 Davis, A.M., and 5 coauthors, MmSAI **72**, 413 (2001).

153 Fujimoto, S., and 5 coauthors, Astrophys. J. **585**, 418 (2003).

The reliable modeling of the p-process involves consideration of an extended network of some 20,000 reactions linking about 2,000 nuclei in the $A \leq 210$ mass range. The basic reactions are photon absorption and emission of a neutron, or a proton, or an α particle, or a combination of all of the above.

The largest fraction of p-isotopes is probably created by sequences of photodissociations and β^+-decays.[154] This may occur in explosive O-Ne burning during SN type II explosions and reproduces the solar abundances for the bulk of p-isotopes[155] within a factor of $\approx$3. The SN shock wave induces temperatures of 2–3 $\times 10^9$ K in the outer layers composed of (C, Ne, O), temperatures that are sufficient to trigger the required photodisintegrations. More massive stellar models with masses greater than $20M_\odot$ seem to reach the required temperatures for efficient photodisintegration even by the end of hydrostatic O-Ne burning. The decrease in temperature after passage of the shock leads to a freeze-out via neutron captures and mainly β^+-decays, resulting in the typical p-process abundance pattern with maxima at ^{92}Mo and ^{144}Sm.

Is this too good to be true? This scenario suffers from a strong underproduction of the most abundant p-isotopes, ^{92,94}Mo and ^{96,98}Ru, due to a lack of seed nuclei with $A \geq 90$. For these missing abundances, alternative processes and sites have been proposed, either using strong neutrino fluxes in the deepest ejected layers of a SNII (ν-p process[156]), or rapid proton captures in hot proton-rich matter accreted on the surface of a neutron star (r-p-process[157]). A few p-nuclides may also be produced by neutrino-induced reactions during the γ-process.

In summary, most recent studies of the p-process still have problems to synthesize p-nuclei in the regions $A \leq 124$ and $150 \leq A \leq 165$. It is not yet clear whether the observed underproductions are only due to a problem with astrophysical models or whether there is also a problem with the nuclear physics input. The reduction of uncertainties in nuclear data is therefore strictly necessary for a consistent understanding of the p-process.

13.19.1 Spallation

The existence of strong magnetic fields in the Sun and other stars led Riddiford and Butler[158] to suggest that growing or shrinking magnetic fields might accelerate ionized particles in sunspots. Similar ideas were proposed by Bagge and Biermann,[159] who estimated that the energy of the particles accelerated by changing solar magnetic

---

154 Woosley, S.E. and Howard, W., Astrophys. J. Suppl. **36**, 285 (1978); Woosley, S.E. and Howard, W., Astrophys. J. **354**, L21 (1990); Rayet, M., Prantzos, N., and Arnould, M., Astron. Astrophys. **227**, 271 (1990).

155 M. Rayet, M., and 4 coauthors, Astron. Astrophys. **298**, 517 (1995).

156 Fröhlich, G., and 7 coauthors, Phys. Rev. Lett. **96**, 142502 (2006).

157 Schatz, H., and 8 coauthors, Phys. Rep. **294**, 167 (1998); ibid. PRL **86**, 3471 (2001).

158 Riddiford, L., and Butler, S.T., Phil. Mag. 7, **43**, 447 (1952).

159 Bagge, E. and Biermann, L., Naturwissenschaften **35**, 120 (1948).

fields could reach 10^9 eV. The purpose of these papers was to explain the source of high energy particles. Burbidge and Burbidge[160] were the first to suggest that these high energy particles could then induce nuclear reactions and in this way synthesize specific isotopes which were not produced in the regular way. In addition, the collisions between the accelerated particles and various elements which serve as targets would take place on the surface of stars, which means that the products, even if produced in small amounts, would be right where they could be most easily observed. The most abundant targets are naturally the α nuclei C, O, Ne, Mg (and N which is not an α nucleus). The energy of the accelerated particles can reach several tens of MeV, and when such particles collide with the targets, they mostly spallate them. Hence the name spallation reactions.

A classic example of such elements is provided by the light elements Li, Be, and B, which have very low abundances relative to their neighboring elements. Yet they cannot be explained by the classical theory of thermonuclear reactions in stars. To explain their formation Fowler, Burbidge, and Burbidge[161] elaborated on the work of Burbidge and Burbidge, and suggested spallation reactions on CNO elements as targets. A year later, Hayakawa[162] independently suggested the idea in a short paper, but did not elaborate on any of the consequences, abundance predictions, etc. Hayakawa, however, assumed the reaction to take place when cosmic rays impinge on the surface of stars.

The spallation theory was expanded upon by many authors.[163] Strong magnetic fields and unique abundances are observed on magnetic Ap stars. These are stars in the spectral class range B9–A0 which exhibit unusually large magnetic dipole fields. Such stars show anomalous abundances of the elements Si, Cr, Fe, and rare earth elements that appear to be non-uniformly distributed across the stellar surface. There are difficulties in explaining these abundances as a result of classical thermonuclear reactions, so several authors have resorted to proton acceleration and spallation. It is not clear to what extent the spallation is restricted to the stellar surfaces of these unique stars. It was Audouze[164] who extended the spallation idea to the *p*-process elements formed in flares of high energy protons which occur on the surface of stars.

Frank-Kamenetski[165] investigated (p,n) and (p,2n) reactions as a source of neutrons. In particular, he searched for ways to synthesize what he called bypassed nuclei, i.e., nuclei bypassed by traditional neutron capture. So the idea was that (p,n) and even (p,2n) reactions might release the required neutrons. However, no calculation of the yield was carried out and nor was there any comparison with observation.

---

[160] Burbidge, G.R. and Burbidge, E.M., Astrophys. J. Suppl. **1**, 431 (1955).

[161] Fowler, W.A., Burbidge, G.R., and Burbidge, E.M., Astrophys. J. Suppl. **2**, 167 (1955).

[162] Hayakawa, S., Prog. Theor. Phys. **13**, 464 (1956).

[163] Bashkin, S., and Peaslee, D.C., Astrophys. J. **134**, 981 (1961); Fowler, W.A., Greenstein, J.L. and Hoyle, F., Geophys. J. R.A.S. **6**, 148 (1962); Yiou, F., and 6 coauthors, Phys, Rev. **166**, 968 (1968); Gradsztajn, E., Ann. Phys. (Paris) **10**, 791 (1965); Bernas, R., Gradsztajn, E., Reeves, H. and Schatzman, E., Ann. Phys. **44**, 426 (1967); Epherre, M. and Reeves, H., Astrophys. J. **3**, 145L (1969).

[164] Audouze, J., Astron. Astrophys. **8**, 436 (1970).

[165] Frank-Kamenetski, D.A., Soviet Astron. (AJ) **5**, 66 (1961).

He just stated that *the necessary protons with energies of several MeV should be formed in cold-acceleration processes*, without any further specification as to where this should take place, let alone what the yield might be.

Ito (1961) showed that the original idea of B2FH could work in regions where the temperatures are of the order of 10^9 K. The problem is that the proton reactions are extremely slow, even to produce the rare p-species, because of the strong Coulomb barrier presented by the HTI elements.

13.20 Astrophysically Relevant Progress Expected in Nuclear Physics

The nuclear physics required for an r-process calculation spans almost all our knowledge of nuclear physics, but in a mass region that is much more difficult for analysis. A major effort has been devoted in recent years to the measurement of nuclear data of relevance to the r-process. Still, a large body of information remains to be acquired, and will remain out of reach of our experimental capabilities for the foreseeable future. This is, of course, a direct consequence of the huge number of nuclear species that may be involved in one r-process or another, along with the fact that nuclei far from the valley of stability are likely to enter the process. Theory must therefore complement laboratory measurements.

In order to meet the demanding nuclear physics specifications of the r-process, nuclear models of choice have to satisfy two basic requirements: they have to be microscopic, meaning correct for a set of isotopes of a given element, and universal, i.e., good over as wide a range of atomic weights as possible. Two characteristics of the nuclear theories have to be considered in these situations. The first is the accuracy of the model. In most nuclear applications, this criterion has been the main, if not the only one for selecting a model. The second is the reliability of the predictions in regions where no data is available. It is common belief that a physically sound model that is as close as possible to a microscopic description of the nuclear systems will provide the most reliable extrapolation. Of course, the accuracy of such microscopic models in reproducing experimental data may be poorer than can be obtained from more phenomenological models, in which a sufficiently large number of free parameters can guarantee a satisfactory reproduction of the data at the expense of the quality of the input physics and overall reliability.

13.20.1 Nuclear Ground State Properties

Impressive progress has been made recently in measuring the masses of unstable nuclei. The 2003 Atomic Mass Evaluation[166] contains 2,228 measured masses, while the calculation of the r-process requires data on about 20,000 nuclei.

[166] Audi, G., Atomic Mass Data Center, AMDC, 2003.

13.20.2 Peculiarities of High N/Z Ratios: Halo Nuclei

Stable nuclei have $N/Z \sim 1$–1.5 and the protons and neutrons are well mixed in the nucleus. On the other hand, when $N/Z \sim 1.5$–4, the nuclei become unstable and the neutrons form a halo around a compact core. As a matter of fact, there is a decoupling between the proton distribution and the neutron distribution. Despite the multiparticle problem, the fundamental dynamics of halo nuclei can be conceived of as a halo that is decoupled to a large extent and a nuclear 'core'. In essence, the entire nuclear physics changes.

In 1985, Tanihata et al.[167] measured the effective radii of all known isotopes of lithium (^{6}Li to ^{11}Li), and discovered that ^{11}Li *showed a remarkably large radius, suggesting a large deformation or a long tail in the matter distribution* inside the nucleus. The result was explained by Hansen and Jonson,[168] who argued that the last two neutrons are weakly bound, in contrast with all other protons and neutrons, and form a cloud around a central core, which is the nucleus of ^{9}Li, as illustrated in Fig. 13.26. While ^{10}Li is a nucleus with 'normal' mass distribution, addition of a single neutron caused this dramatic change. Moreover, if one of the halo neutrons is removed, the other comes away too. The ^{9}Li nucleus plus the two neutrons is a three-body system which disintegrates if any one of the three components is removed. For this reason, Zhukov et al.[169] called such models Borromean nuclei.[170]

So far only light halo elements have been investigated, but there is no a priori reason to think that the same phenomenon would not exist in heavy halo nuclei. In fact, the path of the r-process traverses a territory full of halo nuclei (Fig. 13.27).

A question that remains open is: why are ^{26}O and ^{28}O unbound? The addition of so many neutrons cannot bind the nucleus. But if a single proton is added to ^{24}O, then the nucleus ^{31}F with 6 extra neutrons is stable. This question is related to the fact that, far from the stability line, $N = 16$ appears as a new magic number, while the $N = 20$ magic number disappears.

What about superheavy elements or islands of closed stable shells? This possibility was already mentioned by Myers and Swiatecki.[171] All nuclei in this range have a non-spherical shape. The liquid drop model favors spherical nuclei and the shell effect favors deviations from spherical symmetry. Hence, shell effects are very important at high A values.[172]

---

167 Tanihata, I., and 8 coauthors, PRL **55**, 2676 (1985).

168 Hansen, P.G. and Jonson, B., Europhysics. Lett. **4**, 409 (1987).

169 Zhukov, M., and 5 coauthors, Phys. Rep. **231**, 151 (1993).

170 Borromean rings consist of three circles linked in such a way that removal of any ring involves unlinking the other two.

171 Myers, W.D. and Swiatecki, W.J., Nucl. Phys. **81**, 1 (1966).

172 For a review see Sobiezewski, A., and Pomorski, K., Prog. Part. Nucl. Phys. **58**, 292 (2007).

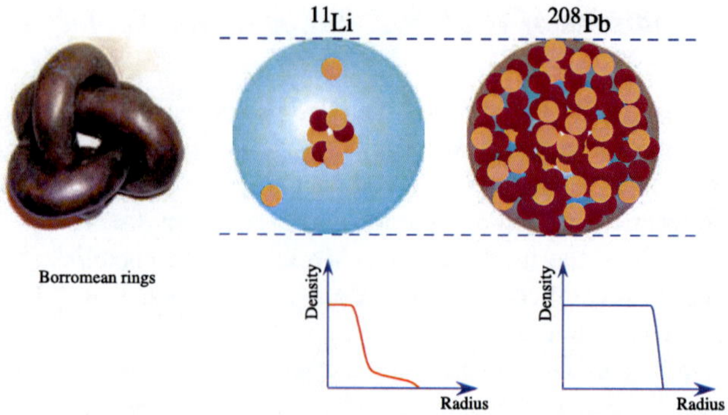

Fig. 13.26 Comparison between the halo nucleus ^{11}Li and the very heavy 'regular' nucleus ^{208}Pb. The extra neutrons in ^{9}Li move in a large volume, much larger than the prediction of the classical theory. The lower graphs show the matter density as a function of radius in the nucleus. After Nörtershäuser, W., The TRIUMF Newsletter, April 2005. *On the left* is an image of the Borromean rings

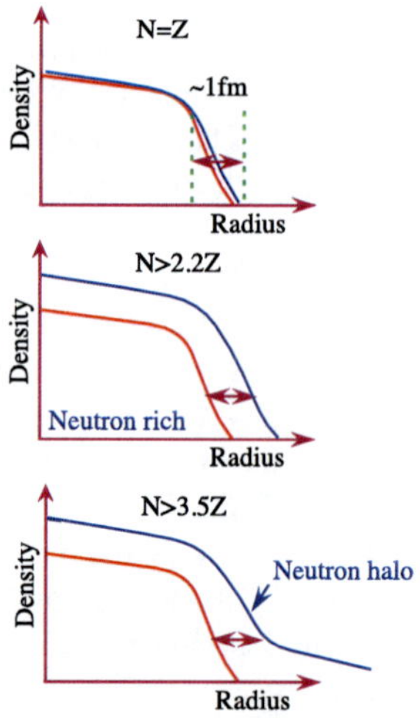

Fig. 13.27 Difference in structure and distribution of protons and neutrons as the number of neutron increases relative to the number of protons

13.20.3 Chaos?

The phenomenological mass binding energy formulae have developed immensely since the idea was first conceived by Bethe and von Weizsäcker in 1935. The evolution of the mass formula is shown schematically in Fig. 13.28. But where do we stand today?

Current empirical formulae for the masses of nuclei contain up to 30 parameters, fitted to a large number of data points, although not those of the nuclei important for the r-process. Despite the efforts of many nuclear physicists, the overall accuracy in

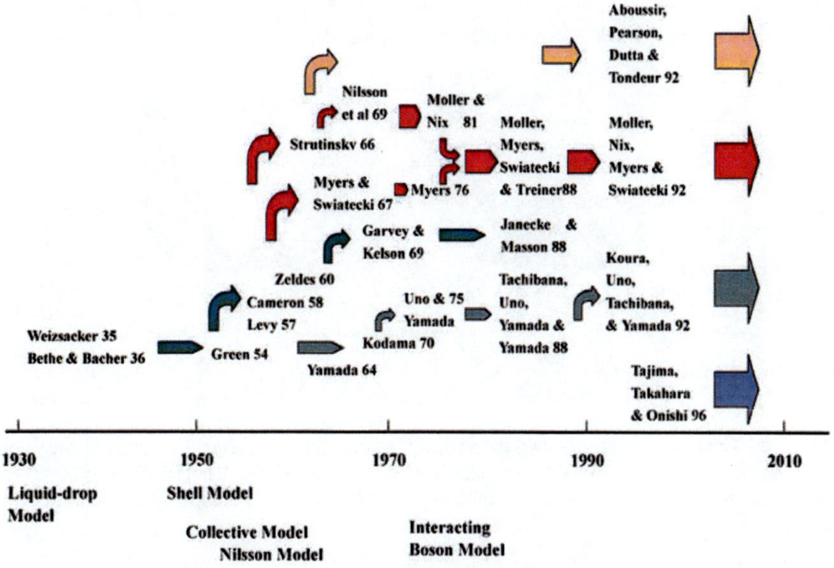

Fig. 13.28 Evolution of the mass formula and the present zoo of formulas. After Uno, Riken Review, 28, 2000

the predicted mass remains worse than 0.5 MeV. A comparison between two such theoretical formulae in regions where no experimental data is available is shown in Fig. 13.29. This may sound impressive because the differences in energy are small compared to the total binding energy of the heaviest nuclei. But what is important for the r-process is the difference in mass between two neighboring nuclei, and here the situation is much worse, mainly because nuclear effects are not a smooth function of the number of neutrons or protons.

This situation led Bohigas and Leboeuf[173] to suggest that the discrepancy can be attributed to the fact that nuclear dynamics is partly chaotic. If this is indeed the case, as these authors have argued, than the chaotic energy is of the order of $2.78/A^{1/3}$ MeV, which is of the same order as the correction for the nuclear shell in the best mass formula. Why is this so? The shell model treats the nucleons as individual particles moving under the influence of all other particles. But this is only an approximation, and a more accurate interaction expression should contain a fluctuating part, thereby appearing random or chaotic.

The heretic suggestion of Bohigas and Leboeuf caused a stir in the nuclear physics community, and Molinari and Weidenmüller[174] gave the idea another twist by replacing the term 'chaotic', which had caused some nuclear physicists displeasure.

---

[173] Bohigas, O. and Leboeuf, P., Phys. Rev. Lett. **88**, 129903 (2002).

[174] Molinari, A. and Weidenmüller, H.A., Phys. Lett. B **637**, 48 (2006).

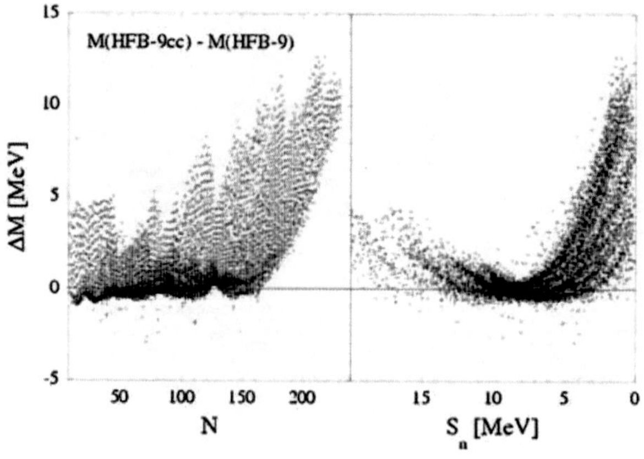

Fig. 13.29 Comparison between two theoretical formulae for the masses (expressed in MeV) of isotopes (*left panel*) and neutron separation energies S_n (*right panel*) for all nuclei with $8 \leq Z \leq 110$ lying between the proton- and neutron-drip lines. Note that this is not a comparison with (available) measured data. After Arnould et al. 2008

These authors treated the discrepancy as the residual effect of the many nucleons moving in the nucleus, a residual not accounted for by the mass formula because it is not smooth, but fluctuating. However, the difference between the first pair of authors and the second is not really fundamental. The consequence for our problem of element synthesis is that we will probably have to abandon any hope of finding a mass formula that can be used for the prediction of the r-process and wait until the masses of all the relevant isotopes are measured in the laboratory.[175]

13.20.4 Are the Magic Numbers Universal?

Recent research has demonstrated that magic numbers are not constants of nature, as was once envisioned. So far, the research has focused mainly on light nuclei with proton and neutron numbers that differ significantly from those of stable nuclei. Some nuclei that were expected to be magic turned out not to be particularly tightly bound. On the other hand, there are indications of new magic numbers for nuclei that have many more neutrons than protons. These experimental observations suggest that the gaps in the sequence of nuclear levels that give rise to magic numbers depend on the balance between protons and neutrons in ways that will require further exploration. Does this make sense? If the nuclear force acting between two neutrons is slightly different from the force acting between two protons (in contrast with the standard assumption that the two forces are identical), then it is to be expected that the behavior

[175] For a review on the potential of measuring the important masses see Lunney, D., Pearson, J.M. and Thibault, C., Rev. Mod. Phys. **75**, 1021 (2003).

will change as the number of neutrons relative to the number of protons increases to the extreme.

As an example, consider recent research on Si nuclei. Silicon ($Z = 14$) has three stable isotopes with 14, 15, and 16 neutrons, respectively. ^{42}Si has 28 neutrons, or 12 more than the most massive stable isotope. Fridmann et al.[176] produced ^{42}Si nuclei that can be identified in a magnetic spectrograph. The studies of ^{42}Si showed that the magic number $N = 28$ remains valid, despite the large excess of neutrons in this nucleus. However, a full understanding of the experimental observations requires the proton number $Z = 14$ to be magic as well (previously, only $N = 14$ had been shown to be magic in exotic nuclei). This makes the silicon isotope doubly magic, and results in an almost spherical shape for the nucleus.

An international team of physicists[177] has found strong evidence for the existence of a new 'doubly magic' nucleus with an unconventional configuration of neutrons. The ^{24}O nucleus ($N = 16$, $Z = 8$) is also the lightest doubly-magic nucleus that is very unstable against radioactive decay.

As a further example, consider the possible new sub-shell closure[178] at $N = 58$ emerging in neutron-rich nuclei beyond ^{78}Ni. The ^{78}Ni nucleus is unique, as it is the most neutron rich ($N/Z \approx 1.8$) and a doubly magic nucleus. Similarly, Motobayashi et al.[179] concluded that the neutron rich ^{32}Mg with $N = 20$ is not a magic number when N/Z is very large.

Are the magic numbers a simple extrapolation or are there new ones? This and many other questions are the subject of present day nuclear research. See, for example, Sorkin and Porquet[180] and Janssens.[181]

13.20.5 More Than One Ground State

An amazing experiment was carried out recently by Gaudefroy et al.[182] who discovered the coexistence of two ground states of the nucleus ^{43}S, one with a spherical shape and the other with a deformed shape (see Fig. 13.30). Strictly speaking, the deformed state has a slightly lower energy than the spherical one. What is surprising is that nuclei with magic numbers of neutrons are spherical and the non-spherical shape has just a trifle less energy than the spherical shape. The small difference in energy between the two shapes, small as compared with the classical energy difference between nuclear energy states, has led to the definition of coexisting ground

[176] Fridmann, J., and 21 coauthors, Nature **435**, 922 (2005).

[177] Morrissey, D., CERN Courier, 20 November 2007.

[178] Winger, J.A., and 17 coauthors, Phys. Rev. C **81**, 044303 (2010).

[179] Motobayashi, T., and 18 coauthors, Phys. Lett. B **346**, 6 (1995).

[180] Sorlin, O. and Porquet, M.-G., *Nuclear magic numbers: New features far from stability*, Prog. Particle and Nucl. Phys. **61**, 602 (2008).

[181] Janssens, R.V.F., Nature **435**, 897 (2005).

[182] Gaudefroy, L., and 22 coauthors, PRL **102**, 092501 (2009).

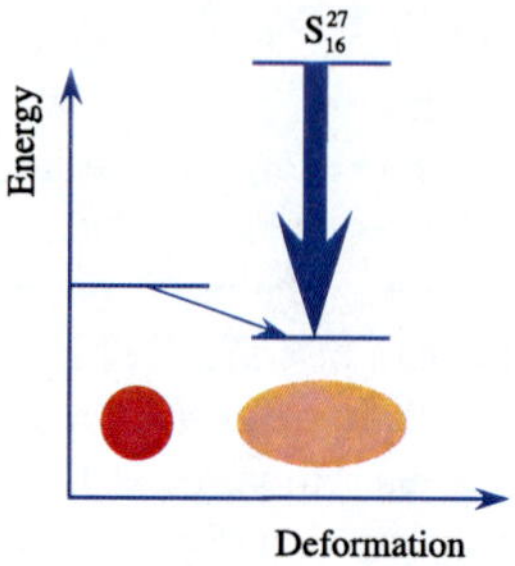

Fig. 13.30 Shape of the ground states of S^{27}_{16}, after Gaudefroy et al. (2009). The energies of the spherical and the deformed ground states are extremely close relative to the mean separation between nuclear energy levels. Hence the expression 'two ground states with different shapes'

states. The upper level of the deformed nucleus decays mostly to the deformed ground state. The spherical state hardly ever decays into the deformed ground state.

Such results indicate that some aspects of the interactions responsible for the shell structure in nuclei are not readily apparent from the properties of the stable nuclei. If these aspects of the forces that bind nuclei are more important in exotic nuclei far from the bottom of the stability valley, then clearly the nuclear physics prevailing along the path of the r-process is not identical to that near the bottom.

13.20.6 Half-Lives Against β-Decay

There are several theories for the β-decay of heavy nuclei. Evidence has accumulated that measured β-decay half-lives of neutron-rich nuclei are often significantly shorter than the predicted values. It has been argued[183] that in some cases this tendency comes simply from the use of the relatively low energy difference values predicted by the adopted liquid-drop-type mass formula of Myers and Swiatecki.[184] This attests to the fact that the large-scale microscopic-type calculations of β-decay rates, as needed for r-process simulations, have not reached a satisfactory level of accuracy and predictive power.

13.20.7 Clustering in Neutron-Rich Nuclei

Clustering has long been known[185] to be influential in the structure of the ground state, as well as excited states, of nuclei with $N = Z$. The first discussions of

---

[183] Takahashi, K., in: *Origin and Distribution of the Elements*, ed. by Mathews, World Scientific, Singapore, 1987.

[184] Myers, W.D., and Swiatecki, W.J., Nucl. Phys. **81**, 1 (1966).

[185] von Oertzen, W., Freer, M. and Kanada-En'Yo, Y., *Nuclear clusters and nuclear molecules*, Phys. Rep. **432**, 43 (2006).

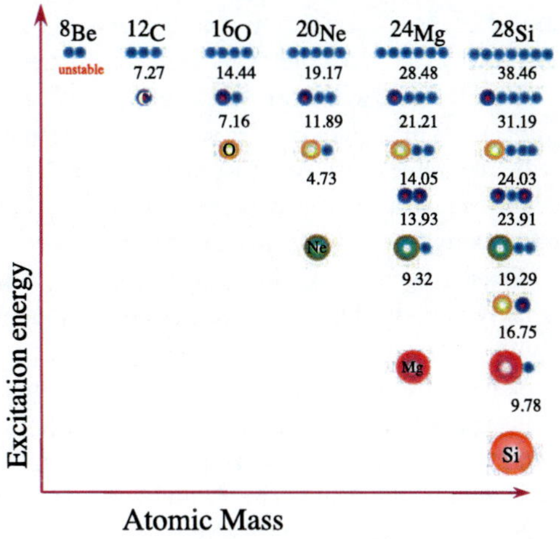

Fig. 13.31 The Ikeda table, which gives the threshold for nuclei with α-clustering. The table and the energies of disintegration of the clusters are important for the α-process. After Ikeda et al. 1968

nuclear clustering[186] took place rather early on, but the idea of clustering was pushed into a corner by the independent particle model and the shell model. However, in recent years we have seen a renaissance of the ideas of nuclear clustering, the reason being the discovery of complex structures composed of α particles in what are called α-conjugate nuclei, that is, nuclei with $N = Z$ which have an even and equal number of protons and neutrons.[187] These nuclei are extra stable. The best example of such a model is the α-particle model, which is rather robust against collapse into more compact shell-model types of configurations. It was this fact that led to the revival of the α-clustering and α-particle models in the 1960s.[188] This is extremely important for the s- and r-processes.

An early product of this research was the Ikeda diagram,[189] which depicts the structure of the most stable and most abundant elements (see Fig. 13.31). The table links the energy required to liberate the cluster constituents to the excitation energies at which the cluster structure prevails in the host nucleus.

The renewed interest in clustering arose from studies of neutron-rich nuclei and exotic weakly bound nuclei. The strong clustering in weakly bound systems can give rise to two-center and multi-center nuclear configurations, whose structure can be described by concepts of molecular physics. Valence neutrons can exist in molecular orbitals. The neutrons establish a covalence bond with the core.[190] The shell

[186] von Weizsäcker, C.F., *Die Atomkerne*, Akadem. Vedagsantalt, Leipzig, 1937; ibid., Naturwiss. **26**, 209 (1938); Wefelmeier, W., Zeit. f. Phys. **107**, 332 (1937).

[187] Freer, M. and Merchant, A.C., J. de Phys. G **23**, 261 (1997).

[188] Wildermuth, K. and Kanellopoulos, Th., Nucl. Phys. **7**, 150 (1958); Wildermuth, K. and McClure, W., Springer Tracts in Modern Physics **41**, Springer, Berlin, 1966; Hiura, J. and Shimodaya, I., Prog. Theor. Phys. **30**, 585 (1963); ibid. **36**, 977 (1966); Tamagaki, R. and Tanaka, H., Prog. Theor. Phys. **34**, 191 (1965).

[189] Ikeda, K., Tagikawa, N., and Horiuchi, H., Prog. Theor. Phys., Suppl., Special Issue 464 (1968).

[190] von Oertzen, W., Zeit. f. Phys. **357**, 355 (1997).

model cannot describe non-spherically-symmetric multi-center configurations. New methods will have to be developed for the treatment of such configurations.

Recall the low energy scattering of $^{12}C + {}^{12}C$ and $^{16}O + {}^{16}O$, which revealed indications of molecular-like structure. Indeed the researchers described the resonant structure in the clustering assuming molecular-like potentials.[191] In a molecule composed of two atoms, the outer electrons move in a two-center field. As a consequence, one can find similar discussions of the two-center shell model.[192]

Clustering becomes important along the drip line, because the systems become weakly bound. There are speculations[193] that clustering may even be the dominant configuration.

α particles have zero spin and hence do not obey the Pauli exclusion principle. Consequently, unlimited numbers of α particles can occupy the same nuclear state and create what is known as a condensed state. The possibility of such a state was discovered by Tohsaki et al.[194] and has gained momentum since then. It appears that such a state may be common in heavy nuclei.[195]

13.20.8 And Yet There Is Life in the Old α Model

From time to time a new feature of the old α model turns up. Indeed, among the light nuclei, 4n nuclei such as ^{12}C, ^{16}O, ^{20}Ne, ^{24}Mg, and so on, have higher binding energies per particle than any of their neighbors.

In 1938, a year after Bohr's famous address, Hafstad and Teller[196] attempted a description of $A = 4n$ nuclei based on individual α particles. To obtain the structure, they developed an α–α interaction which was repulsive at short distance and attractive over longer distances, something like the van der Waals force which acts between molecules. Hafstadt and Teller proposed a cluster model to estimate the binding energies of the $4n$, $4n-1$, and $4n+1$ nuclei, assuming that the $4n$ core was composed of α particles which interact only with the nearest α particle and a satellite particle moving around. The agreement Hafstadt and Teller got was quite nice, except for the simplest case of two α particles, the unbound ^8Be. The theory was improved by Brown and Inglis[197] by including the Coulomb energy.

---

191 Haas, F. and Abe, Y., PRL **46**, 1667 (1981).

192 Braun-Munzinger, P. and Barrette, J., Phys. Rep. **87**, 209 (1982); Greiner, W., Park, J.Y. and Scheid, W., *Nuclear Molecules*, World Scientific, Singapore, 1995; von Oertzen, W., Prog. Theor. Phys., Suppl. **146**, 169 (2002).

193 Horiuchi, H., Proc. Seventh Int. Conf. Clustering Aspects of Nuclear Structure and Dynamics, ed. by Korolija, Basrak, and Caplar, World Scientific, Singapore, 1999.

194 Tohsaki, A., Horiuchi, H., Schuck, P. and Röpke, G., PRL **87**, 192501 (2001).

195 Itagaki, N., Kimura, M., Kurokawa, C., Ito, M. and von Oertzen, W., Phys. Rev. C **75**, Issue 3, id. 037303 (2007).

196 Hafstad, L.R. and Teller, E., Phys. Rev. **54**, 681 (1938).

197 Brown, H., Inglis, D.R., Phys. Rev. **55**, 1182 (1939).

In addition Wheeler,[198] and years later Ikeda, Takigawa, and Horiuchi,[199] attempted to calculate the binding energy of $4n$ nuclei assuming the α model. Wheeler treated the α particles as atoms in a molecule and calculated the relative motion of each α particle relative to the others. The advantage of Wheeler's scheme is that it hardly depends on the nature of the poorly known nuclear force.

The simplest two-center system is ^{8}Be. However, it is unstable. It is not clear why a three-body system like the very stable ^{12}C would hold itself together so well, while a two α system would not.

In 1940, Dennison[200] proposed a model of ^{16}O with four α particles arranged at the corners of a regular tetrahedron, a structure reminiscent of molecules composed of 4 atoms. Energy levels of such a structure should include levels due to rotation and vibration, and should be relatively easy to identify. The observed level at 6.06 MeV was predicted to be a breathing mode (expansion and contraction of the entire nucleus). By 1954, more information had accumulated, and Dennison was able to fit and predict many excited states of the nucleus. The model was eventually rejected when the measurements showed that the 6.06 MeV level had a small monopole matrix element in the ground state, which means a deviation from spherical symmetry, and that the breathing mode was much higher in energy.

Bethe and Bacher claimed in Bethe's Bible that the existence of an α particle in the nucleus was not conclusive at all. A simple calculation shows that the heavy nuclei are unstable against the emission of an α particle, but stable against the emission of a proton, for example. It is energetically more favorable for the nucleus to emit an α particle than any other particle. The argument is not that simple, because if the nucleus is unstable against emission of an α particle, it is also unstable against the emission of a ^{12}C nucleus, for example, as Bethe and Bacher showed. But why does this not happen? It was a question of probability, replied Bethe and Bacher. The chances of an α emission, namely 4 particles escaping together, is already small. So the chances of 12 particles coming out together should be significantly smaller still. Bethe and Bacher summarized as follows:

> [...] we must say that it can at present not be decided how much truth is in the assumption of α particles as nuclear subunits. Certainly, this assumption must not be taken literally.

Of course, the α particle inside the nucleus feels various forces that it does not feel when it is free, so the α particle inside the nucleus looks quite different from one that is free. This harks back to what Schrödinger had to say about α emission in Sect. 8.2.

Recall the 7.65 MeV Hoyle state in ^{12}C. If it were a genuine 3α model, then it should show a nuclear spectrum that resembles that of rotation. A $J = 2^+$ level, which belongs to the rotational band, is thus predicted to lie at ∼8.4 MeV.[201] The closest $J = 2^+$ candidate is about 3 MeV higher in energy, and thus cannot reasonably be rotationally linked with the Hoyle state. We conclude that the Hoyle state is not

---

198 Wheeler, J.A., Phys. Rev. **52**, 1083 (1937).

199 Ikeda, K., Takigawa, N. and Horiuchi, H., Prog. Theor. Phys. Suppl., Extra Number, 464 (1938).

200 Dennison, D.M., Phys. Rev. **57**, 454 (1940).

201 Freer, M. and Merchant, A.C., J. of Phys. G: Nucl. Particle Phys. **23**, 261 (1997).

Fig. 13.32 The nuclear density contours as calculated for the ^{8}Be nucleus. The scale is about 4 fm

a rotational state in a 3α chain which makes up the ^{12}C nucleus. It is generally accepted that the state possesses a rather different character, which has overlaps with an extended α-particle arrangement.[202] In simple terms, the evidence for the ^{12}C nucleus to be composed of three α particles is weak (Fig. 13.32).

On the other hand, Ali and Bodmer[203] scattered α on α and found the mutual interaction and potential. Once the interaction between two α particles became known, several groups calculated the structure of its states with Ali and Bodmer potentials. There was particular interest in the excited 0^+ state with an energy of 0.38 MeV above the threshold for breakup into the 3α channel, which was predicted by Hoyle in 1954. Visschers and van Wageningen[204] found that the Ali and Bodmer potentials gave a ground state binding energy which was too small, and failed to give a reasonable energy for the Hoyle state. Many authors argued that there were fundamental weaknesses in the elementary α-model and that the microscopic structure of the α particle was important. And this may well be so!

Fedorov and Jensen[205] added a 3-body force to one of the Ali and Bodmer potentials and were able to choose the parameters so that the energies of the ground state of the Hoyle state agreed with the prediction. In this way they were able to predict some of the properties.

Nuclei have practically constant density. If the particles in the nucleus are all fermions, then it is clear from the Pauli exclusion principle that there is saturation and constant density. But if there are clusters of α particles, then these are bosons that can condense to the lowest energy level and reach extreme densities. The higher the atomic mass, the higher the density can be, if the basic particles are bosons.

13.21 No Further Neutron Build-up: The Fission Barrier

A natural question to ask is whether irradiation by neutrons can produce heavier and heavier nuclei without limit, or whether at a certain point the newly synthesized nuclei decay or even fission. It seems that the latter is the case.

[202] Fedorov, D.V. and Jensen, A.S., Phys. Lett. A **586**, 493 (1995).

[203] Ali, S. and Bodmer, A.R., Nucl. Phys. **80**, 99 (1966).

[204] Visschers, L.L. and van Wageningen, R., Phys. Lett. B **34**, 455 (1971).

[205] Fedorov, D.V. and Jensen, A.S., PRL **79**, 2411 (1997).

Table 13.9 The fission barrier (in MeV) predicted for neutron-rich nuclei by four different fission theories

Element	*Z*	*N*	BSk8	ETFSI	MS	HM
Po	84	170	28.27	26.85	15.40	6.84
Po	84	184	37.14	39.01	20.08	6.21
U	92	170	11.48	5.25	4.55	3.36
U	92	184	16.77	17.67	7.12	3.80
U	92	194	15.39	10.89	5.06	
Fm	100	179	4.24	2.18	1.72	2.61
Fm	100	184	5.34	5.97	1.86	2.58
Fm	100	194	3.24	1.63	1.36	
Fm	100	210	6.42	7.30	3.59	

(1) BSk8—Samyn, M., Goriely, S. and Pearson, J.M., Phys. Rev. C **72**, 044316 (2005), (2) ETFSI (extended Thomas–Fermi plus Strutinsky integral)—Mamdouh, A., and 3 coauthors, Nucl. Phys. A **679**, 337 (2001), (3) MS—Myers, W.D. and Swiatecki, W.J., Phys. Rev. C **60**, 014606 (1999), (4) HM—Howard, W.M. and Möller, P., Atomic Data Nucl. Data Tables **25**, 219 (1980)

So how well can we predict fission? The fission process is mimicked by a sufficiently strong deformation of the nuclear drop that the nucleus splits into two fragments (and not more).[206] The predicted fission potential barriers are still highly uncertain, particularly for neutron-rich nuclei, as can be seen from Table 13.9. In contrast to the simple classical double-humped picture along the bottom of the stability valley, modern theories exhibit three or even four barriers.[207]

13.22 Some Reflections on Present Day Nuclear Physics

Nuclear physics is almost a century old, and yet a *genuine nuclear theory* of low energy phenomena is still wishful thinking. What we have at our disposal is a large number of ad hoc models, each valid in a small neighborhood but none of which can really be trusted under the conditions in which the heavy elements are supposed to be synthesized.

So why is this? Newton was lucky that the gravitational force of the Sun is the dominant one in the Solar System so that a very good approximation can be obtained by ignoring the interactions of any given planet with all the others. The effect of all the other planets on Mercury, for example, amounts to a rotation of about 530 arcsecond/century in the orbit of the planet, so that ignoring the gravitational influence of all the planets provided a superb first approximation. But the opposite is true for the nucleus, at least for low energy phenomena. The nucleus is a system of many strongly interacting particles without any dominant interaction or any significantly more massive particle. The misleading partial success of the shell model

[206] Bohr, N. and Wheeler, J.A., Phys. Rev. **56**, 426 (1939).

[207] Minato, F. and Hagino, K., Phys. Rev. C **77**, 044308 (2008).

is probably due to the cancelation of many particle interactions leaving some quasi-smooth force between the nucleons. Hence, the nucleus is a horribly complicated many-body problem.

While classical physics had simple problems to solve, and so does quantum theory, these theories were soon accompanied by philosophical implications. In contrast, the 'theory' of nuclear physics is a collection of ad hoc phenomenological models without any new philosophy as yet.

The nucleon–nucleon interaction is very peculiar, and we do not know of any similar behavior in other systems. The interaction between free protons or neutrons is not the same as between bound protons and neutrons. Furthermore, the interaction between a collection of protons and neutrons is not the simple sum of the binary interactions. Thus, there is a genuine conceptual and philosophical difficulty in nuclear physics.

Chapter 14
The Elusive First Stars

14.1 The Need for Population III

All stars observed today show various amounts of heavy elements. But we know that heavy elements could not have been formed in the Big Bang. So what about the heavy elements seen in the most heavy-element-poor stars? Where were they formed? While the amount of heavy elements in the most metal-poor stars exceeds by many orders of magnitude what can be synthesized in the standard Big Bang, no star completely lacking in heavy elements has ever been observed. Typical results predicted by the Big Bang are shown in Fig. 14.1. The data is taken from Wagoner et al.[1] The present day amount of ^{4}He predicted by the Big Bang, which is about 0.24, was considered unrealistic in 1967 when Wagoner et al. carried out the calculations, and so was off the scale. Big Bang models with present day values of the cosmological parameters and which yield a helium abundance (by mass) of $\sim$0.24–0.25 He produce negligible amounts of ^{12}C.

Figure 14.1 (right) shows the abundances of the elements between ^{12}C and ^{24}Mg. The value inserted for $A = 24$ is the sum of the abundances of all elements beginning with magnesium. The figure was plotted for the case in which the carbon abundance is the highest (He abundance of 0.34). This is not an acceptable value today. The sum of all the heavy elements is at most 10^{-7}, and in most cases significantly less. In summary, the amount of elements heavier than carbon which formed in the standard Big Bang is negligible.

Since the Big Bang does not produce any heavy elements and the oldest stars show significant amounts of heavy elements, astrophysicists were led to the hypothesis that the observed generations of stars (Pop I and Pop II) were preceded by a third generation of stars, the so-called Pop III stars, or primordial stars. Here we discuss the properties of this not yet discovered, but badly needed and therefore expectantly hypothesized first generation of stars in which the first synthesis of the heavy elements must have occurred.

---

[1] Wagoner, R.V., Fowler, W.A. and Hoyle, F., Astrophys. J. **148**, 3 (1967).

G. Shaviv, *The Synthesis of the Elements*, Astrophysics and Space Science Library 387,
DOI: 10.1007/978-3-642-28385-7_14, © Springer-Verlag Berlin Heidelberg 2012

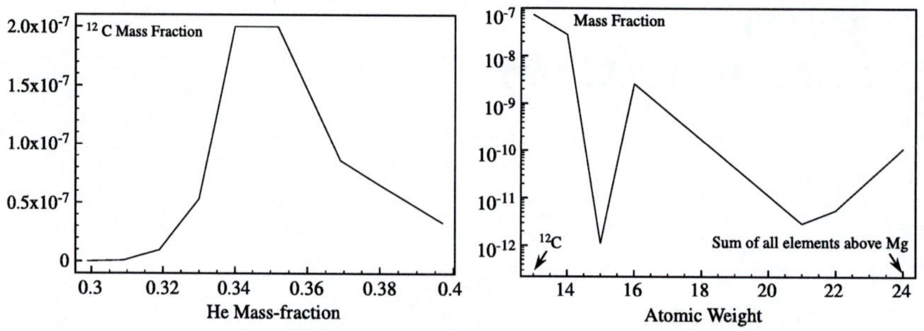

Fig. 14.1 *Left* abundance of ^{12}C as a function of ^{4}He, as predicted by the Big Bang. Wagoner et al. (1967). The present day accepted primordial value of He (0.24–0.25) is off the scale. *Right* abundance of the elements between ^{12}C and ^{24}Mg, as predicted by the Big Bang. Wagoner et al. (1967)

Table 14.1 Cosmic dates. Times since the Big Bang

	Time	T (K)	E (eV)
Today	13.7 Gyrs	2.7	2.349×10^{-4}
Recombination	300,000 years	3,000	0.26
First cosmic structure	~1 Gyrs	10	10^{-3}
Nucleosynthesis	3 min	7×10^{8}	6×10^{4}
Formation of protons and neutrons	$<4 \times 10^{-6}$ s	$>10^{13}$	$>10^{9}$

14.2 Cosmic Timetable: When Did It Happen?

Table 14.1 summarizes the phases of the Universe since the Big Bang. The table also provides the mean temperature of the Universe, as well as the mean energy of the particles it contained in eV. About 300,000 years after the Big Bang, the Universe was already cool enough for electrons to combine with the primordial nuclei, mostly hydrogen and helium and very small amounts of deuterium and lithium, and form neutral atoms for the first time. This is known as the recombination time, although it should rather be called the combination time, since there had never been any neutral atoms prior to that! From this time on, the Universe was in a more familiar cool gaseous state, except for the fact that there were no stars or galaxies.

The Universe was filled almost uniformly with a nearly neutral gas of H and He. There was nothing else to 'see' save an essentially perfect background. As the starless Universe expanded further, it became darker (because the background radiation became fainter) and cooler. This era in the Big Bang theory is known as the Dark Ages, because there were no radiation sources, just a uniform background. It would have seemed something like being in a heavy fog.

The Dark Ages lasted for about 400,000 years until the first stars formed and emitted radiation, thus lighting up the Universe. This was the era of primordial star formation. An artist's concept of the major features of cosmic evolution is shown in Fig. 14.2.

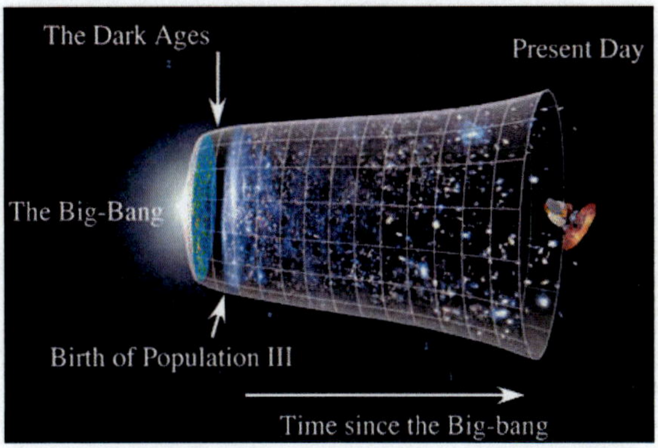

Fig. 14.2 Artist's view of bright structure evolution in the Universe. Credit: NASA/WMAP Science Team

14.3 Some General Properties Expected of Population III

A few general properties can be expected of population III stars. Since the time span between the oldest known Pop II stars and the age of the Big Bang is small, those Pop III stars which synthesized the elements must have been short-lived, no more than a few million years. If so they must have been very massive stars, evolved fast, exploded, and spread the heavy elements in the space around them, so that the next generation of stars would already have contained small amounts of heavy elements. As no such stars have been found at the distances we can observe today, we can limit their lifetime from above, and hence limit their mass from below. To observe these massive stars, we have to look at remote distances, a long way back in time, to the era in which the galaxies and stars had just formed. Present day telescopes are just beginning to reach the relevant distances. But if stars with all possible masses were formed, then stars with mass say $0.5M_{\odot}$ would also have formed, and these stars have a lifetime which exceeds the age of the Universe. So if they were formed, they will still be alive and should also be found in our Galaxy. But are there any such stars?

Since the heavy elements must, by this hypothesis, be generated quickly, they can only be produced in very massive stars. Looking around our own environment, we infer from observation that the probability of forming a massive star decreases with mass and varies, in our neighborhood today, as $M^{-2.35}$. This means (a) that the probability of forming $100M_{\odot}$ stars is only 2×10^{-5} times that of forming a $1M_{\odot}$ star and (b) that most stellar mass resides in low mass stars. The amount of mass in stars with masses in the range $10–100M_{\odot}$ is only 0.044 of the mass in stars with masses in the range $1–10M_{\odot}$. Thus, if the birth of Pop III stars followed the patterns of present day star formation, there would be (a) too few massive stars to generate the required amount of heavy elements and (b) many Pop III stars of low mass ($0.8M_{\odot}$) that would still be alive in our own Galaxy. It is therefore reasonable to venture that the conditions for stellar formation have changed significantly from those prevailing when the Pop III stars were formed to those prevailing when Pop II and Pop I stars were formed. The probability of forming massive stars must have been significantly higher at the time when Pop III stars were formed.

14.4 Ideas Leading to Population III: Why Do We Need them?

The need for Pop III stars was already suggested by Tayler and Hoyle in 1964, but the details were not worked out. In 1970, Unsöld[2] began his review of the origin of the elements with the words:

> After the decline of the Lemaitre–Gamow 'big bang' theory, the origin of the elements and their abundance distributions was discussed principally with regard to our Galaxy. The Galaxy consists mainly of three different stellar populations [...]

By this time, Unsöld accepted the existence of stellar populations, although not classified exactly as Baade had done. Unsöld had a halo population II with an age of $(10 \pm 4) \times 10^9$ years and with metal abundances of 1/200 to 1/5 relative to the Sun, a disk population with age less than $(10 \pm 4) \times 10^9$ years and with mostly normal metal abundances, and a spiral arm population I with an age of 10^5–10^{10} years and with normal metal abundance (relative to the Sun).

Unsöld made the important point that:

> Even the oldest stars of the disk population have, in the main, a metal abundance which cannot be distinguished from that of the Sun or the younger star, spiral arm population, that is, all abundances in the disk and the spiral arm populations are the same, to within an error of at most 50%. That the average composition of the matter in the Galaxy has not changed within the past $\sim 10^{10}$ years is not astonishing, since, with its present luminosity, the Galaxy would have burnt in this time only about 2% of its total hydrogen.

The reader may recognize in Unsöld's words the argument given by Tayler and Hoyle, but returning to the old problem does not solve it. So when did the heavy elements form? By observations of the oldest GC and their metal abundances and the oldest open clusters like M67 and NGC188,[3] which do of course contain metals, Unsöld estimated that the heavy elements must have formed in less than 1 Gyrs after the Big Bang.

---

[2] Unsöld, A., ASPL **10**, 329 (1970).

[3] Messier 67 or NGC 2682 is an open cluster which contains about 500 stars in the constellation of Cancer. It was discovered by Koehler in 1779. Theoretical age estimates vary between 3.2 and 5 Gyrs. NGC 188 is an open cluster in the constellation of Cepheus. It was discovered by Herschel in 1825. Most open clusters disintegrate after a few million years because of the gravitational interaction of our galaxy. However, NGC 188 lies far above the plane of the Galaxy and survived its perpetual action rather well. It is one of the most ancient open clusters.

Has all nucleosynthesis occurred in normal stars? Fowler and Hoyle[4] suggested that massive objects of up to $10^8 M_\odot$, which are not seen today, were responsible for the synthesis of the heavy elements. As they wrote:

> The concept of stellar type objects with masses up to $\sim 10^8 M_\odot$ is of course strange, but the very nature of the case demands an unusual physical situation.

Moreover, if the present rate of conversion of hydrogen into helium has always prevailed in the Galaxy, the total helium produced falls short of what is observed. Add to that the indication that no substantial change in the amount of heavy elements was observed in the past 5×10^9 years and one is forced to conclude that the nuclear evolution in the past must have been much more intense. The early phases of the Galaxy must have been much more active and violent. We remark that the above was written when the Big Bang did not seem to produce the right amount of helium, so that the default solution was fast He generation in stars. In retrospect, the massive primordial stars were conceived to produce, à la Hoyle, the He as well as the heavy elements. As the energy released per unit mass in the conversion of hydrogen into helium is the largest nuclear energy release, no wonder the result was a very active and violent initial galactic phase.

The time constraint makes it plausible that the major part of nucleosynthesis took place in 'massive objects', even with present day established information that He is largely a Big Bang product. Clearly, the stars must be massive so as to evolve quickly, generate the elements, and explode. Consequently, if we consider very massive stars, say, $M > 100 M_\odot$, that explode via a pair creation SN, we find[5] that the effect of these Pop III stars on the chemical enrichment is negligible if only one or two generations of such stars occurred, whereas they produce quite different results from the standard model if they formed continuously for a period not shorter than 0.1 Gyrs. Even then, the results are at variance with the main observational contraints of elliptical galaxies, such as the average abundance ratios observed in stars. Hence, leaving all the eggs in one Pop III basket does not secure a solution.

If the galactic disk was formed by a collapse out of the halo, which is the picture proffered by Eggen, Lynden-Bell, and Sandage, then simple mechanical considerations give a much narrower limit: the time of free fall must be of the same order as the time of revolution, i.e., about $1\text{–}3 \times 10^8$ years. If further the original matter was not only metal deficient but also helium deficient, than it follows that, during the initial 10^8 years, the Galaxy must have been 1,000 times brighter than today to generate the present day levels of helium. But if the helium was formed in the Big Bang and the energy for the helium formation is taken care of, the energy needed to form the heavy elements is so small (because the amount of heavy elements is small) that nothing can be inferred about the original luminosity of the Galaxy.

---

4 Hoyle, F. and Fowler, W.A., MNRAS **125**, 169 (1963); Fowler, W.A., RMP **36**, 545 (1964).

5 Pipino, A., and Matteucci, F., Chemical Abundances and Mixing in Stars in the Milky Way and its Satellites. Proceedings of the ESO Arcetrie Workshop held in Castiglione della Pescaia, Italy, 13–17 September, 2004. ESO Astrophysics Symposia, ed. by S. Randich and L. Pasquini. Series Editor: Bruno Leibundgut, ESO. ISBN: 978-3-540-34135-2. Published by Springer-Verlag, Berlin and Heidelberg, Germany, 373, 2006.

Massive or giant galaxies may form through mergers of smaller objects, where massive Pop III stars have already enriched the gas with heavy elements, and then no stars with vanishing heavy elements would be seen. The oldest stars would not be in a spherical halo, but scattered according to the most recent merger.

14.5 Alternative Ideas

The mystery of the powerful radio sources was one of the most nagging problems in the early 1960s. Extragalactic radio sources are among the most unusual and spectacular objects in the Universe, with sizes in excess of millions of light years, radiating energies over ten times those of normal galaxies, and with a unique morphology. They reveal some of the most dramatic physical events ever seen and provide essential clues to the basic evolutionary tracks followed by many galaxies. These objects thus cause imaginations to run wild.

A prominent example is the galaxy NGC 4486, better known as M87. The galaxy was discovered by Messier in 1781, although he did not notice anything special about it, merely observing a luminous patch in the sky. In fact, a long list of distinguished astronomers observed the nebula and missed the exciting feature. The first to note something special there was Curtis,[6] who noted a 'ray' in the photograph. This note itself went unnoticed for a further 30 years. In 1949, Bolton, Stanley, and Slee[7] identified the NGC 4486 galaxy with the Virgo radio source. The description they gave of the observed nebula was: *Spherical nebula, unresolved.* They did not know about Curtis' remark 30 years earlier. In 1954 Baade and Minkowski[8] tackled the problem of identifying radio sources with optical ones and confirmed the identification of Bolton et al. Baade and Minkowski, who realized that the regular overexposed photograph they took (in an attempt to discover faint features) hid the interesting details, proceeded to take a photograph with a blue filter and Fig. 14.3 is what they got. It clearly shows a jet emerging from the center of the galaxy. Baade and Minkowski reported that Humason[9] had taken a spectrum of the jet and discovered that it had a featureless continuous spectrum.

As Baade and Minkowski wrote:

> No possibility exists at this time of forming any hypothesis on the formation of the jet, the physical state of its material, and the mechanism which connects the existence of the jet with the observed radio emission.

The two observers Baade and Minkowski did not know (probably because it was published in Russian) that a year earlier Shklovskii[10] had suggested that the observed

---

[6] Curtis, H.D., Pub. Lick Obs. **13**, 31 (1918).

[7] Bolton, J.G., Stanley, G.J. and Slee, O.B., Nature **164**, 101 (1949).

[8] Baade, W. and Minkowski, R., Astrophys. J. **119**, 215 (1954).

[9] The information was provided in the paper by Baade and Minkowski. No reference was given. We may guess that, for a spectroscopist, a spectrum without any features was not terribly interesting.

[10] Shklovskii, I.S., Doklay Akad. Nauk. USSR **90**, 983 (1953).

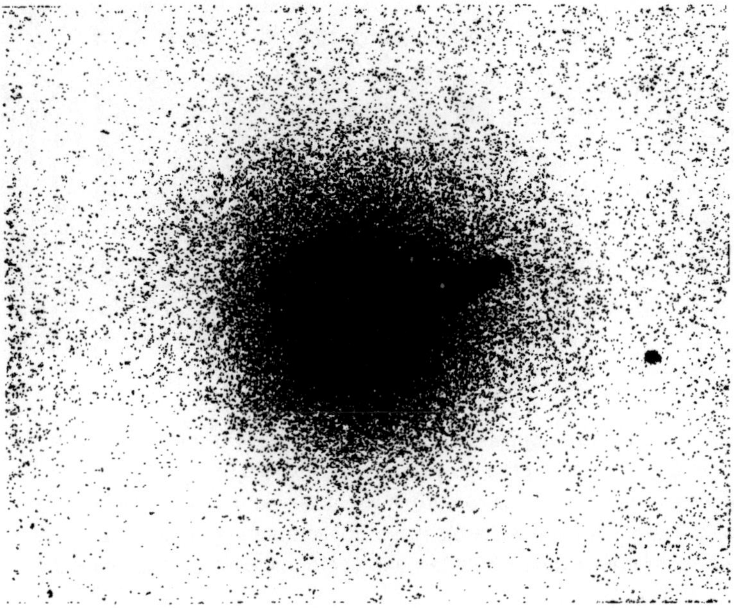

Fig. 14.3 The first photograph of the galaxy M87, indicating its peculiarity. This photograph was taken by Baade and Minkowski in blue light in 1954

featureless radiation might be synchrotron radiation.[11] Several investigators at the 1957 Manchester Symposium on Radio Sources[12] suggested that the synchrotron mechanism, which explains the radiation from the Crab Nebula, might also be responsible for the radiation coming from the jet in M87.[13] Baade[14] decided to test this hypothesis and observed the jet with a polarimeter, to discover that the radiation was, as expected, highly polarized. Baade did not measure the degree of polarization, but guesstimated that it was 30% polarized, which is a very high value. Three years later, Hiltner[15] confirmed Baade's observation when he found that only the jet emitted polarized radiation. Hiltner was unable to determine a lower limit to the degree of polarization. Polarization is one of the clearest signatures of synchrotron radiation.

In Fig. 14.4 we have an excellent example of the typical morphology of these objects. This radio galaxy was discovered in 1977 by Burch[16] and shows two jets

[11] Synchrotron radiation is electromagnetic radiation generated by accelerating charged particles moving through magnetic fields at speeds close to the speed of light.

[12] Proc. Manchester Symposium on Radio Astronomy, Cambridge, England, 1957.

[13] Oort, J.H. and Walraven, J.H., BAN 462 (1956).

[14] Baade, W., Astrophys. J. **123**, 550 (1956).

[15] Hiltner, W.A., Astrophys. J. **130**, 340 (1959).

[16] Burch, S.F., MNRAS **181**, 599 (1977).

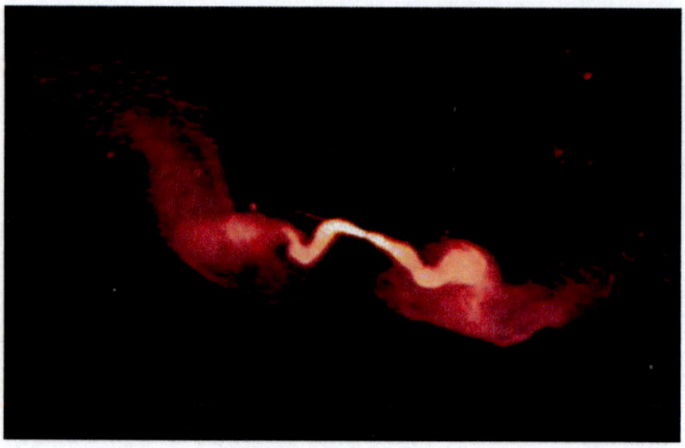

Fig. 14.4 The radio image of the galaxy 3C31. The jet, which changes its shape, extends to about 1 million light years. The image was created by the National Radio Astronomy Observatory/Associated Universities, Inc./National Science Foundation, by permission

coming out in roughly opposite directions. The length of each jet is about 60,000 light years and extends much further than the size of our own galaxy. The jet is hardly seen in visible light[17] and has recently been observed in X rays.[18] The galaxy itself is a bright, elliptical galaxy. As the radio image may give a false impression of its size, radio astronomers at the NRAO made a composite image with the visual galaxy, as shown in Fig. 14.5.

In 1959, Burbidge[19] carried out an estimate of the total energy in the large radio blobs observed at the end of the jets seen in radio sources. Burbidge took the theory of synchrotron emission and applied it to galaxies. The minimum energies required for the synchrotron emission were staggering. The jet in NGC 4486 (M87) has an energy of 2.4×10^{55} erg and the central radio source must contain at least 2.4×10^{58} erg. The estimate for the minimum total energy of the source Cygnus A[20] came out to be 3.9×10^{60} erg. The gravitational energy, even in a large galaxy, is no more than 10^{15} erg/g, so that the total potential energy of a galaxy of mass $10^{11} M_\odot$ is only 10^{59} erg. The energy requirement for a radio source is therefore comparable with the total potential energy of a galaxy. This accordingly raised the question as to how such a gigantic amount of energy could be produced, if not by nuclear explosion, which of course produces many elements.

Two important papers appeared back to back in 1978. In the first paper, Young[21] showed that the density of stars at the core becomes extremely high and causes a deviation from what is seen in regular galaxies (see Fig. 14.6). The authors inferred that:

> The nucleus of M87 contains a compact mass of low luminosity, with $M = 5 \times 10^9 M_\odot$.

---

[17] Croston, J.H., Birkinshaw, M., Conway, E., and Davies, R.L., MNRAS **339**, 82 (2003).

[18] Hardcastle, M.J., Worral, D.M., Birkinshaw, M., Liang, R.A., and Bridle, A.H., MNRAS **334**, 182 (2002).

[19] Burbidge, G., Astrophys. J. **129**, 849 (1959); ibid. **124**, 416 (1956).

[20] Cygnus A is a powerful radio galaxy which exhibits a jet similar to the one seen in M87. It is at a distance of 7×10^8 light years.

[21] Young, P.J., and 4 coauthors, Astrophys. J. **221**, 721 (1978).

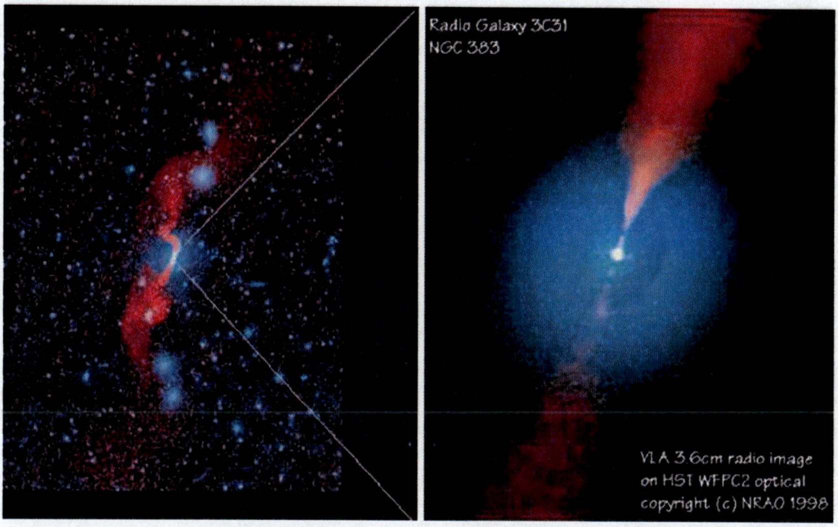

Fig. 14.5 Radio image of the galaxy 3C31 on top of the visual image, created by National Radio Astronomy Observatory/Associated Universities, Inc./National Science Foundation, by permission

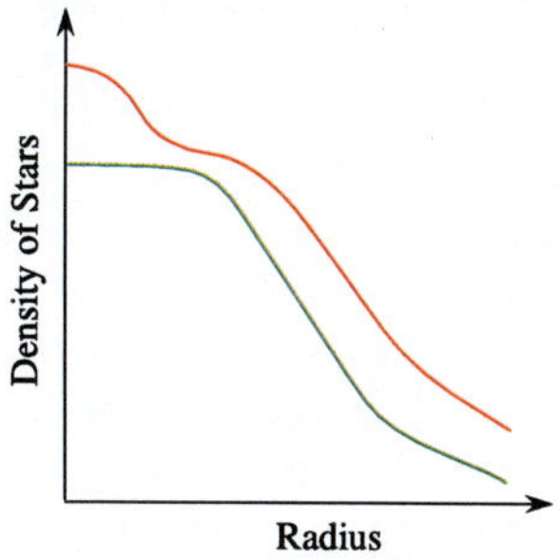

Fig. 14.6 Schematic showing the observed effect of a black hole on the number of stars per unit volume in the core of M87, after Young et al. (1978)

And this with a radius of $r < 100$ parsecs. The data fitted with the standard model of a galaxy plus a black hole. In the second paper, Sargent[22] inferred from the rotational velocities of stars in the core of M87 that the dynamics of the stars was controlled by a mass of $5 \times 10^9 M_\odot$ which had to reside within a radius of 110 parsecs. The only way to fit this kind of a mass into such a relatively small radius was by assuming that the mass was in the form of a giant black hole. At that time it was the best case for *a hypothetical black hole in a galaxy nucleus*.

A year later Ford and Butcher[23] discovered that the gas in the filaments observed near the core was falling into the center of M87. No outflow, which would imply an explosion, was observed. Many papers have since then confirmed the existence of a

[22] Sargent, W.L.W., and 5 coauthors, Astrophys. J. **221**, 731 (1978).

[23] Ford, H.C. and Butcher, H. Astrophys. J. Suppl. **41**, 147 (1979).

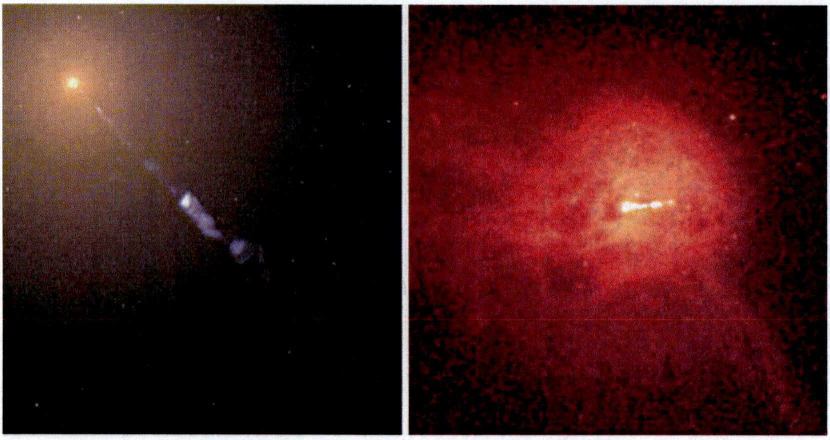

Fig. 14.7 *Left* the radio source M87 with its jet a million light years long. Hubble/NASA. *Right* the innermost part of the galaxy in X rays, as discovered by the Chandra X-ray satellite. Credit: NASA/CXC/W, Forman et al. Chandra X-ray observatory photo album

massive black hole in the cores of active galactic nuclei, like M87. The black hole located in the center of M87 is one of the most massive black holes known in the Universe.

The energy of the black hole comes from mass that falls into the deep gravitational potential created by the black hole. Of course, the radiation must come from a radius which is larger than the Schwarzschild radius. In a way, this extraction of energy from the gravitational potential of a black hole is a modern variant of the old Helmholtz idea of asteroids and comets falling into a potential well. What was apparently overlooked by Hoyle and Fowler is that a black hole can produce more energy per gram of matter than nuclear energy can. In the conversion of hydrogen into helium, the most powerful nuclear source, about 0.007 of the rest mass is converted into energy. However, the definition of the Schwarzschild radius implies that the gravitational energy a mass reaching the Schwarzschild radius has to release is just mc^2. In other words, the entire mass is converted into energy. More accurate calculation shows that the actual amount is about $0.1mc^2$, but still far more than nuclear energy release. Thus, the old idea of gravitational energy release, which was pushed aside when nuclear energy was discovered, regains its dominant role: gravitational energy release by black holes exceeds nuclear energy release (Fig. 14.7).

The connection with element synthesis was made by Hoyle and Fowler,[24] who argued that nuclear energy was the most promising source of energy for the giant radio sources. To produce the energy observed, the fantastic amount of $10^8 M_\odot$ must be quickly converted into energy. It is also obvious that such a phase of intense energy production cannot last for long, and Hoyle and Fowler estimated that the timescale was no longer than 10^6–10^7 years. A long list of distinguished astrophysi-

[24] Hoyle, F. and Fowler, W.A., MNRAS **125**, 169 (1963).

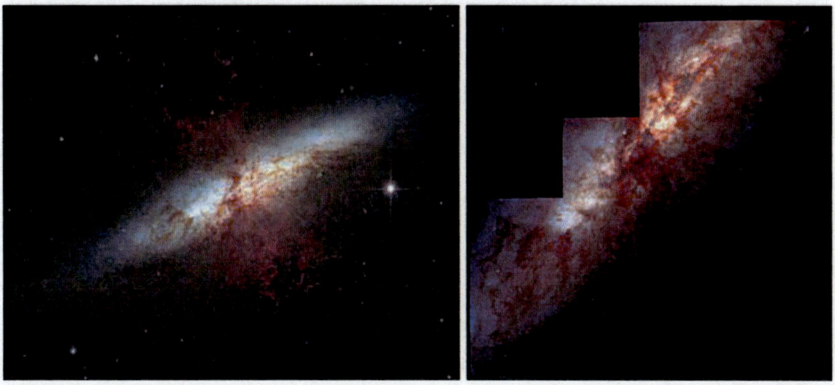

Fig. 14.8 *Left* the M82 galaxy known as a starburst galaxy, in which the burst of stars was triggered by a collision with the M81 galaxy. Credit: Hubble Space Telescope, NASA. *Right* interior parts of the M82 galaxy. Credit: Hubble Space Telescope, NASA

cists, including Ambarzumian,[25] Ginzburg, Pickelner, and Shklovskii,[26] and Pickelner and Shklovskii[27] hypothesized similar scenarios, in view of the data, concluding that a sudden gigantic release of energy must be due to some kind of explosion in the center of the galaxy. Something like 10^8 supernovae exploding within 10^6 years. But then how come so many SNe occur over such a short time? Burbidge suggested that one supernova which blew off in a dense environment of stars could trigger the neighboring stars to become supernovae as well, and in this way start a chain reaction of supernovae.

Hoyle and Fowler suggested that the energy release might be via explosive oxygen burning. If a $10^5 M_\odot$ star explodes in a normal galaxy and a $10^8 M_\odot$ star explodes in an abnormally large galaxy, there will be enough heavy elements for Pop II stars to start forming. The question as to how far such massive explosions would drive all the rest of the gas out of the galaxy and thus choke further star formation was not dealt with.

A beautiful example of a starburst galaxy is M82, where the suspicion is that the starburst was triggered by a violent collision with the M81 galaxy (see Fig. 14.8). A SN, however powerful it may be, is not sufficient to induce the collapse of a star. What happens is the following. Consider a large gas cloud in which a very massive star is born. The star lives for a relatively short time and explodes as a SN. The expanding gases of the explosion shock the remains of the gas cloud which did not have the time to expand and disperse before the massive star has lived out its life and exploded. The shocked gas cloud gives rise to new massive condensation, which evolves into a new generation of massive stars. These stars explode after a short time and, if there is any gas left, shock the gas and repeat the process. In this way we have

[25] Ambarzumian, V.A., *11th Solvay Conference, 241, 1958.*

[26] Ginzburg, V.L., Pickelner, S.B. and Shklovskii, I.S., AJ USSR **32**, 503 (1956).

[27] Shklovsky, I.S. and Pickelner, S.B., Astron. Astrophys. **22**, 913 (1959).

a chain reaction where the number of SN can grow from one generation of stars to the next. The basic physical fact is that the time it takes for a massive star to remove the gas cloud by means of its wind and by mass loss is of the order of, or only slightly longer than, the lifetime of the stars, i.e., just a few million years.

This fact promoted the idea that radio sources originate from the collision between two giant galaxies. The collision shocks the gas in the galaxies, triggering the SN chain reaction. However, evidence accumulated to show that the radio sources were locations of extreme activity, which indicated gigantic explosions rather then a powerful collision between galaxies.

Hoyle and Fowler had difficulty accepting Burbidge's idea, as they estimated that the exploding stars must have been so close as to touch each other, which seemed implausible. Hoyle and Fowler considered the first version of the SN chain reaction in which no gas was considered, only stars. As an alternative idea they hypothesized a single star with a mass somewhere in the range 10^5–$10^8 M_\odot$ at the center of an 'abnormal galaxy'. There were at that time valid theoretical arguments that such a star would be unstable. All the same, as Hoyle and Fowler put it, they turned a blind eye to these arguments and assumed the existence of such objects. It is straightforward to calculate the structure of such a star because, as Eddington showed long ago, it is governed by radiation, so the equations are relatively simple.[28] Was the idea completely farfetched? Not quite. Lallemand, Duchesne, and Walker[29] had already observed the motion of stars near the center of the M31 galaxy in 1960 and found that the rotational velocities could be explained by assuming a central mass of $\sim 10^7 M_\odot$. Similar results were found later for other galaxies. These observations preceded the recognition that massive black holes dominate the gravitational field in the cores of practically all galaxies.

What is the ultimate evolution of such a massive star? Obviously it should become a supernova. Because of the immense dimensions of the star, the density is very low and the collapse takes a long time. The collapse would go, estimated the authors, to the point of oxygen burning in an explosive way, giving rise to a nuclear detonated SN. Independent support for the idea came from Lynds and Sandage,[30] who published in 1963 a paper entitled *Evidence for an Explosion in the Center of the Galaxy M82*, but also from de Vaucouleurs, de Vaucouleurs, and Pence,[31] who provided observational evidence for an explosive event in the peculiar galaxy NGC 1569.

Regarding our point of interest here, Fowler and Hoyle suggested that these explosions might generate the heavy elements we observe today. This hypothesis also explains the observed gradient in composition as a function of the distance from the galactic core. All the heavy elements were synthesized, according to this theory,

[28] In such a star, radiation supplies the pressure to counterbalance the gravity which is produced by the nuclei. The pressure of the gas is completely negligible. In a normal star, the electrons and nuclei supply the pressure of the gas and the nuclei supply the gravity.

[29] Lallemand, A., Duchesne, M. and Walker, M.F., PASP **72**, 76 (1960).

[30] Lynds, C.R. and Sandage, A.R., AJ **68**, 285 (1963); ibid., Astrophys. J. **137**, 1005 (1963).

[31] de Vaucouleurs, G., de Vaucouleurs, A. and Pence, W., Astrophys. J. **194**, 119 (1974).

in the cores of active galaxies and subsequently spread out. The stars with the lowest abundances of heavy elements should be in the outskirts of the galaxies.

There were other proposals to explain the energetic event. For example, supermassive disks.[32] The accretion onto the black hole need not be in the form of a radial flow, but could be circular. The falling gas would create a huge disk around the central accreting object. Viscosity in the disk would cause the matter to flow gradually into the black hole, losing energy which it would radiate out into space. This is again a variant of the Helmholtz idea, except for the fact that the flow is not radial but a gradual circular flow and the gas spirals into the black hole. Under extreme conditions, nuclear reactions can take place in disks.

Burbidge's idea of triggered SNe later became the model for a starburst galaxy. This is a galaxy which, for some reason, experiences an excessive rate of star formation. A trigger for such a phenomenon could be a collision between galaxies. So might the chain SN phenomenon synthesize the heavy elements so quickly and end our story? As a matter of fact, even in 1960, Shklovskii[33] had pointed out that just-formed galaxies must be very bright, i.e., that there should be many bright massive stars. Shklovskii was interested in explaining the phenomenon of radio galaxies and was not in the least bit bothered by the problem of the synthesis of the heavy elements. Radio galaxies are galaxies which are extremely luminous in the radio band, emitting about 10^{45} erg/s or $10^{12}L_\odot$ in the range 10 MHz to 100 GHz. The radio emission is generated by very energetic particles moving in a magnetic field. Only SNe create such energetic particles, and Shklovskii's conclusion was that these objects, the most famous of which was the object known as Cygnus A, had experienced a process of SN bursts. The ongoing hypothesis at the time was that these galaxies had undergone a collision with another galaxy. Shklovskii came up with the alternative explanation, viz., what we see is a young galaxy in which a huge number of SNe explode over a short period of time. In a way, Burbidge's and Shklovskii's hypotheses converged into the same. Shklovskii spoke about the consequence of the process that Burbidge had envisaged.

Does this mean that we should write off M87 as a source of heavy elements or SNe? Not quite. Recently, Simionescu et al.[34] claimed to have discovered, in the core of M87, a cloud rich in heavy elements, and in some locations, even richer than the Sun in heavy elements. The rest of the galaxy has lower abundances of heavy elements. It seems, therefore, that the enrichment is governed by SNe in the core.

---

[32] Kundt, W., Astrophys. Space Sci. **62**, 335 (1979).

[33] Shklovskii, I.S., SvA **4**, 885 (1961). Russian original: Ast. Zhurnal **37**, 945 (1960).

[34] Simionescu, A., Werner, N., Finoguenov, A., Böhringer, H. and Brüggen, M., Astron. Astrophys. **482**, 97 (2008).

14.6 Can a Non-Standard Big Bang Do the Job?

The standard model of the Big Bang assumes that the same physical conditions prevailed everywhere. All locations in space exploded simultaneously and with identical conditions. Recently, deviations from this simplified picture have been considered, supposing that the Universe started its expansion from a non-homogeneous state. The outcome depends on how large the non-homogeneities were, and on what scale. The problem of predicting the observed abundances in such a model is relaxed as compared to the standard Big Bang model, because one no longer has to explain all the abundances with a single set of parameters of the Universe. Appelgate et al.[35] Mathews et al.[36] and Jedamzik et al.[37] considered a situation where the regions with high density are late to expand. Such models tend to produce larger amounts of helium, deuterium, and lithium.

Other authors suggested the existence of matter–antimatter regions (for a review see Steigman 1976[38]). The conclusion is that there is no evidence for the presence of any antimatter in the Universe. From time to time, a hypothesis about non-standard neutrino properties has been raised (see the review by Jedamzik[39]). However, as Jedamzik puts it, the majority of the variants of the Big Bang are implemented to put limits on various effects that might have taken place in the early Universe, so for example the observed abundances of the light elements are used to infer the properties of the Universe. On the other hand, none of the variants suggested so far produces any non-negligible amounts of heavy elements, and nor do they eliminate the need for population III stars. For more ideas concerning non-standard Big Bang models, see Sect. 14.21, where the lithium problem is discussed.

Having said all that, the chain of cosmic puzzles seems endless. Recent observations[40] appear to detect significant amounts of heavy elements in intergalactic space. These are discovered by means of absorption lines due to the intergalactic medium, as imprinted on the light from very distant objects. What these lines tell us is that intergalactic space contains non-negligible amounts of heavy elements, and in some cases the total amount of heavy elements Z may reach $Z = 3 \times 10^{-3} Z_{\odot}$, where $Z_{\odot}$ is the total amount of heavy elements in the Sun. This discovery by itself may not be a surprise. What is amazing is that these heavy elements are discovered at a distance which corresponds to a very young age of the Universe, implying that the heavy element synthesis was very intensive when the Universe was young.

---

[35] Appelgate, J.H., Hogan, C.J., and Scherrer, R.J., Phys. Rev. D **35**, 1151 (1987).

[36] Mathews, G.J., Meyer, B.S., Alcock, C.R., and Fuller, G.M., Astrophys. J. **358**, 36 (1990).

[37] Jedamzik, K., Fuller, G.M. and Mathews, G.J., Astrophys. J. **423**, 50 (1994).

[38] Steigman, G., Ann. Rev. Astron. Astrophys. **14**, 339 (1976).

[39] Jedamzik, K., astro-ph/9805156v1 12 May 1998.

[40] Pichon, C., and 6 coauthors, Astrophys. J. **597**, L97 (2003).

14.7 Why Do Stars Form?

In 1902, Jeans[41] was interested in the stability of spherical nebulae, and in particular in planetary nebulae, which were at that time hypothesized to be the progenitors of planetary systems. So the question was, how do such nebulae hold themselves together and when do they become unstable and shed a ring of mass?[42] But it soon became clear that Jeans' analysis applied to the interstellar medium and provided the basic physics for the collapse of a gas cloud to form a star. The interstellar gas clouds are apparently stable. If they were not stable over a long time, the rate of star formation would have been much higher and we would not have a chance to see gas clouds.

Consider a very large interstellar gas cloud with uniform temperature and density, and in which a very small density enhancement occurs for some reason. Two forces act on this density enhancement: self-gravity which tends to attract all pieces together,[43] and gas pressure which resists gravity, exactly as in a star. The gas pressure depends only on the temperature and density and does not depend on how much mass there is in the density enhancement. On the other hand, self-gravity increases with the total mass of the enhancement, irrespective of its density and temperature. Hence, for every pressure of the gas there exists a radius, or equivalently a mass, beyond which gravity wins over gas pressure and the mass element collapses. The critical mass and radius above which the collapse happens are called the Jeans mass and the Jeans radius/scale, to commemorate their discoverer. From the nature of the situation it follows that every mass bigger than the Jeans mass collapses. Moreover, the bigger the mass, the faster it collapses. The same logic works whenever the gravity resisting force is 'local', e.g., turbulence or a magnetic field. The reason is that gravity is a long range force and acts to infinity.[44]

An estimate of the magnitude of the Jeans mass and scale can be obtained from simple dimensional analysis. After the lengthy and complicated mathematics used to derive his result, Jeans[45] himself posed the question:

> What may determine the linear scale on which a nebula is formed? Three quantities only can be concerned: G the gravitational constant, ρ the mean density, and $R_{\text{gas}}T/\mu$ [which Jeans

[41] Jeans, J.H., *The stability of spherical nebulae*, Phil. Trans. Roy. Soc. Lond. Ser. A **199**, 1 (1902).

[42] There was question of shedding a ring of mass, rather than collapse and formation of a planet.

[43] One assumes an infinite uniform medium. In such a medium all gravitational forces cancel each other. So if there is a perturbation, only the gravity of the perturbation matters.

[44] Consider a sphere of matter with radius R_{Jeans} immersed in a much larger space filled with uniformly distributed matter. The total energy is E = thermal + gravitational. The first term is positive while the second term is negative. The total thermal energy of a sphere of radius R_{Jeans} is $\sim MR_{\text{gas}}T \propto (\rho T)R^3_{\text{Jeans}}$. The gravitational energy is given by $E_{\text{grav}} \sim GM^2/R_{\text{Jeans}} \propto (\rho)R^5_{\text{Jeans}}$. If the radius of the sphere increases, the first term increases as R^3_{Jeans}, while the gravitational energy increases as R^5_{Jeans}. Hence, as R_{Jeans} increases, gravity always wins. The marginal configuration which has vanishing binding energy, $E_{\text{grav}} + E_{\text{therm}} = 0$, has the Jeans radius. If we now equate the thermal energy with the gravitational energy, we get the Jeans radius.

[45] For the dimensional analysis, see p. 51 Sect. 48 of Jeans' 1902 paper.

called the mean elasticity]. Now these quantities can combine in only one way so as to form a length, namely through the expression

$$R_{\mathrm{Jeans}} = \sqrt{\frac{R_{\mathrm{gas}}T}{\mu G\rho}} .$$

Here we have rewritten Jeans' argument using present day notation, so R_{gas} is the gas constant, T the temperature, and ρ the density. μ is the molecular weight. This is exactly the Jeans length, and $4\pi R_{\mathrm{Jeans}}^3\rho/3$ is the Jeans mass. Jeans concluded therefore that the distance between two adjacent star-forming nebulae should be comparable with the above distance. Physically, what we have is that

$$R_{\mathrm{Jeans}} = \text{typical velocity} \times \text{typical time} ,$$

where

$$\text{typical velocity} = \text{speed of sound} \equiv \sqrt{\frac{R_{\mathrm{gas}}T}{\mu}} ,$$

$$\text{typical time} = \text{gravitational time} = \frac{1}{\sqrt{G\rho}} ,$$

where the gravitational time is the collapse time when there is no gas pressure to resist the collapse.

Jeans considered the stability of a cloud with uniform temperature and density extending to infinity. However, it was proven long before Jeans carried out his analysis[46] that such a configuration does not exist. When one checks the stability, the first step is to find the configuration to be checked. But in this case, there is no such configuration and Jeans, who realized this point, wrote explicitly:

> Unfortunately, the stability of a gaseous nebula of finite size is not a subject which lends itself to mathematical treatment. The principal difficulty lies in finding a system which shall satisfy the ordinary assumed gas equations, and shall at the same time give an adequate representation of the primitive nebula of astronomy.

Thus, Jeans chose to examine the stability of a simplified configuration which did not and cannot exist. Jeans was aware of the problem and right through the paper he repeated time and again the 'unacceptable situation'. In view of the difficulty he relied on Darwin.[47] The assumption of an infinite medium was made by Darwin and adopted by the young Jeans.[48] Last but not least, Jeans calculated the stability of

---

[46] Independently by Lane and Ritter.

[47] Darwin, G.H., *On the mechanical condition of a swarm of meteorites, and on theories of cosmogony*, Phil. Trans. A **180**, 1 (1888). Darwin was also the fellow who communicated Jeans' paper to the society. As a matter of fact, Darwin assumed that the isothermal core of the nebula was surrounded by an adiabatic part, so that the problem of uniform temperature was overcome.

[48] Jeans explained on p. 49 of the paper what many others 'rediscovered' time and again. The only solution to the Poisson equation for the gravitational potential in an infinite medium is with vanishing density. However, this was not what Jeans assumed.

a non-uniform medium, but the expression was so complicated that he simplified it by substituting the conditions of uniform matter. Of course, if any bigger mass collapses, one may ask how the bigger mass assembles in the first place. There are two answers:

- When we try to assemble a bigger and bigger mass, it eventually collapses when the mass exceeds the Jeans mass.
- Suppose we assemble a large mass and the temperature is very high, so that the Jeans mass is greater than the mass assembled. If we let the mass cool, the critical Jeans mass decreases and, at the moment it has decreased to the size of the mass assembled, it will collapse.

In 1928, Jeans wrote his famous book *Astronomy and Cosmogony* and returned to the problem of stability. This time he applied the analysis to star formation, but modified most of the mathematics. Jeans refrained from using the gravitational potential, which was ill-defined for the case of an infinite medium. His notes are of interest here[49]:

> The general conception of the stars having been formed out of a homogeneous medium by a process of condensation under gravity is of course very old, being indeed almost as old as the law of gravitation itself. We find Sir Isaac Newton in his first letter to Dr. Bentley (10 Dec. 1692) writing as follows: "If all matter were evenly scattered throughout all the heavens, and every particle had an innate gravity towards all the rest, and the whole space throughout which this matter was scattered, was finite, the matter outside of this space would by its gravity tend towards all the matter on the inside, and by consequence fall down into the middle of the whole space [...] but if the matter were evenly disposed throughout an infinite space, it could never convene into one mass; but some of it would convene into one mass and some into another, so as to make an infinite number of great masses, scattered great distances from one another throughout all that infinite space. And thus might the Sun and fixed stars have formed, supposing the matter were of a lucid nature.

Newton had the idea but could not carry out the analysis, as Jeans had done, for lack of the relevant physics and mathematics.

The problem was of no interest to astrophysicists until the early 1950s when Ledoux and Chandrasekhar reanalyzed the problem. Ledoux[50] showed that the Jeans criterion was valid even when the matter was not uniform. In 1951, Chandrasekhar[51] extended Jeans' stability criterion to include turbulent media and rederived Jeans' analysis and result. However, Chandrasekhar, who had the reputation of being very rigorous in his mathematical analysis, did not find Jeans' analysis faulty in any way. His explanation as to why turbulence must be included had nothing to do with reservations about the rigorousness of Jeans' derivation, as expressed later by many commentators.

Finally, in a rarely cited paper from 1914, Jeans[52] returned to the problem of the stability of spherical nebulae and compared the results obtained for gaseous objects with those obtained assuming liquids. But most important for us, he assumed

---

[49] Jeans, J., *Astronomy and Cosmogony*, Cambridge University Press, London, 1929, p. 352.

[50] Ledoux, P., AnAp **14**, 438 (1951).

[51] Chandrasekhar, S., Proc. Roy. Soc. London A **210**, 26 (1951).

[52] Jeans, J.H., Phil. Trans. Roy. Soc. London A **213**, 457 (1914).

the existence of rotation and showed that in this case one once again retrieves the original stability criterion.

Sometime in the 1960s, the mathematical problem in Jeans' derivation resurfaced and Jeans' result was referred to as the 'Jeans swindle'. Binney and Tremaine[53] wrote in 1987:

> We construct our fictitious homogeneous equilibrium by perpetrating what we shall call the Jeans swindle after Sir James Jeans, who studied this problem.

But it was not Binney and Tremaine who coined the expression. It had already appeared in Stewart 1972,[54] and without reference nor claim to copyright. But Jeans had laid all his cards on the table, and to my mind there is no justification for using a term of abuse like 'swindle'.

Recently, in 1999, Kiessling[55] considered the problem of the Jeans swindle as the limit of a well-behaved mathematical problem and demonstrated that Jeans' result was perfectly valid in the limit of an infinite medium, whence there was no physical justification for the word 'swindle'. Kiessling's argument is basically a repetition of Layzer's argument[56] from 1964. A similar derivation of the Jeans criterion as a limiting process can be found in Chu[57] eight years later. In short, Jeans' physical arguments were absolutely right, and one only had to find the right mathematical treatment to prove it, if so desired. It is appropriate at this point to cite the second of the three pieces of advice Bethe gave to Salpeter,[58] namely:

> (b) Use only the minimum mathematical techniques necessary.

Jeans, unaware of this advice to be given years later, nevertheless followed the same suggestion towards the end of his first paper.

The Jeans[59] length and mass can be simply expressed by

$$R_{\text{Jeans}} = 12.5\sqrt{\frac{T}{n}}\,\text{l years}\,, \quad M_{\text{Jeans}} = 3\sqrt{\frac{T^3}{n}}\text{M}_\odot\,,$$

where T is the temperature of the gas and n the number density in particles per cm^3. The criterion tells us when a mass element becomes unstable. It does not tell us

---

53 Binney, J. and Tremaine, S., *Galactic Dynamics*, Princeton University Press, 1987, p. 287.

54 Stewart, J.M., Astrophys. J. **176**, 323 (1972).

55 Kiessling, M.K-H., arXiv:astro-ph/9910247v1 13 Oct 1999. The idea is to use a screened gravitational potential, where the gravitational potential varies as $V = \exp(-r/l)/r$ and send the screening distance l to infinity.

56 Layzer, D., Ann. Rev. Astron. Astrophys. **2**, 341 (1964).

57 Chu, K-H.W., Eur. J. Phys. **28**, 501 (2007).

58 Salpeter, E.E., Ann. Rev. Ast. Astrophys. **40**, 1 (2002). The two others were: (a) Be prepared to switch fields and (c) in the case of uncertainty, be prepared to use conjecture and shortcuts and take risks, in other words, have CHUTZPAH.

59 Jeans, J.H., *The stability of spherical nebulae*, Phil. Trans. Roy. Soc. London Ser. A **199**, 1 (1902). A summary of the results is given in: *Astronomy and Cosmogony*, Cambridge University Press, 1929.

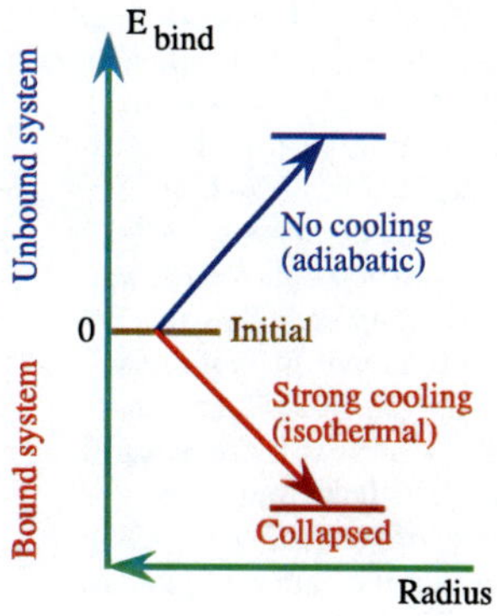

Fig. 14.9 The crucial role of cooling. Strong cooling leads to bound systems. No cooling cannot produce bound systems

what the end product of the collapse will be. As shown before in a footnote, a mass element with mass greater than the Jeans mass has negative energy and hence can in principle be bound, this determining the Jeans critical distance. The mass is given by $M_{\rm Jeans} \sim 4\pi\rho R^3_{\rm Jeans}/3$. The more the radius exceeds the Jeans radius, the more negative the total energy, and vice versa. The basic Jeans analysis can be repeated in various situations. Different mechanisms operate on different scales. It all depends on what force balances gravity, e.g., turbulence or magnetic fields.

How does the collapse proceed? We have seen that mass elements with masses larger than the Jeans mass have a negative binding energy and can in principle be bound. The question now is: what happens during the collapse and to what extent is energy exchanged with the surroundings (see Fig. 14.9)? If cooling is very effective (or in Jeans' language, thermal conductivity is high so that heat flows quickly out of the blob), the temperature stays uniform and isothermal, and the rising gas pressure (due to the rise in density) will not be sufficient to counterbalance gravity and halt the collapse. In this case the system loses the energy which contributed to the gas pressure and becomes unstable. Actually, an isothermal system sinks deeper into the gravitational potential well the more it contracts.

On the other hand, if there is no cooling, say the blob does not lose energy at all, the pressure will increase very fast and soon resist further collapse. In this case the binding energy becomes positive and the system is no longer bound.[60] Hence the detailed physics and thermodynamic properties of the gas are crucial for the fate of the collapse. Without cooling, which Jeans did not allude to, no star can form a bound system. The time evolution of the collapsing protostar is driven by the energy loss via cooling, exactly as stellar evolution is driven by the energy loss via the luminosity of the star. The isothermal collapse continues until the density becomes high enough to make the radiation absorption coefficient significantly high, and the center of

[60] If the collapse is isothermal, the gas pressure P goes as $P \sim \rho$. On the other hand, if the collapse is adiabatic, the pressure rises as $P \sim \rho^\gamma$, where $\gamma = 5/3$ is the adiabatic exponent. The rise in pressure in adiabatic collapse is faster because the energy that the gas did not lose in the cooling adds to the pressure and counterbalances gravity.

Fig. 14.10 The Eagle Nebula, where star formation takes place today, was discovered by Jean-Philippe de Cheseaux in 1745–1746. This nebula, also known as M16, is about 6,500 light years away. It is composed of relatively dense clouds of molecular hydrogen gas, H_2, and dust. The nebula is illuminated by UV light from massive newborn stars. The incredibly beautiful columns of gas have won it the imaginative name the *Pillars of Creation*. Credit: NASA, ESA, STScI, J. Hester and P. Scowen (Arizona State University)

the collapsing blob then becomes opaque. The flow of radiation from the center is hindered, the collapse slows down, and in order to overcome the rising radiation absorption coefficient, a temperature gradient develops. The increase in temperature causes an increase in the pressure of the gas to the point where the gas pressure can balance gravity and stop the collapse. This is the 'official' moment of birth of the star. From now on it will be driven by the surface luminosity and its temperature will rise until it reaches the hydrogen burning phase, i.e., the main sequence (Figs. 14.10, 14.11, 14.12).

In 1930, Eddington suggested that the expansion of the Universe from an equilibrium state might have started with the formation of condensation.[61] McCrea and McVittie[62] found that such condensations have an effect in the opposite sense. Lemaitre[63] thus had a cosmological problem. He had found that:

> The formation of condensations and the degree of concentration of these condensations have no effect whatever on the equilibrium of the Universe. Nevertheless, the expansion of the Universe is due to an effect very closely related to the formation of condensations, which may be named the 'stagnation' of the Universe. When there is no condensation, the energy,

---

[61] Eddington, S.A., MNRAS **90**, 668 (1930). In this paper Eddington wrote that *on general philosophical grounds there can be little doubt* that Einstein's equation with non-vanishing cosmological constant is the correct one and not the earlier form (with no constant). This was written after Hubble's discovery of the expansion of the Universe, which actually eliminated Einstein's reason for introducing the constant. Sixty years later it was discovered that the expansion is in fact accelerating, whence resurrection of the cosmological constant became unavoidable. Could Eddington have had such foresight?

[62] McCrea, W.H. and McVittie, G.C., MNRAS **91**, 128 (1930).

[63] Lemaitre, G., MNRAS **91**, 483 (1931).

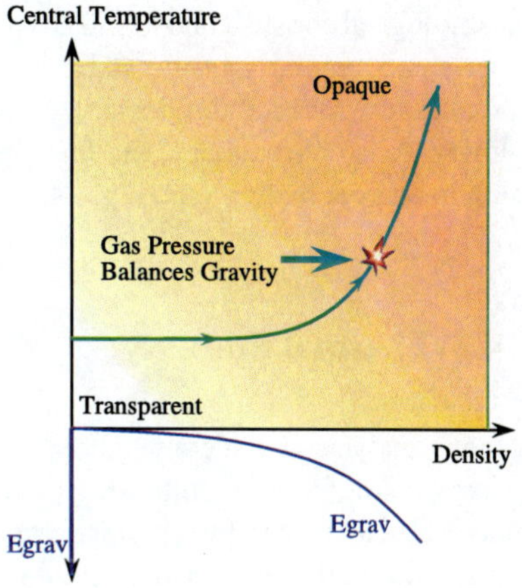

Fig. 14.11 The unstable gas cloud starts out transparent and collapses at constant temperature. Gradually, the gas becomes opaque to the radiation and the central temperature rises. Once most of the cloud has become opaque, we have a new born star which continues to lose energy, sinking more deeply into the gravitational potential well and raising its central temperature until nuclear reactions are turned on

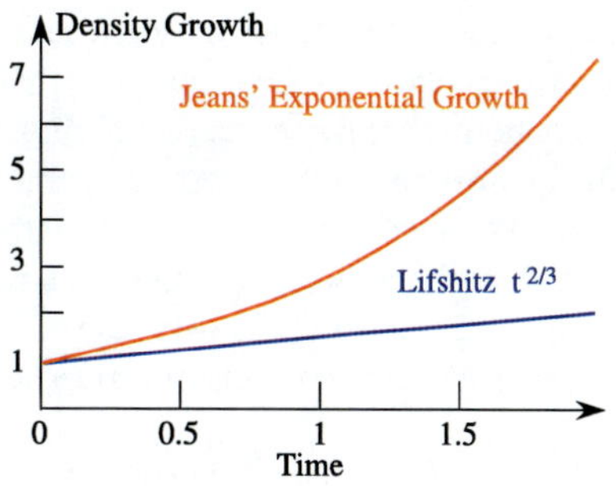

Fig. 14.12 The difference between Jeans' exponential growth and Lifshitz's $t^{2/3}$. When the expansion of the Universe is taken into account, the rate of growth decreases

> or at least a notable part of it, may be able to wander freely through the Universe. When condensations form, this free kinetic energy has a chance to be captured by the condensations and then to remain bound to them. That is what I mean by a 'stagnation' of the world—a diminution of the exchange of energy between distant parts of it.

In other words, the forming stars exchange energy with the rest of the Universe, but they are not responsible for the expansion. Tolman[64] suggested that the radiation emitted by stars as they fall deeper and deeper into their gravitational potential wells could drive the expansion of the Universe.

The radiation emitted by the cooling condensation remains in the Universe and becomes part of the background radiation. Can it be detected? Can we look sufficiently far back in time?

Lemaitre's analysis was totally different from that of Jeans, whose results were not even mentioned, and he ruled out any such effect. In 1946, Lifshitz[65] solved the problem of how a density enhancement evolves in the Universe without the

[64] Tolman, R.C., PNAS **16**, 320 (1930).

[65] Lifshitz, E.M., J. Phys. USSR **10**, 116 (1946). This is the paper that Gamow got while his student Alpher was trying to solve the same problem. See Sect. 7.16.

cosmological constant and found an equivalent instability. However, while the density grows exponentially in the static case solved by Jeans, Lifshitz found that, in an expanding Universe, the density grows only as $t^{2/3}$, where t is the time, i.e., very slowly. As a consequence, the Jeans mass grows to $10^{15}M_\odot$ before thermal effects step in and decrease it (see Fig. 14.26).

14.8 Thermal Stability

Jeans' analysis completely ignored the thermodynamics of the collapse, i.e., the energy losses from the collapsing blob. The original idea that thermodynamics might play a crucial role in the collapse was due to Parker.[66] He reasoned that, if there is a blob which both cools and heats, and which is in thermal balance, the stability depends on how the heating and cooling respond to a change in temperature. Consider a small region in space where for some reason the temperature decreases. If the cooling decreases less than the heating, so that the net result is cooling, the region will cool even more, giving rise to an instability. Parker's reasoning did not include the reaction of the medium to the lower temperature. As the temperature changes, so does the pressure, and the tendency of the matter around is to compress the region and in this way raise the temperature, in such a way as to return to its original state.

In 1955, Zanstra (1894–1972m)[67] included the change in pressure in the analysis, but both Parker and Zanstra arrived at incorrect criteria because their physics was incomplete.

In 1960, Weymann[68] investigated the structure of the outer layers of the Sun, at the base of the corona, and was interested in how heating takes place. In particular, he focused on the thermal stability of the corona. The analysis was general, without reference to any particular heating or cooling process. Weymann's main result was that, if the heating depends on the temperature to a power greater than 1, the heating will be unstable and abrupt temperature rises can be expected. Weymann did take into account the response of the medium.

Field[69] extended Weymann's analysis and generalized it to the interstellar medium. Field systematically analyzed a large number of different cases where the same basic physics applies. In particular for our case here, Field showed that, under a wide range of conditions obtaining in the interstellar medium, thermal equilibrium is unstable and can result in the formation of condensations. Field's results for the condition of thermal stability differed from what was found earlier because he included the finite speed at which the perturbations would propagate in the gas. Field's goal was to discuss those cases in which self-gravity, unlike the Jeans instability, was unimportant.

---

[66] Parker, E.N., Astrophys. J. **117**, 431 (1953).

[67] Zanstra, H., in *Gas Dynamics of Cosmic Clouds*, ed. by Burger and van den Hulst, North-Holland Pub., Chap. XIII, 1955.

[68] Weymann, R., Astrophys. J. **132**, 452 (1960).

[69] Field, G.B., Astrophys. J. **142**, 531 (1965).

He was interested in interstellar clouds, planetary nebulae, the outer layers of stars, etc. Consequently, he ignored the possible role of gravitation.

The conditions found by Field were quite complicated, but the basic result can be understood in the following way. When the gas can cool, a new timescale comes in, namely, the cooling time. The basic length that becomes unstable is then given by

$$R_{\text{thermal}} = \text{speed of sound} \times \text{cooling time} .$$

If the cooling time becomes shorter than the gravitational time, masses smaller than the Jeans mass become unstable. One can summarize as follows:

$$\text{instability scale} = \text{speed of sound} \times \text{Min(cooling time, gravitation time)} .$$

By the way, Field returned to the old problem of Jeans, the stability of planetary nebulae, but of course took into account the cooling. He examined the results of Zanstra, who had found using an incorrect criterion that local fluctuations in the gas give rise to condensations. On the other hand, Daub[70] carried out numerical calculations on the basis of the incorrect Zanstra criterion and concluded that the nebulae were stable.[71]

Field predicted that:

> One might find that the instability sets in as the temperature and densities corresponding to definite radii in the diverging flow within a planetary nebula, and that this could account for the apparent onset of the condensation phenomenon at a certain radius in NGC 7293 and its subsequent growth at greater radii.

Indeed, condensations are observed in the Helix Nebula (see Fig. 14.13 left). However, note that the photograph of the condensations was taken by the Hubble Space Telescope many years after Field's analysis. Some thirty years later, Field's prediction was confirmed.[72]

Hoyle[73] was interested in the formation of galaxies and Pop II stars. The idea of Pop III stars was not yet mooted at the time of the Vatican meeting on stellar populations. Hoyle realized what Fig. 14.9 depicts, namely that the fate of the collapse depends on the degree of cooling. Hoyle did not discuss instabilities, but by simple dimensional analysis, he arrived at the Jeans critical mass, although the identity

---

[70] Daub, C.T., Astrophys. J. **137**, 184 (1963).

[71] Field complained that Daub did not present his basic data.

[72] The Helix Nebula is the largest apparent planetary nebula in the sky. The apparent diameter is about a quarter of a degree, i.e., the apparent size of half the full Moon. The nebula can be seen with a small pair of binoculars, provided the night sky is dark. Sometimes called the Sunflower Nebula, it is a shell of expanding gas blown off by the hot central star. Almost a century after Jeans' analysis of the stability of planetary nebulae, the Hubble Space Telescope confirmed the suspicion that the nebula is unstable. However, the condensations have nothing to do with planets.

[73] Hoyle, F., Proc. Conf. Vatican Observatory, Castel Gandolfo, 1957, North-Holland, and New York: Interscience, 1958, ed. by O'Connell, p. 223.

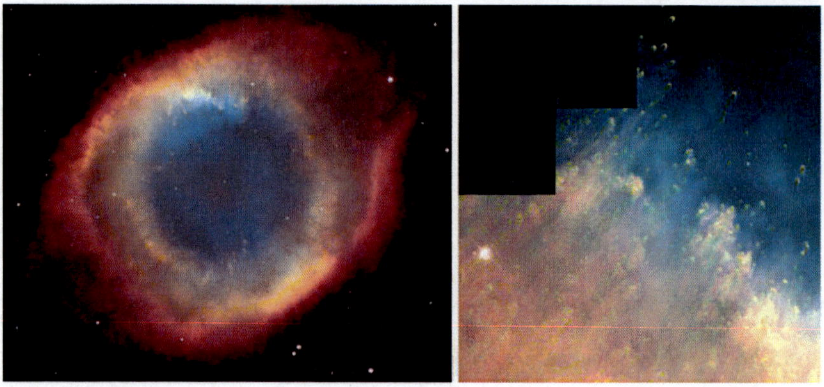

Fig. 14.13 *Left* a classic round planetary nebula: the Helix planetary nebula. Credit: NASA, NOAO, ESA, Hubble Space Telescope. *Right* a magnification of a section of the Helix nebula showing the condensations in the expanding nebula. Credit: NASA, NOAO, ESA, Hubble Space Telescope

between his result and the Jeans mass was not recognized. Gold and Hoyle[74] realized that:

> Attempts to explain both the expansion of the Universe and the condensation of galaxies must be very largely contradictory so long as gravitation is the only force under consideration.

So they hypothesized that the galaxies form by cooling. They did not discuss an instability, and nor did they refer to the earlier work by Parker or Zanstra.

Harwit[75] analyzed the conditions for galaxy formation in the steady state theory and reached the conclusion that gravitational instabilities would be too weak for galaxy formation.

So far the Jeans mass had been found to be the minimal mass able to collapse. Shaviv and Kovetz[76] discussed the maximum mass of an object that can be formed in the Universe. The mass is limited by general relativistic effects. In classical Newtonian gravitational theory, an isothermal configuration extends to infinity. In general relativity, an isothermal mass can have a finite size and if larger than this size, it will collapse. Shaviv and Kovetz showed that the maximal mass is actually given by Eddington's equation

$$M_{\rm max} = 18.0\frac{\sqrt{1-\beta}}{\mu^2\beta^2}{\rm M}_\odot\,,$$

where β is the ratio of gas pressure to total pressure and μ is the molecular weight. What this means is that a higher mass cannot reach hydrostatic equilibrium. The amazing fact is that the Jeans mass depends on β and μ in exactly the same way,

---

74 Gold, T. and Hoyle, F., IAUS **9**, 583 (1959).

75 Harwit, M., MNRAS **122**, 47 (1961); ibid. **123**, 257 (1961).

76 Shaviv, G. and Kovetz, A., Astrophys. Lett. **6**, 129 (1970).

and consequently M_{max} is of the same order as the Jeans mass. Thus, under such conditions, the collapsing mass would result only in objects having a mass of the order of the Jeans mass.

14.9 Primordial Cooling and the Hydrogen Molecule

As the ability to cool is the dominant factor in the formation of gas condensations which become stars, we consider the evolution of the conditions in the Universe. The story is contingent upon the chemistry of the formation of the hydrogen molecule H_2 at low temperatures.

It all started with the question raised by McCrea[77]: how could the observed hydrogen molecules form in interstellar clouds. Simple considerations by McCrea indicated that, at the low temperatures and densities found in such clouds, the formation of molecules would be unlikely. Consequently, McCrea and McNally[78] suggested that the building of molecules could take place on dust grains. Now these grains are extremely big relative to the size of a molecule, so the collision between a grain and a hydrogen atom is much more frequent than the collision between two atoms. Therefore the idea of McCrea and McNally was that the dust grains might act as vacuum cleaners by moving through space and collecting hydrogen atoms which would stick to them. Once on the grain, the atoms would attract each other and perform what is called a surface reaction, thereby creating the hydrogen molecules.[79]

McDowell[80] disagreed with McCrea and McNally and wanted to prove that hydrogen molecules could form even without grains, provided that there were free electrons in the gas. The reactions McDowell envisaged were:

$$\mathrm{H} + \mathrm{e} \rightleftharpoons \mathrm{H}^- + \gamma\,, \quad \text{followed by } \mathrm{H}^- + \mathrm{H} \rightarrow \mathrm{H}_2 + \mathrm{e}\,.$$

The H^- ion is the same familiar ion that controls the radiative absorption coefficient on the surface of the Sun. However, McDowell had to suggest where the electrons might have come from. This was relatively easy since the clouds he discussed contained heavy elements. Hence McDowell assumed (!) that the background radiation was sufficient to ionize the metals and produce an electron density of $n_e \approx 5 \times 10^{-4} n_H$. Once this electron density had been assumed, the road was open.

In his first treatment, McDowell assumed that the radiation available in space would be sufficient to knock an electron out from the heavy metals in which the electron has a small binding energy. The reaction was also considered by Pagel[81] in very

[77] McCrea, W.H., Proc. Roy. Soc. A **256**, 245 (1960).

[78] McCrea, W.H. and McNally, D., MNRAS **121**, 238 (1960).

[79] The dust grains are made of SiO (sand) and C (graphite).

[80] McDowell, M.R.C., The Obs. **81**, 240 (1961).

[81] Pagel, B., MNRAS **119**, 668 (1960).

cool stellar atmospheres. Collecting the physical data for the reactions, McDowell was able to show that, if the densities of atomic hydrogen exceed 10^4 atoms/cm^3, the hydrogen will convert rapidly into H_2.

Six years later, the idea of primordial stars was in the air, proposed by Tayler and Hoyle, for example, and the question was: how could stars form from a pure hydrogen/helium gas? Saslaw and Zipoy[82] were the first to realize that, when a primordial gas devoid of metals collapses, by the time the density reaches about 10^4 particles/cm^3, hydrogen molecules will form and start to dominate the cooling. A density of 10^4 particles/cm^3 is about a factor of 10 to 100 higher than the density in interstellar gas. The best vacuum that can be produced on Earth contains about 5,000 particles/cm^3. It is amazing to think that the Universe started from extremely high temperature and density, expanded to extremely low temperature and density, and then collapsed back to form bound systems which separated from the overall expansion of the Universe, contracted, and reached extremely high temperatures and densities before exploding as supernovae.

The reactions Saslaw and Zipoy considered were

$$H^+ + H \rightarrow H_2^+ + \gamma\,, \quad H_2^+ + H \rightarrow H_2 + H^+\,, \quad H_2 + H \rightarrow 3H\,,$$

whence the free electrons did not participate in the process, which was basically an exchange of charge between atomic and molecular hydrogen. Only one photon is emitted. Saslaw and Zipoy did not calculate the amount of H^+, but took it as a free parameter. They found that the rate of cooling and the radiation emitted were quite independent of the initial assumed concentration of H^+.

Saslaw and Zipoy were the first to discuss the possible detection of the photons released during the decay of the excited hydrogen molecule. This should be the signature of the way primordial cooling took place. The signature should add to the background radiation in the infrared domain.

Once the temperature of the gas has reached 10–100 K and hydrogen molecules have formed, the collapse of the Jeans unstable mass can proceed almost at constant temperature, leading to bound luminous objects and the end of the Dark Ages. Peebles and Dicke[83] calculated that, at that moment in time, the minimal Jeans mass would have been 10^5–$10^6 M_\odot$, which is the mass of a typical globular cluster. The collapsing blob then starts to dissociate itself from the rest of the matter and evolves towards a bound system. At some point during the collapse, fragmentation takes place. The fragments may still be too massive and need a process to remove angular momentum, for example, by forming binary stars and converting the spin rotation into the rotation of two large bodies, one around the other.

Peebles and Dicke followed the chemistry of McDowell. For the reactions to go ahead, one requires free electrons. In the same year, Peebles calculated that the recombination processes were not complete. The situation with recombination is similar to that with nuclear reactions in an expanding Universe. The expansion decreases the

---

[82] Saslaw, W.C. and Zipoy, D., Nature **216**, 976 (1967).

[83] Peebles, P.J.E. and Dicke, R.H., Astrophys. J. **154**, 891 (1968).

rate of electron–proton recombination, and if the expansion is very fast, very little recombination will take place, and vice versa. So Peebles[84] calculated the fraction of the electrons that would remain free given the rate of expansion of the Universe. The fraction he found was 3×10^{-5}. Note how close this number is to the one assumed by McDowell. This result allows us to hypothesize about what the Universe would have looked like if the probability for the protons to capture electrons had been, say, a factor of 1,000 higher, or the rate of expansion of the Universe slower. The result would have been complete recombination with no free electrons left over. Would that mean no formation of H_2 and no possible cooling of clouds and star formation?

Peebles and Dicke argued that the first bound systems to have formed in the expanding Universe would have been gas clouds with mass and shape quite similar to the globular star clusters. They pointed to the fact that, although the globular clusters show some variation, they are to a large extent quite uniform in their properties. Indeed, in an expanding Universe with the cosmological parameters that lead to a 2.7 K background radiation, the first objects to become Jeans unstable would have been gas masses with the size of globular clusters (a few million solar masses).

This scenario for galaxy formation deviates from the conventional one, which claims that the globular clusters formed along with the halo field population stars during the initial collapse of the protogalaxy, and it generates new problems like why the heavy element content of globular cluster stars is correlated with the cluster position in the Galaxy, or why the stars in GCs are always polluted with at least a trace of heavy elements, and so on. In short, there is much work to do before the process is fully understood.

14.10 Must the First Star Have Been Massive?

Palla, Salpeter, and Stahler[85] posed the vexing question:

> Must the first stars, forming out of pure hydrogen and helium gas, have been very massive?

This was what several authors had concluded[86] following Peebles and Dicke. The gas coming out of the Big Bang expands from high densities. The densities are sufficiently high to give rise to three-body reactions[87] of the type $H + H + H \to H_2 + H$ and $H + H + H_2 \to 2H_2$. Such reactions are important all the time because of the high

[84] Peebles, P.J.E., Astrophys. J. **153**, 1 (1968).

[85] Palla, F., Salpeter, E.E. and Stahler, S.W., Astrophys. J. **271**, 632 (1983).

[86] Yoneyama, T., PAS Japan **24**, 87 (1971); Hutchins, J.B., Astrophys. J. **205**, 103 (1976); Silk, J., Astrophys. J. **211**, 638 (1977); Carlberg, R.G., MNRAS **197**, 1021 (1981).

[87] The reader may be surprised to hear of these three-body reactions. We start with two free particles and we want to get a bound system. Hence energy must be removed from the system. So when two particles become bound and they cannot radiate away the extra energy, one possible solution is to deliver the energy removed from the pair to the third particle. In a similar way, a free star cannot capture another free star in a stellar collision unless a third star participates in the collision and takes away the extra energy in the form of kinetic energy.

density. However, as long as the temperature is high, the newly formed bound system will disintegrate. This situation prevails until the temperature decreases sufficiently to allow for the survival of the molecules. In this way Palla et al. argued that significant cooling starts when the density is relatively high and stars with masses as low as $0.1M_\odot$ can form. This result has a profound impact. If all primordial stars are massive, the only hope of discovering such stars is to look back in time to very large distances. But if Palla, Salpeter, and Stahler are right, then the lifetime of $0.1M_\odot$ stars is longer than the age of the Universe and we should be able to discover such stars, and slightly more massive ones, in our own galaxy.

What is the physical basis for this claim? Atomic hydrogen is a poor radiator at low temperatures, so the collapse of a pure atomic cloud of hydrogen would be adiabatic. But if the collapse is adiabatic, we know that no bound system will form. As a matter of fact, another expression for this physical result is that the Jeans mass increases, so what was unstable becomes stable. Palla et al. showed that, at densities above $n \geq 10^8$ cm^{-3}, the three-body reaction is very fast and dominates the formation of H_2.

As a consequence, Stahler, Palla, and Salpeter[88] even calculated the evolution of a $5M_\odot$ star devoid of metals. They found that the lack of metals in the atmosphere renders such stars bluer. When SN 1987A exploded a year later and the progenitor was found to be a blue star and not a red one, this fact was rediscovered several times. As a matter of fact, in 1960, Wallerstein and Carlson[89] and in 1962, Wallerstein[90] showed that a rather close correlation exists between the metal abundance and the extra blue color of main sequence stars. In short, the lower the abundance of heavy elements, the bluer the star. This phenomenon was later applied in the search for stars with low metal abundances.

In 1997, Tegmark et al.[91] took up the problem of the smallest mass that can be formed in clouds devoid of metal and applied detailed chemistry to it. After a long discussion, the bottom line was that the outcome depends on an unknown initial mass function, with the possibility that even as much as half of the visible mass condensed into faint low mass stars, known today as MACHOs (acronym for MAssive Compact Halo Objects).

14.11 How Stars Form Today

It may be instructive to examine the way stars are formed today in our Milky Way. The contrast with the conditions when population III stars were supposed to have been born will then become clearer.

Present day stars are formed in giant molecular clouds (hereafter, GMC), as exemplified in Fig. 14.10. These are huge complexes of interstellar gas and dust.

---

[88] Stahler, S.W., Palla, F. and Salpeter, E.E., Astrophys. J. **308**, 697 (1986).

[89] Wallerstein, G. and Carlson, M., Astrophys. J. **132**, 276 (1960).

[90] Wallerstein, G., Astrophys. J. Suppl. **6**, 407 (1962).

[91] Tegmark, M., and 5 coauthors, Astrophys. J. **474**, 1 (1997).

As the temperature is as low as 10–20 K, the hydrogen in the gas is mainly in the molecular form H_2. Traces of other molecules exist as well. The GMCs have relatively high densities (in the range 10^6–10^{10} particles/cm^3). A typical cloud contains about $10^5 M_\odot$, and this is bound, so the entire cloud has negative total energy. GMCs are the only places where star formation is observed to take place. The other types of interstellar clouds, in which hydrogen is in atomic rather than molecular form, are too warm and diffuse for stars to form. In the absence of molecules, dust serves as one of the main coolants. Consequently, those places where an instability has taken place and a blob has formed are as a rule embedded in heavily obscured regions.

The fate of a GMC depends on its environment since this may trigger the initial perturbation which becomes unstable and grows into a star. Once several massive blobs have collapsed and become bright stars, these stars disperse what remains of the cloud by extensive radiation pressure and winds, whereupon the young stars become visible in the visible part of the spectrum.

14.12 The Virial Theorem in Clusters of Galaxies

In 1933, a few years after Hubble's discovery of the recession velocities of galaxies, Zwicky,[92] the perpetual iconoclast, was interested in the physical meaning of the redshift–distance relation. In particular he wanted to see if gravitational redshift could explain Hubble's discovery. Recall that, in 1929,[93] Zwicky had suggested the tired light theory as an explanation for the redshift–distance relation.[94]

Zwicky was the first astronomer to apply the virial theorem to observations (after the mathematician Poincaré) (see Fig. 14.14). Zwicky investigated the Coma cluster of galaxies, which contains about 1,000 galaxies. For the present discussion, one can consider the galaxies in the cluster as point masses moving under their mutual gravitational potential and assume the system to be in a steady state. Then it becomes a perfect case for applying the virial theorem. Zwicky found from the Doppler shift of just 8 individual galaxies[95] that the typical velocities of the galaxies in the cluster are 1500–2000 km/s. On the other hand, the total mass of galaxies estimated on the basis

---

[92] Zwicky, F., Helv. Phys. Acta **6**, 110 (1933).

[93] Zwicky, F., PNAS **15**, 773 (1929). See also, Phys. Rev. **34**, 1623 (1929).

[94] Zwicky estimated the effect of the matter in the Universe on the energy of the photons, assuming a mean density of 10^{-26} g/cm^3. He found that, with this assumption, the tired light effect would yield a change of $\Delta\lambda/\lambda > 3 \times 10^{-7}$ for a distance of 10^6 parsec $= 3.2 \times 10^6$ light years. On the other hand, in those days, Hubble's law gave $\Delta\lambda/\lambda \sim 1/600$. With these two numbers in hand, Zwicky concluded that:

> In view of this agreement in order of magnitude, a further elaboration of the theory seems to be worthwhile.

About a decade and a half later, Hubble found that the 'Hubble constant' had to be revised downward, making the 'agreement' between the two numbers even better.

[95] The mean redshift is $\Delta\lambda/\lambda = 0.023$.

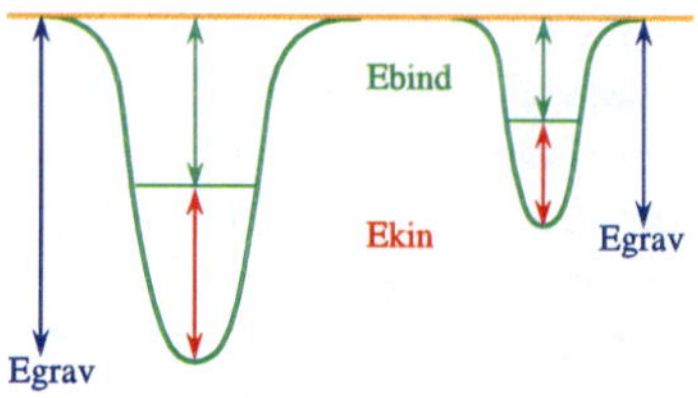

Fig. 14.14 The greater the kinetic energy, the deeper the gravitational potential well must be to keep the system bound (negative total energy). The virial theorem states that $E_{\text{bind}} = -E_{\text{kin}}$

of the individual values[96] of the mass-to-light-ratio M/L obtained from the particular class of each galaxy was $8 \times 10^{11} M_{\odot}$. If such a collection of masses generates the gravitational potential well inside which all the observed galaxies move, the mean kinetic energy of a galaxy in the cluster should be, as estimated correctly by Zwicky, only 80 km/s. This analysis therefore revealed a discrepancy (see Fig. 14.14).

How come the galaxies move so fast when there is no observed matter to hold them together? Moreover, the observed mass of the cluster would yield a gravitational redshift of $\Delta\lambda/\lambda \sim E_{\text{grav}}/c^2 \sim 3.5 \times 10^{-8}$, which is much too low compared with the observed value of $z = 2.3 \times 10^{-2}$ attributed to the expansion of the Universe. Consequently, Zwicky hypothesized that his mass estimate was off and there was what he called *dunkler materie*, or *dark material*.[97] Parenthetically, the mass needed to solve the redshift–distance relation according to Zwicky's tired light theory was much too high, according to what was found years later.

From that time on, the 'missing' mass was called dark matter. Zwicky did not make a big fuss over the idea of dark matter, because he assumed that the dark matter was simply cold matter, shining at a luminosity well below contemporary detection levels.

In 1936, Smith analyzed the Virgo cluster of galaxies.[98] Smith's aim was to determine the nature of the material between galaxies and to understand why space is so transparent as to let us see distant objects. This is not a trivial question. Has all matter condensed into galaxies leaving nothing in-between the galaxies, or is intergalactic matter in some special form that makes its detection difficult? Smith made a critical point: we can measure velocities, but we cannot measure accelerations directly. Accelerations enter Newton's second law of motion, while velocities enter the energy equation. Smith used the data of 25 galaxies measured by Humason and 5 galaxies measured by Slipher when he first noted that the speed of the galaxies did not change with location in the cluster. For example, the galaxies near

[96] Different types of galaxies have typical M/L ratios depending on the type of the majority of stars. Since the luminosity varies as the mass cubed, the M/L value of a star varies as the inverse mass squared, i.e., low mass stars produce less energy per gram and have a high M/L, while massive stars produce a lot of energy per gram and have a low M/L. Elliptical galaxies are dominated by low mass stars and hence show a high value of M/L, while spirals, in which the formation of massive stars continues today, show a low value of M/L.

[97] The paper was published in German, which appears to be a problem for many astronomers in the USA. This may explain the late 'rediscovery' of this paper.

[98] Smith, S., Astrophys. J. **83**, 23 (1936).

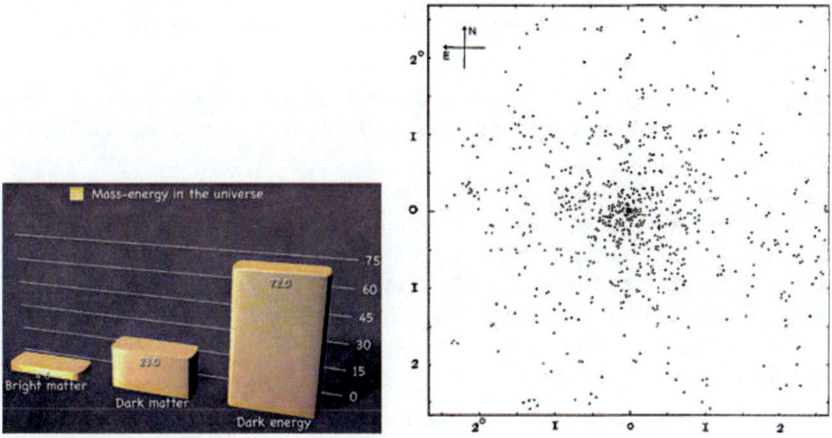

Fig. 14.15 *Left* relative amounts of bright and dark matter in the Universe. The bright matter is only 4.6% of the total. *Right* distribution of galaxies in the Coma cluster of galaxies, as given by Zwicky (1937)

the center did not have higher velocities. Smith interpreted this result as proving that the cluster was in steady state. If the Virgo cluster was collapsing, argued Smith, we should have seen the high velocity galaxies near the center, and conversely if the cluster was evaporating. The analysis of the missing mass depends crucially on the assumption that the cluster is in a steady state. Once he had 'proved' that the cluster was in a steady state, Smith found that the mass needed to generate the gravitational well to keep the galaxies bound was $10^{14}M_{\odot}$, while assuming the cluster to have 500 galaxies yielded a mass of only $2 \times 10^{11}M_{\odot}$. It should be noted that at that time Hubble estimated the mass of the typical galaxy to be a factor of about 200 smaller. Hence, had Smith taken Hubble's estimate, the discrepancy would have been even greater.

Zwicky and Smith showed that, if clusters of galaxies are bound systems, the total mass must considerably exceed the sum of the masses of the individual member galaxies. The excess was a factor of 500 for the Virgo cluster and a factor of 400 for Coma (see Fig. 14.15 right).

Zwicky returned to the problem of the masses of nebulae in 1937.[99] He examined various methods to determine the masses of galaxies and found most of the observational based methods to be inaccurate, while the one he most relied on, the virial theorem for the cluster, was deemed to be satisfactory. On top of this, Zwicky criticized all astronomers who used other methods with contempt. This was naturally the method which yielded the mass discrepancy. Such criticism later became Zwicky's hallmark.[100] Here and there he mentioned that:

---

[99] Zwicky, F., Astrophys. J. **86**, 217 (1937).

[100] Zwicky used to alienate his colleagues on any possible occasion, especially if he felt that his priority for a discovery had been ignored. He used to call his fellow astronomers 'spherical bastards'.

> The luminosity of the nebula is diminished by the absorption of radiation due to the presence of dark matter.

Later, in 1942, Zwicky claimed[101] that all known distributions of galaxies were wrong and the Universe contained half of its mass in the form of faint, as yet unobserved, galaxies.

In December 1951, Oort delivered the distinguished Henry Norris Russell lecture[102] on *Problems of Galactic Structure*, and discussed the variation of the M/L ratio[103] in galaxies and in the Andromeda galaxy in particular. Oort noticed that the M/L ratio varies from 13 at 600 parsecs from the center to about 70 at 5,500 parsecs. These results for M/L were much less alarming then the values found by Zwicky and Smith in the only two clusters of galaxies that had been examined at that time. The explanation Oort gave was that these high M/L values:

> [...] indicate either that the amount of interstellar gas must be some hundred times higher than in our surroundings or else that practically all the mass is due to stars of extremely low intrinsic brightness.

Similar results were obtained for other galaxies. In the galaxy NGC 3115, Oort[104] found that M/L reached the value of 250. It was improbable that such a high ratio would be the result of large amounts of dark gas, and by elimination the only possibility was a large number of very faint stars with M/L of order 1,000. But no low mass stars exist with such a value of M/L. The lowest mass star which ignites hydrogen has $M/L \sim 100$. Hence, the dark matter should be in the form of large planets like Jupiter rather than very low mass main sequence stars.

Oort inferred that (a) while the bright stars concentrate near the center, the faint stars are spread uniformly in the galaxy, and (b) the bright stars are embedded in the gravitational well produced by the faint stars. This important conclusion was not fully appreciated for many years. The implication is that the structure of the Galaxy depends on the dark matter, and the bright matter just follows. The two kinds of stars have different distributions. No hypothesis was put forward as to what distinguishes between the two kinds of stars. The extreme values of M/L were not sufficiently extreme to alarm astronomers that this faint matter was not and could not be in the form of low mass main sequence stars.

For several decades it was considered that Zwicky's and Smith's results for clusters were not worth checking or following up. The question of large scale structure in the Universe was not yet on the astronomers' desk, despite Zwicky's attempts to set off the alarm bells. As a matter of fact, Oort made no connection with the problem of dark

---

When asked to explain why, he retorted: *They look like bastards from every side*. Zwicky's law was: *The more irrelevant garbage you put into a sentence, the better it sounds*. He implemented his law by writing concise papers.

[101] Zwicky, F., Phys. Rev. **61**, 489 (1942).

[102] Oort, J.H., Astrophys. J. **116**, 233 (1952).

[103] The mass-to-light ratio is usually written as M/L, but it is in fact $(M/M_\odot)/(L/L_\odot)$, i.e., expressed in solar units.

[104] Oort, J.H., Astrophys. J. **91**, 302 (1940).

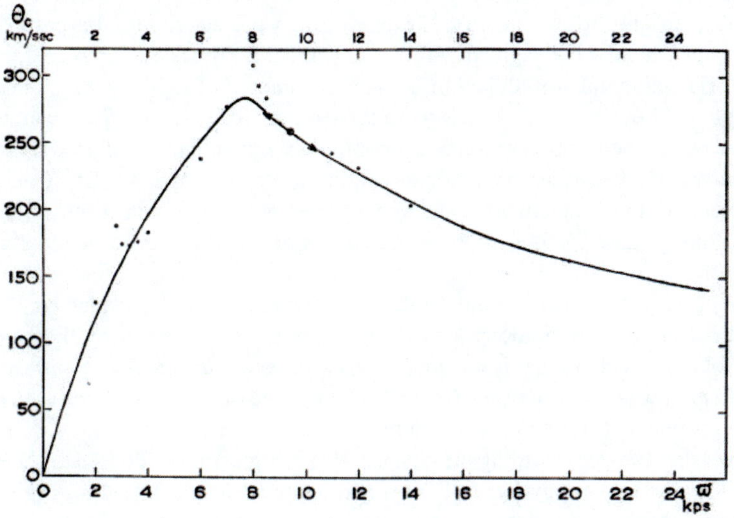

Fig. 14.16 Rotation curve of the Milky Way, as presented by Oort in his Russell lecture 1951

matter in clusters of galaxies, and Zwicky's and Smith's results were not mentioned. The only figure of galaxy rotation shown was that of the Milky Way, where we can see that the velocity of rotation decreases nicely as one moves outwards (see Fig. 14.16). The discovery of dark matter came too early to be properly appreciated.

In 1956, Schmidt[105] published a model for the distribution of mass in the galactic system. Schmidt used the observed rotation of the Galaxy to derive the mass by assuming that the centrifugal force balances the gravitational pull of the mass inside a given radius. There was no hint about problems with the mass that provides the gravitational pull. The rotation was determined up to a distance of 8.2 kpc from the center by means of 12 cm wavelength radio waves emitted by neutral hydrogen. There was not a word about the M/L ratio of our galaxy. A year later, Schmidt[106] repeated the procedure when he applied the model of our galaxy to M31 and found that the M/L value was practically constant at 27 for the inner part, while the mean value for the entire visible galaxy was 23. This value appeared very high for the type of stars seen in M31, i.e., the type of stars observed in this galaxy actually have much lower M/L ratios. Some inconsistency thus began to surface.

Holmberg (1908–2000)[107] was a well known galaxy researcher, and like Zwicky doggedly refused to accept the Hubble discovery as implying an expansion of the

---

[105] Schmidt, M., BAN **13**, 13 (1956).

[106] Schmidt, M., BAN **14**, 17 (1957).

[107] The Holmberg IX galaxy is named after its discoverer. This is an irregular dwarf galaxy and is a satellite of the giant galaxy M81. Holmberg found a correlation between galaxy magnitudes and colors (known as the Holmberg relation) and showed that elliptical galaxies were older than spiral galaxies. Holmberg, E., Arkiv für Astronomi **2**, 559 (1961).

Universe, devising an alternative to Einstein's theory of relativity for this purpose.[108] On the other hand, he disagreed with Zwicky and did not accept the need for dark matter. He claimed that it all had to do with the amount of light reaching the spectrograph and how the observations were carried out, and that the effect, which was interpreted as due to observational errors of a systematic nature, offered *a natural explanation* for the excessive velocities found in clusters and groups of nebulae. However, it did not explain what had already been observed by this time in individual rotating galaxies, namely, that dark matter was already needed on the scale of galaxies.

Holmberg's 'new explanation' for the recession of the nebulae, which could have turned cosmology upside-down had it been correct, was rejected by the community. Holmberg's claim that it was *an illusion inherent in the method of observation allowed to us* was attacked by McCrea,[109] who pointed out that Holmberg used many seemingly plausible reinterpretive arguments which *eliminate any element of deduction from the theory*. Regarding the physics, Holmberg's proposal was *that the space appropriate to cosmology has three, not four, dimensions*. He based his arguments on the idea that:

> Our experience of events outside [...] our own Galaxy is [...] effectively confined to one observation at one instant of time [...] these events can be considered as lying on a light cone.

All this explanation and theory induced on the basis of personal feelings were critically analyzed and rejected by McCrea. Holmberg replied[110] to McCrea, and it seems that McCrea gave up trying to convince him. In 1958, Holmberg defined what is known today as the Holmberg radius, namely the radius of an external galaxy at which the surface brightness has a given a priori value. The idea of such a radius was developed by Holmberg in order to estimate the actual dimensions of the major and minor axes of a galaxy, without regard to its orientation in space, and eventually eliminate the need for dark matter.

In 1961, Holmberg[111] was more explicit. He carried out an analysis of the Virgo cluster and concluded that:

> The observational material available for the Virgo cluster of galaxies leads to the conclusion that the high redshift dispersion can be satisfactorily explained as a result of orbital motions, systematic errors in the measures of redshift, and disturbances by optical members. [...] Accordingly there is no need to introduce any additional [...] strong gravitational field due to nonluminous intergalactic matter.

In other words, there was no dark matter and no expansion of the Universe.

Attempts to find the 'missing mass' in individual galaxies were carried out by Burbidge, Burbidge, and Prendergast,[112] who used Schmidt's method. But Burbidge,

---

[108] Holmberg, E., MNRAS **116**, 691 (1956).

[109] McCrea, W.H., Obs. **77**, 208 (1957).

[110] Holmberg, E.R.R., Obs. **77**, 210 (1957).

[111] Holmberg, E., AJ **66**, 620 (1961).

[112] Burbidge, E.M., Burbidge, G.R. and Prendergast, K.H., Astrophys. J. **130**, 739 (1959); ibid. **142**, 649 (1965).

Burbidge, and Prendergast, who got a slightly higher M/L than observed previously, clearly expressed the suspicion that the assumption of an exact balance between gravity and centrifugal forces might not apply.

In 1960, van den Bergh[113] repeated the observation of the Coma cluster of galaxies and got $M/L = 900$, but he did not really accept the result and noted that Burbidge and Burbidge[114] and de Vacouleurs[115] had expressed doubts about the applicability of the virial theorem to clusters of galaxies.[116] However, de Vaucouleurs examined a group of galaxies, not a cluster, in which the number of galaxies was small (21 in his case), unlike the cluster examined by Zwicky, which contained about 1,000 galaxies (see Fig. 14.18). In the Stephan Quintet examined by the Burbidges, there are only 6 galaxies, yet the virial theorem was implemented (see Fig. 14.17).

The Burbidges calculated the gravitational and kinetic energies to see whether the virial theorem was satisfied. So with careful mass-to-light estimates, they found that $2T + W > 0$, where T was the total kinetic energy of the galaxies and W their total potential energy. They concluded like de Vaucouleurs that the group was expanding. They then inverted the problem, arguing that only if the masses of the individual galaxies were enhanced by a factor of 'many hundreds' could the virial condition be met. Unfortunately, this particular case was the wrong one to use to make inferences about dark matter because, as was realized later, it was not a steady-state system, but a system in collision.

The basic problem is that the virial theorem does not hold if the system is not bound. Even when the system is bound, the correct expression for the virial theorem is that the long time average of the kinetic and gravitational energies will satisfy the relation $2\overline{T} + \overline{W} = 0$, where the bar indicates a time average. But we can only take snapshots, because we are unable to observe the cluster for a few billion years in order to obtain a proper time average for it.

On the other hand, if the clusters are not bound, then examining a large number of clusters will reveal that the instantaneous value of $2T + W$ is sometimes positive and sometimes negative, and is not always the same.

Consider the problem from another angle. Is the group or cluster of galaxies in a bound state or a transient state? Is the accumulation of galaxies in different locations in space a fluctuation or does it represent a real bound system? Compare the Coma cluster (Fig. 14.15 right) and the small group (Fig. 14.18) studied by de Vaucouleurs. It is highly improbable that the Coma cluster is a random fluctuation, while it is not

---

[113] van den Bergh, S., Astrophys. J. **131**, 558 (1960).

[114] Burbidge, G.R. and Burbidge, E.M., Astrophys. J. **130**, 15 (1959).

[115] de Vaucouleurs, G., Obs. **79**, 113 (1959); ibid., Astrophys. J. **130**, 718 (1959).

[116] When the application of a physical law yields what appear to be hallucinating results, the physicist/astronomer has two options: throw away the physical law which for some reason does not apply, or continue to believe in the law but invent a reason for its failure to yield 'reasonable results'. In the early 1950s, before the new theory of weak interactions was invented, there were serious problems in believing that so much unseen matter could be hidden in galaxies. A decade later, when a zoo of new particles had been invented/discovered, the views changed, and people were more inclined to accept the idea of a weakly interacting particle as the main (or sole?) component of dark matter.

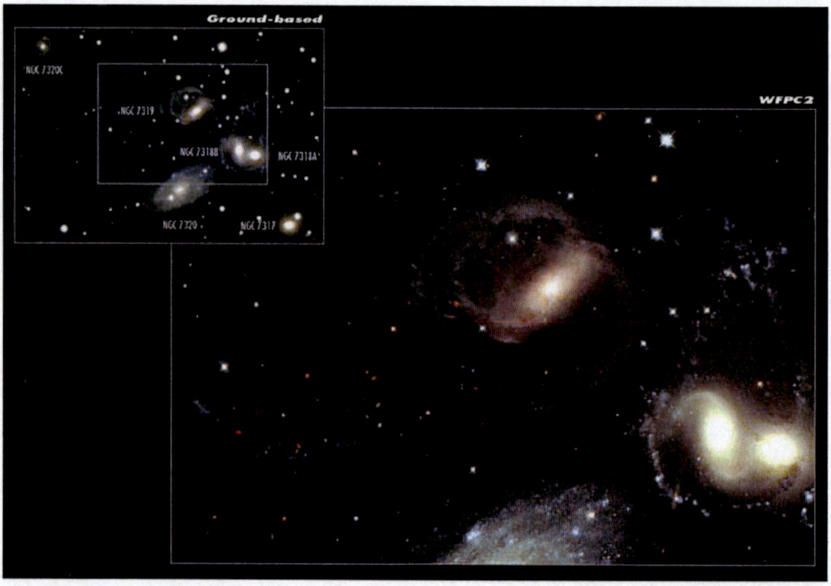

Fig. 14.17 The Stephan quintet used by the Burbidges to infer a large, rather dark mass of galaxies. However, these are colliding galaxies and hence out of equilibrium, so that the dynamic mass estimates are unjustified. Credit: Hubble Space Telescope, NASA

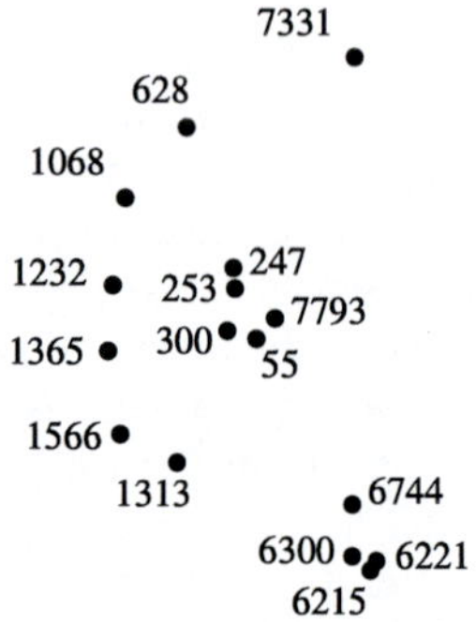

Fig. 14.18 The small group of galaxies examined by de Vaucouleurs (1959). Note the two pairs of very close galaxies which may be in collision (247–253 and 6221–6215)

obvious that the small group is a bound system. Indeed, de Vaucouleurs found that it is expanding and dispersing.

In the same year, van den Bergh[117] tried to check the assumption that the Virgo cluster of galaxies is well balanced, or 'relaxed', to use physicists' jargon, meaning that the galaxies have exchanged energies with one another like molecules in gas, and that the velocities correspond to a state in which a sufficiently large number of energy exchanges has taken place. But van den Bergh found that the cluster is not

[117] van den Bergh, S., MNRAS **121**, 387 (1960).

'relaxed', implying that it is young, whence application of the virial theorem may be incorrect and lead to an erroneous M/L value. Much of the mystery could have been eliminated, but van den Bergh's logic was circular. He assumed mass-to-light ratios of 40, 30, 20, 10, and 5 for E, Sa, Sb, Sc, and Ir galaxies, respectively,[118] and then inferred that the virial theorem was not satisfied, whence the Virgo cluster was not in equilibrium. Had more realistic M/L values been assumed for the galaxies, the opposite result would have been obtained.

Similar doubts about the applicability of the virial theorem were expressed by Greenstein[119] when he examined the extreme case of a pair of galaxies in the 3C 278 radio source. It is known today that these two galaxies are colliding with one another.[120]

Even with such large discrepancies between observed and inferred masses, the suspicion that something fundamental was buried here was not awoken. In 1962,[121] Layzer summarized:

> The correct conclusion to be drawn from the published data appears to be that only a small fraction of the matter in the Universe occurs in the form of bright galaxies.

However, he warned that:

> The data in question are not of high quality and are consequently incapable of providing strong support for any hypothesis.

Finzi realized the fundamental difficulty and suggested in 1963[122] that Newton's law of gravitation might be at fault over large distances. He proposed that the force of gravity might decline more slowly than the distance squared beyond a certain critical distance. Finzi assumed the critical distance to be about 16,000 light years. Gravity at large distances, according to Finzi, was stronger than predicted by Newton's law. At such a distance, the local gravitational acceleration of a typical galaxy is of the order of 5×10^{-8} cm/s^2.

In 1983, Milgrom suggested[123] that Newton's law of gravity should be changed for very low accelerations. Indeed, the accelerations of galaxies, assuming the visible mass only, are about 2×10^{-8} cm/s^2 and are of the same order as $cH = 5 \times 10^{-8}$ cm/s^2, which is the acceleration due to the expanding Universe. Here c is the speed of light and H is the Hubble constant.[124] This observation was called Milgrom's law by Kaplinghat and Turner,[125] and they demonstrated how the supposition of cold

[118] E means elliptical galaxy and S refers to a spiral galaxy. The subdivisions a, b, c indicate as follows: Sa are tightly wound, with poorly defined arms and a relatively large core region. An Sc galaxy has open, well-defined arms and a small core region. Sb means that the arms start as a bar extending through the core of the galaxy.

[119] Greenstein, J.L., Astrophys. J. **133**, 335 (1961).

[120] Machacek, M.E., and 5 coauthors, AAS, 210.0801, 2007.

[121] Layzer, D., Astrophys. J. **136**, 138 (1962).

[122] Finzi, A., MNRAS **127**, 21 (1963).

[123] Milgrom, M., Astrophys. J. **270**, 365 (1983).

[124] The cosmic acceleration was calculated assuming $H = 50$ (km/s)/Mpc.

[125] Kaplinghat, M. and Turner, M., Astrophys. J. **569**, 19 (2002).

dark matter brings about this correspondence. Note that, had Milgrom assumed the existence of dark matter, his number for the acceleration of the galaxies in the cluster would have exceeded the cosmic acceleration. Finzi's earlier alternative suggestion, which treated the same accelerations, was not mentioned. So what kind of accelerations are we talking about here? Consider the gravitational acceleration between two protons touching each other. It has a value of roughly 10^{-5} cm/s^2, about a factor of 1,000 bigger than Milgrom's critical acceleration. For comparison, the Coulomb repulsion between the two touching protons is 10^{36} times bigger.[126]

14.13 Dark Matter in Galaxies. Tracers of a Larger Mass

All spiral galaxies, and to a large extent irregular galaxies, exhibit ordered rotational motions. The best available method for the determination of the mass of a galaxy is from the rotation, assuming a balance between gravity and the centrifugal force. The situation with regard to galaxies changed significantly when radio observations in the 21 cm wavelength became possible.[127] It soon became clear[128] that:

- Galaxies contain hydrogen gas way past the 'edge of the visible galaxy'. This gas has a low temperature and hence does not shine in the visible, but rather in the radio domain.
- The total amount of gas is small relative to the mass of the galaxy.
- The hydrogen gas rotates around the center of the galaxy and the rotation velocity hardly varies with distance from the galactic center beyond a certain radius (see the schematic Fig. 14.19 left).

If the matter in the galaxy ended where the luminosity of the galaxy declines to zero, namely where we see the 'edge of the galaxy', then any gas outside this region should follow the classical Kepler law in its rotation around the center of the galaxy. This classical Kepler law is based on the assumption that the entire mass acting on the gas lies inside the radius of rotation, and it leads to a rotation velocity (the blue curve) which decreases with distance, in contrast to the observed curve. Parenthetically, if we consider cosmic rotation, say the Earth going around the Sun, we also find that

---

[126] If you stand on the surface of the Earth and the Moon is above your head, the Moon-induced acceleration is 0.3 cm/s^2, so your weight is reduced by a factor of $1{-}3 \times 10^{-4}$. The equivalence principle has been tested with a relative accuracy of 7×10^{-10} (Carusotto, S. and 6 coauthors, PRL **69**, 1722, 1992), but testing Milgrom's suggestion in the laboratory would require a relative accuracy of 5×10^{-11}.

[127] The 21 centimeter line or HI line refers to a spectral line created by an atomic transition between the two hyperfine levels of neutral hydrogen from a state of parallel spins of the proton and electron to an antiparallel spin configuration. The transition is strongly forbidden, with an extremely small probability of 2.9×10^{-15} s^{-1}. An isolated, unperturbed atom of neutral hydrogen would undergo this transition once in about 10 million years, so it is unlikely to be seen in a laboratory here on Earth. However, the huge amounts of neutral hydrogen in space are sufficient to produce a detectable signal in radio telescopes.

[128] Huchtmeier, W.K., Astron. Astrophys. **45**, 259 (1975).

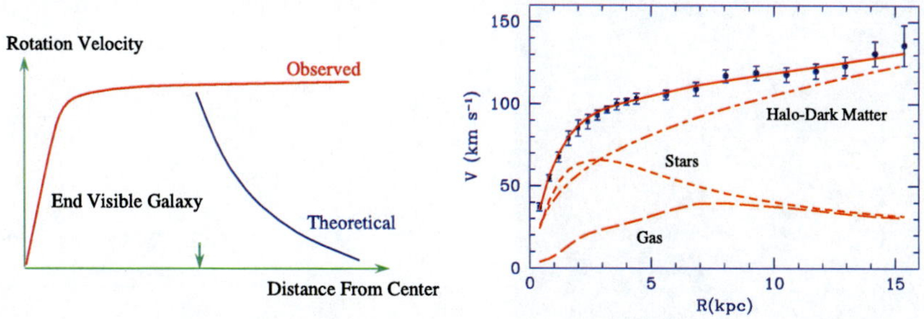

Fig. 14.19 *Left* schematic description of the rotation curve of spiral galaxies. The theoretical curve is the prediction had the actual mass of the galaxy ends where the visible mass ends (marked by an *arrow*). *Right points* indicate the measured rotation curve for M33. Theoretical fits to the different components according to Corbelli and Salucci (2000) are shown with *broken lines*

the kinetic energy is twice the absolute value of the gravitational energy, the result given by the virial theorem.

Let us present one detailed example so as to appreciate the properties of galaxies. Consider then the rotation curve for the nearby galaxy M33 (see Fig. 14.20 left). M33 is a spiral galaxy which belongs to our own Local Group of galaxies. It is frequently called the Triangulum galaxy. This galaxy may be gravitationally bound in the same gravitational potential well of dark matter as M31 (the Andromeda galaxy), as can be seen from Fig. 14.20 (right). Corbelli and Salucci[129] attempted to reproduce the rotation curve by assuming a distribution of bright mass (stars and gas), based on the best available observations, and dark matter. The result, given in Fig. 14.19 (right), shows how bright matter declines towards the edge of the galaxy, while dark matter continues to grow.

By 1978, it had become clear that the phenomenon was almost universal. Thonnard et al.[130] compiled a list of 300 galaxies which was the basis for Rubin's[131] review (see Fig. 14.21). Rubin pointed out that no spiral galaxy with declining rotation curve had ever been found:

> It is impossible to identify a galaxy with a falling optical rotation curve.

The phenomenon appears to be universal.

Elliptical galaxies have shapes ranging from spherical to elliptical. The classical wisdom in the late 1970s was that they were flattened due to rotation. But in 1977, Illinworth[132] examined 13 elliptical galaxies and compared them with existing models of such rotating galaxies. What he found was that, on the average, most elliptical galaxies have one-third the peak rotational velocity required by these models, with an estimated upper limit of two-thirds of the expected rotation:

---

[129] Corbelli, E., and Salucci, P., MNRAS **311**, 441 (2000).

[130] Thonnard, N., Rubin, V.C., Ford, W.K. Jr., and Roberts, M.S., AJ **83**, 156 (1978).

[131] Rubin, V.C., IAUS **84**, 211 (1979).

[132] Illingworth, G., Astrophys. J. **218**, 43 (1977).

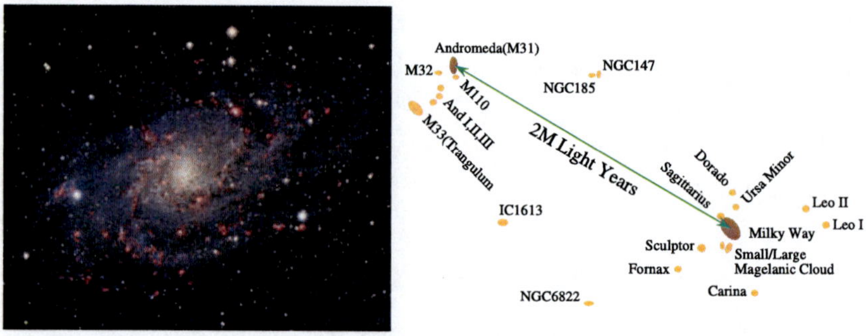

Fig. 14.20 *Left* the M33 spiral galaxy. With a diameter of about 5×10^4 light years, it is the third largest in the Local Group. One can identify the blue star clusters and pink star-forming regions which trace the galaxy's spiral arms. The nebula NGC 604, which belongs to a spiral arm of this galaxy, is the brightest star-forming region. Credit SAO/NASA. *Right* map of the Local Group of galaxies. The two big galaxies, the Milky Way and Andromeda, each have several smaller galaxies around them. It is plausible that the smaller galaxies are moving in the potential well of the big ones

Fig. 14.21 A couple of examples of rotation curves, after Rubin (1979)

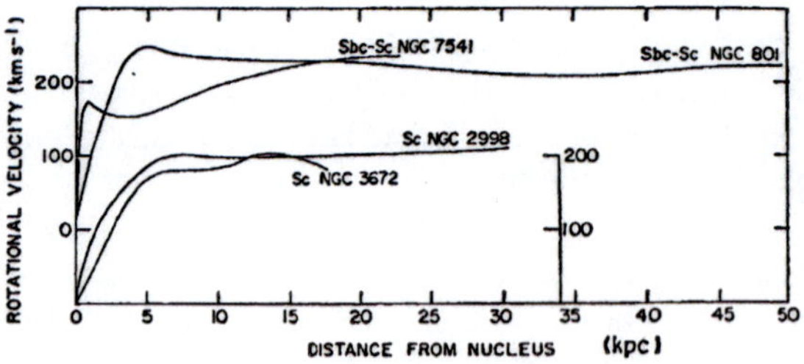

Oblate models in which the flattening is a result of rotation appear to be precluded by these data.

Only one giant elliptical galaxy rotates as predicted by the theory. It thus looks risky to rely on the slow rotation to infer the mass.

Various attempts to relate the distribution of stars to mass in galaxies were made by Dressler,[133] for example, who considered the large elliptical galaxy at the center of the Abell 2029 cluster of galaxies. Dressler observed the luminosity of the galaxy in certain common spectral lines and built a model of the galaxy to account for the distribution of the inferred number of stars. He found that three different components were needed to obtain good agreement with the observations:

- A normal elliptical galaxy with $M/L \approx 10$.
- A halo with $M/L \approx 35$.
- Dark cluster-filling and cluster-binding matter with $M/L > 500$.

[133] Dressler, A., Astrophys. J. **231**, 659 (1979).

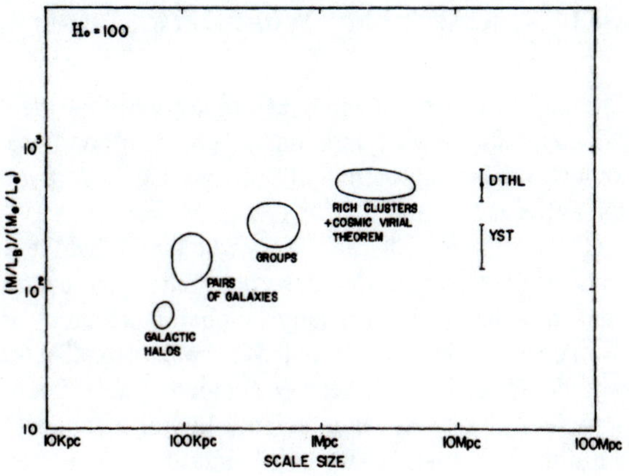

Fig. 14.22 Mass-to-luminosity ratio as a function of scale in the Universe. The larger the scale, the greater the ratio, indicating that the ratio of dark to bright matter increases with scale. After Hoffman et al. (1982)

In 1977, Gallagher et al.[134] discovered for the first time the existence of neutral hydrogen in a normal elliptical galaxy (NGC4278, a type E1 galaxy). Following this discovery, Knapp, Kerr, and Williams[135] investigated a collection of elliptical galaxies and found emission from neutral hydrogen from most of them. Furthermore, they discovered that the gas is not distributed all over the galaxy but resides *in a rotating disk which extends well beyond the visible body of the galaxy*. The M/L ratio they found was greater than 20. Hence, their conclusion was:

> This observation is consistent with the view that this elliptical galaxy, like many spiral galaxies, is embedded in a massive, low luminosity halo.

In 1982, Hoffman, Shaham, and Shaviv[136] considered how M/L varies with the scale of the object. The results are displayed in Fig. 14.22 and show that the larger the object, the greater the discrepancy. This implies that the effect of dark matter on the scale of a star is negligible. The dark matter exists everywhere, but as long as it has a uniform density (and temperature) and extends significantly beyond the size of the system, there are no noticeable effects. On the other hand, the mass-to-light ratio appears to level off at large scales.[137]

---

[134] Gallagher, J.S., Knapp, G.R., Faber, S.M. and Balick, B., Astrophys. J. **215**, 463 (1977).

[135] Knapp, G.R., Kerr, F.J. and Williams, B.A., Astrophys. J. **222**, 800 (1978).

[136] Hoffman, Y., Shaham, J. and Shaviv, G., Astrophys. J. **262**, 413 (1982).

[137] The dependence of the mass-to-light ratio on the scale was rediscovered by Bahcall, N.A., Nucl. Phys. B **138**, 16 (2005); Bahcall, N.A., Lubin, L.M. and Dorman, V., Astrophys. J. **447**, 81 (1995), who failed to discover the paper by Hoffman et al. including the apparent flattening for large scales.

14.14 Guesses about What Dark Matter Is Made of?

The nature of dark matter presents today one of the greatest mysteries of astrophysics. We do not know what dark matter is made of, but we can make several guesses. We know that it is not protons and electrons. Could dark matter be explained by Zwicky's faint objects?

Palla, Salpeter, and Stahler,[138] who found that the first stars could have masses as small as $0.1M_{\odot}$, considered the possibility that low mass stars with zero metallicity could have been formed in large numbers and thereby provide the dark matter inferred from observations. They reached the conclusion that the total luminosity of such stars would be very low and hence difficult to detect. Such low mass stars never become sufficiently hot at the center to ignite hydrogen, and all the energy they radiate away is the Kelvin–Helmholtz–Ritter gravitational energy. So assuming that the entire dark mass is due to such objects, the predicted M/L ratio for such 'dark matter' would be about 60. Hence Palla et al. predicted that it would be difficult to detect these faint objects, even in nearby galaxies.

One of the latest ideas for dark matter along the lines suggested by Zwicky was proposed in 1984 by Carr, Bond, and Arnett,[139] who calculated the evolution of very massive Pop III stars and found that stars more massive then $200M_{\odot}$ should end up as black holes. While being able to produce the helium seen in stars, the black hole remnants of these stars could provide the 'missing mass'.

Most recent attempts to look for massive compact halo objects, or MACHOs for short, are based on the idea that these objects may be brown dwarfs or giant planets like Jupiter. A truly fantastic idea was to look for them using gravitational lensing. A gravitational lens is formed when the light from a distant source is deflected by a compact object, placed along the path of the light to the Earth, mimicking the solar deflection of stellar light which served to confirm Einstein's general theory of relativity back in 1919.[140] The idea is that faint unseen moving objects can cross

---

[138] Palla, F., Salpeter, E.E. and Stahler, S.W., Astrophys. J. **271**, 632 (1983).

[139] Carr, B.J., Bond, J.R. and Arnett, W.D., Astrophys. J. **277**, 445 (1984).

[140] In this solar eclipse, which took place in Sorbal, Brazil, a delegation of the Royal Astronomical Society headed by Eddington confirmed the effect as predicted by the general theory of relativity. This made headline news around the world and tremendously increased Eddington's, and of course Einstein's, scientific reputation in the public eye. When years later, in 1935, Chandrasekhar expressed his admiration for Eddington for planning a scientific expedition during *the darkest days of the war*, Eddington responded in a typical Eddingtonian manner:

> I would not have planned the expedition since I was fully convinced of the truth of the general theory of relativity.

In 1917, Eddington was called for military service, to which, as a declared quaker, he objected. It was Dyson, the then astronomer royal, who organized the expedition and who testified to the deferment committee which gave the quaker Eddington a temporary release to carry out the required measurement, even if the aim was to confirm the theory of a German enemy physicist. In this way Eddington escaped an English prison for refusing to be drafted. The results of the expedition were published by Dyson, F.W., Eddington, A.S., and Davidson, C.R., *A Determination of the deflection of light by the Sun's gravitational field, from observations made at the total eclipse of May 29,* the line of sight between the observer and a distant source and cause the light of the source to brighten (concentrate) in a very peculiar way. The original idea of using this effect to discover distant faint objects was due to Zwicky in 1937,[141] and it is a widely applied technique today. The most up-to-date result today[142] indicates that the galactic halo may contain about 20% MACHOs. So using Zwicky's suggested technique some seventy years on, it became clear that he was about 20% right.

This result also kills the idea that dark matter might be made of primordial black holes, or black holes which formed before recombination, or that these objects are the burial locations of synthesized heavy elements. Consequently, the failure to identify dark matter with cosmic macroscopic bodies seems to indicate that dark matter is predominantly some kind of elementary particle with unique properties which make its detection particularly difficult. It has to be some relic of the early Universe. But is this particle stable and long-lived, or does it have a middlingly long lifetime, say 10^{10} years, and send some kind of signal from its decay?

Different particles have been considered at different times as more and more data has accumulated. The theory of elementary particles has made progress and various observational constraints have been established. All sorts of particles have been suggested, including photinos, neutrinos, weakly interacting massive particles (WIMP), gravitinos, right-handed neutrinos, monopoles, and massive neutrinos.

With the progress in supersymmetric theories, new ideas for constituents of dark matter have been suggested. Past experience[143] has shown that when we combine different theories, we may discover something new. For example, when Dirac devised relativistic quantum mechanics, which combined the theories of quantum mechanics and relativity, he discovered antimatter. Each particle has its antiparticle. The merging of the two theories resulted in, or rather required, the doubling of all known particles. Would a similar process take place when quantum theory merges with general relativity? So far such merging attempts have failed and we can only hypothesize about the successful one. Could it be that the merge will be supersymmetric and involve once again a doubling of all particles? In supersymmetric theories each particle has a partner with a spin that differs by 1/2. And the best (whatever the best means when we do not know the answer) candidate for dark matter may be the lightest *hypothetical* supersymmetric partner particle, called the neutralino.

In many models, the lightest neutralino can be produced thermally in the hot early Universe and leave approximately the right relic abundance to account for the observed dark matter. The lightest neutralino with a mass of roughly $10–10^4$ GeV is

---

(Footnote 140 continued)
1919, Mem. Roy. Astron. Soc. **220**, 291 (1920). For more details, see Chandrasekhar, S., *Verifying the theory of relativity*, Notes and Records of the Royal Society of London **30**, 249 (1976). Dyson was awarded the Gold Medal of the Royal Astronomical Society for organizing the expedition and Jeans presented him with the medal. See Jeans, J.H., MNRAS **83**, 672 (1924).

[141] Zwicky, F., Phys. Rev. **51**, 290 (1937); ibid. 679.

[142] Mancini, L., *1st Workshop of Astronomy and Astrophysics for Students*, Naples, 19–20 April 2006. Published by INFN-Naples, eds. Napolitano and Paolillo, p. 85. See also Freese, K., astro-ph/0508279 11 Aug 2008.

[143] Murayama, H., hep-ph/0002232v2 4 March 2000.

the leading weakly interacting massive particle (WIMP) dark matter candidate. The idea of the neutralino was due to Greene and Miron[144] in 1986. In 1988, Barbieri and Berezinsky[145] came up with the same idea of the neutralino. If the energy of the neutralino is 10 GeV, this means that the density of these elementary particles in the Universe is about 1,000 particles per kilometer cubed.

The recent discovery that the expansion of the Universe is accelerating has opened new, bizarre, and previously inconceivable possibilities. The mysterious dark energy and/or matter behave as though they have negative pressure. In 1902, Chaplygin[146] was the first to examine the properties of such a unique class of media, where many phenomena contrast starkly with our daily experience. The particular medium he investigated was matter in which the pressure is inversely proportional to the density, i.e., the higher the density, the lower the pressure. If this medium is compressed, the pressure does not increase as it would in any normal medium. Instead, the medium responds by decreasing its pressure. Since Chaplygin's first suggestion, similar ideas have appeared in different situations. The case of gravity was considered by Polyachenko and Fridman.[147] Mechanical situations where such behavior is seen are described by Book, Ott, and Sultan.[148] In the case of cosmology, it turns out that the pressure depends on minus some power of the density. The pressure is negative.

In view of the relatively new discovery of the accelerating Universe and the resuscitation of the cosmological constant, the hypothesis that dark matter and dark energy behave like Chaplygin's medium have become bona fide possibilities. Extensive research is going on in this area on the front line of science, and it would be premature to predict the outcome, save for one point: whatever happens, it will be exciting.

14.15 The Role of Dark Matter in Structure formation

The acceptance of the idea that the dark matter is due to some elementary particles, which predominantly exert a gravitational force and may in addition involve some very weak interaction but do not emit any photons for us to detect, led to a revolution in the concepts of structure formation in the Universe. As these particles contribute the lion's share of the mass, it is they who control the gravitational attraction. Moreover, dark matter appears to be collisionless and dissipation-free, in the sense that particles of dark matter do not collide with one another and do not lose energy. The only interaction between particles of dark matter is therefore via the gravitational force between them, and most importantly, between them and the mean force generated

[144] Greene, B.R., and Miron, P.J., Phys. Lett. B **168**, 226 (1986).

[145] Barbieri, R. and Berezinsky, V., Phys. Lett. **205**, 559 (1988).

[146] Chaplygin, S.A., Izbrannye trudy (Selected Works), Nauka, Moscow, 1976, p. 94.

[147] Polyachenko, V.L. and Fridman, A.M., Equilibrium and stability of gravitating systems, Nauka, Moscow, 1986, p. 347.

[148] Book, D.L., Ott, E. and Sultan, A.L., Phys. Fluids, **17**, 676 (1974).

by fluctuations in their density. So if there is no energy loss in the usual sense of energy loss by radiation, how can dark matter, or any pure gravitational system, form gravitational potential wells?

One possible answer was given by Lynden-Bell in 1967,[149] considering the dynamics of stars in the Galaxy. Collision between stars is a rare event and can be neglected for most purposes. The energy of a particle moving in a time-constant gravitational field is conserved. But if the gravitational field changes in time, this is no longer true. Consider the simplest case of many concentric shells composed of individual particles which interact only through the gravitational field. If the concentric layers are for some reason unstable, as in the Jeans instability, then each concentric shell will fall inward, converge to the center, become a 'point', and then expand to the original radius if the energy is conserved. The gravitational force acting on each particle depends only on the mass inside its radius. If this mass changes, then the force will change and consequently it will change its energy. It all depends, in the spherical case, on how the concentric shells move. If an outer shell can overtake another shell, there will be a change in the mass enclosed within the radius of the overtaken shell, and its energy will change. This is the principle of what Lynden-Bell called violent relaxation of gravitational systems.

Shortly after the Big Bang, the Universe was opaque to electromagnetic radiation. Then, at a certain moment after recombination, it became transparent. At this point, existing photons stopped scattering and continued to propagate almost freely in the Universe, until they reached the satellite WMAP, where their intensity was measured.[150] Hence, there is a rough surface, at a distance equal to the age of the Universe minus the age of recombination (approximately), on which the photons we see today were born. This is the so-called last scattering surface. If the Universe were strictly homogeneous, we would see no structure. But the existence of fluctuations in density and temperature cause the surface to look as shown in Fig. 14.23. The relative difference between the maximum and minimum temperatures is as small as 2×10^{-4} K.

Assuming a distribution of initial fluctuations as measured by the WMAP satellite, these fluctuations will be subject to mutual gravitational interaction. Present day theoretical modeling of structure formation in the Universe therefore starts with simulation of fluctuations in the dark matter and how they evolve to form a cosmic network of gravitational potential wells connected to one another by 'filaments' or 'bridges'. A typical example is shown in Fig. 14.24. This structure is the underlying dark matter structure determining what we see in the Universe today. In this view, the bright matter, which emits photons which we can observe, had only a minor effect

[149] Lynden-Bell, D., MNRAS **136**, 101 (1967).

[150] WMAP is an acronym for Wilkinson Microwave Anisotropy Probe, which measures cosmic microwave temperature fluctuations. WMAP produced the first full-sky map of the microwave sky with a resolution of under a degree, about the angular size of the Moon. The search is made in the microwave range because the radiation is light left over from the Big Bang, redshifted to microwave wavelengths due to the expansion of the Universe. The WMAP mission was proposed to NASA in 1995 and launched in 2001. The instruments are capable of measuring the temperature of the sky's microwave background to the incredible accuracy of one part in a million.

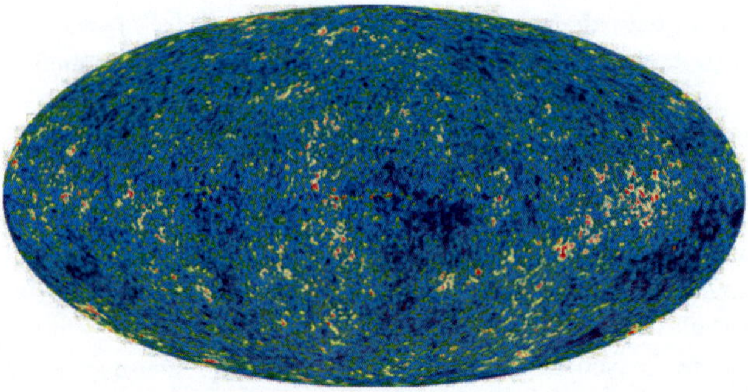

Fig. 14.23 Image of the Universe when the photons interacted for the last time with the matter, known as the last scattering surface. The average temperature is 2.725 K. *Colors* represent the temperature fluctuations. The maximum temperature difference is 2×10^{-4} K. Credit: NASA/WMAP Science Team

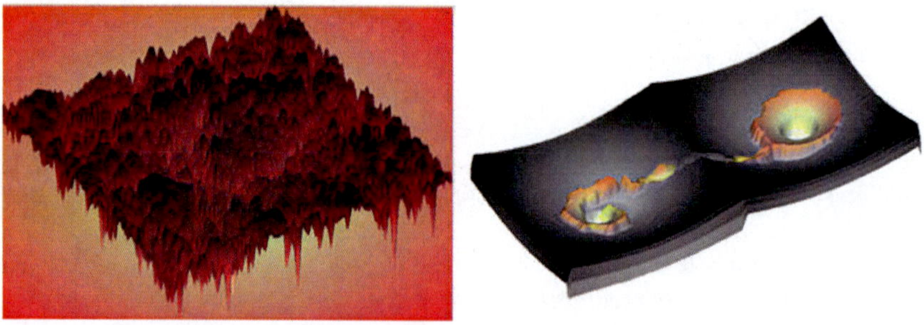

Fig. 14.24 *Left* the distribution of gravitational potential wells of dark matter as formed from the time evolution of the initial fluctuations in the Universe. Calculated by Shaviv (unpublished). *Right* the bright matter, mostly hydrogen–helium gas produced in Big Bang nucleosynthesis, falls into the gravitational potential wells created by the dark matter. Each galaxy resides in such a potential well. Empty wells may exist

on the formation of the large structure of the Universe, simply falling into the 'dark matter potential wells', cooling, and condensing into clusters of stars, galaxies, and massive stars. In this way a nested system of gravitational potential wells was formed. The shape of the galaxy distribution and the large scale structure observed today in the Universe was determined by the primordial perturbations in the Big Bang!

Due to the properties of the dark matter, as seen from its inferred structure, it seems that dark matter did not play a role in the condensation of masses much smaller than about $10^8 M_\odot$, which in turn are much smaller than the masses of galaxies. If so, all the physics of star formation, cooling, fragmentation, and so on, which we discussed before, is at least valid in the first approximation. The dark matter varies only on scales of the order of the scale of a galaxy or larger, and hence it may plausibly have no effect on smaller scales. The stellar structure and evolution theory we have learnt so far would then remain valid.

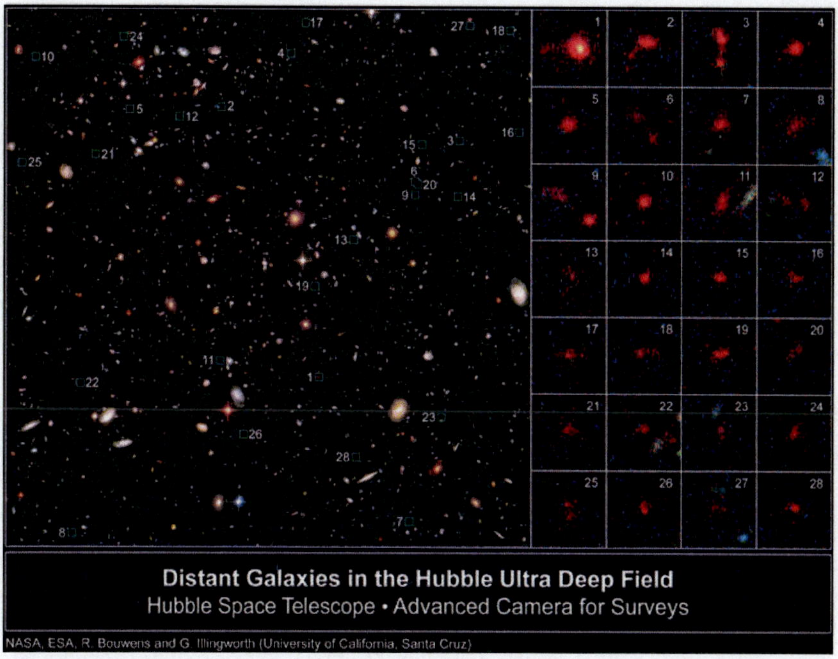

Fig. 14.25 One of the most spectacular images. Distant galaxies discovered by the Hubble telescope. The table on the right is a magnification of the details of the emerging galaxies. At this early phase, the galaxies do not yet have the shapes we know so well today. Credit: HST, NASA

14.16 Almost Seeing the Beginning

One of the thrilling moments in modern cosmology happened when the Hubble Space Telescope was dedicated to a very long exposure in an attempt to discover objects from the era of galaxy formation. The Hubble Deep Field (HDF) is an image of a small region in the constellation Ursa Major, based on the results of a series of observations by the Hubble Space Telescope. The image seen in Fig. 14.25 was assembled from 342 frames. The field is so small that only a few foreground Milky Way stars are observed. Almost all of the 3,000 objects seen in the image are galaxies in different phases of their evolution. Some of the galaxies are in their earliest phase of evolution, and hence the furthest away. The young galaxies have amorphous shapes, showing us the building blocks of galaxies. Thus, at least in that phase, galaxies were assembled from smaller pieces. Each great galaxy does not therefore result from the collapse of a single large object. The galaxies are said to be built bottom-up from smaller pieces that merge together. However, the pieces were bigger than stars.

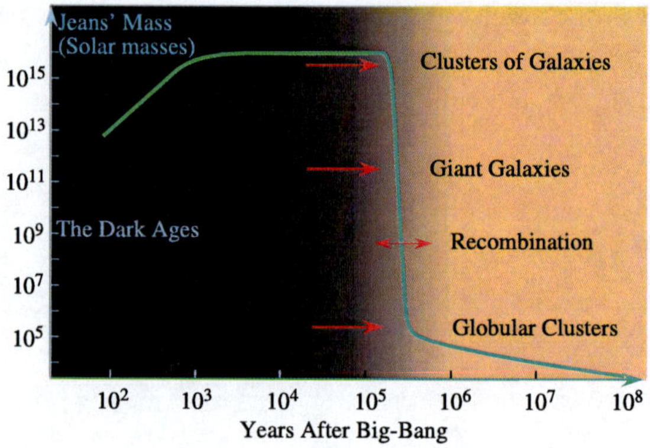

Fig. 14.26 Evolution of the Jeans mass in the expanding Universe

14.17 When Did the First Stars Form?

The time evolution of the Jeans mass is shown in Fig. 14.26. The Jeans mass increased at the beginning, then reached a plateau and subsequently decreased quickly when recombination took place. As we saw earlier, cooling must be efficient for the instability to end with a bound system. Hence we have to follow the curve to the point where cooling wins. This happens when the temperature is so low that hydrogen molecules form. At this moment, the first stars form. The formation of the massive stars caused the end of the Dark Ages because they illuminated the Universe. Moreover, the copious UV radiation emitted by these stars ionized the rest of the matter which had not condensed into stars. It was this re-ionization which turned the Universe into a transparent cosmos where we can see so far back in time.

14.18 Where Are the First Heavy Elements Expected to Have Been Synthesized?

In 1954, when Salpeter calculated the initial mass function for the first time, he assumed a constant rate of star formation. But the need for improvement was quick to appear. Likewise, in 1957, von Hoerner (1919–2003)[151] attempted to simulate the evolution of the Galaxy and assumed that the rate of star formation was constant, but ignored the gas in his calculations and got no resemblance to the observed patterns. In 1959, Mathis[152] repeated von Hoerner's calculation and tried to predict the stellar population in the Galaxy. To this end, he also assumed that the rate of star formation was constant in the Galaxy. The result was that he could not fit the observations. His explanation for the failure was mixing between local stars and stars from other regions far away in the disk. This explanation turned out to be wrong. In the same year,

[151] von Hoerner, S., Astrophys. J. **126**, 592 (1957).

[152] Mathis, J.S., Astrophys. J. **129**, 259 (1959).

Schmidt[153] assumed that (a) the initial mass function for Pop I was time-independent and (b) the rate of formation was proportional to the density of the gas squared. He found that the present rate of star formation would be 5 times lower than the average over the lifetime of the Galaxy. This of course implies massive element synthesis at the beginning of the Galaxy or during the collapse to form the Galaxy.

In parallel, Salpeter[154] found that a better approximation to the rate of star formation was needed. He abandoned the assumption that the rate of star formation was constant, taking it rather to be proportional to the amount of gas in the Galaxy. He then found a high rate of star formation at the beginning, while the first billion years was followed by a constant decay. Such behaviour is to be expected, because there are stars which live longer than the Galaxy and any matter condensing into such stars is removed from the star–gas–star cycle. Accordingly, extensive synthesis of heavy elements would have taken place in the beginning stages of the Galaxy and the abundance of the heavy elements must have risen very quickly.

In 1962, Wallerstein examined 21 G-dwarf stars (in layman's terms, main sequence stars like the Sun) in the solar neighborhood. According to the simple model of stellar populations, they should all be Pop I stars with high abundances of heavy elements. Seven of these stars were located above the main sequence (on their way to becoming giants) and six of them, concluded Wallerstein, were as old as the cluster NGC 188, which he claimed was 15 Gyrs old (this was the number Wallerstein quoted, but more accurate determinations fix the age at 6.8 Gyrs[155]). Among the old stars, most were metal-rich and a few were metal-poor. Consequently, Wallerstein claimed that:

> The rate of metal enrichment of the interstellar medium has been low for the last 15 Gyrs. Such a low rate of metal enrichment supports the idea of Schmidt and Mathis that the present rate of star formation is much lower than the rate during the very early history of the Galaxy.

His conclusion today would have referred to the last 6.8 Gyrs. Wallerstein was not looking for Pop III stars. He just examined the nature of stars like the Sun in the solar neighborhood. Hence his results implied that most of the heavy elements were synthesized before Pop I stars were formed.

On the other hand, as Wallerstein found, there is a very strong correlation between the amount of heavy elements and the velocity of the stars, viz., the faster the star, the lower the amount of heavy elements. So far so good.

The first to notice a problem with this conclusion was van den Bergh who, in 1962,[156] discovered that the rate of heavy element creation in the Galaxy decreased much faster than the rate of star formation. He obtained another interesting result when he took Wallerstein's results and plotted them differently, finding that the abundance of manganese (Mn) varies with the abundance of iron. This in itself is not surprising. The lower the iron, the lower the abundance of Mn. This was true for

---

153 Schmidt, M., Astrophys. J. **129**, 243 (1959).

154 Salpeter, E.E., Astrophys. J. **129**, 608 (1959).

155 VandenBerg, D.A. and Stetson, P.B., PASP **116**, 997 (2004).

156 van den Bergh, S., AJ **67**, 486 (1962).

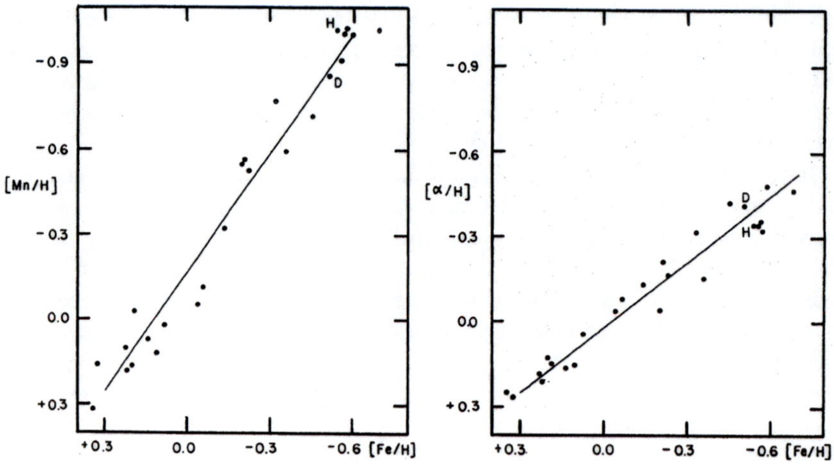

Fig. 14.27 Correlation between the abundance of Mn and Fe (*left panel*) and between α nuclei and iron (*right panel*). The units are the logarithm of the abundance ratio in the star relative to the ratio in the Sun. Note that, while the depletion of all elements is proportional to that of iron, different nuclei are depleted differently (the slope of the two curves is different). From van den Bergh (1962)

practically all tested elements (see Fig. 14.27). The surprise was that the slopes of the curves were different. This implied that there were differences in the synthesis and its relative outcome as a function of the heavy element abundance. van den Bergh discovered that stellar evolution depends on the heavy element abundance and vice versa. This important fact had not really been predicted before, and was not sufficiently stressed by van den Bergh. Indeed, it was clearly not properly appreciated by researchers until the first years of the twenty-first century, when it was rediscovered.

If the initial rate of star formation was very high and stars of all masses were formed, then one should expect to see many G-dwarf stars (with masses less than $0.8M_\odot$) with very low metal abundances. But we do not see these! In fact, there should be more stars with low metal abundances than stars with any other metal abundance. This was later defined as the *G-dwarf problem*, formulated by van den Bergh and Schmidt.

In 1970, Bond[157] observed that the number of metal-poor stars seen today is below the predicted number. Consequently, he suggested that it provides support for the idea that the Pop III stars were massive. As he put it:

> Massive objects played an important role in the earliest stellar generations.

He stated that his results were *at some variance* with those of von Hoerner, van den Bergh, and Dixon[158] in 1966, who concluded that they found roughly comparable numbers of stars throughout the range of observed metal content.

---

[157] Bond, H.E., Astrophys. J. Suppl. **22**, 117 (1970).

[158] Dixon, M.E., MNRAS **129**, 51 (1965); ibid. **131**, 325 (1966).

Bond[159] returned in 1980 to the search for low metallicity stars, and a year later, in view of his failure to discover many of these, he posed the question[160]:

> Where is population III?

Bond was looking for stars with metal content less than 1,000 times that of the Sun.[161] He searched over half the sky, examined several hundred thousand stars, and discovered only two stars with metal abundance 1/1000 that of the Sun. In 1981, Bond concluded that:

> The rarity of metal-poor disk stars (van den Bergh 1963, Schmidt 1963, Bond 1970) is often reffered to as the 'G-dwarf problem'. The main result of this paper has been the demonstration that there is a significant paucity of halo stars with $Fe \leq Fe_{\odot}/400$, that is, a 'G-dwarf problem' also exists in the galactic halo.

So this problem is not restricted to the solar neighborhood.

Major progress was initiated in 1985 when Beers, Preston, and Shectman[162] started a systematic search for very low metal abundance stars. The main new feature in the search was that Beers et al. observed fainter objects and looked into the halo of the Galaxy. The newly discovered stars were therefore fainter than those sought by Bond. Recently, a new survey which goes to still fainter stars, the Hamburg/ESO survey (HES)[163] offers further possibilities.

Today astronomers define extremely metal-poor stars to be ones with 1/1000 the solar amount of heavy elements. The lowest metallicity observed in the halo of the Galaxy is $Z \sim Z_{\odot}/10{,}000$, so the first generation stars must have produced at least these amounts of heavy elements. There are about 200 confirmed extremely metal poor (EMP) stars and about 2,000 suspected.[164]

A few exceptions should be noted. In 1984, Bessel and Norris[165] serendipitously discovered the unique star CD-38° 245, which exhibited a record low Fe content of $[Fe/H] = -4.5$. They concluded that the star was an extreme Pop II, although there was some speculation that it might be a true Pop III star. A more accurate measurement by Norris et al.[166] found that the iron content was lower by a factor of 10,000 relative to the Sun. Consequently, this star held the record low Fe level for 20 years.

---

[159] Bond, H.E., Astrophys. J. Suppl. **44**, 517 (1980).

[160] Bond, H.E., Astrophys. J. **248**, 606 (1981).

[161] A similar attempt was carried out by Bidelman, W.P. and MacConnell, D.J., AJ **78**, 687 (1973), and yielded similar results.

[162] Beers, T.C., Preston, G.W., Shectman, S.A., AJ **90**, 2089 (1985).

[163] Wisotzki, L., and 6 coauthors, Astron. Astrophys. **358**, 77 (2000).

[164] Christlieb, N., Review Modern Astronomy **16**, 191 (2003).

[165] Bessel, M.S. and Norris, J., Astrophys. J. **285**, 622 (1984).

[166] Norris, J.E., Ryan, S.G., and Beers, T.C., Astrophys. J. **561**, 1034 (2001).

The record in low iron abundances is now held by the stars HE 0107-5240[167] and HE 1327-2326.[168] The two stars differ in their abundance patterns. HE0107-5240 is a giant star of the galactic halo with iron abundance 1/200,000 of the solar iron abundance. Its mass is $M = 0.8M_{\odot}$. Another interesting feature is that carbon is enhanced relative to iron by a factor of 10,000, while nitrogen is enhanced by only 200, sodium is enhanced by 6, and magnesium is enhanced by 1.5, and the strontium content is lower by a factor of 0.3.

HE 1327-2326 has an iron abundance which is only 1/250,000 of the solar value. Another peculiar feature is the extremely low Li level. Such a Li abundance is about a factor of 10 below the Big Bang value. C and O are enhanced relative to the Sun by about 12,000 and 30,000 respectively. O is enhanced by less then 10,000, Na and Mg are only enhanced by about 100, and strontium is enhanced by 10. At present there are no model predictions leading to such abundances.

14.19 What Has Been Found So Far?

About 70 out of 94 elements have been discovered in the metal-poor stars. The synthesis site has not yet been identified with certainty, but the most likely suggestion is primordial supernovae of very massive stars which started with vanishing amounts of heavy elements.

Usually, the α nuclei (Mg, Si, Ca, and Ti) are enhanced by a factor of about 2.5 relative to iron. But exceptions exist. For example, in 2003, Ivans et al.[169] discovered several stars in which the α elements were depleted relative to iron.

In 1999, Rossi et al.[170] discovered that many extremely metal-poor stars have a relatively high carbon abundance. This was later confirmed by Cohen et al. in 2005[171] and Lucatello et al. in 2006.[172] It appears that the enhancement of C relative to Fe increases as Fe decreases relative to the Sun. Many of the carbon-enriched stars also show neutron-capture-enhanced elements. The situation updated to 2007 is shown in Fig. 14.28 (after Norris et al.[173]). The enhancement of carbon can be as high as a factor of 10,000.

The observation of stars (but too few of them) with low metallicity showed a rather large variation, and hence people concluded that the interstellar medium was still inhomogeneous and not well mixed when the cosmic abundance of iron was

---

167 Chrislieb, N., and 8 coauthors, Nature **419**, 905 (2002). HE refers to the Hamburg/ESO survey covering the whole southern extragalactic sky.

168 Frebel, A., and 18 coauthors, Nature **434**, 871 (2005).

169 Ivans, I.I., and 7 coauthors, Astrophys. J. **592**, 906 (2003).

170 Rossi, S.V., Beers, T.C., and Sneden, C., in *The 3rd Stromolo Symposium: The galactic halo*, ASP Conf. Ser. **165**, 264 (1999).

171 Cohen, J.G., and 8 coauthors, Astrophys. J. **633**, 109 (2005).

172 Lucatello, S., et al. Astrophys. J. **625**, 37 (2006).

173 Norris, J.E., and 7 coauthors, Astrophys. J. **670**, 774 (2007).

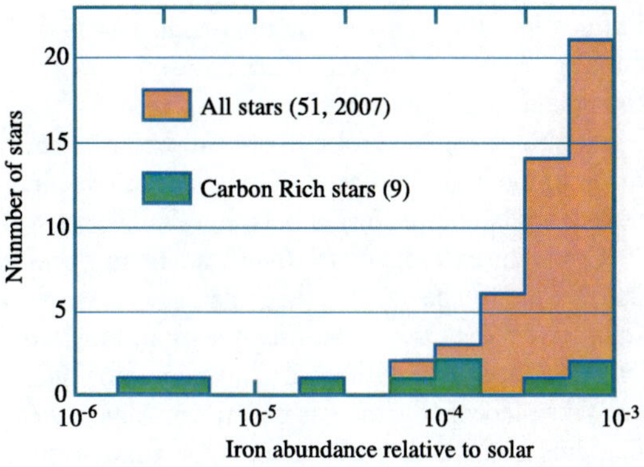

Fig. 14.28 The statistics of metal-poor stars in the halo of our Galaxy, as of 2007. From Norris et al. (2007)

[Fe/H] = −3.0 (e.g., Argast et al.[174]). This conclusion was turned upside-down when Cayrel et al.[175] compared 35 stars and found quite a high level of regularity, indicating that elements with atomic number less than 30 were well mixed when [Fe/H] ∼ −4.0. Thus, at this time the interstellar medium was well mixed, as all stars at different locations in the halo had the same composition.

The first attempts to reproduce the observed abundances by means of theoretical models[176] show that[177]:

> The results are clearly incompatible with the predicted yields of pair-instability supernovae/hypernovae.

Observations of other galaxies indicate that our galaxy is not unique. In 1996, Worthey, Dorman, and Jones[178] showed that the number of metal-poor stars in other galaxies is at least a factor of 2 less than predicted by the 'simple' model, much like the situation in our own galaxy.

It seems[179] that, in dealing with the chemical evolution of the Galaxy, the hypothesis of instantaneous mixing (which means that any ejecta from stars is mixed with the existing gas on a very short time scale) is more severe than instantaneous recycling (which means that the ejected gas condenses quickly into new stars), and in fact allows no fit to the G-dwarf metallicity distribution.

In 1996, the first *r*-process star was discovered by Sneden et al.[180] The abundance ratios were found to be similar to those in the Sun, and Frebel concluded that this

---

[174] Argast, D., Samland, M., Gerhard, O.E. and Thielemann, F.-K., Astron. Astrophys. **356**, 873 (2000).

[175] Cayrel, R., and 13 coauthors, Astron. Astrophys. **416**, 1117 (2004).

[176] Francois, F., and 5 coauthors, Astron. Astrophys. **421**, 613 (2004).

[177] Umeda, H. and Nomoto, K., astro-ph/0205365v1 22 May 2002; Heger, A. and Woosley, S.E., Astrophys. J. **567**, 532 (2002).

[178] Worthey, G., Dorman, B. and Jones, L.A., AJ **112**, 948 (1996).

[179] Caimmi, R., AN **321**, 323 (2000).

[180] Sneden, C., and 5 coauthors, Astrophys. J. **467**, 819 (1996).

implies that the r-process is universal. But two similar stars do not announce the coming of spring. Indeed, in 2005, Aoki et al.[181] required more than one r-process to explain the abundances.

Complicating the problem once more, in 2005, Lucatello et al.[182] discovered that many of the high s-process elements are close binary stars, where nuclear processed matter has been transferred between them. The evolution is therefore more intricate.

Could the extreme metal-poor stars be just peculiar stars which have unique abundances due to special rare circumstances? Are they really old? Recently, Venn and Lambert[183] examined just this question. How could such a situation come about? The gas out of which the stars form contains dust, and various separation processes can take place inside the gas before condensation begins. For example, it is plausible that dust and metal, the best coolants, condensed first to form a low mass protostar, which later accreted metal-free gas to form the bigger star.

A temporary summary is therefore that too few EMP stars have been discovered and it is premature to draw conclusions about major and minor nucleosynthesis processes. In any case, Bond's conclusion was vindicated: there are too few metal-poor stars. Since the heavy element abundance of these stars is still higher than could be produced in the Big Bang, they should be second generation stars, but not the very first ones.

What about massive Pop III stars in distant galaxies? Various attempts to detect Pop III stars have been made. For example, Nagao et al.[184] declared that:

> Pop III stars have not yet been discovered.

As the Pop III stars are high mass stars, the expectation is that they should have a high surface temperature and hence radiate copiously in the UV, as expected for surface temperatures of 10^5 K and above. There are suggestions that certain helium lines (the He II λ 1640) could be a signature of such stars. But all results today are upper limits. The null result applies to galaxies within a certain distance from us, and so far the result is not universally confirmed.

14.20 The Most Massive Stars Known Today

It is easier to determine the most luminous stars than the most massive stars. To determine luminosity it is sufficient to find the distance by some other means, and there are many possibilities. On the other hand, to determine the mass directly and without relying on theory, the star must be a member of a binary system, and not all massive stars are binaries. About 20 stars in the neighborhood of our galaxy are known with luminosities in excess of $2 \times 10^6 L_\odot$. Each of these luminous stars has

[181] Aoki, W., and 9 coauthors, Astrophys. J. **632**, 611 (2005).

[182] Lucatello, S., and 5 coauthors, Astrophys. J. **625**, 825 (2005).

[183] Venn, K.A. and Lambert, D.L., Astrophys. J. **677**, 572 (2008).

[184] Nagao, T., and 11 coauthors, arXiv/astro-ph/arXiv:0802.4123v1 28 Feb 2008.

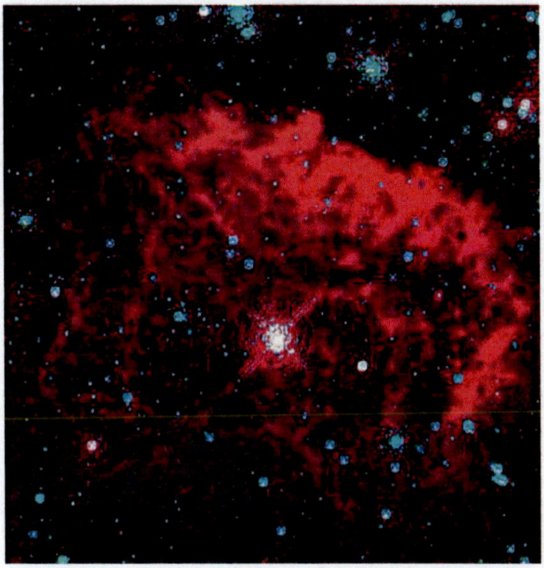

Fig. 14.29 One of the most luminous stars in our galaxy, the Pistol star and the nebula formed by its mass loss. Credit: NICMOS, STScI, NASA (Figer et al. 1995)

Table 14.2 The most luminous stars

	η Car	P Cyg	R136a	Var 33 in M33	Pistol star
Mass ($M_\odot$)	~200	~80–100	$\leq 2,000$	~100–150	200–250
Luminosity ($L_\odot$)	5×10^6	$\leq 1.5 \times 10^6$	$\leq 6 \times 10^7$	2×10^6	4×10^6
Estimated initial luminosity					1.5×10^7
Rate of mass loss [$M_\odot$/year]	10^{-1}–10^{-3}	10^{-4}–10^{-5}	$\sim 5 \times 10^{-4}$	$\sim \times 10^{-5}$	?

Based on Humphreys, R.M. and Davidson, K., Science **223**, 243 (2008) and Figer, D.F., and 6 coauthors, Astrophys. J. **506**, 384 (1998)

an interesting story. The Pistol star, for example, is hidden behind a heavy cloud of dust. Consequently, Hubble's Near Infrared Camera and Multi-Object Spectrometer were needed to take its false-color image (see Fig. 14.29), which is a composite of two separately filtered images.

The powerful luminosity of the star causes extensive mass loss, and the expanding red nebula is the ejected mass. The fact that we see a non-spherical nebula and not a continuous spherical gas cloud implies that the mass loss is via eruptions which are not necessarily spherical. The star is at a distance of 25,000 light years in the direction of the center of our galaxy. Despite its great distance, the star would have been visible to the naked eye were it not for the dust cloud inside which it is embedded.

The masses listed in Table 14.2 are inferred from theory, using difficult measurements of the stars' temperatures and absolute brightnesses. All the listed masses are uncertain: both the theory and the measurements are pushed to the limit.

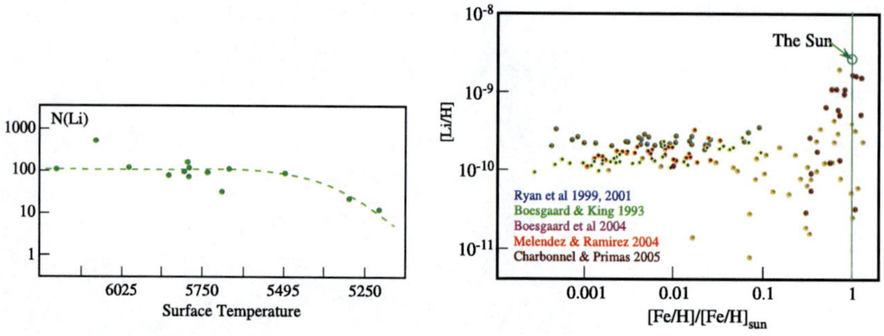

Fig. 14.30 *Left* the original way the Spite plateau was presented (Spite and Spite 1982). Lithium abundance as a function of the surface temperature of the main sequence star (equivalent to the mass of the star). *Right* spite plateau converted into lithium abundance as a function of the iron depletion in the star

14.21 The Intriguing Abundance of Lithium in Metal-Poor Stars

The synthesis of lithium is one of the most important consequences of Big Bang nucleosynthesis, as was shown by Wagoner et al. in 1967, and it demonstrates the complexity of the conundrum. Consequently, the determination of the abundance of lithium in population III stars was considered for many years a good tool to constrain the assumed parameters of Big Bang nucleosynthesis and subsequent cosmological models.

In 1982, Spite and Spite[185] discovered a unique behavior of lithium in Pop II metal-poor stars. All stars with iron depletion by a factor of 0.04–0.004 relative to the Sun and surface temperatures in the range 5,700–6,250 K show the same lithium abundance, i.e., while the iron abundance decreases, and with it the abundances of all the heavy elements, the abundance of lithium remains constant (see Fig. 14.30 left). No other element exhibits similar behavior. Today, this unique behavior is called the Spite plateau.[186] It was interpreted as a signature of Big Bang nucleosynthesis, because these observations agreed with the predictions of Wagoner et al. (1967). It was a very nice and reassuring result.

Note that the original figure plotted by Spite and Spite (see Fig. 14.30 left) showed lithium as a function of the surface temperature for old metal-depleted stars. The first time the lithium abundance was plotted as a function of the iron abundance, as shown in Fig. 14.30 (right), was by Thorburn,[187] who noticed *a slight trend towards lower*

[185] Spite, M. and Spite, F., Nature **297**, 483 (1982); Astron. Astrophys. **115**, 357 (1982).

[186] The word 'plateau' appeared for the first time in Spite, M., Maillard, J.P. and Spite, F., Astron. Astrophys. **141**, 56 (1984), where they added an observational point at an even higher surface temperature.

[187] Thorburn, J.A., Astrophys. J. **421**, 318 (1994).

lithium abundance as metallicity declines. However, the data was not sufficient to nail this trend down. The most important result of Thorburn was:

> A large proportion of the scatter about the Spite plateau is removed when the [Fe/H] dependence of N(Li) is taken into account.

In other words, the plateau emerges more clearly and more sharply (with less scatter of points) when the lithium abundance is plotted against the abundance of iron. Since then, this is the way the so-called Spite plateau has appeared in the literature.

Lithium differs from all other metals in that (a) it is produced in the Big Bang[188] and (b) it burns easily and fast at relatively low temperatures of 2.5×10^6 K. Along the lower part of the main sequence, stars have extensive convective envelopes which extend to a sufficiently high temperature to burn lithium. The more massive the star, the shallower the convective region below the surface, and the lower the temperature at the bottom of the convective region (until the convective zone disappears for massive stars). When the bottom of the convective region extends to the burning temperature of lithium, lithium depletion takes place. All the inner parts of the star become devoid of lithium, but the primordial lithium plays no role in the nuclear reaction and energy generation.

Even when the lithium on the surface is mixed into the convective region, dragged down to high temperatures, and consequently depleted, there is no effect on stellar evolution and structure. The importance of lithium is, as mentioned above, only for cosmological reasons. The extent of the convective zone depends on the total heavy element abundance,[189] and hence changes with stellar population.

Cracks started to appear in this picture with the discovery of stars that had even lower than Big Bang predicted amounts of lithium. Ryan et al.[190] noted the existence of at least 8 stars with lithium abundances lower than the Spite plateau by a factor of 3. A further examination of these stars[191] provided more surprising results. Two stars had element compositions very similar to stars on the Spite plateau. While two other stars showed clear evidence of neutron capture elements, like excess of Sr, Y, and Ba relative to iron. The rest of the stars showed other peculiarities. The confusion was that no general pattern could be found. It looked as if the lithium abundance was erratic. So does the existence of such compositional anomalies in the lithium abundance indicate the existence of several different evolutionary routes?

Bonifacio et al.[192] extended the lithium abundance observations to stars with even lower iron abundances and confirmed Ryan et al.'s basic results (Fig. 14.31). However, the state of affairs is more complex. The present day updated Big Bang

---

[188] The suggestion that lithium is produced in the B2FH x-process, like spallation of heavier nuclei, was proven not to be feasible on account of the required energy and the low yield.

[189] This happens through the effect of the heavy elements on the radiation absorption coefficients and hence on the run of the temperature in the star. The heavy elements significantly increase the radiation absorption coefficient.

[190] Ryan, S.G., Norris, J.E. and Beers, T.C., Astrophys. J. **523**, 654 (1999).

[191] Norris, J.E., and 4 coauthors, Astrophys. J. **485**, 370 (1997); Ryan, S.G., Norris, J.E. and Beers, T.C., Astrophys. J. **506**, 892 (1998).

[192] Bonifacio, P., and 14 coauthors, Astron. Astrophys. **462**, 851 (2007).

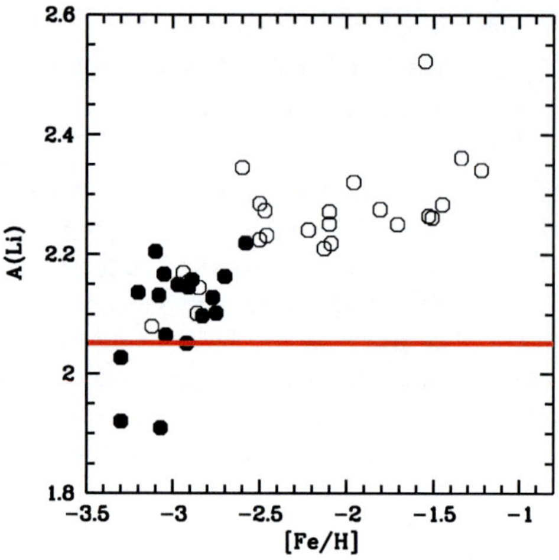

Fig. 14.31 The lithium abundance as a function of the abundance of iron in metal-poor stars, as found by Bonifacio et al. (2007). The quantities depicted are $A(\mathrm{Li}) = \mathrm{Log}\,(N(\mathrm{Li})/N(\mathrm{H})) - 12$ and $[\mathrm{Fe/H}] = \mathrm{Log}\,(N(\mathrm{Fe})/N(\mathrm{H})) - \mathrm{Log}\,(N(\mathrm{Fe})/N(\mathrm{H}))_{\odot}$. The *red line* is the present day Big Bang prediction of lithium abundance

calculations yield a lithium abundance of 4.4×10^{-10} relative to hydrogen, while the highest number on the Spite plateau is a factor of two lower, i.e., well below the primeval value. If lithium could not be depleted in these stars, we have a problem.

What could be the reasons? Bonifacio et al. were ready to conclude that new physics is needed in Big Bang nucleosythesis to explain the discrepancy. A number of esoteric ideas were tossed up, such as the formation of lithium in shock waves just before galaxy formation takes place,[193] or the decay of a strange particle after the Big Bang nucleosynthesis ended.[194] Other ideas were taken from the shelf and dusted off, such as variations of the fine structure constant[195] $\alpha \approx 1/137$, massive particle decay[196] which destroys ^{7}Li, and pre-galactic ^{6}Li production.[197] If dark matter is composed of a particle which has a bound state, it can affect the abundance of lithium.[198]

We can conclude, this time with some assurance, that we still do not know what happens to lithium after its formation in the Big Bang. Nor do we know the extent to

193 Suzuki, T.K. and Inoue, S., Astrophys. J. **573**, 168 (2002); Fields, B.D. and Prodanovic, T., Astrophys. J. **623**, 877 (2005).

194 Jedamzik, K., Choi, K.-Y., Roszkowski, L., and Ruiz de Austri, R.J., Cosmology and Astroparticle Phys. **7**, 7 (2006).

195 Ichikawa, K. and Kawasaki, M., Phys. Rev. D **69**, 123506 (2004); Landau, S.J., Mosquera, M.E. and Vucetich, H., Astrophys. J. **637**, 38 (2006).

196 Jedamzik, K., Phys. Rev. D **70**, 083510 (2004); Kawasaki, M., Kohri, K. and Moroi, T., Phys. Lett. B **625**, 7 (2005); Ellis, J., Olive, K.A. and Vangioni, E., Phys. Lett. B **619**, 30 (2005).

197 Suzuki, T.K. and Inoue, S., Astrophys. J. **573**, 168 (2002); Rollinde, E., Vangioni, E. and Olive, K., Astrophys. J. **627**, 666 (2005).

198 Cyburt, R.H., and 5 coauthors, J. Cos. and Astroparticle **11**, 14 (2006).

which its abundance is affected by other mechanisms, or where exactly it is destroyed or hidden.[199] In short, all options are still available.

14.22 The First Direct Age Measurement of a Star

The measurable range of ages depends on the half-life of the radioactive nucleus used for dating. Since we are interested in ages of the order of the age of the Universe, our attention is directed to uranium–thorium dating. The half-life of thorium is 14.05 Gyrs, while the half-life of uranium is 4.47 Gyrs. Could the idea be implemented to measure the age of stars directly, rather than using the theory of stellar evolution?

In Sect. 13.18, we discussed the problems of dating by means of r-process-produced elements. Along the way the following stable isotopes are synthesized: Eu, Os, and Ir. On the other hand, Th and U are among the unstable isotopes formed by the r-process. Now, if we know the abundance ratios of Th/Os and Th/Ir at the end of neutron irradiation, it is a straightforward matter to measure the present day value and get the moment at which the decay started. The intriguing theoretical fact is that the three abundance ratios Th/Eu, Th/Os, and Th/Ir are quite invariant, and hence can be safely assumed to be the initial ratio from which the decay of U and Th starts. Beside being stable nuclei, these atoms also have spectral lines in the observed range, so the abundance of Th, Eu, Os, and Ir can be measured with significant accuracy.[200] Before the discovery of uranium, only the thorium ratios were known and they were not very accurate, as was shown by Goriely and Clerbaux.[201] The discovery of U added a new dimension, and the U/Th ratio could be measured directly and compared with theory. Also U and Th have very close atomic numbers and weights, so the initial difference in abundances is much smaller than in all other suggested isotopes.

Even in 2001, Cayrel et al.[202] published the first detection of uranium in the r-process-enhanced metal-poor star CS 31082-001.[203] Hill et al.[204] used the U/Th ratio to find an age of 14.0 ± 2.4 Gyrs. However, they could not assign a reliable error estimate to the result. Moreover, they got a contradictory result from the Th/Eu ratio, indicating some basic problem either in the entire method or in the observational results and measurements.

---

[199] Ryan, S.G. and Elliott, L., in *Chemical Abundances and Mixing in Stars in the Milky Way and its Satellites*, ESO Astrophysics Symposia. Springer-Verlag, 2006, p. 185.

[200] Note that the dating method discussed here is somewhat different from the one discussed in Sect. 13.18.

[201] Goriely, S., and Clerbaux, B., Astron. Astrophys. **346**, 798 (1999).

[202] Cayrel, R., et al. Nature **409**, 691 (2001).

[203] In this star, the heavy elements are reduced by a factor of 1.2×10^{-3} relative to the Sun.

[204] Hill, V., and 12 coauthors, Astron. Astrophys. **387**, 560 (2002).

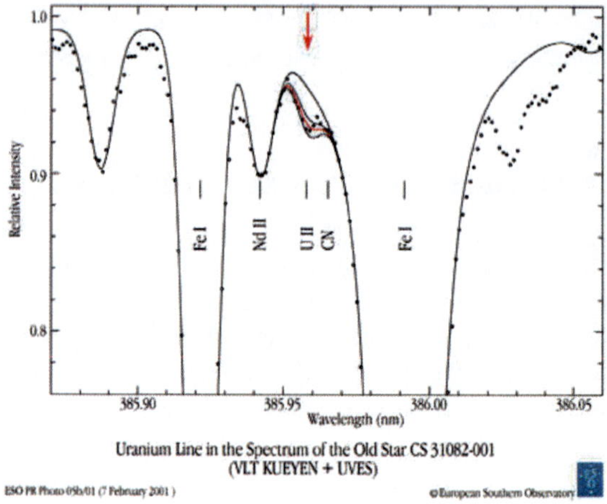

Fig. 14.32 The uranium line (marked by the *red arrow*) near the CN lines. Note that the *vertical scale* does not start from zero, and the minimum of the iron lines is off the scale. Credit: ESO

A major step forward was taken in 2007, when the age determination was carried out for the star HE 1523-0901 by Frebel et al.[205] This star is known to be very metal poor, the heavy elements being reduced by a factor of 1,000 relative to the Sun. The interesting feature is that the radioactive elements Th and U were discovered on the surface of this star. To appreciate the result, Fig. 14.32 shows the spectrum in which the relevant uranium line was discovered. The question as to how it got to the surface is another issue. High expertise in spectroscopy is required, as well as reliable modeling of the stellar atmosphere. The once ionized uranium (denoted by UII) line is marked in the figure with a red arrow. As can be seen, the uranium line appears on the shoulder of the CN line, which is relatively strong. Note that the iron line, for example, reaches out of the figure and the y-axis begins at 0.75.

The problems of conflicting data did not exist at this time, and Frebel et al. were able to get a good estimate with an acceptably small error. The weighted average of all the ages found from the above abundance ratios yielded an age of 13.2 ± 0.7 Gyrs. The most recent and accurate age of the Universe was determined by the WMAP satellite,[206] and is 13.7 ± 0.2 Gyrs.

Frebel et al. were successful because the star they discovered contained relatively large amounts of uranium. An additional coincidence which allowed this result is the low abundances of N and C in the star. These elements form the CN molecule which has a spectral line very close to the uranium line, and if these elements have a normal abundance, this particular line of CN becomes so strong and broad as to completely cover the uranium line. Finally, the star is a giant, so it is bright and relatively easy to observe.

---

[205] Frebel, A., Christlieb, N., Norris, J., Thom, C., Beers, T.C., and Rhee, J., Astrophys. J. **660**, L117 (2007).

[206] Spergel, D.N., Frebel, and 15 coauthors, Astrophys. J. Suppl. **148**, 175 (2003).

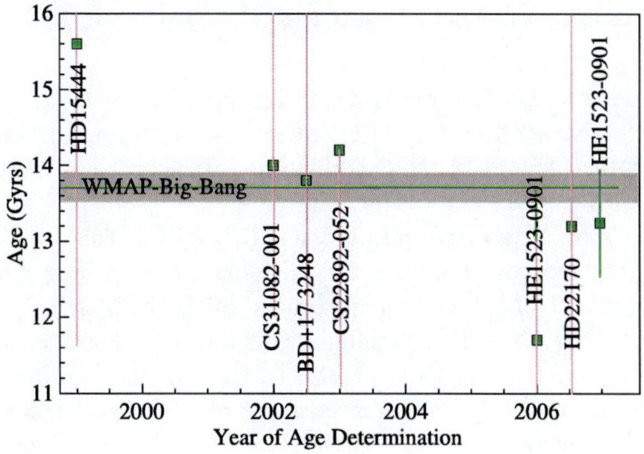

Fig. 14.33 Radioactive age determination of metal-poor stars as a function of the time the measurement was made

In this way, the authenticity of the cosmological age was confirmed. The low metallicity stars are indeed very old, and in this particular case the age agrees with the age of the Universe. These stars were formed soon after the formation of the Universe itself.

The method was applied to other galactic objects. Sneden et al.[207] detected Th in the globular cluster M15. From the Th/Eu ratios, an age estimate of 14 ± 3 Gyr was obtained. This is within acceptable agreement with the most recent standard estimates of the mean age of the globular clusters of 12.9 ± 2.9 Gyr by Carretta et al.[208]

Figure 14.33 summarizes recent age determinations by means of radioactive decay.[209] Also marked is the most up-to-date age of the Universe as inferred from the recent WMAP satellite experiment determinations. The band denotes the error in that determination. It is evident that the accuracy of the two independent determinations has improved over the past years.

---

207 Sneden, C., and 5 coauthors, Astrophys. J. **536**, 85 (2000).

208 Carretta, E., Gratton, R.G., Clementini, G. and Fusi Pecci, F., Astrophys. J. **533**, 215 (2000).

209 HD15444: Cowan, J.J., and 7 coauthors, Astrophys. J. **521**, 194 (1999). CS 31082-001: Hill, V., and 11 coauthors, Astron. Astrophys. **387**, 560 (2002). BD +17° 3248: Cowan, J.J., and 10 coauthors, Astrophys. J. **572**, 861 (2002). CS 22892-052: Sneden, C., and 11 coauthors, Astrophys. J. **591**, 936 (2003). HE 1523-0901: Frebel, A., and 5 coauthors, Astrophys. J. **652**, 117 (2007). HD 22170: Ivans, I.I., and 6 coauthors, Astrophys. J. **645**, 613 (2006).

14.23 The Tale of G77-61: What an Old Star Can Tell Us

Astronomers were looking for metal-poor stars, but if a star is really metal poor, would they recognize it? The story of G77-61[210] is very instructive in this respect. It demonstrates the complexity and the difficulties of the problem, and above all the steady and healthy skepticism of astronomers.

In 1977, Dahn et al. rediscovered the star G77-61.[211] The original discovery was made during the course of the Lowell Observatory proper motion survey. The star attracted the attention of astronomers due to its high proper motion of 0.76 arcsec/year.[212] As a rule, such a high value indicates that the star is a nearby object. But regular monitoring soon revealed that the parallax was quite small, only 0.013 arcsec/year, which implies that the star is far away. A star with a large apparent motion and far away should be something exceptional because it implies a very high spatial velocity. But the unexpected occurred: astronomers lost interest in this distant star and it was not monitored any longer. However, a discrepancy emerged between the distance implied by the parallax, the apparent magnitude (which yields the absolute luminosity when the distance is known), and the colors of the star. In this case, the colors implied that it should have been much fainter. All possible reasons for an error were checked, leaving the contradiction unsolved. If it were a dwarf star, it would have been much cooler than the coolest known white dwarf, and hence older than any known white dwarf. Dahn et al. realized that the star is similar to a carbon star, since the atmosphere of the star contains more carbon than oxygen. The high carbon abundance on the one hand and the very low amount of iron on the other led Dahn and his associates to suspect that the star might be a close binary star in which mass and composition were exchanged between the two stars, so that the more evolved star synthesized the elements and transferred them to the less evolved companion. Such complicated evolution is known in other well established cases. Yet, the star was so strange that astronomers did not believe these results (Fig. 14.34).

In 1986, Eggen[213] examined the kinematics and color properties of several large groups of stars and concluded that G77-61 must be 19 Gyrs old! This fantastic age was found by comparing the theoretical evolution of stars with heavy element abundance 0.0003 by mass and helium abundance 0.2 by mass. Eggen was indeed worried about the incredible age derived in this way, and cautioned that the star's strange composition could have affected the age determination. However, the obvious conflict with the age of the Universe was not mentioned.

In 1986, Dearborn et al.[214] were able to prove that Dahn et al. were right in their hypothesis that the star was a binary system. The researchers had at their disposal 25 photographs taken at the US Naval Observatory over 6 years, and carefully mea-

---

[210] The G stands for Giclas, H.L. (1910–2007) from the Lowel Observatory, who prepared the catalogue of parallaxes.

[211] Dahn, C.C., and 4 coauthors, Astrophys. J. **216**, 757 (1977).

[212] Giclas, H.L., Burnham, R., Jr., and Thomas, N.G., Bull. Lowel. Obs. **112**, 611 (1961).

[213] Eggen, O., AJ **92**, 910 (1986).

[214] Dearborn, D.S.P., and 6 coauthors, Astrophys. J. **300**, 314 (1986).

Fig. 14.34 The metal-poor star CS22892-052 located at a distance of 1.5×10^4 light years in the halo of our Galaxy. The estimated mass is $0.8M_\odot$ and the iron to hydrogen ratio is 1/1000 the ratio in the Sun. A simple eye observation would not reveal how special this star is, so finding it was really like searching for a needle in a haystack

sured the location of the star. The result was that it actually oscillates in space with a period of 245.5 days. But the companion had never been observed! Yet, people were inclined to hypothesize that the now unseen companion was the star which synthesized the carbon-enriched material and transferred it to the metal-poor star. If the mass of G77-61 is assumed to be 0.30–$0.35M_\odot$ (an estimate based on the colors of the star), then it follows that the unseen companion must have a mass of $0.55M_\odot$, so it could be a very old white dwarf which has cooled so much as not to be observed.

We are in fact compelled to assume that this is an old binary star, so the authors tried to fit the star with a theoretical model and found that, unless hydrogen is significantly less abundant than the primordial value, even reducing the heavy elements to 0.000 01 does not generate an agreement between theory and observation. In other words, the star must contain an unusually large amount of helium in its atmosphere. Another possibility considered was that the helium might have been transferred from the unseen companion. But this scenario is not free from problems either. For example, when the binary period is as long as 245.5 days, the chances of mass transfer are extremely low. Could the binary have been much closer in the past, transferring mass to the companion and then moving away from it? This is in principle possible, but its history becomes much more complicated.

If it were so, could the unique composition be pristine, i.e., cosmological, dating from the beginning of the Galaxy? This alternative scenario was rejected because no other star which is so metal poor was known to have such a high carbon-to-oxygen ratio (even though the absolute abundances of C and O are much lower than those in the Sun), and because it ignores the fact that the star is a binary with an unobserved companion.

In 1988, Gass, Liebert, and Wehrse[215] improved the analysis of the spectrum of the star using stellar atmosphere models. They confirmed the findings that this object is a star with extremely low heavy element abundance. In fact, the amount of heavy

[215] Gass, H., Liebert, J. and Wehrse, R., Astron. Astrophys. **189**, 194 (1988).

Table 14.3 The unique abundance ratios of G77-61, based on Plez and Cohen 2005

Species	Ratio
He/H	0.1
C/H	2×10^{-5}
N/H	1.6×10^{-7}
O/H	$<10^{-7}$
Na/H	1.2×10^{-9}
Mg/H	2×10^{-9}
K/H	$<1.6 \times 10^{-9}$
Ca/H	$<3.2 \times 10^{-11}$
Fe/H	$1. \times 10^{-10}$

elements in G77-61 is lower by a factor of $1/400{,}000$ than the amount of heavy elements in the Sun! So it was suggested that the unseen companion transferred a modest amount of carbon-enriched material at the time when the object was an evolved star near its maximum radius.

Gass et al. reached the conclusion that both scenarios were required, i.e., that it is indeed a very old star with composition typical of the product of Pop III stars and that it has a special evolution history as a binary star. The only 'normal' fact about this star was that the abundances of the light elements in the range of atomic number $1 < Z < 12$ agreed well with the Big Bang predictions of Wagoner et al. 1967.

The most recent investigation of this star was carried out by Plez and Cohen[216] (see Table 14.3). Using the largest available telescope, they discovered that G77-61 is indeed metal poor, although not to the extreme extent previously measured. The heavy elements are in fact reduced by a factor of 1/10,700 relative to the Sun. The star is among the most C- and N-rich metal-poor stars. No overabundance of s- or r-process elements was detected. If indeed the amounts of these elements are below detection level, we will have to look for younger stars to produce them, since the unavoidable implication would be that Pop III did not synthesize them.

Finally, Plez and Cohen's more accurate determination of the amount of heavy elements and the recent discovery of more heavy-element-poor stars displaced G77-61 from its position as the most metal-poor star, leaving first place to the star HE-0107-5240, which has a heavy element abundance of $1/200{,}000$ relative to the Sun.[217] The record was recently broken once more when Frebel et al.[218] discovered that the heavy element abundance in HE 1327-2326 is lower by a factor of $1/250,000$ relative to the Sun.

Comparison with other galaxies also shows differences between their abundance histories. For example, in 2004, Burstein et al.[219] compared the globular clusters in the Milky Way with those in Andromeda and found marked differences. First, M31

[216] Plez, B., and Cohen, J.G., Astron. Astrophys. **434**, 1117 (2005).

[217] Christlieb, N., and 8 coauthors, Nature **419**, 904 (2002).

[218] Frebel, A., and 19 coauthors, Nature **434**, 871 (2005).

[219] Burstein, D., and 15 coauthors, Astrophys. J. **614**, 158 (2004).

contains GCs with a wide age range, while the Milky Way does not. Next, while both Milky Way and M31 GCs show enhanced nitrogen abundances, the nitrogen abundance of the M31 GCs is clearly greatly enhanced relative to what is seen in Milky Way GCs. It is still not known why this should be so.

14.24 Metal-Rich Stars: A Star Loaded with Puzzles

Up to now we have been discussing the riddle of metal-poor stars. However, there are mysteries on the other side as well, that is, concerning stars with unusually high metal abundances. Such a star was discovered by Przybylski (1913–1984) in 1961.[220] The star had been known previously as HD101065 and was classified as a B5 star. However, Przybylski discovered that it is not a B star at all. Following the rules of spectral classification, Przybylski claimed that there were indications that it might be a K0, an F, or even a G0 star. In short, the star showed confusing signs that did not fit any classification scheme.

The standard stellar classification schemes were established on the basis of 'well-behaved' Pop I stars in the solar neighborhood, and when this star refused to fit the general pattern of Pop I stars, it caused havoc. Przybylski achieved the honor that this very strange star was named after him by Wegner and Petford in 1974, although Przybylski continued to refer to it as HD101065. The original spectroscopic observation was carried out without great accuracy, and for this reason the full peculiarity of the star was not recognized by its discoverer. In 1963, Przybylski and Kennedy[221] carried out a much more accurate observation and discovered a very baffling star indeed.

Its strangeness was expressed by an unusual amount of elements belonging to the lanthanide group, including lanthanum. The lanthanides are the elements with atomic numbers between $Z = 57$ and $Z = 71$. There are 15 elements in this group altogether, and since they are all very rare on the Earth, they are also called the rare earth elements. As the chemical properties of the elements of this group are extremely similar, they appear in the periodic table as an appendix to the first element of the group. The lanthanide abundances obey the Oddo–Harkins rule.[222] In 1974, Wegner and Petford[223] discovered that (Fig. 14.35):

- The abundance of the lanthanides in HD101065 is about 10^5 times higher than what we find in the Sun.
- At the same time, the abundance of iron seemed to be normal.

---

[220] Przybylski, A., Nature **189**, 739 (1961).

[221] Przybylski, A. and Kennedy, P.M., PASP **75**, 349 (1963).

[222] See Shaviv, G., *The Life of the Stars*, Springer & Magnes Pub., Heidelberg, 2009. This rule implies, for example, that carbon should always be more abundant than nitrogen. However, stars do not obey the rule, and we find stars with $C/N > 1$ and stars with $C/N < 1$, probably implying two means of formation or a variation of the ratio with the mass of the star.

[223] Wegner, G., and Petford, A.D., MNRAS **168**, 557 (1974).

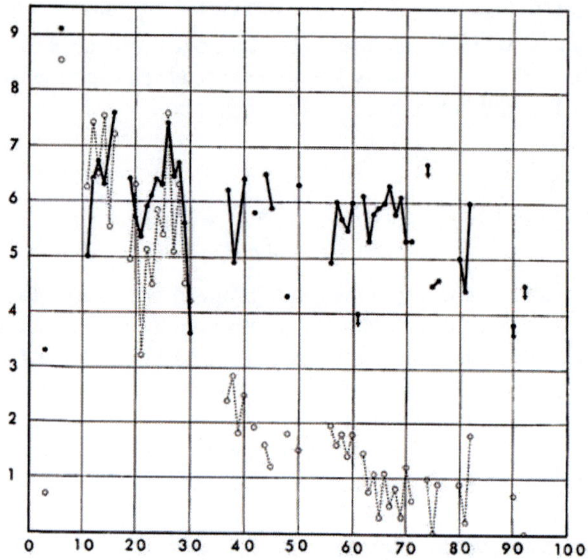

Fig. 14.35 Relative abundance of the lanthanides according to Wegner and Petford (1974). The *horizontal axis* is the atomic weight and the *vertical axis* is the logarithm of the abundance relative to hydrogen, which is taken as 12. *Full black points* are the results of Wegner and Petford. *Open circles* are solar abundances as given by Allen (1973)

- The star was a relatively cool Ap star.[224]

What could be the source of such a bizarre abundance pattern? B2FH suggested the x-process. This contained everything that did not agree with the canonical theory, like reactions on the surface of stars. Another suggestion was put forward by Michaud in 1970,[225] namely, the action of radiation pressure. What distinguishes this group of elements from the rest is the large number of spectral lines, and radiation pressure in those lines could lift elements to the surface of these stars. In other words, radiation pressure purges the star of the lanthanides and causes their accumulation on the surface. But if this is so, why do we only witness such abundances in the Przybylski star?

A great battle took place over the nature of this star during the IAU Colloquium 32 in 1975, entitled *The Physics of Ap Stars*. It was here that Przybylski claimed for the first time that, while the lanthanides were very overabundant, iron was deficient

[224] An Ap star is an A-type star but with high abundances of most metals, including the rare earth elements. The overabundance can reach a factor of 10^3 to 10^6 of the solar values. Ap stars typically have surface temperatures of 8,000 to 15,000 K, strong magnetic fields, and low rotational velocities. The mass of these stars lies in the range 2–4$M_{\odot}$.

[225] Michaud, G.J., Astrophys. J. **160**, 641 (1970).

by a factor of 300 relative to the Sun.[226] This anomaly was too much for the audience to swallow, and a heated debate erupted.

By 1973, it was clear that the star escaped any explanation of its peculiar abundances,[227] and with time the problem only became worse, as more detailed information accumulated. In 1976, Wolff and Hagen[228] discovered that HD101065 had a magnetic field of several thousand gauss.[229] Towards the end of this decade, Kurtz discovered[230] that HD101065 is a rapidly pulsating star. It was in fact the first star of this class to be discovered. It was not clear what the pulsation had to do with the peculiar abundances, save to point to some additional unique property of the star.

Przybylski himself did not agree with those who claimed that the iron abundance in the star was normal because the spectra of this star differed very much from the spectra shown by typical Ap stars. He maintained that there was no proof at all for the presence of iron. The confusion over the spectral lines was so great as to make such a claim possible.

In 1975, Przybylski suggested[231] that the abundances could be explained by spontaneous fission of transuranium elements.[232] The idea was basically that the r-process was extremely effective and that all the iron had been depleted by neutron captures to form the transuranium elements, which would require a huge neutron flux. These elements subsequently decayed by spontaneous fission. However, this would not be sufficient to explain the peculiarities. As all the elements between iron and lanthanum (the first element of the group) are also depleted, the r-process had to continue and deplete these elements, too! As Przybylski realized, the problem with this idea is just to explain where the gigantic neutron flux could have come from? In 1977, Cowley et al.[233] came to the conclusion that there was very little iron in the star, in contrast to Wegner, who maintained that the iron abundance was normal, but that the iron lines were overshadowed by the overabundance of the lanthanides. When spectral lines overlap and there is too much information, such arguments are unavoidable.

In 1998, Cowley and Mathys[234] observed the star and concluded that Wegner, Petford, and Kurtz were right. While the lanthanides show an overabundance by a factor of 10,000 relative to the Sun, the iron content is quite normal relative to the

226 Przybylski, A., PASAu **2**, 352 (1975).

227 Adelman, S.J., Astrophys. J. **183**, 95 (1973).

228 Wolff, S.C. and Hagen, W., PASP **88**, 119 (1976).

229 Recall again that the magnetic field of the Sun is about 4,000 gauss inside sunspots, while the average field, averaging the sunspots over the entire Sun, is only about 50 gauss.

230 Kurtz, D.W., ASPC **279**, 351 (2002); ibid. MNRAS **242**, 636 (1990); ibid. MNRAS **191**, 115 (1980).

231 Przybylski, A., PAS Australia **2**, 352 (1975). The sensational idea had been suggested to the author by Neiwodniczanski ten years earlier. It was independently proposed by Kuchowicz (Kuchowicz, B., Nature **232**, 552, 1971; Quart. J.R. Astron. Soc. **14**, 121, 1973).

232 The physical criterion for the occurrence of spontaneous fission is approximately $Z^2/A \geq 45$.

233 Cowley, C.R., Cowley, A.P., Aikman, G.C.L. and Crosswhite, H.M., Astrophys. J. **216**, 37 (1977).

234 Cowley, C.R. and Mathys, G., Astron. Astrophys. **339**, 165 (1998).

Sun. At the same time, the measurement of the strength of the magnetic field yielded 1,400 gauss (average for the entire star!), which is a very high value.

Recently, lines of the radioactive elements technetium and promethium, as well as elements with $84 < Z < 99$, were discovered. The slow diffusion of the elements from the core to the surface cannot account for the abundances of the radioactive elements. Hence a logical possibility is a reaction on the surface of the star. Calculations by Goriely[235] show that the elements can be produced by charged particles accelerated by the magnetic field of the star. The problem is that the fluxes needed are extremely large because of the low yield, and it is not clear whether such fluxes could exist on the surface of the star.

In 2004, Gopka et al.[236] discovered a sequence of radioactive elements. The only way they could explain their presence and relative abundances was by assuming that they emerged from the radioactive decay of Th and U. How the Th and the U reached the surface was another puzzle. They hypothesized that a strong r-process operated on the star, followed by the decay of Th and U. If correct, it will be the first time that fast neutron capture has been found to operate on the surface of stars. It appears that Przybylski's claim for fission of U and Th was vindicated.

Let us make a few last comments. Despite the discovery of 57 elements in this unusual star, there are still many unidentified spectral lines. Moreover, nobody understands the appearance of the spectrum. It should be noted that the temperature of the atmosphere cannot be determined accurately, let alone its distribution with depth. Thus, a large uncertainty is expected in the abundance determinations. So far, no quantitative theory can predict the observed abundances. For over 40 years, this star has remained a mystery and presents a standing challenge to the theory of stellar structure and evolution.

The Przybylski star is in our Galaxy and indicates the local peculiarities. Recently, Hasinger, Kossoma, and Schartel[237] have observed the quasar APM 8279-5255 and discovered that the iron abundance in this young object is a factor of 3 higher than the solar abundance. Quasars are very luminous objects and this one is exceptionally luminous compared with other quasars. Its luminosity reaches $10^{13}L_{\odot}$, which is a factor of 10 above the most luminous galaxy. This quasar is at such a distance that we see it today as it was when the age of the Universe was only 1.5 billion years. The puzzle is therefore to understand how an object with the size of a galaxy that is so much younger than our own galaxy and the Sun can exhibit such a high level of iron. This should be contrasted with the fact that, in most quasars, we see a more or less solar abundance, another fact that has not been explained. Could it be that star formation was accelerated in the typical quasar environment?

---

[235] Goriely, S., Astron. Astrophys. **466**, 619 (2007).

[236] Gopka, V.F., and 6 coauthors, Proc. IAU Symp. 224, ed. by Zverko, Ziznovsky, Adelman, and Weiss, p. 734.

[237] Hasinger, G., Schartel, N. and Kossoma, S., Astrophys. J. **573**, 77 (2002)

14.25 How Pristine Can Stars Be?

We observe stars with ages that approach the age of the Universe and measure the abundances of various elements. We assume that what we see is the composition that the star was born with. This is, of course, a far-reaching and courageous assumption. As a matter of fact, this is one of the most uncertain theoretical questions concerning old stars. The stars, being gaseous, are expected to experience diffusion or sedimentation, whence the heavy elements would sink and the light elements float. The existence of sedimentation has been proven to operate in the Sun, which is younger than most stars we have discussed here. While the process is very slow, it has a long time in which to take effect.

Sedimentation causes two consequences. Firstly, when the ages of globular clusters are calculated assuming sedimentation, the ages found are about 10^9 years greater than ages found when the phenomenon is ignored. Secondly, one expects the ratio of metals to hydrogen to be affected and indeed reduced if the effect is included (the hydrogen floats and the heavy elements sink). Richard et al.[238] considered that effect of sedimentation, but included in the calculation the effect of radiation pressure and the fact that the metals are easily ionized. The effect of radiation pressure is to lift the ions which have many absorption lines in the wavelength range where most of the radiation comes out. The lines themselves depend on the degree of ionization. Hence, radiation pressure has a tendency to counteract sedimentation and it is not obvious which will win out, sedimentation or lifting, and in any case the result will depend on the composition.

Richard et al. found that the abundance of iron, which has in particular a low ionization energy and many spectral lines, increases with time, while the abundance of lithium decreases. However, in 2003, Gratton found[239] no traces of such a change in the abundances when they observed globular clusters. While this is good news, because it allows many fine measurements, it is a surprise. A possible explanation is that, below the radiative layer, where the above effects take place, there is a convective layer where turbulent currents carry the heat and at the same time thoroughly mix the layer. The sunk or lifted elements move into and out of this turbulent layer, but the amounts of them does not change. As the convection zone expands towards the surface, the original abundance is reconstituted. This explanation requires very fine tuning between many factors and it is not obvious that such tuning exists. The stellar models should exhibit such fine tuning if it exists, but the predictions of Richard et al. as mentioned above do not attest to it.

---

[238] Richard, O., and 5 coauthors, Astrophys. J. **568**, 979 (2002).

[239] Gratton, R.G., and 19 coauthors, Msngr **112**, 31 (2003).

14.26 Open Questions

We see in the cosmos two basic types of object which emerged from the Big Bang: galaxies and stars. The two fundamental differences between these two types of object concern size and shape:

- Galaxies: rotation balances gravitation. The basic shape is flat. Dark matter is needed for the balance.
- Stars: gas pressure balances gravitation. The basic shape is spherical. Dark matter has no effect.

The rotation in stars has a small effect on the global structure—stars are good spheres. However, rotation affects slow mixing processes and circulations, and research into these subjects is in its infancy. It is clear that there must be basic differences between the dominant processes forming these two types of object. In particular, the original gas cloud which formed a galaxy must have undergone fragmentation, while stars, once they have formed, do not fragment.

Star formation is still a mystery. Over 50 years after Salpeter wrote his famous paper about the initial mass function (IMF), which gives the empirical probability for the formation of stars with different masses, we still do not know how to derive it from basic physical principles. Salpeter did what is known as reverse engineering in that he found what the IMF should be in order to yield the present day mass distributions of main sequence stars. As a first approximation, Salpeter considered all stars in the disk of the Galaxy, ignoring for example the distribution with height above the disk.

Celebrating 50 years of the introduction of the IMF idea,[240] Zinnecker summarized the outstanding fundamental problems in star formation:

- What theory predicts the observed IMF? At the moment there is no working theory.
- What is the physics of the maximum and minimum mass and how does it depend on the heavy element abundance? What is the physics of gravitational dynamics, collapse, collisions between blobs, mergers, etc. Big computers crank out the answer through heavy numerical simulations, but unfortunately, simple solutions do not abound.
- Is the formation top-down, in the sense that the giant cloud fragments in many steps to smaller and smaller blobs, or is it bottom-up, wherein very small blobs merge to form bigger and bigger objects? What is the nature of the collapse/instability of the medium out of which the blobs form? Is it an instability of small blobs or big ones? It is plausible that both processes exist, but on a different scales, galaxies undergoing a bottom-up process and stars forming top-down. Is the dividing mass the mass of the globular clusters?
- Does stellar formation arise due to an outside trigger? A shock wave that passes through the medium might trigger instabilities that lead to the formation of blobs

[240] *The Initial Mass Function 50 Years Later*, ed. by Corbelli, E., Palla, F. and Zinnecker, H., Springer, 2005.

which then go on to form stars. How violent should the means be to trigger star formation? What do we know about the universality of the IMF? Do different physical conditions lead to the same IMF whenever the heavy element abundance is above some critical value?

- What fraction of stars form in clusters of stars? All of them, perhaps? What fraction forms in binary systems? Is the binarity of stars a prerequisite for star formation?
- The observation of giant molecular clouds, where we believe stars continue to form today, reveals very broad spectral lines meaning a large spread of velocities that is typical of turbulent flows. It thus appears that the ill-understood phenomenon of turbulence will have to be tackled in order to understand the state of the star-forming cloud. How turbulent was the gas coming out of the Big Bang?

What can be said today with some degree of confidence[241] is the following:

- The formation process creates some objects with masses that are too low ever to ignite hydrogen. These objects will never participate in the synthesis of the elements. They will heat up, reach a temperature below the hydrogen burning temperature, and then cool to become what is known as brown dwarfs. The total mass in the form of brown dwarfs is too small to provide the missing dark matter.
- The shape of the IMF seems to be very similar under very diverse environments checked for Pop I and Pop II, but it has not yet been possible to check the case of Pop III stars. It seems that, at a heavy element abundance of $Z \sim 10^{-5} Z_\odot$, there is a major change in the physics, and consequently in the star formation and probably also in the synthesis of the heavy elements.

So that may be the end of this book, but it is not the end of the story!

[241] See also Chabrier, G., in *The Initial Mass Function 50 Years Later*, ed. by Corbelli, E., Palla, F. and Zinnecker, H., Springer, 2005, p. 41.

Index

G. Shaviv, *The Synthesis of the Elements*, Astrophysics and Space Science Library 387,
DOI: 10.1007/978-3-642-28385-7,